WATER QUALITY MONITORING AND ASSESSMENT

Edited by **Kostas Voudouris** and **Dimitra Voutsa**

INTECHOPEN.COM

Water Quality Monitoring and Assessment
Edited by Kostas Voudouris and Dimitra Voutsa

CBS, Edition **2017**

Published by InTech
Janeza Trdine 9, 51000 Rijeka, Croatia

Publishing Process Manager Marija Radja

Technical Editor Teodora Smiljanic

Cover Designer InTech Design Team

Additional hard copies can be obtained from orders@intechopen.com

Water Quality Monitoring and Assessment,
Edited by Kostas Voudouris and Dimitra Voutsa
p. cm.
ISBN 978-953-51-0486-5

Contents

Preface

Water is a valuable and finite resource on Earth. Both water quantity and quality are becoming dominant issues in many countries. European Environment Agency notes that except in some northern countries that possess abundant water resources, water scarcity occurs in many countries, particularly in the Mediterranean, Middle East, Africa etc, confronted with a crucial combination of a severe lack of and high demand for water. The growth of world population, up to 9 billion by 2050, leading to increase demands of water, growing urbanization and high living standards, intensive agricultural activities and industrial demands as well as climate change with droughts and floods episodes are significant pressures for the available water resources. Consequently, many countries have significant problems concerning both severe water scarcity and poor water quality.

Surface water and groundwater that are the main sources of fresh water for drinking purposes, irrigation and various other uses, represent as small fraction of water burden on earth. It is pointed out that only 30% of the freshwater (3% of the total volume of water) on Earth is groundwater. In many areas, water needs are mainly covered by groundwater abstracted from the aquifers via numerous wells and boreholes. As a result, a negative water balance is established in the aquifer systems around the world and water levels are dropping rapidly.

Point and non-point sources such as sewage effluents, wastewater discharges, agricultural runoff, industrial and mining activities, atmospheric deposition may seriously affect these water resources. As a consequence various pollutants such as pathogen microorganisms, nutrients, heavy metals, toxic elements, pesticides, pharmaceuticals and various other organic micropollutants may occur in water resulting in degradation of water quality. Another, severe problem, especially in coastal areas is the increase salinity of groundwater, due to seawater intrusion in coastal aquifers as a cause of high water demands and overexploitation.

The access to good quality freshwater is a decisive factor for socio-economic development of the countries. Recently, the European Community through Water Directive 2000/60/EC, established the framework for actions in the field of water policy for the protection of inland surface waters, transitional waters, coastal waters and groundwater. This Directive aims at the protection and enhancement of the aquatic ecosystems, promotion of sustainable water use based on a long-term protection of available water resources, progressive reduction or cessation of discharges of hazardous substances into aquatic environment and mitigation the effects of floods and droughts. These actions contribute to the provision of sufficient supply of good quality surface water and groundwater as needed for sustainable as well as to balanced and equitable water use.

This book entitled "**Water Quality Monitoring and Assessment**" attempts to covers the main fields of water quality issues presenting case studies in various countries concerning the physicochemical characteristics of surface and groundwaters and possible pollution sources as well as methods and tools for the evaluation of water quality status. Particularly, this book is divided **into** two sections:

1) Statistical analysis of water quality data

The first ten chapters focus on the evaluation of water quality data by employing conventional hydrochemical techniques and statistical analysis (e.g. cluster, factor and trend analysis, risk analysis and decision support systems).

2) Water quality monitoring studies

This section includes seventeen chapters related to the water quality and the assessment of water pollution. These chapters represent case studies from different countries of the world regarding the quality of surface and groundwater.

We would like to express our thanks to the authors who contributed to this volume, to the reviewers for their valuable assistance, as well as to the organizers and the staff of the INTECH Open Access Publisher, especially **Marija Radja**, for their efforts to publish this book.

Dr. Kostas Voudouris
Laboratory of Engineering Geology & Hydrogeology,
Department of Geology,
Aristotle University of Thessaloniki,
Greece

Dr. Dimitra Voutsa
Department of Chemistry,
Laboratory of Environmental Pollution Control,
Aristotle University of Thessaloniki,
Greece

List of reviewers

Bobori Dimitra, Lab. of Ichthyology, School of Biology, Aristotle University of Thessaloniki
Genitsaris Savvas, Dep. of Botany, School of Biology, Aristotle University of Thessaloniki
Georgiou Pantazis, School of Agriculture, Aristotle University of Thessaloniki
Kaklis Akis, Dr. of Hydrogeology, Aristotle University of Thessaloniki
Karayanni Hera, Dep. of Biological Applications and Technology, University of Ioannina
Katsiapi Maria, Dep. of Botany, School of Biology, Aristotle University of Thessaloniki
Kormas Kostas, Dep. of Ichthyology and Aquatic Environment, University of Thessaly
Lazaridou Maria, Professor of Biology, Aristotle University of Thessaloniki
Mattas Christos, Dr. of Hydrogeology, Aristotle University of Thessaloniki
Melfos Basil, Lecturer of Geology, Aristotle University of Thessaloniki
Michaloudi Evagelia, Lab. of Ichthyology, School of Biology, Aristotle University of Thessaloniki
Moustaka-Gouni Maria, Dept. of Botany, School of Biology, Aristotle University of Thessaloniki
Polemio Maurizio, Istituto di Ricerca per la Protezione Idrogeologica, Bari, Italy
Theodosiou Nikolaos, Assistant Professor, Civil Engineering, Aristotle University of Thessaloniki
Voudouris Kostas, Ass. Professor of Hydrogeology, Aristotle University of Thessaloniki, Greece
Voutsa Dimitra, Assistant Professor, Dep. of Chemistry, Aristotle University of Thessaloniki, Greece

Part 1

Statistical Analysis of Water Quality Data

1

Spatial Decision Support System (SDSS) for Stormwater Management and Water Quality Assessment

Nally Kaunda-Bukenya, Wubishet Tadesse, Yujian Fu, Teferi Tsegaye and Mezemir Wagaw
Alabama A&M University & City of Huntsville, AL, Planning Division
USA

1. Introduction

Land use policy in the United States is a predominantly local issue (Giannotti & Arnold, 2002). The challenge is that land use policies and decisions are made by elected and appointed municipal officials (Stocker et al., 1999) whose training may not necessarily be in environmental management. Because of the critical importance of their work, and because they deal with land-use planning and regulation on a daily basis, local officials need decision tools that can allow them to place case-by-case land-use decisions within the broader context of the watershed. These land use managers need tools for assisting them to evaluate environmental impacts of their land-use decisions, visualize alternative scenarios, and educate their constituency (Arnold, 2000). Historically, decision makers have indicated that inaccessibility of required geographic data and difficulties in synthesizing various recommendations are primary obstacles to spatial problem solving (Ascough et al., 2002). Indeed studies have shown that the ability to produce meaningful solutions can be improved if these obstacles are lessened or removed through an integrated systems approach, such as a Spatial Decision Support System (SDSS). As Ascough et al. (2002) have observed, a SDSS makes a positive contribution to decision makers' task if it enables them to reach: (i) a more accurate solution, (ii) a faster solution to a given problem, or (iii) both of these.

The driving force for developing this SDSS is the limited use of Geographic Information Systems (GIS) for environmental planning in municipalities. This limitation is due to the fact that, even though GIS software is available in most municipal land management data centers, it is too complex for policy makers and environmental officials to use "out of the box" without acquiring expertise in GIS. Thus, there is a need to develop custom tools that are less intimidating to non-GIS audience or users, but robust enough to perform complex geoprocessing tasks and hydrological models in the background. The goal of this Chapter is to develop an adaptive and customer-driven environmental SDSS to assist municipal officials fulfill environmental legislations and minimize the impact of pollution resulting from urban development. There are two specific objectives that we will address. The first objective is to develop a front-end graphical user interface (GUI) that is robust enough to quantify and geolocate pollution hot-spots in an urban area, but simple enough for use by land use decision-makers whose expertise is neither GIS nor water quality assessment. The second objective is to demonstrate the use of the SDSS for generating environmental compliance reports and for assessing pre-development and post-development conditions of a land use change. The rest of the chapter is organized into four sections. First, a review of previous studies is presented in section 2, the methods for developing the SDSS components in section 3. and the last two sections are the results and summary.

2. Literature review

A spatial decision support system (SDSS) is an interactive, computer-based system designed to support a user or group of users in achieving a higher effectiveness of decision making while solving a spatial problem (Sprague, 1982). According to Sprague (1982), a SDSS has three primary components: a geographic database management system for handling geographic data; a number of potential models that can be used to forecast the possible outcomes of decisions; and a user interface to provide interaction of the user to model scenarios. Similarly, Armstrong and Densham (1990) suggest that five key modules are needed in a SDSS: (i) a database management system (DBMS), (ii) analysis procedures in a model base management system (MBMS), (iii) a display generator, (iv) a report generator, and (v) a user interface.

Purdue Research Foundation (2010) developed a model called Long Term Hydrologic Impact Analysis (L-THIA). This model takes land use, soil, and long-term precipitation data as input and computes changes in recharge, runoff, and nonpoint source pollution resulting from past or proposed development as an output. The L-THIA modeling program is available in three forms as an online spreadsheet, as Avenue scripts that run as an extension of ArcView 3.x, or as an interactive mapping application developed using Java programming (Purdue Research Foundation, 2010). Although L-THIA was originally developed for municipal planners, its main focus seems to be non-point source (NPS) pollution. A similar tool is needed for addressing point source pollution in addition to NPS pollution, and for assisting municipalities with tools for environmental legislation compliance. The current research fills this gap by providing the ability to quantify point source pollution and use the newly developed user interface to generate reports for environmental compliance. The idea is that, having an environmental compliance tool that produces qualitative pollution hotspot maps or charts in addition to quantitative outputs (such as tabular data), enables decision-makers to track the environmental status of their watersheds and monitor the long-term effect of land use on the environment.

Wilkerson et al. (2010) developed a SDSS that allows users to balance watershed protection with smart growth/low-impact site development strategies. The SDSS was developed to calculate: time-varying runoff and water quality as a function of rainfall, site characteristics, and BMPs for development sites within the Southeastern U.S; and BMP cost, and compare various scenarios for effectiveness and cost. The authors used an existing Hydrological Simulation Program-FORTRAN (HSPF) for computing movement of water through a complete hydrologic cycle—rainfall, interception, evapo-transpiration, runoff, infiltration, and flow through the ground. HSPF runs on Better Assessment Science Integrating point and Nonpoint Sources (BASINS) interface called WinHSPF. A simplified windows interface called Latis was developed by the authors as a simpler replacement of BASINS. (Wilkerson et al., 2010) The Latis model involves using specific rain event to model and compare scenarios under pre-development, and "as-built" with BMP options, and even worst case scenario (100 percent impervious). Later on, Latis, was further improved into Latis-LIDIA to estimate runoff based on pre- and post-developed site conditions using the widely-used Soil Conservation Service (SCS) runoff curve number (CN) method. The first step in the model requires user input of project information, site dimensions, and precipitation data. Precipitation data are automatically generated by selecting state and county, or manually entered by user-defined values. The precipitation database is tailored for sites within Alabama, Louisiana, and Mississippi (Wilkerson et al., 2010). The user then characterizes land use and land cover for each respective hydrologic soil group (HSG), cover type, and size. The model then generates runoff coefficients. Although this is significant contribution to the existing body of knowledge, the authors admit that the model needs further development to accommodate pollutant loading computation. This current research addresses this deficiency.

In another instance, the Decision Evaluation in Complex Risk Network Systems (DECERNS) is a similar SDSS tool which focuses on land use planning and management (DECERNS - Team, 2006). DECERNS provides spatial data visualization for vector and raster models. It is used in the development of alternatives and criteria specification; implementation of the basic and advanced multi-criteria decision; and generation of various reports, including text descriptions, tables, diagrams, and maps (DECERNS -Team, 2006). It is a powerful commercial SDSS that caters for a larger community of state and regional officials, educators, and researchers. However, local governments who make critical land-use decisions are not direct beneficiaries because of the costs associated with this commercial software and also because this system lacks specific direct benefits to municipalities such as environmental legislation compliance tools. Thus, justifying the need to develop a municipal SDSS targeted for local communities to visualize the impacts of their decisions and simultaneously fulfill environmental legislation.

In another example, Rodman and Jackson (2006) used Python programming and the ArcGIS geoprocessor to develop a standalone spatial application, for the US Army Corps of Engineers (USACE), which performs data mining in geographic datasets. Python was selected as the language of choice because it is a powerful open source cross-platform programming language, that can run on Windows, Mac, or Unix and has a wealth of available code and tools that connect to databases for developing graphical user interfaces (GUI). The authors' goal was to create an application that relies on ArcGIS for handling spatial data formats, geographic coordinate system transformation, mapping, and geoprocessing. The resulting data mining application known as *Aspect*, is used to discover association rules that describe spatial relationships between geographic features. This allows decision makers to have tools that combine knowledge of terrain, travel routes, structure, land use/cover to improve situational awareness (Rodman & Jackson, 2006).

In this research, several implementation options were examined to determine the best approach to develop the graphical user interface. The option selected would need to take advantage of the already established ArcGIS geoprocessor for handling spatial data processing, coordinate system, and data editing. For this reason, some of the options examined for use include Esri's ArcGIS Add-ins, Tkinter/python-based GUI, a web-mapping application, and a Java-based custom desktop application. Each of these applications is briefly described below while highlighting possible shortcomings as they relate to the current research.

2.1 ArcGIS Add-ins

In the 2011 version of ArcGIS 10 software, the concept of Add-ins was introduced to expand ArcGIS desktop's functionality and extend the interface (Burke and Elkins, 2010). Using Esri's ArcObjects, one can create new custom functionality using Add-ins to create a button or tool that a user interacts with to do something with the map or with GIS data. There are several types of Add-ins ranging from menus, buttons, toolbars, dockable windows, tool palettes, or applications and extensions. Add-in extensions are invisible to the user, but are event listeners that react to events by running code attached to them. The advantage here is that the complexity of the code is hidden from the end user, making complex processes more user-friendly. Add-ins can be built with C++ programming and often require a lot less code since the programmer simply adds more functionality instead of creating a new software package. They are also portable as they can be easily shared by email or file transfer from one user to another using a few steps to install (Burke & Elkins, 2010).

There are several ways to create Add-ins. Burke and Elkins (2010) use Microsoft Visual Studio 2008 and .NET to create a button Add-in using a wizard, assign an image icon, a name, and add reference to it. Microsoft Visual Studio 2008 Express is a free version that can be used to get started with Add-in development. Add-Ins can also be built with Java, for example using Eclipse development environment. To do any Add-in development, ArcObjects Software Development Kit (for .NET if using Visual Studio) needs to be installed. An Add-in is just one file with *EsriAddin* file extension, a folder-list container that contains everything needed for it to work without much setup. Add-ins can be placed in a specific directory where ArcGIS checks every time the software is launched. In summary, the Add-in creation process in Visual Studio involves making a Visual Studio project, creating an add-in, adding type to it, writing the business logic (what it does, and code behind it), and then testing it (Burke & Elkins, 2010).

Although Add-Ins seems straightforward, a Java or C++ programming skill is required to get the full effects (Burke & Elkins, 2010). Also, they are relatively new in ArcGIS and do not have a large user community yet, thus they were not selected for implementation in this research. Alternatively, Esri's ArcGIS Engine software can be used with ArcObjects to add dynamic mapping and GIS capabilities to existing applications, build custom mapping applications, or add geoprocessing scripts using application programming interfaces (APIs) for COM, .NET, Java, and C++ (Burke & Elkins, 2010). This option was not utilized in the current research because ArcGIS Engine software package was not available. The second option that was considered using a python-based module called Tkinter.

2.2 Tkinter/python-based graphical user interface

Tkinter Tool Command Language (Tk/Tcl) is an integral part of Python that provides a platform-independent windowing toolkit that is available to Python programmers using the Tkinter module (Official Tk/Tcl website, 2010). The Tkinter module (renamed to tkinter in Python version 3.x) is the standard Python interface to the Tk GUI toolkit (Official Python Website, 2011). Tkinter is basically a set of wrappers that implement the Tk widgets as Python classes. Tcl (Tool Command Language) is a dynamic programming language that is suitable for a very wide range of uses, including Web and desktop applications, network programming, embedded development, testing, general purpose programming, system administration, database work, and many more (Official Tk/Tcl website, 2010). (Official Tk/Tcl website, 2010).

Tk, a graphical user interface toolkit can be used in Tcl or in Perl language to create a number of GUI components such as buttons, labels and canvas. These components are known as widgets (Official Tk/Tcl website, 2010). Once these widgets have been created, three geometry managers are used to display them in relation to each other: pack, grid, and place. The "pack" geometry manager allows one to place your widgets in a row or column. "Grid" geometry manager allows the placement of widgets in a matrix. The third geometry manager, "place" provides the ability to place widgets by pixel or by the proportion of the way across the window that the programmer wants them to appear. More complex windows can be built using frames or nested frames (Well House Consultants, 2006). The python-based Tkinter appears to be a good option since the underlying geoprocessing scripts developed for this research are also Python-based. The limitation with this application is that the spatial components of the GUI such as coordinate system handling and spatial analysis would need to be programmed, which could be time-consuming. The third option examined for the GUI was a web-mapping application.

2.3 Web mapping application

Esri's software, ArcGIS Server enables the user to create, manage, and distribute GIS services over the Web (Esri, 2011). Different Application Programming Interfaces (API) are available for web application development on various platforms. The Esri APIs include JavaScript, Flex, Silverlight, and Java Web Application Development Framework (ADF), and the NET Web ADF (Esri, 2011). ArcGIS Explorer Online is another option to present web map services, add other content to it, navigate, present and share the map. Explorer Online makes it possible to disseminate work on the Web and integrate map services from various sources (Esri, 2011). ArcGIS Explorer Online allows one to save maps to ArcGIS.com and choose to save them privately, share with a group or share over the Internet (ArcGIS Explorer Online Team, 2011). While the web-mapping approach is not emphasized in this research, a prototype web-mapping application was developed for spatial data editing and for interactive mapping. The next section further describes the SDSS development methodology.

3. Spatial Decision Support System methodology

The overall methodology for this research encompasses a three-tier approach that leads to the development of a spatial decision support system. The three levels are the data level, the model development level, and the development of a graphical user interface. The data level was implemented primarily for creating a comprehensive database that stores all the data needed for water quality evaluation. Modeling the effects of land use change on water quantity and quality requires a multidisciplinary approach which incorporates many different data types. Therefore, several data sets such as sewer infrastructure data, rainfall data, soils/curve number, and pollutant sampling data were compiled into a central geodatabase (geographic geodatabase).

3.1 Database component

Geodatabase development for stormwater management in particular can be challenging due to the multitude of different types of stormwater features involved, and the complex topological relationships that exist between them. In this research a geodatabase schema was developed that can be adopted by other municipalities as a template for stormwater mapping. Specifically, ArcGIS software suite by Esri was used to develop an enterprise level geodatabase for managing the watershed and subsurface infrastructure data. Geodatabases are needed for the successful storage, access, retrieval, manipulation, and management of massive data sets typical of municipalities. The most critical data set emphasized in this research was the sewer infrastructure data. Specifically, two separate datasets were developed for stormwater networks and sanitary sewer (wastewater) networks. A dataset is a set of georeferenced data layers that are topologically related and are in the same spatial extent (Esri, 2010). The stormwater dataset includes locations of stormwater inlets, pipes, headwalls, and culverts captured using mapping-grade Global Positioning Systems (GPS), as well as creeks, rivers, and delineated drainage basins from topographic mapping. Similarly, the sanitary sewer database consists of GPS locations and dimensions of manholes, sewer lines, pump stations, wastewater treatment plants, and sewer drainage basins. Network topology rules were developed to enforce the connectivity of the sewer features such that all features that participate in a network are topologically connected. For example, if an inlet is not connected to any pipe or waterway, it would be marked as an error. Similarly, attribute validation rules are also established for the sewer features to minimize errors when editing the sewer data. After the database design the next step was to populate the "spatial container" with data from multiple sources including GPS, aerial photos, engineering design drawings, and digital elevation models. Once all the data had been collected, the next step in the SDSS development was model development.

3.2 Model development component

The model development stage is critical to the full-functioning of any SDSS because data has to be processed to make it meaningful for decision-making. The general methodology of the model development stage is as follows:

- Stormwater outfalls were extracted by querying for stormwater pipes (diameter of at least 12 inches in industrial areas and at least 36 inches in all other land uses) that empty into major rivers and creeks. This query is based on the definition of an outfall by the United States Environmental Protection Agency (EPA, 1992).
- Drainage boundaries were delineated using topographic elevation and the spatial and topologic locations of underground and above-ground stormwater infrastructure. The

result was a delineation of basins, subbasins, and mini-basins. It is important to note that mini-basins are categorized into outfall basins (point source of pollution) and diffuse basins (non- point source of pollution).

- Land use characterization was achieved by performing a spatial union of an existing land use layer with the mini-basin polygons. This resulted in one of the parameters in the hydrological model; area values of each land use type in each mini-basin.
- A geospatial approach for hydrological modeling was developed using python programming to determine the water quality and runoff effect of land use change.

The uniqueness of the model component of the SDSS is in the integration of Java and Python programming languages. The user interface was written using Java, but an existing hydrological model was programmed into a geospatially-enabled hydrological model using Python programming. The hydrological model adopted is a series of three equations based on "The Simple Method" by Schueler (1987), often called the Curve Number Method. The equations were spatially enabled by encoding them into ArcGIS scripting environment called arcpy (python for ArcGIS). First, an area-weighted runoff coefficient (weighted C-value) script was written using Python's mathematical operations and program looping to calculate weighted runoff coefficient (Rvi) for each outfall basin (**equation 1**).

$$Rv_i=\sum(A_i*Rv)/\sum A_i \tag{1}$$

Where *Rv*= Runoff coefficient for each land use within the outfall drainage area.

A_i=Land area of each land use within the drainage area of the major outfall. Next, Event Mean Concentration (EMC) of each pollutant was calculated using the result from equation 1 above.

$$EMC=\sum(A_i/A_T)*(Rv/Rv_i)*C_i \tag{2}$$

Where A_i=Area of a specific land use within the outfall drainage basin

- A_T=Total land area within the drainage area of the major outfall
- Rv=Runoff coefficient (C-value) for a particular land use
- Rv_i=weighted C-value for the drainage area of the major outfall
- C_i=measured pollutant concentration in each land use

Finally, the pollutant loadings (weight of pollutant per season) were calculated using the following equation:

$$Li=[(P*CF*Rvi)/12](EMC*Ai) \tag{3}$$

Where *Li*= Seasonal pollutant load in pounds per outfall/season

- *P*= Precipitation (inches/season)
- *CF*=Correction factor that adjusts for storms which produce no runoff
- *Rvi*=Weighted runoff coefficient for the area drained by each outfall
- *EMC*=Event Mean Concentration of pollutant in milligram/liter
- *Ai*=Land area drained by each major outfall (acres)
- 12 and 2.72 are unit conversion factors.

These three equations have been sequentially executed and the resulting table would have the output that is used to create thematic maps that show pollutant hotspots in the study area. The equations would be encapsulated in the user interface and simplified as a one-button click. The next section describes the user interface in more detail.

3.3 SDSS GUI implementation

The Java-based custom desktop application was selected for the development of the SDSS user interface. GUI programming is a complicated task involving multiple steps, because many interface widgets need to be generated and specified. Furthermore, the alignment and appearance of these interface widgets also required a lot of programming. The GUI was developed using version 1.0.23 of Java Software Development Kit (JDK) and version 6.9.1 of NetBeans Integrated Development Environment (IDE). NetBeans and JDK are free programs obtained from NetBeans' website (NetBeans official website, 2011). NetBeans is an IDE that simplifies the programmer's task and ease the burden of GUI programming. To make the GUI components produce results that react to user input, many kinds of events must be handled properly and programmed to make the widgets functional. Almost every GUI widget is a generator that is capable of initiating multiple kinds of events. Java uses "event listener" to capture events that are generated from user interactions. A "Listener" is a an interface that a programmer can put all information that is needed to make the buttons functional. An event object is a holder of generated events. The "Listener" can only capture the registered event handler. By this observer pattern, a Java GUI program can properly handle and process user interactions, providing a flexible and extendable programming strategy.

In our GUI development, the typical events generated were action events and mouse events. An action event is generated when a button is pressed, or by pressing *Enter* in a text field, or when a menu item is selected. (Schildt, 2001). The programmer has to implement an action listener to define what methods are invoked when a user performs certain operations. With registered necessary event handlers, an "actionPerformed" message is implemented to handle all generated events on the relevant component. For example, the *Save/Add* button on the user interface listens for the user to click the button, then performs the save operation to save the user input from the text field into the corresponding table in the basins ArGIS personal geodatabase (Microsoft Access database).

Swing components were developed by SUN Microsystem's, to provide a more sophisticated and user-friendly GUI programming paradigm, including frames, buttons, panels, text fields, and labels (NetBeans official website, 2011). **Figure 1** shows the component hierarchy of the Java Swing as illustrated by Reddy (2007). Other modules used include input/output (java.io), abstract windowing toolkit event (java.awt.event), and java.lang.reflect methods. Due to the feature of not invoking OS resource, Swing is considered as lightweight component that is extended on top of many widgets of an AWT packet.

Figure 2 illustrates a use-case scenario that shows how a user would interact with the system and how the system responds to the user's interaction. In general, the SDSS allows the users to:

1. input data into simplified forms on the interface and save edits at the click of a button,
2. use one-button click to run complex geoprocessing in the background, and
3. use one-button click to generate maps and output reports that are needed for informed decision-making and for environmental legislation compliance.

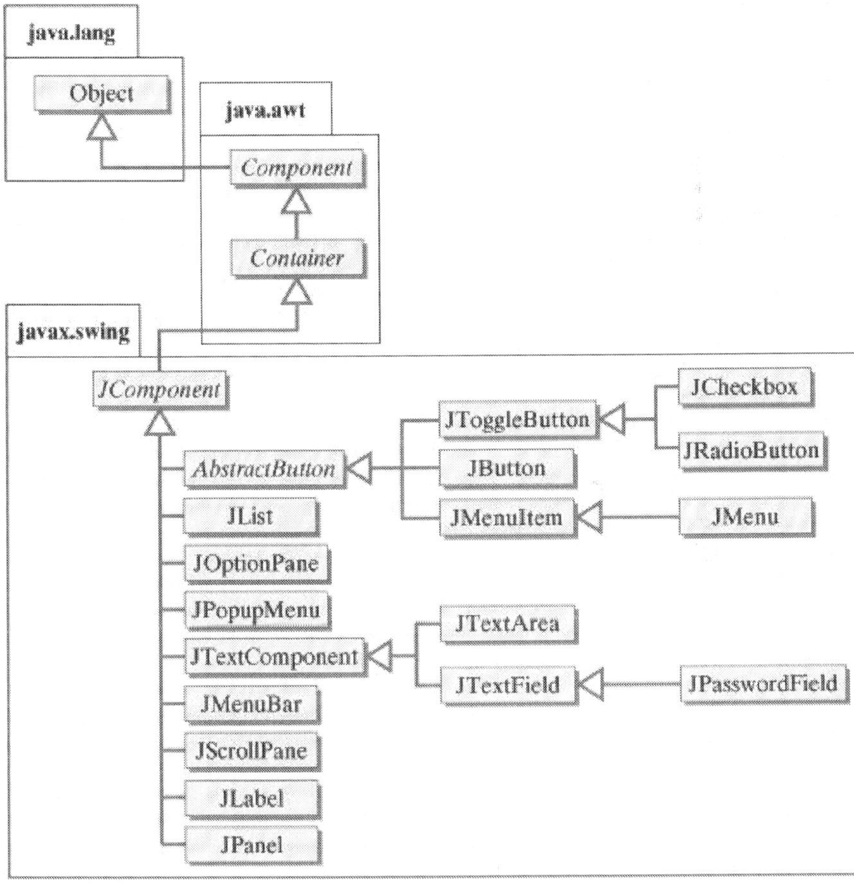

Fig. 1. The hierarchy of Java Swing Components

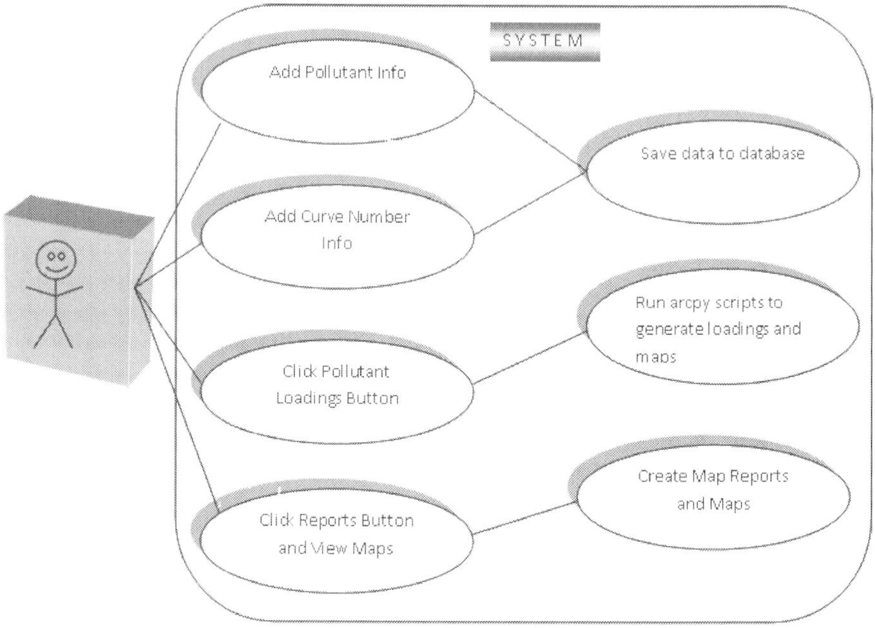

Fig. 2. Use Case Diagram showing how a user interacts with the interface

4. Results

A graphical user interface for the SDSS was developed for non-GIS professionals as a front-end for data collection as well as for executing hydrological modeling scripts. The developed geospatially-enabled hydrological model can optionally be executed in ArcGIS as a Python script tool, but the complexity of the ArcGIS interface **(Figure 3a)** has been simplified for non-GIS users into a simple, straightforward interface as shown in **Figure 3b.** The resulting GUI allows the complex pollutant loadings scripts to be executed in a less intimidating environment.

The *Calculate Loadings* in Figure 3b button runs a Java command that executes arcpy scripts (ArcGIS 10 Python scripts), which then progress by invoking classes and methods on the ArcGIS geoprocessor object. Specifically, the *Generate Loading* scripts fetch pollutant data such as pollutant concentration and curve number information entered by the user through the Java interface and use that as input for calculating the Estimated Mean Concentrations and pollutants loadings. Pollutant hotspot maps indicating the spatial distribution of outfall basin pollutant loadings are also generated upon pressing the *Calculate Loadings* button. The *Loadings Maps* button fetches the generated maps and displays them using a PDF reader such as Adobe Reader, ready for inclusion in reports, and for immediate decision-making. The user also has the option to press the *Generate Report* function on the file menu executes Java commands to generate water quality reports based on EMC's and Pollutant loadings fields from the ArcGIS geodatabase. When clicked, the *Open Online Maps* button opens a web browser showing a FlexViewer web map developed by the authors for viewing and editing geographic data associated with the SDSS. Finally, the *Open GTViewer* button opens an internal GIS desktop application (by Graphics Technologies, Inc.) that the City of Huntsville uses.

With this application, non-GIS-expert users can collect pollutant data such as pollutant name, concentrations, etc., and save into the master database tables via the *Pollutant Editor* form. The data entry is developed in such a way that the pollutant data can be edited by adding new records or editing existing records, **Figure 4** shows the pollutant editor form for this purpose.

Since different municipalities use different land use classification schemes, they may find it easier to manually enter literature-based runoff coefficients from a land use lookup table. For this reason, a runoff coefficient editor was created for manually populating C-Values/runoff coefficients; **Figure 5** shows the *Curve Number Editor* form.

The core of the model component is encapsulated in the "*Calculate Loadings*" Button on the GUI. When this button is pressed, the hydrological modeling equations (1-3) described in the model development stage would be executed. **Figure 6** shows the two buttons for processing pollutant loading calculations and for viewing output maps. To demonstrate the SDSS application, two applications were illustrated: a report generation function and a comparison of pre-development and post-development pollution contributions.

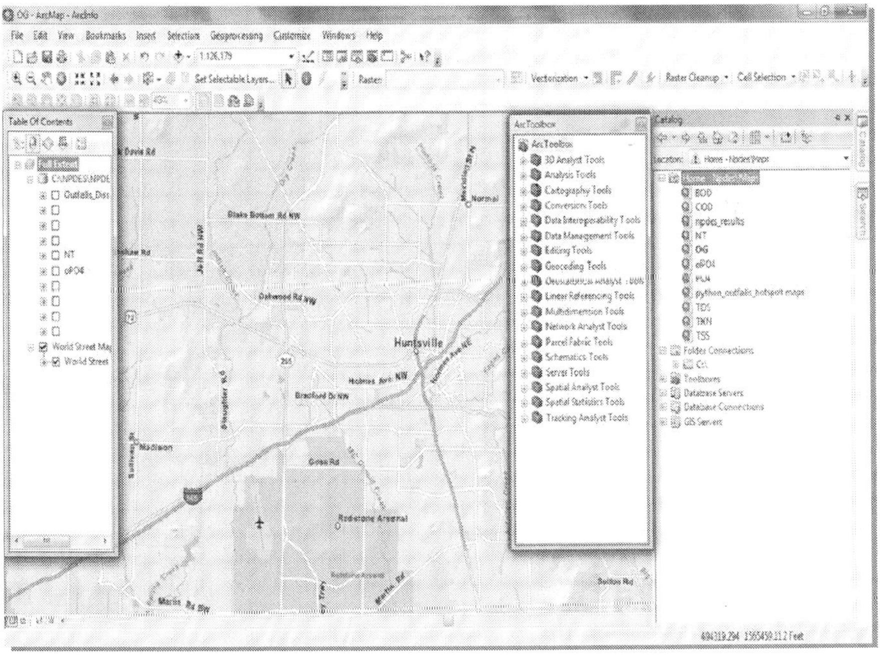

Fig. 3a. ArcGIS 10 interface

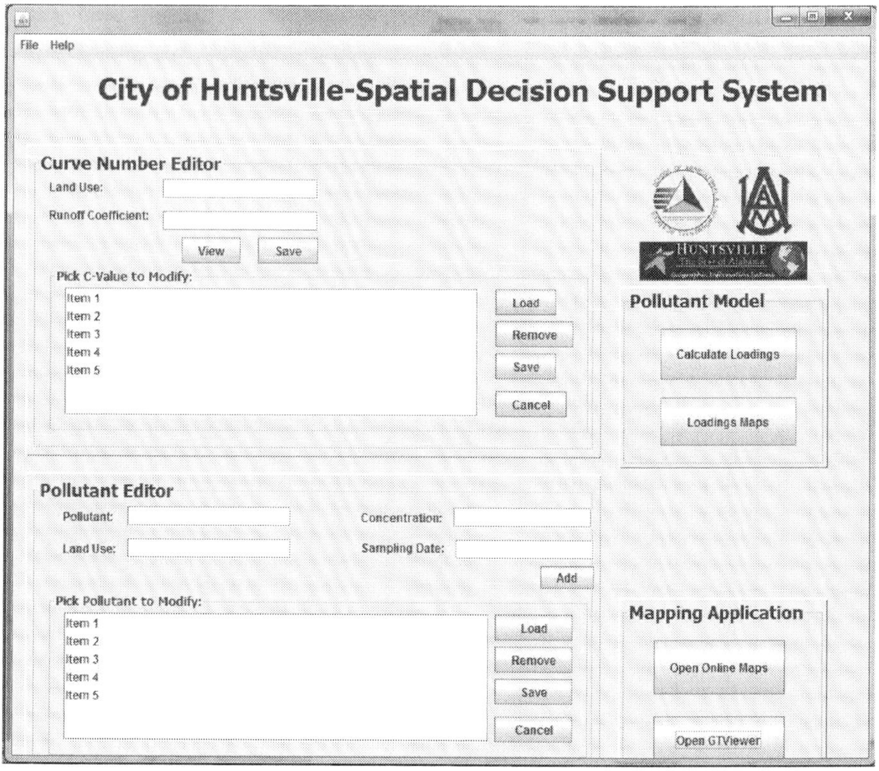

Fig. 3b. The newly developed, less intimidating graphical user interface

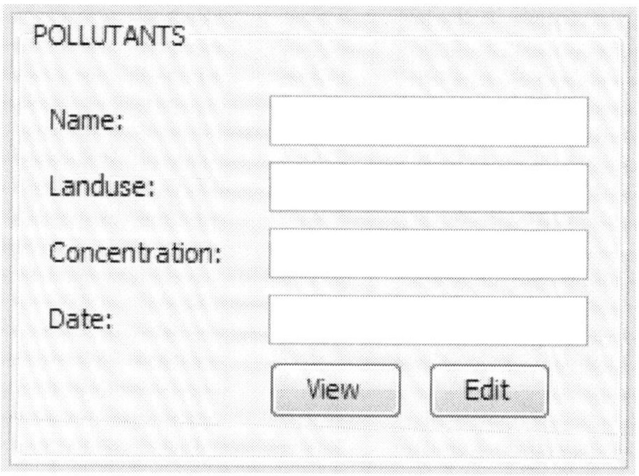

Fig. 4. Pollutant Editor Form

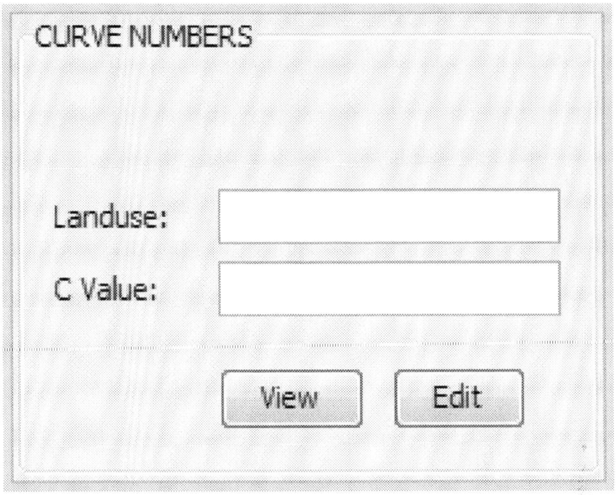

Fig. 5. The Curve Number Editor

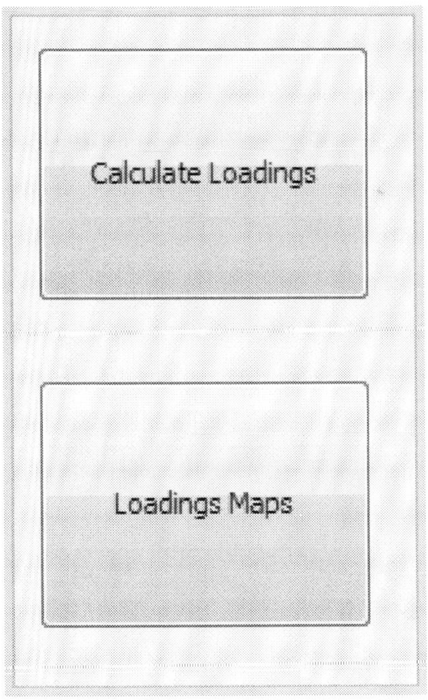

Fig. 6. Pollutant loadings and output map button

4.1 SDSS Demonstration 1: Report generation and pollution contributions of land use change (Bridge Street Town Center)

First, a report generation function for municipal environmental compliance and pollution contributions of a land use change are illustrated. To test the report function, a report was generated for the City of Huntsville stormwater outfalls using 2004 data. The *Generate Report* function on the file menu generates a report ready for submission to the United States Environmental Protection Agency (US EPA) for stormwater regulation compliance. Stormwater pollution is regulated under the Clean Water Act (CWA) of 1972. Under the CWA, the US EPA has implemented pollution control programs and set standards that make it unlawful for industries, municipalities, and other facilities to discharge any pollutant into navigable waters, without a permit (US Congress, 1972). US EPA's National Pollutant Discharge Elimination System (NPDES) permit program controls these point source discharges and has put specific regulation in place as a guide for NPDES permit *applicants* (US Congress, 1972). **Figure 7** illustrates the report generation workflow. The user interface simplifies this process for land use decision-makers by using just a few clicks to quantify pollution and generate maps for decision support or reports for environmental legislation programs such as NPDES permit compliance. Instant report generation saves time for municipal officials so that they focus more on decision-making instead of technical setbacks. The added value of the geospatially-enabled hydrological model is the ability to produce pollutant hotspot maps that unveil spatial trends, allowing land use policy makers to visualize the environmental impact of their decisions.

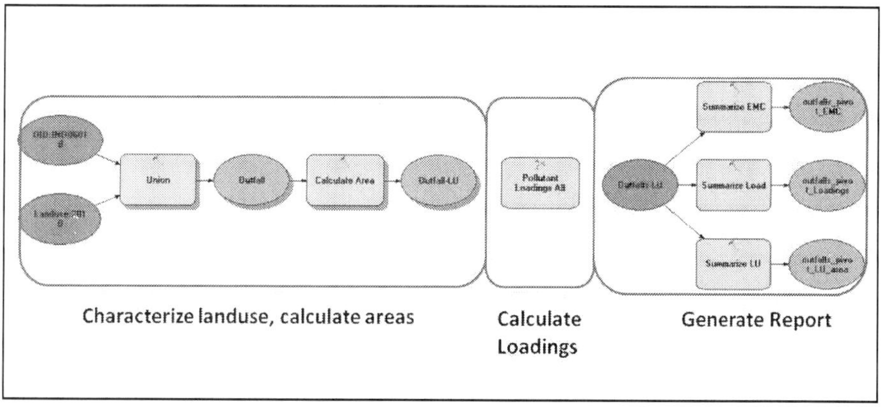

Fig. 7. Complete workflow showing land use characterization, model execution, and report generation

The first set of processes in **Figure 7** calculate the land use acreage in each mini-basin, the second calculates EMC's and pollutant loadings, and the last step selects outfall polygons and generates summarized Pivot tables with a report output. The report includes the stormwater outfall ID, weighted runoff coefficients, event mean concentration of pollutants, and pollutant loading in each outfall basin. **Tables 1-3** are respectively, land use summaries, Event Mean Concentrations, pollutant loadings generated as part of the report. **Figure 8** is an example of pollutant hotspot maps generated using the tools from the user interface.

OUTFALL_ID	Tot_Area	Rvi	Low_Density	Forest	Recreation	Transportation	Urban_Open
ALD03001	6.828	0.40	6.83	0.00	0.00	0.00	0.00
ALD03002	25.045	0.39	23.19	1.85	0.00	0.00	0.00
ALD04001	44.683	0.37	34.44	10.23	0.01	0.00	0.00
ALD04002	27.586	0.39	25.26	0.36	0.00	0.49	1.48
ALD05001	101.876	0.27	16.90	84.97	0.00	0.00	0.00
ALD05002	44.346	0.27	7.26	37.09	0.00	0.00	0.00
ALD06001	27.607	0.52	0.75	0.00	0.01	2.06	10.38
ALD06002	179.886	0.46	97.83	16.47	4.02	18.06	8.28
ALD06003	6.501	0.60	0.08	0.00	0.00	0.00	0.00
ALD06004	8.751	0.53	0.01	0.00	1.67	0.00	0.00
ALD06005	5.738	0.59	0.00	0.00	0.20	0.00	0.00
ALD07001	14.053	0.39	13.57	0.00	0.00	0.00	0.00
ALD07002	53.039	0.42	41.64	2.00	0.00	0.00	0.00
ALD07003	141.464	0.44	76.32	24.35	0.06	0.00	0.00
ALD07004	36.601	0.49	18.33	0.00	0.31	4.90	0.00
ALD07005	95.606	0.39	86.50	5.65	0.00	0.00	1.17
ALD07006	79.904	0.39	75.48	0.00	0.00	0.00	2.36
ALD07007	20.849	0.39	19.51	0.00	0.00	0.00	1.22
ALD07008	11.493	0.57	0.00	3.11	0.00	1.27	0.00
ALD07009	68.361	0.40	51.92	1.29	4.69	1.43	3.25
ALD07010	136.089	0.38	67.64	33.98	0.00	2.36	16.01
ALD07011	20.149	0.52	12.01	0.00	0.00	0.17	0.00

Table 1. Land use area of each mini-basin in acres. (Rvi = weighted runoff coefficient)

OUTFALL_ID	Tot_Area	Rvi	EMC_BOD	EMC_COD	EMC_TSS	EMC_TDS	EMC_PO4
ALD03001	6.83	0.40	13.50	19.38	42.25	89.50	0.87
ALD03002	25.05	0.39	13.13	20.36	43.57	88.93	0.84
ALD04001	44.68	0.37	12.27	22.61	46.60	87.63	0.75
ALD04002	27.59	0.39	12.89	20.45	42.41	90.48	0.82
ALD05001	101.88	0.27	7.56	35.02	63.30	80.47	0.30
ALD05002	44.35	0.27	7.54	35.08	63.39	80.44	0.30
ALD06001	27.61	0.52	7.69	35.27	63.12	76.09	0.40
ALD06002	179.89	0.46	9.95	26.71	46.96	88.68	0.58
ALD06003	6.50	0.60	8.37	18.51	13.90	64.87	0.42
ALD06004	8.75	0.53	8.10	20.42	18.74	65.86	0.39
ALD06005	5.74	0.59	8.29	18.82	14.51	64.86	0.42
ALD07001	14.05	0.39	13.50	19.38	42.25	89.50	0.87
ALD07002	53.04	0.42	12.19	20.57	39.51	83.85	0.76
ALD07003	141.46	0.44	10.49	25.24	47.18	77.43	0.64
ALD07004	36.60	0.49	9.52	22.42	31.59	87.32	0.55
ALD07005	95.61	0.39	13.00	20.41	43.18	88.28	0.83
ALD07006	79.90	0.39	13.35	19.77	42.78	89.27	0.86
ALD07007	20.85	0.39	13.21	20.15	43.29	89.05	0.84
ALD07008	11.49	0.57	7.40	36.18	65.65	76.01	0.42
ALD07009	68.36	0.40	11.88	24.62	50.71	87.34	0.74
ALD07010	136.09	0.38	10.28	27.33	51.52	83.29	0.59
ALD07011	20.15	0.52	10.82	34.49	76.71	78.69	0.70

(Rvi = weighted runoff coefficient, EMC= Event Mean Concentration, BOD=Biological Oxygen Demand, COD= Chemical Oxygen Demand, TSS= Total Suspended Solids, TDS = Total Dissolved Solids, PO4 = Phosphates)

Table 2. Annual Event Mean Concentration of each pollutant in each mini-basin

OUTFALL_ID	Tot_Area	Rvi	BOD_Load	COD_Load	TSS_Load	TDS_Load	PO4_Load	oPO4_Load
ALD03001	6.8	0.4	424.6	609.4	1328.9	2815.0	27.4	23.1
ALD03002	25.0	0.4	1472.4	2283.3	4887.1	9975.3	93.9	78.7
ALD04001	44.7	0.4	2308.9	4253.3	8766.9	16486.9	142.0	117.4
ALD04002	27.6	0.4	1615.2	2562.5	5315.1	11339.0	102.3	85.8
ALD05001	101.9	0.3	2438.3	11293.9	20414.0	25952.4	97.5	63.6
ALD05002	44.3	0.3	1056.6	4918.8	8886.8	11277.4	42.1	27.3
ALD06001	27.6	0.5	1272.8	5839.1	10449.3	12595.8	66.8	69.8
ALD06002	179.9	0.5	9417.9	25294.8	44464.3	83974.1	551.2	486.9
ALD06003	6.5	0.6	374.5	827.9	621.9	2902.0	19.0	23.8
ALD06004	8.8	0.5	434.9	1096.9	1006.5	3536.8	21.1	26.1
ALD06005	5.7	0.6	321.9	730.9	563.6	2518.8	16.1	20.3
ALD07001	14.1	0.4	844.0	1211.3	2641.5	5595.5	54.5	46.0
ALD07002	53.0	0.4	3092.6	5218.3	10025.7	21275.2	192.8	171.4
ALD07003	141.5	0.4	7466.8	17965.6	33583.7	55109.0	456.6	395.0
ALD07004	36.6	0.5	1969.3	4641.0	6537.6	18070.8	114.5	102.7
ALD07005	95.6	0.4	5577.3	8753.6	18519.9	37862.9	354.9	298.5
ALD07006	79.9	0.4	4732.6	7009.2	15167.2	31648.1	304.2	255.7
ALD07007	20.8	0.4	1233.3	1881.9	4043.2	8316.6	78.8	66.2
ALD07008	11.5	0.6	554.3	2709.7	4916.4	5692.8	31.2	29.4
ALD07009	68.4	0.4	3774.8	7823.7	16111.5	27751.5	235.4	200.3
ALD07010	136.1	0.4	6155.1	16354.0	30834.3	49845.4	352.0	303.1
ALD07011	20.1	0.5	1316.2	4196.0	9332.4	9573.9	85.4	79.1
ALD07012	73.7	0.4	3852.1	7301.4	13761.6	29609.7	233.1	196.4
ALD07013	36.9	0.4	2252.1	3232.2	7048.4	14930.9	145.6	122.6
ALD07014	15.3	0.4	933.1	1339.2	2920.2	6186.0	60.3	50.8
ALD08001	104.7	0.4	4847.8	10505.1	20265.3	38260.8	285.1	232.4
ALD08002	140.0	0.4	7269.9	13911.8	25757.2	52707.2	435.1	380.6
ALD08003	63.6	0.4	3952.2	5672.1	12368.8	26201.3	255.4	215.2
ALD08004	31.7	0.4	1747.0	3057.1	6056.4	12992.0	108.3	90.2

Table 3. Annual pollutant loadings for each pollutant in each mini-basin in pounds/year. (oPO4_Load = orthophosphates, other pollutant acronyms as described in Table 2)

4.2 SDSS Demonstration 2: Pollution contribution of land use change

The second application of the SDSS is the assessment of pre-development (2004) and post-development (2010) conditions of an existing commercial establishment to evaluate the pollution contribution of Bridge Street Town Center in Huntsville, Alabama. Bridge Street is located in Cummings Research Park and is mostly a commercial development that also has condominiums, hotels and recreational facilities. TThis section demonstrates the use of the SDSS by comparing Pre and Post Development conditions of mini-basin IND06018 which encompasses Bridge Street Town Center. In 2004 Bridge Street did not exist, but in 2010 the land was highly developed as shown in **Figure 9.**

The land use changes shown in **Figure 10** indicate that cropland (-38%) and campus/institutional (-30) land use areas decreased while all other land uses increased. The highest land use change after the development was commercial, at 41% increase. **Figure 11** shows that runoff coefficients increased by 12%, and the event mean concentration for all pollutants increased, with Oil and grease showing the highest increase (50%). The increase in runoff coefficients can be attributed to an increase in impervious surface as the area was mostly cropland in 2004 and in 2011 it is mostly urban. The higher oil and grease increase can be attributed to the increase in parking areas where oil leaks are possible from parked vehicles, and grease from the commercial establishments in the town center are inevitable. The percent change in pollutant loadings for oil and grease were also analyzed and shown in **Figure 12.** Consequently, the loadings for oil and grease also increased for the dry, wet, transitional and annual seasons.

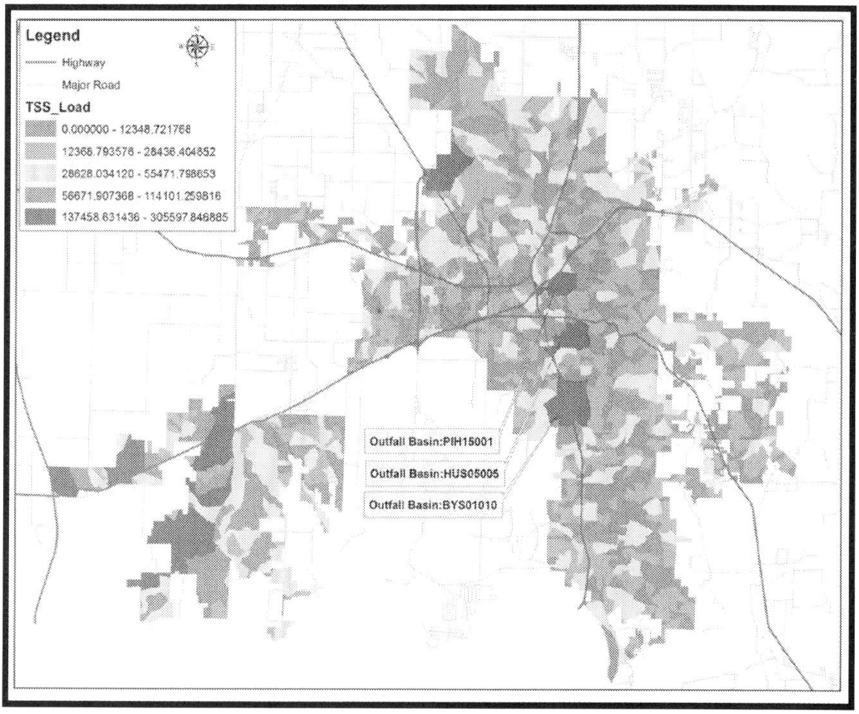

Fig. 8. Example of hotspot map for Total Suspended Solids (TSS) loads in Huntsville, Alabama (units in pounds/year) The map is presented in State Plane Coordinate System, North American Datum of 1983, Alabama East FIPS 0101.

Fig. 9. 2004 and 2010 Orthophotos at the Bridge Street Town Center location: pre and post development

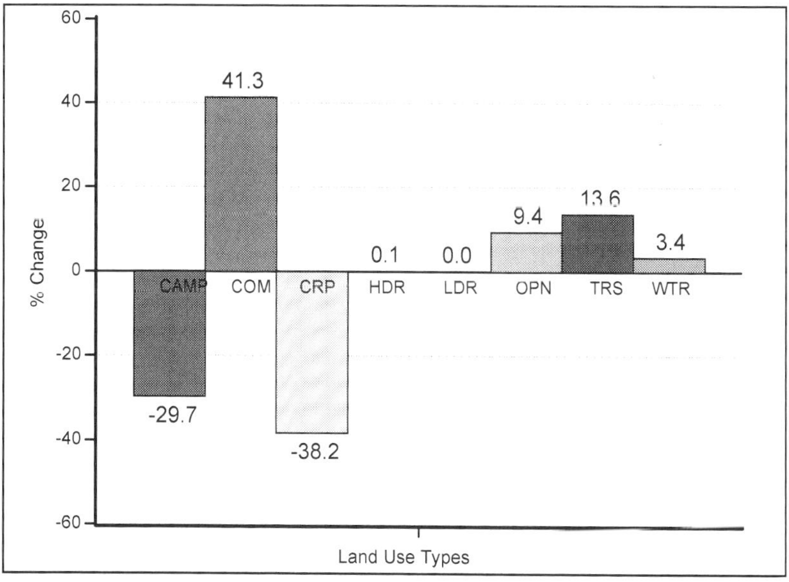

Fig. 10. Percent Change in Land use Acreage after Bridge Street Development

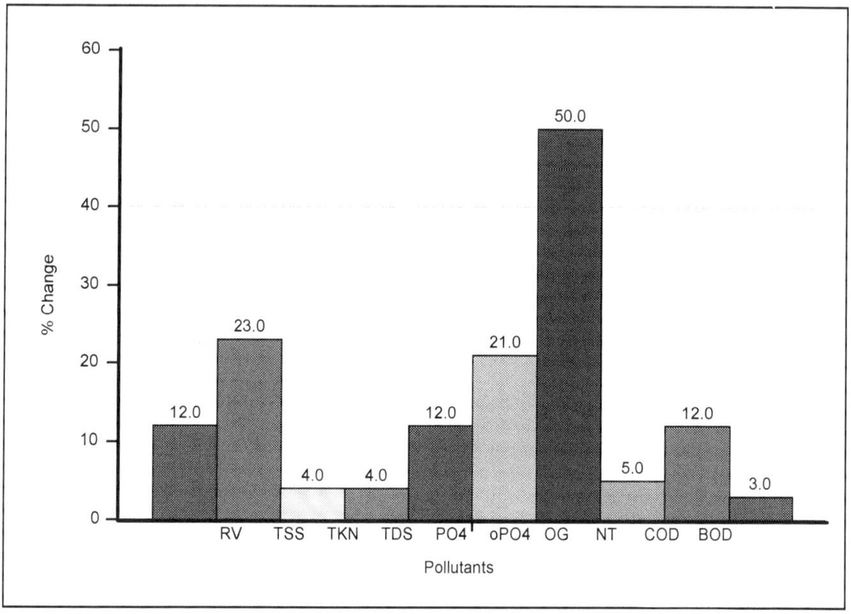

Fig. 11. Percent change in runoff coefficients and Event Mean Concentrations

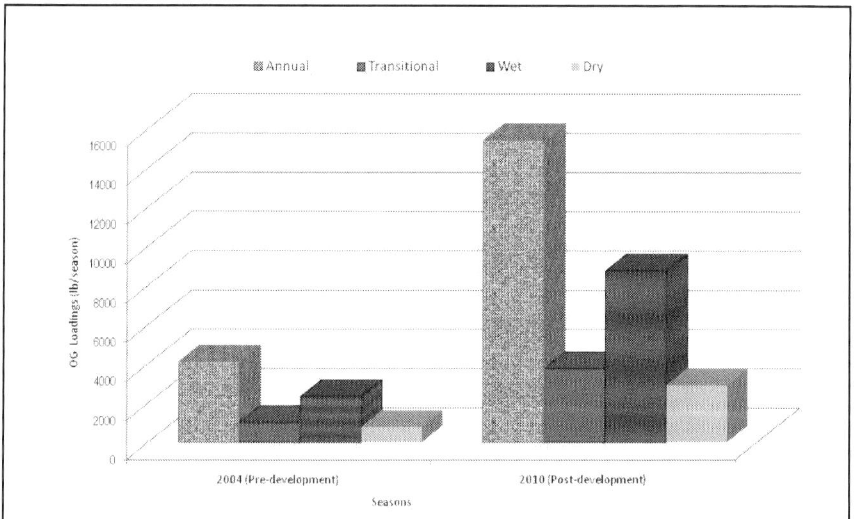

Dry: August-October, Wet: December-May, Transitional (in between): November, June-July, (Used seasonal designations per City of Huntsville, AL, (2005).

Fig. 12. Change in Seasonal Pollutant Loadings for Bridge Street area

5. Summary and conclusions

The first objective of this study was to develop a front-end graphical user interface (GUI) that is robust enough to facilitate data collection, quantify and geolocate urban pollution, but simple enough for land use decision-makers, whose expertise is neither GIS nor water quality assessment. The second objective was to demonstrate the use of the SDSS for generating reports and assessing pre-development and post-development conditions of a land use change. A desktop application has been designed and implemented using Java programming in NetBeans IDE. The GUI is a user-friendly interface that conceals program details, saving the user valuable time from focusing on technical complications, while still getting a powerful tool for the intended needs. The GUI provides custom tools for quick data input, spatial analysis, report generation, and environmental regulation compliance. The output is a graphical user interface for municipal officials or other land-use decision-makers and watershed managers for visualizing and quantifying the effects of land use on the environment. A robust, but user-friendly custom interface for local land-use officials is necessary for decision-makers to be more environmentally aware and to channel resources where they are needed the most.

One shortcoming of the desktop application is that it only allows for pollutant and curve number editing, but does not support editing GIS data. As a result, a web-mapping application is under development to address this limitation. A link to the website was established in the user interface using the *Open Online Maps* button. The web mapping application was developed using Esri's FlexViewer API and hosted on the City of Huntsville's ArcGIS server as a prototype that is not yet available for public view. The web map can be loaded by field inspectors onto a mobile device or compatible smart phones to edit and modify GIS data, sending data back to the master database via a cell phone network or the Internet. Similarly, it can be used by local citizens to pinpoint incidents such as locations of illegal pollutant discharge, sanitary sewer overflows, or flooding complaints that may result in pollution.

6. References

ArcGIS Explorer Online Team. (2011). Online Electronic Documentation. Accessed February 4, 2011. Available from http://explorer.arcgis.com/

DECERNS-Team. (2006). Decision Evaluation in Complex Risk Network Systems. Accessed 2/4/11. Available from http://195.112.127.216/?q=rd

EPA. (1992). Guidance Manual for the Preparation of Part 2 of the NPDES Permit Applications for Discharge from Municipal Separate Storm Systems. *EPA Office of Water*. Washington, DC.

Esri. (2010). ArcGIS Server Technology. Accessed February 10, 2010. Available from http://www.Esri.com/

Esri. (2011). Add-ins for ArcGIS Desktop. Accessed March 26, 2011. Available from http://blogs.esri.com/Dev/blogs/arcgisdesktop/archive/2010/05/05/Add_2D00_ins-for-ArcGIS-Desktop-10.aspx

Giannotti, L A. & Arnold, C L. (2011). Changing Land use Decision Making One Town at a Time. *North Carolina State University*. Accessed March 1, 2011. Available from http://nemo.uconn.edu/publications/about_nemo/changing_land use_10year.pdf

NetBeans Official Website. (2011). NetBeans Integrated Development Environment. Accessed January 25, 2011. Available from http://NetBeans.org/

Official Python Website. (2011). Python Programming Language. Accessed January 25, 2011. Available from http:// www.python.org

Official Tk/Tcl website. (2011). The Tcl Developer Exchange. Accessed January 25, 2011. Available from http://www.tcl.tk/

Purdue Research Foundation. (2010). Long Term Hydrologic Impact Analysis. Accessed December 8, 2011. Available from https://engineering.purdue.edu/~lthia/

Reddy, D. (2007). UI's and Java Swing -Java Swing Components Hierarchy. Accessed Feb. 4, 2011. Available from http://www.google.com/imgres?imgurl=http://jroller.com/DhilshukReddy/resource/JavaSwing/JavaSwingCompoentsList.PNG

Rodman L C. & Jackson, J. (2006). Creating Standalone Spatially Enabled Python Applications using the Arcgis Geoprocessor. Accessed Dec. 8, 2010. Available from http://proceedings.esri.com/library/userconf/proc06/papers/papers/pap_1091.pdf

Schildt, H. (2001). *The Complete Java Reference: Java 2 Fourth Edition*. New York: Osborne/McGraw-Hill.

Singh,V P. & Frevert, D.K. (2010). Hydrologic Modeling Inventory. Texas A&M University and the Bureau of Reclamation. Accessed December 8, 2010. Available from http://hydrologicmodels.tamu.edu/

Sprague, R. H. & Carlson, E.D. (1982). Building effective Decision Support Systems. Englewood Cliffs, N.J.:Prentice-Hall, Inc.

Stocker, J., C., Prisloe, A.S. & Civco, D. (1999). Putting Geospatial Information into the Hands of the Real Natural Resource Managers. *Proceedings of the 1999 ASPRS Annual Convention*, Portland, Oregon. 1070-1076.

U.S. Congress. (1972). Federal Water Pollution Control Act (33 U.S.C. §1251 et seq. Accessed September 18, 2010. Available from http://epw.senate.gov/water.pdf

Well House Consultants. (2006). Tk - Laying Out Your GUI with Frames, Pack And Grid. Accessed Feb. 1, 2011. Available from http://www.wellho.net/mouth/787_Tk-laying-out-your-GUI-with-frames-pack-and-grid.html

Wilkerson, G W., McAnally, W.H., Martin, J.L. Ballweber, J.A., Pevey, K.C., Diaz-Ramirez, J. & Moore, A. (2010). Latis: a Spatial Decision Support System to Assess Low-Impact Site Development Strategies." *Advances in Civil Engineering* Volume 2010. Accessed March 18, 2011. Available from http://www.hindawi.com/journals/ace/2010/810402/

2

Water Quality Monitoring and Associated Distributed Measurement Systems: An Overview

Octavian Postolache[1,2], Pedro Silva Girão[2] and José Miguel Dias Pereira[1,2]
[1]*ESTSetúbal-LabIM/IPS, Setúbal,*
[2]*Instituto de Telecomunicações/Instituto Superior Técnico, Lisboa,*
Portugal

1. Introduction

Water is essential to life, as we know it. However, statistics reveal that, in 2000, one billion people lacked access to safe drinking water and 2.4 billion to adequate sanitation. To achieve United Nations target of reducing by half the proportion of people without sustainable access to safe drinking water by 2015, an additional 1.5 billion people would require access to some form of improved water supply by 2015, that is an additional 100 million people each year (or 274,000/day) until 2015.

Because water sources are limited, it is of paramount importance to keep its quality at the highest level possible. Threats to water are manifold, from industry to natural phenomena, and water quality assurance is a basic environmental issue involving from political to technical aspects and options, but it is obvious that no assessment of water quality is possible without a quantitative identification of some characteristics, a process commonly called water quality monitoring.

This chapter is an overview on water quality and on its monitoring. The text reflects the experience of the authors on the subject, presents some research and development results they obtained in the last decade and includes data gathered from different sources, namely from USEPA reports and North Caroline State University Water Quality Group documents. The text includes remarks about measuring techniques for different water quality parameters that result from the experience acquired by the authors in the implementation of several water quality measuring units. The last part of the chapter proposes architectures and intelligent signal processing techniques for distributed water quality monitoring networks.

2. Water quality

Water quality is commonly defined by its physical, chemical, biological and aesthetic (appearance and smell) characteristics. Water may be used for drinking, irrigating crops and watering stock, industrial processes, production of fish, shellfish and crustaceans, wildlife habitats, protection of aquatic ecosystems, navigation and shipping, recreation (swimming, boating), and scientific study and education.

2.1 Factors influencing water quality

Water quality is closely linked to the surrounding environment and land use. Liquid water is never pure and is affected by agriculture, urban, industrial and recreation uses. The modification of natural stream flows and the weather can also have a major impact on water quality.

Groundwater is a major source of water and, when close to urban or industrial development, is vulnerable to contamination.

Generally, water quality of rivers is best in the headwaters, where rainfall is often abundant, declining as rivers flow through regions where land use and water use are intense and pollution from intensive agriculture, large towns, industry and recreation areas increases. There are of course exceptions to the rule and water quality may improve downstream, behind dams and weirs, at points where tributaries or better quality groundwater enter the mainstream, and in wetlands.

Rivers frequently act as conduits for pollutants by collecting and carrying wastewater from catchments and, ultimately, discharging it into the ocean. Storm water, which can also be rich in nutrients, organic matter and pollutants, finds its way into rivers and oceans mostly via the storm water drain network.

2.2 Water quality and ecosystems

An ecosystem is a community of organisms - plants, animals, fungi and bacteria - interacting with one another and with the environment in which they live. Protecting aquatic ecosystems is in many ways as important as maintaining water quality, for the following reasons:

- Aquatic ecosystems are an integral part of our environment. They need to be maintained if the environment is to continue to support people. World conservation strategies stress the importance of maintaining healthy ecosystems and genetic diversity.
- Aquatic ecosystems play an important role in maintaining water quality and are a valuable indicator of water quality and the suitability of the water for other uses.
- Aquatic ecosystems are valuable resources. Aquatic life is a major source of protein for humans. In most countries, like Portugal, commercial and sport fishing is economically important.

2.3 Water quality assessment

The presence of contaminants and the characteristics of water are used to indicate the quality of water. These water quality indicators can be categorized as:

Biological: algae, bacteria

Physical: temperature, turbidity and clarity, color, salinity, suspended solids, dissolved solids, sediment

Chemical: pH, dissolved oxygen, biological oxygen demand, nutrients (including nitrogen and phosphorus), organic and inorganic compounds (including toxicants)

Aesthetic: odors, taints, color, floating matter

Radioactive: alpha, beta and gamma radiation emitters.

Measurements of these indicators can be used to determine and monitor changes in water quality and to determine whether the quality of the water is suitable for the health of the natural environment and the uses for which the water is required.

The design of water quality monitoring systems is a complex and specialized field. The range of indicators that can be measured is wide and other indicators may be adopted in the future. The cost of a monitoring system to assess them all would be prohibitive, so resources are usually directed towards assessing contaminants that are important for the local environment or for a specific use of the water.

The paragraphs that follow detail several aspects of these quantities, algae, bacteria and radiations excluded. The paper includes a short reference to systems for on-line, in-situ water quality monitoring and ends with a list of references.

3. Water quality parameters and measuring techniques

3.1 Temperature

Temperature is an important water parameter because it is an influence quantity for the generality of other water parameters and also because it determines many physical characteristics of a water body. In the winter, water's temperature-dependent density allows aquatic life to survive. Ice is formed at 0 °C and thus remains at the top of the water body. Sun shining through the ice will serve to warm the water below slightly, keeping the temperature just above freezing. Water at 4 °C is the densest, and will sink to the bottom and be replaced by lighter 1 - 3.9 °C water. The continual process of heating and sinking keeps the water body from freezing entirely [1].

In addition, temperate lakes stratify during the summer because of water's temperature-dependent density. Stratification prevents the mixing of oxygen and nutrients in the water body, and often encourages dissolved oxygen depletion. During the spring, stratification will break down allowing mixing of oxygen and nutrients. During the fall, the water body loses heat until its temperature is uniform at 4 °C. Wind creates circulation, which distributes oxygen and nutrients throughout the water body (fall overturn). Eventually, the surface water layer falls below 4 °C, becomes less dense, and remains at the surface. Ice will form if temperatures are low enough; otherwise, this upper layer will remain just above 0 °C. Deeper water will remain roughly at 4 °C until spring [1].

Higher temperatures often exacerbate low dissolved oxygen level problems in lakes and reservoirs. High temperatures encourage the microbial breakdown of organic matter, a process that requires dissolved oxygen. Unfortunately, warm water naturally holds less dissolved oxygen. Thus, persistent warm conditions may lead to a depletion of dissolved oxygen in the water body.

3.1.1 Measuring techniques

3.1.1.1 Temperature probes

Temperature range is usually from 0 to 30 °C. Thus thermistor, platinum or even electronic based probes are adequate. Some manufacturers, like Quanta, commercializes probes that can measure several water parameters, including temperature (multi-parameter probes).

3.2 Turbidity

Turbidity is a quantity quantifying the degree to which light traveling through a water column is scattered by the suspended organic (including algae) and inorganic particles. Light scattering increases with the quantity of solids suspended in water. According to the research work developed by Campbell Scientific, usually the values of turbidity are correlated with the suspended solids concentration -SCC (Fig. 1); however, cases are also reported where no correlation between these two quantities is registered. Turbidity is commonly measured in Nephelometric Turbidity Units (NTU).

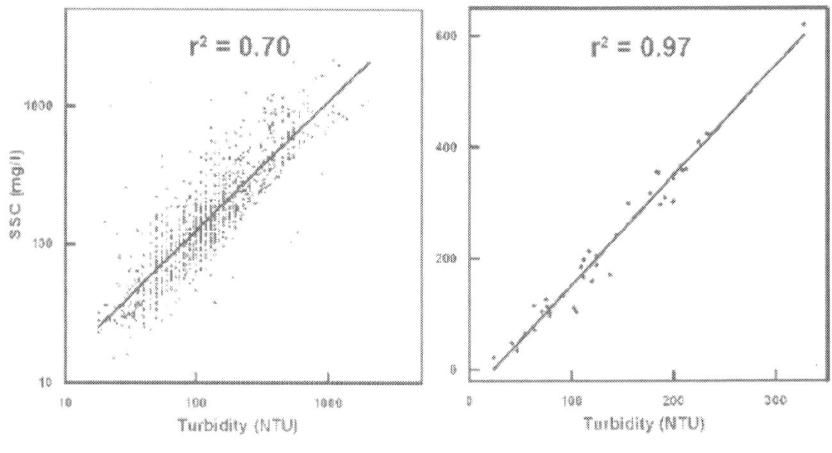

Fig. 1. The graph on the left provides measurements of runoff from a freeway, which indicates a bad correlation between SSC and turbidity. The graph on the right provides measurements from San Francisco Bay that indicates a good correlation between SSC and turbidity (Campbell Scientific document)

The velocity of the water resource largely determines the composition of the suspended load. Suspended loads are carried in both the gentle currents of lentic (lake) waters and the fast currents of lotic (flowing) waters. Even in flowing waters, the suspended load usually consists of grains less than 0.5 mm in diameter (Table 1). Suspended loads in lentic waters usually consist of the smallest sediment fractions, such as silt and clay [2].

Sediment class	Size (mm)
Sand	
V. Coarse	1.5
Medium	0.375
V. Fine	0.094
Silt	
V. Coarse	0.047
Medium	0.0117 (not visible to the human eye)
V. Fine	0.0049
Clay	< 0.00195

Table 1. Size classification of sediments (adapted from [3]).

3.2.1 Numerical categories

Designated Use	**Acceptable Ranges**
Recreation	5 NTU [4]
Aquatic Life	< 50 NTU instantaneously or
	< 25 NTU for a 10 day average [5]
	< 10 NTU for trout waters or
	< 25 NTU for streams (non-trout waters) or
	< 50 NTU for lakes and reservoirs (non-trout waters) [6]

Human Consumption 1 to 5 NTU (up to 5 NTU is allowed if the water supplier can demonstrate that this level does not interfere with:

1. disinfection
2. maintenance of a disinfecting agent
3. microbiological determination [7]

Turbidity may be due to organic and/or inorganic constituents. Organic particulates may harbor micro organisms. Thus, turbid conditions may increase the possibility for waterborne disease. Inorganic constituents have no notable health effects.

If turbidity is largely due to organic particles, dissolved oxygen depletion may occur in the water body. The excess nutrients available will encourage microbial breakdown, a process that requires dissolved oxygen. In addition, excess nutrients may result in algal growth. Although photosynthetic by day, algae respire at night, using valuable dissolved oxygen. Fish kills often result from extensive oxygen depletion.

3.2.2 Measuring techniques [8]

Nephelometric Method: Comparison of the light scattered by the sample and the light scattered by a reference solution [9].

- *Detection limits:* Should be able to detect turbidity differences of 0.02 NTU with a range of 0 to 100 NTU.

- *Interferences:* Rapidly settling coarse debris, dirty glassware, presence of air bubbles, and surface vibrations.

It is important to underline that turbidity is a measurement of the light scattering intensity relatively to the one that is obtained with the turbidity calibration standards. Visual clarity, measured as Secchi [10][11] or black disc visibility, is a direct measurement of the amount of suspended solids in water but its measurement requires more expensive equipments. Nevertheless, clarity measurements are more precise than turbidity measurements.

3.3 Salinity

The total dissolved solids (TDS) in water consist of inorganic salts and dissolved materials. In natural waters, salts are chemical compounds comprised of anions such as carbonates, chlorides, sulphates, and nitrates (primarily in ground water), and cations such as potassium (K), magnesium (Mg), calcium (Ca), and sodium (Na) [12]. In ambient conditions, these compounds are present in proportions that create a balanced solution. If there are additional inputs of dissolved solids to the system, the balance is altered and detrimental effects may be seen. Inputs include both natural and anthropogenic sources.

3.3.1 Numerical categories

Designated Use	(mg/l)
Human Consumption	500 TDS
	250 chloride
	250 sulphate
Irrigation	500-1 000 TDS (dependent upon crop sensitivity)
	250 mg/1 chloride
Industry *	
Brewing	
Light beer	500 TDS
Dark beer	1000 TDS
Pulp and paper	
Fine paper	200 TDS
Ground wood paper	500 TDS
Boiler feed water	50 to 3000 TDS depending on pressure
Canning/Freezing	850 TDS
Aquatic Life	Varies, depending on natural conditions

*Industry can de-ionize water to meet requirements; economics is the limiting factor [12].

Table 2. Salinity: designated use limits [13].

3.3.2 Total dissolved solids and conductivity

The presence in water of different anions and cations in different proportions leads to different values of water electric conductivity. However, even if it exists almost a linear relation between salinity and conductivity, that relation depends on the type of the dissolved salt. Moreover, conductivity is a non-selective measurement because instead of salinity, it gives the contribution of all charge carriers and not of a specific one. Notwithstanding, commonly, salinity is indirectly measured using conductivity meters.

As an example, Fig. 2 represents the experiment results of the relation that is obtained between TDS and conductivity for a variable amount of NaCl dissolved in water. In this case, the correlation coefficient between both variables is almost equal to 1, meaning that the relation between both variables is practically linear.

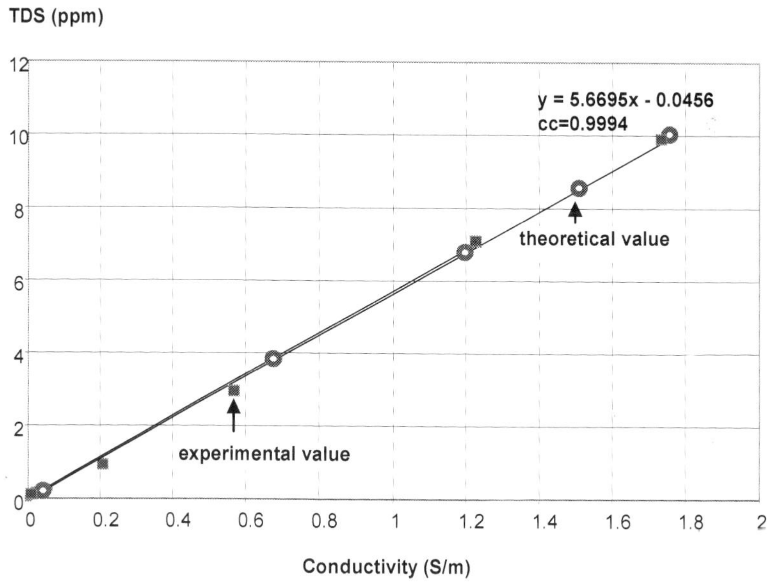

Fig. 2. Experiment results of the relation that is obtained between TDS and conductivity for different values of NaCl dissolved in water (square symbol: experimental data; circle symbol: theoretical data).

It is important to underline that the equation of the straight line that is represented in the graph, namely its slope, depends mainly on the type of salt and on the ionic activity of the elements that are dissolved in the water and on the temperature of the solution.

3.3.3 Measuring techniques [14][15]

1. **Electric Conductivity (EC):** Uses a conductivity bridge calibrated with standard seawater solution. The solution's ability to transmit electricity is facilitated by increasing

salt content. The EC is normally measured in mhos/cm, mmho/cm, or umho/cm (non-SI units) or siemens per meter (S/m) (SI units), depending on dissolved salt concentration.

The conversion of EC values into the total quantity of dissolved salts depends on the dissolved salts and manufacturers of this type of meter use conversion factors such that:

TDS (mg/l) = 50 to 70*EC (S/m) = 500 to 700*EC (mmho/cm).

Note that, in this context, 1 mg/l is approximately equal to 1 ppm and that these conversion factors are some ten times higher than the value obtained from Fig. 2.

A simple sensor for water (or any liquid) electrical conductivity measurement is the two electrode cell [16][17]. The current that flows between two electrodes immersed in the water, when a voltage is applied between them, is a function of all dissolved ionized solids in the solution. Theoretical results of conductivity, σ, are determined from the sensor measured conductance (1/R), using a geometric coefficient or "cell constant" (K_C) that reflects the ratio between the length (d) and the cross-section area (A) of the sampled water volume in which the electrical current actually flows:

$$\sigma = \frac{d}{A}\frac{1}{R} = K_C \frac{1}{R} \tag{1}$$

The cell and associated conditioning circuits are projected to reduce electrode-water interface contributions to the measurement. For instance, almost invariably, alternating current is used in order to minimize the polarization effects [18]. Notwithstanding, these effects cause a nonlinear variation of the cell constant K_C, that can only be compensated by calibration. Thus, it is not convenient to consider K_C as the geometric value of the ratio d/A for the cell, but the value obtained experimentally using solutions of accurate known conductivity.

For high conductance measurements, when coating and electrodes fouling are a concern, the four-electrode based conductivity measuring circuit, as the one presented in Fig. 3, is a more suitable solution [19]. Current is imposed across two drive electrodes and the other two sense electrodes are used for a null-current voltage measurement. Polarization at the drive electrodes has no effect on the measurement, provided the drive voltage is able to maintain the control current through the cell.

Another way to eliminate polarization effects and to obtain contact less measurements is to use an inductive sensing structure [20]. The basic principle consists in the presence of eddy currents induced in the interior of the body under test. These currents are due to the time variation of the magnetic flux originated by the primary coil current. The intensity of the induced currents is related to the electric conductivity of the salty water and there is a correlation between the electrical parameters that can be assessed and the medium conductivity to be determined. In Fig. 4 the simplest form of inductive conductivity sensor is depicted. The sensor is shunted with a capacitance C. It is a transformer whose secondary coil is the surrounding liquid. The primary current phasor, $\bar{I}$, is the sensor signal output.

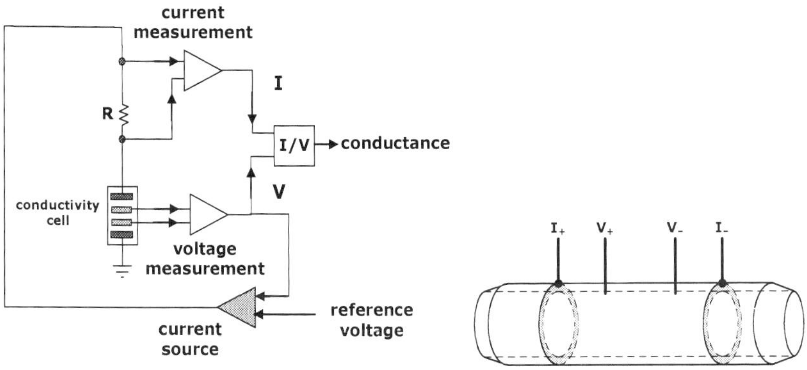

(a) (b)

Fig. 3. Four-electrode conductivity cell: (a) measuring circuit and (b) cell structure.

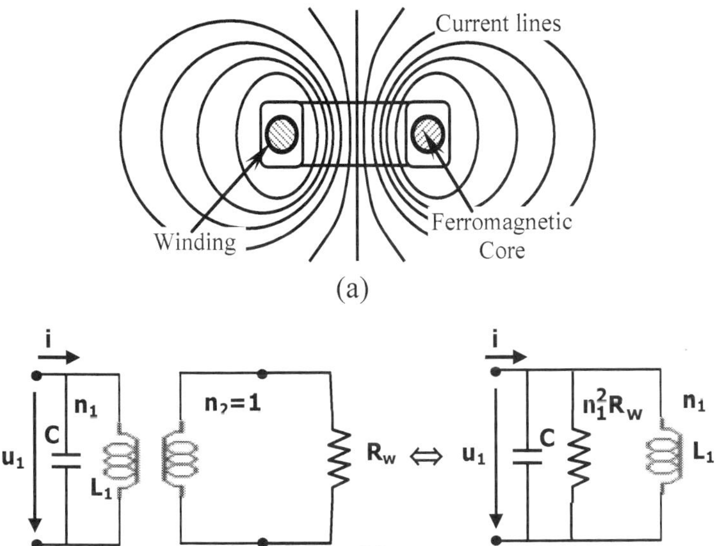

(b)

Fig. 4. (a) Inductive sensor with a single ferromagnetic core; (b) Equivalent electrical circuit (Rw- water resistance).

Based on the equivalent circuit of the sensor, presented in Fig.4 (b), the relationship between the applied voltage, u_1, and the sensor output current, i, is:

$$\bar{I} = \left(\frac{1}{n_1^2 R_w} + \frac{1}{j\omega L_{11}} + j\omega C \right) \overline{U}_1 \tag{2}$$

where n_1 is the number of windings, L_{11} the inductance of the coil, C is the capacitance of the shunt capacitor, $\omega=2\pi f$ and f the frequency. By choosing C so that $\omega C=1/\omega L$, the quadrature component of the current is null and the primary current phasor becomes an exact measure of R_W $\bar{I}=\left(\bar{U}_1 / n_1^2 R_W\right)$.

Finally, it is important to underline that conductivity depends on temperature, exhibiting a dependence coefficient approximately equal to 2 %/°C for salt waters. Hence, it is necessary to normalize the conductivity measurements to a reference temperature, usually equal to 25 °C.

2. **Density Method:** Uses a precise vibrating flow densimeter.
3. **Gravimetric Method** [8]: As an example, magnesium is measured using the gravimetric method. Diammonium hydrogen phosphate precipitates magnesium in ammonical solution as magnesium ammonium phosphate. The test can be performed two ways. First, the ammonium salts and oxalate can be destroyed, followed by precipitation of magnesium ammonium phosphate. Second, the diammonium hydrogen phosphate can undergo double precipitation without pre-treatment (preferable option). Sample is then dried and weighed.

   Interferences: Presence of aluminium, calcium, iron, manganese, silica, strontium, and suspended matter might interfere with test. Solution should not contain more than 3.5 g NH_4Cl.

3.4 Sediment

Sediment is composed of organic and inorganic particles of various sizes. The major classes of sediment, from largest to smallest, are boulders, cobbles, pebbles, sand, silt, and clay [3],

Class	Size (mm)	Approx size
Boulders	> 256	> Volleyball
Cobbles	> 64	> Tennis ball
Pebbles	> 2	> Match head
Sand		
V. Coarse	1.5	
Medium	0.375	
V.Fine	0.094	
Silt		
V. Coarse	0.047	
Medium	0.0117 (no longer visible to the human eye)	
V.Fine	0.0049	
Clay	< 0.00195	

(Adapted from [3]).

Table 3. Sediment classes.

Sediments are classified into four broad categories, according to their origin in relation to the basin of water in which they are deposited: extrabasinal, carbonaceous, pyroclastic, and intrabasinal.

1. **Extrabasinal (terrigenous) particles:** Terrigenous particles have been eroded from the land outside the body of water experiencing the deposition. The particles either retain their chemical make-up, or become chemically altered to clays and iron oxides.
2. **Carbonaceous particles:** These particles are organic in nature and are derived from either solid carbonaceous material (coal, amber, wax, and kerogen), reworked from other geologic formations, or from modern plant detritus.
3. **Pyroclastic particles:** These particles are derived during the explosive action of a volcano. Particles include rock fragment, single crystals, or bits of volcanic glass.
4. **Intrabasinal particles:** These particles grow biochemically or chemically in the waters experiencing deposition. These particles include carbonate biocrystals, silica biocrystals, particles composed of evaporated minerals, and minerals that grow at the water/sediment interface.

Sediment introduced into surface water is either deposited on the bed of the stream or lake or suspended in the water column (suspended load). Bed load is large sediment particles that move by bouncing along the bottom. Generally, the suspended loads in lotic (flowing) water consist of grains less than 0.5 mm in diameter [2]. Lentic (lake) suspended loads usually consist of the smallest sediment fractions, such as silt and clays [2]. A water body's suspended load is a component of the total turbidity. Any sediment transported by water is subject to deposition as flow velocity decreases [21].

The amount of sediment deposited on a rocky substrate can be quantitatively defined by an estimation of the percent embeddedness. The percent embeddedness is the degree to which fine sediments such as sand, silt, and clay fill the interstitial spaces between rocks on a substrate.

3.4.1 Numerical categories

Acceptable Ranges to Maintain Designated Use

Optimal Ranges	**Designated Use**
	Aquatic life [22]
< 25% embeddedness	Excellent Conditions
25 - 50% embeddedness	Good Conditions
50 - 75% embeddedness	Fair Conditions
> 75% embeddedness	Poor Conditions
	Industry (total solids) [4]
	Boiler Feedwater
500 - 3000 mg/l	0 - 150 psi
500 - 2500 mg/l	150 - 250 psi
100 - 1500 mg/l	250 - 400 psi
50 mg/l	> 400 psi
200 mg/l	Photographic Processing
200 mg/l	Clear Plastic Production
100 mg/l	Pulp Production

Table 4. Sediment: designated use and optimal ranges.

3.4.2 Measuring techniques

3.4.2.1 Total Suspended Solids (TSS) sampling technique

A number of different methods are available for sampling suspended sediment in streams. Two types of suspended sediment samplers are available for perennial streams: depth-integrating and point-integrating. Both samplers are usually made from cast aluminium or bronze and have a tail fin to orient the sampler's intake nozzle upstream. The depth-integrating sampler is designed to sample continuously as it is lowered at a constant speed from the water surface to the stream bed and back. The point-integrating sampler is equipped with a mechanism at the end of the sampler that can open and collect a sample at a specified depth in the stream [23].

Ephemeral streams are sampled using rising-stage samplers. These samplers have a number of bottles arranged on top of one another in a frame. Each bottle is equipped with kinked tubing pointed into the flow and will collect a sample as the water rises. This sampler is better suited for sampling silts and clays because the intake flow velocity is slower than the stream velocity and larger particles might settle before entering the sampler [23].

Many automated samplers are also available. They differ in cost, ease of maintenance, run time, and ability to extract a representative sample. The samplers extract samples by pumping water from the stream and retaining some in a sample bottle. The samplers can be set to extract samples at set intervals or can be attached to a float that will automatically trigger the sampling process when a pre-selected stream stage is reached [23].

Finally, suspended sediment concentration is often monitored with a turbidity meter. Although turbidity can be influenced by other factors such as size distribution, shape, and absorptivity of the sediment, and the color of the water, turbidity meters give satisfactory estimates [23].

3.4.2.2 Percent embeddedness technique

The amount of sediment deposited on a rocky substrate can be quantitatively defined by an estimation of the percent embeddedness. This method is only applicable on substrates of coarse pebbles, cobbles, or rubble. Percent embeddedness is measured at transect points of an area of known size (e.g. a square 0.5m X 0.5m). The percentages normally assigned to various substrate conditions are [24]:

- *100% embeddedness*= Rocks are completely surrounded by sediment and completely covered by sediment.
- *75% embeddedness*= Rocks are completely surrounded by sediment and half covered by sediment.
- *50% embeddedness*= Rocks are completely surrounded by sediment but are not covered by sediment.
- *25% embeddedness*= Rocks are half surrounded by sediment and are not covered by sediment.
- *0% embeddedness*= No fine sediments on substrate.

3.4.2.3 Assessment of capacity loss due to sedimentation

The rate of sediment-induced storage loss in lakes and reservoirs can be measured with a series of sedimentation or bathymetric surveys. Transects and sampling points are often established perpendicular to the main axis of tributary inflow or the main axis of the lake. Some base strata (original lake or reservoir bottom) should be established to track the rate (cm/yr) of deposition. A long pole may be used in shallow areas to measure sediment depth. For deeper areas, sonar or a SCUBA diver can record sediment depth measurements. Although measurements every year may not be needed, the same transects and stations should be monitored periodically. Major changes in land use, best management practices, or stream bank erosion that could increase sedimentation should also be monitored [25].

Sediment accumulation (deposition rates) can also be determined by measuring radionuclides in the sediment [26]. Cesium-137, a fallout product of nuclear testing, binds tightly to soil particles and can be used to estimate the time of sediment deposition. Measurable levels of Cesium-137 were introduced into the atmosphere during the beginning of the nuclear age and can only be used to estimate deposition after 1957.

Lead-210 may also be used to measure sedimentation rates [26]. Lead-210 is a naturally-occurring uranium isotope that is a decay product of radon. When atmospheric radon decays, the lead-210 is deposited on the earth's surface. Lead-210 will bind to soil particles and can be used to measure sedimentation rates for the past 100 years [26].

3.5 pH

pH is defined as the negative log-base 10 of the hydrogen ion activity:

$$pH = -\log_{10} [a_{H+}] \tag{3}$$

As long as the hydrogen ion concentration is not too high (less than 10^{-3} M), this activity is approximately proportional to the hydrogen ion concentration, more precisely:

$$[a_{H+}] = \gamma \cdot [H^+] \tag{4}$$

where γ represents the activity coefficient.

This coefficient is always lower than one and is almost equal to one for low hydrogen ion concentration.

The pH is a log-base 10 scale that measures acidity of a solution on a scale of 0 to 14. The pH of neutral solutions, such as pure water, is equal to 7. Alkaline solutions will have high pHs (8-14) and acidic solutions will have low pHs (1-6).

Since the pH is a log-base-10 scale, the pH changes 1 unit for every power of ten changes in [H+]. For example, water with a pH of 3 has 100 times the amount of [H+] that is found in a pH 5 water. Because pH = - $\log_{10}$ [H+], the pH will decrease as the [H+] increases [27].

3.5.1 Numerical categories

Although near-neutral pH values are preferred, industry as a whole can tolerate a wide pH range, depending on the intended water use. The Environmental Protection Agency reports that the widest pH range is between 3.0 and 11.7 for process waters, and between 5.0 and 8.9 for cooling waters. Specific industries will require more limited ranges. An industry can usually prepare water of the proper pH to meet its needs [12].

Optimal pH Ranges	Designated Use
6.0 -8.5	General Agriculture [28]
6.8 - 8.5	Dairy Sanitation
4.5 - 9.0	Irrigation water [12]
5.0 - 9.0	Human Consumption [28]
6.5 - 9.0	Freshwater aquatic life
6.5 - 8.5	Marine aquatic life [12]
	Industry [28]
> 8.0	Boiler Feedwater
6.5 - 7.0	Brewery
6.5 - 7.5	Cooling Water
> 7.5	Cannery
6.0 - 6.8	Laundering
> 7.0	Oil Well Flooding
7.8 - 8.3	Rayon Manufacturing
6.8 - 7.0	Steel Manufacturing
6.8 - 8.0	Tanning

Table 5. pH ranges for different designated use.

A reduction in pH (more acidic) may allow the release of toxic metals that would otherwise be absorbed to sediment and essentially removed from the water system. Once mobilized, these metals are available for uptake by organisms. For many metals, the rate of uptake is directly proportional to the levels of metal availability in the environment. Thus, a decrease in pH increases metal availability, lending itself to greater metal uptake by organisms. Metal uptake can cause extreme physiological damage to aquatic life [29].

An increase in pH may cause heightened ammonia concentrations [12]. At low pH, ammonia combines with water ($H2O$) to produce an ammonium ion ($NH4+$) and a hydroxide ion ($OH-$). The ammonium ion is non-toxic and not of concern to organisms. Above a pH of 9, ammonia (un-ionized) is the predominant species [30]. The un-ionized ammonia ($NH3$) is very toxic to organisms. Thus, organisms experience ammonia toxicity more readily at higher pH [31].

Experiments have shown that a pH decrease of 1.4 units of pH can disturb the aquatic community.

3.5.2 Measuring techniques [27]

1. **Electronic pH Meter:** A probe containing an acidic aqueous solution encased in a special glass membrane allows migration of hydrogen ions (H+). If the water has a pH different from that of the solution within the probe, an electric potential results.

   Usually, electrochemical sensors are used to measure pH. According to the Nernst equation, the voltage difference (E) that is obtained from these sensors is given by:

$$E = -\frac{2{,}3R(T + 273{,}15)}{nF}(7 - pH) \tag{5}$$

   where E represents the voltage difference between measuring and reference electrodes, T represents temperature in °C, F represents the Faraday constant and n represents the number of electrodes that participate in the REDOX reaction. This value is equal to one for the auto-ionization reaction of water:

$$2H_2O \leftrightarrow OH^- + H_3O^+ \tag{6}$$

   Fig. 5 represents a pH measuring diagram based on a glass membrane pH sensor and its equivalent electrical circuit. The electrical circuit contains multiple resistors and voltage sources that are associated with all resistive and polarization effects existing inside the measuring cell.

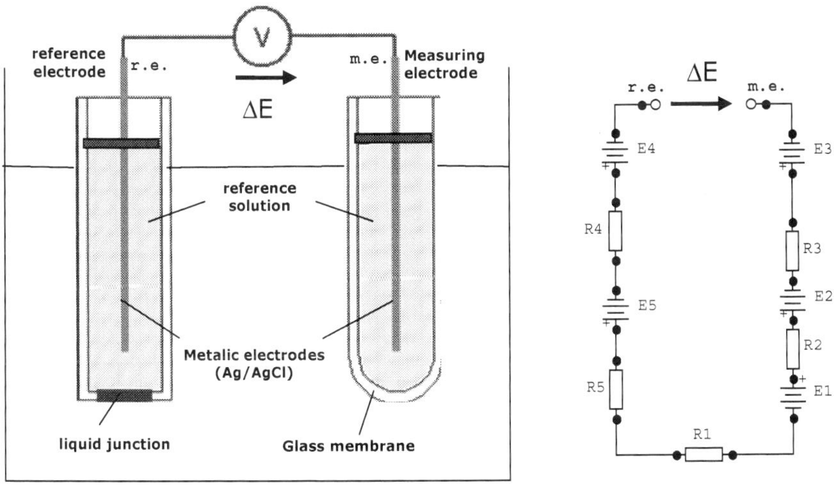

Fig. 5. Structure of a glass membrane pH sensor and its equivalent electrical circuit.

   Regarding temperature effects, pH measurements must be compensated for the explicit effect of temperature defined in Nernst equation and also for the variation of the ionic product of the water [32][33] with temperature that is approximately given by:

$$K_W = 10^{-14.00-0.0331(25-T)+0.00017(25-T)^2}\ mol^2.l^{-2} \tag{7}$$

where T represents the temperature in ° C.

Finally, it is important to underline that because the internal resistance (R_S) is very high, sometimes higher than 500 MΩ, an accurate measurement of the low voltage delivered by the pH transducer requires an amplifier with a very high input resistance. Usually the maximum bias currents of the amplifier must be lower than a few pA. As an example, Fig. 6 represents an equivalent electrical circuit of a signal conditioner that can be used for such a pH sensor.

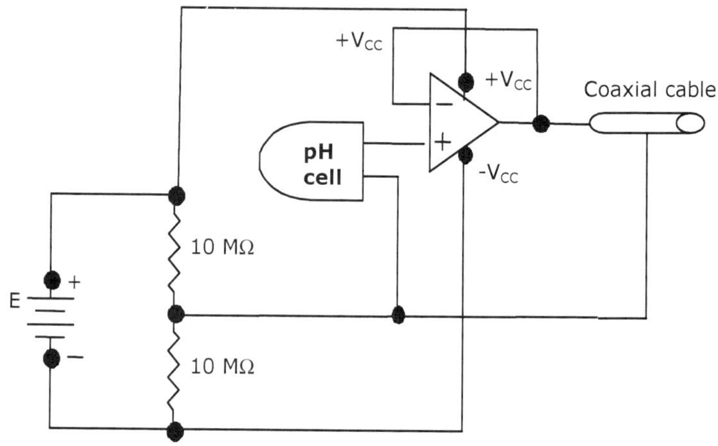

Fig. 6. Signal conditioning for a glass membrane pH sensor.

2. **Electronic devices:** a semiconductor device, usually a JFET or MOSFET, whose electrical characteristics change with the hydrogen ion concentration.

3.6 Dissolved oxygen

Dissolved oxygen (DO) refers to the volume of oxygen that is contained in water. Oxygen enters the water by photosynthesis of aquatic biota and by the transfer of oxygen across the air-water interface. The amount of oxygen that can be held by the water depends on the water temperature, salinity, and pressure. Gas solubility increases with decreasing temperature (colder water holds more oxygen). Gas solubility increases with decreasing salinity (freshwater holds more oxygen than does saltwater). Both the partial pressure and the degree of saturation of oxygen will change with altitude. Finally, gas solubility decreases as pressure decreases. Thus, the amount of oxygen absorbed in water decreases as altitude increases because the atmospheric pressure decreases [1].

Microbes play a key role in the loss of oxygen from surface waters because they use oxygen as energy to break down long-chained organic molecules into simpler, more stable end-products such as carbon dioxide, water, phosphate and nitrate [2]. If high levels of organic matter are present in water, microbes may use all available oxygen.

3.6.1 Numerical categories

Criteria to maintain designated use	
Designated Use	**Lowest acceptable DO levels (mg/l)***
Aquatic life	
Warm water fish	5.0
Cold water fish	6.0
Spawning season	7.0
Estuarine biota	5.0
Recreation	
Primary Contact	3.0
Secondary Contact	3.0
* Summary of state standards.	
Preferred ranges for designated use	
Designated Use	**Ranges (mg/l DO)**
Industry	
Boiler Feed Water	
High Pressure	0
Low Pressure	0.1 - 1.4

Table 6. Dissolved oxygen designated use and lowest acceptable levels.

3.6.2 Measuring techniques

1. **Iodometric Method:** Most reliable method. Requires the addition of divalent manganese solution, followed by a strong alkali, in a stoppered glass bottle. The dissolved oxygen oxidizes to manganous hydroxide precipitate. With the addition of iodide, the oxidized manganese reverts back to divalent state, releasing iodide equivalent to original dissolved oxygen content. Iodide is titrated with thiosulphate.
   - *Interferences:* Oxidizing agents may release iodines from iodides (false positive). Reducing agents may reduce iodine to iodide (false negative). Air entrapped in the sample bottle will also interfere.
2. **Azide Method:** This method is suitable for samples containing more than 50 ug/l nitrite and not more than 1 mg/l of ferrous iron.
   - *Interferences:* Oxidizing and reducing agents may provide false positives or negatives.
3. **Permanganate Modification:** This method is suitable for samples containing ferrous iron.
   - *Interferences:* High concentrations of ferric iron will interfere. Can be overcome by the addition of 1 ml KF.

4. **Alum Flocculation Modification:**
   - *Interferences:* High suspended solids may consume iodide in acid solution. This interference is removable by alum flocculation.
5. **Membrane Electrode Method:** This method is ideal for field testing. Not applicable for industrial and domestic wastewater. Polarographic or galvanic oxygen-sensitive membrane electrodes are composed of two metal electrodes in contact with a supporting electrolyte that is separated from the test solution by a selective membrane.
   - *Interferences:* The presence of hydrogen sulphide gas will desensitize the electrode cells.

3.7 Nitrogen

Nitrogen makes up 78% of the atmosphere as gaseous molecular nitrogen, but most plants can use it only in the fixed forms of nitrate and ammonium. Nitrate and nitrite are inorganic ions occurring naturally as part of the nitrogen cycle [1].

The nitrogen cycle is composed of four processes. Three of the processes - fixation, ammonification, and nitrification - convert gaseous nitrogen into usable chemical forms. The fourth process, denitrification, converts fixed nitrogen back to the unusable gaseous nitrogen state [1].

- **Nitrogen fixation** is the conversion of nitrogen in its gaseous state to ammonia or nitrate. Nitrate is the product of high-energy fixation of atmospheric nitrogen and oxygen. High-energy fixation accounts for little (10%) of the nitrate entering the nitrogen cycle. In contrast, biological fixation accounts for 90% of the fixed nitrogen in the cycle. In biological fixation, molecular nitrogen (N2) is split into two free N molecules. The N molecules combine with hydrogen (H) molecules to yield ammonia (NH3).
- **Ammonification** is a one-way reaction in which organisms break down amino acids and produce ammonia (NH3).
- **Nitrification** is the process in which ammonia is oxidized to nitrite and nitrate, yielding energy for decomposer organisms.
- **Denitrification** is the process in which nitrates are reduced to gaseous nitrogen. This process is used by facultative anaerobes.

In temperate zones, soil nitrate concentrations will vary seasonally with temperature and moisture levels. Fall and winter rains thoroughly remove all nitrates from the soil. During the spring and summer, the increased nitrogen-fixing activity of organisms and the addition of fertilizer cause the concentration of nitrates in the soil to steadily increase. Most of this nitrate is absorbed by plants. Thus, the removal of crops in the fall increases the chances for large flushes of nitrate from the soil to water bodies. Some leaching may occur in the spring if crops are not well- established enough to absorb the nitrogen [34].

3.7.1 Numerical categories

Water contaminated with nitrate is very difficult and costly to treat. Thus, if contamination affects a large water supply, the best alternative may be a new water source [35].

Generally, phosphorus is the limiting nutrient in freshwater aquatic systems. If all phosphorous is used, plant growth will cease, no matter the amount of nitrogen available. In contrast to freshwater, nitrogen is the primary limiting nutrient in the seaward portions of most estuarine systems [36]. Thus, nitrogen levels control the rate of primary production. If a nitrogen limited system is supplied with high levels of nitrogen, significant increases in phytoplankton (algae) and macrophyte (larger aquatic plants) production may occur. The recommended level of nitrogen in estuaries to avoid algal blooms is 0.1 to 1 mg/l, while the phosphorus concentration is 0.01 to 0.1 mg/l.

Limits suggested to maintain designated use:

Designated Use	Limit (mg/l)(AWWA 1990 [4])
Nitrate (NO3-N):	
Human Consumption	10.0
Aquatic Life	
Warm water fish	90.0
Industry	
Brewing	30.0
Nitrite (NO2-):	
Human Consumption	1.0
Aquatic Life	
Warm water fish	5.0
Nitrate+Nitrite:	
Human Consumption	10.0
Agriculture (Livestock etc.)	100.0
Aquatic life	
Estuaries (recommended)	
maximum diversity	0.1 (and phosphorus 0.01)
moderate diversity	1.0 (and phosphorus 0.1)

Table 7. Nitrogen designated use and corresponding limits.

3.7.2 Measuring techniques

A. Nitrate-Nitrogen [8]

1. **Ultraviolet Spectrophotometer Screening Method:**
   - *Detection limits:* Used to screen non-contaminated samples (low inorganic matter) to determine most suitable method.
   - *Interferences:* Dissolved organic matter, surfactants, NO2(-) and Cr(+).
2. **Ion Chromatography Method:**
   - *Detection limits:* 0.1 mg/l nitrate
   - *Interferences:* N/A
3. **Nitrate Electrode Method:**
   - *Detection limits:* NO3(-) ion activity between 0.00001 and 0.1 M (0.14 to 1400 mg/l)
   - *Interferences:* Chloride and bicarbonate, when their weight ratios to nitrate are >10, or >5, respectively.

4. **Cadmium Reduction Method:** Nitrate is reduced to nitrite in the presence of cadmium. The nitrite concentration is determined by diazotizing with sulphanilamide and coupling with NED dihydrochloride to form a colored azo dye that is measured colorimetrically.
   - *Detection limits:* 0.01 mg/l to 1.0 mg/l nitrate. Recommended especially for nitrate concentrations below 0.1 mg/l, when other methods lack sufficient sensitivity.
   - *Interferences:* Suspended matter in the column will restrict sample flow.
5. **Automated Cadmium Reduction Method:**
   - *Detection limits:* 0.5 mg/l to 10 mg/l nitrate.
   - *Interferences:* Turbidity, color
6. **Titanous Chloride Method:** Nitrate is determined potentiometrically using an NH3 gas-sensing electrode after nitrate is reduced to NH3 by a titanous chloride reagent.
   - *Detection limits:* 0.01 mg/l to 10 mg/l nitrate.
   - *Interferences:* NH3 and NO2(-), if present, are measured with NO3(-). Measure separately and subtract.
7. **Automated Hydrazine Reduction Method:** Nitrate is reduced to nitrite by hydrazine sulphate. The nitrite concentration is determined by diazotizing with sulphanilamide and coupling with NED dihydrochloride to form a colored azo dye that is measured calorimetrically.
   - *Detection limits:* 0.01 mg/l to 10 mg/l nitrate.
   - *Interferences:* Color, sulphide ion concentrations of less than 10 mg/l.

B. Total Kjeldahl Nitrogen [8][37]

1. **Digestion followed by distillation.**
2. **Automated Phenate Colorimetric Method:** Reaction produces indophenol, an intensely blue compound.
   - *Detection limits:* 0.05 mg/l to 2.0 mg/l.
   - *Interferences:* Iron and chromium ions tend to catalyze, while copper ions will inhibit the color reaction.

3.8 Phosphorus

Phosphorus (P) is an essential nutrient for all life forms. It plays a role in deoxyribonucleic acid (DNA), ribonucleic acid (RNA), adenosine diphosphate (ADP), and adenosine triphosphate (ATP). Phosphorus is required for these necessary components of life to occur.

Phosphorus does not exist in a gaseous state. Natural inorganic phosphorus deposits occur primarily as phosphate in the mineral apatite. Apatite is found in igneous and metamorphic rocks, and sedimentary rocks. When released into the environment, phosphates will speciate as orthophosphate according to the pH of the surrounding soil.

Phosphate is usually not readily available for uptake in soils. Phosphate is only freely soluble in acid solutions and under reducing conditions. In the soil it is rapidly immobilized as calcium or iron phosphates. Most of the phosphorus in soils is adsorbed to soil particles or incorporated into organic matter [1][38][39].

Phosphorus in freshwater and marine systems exists in either a particulate phase or a dissolved phase. Particulate matter includes living and dead plankton, precipitates of phosphorus, phosphorus adsorbed to particulates, and amorphous phosphorus. The dissolved phase includes inorganic phosphorus (generally in the soluble orthophosphate form), organic phosphorus excreted by organisms, and macromolecular colloidal phosphorus.

The organic and inorganic particulate and soluble forms of phosphorus undergo continuous transformations. The dissolved phosphorus (usually as orthophosphate) is assimilated by phytoplankton and altered to organic phosphorus. The phytoplankton is then ingested by detritivores or zooplankton. Over half of the organic phosphorus taken up by zooplankton is excreted as inorganic phosphorus. Continuing the cycle, the inorganic P is rapidly assimilated by phytoplankton [1][39].

The EPA water quality criteria state that phosphates should not exceed 0.05 mg/l if streams discharge into lakes or reservoirs, 0.025 mg/l within a lake or reservoir, and 0.1 mg/l in streams or flowing waters not discharging into lakes or reservoirs to control algal growth [12]. Surface waters that are maintained at 0.01 to 0.03 mg/l of total phosphorus tend to remain uncontaminated by algal blooms.

3.8.1 Numerical categories

Designated Use Limit

Freshwater		
Aesthetics		
Federal criteria		
	streams/rivers:	0.1 mg/l
	streams entering lakes:	0.05 mg/l
	lakes/reservoirs:	0.025 mg/l
Example. State criteria used:		
Reservoirs (CO)	chlorophyll a	15 ug/l
	Total P	0.035 mg/l
(Minn.)	Total P	0.015 mg/l
Impoundments (EPA Region 4)		
water supply Total P .015 mg/l		
aquatic life	Total P	0.025 mg/l
Lakes (NC)	chlorophyll a	40 ug/l
	Total P	0.05 mg/l
mountain lakes		0.02 mg/l
(VT)	Total P	0.014 mg/l
Estuaries (recommended)		
Aquatic life support	0.1 ug/l elemental phosphorus	
maximum diversity	0.01* total phosphorus (and nitrogen < 0.1) mg/l	
moderate diversity	0.1* (and nitrogen < 1.0) mg/l	

*These figures are recommended; eutrophication is also dependent on freshwater influx, nutrient cycling, dilution, and flushing of a pollutant load in a particular estuary [40].

Table 8. Phosphorus designated use limits.

Generally, phosphorus (as orthophosphate) is the limiting nutrient in freshwater aquatic systems; if all phosphorus is used, plant growth will cease, no matter how much nitrogen is available. The natural background levels of total phosphorus are generally less than 0.03 mg/l. The natural levels of orthophosphate usually range from 0.005 to 0.05 mg/l [2]. As mentioned before, in contrast to freshwater, nitrogen is generally the primary limiting nutrient in the seaward portions of estuarine systems. Systems may be phosphorus limited, however, or become so when nitrogen concentrations are high and N:P>16:1 [41].

3.8.2 Measuring techniques

Total Phosphorus and Orthophosphate: Analysis involves two procedural steps:

1. conversion of the phosphorus form into dissolved orthophosphate by a digestion method, and
2. colorimetric evaluation of the dissolved orthophosphate concentration [8].

Step 1. Digestion methods

1. **Perchloric Acid Digestion:** Recommended only for extremely difficult-to-analyze samples, such as sediments.
2. **Nitric Acid-Sulphuric Acid Method:** Recommended for most samples.
3. **Persulphate Oxidation Method:** This simple method should be cross-checked with one or more thorough techniques and adopted if results are identical.

Step 2. Colorimetric methods

1. **Ascorbic Acid Method:** Ammonium molybdate and potassium antimonyl tartrate react with orthophosphate to form a heteropoly acid that is reduced to molybdenum blue by ascorbic acid.
   - *Detection limits:* Ranges change with light path used.
   - *Interferences:* Arsenates react with the molybdate to form a similar blue color. Nitrite and hexavalent chromium interfere to yield results 3% less than actual at 1 mg/l and 10% to 15% less than actual at 10 mg/l.

Range (mg/l as P)	Path (cm)
0.3 - 2.0	0.5
0.15 - 1.3	1.0
0.01 - 0.25	5.0

Table 9. Ascorbic acid method detection limits as a function of light path length.

2. **Automated Ascorbic Acid Reduction Method:** Ammonium molybdate and potassium antimonyl tartrate react with orthophosphate in an acid medium to form an antimony-phosphomolybdate complex that forms a blue color suitable for photometric measurements when reduced by ascorbic acid.

- *Detection limits:* 0.001 to 10.0 mg/l as P when photometric measurements are performed at 600 to 650 nm in a 15mm tubular flow cell, or 880 nm in a 50mm tubular flow cell.
- *Interferences:* >50 mg/l $Fe(3+)$, 10 mg/l Cu, and 10 mg/l $SiO2$. Turbidity, color may interfere. Arsenate provides a positive interference.

3. **Vanadomolybdophosphoric Acid Colorimetric Method:** Ammonium molybdate reacts under acid conditions to form a heteropolyacid. In the presence of vanadium, yellow vanadomolybdophosphoric acid is formed, the intensity of which indicates the amount of orthophosphate present.
   - *Detection limits:* 1 to 20 mg/l as P. This method is not good for water samples - best for soils.
   - *Interferences:* Silica and arsenate interfere in heated samples. Blue color is formed by ferrous iron, but does not interfere if iron concentration is < 100 mg/l.
4. **Stannous Chloride Method:** Molybdophosphoric acid is formed and reduced by stannous chloride, forming intensely colored molybdenum blue.
   - *Detection limits:* 0.001 to 6 mg/l as P.
   - *Interferences:* Silica and arsenate interfere in heated samples. Blue color is formed by ferrous iron, but does not interfere if iron concentration is < 100 mg/l.

3.9 Heavy metals

Heavy metals are elements having atomic weights between 63.546 and 200.590, and a specific gravity greater than 4.0. Living organisms require trace amounts of some heavy metals, including cobalt, copper, iron, manganese, molybdenum, vanadium, strontium, and zinc. Excessive levels of essential metals, however, can be detrimental to the organism. Non-essential heavy metals of particular concern to surface water systems are cadmium, chromium, mercury, lead, arsenic, and antimony.

All heavy metals exist in surface waters in colloidal, particulate, and dissolved phases, although dissolved concentrations are generally low. The solubility of trace metals in surface waters is predominately controlled by the water pH, the type and concentration of ligands on which the metal could adsorb, and the oxidation state of the mineral components and the REDOX environment of the system [29].

The behaviour of metals in natural waters is a function of the substrate sediment composition, the suspended sediment composition, and the water chemistry. Sediment composed of fine sand and silt will generally have higher levels of adsorbed metal than will quartz, feldspar, and detrital carbonate-rich sediment. Metals also have a high affinity for humic acids, organo-clays, and oxides coated with organic matter [29].

Heavy metals in surface water systems can be from natural or anthropogenic sources. Currently, anthropogenic inputs of metals exceed natural inputs.

3.9.1 Numerical categories

Numeric aquatic life guideline criteria extracted from several US codes are:

Designated Use	Metal	Water Hardness (mg/l)	Limit (ug/l)
Fresh Water Aquatic Life			
	As		50.0 #
	As(III)		190.0 *
	As(III)		360.0
	Be		130.0
			5.3 *
	Cd	50	0.66 *
			1.80
		150	1.10 *
			3.90
		200	2.00 *
			8.60
	Cu	50	6.50 *
			9.20
		150	12.00 *
			18.00
		200	21.00 *
			34.00
	Hg		0.012 *
	Ni	50	56.00 +
		150	96.00 +
		200	160.00 +
	Pb	50	1.30 *
			34.00
		150	3.20 *
			82.00
		200	7.7
		200	7.70 *
			200.00
	Se		5.0 #
	Zn	50	180.00 ++
		150	320.00 ++
		200	570.00 ++
			47.00 +

* Four-day average concentration
One-hour average concentration
\+ Twenty-four hour average concentration
++ Level not to be exceeded at any time
(Adapted from [42][43][44]).
[45]

Table 10. Limits of heavy metals concentration in fresh water.

Designated Use	Metal	Water Hardness (mg/l)	Limit (ug/l)
Estuarine/ Coastal Aquatic Life		N/A	
	As		50.0 #
	Ag		0.10 #
	Cd		8.00 *
	Cu		2.90
	Hg		0.025 *
	Ni		7.10 +
	Pb		5.80 *
	Se		71.0 #
	Zn		76.6 *

* Four-day average concentration
One-hour average concentration
\+ Twenty-four hour average concentration
(Adapted from [42][43][44][46][47][48]).
\# [45]

Table 11. Limits of heavy metals concentration in estuarine and coastal water.

Designated Use	Metal	Water Hardness (mg/l)	Limit (ug/l)
Human Consumption	As		0.05 (mg/l)
	Ba		1.0 (mg/l)
	Cd		10.0 *
	Cr		0.05 (mg/l)
	Cu		1.0 ++
	Hg		144.0 ng/l *
	Ni		632.0 *
	Pb		50.0 * (adults)
	Zn		5.0 *

^ Ambient water criteria
Maximum contaminant level [42]
++ Level not to be exceeded at any time
(Adapted from [4][42][49]).

Table 12. Limits of heavy metals concentration in water for human consumption.

Designated Use	Metal	Water Hardness (mg/l)	Limit (ug/l)
Irrigation	Cd	N/A	10
	Cu		200
	Pb		5000
	Zn		2000

(Adapted from [49]).

Table 13. Limits of heavy metals concentration in irrigation water.

3.9.2 Criteria for the nine national toxics rule states - Dissolved metals criteria

Freshwater dissolved metals criteria that are not hardness dependent

Metal	Total Recoverable Metals Criteria		Dissolved Metals Criteria	
in ug/l	CMC	CCC	CMC	CCC
Arsenic	359.1	188.9	360	190
Chromium (VI)	15.74	10.80	15	10
Mercury	2.428	0.0122	2.1	N/A

CMC= Criteria Maximum Concentration
CCC= Criteria Continuous Concentration

Table 14. Freshwater dissolved metals criteria (not hardness dependent).

For Hardness-Dependent Freshwater Criteria see the Federal Register notice of May 4, 1995 (60 FR 22230). The hardness-dependent criteria require site specific calculations.

Saltwater dissolved metals criteria

Arsenic	68.55	36.05	69	36
Cadmium	42.54	9.345	42	9.3
Chromium (VI)	1079	49.86	1100	50
Copper	2.916	2.916	2.4	2.4
Lead	217.16	8.468	210	8.1
Mercury	2.062	0.0250	1.8	N/A
Nickel	74.6	8.293	74	8.2
Selenium	293.8	70.69	290	71
Silver	2.3	N/A	1.9	N/A
Zinc	95.1	86.14	90	81

Table 15. Saltwater dissolved metals criteria.

- Total recoverable metals criteria is from EPA National Ambient Water Quality Criteria Documents - Criteria Maximum Concentration (CMC) is the highest concentration of a pollutant to which aquatic life can be exposed for a short period of time (1 hour average) (acute); - Criteria Continuous Concentration (CCC) is the highest concentration of a pollutant to which aquatic life can be exposed for an extended period of time (4 days) without deleterious effects. - A more conservative approach to aquatic life protection may be preferred; in such cases the total recoverable metals criteria may be used.

Human Consumption

	Water and Organisms	Organisms Only
	ug/l	
Arsenic	0.018	0.14
Mercury	0.14	0.15
Nickel	610	4600

Table 16. Arsenic, mercury and nickel concentration in water for human consumption.

Ingestion of metals such as lead (Pb), cadmium (Cd), mercury (Hg), arsenic (As), barium (Ba), and chromium (Cr), may pose great risks to human health. Trace metals such as lead and cadmium will interfere with essential nutrients of similar appearance, such as calcium (Ca^{2+}) and zinc (Zn^{2+}).

Lead can substitute for calcium and included in bone. Lead that is stored in bone is not harmful, but if high levels of calcium are ingested later, the lead in the bone may be replaced by calcium and mobilized. Once free in the system, lead may cause nephrotoxicity, neurotoxicity, and hypertension. Cadmium may interfere with the metallothionein's ability to regulate zinc and copper concentrations in the body. Mercury poses a great risk to humans, especially in the form of methyl mercury. When mercury enters water it is often transformed by micro-organisms into the toxic methyl mercury form. Chronic poisoning is usually a result of industrial exposure or a diet consisting of contaminated fish (mercury is the only metal that will bio accumulate). Arsenic ingestion can cause severe toxicity through ingestion of contaminated food and water. Ingestion causes vomiting, diarrhea, and cardiac abnormalities. The presence of abundant chromium anions in the water is generally a result of industrial waste. The chronic adverse health effects are respiratory and dermatologic.

3.9.3 Measuring techniques

Total Metals: Includes all metals, organically and inorganically bound, both dissolved and particulate [8]. Most samples will require digestion before analysis to reduce organic matter interference and to convert metal to a form that can be analyzed by Atomic Absorption Spectroscopy or Inductively Coupled Plasma Spectroscopy.

General methods:

1. **Direct Atomic Absorption Spectroscopy or Inductively Coupled Plasma Spectroscopy:** Sample must be colorless, transparent, odorless, single phase, and have turbidity <1 Nephelometric Turbidity Unit. Otherwise, sample must first be digested.
2. **Voltammetry:** It is an electrochemical measuring technique where metal ion concentration is evaluated through its REDOX properties. The electrons transfer present in every metal oxidation or reduction reaction originate currents at given potential values that can be used to identify individual metals present in a solution and the correspondent concentrations. Voltammetry is a very sensitive and accurate analysis technique that is particularly suited for the detection of low level concentration of heavy metals (HM) in the lower parts per billion (ppb) region [50][51].
3. **Dedicated solutions:**
   Fig. 7 represents the data acquisition system and the schematic diagram of an electrical circuit that can be used for HM measurements [52][53]. The circuit includes the following main elements: a three electrodes measuring cell (MC), a data acquisition unit (DAQu), and an inverter sum amplifier (OA1) whose input voltage signals come from the DAQu (VDAC) and from a voltage follower circuit (V02). It is important to underline that the linear working condition of OA1 is assured by the negative feedback loop that is closed through the electrolyte solution contained in the MC.

Basically, after a pre-concentration phase, a positive potential, with linear variation, is applied to the counter electrode of the MC and current peaks associated with the REDOX potential of each metal dissolved in the solution are generated and detected by the current to voltage converter that is connected to the working electrode of the MC. Before the measurement the oxygen must be removed from the solution in order to avoid large measurement errors [54].

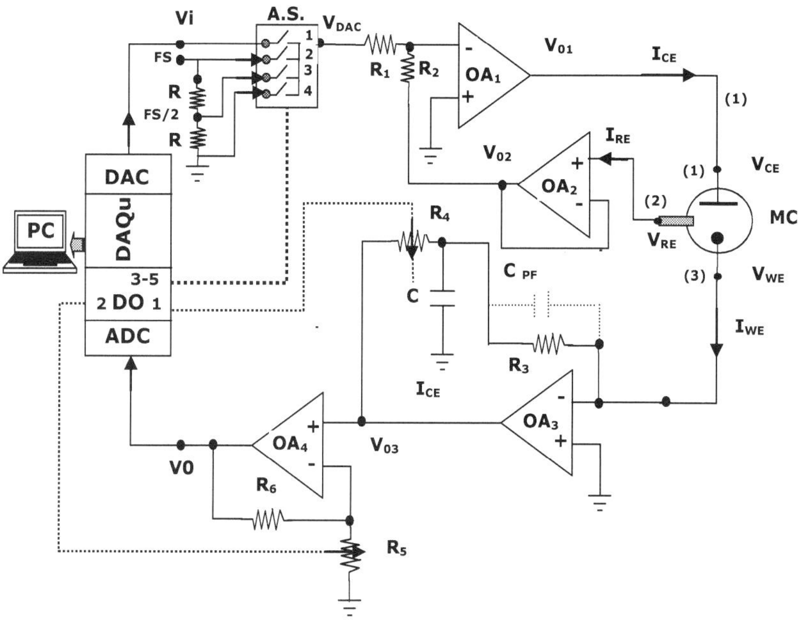

Fig. 7. Data acquisition system and the schematic diagram of an electrical circuit that can be used for HM measurements (MC-measuring cell, DAQu- data acquisition unit, ADC- analog to digital converter, DAC- digital to analog converter, OA- operational amplifier, AS- analog switch).

Digestion methods:

- **Nitric Acid Digestion:** Digestion is complete when solution is clear or light-colored.
- **Nitric Acid - Hydrochloric Acid Digestion:** Digestion is complete when digestate is light in color.
- **Nitric Acid - Sulphuric Acid Digestion:** Digestion is complete when solution is clear.
- **Nitric Acid - Perchloric Acid Digestion:** Digestion is complete when solution is clear and white HClO4 fumes appear.
- **Nitric Acid - Perchloric Acid - Hydrofluoric Acid Digestion:** Digestion is complete when solution is clear and white HClO4 fumes appear.

4. Distributed measuring systems architectures

In paragraph 3 the principal quantities commonly used to evaluate water quality were reviewed and for each quantity, the measuring techniques for its measurement briefly mentioned. To continuous measure the water quality parameters for extended areas, distributed measurement system architectures are usually used [55][56] rather than individual instruments. Generally, a distributed measurement system for water quality monitoring includes a main unit - typically a base unit that can store a server application and also additional software for data processing, and data presentation - and field nodes each with the capability of measuring quantities of water quality. The sensor for each quantity should be able to translate the quantity to the electric domain because the nodes must have several capabilities only possible if data is supported on digital format electric signals. Contrary to the main unit that is land based, most nodes should be placed on the water body to be monitored, which leads to the following considerations:

1. Node's equipment must be placed in a raft or buoy;
2. Even if anchored, the raft or buoy should have geographical position tracking capabilities (e.g. GPS) both for an accurate identification of its localization and to detect when anchorage is lost and the buoy or raft drifts away;
3. The equipment must be rugged and contained in a waterproof enclosure;
4. Node's equipment must include the hardware required to immerge the sensors in the water or to take samples of the water into a container accessible to the sensors;
5. Long term, unattended operation requires provision for periodic calibration of at least some of the sensors. The hardware and the procedure must be implemented within the node's hardware. The authors overcome 4 and 5 by choosing the type of solution depicted in Fig. 8, taken from [57] and also described in [58].
6. Electric power consumption of the whole equipment must be carefully planned. Components design should aim at minimum power consumption and high power capacity of the powering batteries should also be an objective;
7. Long term, unattended operation requires provision for re-chargeable batteries and for means for its recharge (e.g. solar panels or power harvesting solutions);
8. Power considerations also impacts on the management of the network, namely on nodes' measuring rates, on periods when the nodes are in sleeping mode, and on how data goes from each node to the main unit; Issues 6, 7 and 8 were addressed in several publications of the authors included in the references.
9. A decision concerning node's data processing capabilities must be taken beforehand. Each node must have a processing unit embedded but the characteristics and performance of that unit and of its interfaces depend on the algorithms and data they act upon, that is to say, on how distributed the system is in terms of data processing. Network nodes are thus designed as smart sensing nodes. Solutions based on FPGA [59], SDI-12 [60], and FieldPoint [61] technologies have been reported by the authors.
10. The characteristics of the network, namely the area covered by the network and its hosting medium, advise the use of wireless communications. This option means not only that nodes must have the hardware able to implement it, but also that all problems related to wireless data transmission, including data flow, data integrity, data security, access control, and protocol must be addressed before implementation of each node and of the whole network.

The references include several publications of the authors describing topologies and discussing technologies both for node implementation and for wireless data transmission. Regarding distributed systems practical implementations, different architectures were also developed by the authors and already reported in [9] and [57][58][62-96]. The main types of architectures designed and implemented are based on real time sensing nodes and are presented in Fig. 8 and Fig. 9.

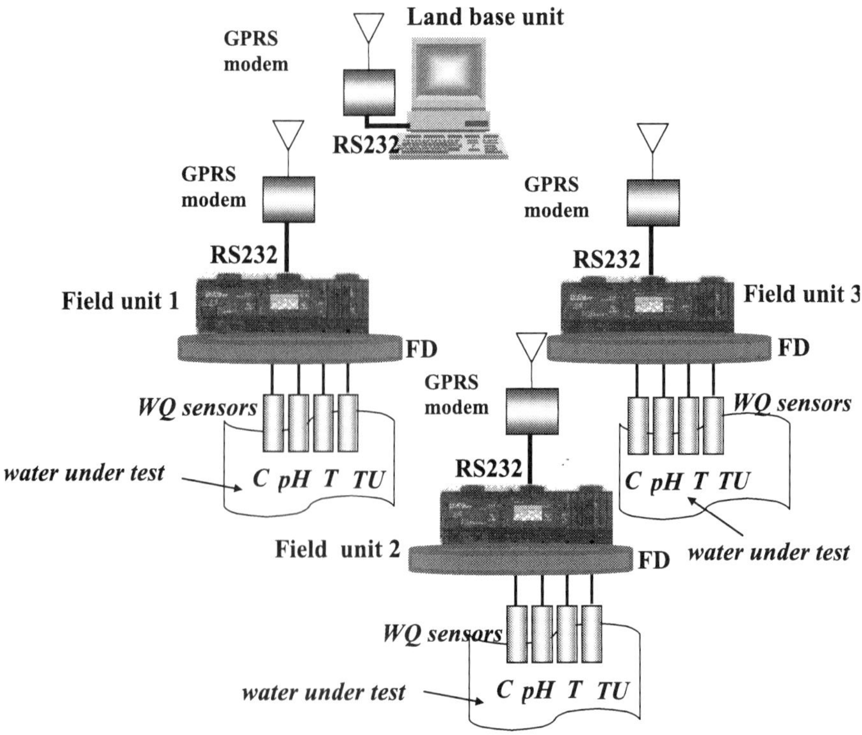

Fig. 8. Water quality (WQ) monitoring architecture based on FieldPoint embedded systems and GPRS modems.

FieldPoint based nodes are presented in Fig. 8 that depicts a wireless local area network (WLAN) with nodes with 3G communication capability and a land based unit expressed by a PC provided with interned connection. Fig. 9 presents an architecture that includes real time water quality monitoring nodes Wi-Fi compatible that are part of a WLAN including a base unit mounted in a ship. The values from the water quality monitoring nodes are uploaded through a 3G/UMTS connection in a web based information system that provides information about the water quality of the monitored area.

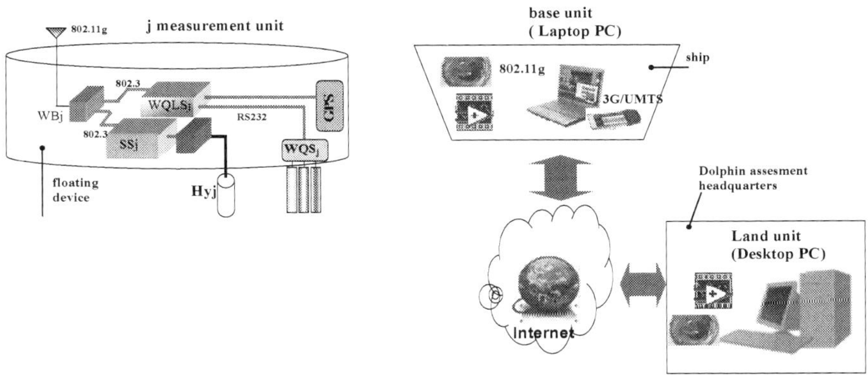

Fig. 9. The WLAN with web based WQ information system architecture (SSj - underwater acoustic signal server, WQLSj - water quality parameter and localization server, WQSj - water quality sensing unit, Hyj- hydrophone, GPS - global positioning system unit, 3G/UMTS - mobile Internet modem, WBj - Ethernet bridge).

As can be observed for the presented architectures, the land-based unit, which must be also compatible with the wireless data transmission solution, is mainly in charge of processing data received from the measuring nodes and of organizing and presenting the information according to the purposes it aims, including web server functionality. The basic processing at node's level should yield the values of the measured quantities (water parameters) but it is common that the node's processing unit implements algorithms for purposes such as sensor's inverse modeling, sensor's characteristics linearization, correction of influence quantities, sensors auto-calibration, node's self-testing, etc. The main unit, usually with better processing resources and data presentation means, is also the interface of the network with the outside world. The type of information output by the whole wireless distributed measuring system assumes many possible forms, depending mainly on the use given to that information. Because the system covers a more or less wide area, it is normal to want to know the water parameters values in the points where the nodes are installed and eventually pollution events. The knowledge of parameter values and the tendency of the evolution of those values can be used not only to foresee pollution events, but also to detect pollution sources. Sensors' data fusion, data mining, artificial neural networks, and Kohonen maps are some of the techniques that can be and have been used and referenced here by the authors for those purposes.

The availability and access to the information is nowadays made easier through the internet. The implementation of data bases allows users with the appropriate privileges to explore stored data and to search for data satisfying specific criteria. A geographic information system (GIS) not only allows data presentation in a user's more friendly way, but also to use an adequate browser to view the data. The integration of a low cost GPS device in each node allows the implementation of a web based geographical information system. Work regarding GIS implementation associated with WQ monitoring is reported by the authors in [97][98].

The authors also lead a team that implemented a geographic information system (GIS - SonicQualSado) designed to monitor not only the evolution of the water quality in Sado Estuary, but also to acquire and process Sado's dolphins population emitted sounds and anthropogenic produced sounds in order to establish possible correlations between dolphins behavior and water quality parameters. Thus, the GIS was designed to receive the data from the base unit that is a part of the architecture presented on Fig. 9. More elements associated with hardware and software architecture of the system are represented in the UML deployment diagram presented in Fig. 10.

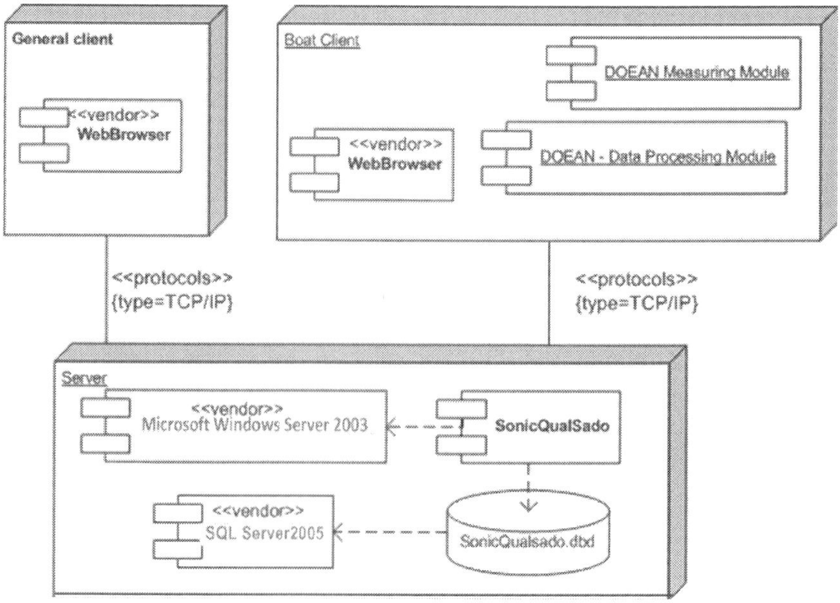

Fig. 10. GIS SonicQualSado software block diagram.

The number of the measured water parameters depends on the used measuring probe. Thus, using a low cost measuring probe, temperature, conductivity and turbidity are measured while with the utilization of a commercial multi-parameter measuring probe (Quanta Hidrolab) additional parameters, such as pH and dissolved oxygen, are also measured.

Concerning the GIS software implementation, the ASP.net was mainly used. Additionally, the Microsoft SQL Server software technology was used to develop the database and to perform the data management tasks. The GIS server implemented in the land computer uses the Microsoft Windows Server 2003 operating system. The GIS interface with the user is made through a browser (e.g. Internet Explorer). The user can verify through the developed SonicQualSado site [98] information about the latest field measurement on a digital map of the Sado Estuary. This data mapping functionalities were implemented using the Google Map APIs and .net software. The implemented GIS assures automatic data storage, map description of the measurement points, data searching, and data extraction for post-processing tasks. Some graphical elements of the implemented GIS are presented in Fig. 11. The figure presents an example of how data of one measurement point is shown to the user. In this example, the water quality measurements have also associated dolphins' pictures and sounds. The access to the pictures and to the sound reproduction is made through the picture tab and the sound tab, respectively.

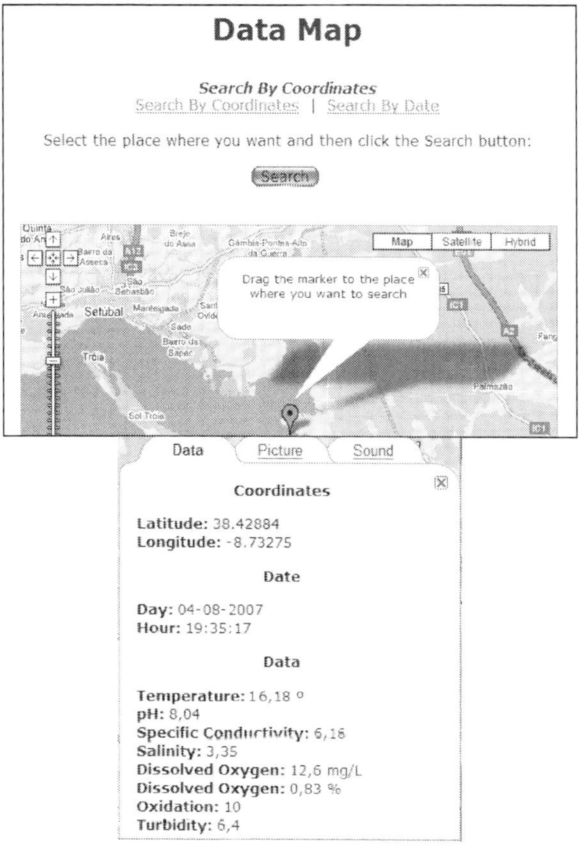

Fig. 11. The GIS SonicQualSado GUI and the data frame example associated with water quality and underwater acoustic sound monitoring.

The GIS uses a digital map as the main element of the user interface (GUI). The map is used both for visualization purposes and to entry data for searching tasks.

5. Conclusion

Water quality monitoring is a complex issue because there are a large number of parameters that can be used to access its quality. Understanding the meaning of each parameter requires a good knowledge of different areas, namely electrical, electrochemical and biological engineering, among others. Measuring techniques is still an open field for some variables, particularly in which concerns in-situ and on-line measurement systems. As referred in the introduction, this chapter contains an initial overview related with water quality parameters based on the information available from different entities accredited in the field. However, some remarks about measuring techniques for different water quality parameters were introduced by the authors according to their experience in the field. The last part of the paper focus different water quality monitoring architectures that were used by the authors for distributed sensing and intelligent signal processing network for water quality monitoring.

6. References

[1] R.L. Smith. Ecology and Field Biology. 4th ed. Harper Collins Publishers, NY., 1990.
[2] T. Dunne, and L.B. Leopold. Water in Environmental Planning. W.H. Freeman and Company, NY., 1978.
[3] G.M. Friedman, J.E. Sanders, and D.C. Kopaska-Merkel. Principle of Sedimentary Deposits. Macmillan Publishing Company, NY., 1992.
[4] American Water Works Association (AWWA). 1990. Water Quality and Treatment: A Handbook of Community Water Supplies. 4th ed. McGraw-Hill.
[5] G.W. Harvey. Technical Review of Sediment Criteria, for Consideration for Inclusion in Idaho Water Quality Standards. Idaho Dept. of Health and Welfare, Water Quality Bureau, Boise, ID., 1989.
[6] North Carolina Administrative Code Section T15A NCAC 02B .0100 & .0200, amended 2002. "RedBook" Surface Water and Wetlands Standards. North Carolina Department of Environment and Natural Resources, Division of Water Quality. Accessed: http://h2o.enr.state.nc.us/admin/rules/rb010102.pdf.
[7] North Carolina Administrative Code Section T15A.18C.0101-.0102. 1992. Rules Governing Public Water Systems. North Carolina Department of Environment, Health, and Natural Resources, Division of Environmental Health, Raleigh, NC.
[8] American Public Health Association (APHA). 1992. Standard Methods of Water and Wastewater. 18th ed. American Public Health Association, American Water Works Association, Water Environment Federation publication. APHA, Washington D.C.
[9] O. Postolache, P.M. Girão, J.M. Dias Pereira, H.G. Ramos. "Multibeam Optical System and Neural Processing for Turbidity Measurement", IEEE Sensors Journal, Vol. 7, No. 5, pp. 677 - 684, May, 2007.
[10] R.E. Carlson, "The Secchi disk and the volunteer monitor". LakeLine. 15(1): 28-29, 35-37, 1995.
[11] R.E. Carlson and J. Simpson, North American Lake Management Society, "A Coordinator's Guide to Volunteer Lake Monitoring Methods", 96 pp, 1996.
[12] USEPA. U.S. Environmental Protection Agency. 1986. *Ambient Water Quality Criteria for Bacteria - 1986*. EPA 440/5-84-002.
[13] J.H. Sherrard, D.R. Moore, and T.A. Dillaha. Total dissolved solids: determinations, sources, effects, and removal. Journal of Environmental Education 18(2):19-24, 1987.
[14] M. Sittig, Handbook of Toxic and Hazardous Chemicals. Noyes Publications. Park Ridge, N.J., 1981.
[15] D. Yaron. Salinity in Irrigation and Water Resources. Marcel Dekker, Inc., NY., 1981.

[16] Belá G. Lipták (editor), Instrument Engineers' Handbook - Process Measurement and Analysis, 4th edition, CRC Press/ISA, 2003.
[17] Thermo Orion, Conductivity Theory, The Technical Edge, 2001.
[18] J. Braunstein and G. D. Robbins, "Electrolytic Conductance Measurements and Capacitive Balance", in Journal of Chemical Education, Vol. 48, No. 1, pp. 52-59, January 1971.
[19] Pallás-Areny, Ramón and Webster, John G, Sensors and Signal Conditioning, John Wiley and Sons, 1991.
[20] K. Striggow and R. Dankert, "The exact Theory of Inductive Conductivity Sensors for Oceanographic Application", IEEE Journal of Oceanic Engineering, Vol. OE-10, No. 2, pp. 175-179, April 1985.
[21] J.M. McCabe, and C.L. Sandretto. Some Aquatic Impacts of Sediment, Nutrients, and Pesticides in Agricultural Runoff. Publication No. 201. Limnological Research Laboratory, Dept. of Fisheries and Wildlife, Michigan State University, 1985.
[22] J.L Plafkin, M.T. Barbour, K.D. Porter, S.K. Gross, and R.M. Hughes. Rapid Bioassessment Protocols for Use in Streams and Rivers: Benthic Macroinvertebrates and Fish. EPA-444/4-89-001. US Environmental Protection Agency. US Gov. Print. Office, Washington, D.C., 1989.
[23] N.D. Gordon, T.A. McMahon, and B.L. Finlayson. Stream Hydrology: An Introduction for Ecologists. John Wiley & Sons, West Sussex, England, 1992.
[24] T.D. Simonson, J. Lyon, and P.D. Kanehl. Guidelines for Evaluating Fish Habitat in Wisconsin Streams. US Department of Agriculture Technical report NC-164. St. Paul, MN., 1994.
[25] S.W. Coffey, J. Spooner, and M.D. Smolen. The Nonpoint Source Manager's Guide to Water Quality and Land Treatment Monitoring. NCSU Water Quality Group, Department of Biological and Agricultural Engineering, North Carolina State University, Raleigh, North Carolina, 1995.
[26] S.C. McIntyre, J.W. Naney, and J.R. McHenry. Quantification of lake sedimentation rates utilizing radioisotopes present in the environment. p. 1-5. In Environmental Protection Agency. Off-site Assessment. North American Lake Management Society and the U.S. Department of Agriculture Soil Conservation Service: National Off-site Assessment Workshop, St. Louis, Missouri. November 15, 1988. U.S. Gov. Print. Office, Washington, D.C., 1989.
[27] S.S. Zumdahl, Chemistry. 2nd ed. D.C. Heath and Company, Lexington, MA., 1989.
[28] C.P. Straub, Practical Handbook of Environmental Control. CRC Press, Inc., Boca Raton, FL., 1989
[29] D.W. Connell, G.J. Miller. Chemistry and Ecotoxicology of Pollution. John Wiley & Sons, NY., 1984.
[30] J.J. Morgan, and W. Stumm, Aquatic Chemistry. John Wiley & Sons, NY., 1981.
[31] National Research Council (NRC), Subcommittee on Ammonia. 1979. Ammonia. University Park Press, Baltimore, MD.
[32] Ville Voipio, Instrument for Measuring pH with Optical Indicator Thin Film - PhD Thesis, Technical University of Helsinki, Finland, 2001.
[33] Daniel C. Harris, Quantitative Chemical Analysis, 6th edition, W. H. Freeman, New York, 2003.

[34] A.M. Gower. Water Quality in Catchment Ecosystems. John Wiley & Sons: NY, New York., 1980.
[35] P.B. Kubek, and P.D. Robillard. Drinking Water Solutions 2.0: Computer Information System. Pennsylvania State University Department of Agricultural and Biological Engineering., University Park, PA., 1990.
[36] H.W. Paerl. “Emerging Role of Atmospheric Nitrogen Deposition in Coastal Eutrophication: Biogeochemical and Trophic Perspectives”. Can J. Fish. Aquat. Sci. Vol. 50:2254-2269, 1993.
[37] USEPA. U.S. Environmental Protection Agency. 1984. Test Methods: Technical Addition to Methods for Chemical Analysis of Water and Wastes (EPA-600/4-79-020). EPA Technical Manual EPA-600/4-84-017. U.S. Gov. Print. Office, Washington, D.C.
[38] J.R. Craig, D.J. Vaughan, and B.J. Skinner. Resources of the Earth. Prentice-Hall: Englewood Cliffs, NJ., 1988..
[39] H. Holtan, L. Kamp-Nielson, and A.O. Stuanes. Phosphorus in sediment, water, and soil: an overview. Hydrobiologia 170:19-34, 1988.
[40] NOAA/EPA. National Oceanic and Atmospheric Administration and Environmental Protection Agency. 1988. Strategic Assessment of Near Coastal Waters, Chapter 3, Susceptibility and Concentration Status of Northeast Estuaries to Nutrient Discharges. NOA A: Washington, D.C.
[41] N.A. Jaworski, O. Villa, Jr. A Suggested Approach for Developing Estuarine Water Quality Criteria for Management of Eutrophication. p. 499 in Estuaries and Nutrients, Nielsen, B.J., and L.E. Cronin (eds.) 1981. Humana Press: Clifton, NJ., 1981.
[42] USEPA. U.S. Environmental Protection Agency. 1987. Quality Criteria for Water. EPA Publication 440/5-86- 001. U.S. Gov. Prin. Office, Washington D.C.
[43] Georgia State Code Chapter 391-3-6. 1993. Rules and Regulations for Water Quality Control. Georgia Dept. of Natural Resources, Environmental Protection Division.
[44] Washington State Code Chapter 173-201A WAC. 1992. Water Quality Standards for Surface Waters of the State of Washington. State of Washington Department of Ecology.
[45] North Carolina Administrative Code Section T15A NCAC 02B .0100 & .0200, amended 2002. "RedBook" Surface Water and Wetlands Standards. North Carolina Department of Environment and Natural Resources, Division of Water Quality.
[46] Florida State Code Chapter 17-302. 1993. Surface Water Quality Standards. Florida Department of Environmental Regulation, Bureau of Surface Water Management.
[47] Texas State Code Chapter 307.2-307.10. 1991. Texas Surface Water Quality Standards. Texas Water Commission.
[48] North Carolina Administrative Code Section 15A NCAC 2B .0200. 1991. Classifications and Water Quality Standards Applicable to Surface Waters of North Carolina. North Carolina Department of Environment, Health, and Natural Resources, Division of Environmental Management, Raleigh, N.C.
[49] Kansas Administrative Code K.A.R. 28-16-28. 1994. Outline of Kansas Surface Water Quality Standards. Kansas Department of Health and Environment Bureau of Water, Topeka, KS.

[50] Kovaleva, V.P. Gladyshev et al., "Methods of Stripping Voltammetry in System of Monitoring Priority Ecotoxicants in Objects of An Environment", Proceedings of Korean International Symposium on Science and Technology", Vol.1, pp.186-187, Korea, 2005.

[51] M. Simion, E.E. Angelescu et al., "Micro and Nanoelectrode Voltammetric Measurements", Semiconductor Conference, Vol.1, pp.123-126, 2002.

[52] J.M. Dias Pereira, O. Postolache, P. Silva Girão, "A Smart and Portable Solution for Heavy Metals Concentration Measurements", 24th IEEE Instrumentation and Measurement Technology Conference (IMTC/2007), Warsaw, Polónia, May, 2007.

[53] J.M. Dias Pereira, O. Postolache, Ricardo Salgado, P. Silva Girão, "Voltammetry Based Automated Instrument for In-situ and Online Measurement of Heavy Metals Concentration in Water", 1st IMEKO TC19 Symposium on Measurement and Instrumentation for Environmental Monitoring, pp. 85-90, Iasi, Roménia, September, 2007.

[54] M. D. Pereira, O. Postolache, P. S. Girão, "Improving Celerity of Heavy Metals Measurements", International Instrumentation & Measurement Technology Conference 2010 (I2MTC'2010), pp. 1073-1077, Austin, May, 2010.

[55] O. Postolache, Helena Ramos, P. Girão, M. Pereira, "A Practical Approach on Water Quality Monitoring Based on Distributed Measurement System and Intelligent Signal Processing", MARTECH'2007, pp. 1-5, Barcelona, Spain, Nov., 2007.

[56] Pedro Girão, Octavian Postolache, Miguel Pereira, Helena Ramos, "Distributed Measurement Systems and Intelligent Processing for Water Quality Assessment", Sensors & Transducers Journal, Vol.38, Issue 12, pp. 74-85, Dec., 2003.

[57] O. Postolache, P.M. Girão, J.M. Dias Pereira, H.G. Ramos. "Auto Calibration of Stand-alone Filed Operating Sensors for Distributed Water Quality Monitoring Systems", Proc IMEKO TC4 Symp., Gdynia/Jurata Poland, Poland, Vol. Vol I, pp. 75 - 80, September, 2005.

[58] H.G. Ramos, O. Postolache, P.M. Girão, J.M. Dias Pereira. "Water Quality Sensors Calibration System Based on Reconfigurable FPGA Technology", Proc IMEKO World Congress, Rio de Janeiro, Brazil, Vol. 00, pp. 00 - 00, September, 2006.

[59] O. Postolache, J.M. Dias Pereira, P. Silva Girão, "Real-Time Sensing Channel Modelling Based on an FPGA and Real-Time Controller", 23th IEEE Instrumentation and Measurement Technology Conference (IMTC/2006), pp. 557-562, Sorrento, Itália, April, 2006.

[60] Helena Ramos, P. Silva Girão, O. Postolache, J.M. Dias Pereira, "Distributed Water Quality Measurement System Based on SDI-12", IEEE Africon'2004, Vol.1, pp. 57-60, Gaborone, Botswana, September, 2004.

[61] O. Postolache, P. Girão, M. Pereira, Helena Ramos, "Self Organizing Maps in a Remote Water Quality Monitoring System", 20th IEEE Instrumentation and Measurement Technology Conference (IMTC/2003), Vol.1, pp. 529-533, Colorado, USA, May,2003.

[62] O. Postolache, P.M. Girão, J.M. Dias Pereira. "Smart Sensors and Intelligent Signal Processing in Water Quality Monitoring Context", Proc International Conf. on Sensing Technology - ICST, Lecce, Italy, June, 2010.

[63] J.M. Dias Pereira, O. Postolache, P.M. Girão. "Improving Celerity of Heavy Metals Measurements", Proc IEEE International Instrumentation and Technology Conf. - I2MTC, Austin, United States, May, 2010.

[64] O. Postolache, P.M. Girão, J.M. Dias Pereira. "Mobile Solutions for Remote Environmental Monitoring Systems", Proc IMEKO TC19 Symp, Budapest, Hungary, Vol. I, pp. 1 - 5, September, 2008.

[65] H.G. Ramos, L.M. Gurriana, O. Postolache, J.M. Dias Pereira, P.M. Girão. "Development and Characterization of a Conductivity Cell for Water Quality Monitoring", Trans. on Systems, Signals and Devices, Vol. 3, No. 4, pp. 483 - 498, April, 2008.

[66] O. Postolache, P.M. Girão, J.M. Dias Pereira, H.G. Ramos. "Intelligent Processing of the Dynamic Response of Sensors for Water Quality Monitoring", Trans. on Systems, Signals and Devices, Vol. 3, No. 4, pp. 539 - 550, April, 2008.

[67] O. Postolache, H.G. Ramos, P.M. Girão, J.M. Dias Pereira. "A Practical Approach on Water Quality Monitoring Based on Distributed Measurement System and Intelligent Signal Processing", Proc UPC Workshop Internacional sobre Tecnologia Marina - 28) MARTECH 2007, Vilanova i la Geltru, Spain, Vol. I, pp. 1 - 5, November, 2007.

[68] O. Postolache, J.M. Dias Pereira, P.M. Girão. "Optical Fiber Based Turbidity Sensing System", Proc IMEKO TC19 Symp, Iasi, Romania, Vol. II, pp. 71 - 75, September, 2007.

[69] J.M. Dias Pereira, O. Postolache, P.M. Girão. "Voltammetry Based Automated Instrument for In-Situ and Online Measurement of Heavy Metals Concentration in Water", Proc IMEKO TC4 Symp., Iasi, Romania, Vol. 1, pp. 85 - 90, September, 2007.

[70] O. Postolache, P.M. Girão, J.M. Dias Pereira, H.G. Ramos. "Multibeam Optical System and Neural Processing for Turbidity Measurement", IEEE Sensors Journal, Vol. 7, No. 5, pp. 677 - 684, May, 2007.

[71] J.M. Dias Pereira, O. Postolache, P.M. Girão. "A Smart and Portable Solution for Heavy Metals Concentration Measurements ", Proc IEEE Instrumentation and Measurement Technology Conf., Warsaw, Poland, Vol. CD, pp. 100 - 105, May, 2007.

[72] O. Postolache, J.M. Dias Pereira, P.M. Girão, H.G. Ramos. "Smart Flexible Turbidity Sensing Based on Embedded Neural Network", Proc IEEE Sensors, Daegu, Korea, South, Vol. I, pp. 658 - 661, October, 2006.

[73] O. Postolache, J.M. Dias Pereira, P.M. Girão. "Embedded FPGA Solution for Water Quality Monitoring System", Proc International Conf. on Informatics in Control, Automation and Robotics - ICINCO, Setubal, Portugal, Vol. 1, pp. 154 - 160, August, 2006.

[74] J.M. Dias Pereira, O. Postolache, P.M. Girão. "A Low-Cost Tide Measurement System for Water Quality Assessment", Proc IEEE Instrumentation and Measurement Technology Conf., Sorrento, Italy, Vol. 1, pp. 2226 - 2230, April, 2006.

[75] O. Postolache, J.M. Dias Pereira, P.M. Girão. "Heavy Metal Concentration Measurement Based on Virtual Instrumentation", Proc Conf. on Telecommunications - ConfTele, Tomar, Portugal, Vol. I, pp. 5 - 9, April, 2005.

[76] H.G. Ramos, P.M. Girão, O. Postolache, J.M. Dias Pereira. "Distributed Water Quality Measurement System Based on SDI-12", Proc IEEE AFRICON Conf. in Africa, Gaborone, Botswana, Vol. 01, pp. 57 - 60, September, 2004.

[77] J.M. Dias Pereira, O. Postolache, P.M. Girão, H.G. Ramos. "Smart Oil and Conductivity Sensor for Water Quality Monitoring", Proc IMEKO TC4 Symp., Athens, Greece, Vol. 2, pp. 417 - 422, September, 2004.

[78] O. Postolache, J.M. Dias Pereira, P.M. Girão, H.G. Ramos. "A Temperature Compensated Conductivity Sensor with Auto-Calibration Capabilities", Proc MultiConf. on Systemics, Cybernetics and Informatics, Orlando, United States, Vol. XIII, pp. 74 - 79, July, 2004.

[79] O. Postolache, J.M. Dias Pereira, P.M. Girão, H.G. Ramos. "Matching Ion Selective Electrode Response Characteristics Using Fuzzy-Neural Network", Proc MultiConf. on Systemics, Cybernetics and Informatics, Orlando, United States, Vol. XV, pp. 135 - 139, July, 2004.

[80] O. Postolache, J.M. Dias Pereira, P.M. Girão, H.G. Ramos. "Water Quality Data Processing Using Fuzzy Neural Networks and Kohonen Self Organizing Maps", Proc IMEKO TC-7 Symp., St. Petersburg, Russia, Vol. 1, pp. 509 - 513, June, 2004.

[81] J.M. Dias Pereira, O. Postolache, P.M. Girão, H.G. Ramos. "SDI-12 Based Turbidity Measurement System With Field Calibration Capability", Proc IEEE Canadian Conf. on Electrical & Computer, Niagara Falls, Canada, Vol. IV, pp. 1975 - 1979, May, 2004.

[82] P.M. Girão, O. Postolache, J.M. Dias Pereira, H.G. Ramos. "Distributed Measurement Systems and Intelligent Processing for Water Quality Assessment", Sensors & Transducers Magazine, Vol. 38, No. 12, pp. 74 - 85, December, 2003.

[83] J.M. Dias Pereira, H.G. Ramos, O. Postolache, P.M. Girão. "Colored Light-to-Voltage Converters Based Absorbance Meter (invited paper)", Journal of Scientific Computing, Vol. 2, No. 1, pp. 58 - 64, October, 2003.

[84] O. Postolache, J.M. Dias Pereira, P.M. Girão, H.G. Ramos. "Smart Wireless System for Pollution of Rivers and Estuaries", Proc Conf. on Electronic Measurement & Instruments - ICEMI, Taiyuan, China, Vol. 2, pp. 1139 - 1143, August, 2003.

[85] O. Postolache, J.M. Dias Pereira, P.M. Girão, H.G. Ramos. "Increasing Ion Selective Electrode Performance Using Neural Networks", Journal of Scientific Computing, Vol. 2, No. 1, pp. 17 - 24, July, 2003.

[86] O. Postolache, P.M. Girão, J.M. Dias Pereira, H.G. Ramos. "A Practical Approach to Ion Selective Electrodes Performance in Static and Dynamic Conditions", Proc IMEKO World Congress, Dubrovnik, Croatia, Vol. 1, pp. 630 - 634, June, 2003.

[87] O. Postolache, P.M. Girão, J.M. Dias Pereira, H.G. Ramos. "Wireless Water Quality Monitoring System Based on FieldPoint Technology and Kohonen Maps", Proc IEEE Canadian Conf. on Electrical & Computer, Montreal, Canada, Vol. 3, pp. 1873 - 1876, May, 2003.

[88] O. Postolache, J.M. Dias Pereira, P.M. Girão. "A Distributed Hybrid Neural Network Based System for Water Quality Monitoring", Buletinul Institutului Politehnic din Iasi, Vol. XLVIII(LII), No. Fasc5, pp. 235 - 241, November, 2002.

[89] O. Postolache, P.M. Girão, J.M. Dias Pereira, H.G. Ramos. "Increasing Ion Selective Electrodes Performances Using Neural Networks", Proc IEEE Sensors for Industry Conf., Houston, United States, Vol. 1, pp. 127 - 132, November, 2002.

[90] P.M. Girão, O. Postolache, J.M. Dias Pereira, H.G. Ramos. "Accuracy Increase of Multi-Sensor Measuring Systems Using Sensors Data Fusion", Proc IMEKO TC4 Symp., Zagreb, Croatia, Vol. Part 2, pp. 442 - 447, September, 2002.

[91] C. Temneanu, O. Postolache, P.M. Girão, J.M. Dias Pereira. "Fuzzy Modeling of Measurement Data in Water Quality Assessment", Proc IMEKO TC4 Symp., Zagreb, Croatia, Vol. 2, pp. 396 - 476, September, 2002.

[92] O. Postolache, P.M. Girão, J.M. Dias Pereira, H.G. Ramos. "An Internet and Microcontroller Based Remote Operation Multisensor System for Water Quality Monitoring", Proc IEEE Sensors, Orlando, United States, Vol. 1, pp. 254 - 259, June, 2002.

[93] O. Postolache, J.M. Dias Pereira, P.M. Girão. "An Intelligent Turbidity and Temperature Sensing Unit for Water Quality Assesment", Proc IEEE Canadian Conf. on Electrical & Computer, Winnipeg, Canada, Vol. 1, pp. 494 - 499, May, 2002.

[94] O. Postolache, P.M. Girão, J.M. Dias Pereira, H.G. Ramos. "An IR Turbidity Sensor Design and Application", Proc IEEE Instrumentation and Measurement Technology Conf., Anchorage, United States, Vol. Vol 1, pp. 535 - 539, May, 2002.

[95] O. Postolache, P.M. Girão, H.G. Ramos. "A Neural Virtual Instrument for Turbidity Measurement", Proc Seminário Anual de Automática, Electrónica Industrial e Instrumentacion, Matanzas, Cuba, Vol. I, pp. 112 - 115, September, 2001.

[96] O. Postolache, J.M. Dias Pereira, P.M. Girão. "Application of Neural Structures in Water Quality Measurements", Proc IMEKO World Congress, Wien, Austria, Vol. IX, pp. 353 - 358, September, 2000.

[97] O. Postolache, P. S. Girão, G. Patricio, J. Sacramento, P. Macedo, M. D. Pereira, "Distributed Instrumentation and Geographic Information System for Dolphins' Environmental Assessment", International Instrumentation & Measurement Tecnology Conference 2008 (I2MTC'2008), Vol. I, pp. 1777-1782, Victoria, Canada, May, 2008.

[98] O. Postolache, P.S. Girão, J.A. Apolonia, N.B. Beirante, P.M. Macedo, J.M. Dias Pereira, J.M.; "Dolphins' Environment Assessment and Knowledge Management using a Distributed Instrumentation and Geographic Information System", Proc IMEKO TC4 Symp., Florence, Italy, Vol. I, pp. 454 - 459, September, 2008.

3

Analysis of Water Quality Data for Scientists

József Kovács[1], Péter Tanos[1], János Korponai[2],
Ilona Kovácsné Székely[3], Károly Gondár[4],
Katalin Gondár-Sőregi[4] and István Gábor Hatvani[1]*
*[1]Eötvös Loránd University,
Department of Physical and Applied Geology, Budapest,
[2]West Transdanubian Water Authority,
Department Kis-Balaton, Keszthely,
[3]Budapest Business School, Institute of Methodology, Budapest,
[4]Smaragd GSH Ltd., Budapest,
Hungary*

1. Introduction

In the last few decades the need for stochastic models and the use of time series and data analysis methods in surface and groundwater research has increased greatly. The reason behind this phenomenon is the increase in the size and number of datasets in which it has become necessary to investigate the connection between random variables, and in the case of time series, their characteristics.

The most often used models are deterministic, although they are prepared from one sampling event. It must be stated that the statistics and model results obtained from this sampling event can significantly change if the sampling is to be reproduced because their results are probability variables (Kovács & Székely, 2006). In the case of deterministic models this problem is solved by means of sensitivity analyses, thus the uncertainty in the applied model remains. This may be the reason why the following can be found in the international literature regarding this question: "The future is stochastic modeling" (Kovács & Szanyi, 2005; Wilkinson, 2006).

This chapter is intended to introduce the application of a few exploratory data analysis techniques, primarily through examples. Exploratory data analysis methods are useful and important tools for obtaining an overview of systems which can be described by many different parameters, for determining the latent and explicit connections between the parameters and for sorting and grouping the data obtained based on mathematics.

The greatest value of this chapter lies in its interdisciplinary character; it casts light on environmental problems originating from a wetland ecosystem a river and a groundwater system as well.

*Corresponding Author

2. Materials and methods

2.1 Data bases and data handling

2.1.1 Data base in four dimensions

Before describing in detail the methods applied, the general properties of the environmental datasets and their handling should be discussed. In most aspects of water research the events analyzed are described by different parameter samples obtained from more than one spatial sampling location. The location of a planar sampling site is determined by two spatial coordinates (x; y) forming the two dimensions, while the parameter types (e.g.: runoff, water level, calcium, height etc.) take place in the third dimension. A hydrogeological process is often described by status parameters sampled only once. If, however, samples are analyzed over time, one is dealing with time series and the three dimensions are extended to four, with time as the fourth axis (Fig. 1).

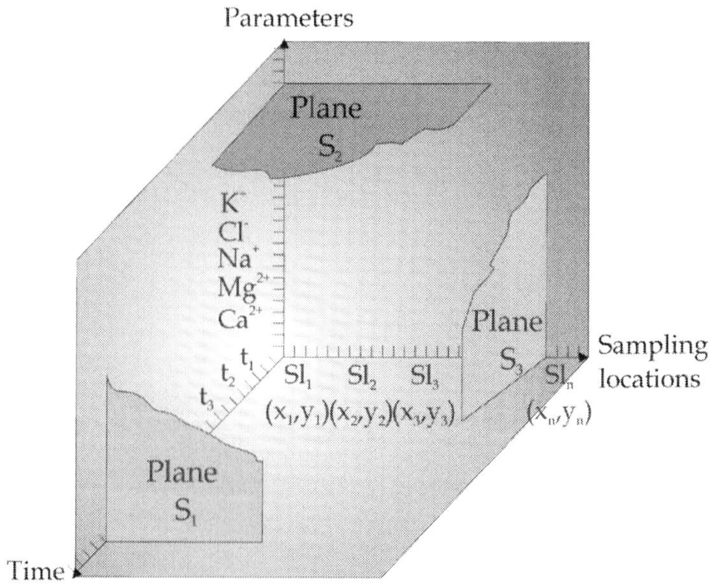

Fig. 1. Data in four dimensions (Sampling locations (x_i; y_i), parameters, time) (based on Kovács et al., 2008)

As an example let us imagine that in a certain area many parameters are sampled from more than one groundwater monitoring well (GMW) at the same time (plane S1). The data obtained is then recorded on worksheets where each row corresponds to a GMW and each column to a parameter. This data matrix is considered to be static. Besides the common uni-variate statistical methods multi-variate ones can be used, such as cluster-, discriminant-, principal component- and factor analysis along with multi-dimensional scaling.

Cluster analysis (CA) and multi-dimensional scaling can be used on sampling locations when there is a need to reveal similarities. Another aim can be to determine the background processes that explain most of the original dataset's variance. This can be achieved using principal component analysis (PCA) or factor analysis (FA). During CA the rows of the data matrix and during PCA of FA its columns are the object and the analyses.

In most cases the datasets obtained contain the parameter determining the fourth dimension, time. In this case the data matrix is not static. Staying with the previous example, if more than one GMW is sampled equidistantly in time one is dealing with problems described in plane S2. In this plane the most frequent question is which background processes drive the sampled parameter's (in our case the water levels) temporal fluctuation. Because consecutive temporal samples are not independent of each other only dynamic factor analysis (DFA) can be applied to answer the question raised (Márkus et al., 1999; Ritter et al., 2007). Its application began only in the last years of the 20th century, and with its promising results its role in solving environmental problems is expected to increase (Kovács et al., 2004).

Plane S3 is where the time series of multiple parameters are examined at only one sampling location. Here "classical time series methods" (Shumway & Davis, 2000; Hans, 2005) can be employed, the use of which implies determining the parameters' trend and periodicity etc. In many cases determining these two characteristics is a key question of the study. If the periodicity and trend of a certain process is extracted it can be used for forecasting, but only if one is certain that the driving processes still exist and will exist in the distant future. However further expansion of this topic is not an aim of this chapter.

Returning to the spatial alignment of the sampling locations, by having two spatial coordinates arranged for every parameter sample one will be able to visualize its spatial distribution (on isoline maps) which could be of great use. However if a parameter's spatiotemporal changes are of interest -for example in non-stationary cases- only a few tools are at hand, and the research to solve four dimensional problems in ongoing. Here we refer the reader to Dryden et al. (2005).

2.1.2 Common problems in data handling

Accurate results can only be expected from multi-variate methods if the datasets used contain the desired information that describes the investigated processes precisely; in other words, the amount of data obtained is sufficient. Determining what amount is sufficient information is the duty and responsibility of the given discipline. As an important requirement the number of sampling locations should exceed the number of analyzed parameters (Füstös et al., 1986), in this way statistical stability would be ensured.

One of the most important criteria regarding the data matrix is that there should be no data missing. Unfortunately in many cases (mostly water quality data) data is missing from the datasets. The solution could be data replacement but this has to be done with caution. In practice it often happens that the missing data is replaced with '0'. This is a huge error and should be corrected at all times. Another frequent mistake is when a measured parameter's values are below detection limit and the analyst therefore sets its value in the matrix to, for example: the detection limit, or half of the detection limit etc. A dataset like this can give misleading results.

Extreme or outlying values can also lead to inaccurate results. To decide precisely which datum is really outlying or extreme and which one is mistyped is a key question. The certain parameters' variability can be of help in deciding this question.

In water quality data it is often observable that a parameter is the linear combination of one or more others. This of course cannot be used in the course of multi-variate data analysis. Every criterion can only be held if the data matrix is checked for these and other kinds of errors before analysis. This is the most annoying and time-consuming part of the research. However, skipping this step will inevitably lead to incorrect results and conclusions.

2.2 Applied methods

The most effective order in which the presented methods should be used is the following. After checking the data matrix for the errors described above it is necessary to examine the data using uni-variate methods like descriptive statistics, distribution analysis, hypothesis testing of some sort and finally determining the stochastic connections with correlation analysis (Helsel & Hirsch, 2002). Out of these methods, correlation analysis is the one that will be discussed.

The next step is the application of multi-variate methods. The first method suggested is cluster analysis. Its results are groups of similar sampling locations. As a verification tool discriminant analysis and as a tool for determining which parameters influenced the formation of the cluster groups the most, Wilks' lambda distribution is suggested, along with the overview of the groups' statistics (Box-and-whiskers plot).

As a last step it is proposed to determine the driving background processes using PCA and if possible visualizing the results on maps for better interpretation.

2.2.1 Correlation (stochastic connections)

A frequent question is what kind of relation can be revealed between two parameters. The connection's strength should be described numerically. The most common is the Pearson correlation coefficient which measures the strength and direction of a linear relationship between two variables. It is calculated as follows:

$$r_{(x,y)} = \frac{\sigma_{(x,y)}}{\sigma_x \sigma_y}$$

Where σ_x, σ_y are the standard deviation of variables X and Y, while the numerator is the covariance.

This means that the correlation coefficient includes the covariance's every good quality and with the division by the standard deviation it will be independent from measurement units, and the upper lower boundary problem will be solved as well. The properties of the correlation coefficients are the following:

- Its value is independent from the measurement unit of X and Y
- if $\sigma_x = 0$ or $\sigma_y = 0$, then $r_{(x,y)} = 0$;
- $-1 \leq r_{(x,y)} \leq 1$
- If the relation between X and Y is positive, then $0 < r_{(x,y)}$, if this relation is negative, then $r_{(x,y)} < 0$
- if $r_{(x,y)} = \pm 1$, then there is a linear functional relation between the two variables
- any variable's correlation with itself is 1

- if the correlation coefficient is zero the two variables are uncorrelated, however this does not mean that they are independent
- if two variables are independent then $r_{(x,y)}=0$

In the studies presented correlation coefficients (in absolute value) higher than 0.71 were considered to represent strong linear relationship (Füst, 1997).

2.2.2 Cluster

Clustering is a kind of coding, in which a certain sampling location -originally described with many parameters (runoff, chemical oxygen demand etc.) is now described with only one value, its group code (cluster number). It is important to note that during clustering not the number of parameters but the number of sampling locations is decreased by placing the similar ones into groups. It is an important criterion that every sampling location has to belong to a group, but only one group. It is obvious that there are many possible group conformations. The main aim is to settle the similar sampling locations into the same group, however this similarity has to be measured by assigning a distance (metrics) to each sampling location which is placed in an *N*† dimensional space. If the distance between two sampling locations is small, then they are highly similar to each other. If the distance is zero they are perfectly similar. From this it should be clear that choosing the right distance is a key question. It needs skill and practice. This means that the verification of cluster results is compulsory.

"Cluster analysis (CA) classifies a set of observations into two or more mutually exclusive "unknown" groups based on combinations of interval variables. The purpose of cluster analysis is to discover a system of organizing observations, usually people, into groups. where members of the groups share properties in common" (Stockburger, 2001).

There are basically two types of clustering, the K-Means CA and the Hierarchical CA. In the former, one has to predetermine how many groups are required and is frequently suggested to be used with large datasets, in latter one only has to determine the groups after obtaining the dendrogram, the graphical output of the CA. In this study divisive Hierarchical CA was applied, where one group is divided to many more and so on. Its opposite is the agglomerative Hierarchical CA when the number of groups is reduced during the analysis.

2.2.3 Discriminant analysis and Wilks' lambda distribution

To verify the accuracy of the results, discriminant analysis can be used. It shows to what extent the planes separating the groups can be distinguished by building a predictive model for group membership. The model is composed of a discriminant function (for more than two groups a set of discriminant functions) based on linear combinations of the predictor variables that provide the best discrimination between the groups. The functions are generated from a sample of cases for which the group membership is known; the functions can then be applied to new cases that have measurements for the predictor variables but their group membership is as yet unknown (Afifi et al., 2004). The result of the discriminant

†In this case the number of dimensions (*N*) is equal to the number of measured parameters.

analysis is often visualized on the surface stretched between the first two discriminating planes (function 1 & function 2, e.g. Fig. 13) (Ketskeméty & Izsó, 2005).

After the verification of the cluster groups the role of each parameter should be analyzed in determining the formation of the cluster groups. Using Wilks' λ distribution a Wilks' λ quotient is assigned to every parameter, where the quotient is:

$$\lambda = \frac{\sum_i \sum_j (x_{ij} - \bar{x}_i)^2}{\sum_i \sum_j (x_{ij} - \bar{x})^2}$$

Where x_{ij} is the jth element of the ith group, $\bar{x}_i$ the ith group's mean, and $\bar{x}$ the total mean.

The value of λ is the ratio of the within-group sum of squares to the total sum of squares. It is a number between 0 and 1. If $\lambda=1$, then the mean of the discriminant scores is the same in all groups and there is no inter-group variability, so, in our case the parameter did not affect the formation of the cluster groups (Afifi et al., op. cit.). On the contrary, if $\lambda=0$, then that particular parameter affected the formation of the cluster groups the most. The smaller the quotient is, the more it determines the formation of the cluster groups.

2.2.4 Box-and-whiskers-plots

Box-and-whiskers plots are great tools for visualizing more than one statistic of a parameter on one graph, making the interpretation clearer. The boxes show the interquartile range and the black line in the box is the median. Two upright lines represent the data within the 1.5 interquartile range. The data between 1.5 and 3 times the interquartile range are indicated with a circle (outliers), and the ones with higher values than 3 times the interquartile range are considered to be extreme values indicated with an asterisk (Norusis, 1993). For example see Fig. 8.

For better interpretation it is necessary in every case possible to visualize the results on maps of some sort.

2.2.5 Principal component analysis compared with factor analysis

The principal component analysis (PCA) and factor analysis (FA) methods are used to analyze multidimensional data. The goal is to determine the background processes, while describing the observed parameters with fewer hypothetical variables without any significant information-loss in the original data.

During PCA the measured chemical and physical parameters are correlated, whereas the hypothetical variables (called principal components) are uncorrelated and are obtained as a linear combination of the original parameters. The PCA decomposes the total variance of the original variables to principal components which explain the original variance in a monotonically decreasing way. The correlation coefficients between the original parameters and the principal components are the factor loadings. They explain the weights of the original parameters in the principal components, however they do not give an exact answer whether a weight has to be considered as significant or not, and how many principal components are important.

During FA the hypothetical variables are called factors. They are classified into three categories: (1) common factors (influencing multiple parameters), (2) specific factors (influencing one parameter) and (3) error factors (arising for example from inaccuracy during the measurement). In this method only the common factors and their factor loadings are determined, because identifying specific-, and error factors along with their factor loadings usually causes mathematical difficulties. As a result, the common factors only explain a part of the total variance of the original parameters. While the PCA uses a correlation matrix, the FA uses an adjusted correlation matrix. In this matrix the elements on the diagonal (commonalities) can be estimated in different ways, as a result different solutions (of the FA) may be acceptable. The basic model of FA suits the conditions of datasets in earth sciences better (Geiger, 2007), this may be the reason for its successful application (Voudouris et al., 1997; 2000). In this chapter only PCA was applied.

In regard to the software used, we suggest using STATISTICA for its good visual output and user friendly interface, SPSS because its more commonly used and for its "syntax system" and last but not least R because it is an open source freeware and the most up-to-date software available.

3. Case studies giving an example on the data analysis methods

In sections 3.1-3.3 three cases studies are presented to give an example of the most efficient application of the basic data analysis methods (section 2) using the datasets of a mitigation wetland, a river and a groundwater area (Fig. 2).

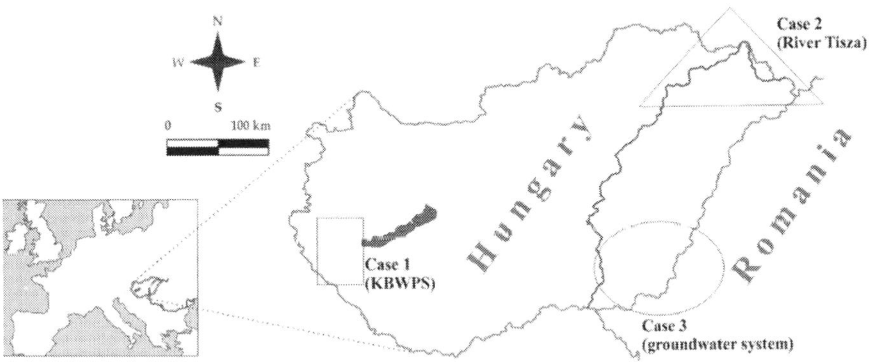

Fig. 2. The location of the three case studies numbered in order of appearance

3.1 The Kis-Balaton Water Protection System (KBWPS)

The KBWPS is a mitigation wetland located at the mouth of the River Zala at Lake Balaton, the largest shallow freshwater lake in Central Europe (Padisák & Reynolds, 2003). It was settled on the remains of the original Kis-Balaton Wetland (KBW), which due to artificial water level modifications decreased in area and functionality in the late 19th century. The original KBW used to naturally filter the waters of the River Zala which supplies 45% of Lake Balaton's water and 35-40% of its nutrient input (Lotz, 1988; Kovács et al., 2010).

During the time period between the function loss of the KBW and the construction of the KBWPS the Zala's waters were less filtered. This resulted in the water quality deterioration of Lake Balaton. As a solution to the problem of nutrient retention at the mouth of the River Zala and stop the degradation of Balaton's water quality the KBWPS was constructed (Tátrai et al., 2000; Kovács et al., 2010).

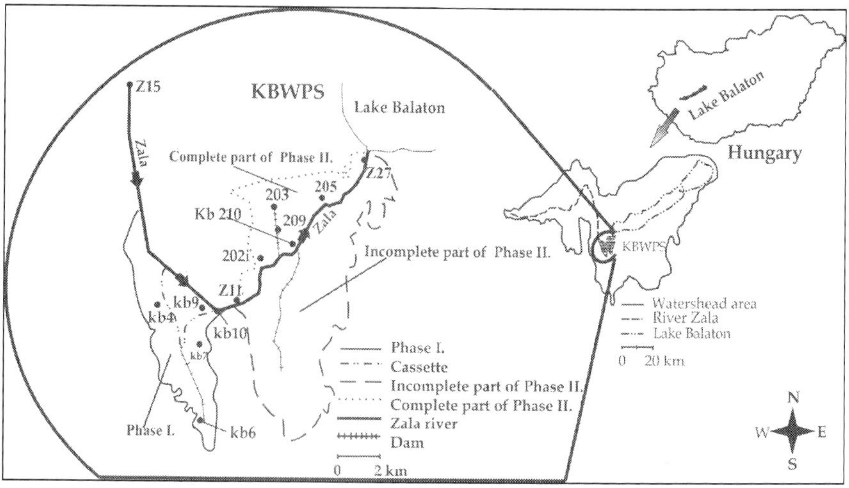

Fig. 3. The KBWPS and its sampling locations (based on Hatvani et al., 2011)

3.1.1 The structure of the KBWPS

The construction of the KBWPS was planned to take place in two phases (Fig. 3) as an extended wetland. Phase I was finished in 1985 after a five step flooding (Korponai et al., 2010). It resulted in a eutrophic pond, which gives ideal conditions for algae to reproduce. Phase II remains incomplete. Only a 16 km^2 area has been functioning since 1992. Its habitat could be described as a "classic" wetland, with 95% macrophyte coverage, primarily reeds (Nguyen et al., 2005; Tátrai et al., op. cit.). It is a highly protected nature conservation area and under the de-jure protection of the Ramsar Convention (1971).

3.1.2 Sampling locations and examined parameters on the KBWPS

Since the installation of Phase I weekly sampling has been carried out by the laboratory of the West Transdanubian Water Authority's Kis-Balaton Department on thirteen sampling locations. In the following the most important locations are described in detail heading downstream.

- "Z15", the inlet of Phase I, which typifies the water of the River Zala
- "Kb4; Kb6; Kb7"
- "Kb9", the sampling location that typifies the waters of the Cassette, which is the site of biological experiments, and is the most eutrophic place in the system
- "Kb10"

- "Z11", the interface between Phase I and Phase II, which typifies the water generated by Phase I, the eutrophic pond
- "202i; 203; 209; 210"
- "205" is the location where the Combined Belt Canal and drainage pipes from fishing lakes coming from Somogy County enter the KBWPS
- "Z27", the outlet of KBWPS, which typifies the water output of Phase II, the "classic" wetland area of the KBWPS

In this study the following parameters were examined for the time interval (1984-2008) with the methods described in section 2.2.

Runoff (m^3 s^{-1}); Chemical oxygen demand-potassium-dichromate original (COD^C); biological oxygen demand (BOD-5); dissolved O_2 (mg l^{-1}); chlorophyll-a (mg m^{-3}); Cl^-; SO_4^{2-}; HCO_3^-; CO_3^{2-}; Fe^{2+}; Mn^{2+}; Mg^{2+}; Ca^{2+}; Na^+; K^+; NH_4-N; NO_2-N, NO_3-N; total nitrogen (TN); total phosphorus (TP); dissolved phosphorus; soluble reactive phosphorous (SRP); suspended solids (mg l^{-1}); pH; conductivity (μS cm^{-1}). This resulted in around ~34,000 data per sampling location.

The sampling was conducted and the samples were prepared according to the current Hungarian standards (MSZ 12749 & MSZ 12750).

In case of the KBWPS the dataset was prepared by the same authority during the whole investigated time period. However in most cases the scientist is not this fortunate, he has to deal with the problems that originate from different sampling methods, and even standards (in case of an international research). To explore these problems and maybe even solve them a preanalysis and probably cross-verification of the data must be carried out before its analysis. Even in our case -the KBWPS- each-and-every data was checked whether it is an extreme-, an outlier- or a mistyped value or not. An x-y scatterplot was created for each parameter for the whole time interval and analyzed manually. It would not have been sufficient to set a high and low margin or even apply the three-sigma rule (Pukelsheim, 1994) because there were certain events that resulted in extremely high values (e.g. floods) that can easily be recognized from their surroundings in the graphs, and these would have been falsely discarded.

3.1.3 Results of the analyses conducted on the KBWPS

3.1.3.1 Cluster-, discriminant and Wilks' lambda analyses

Using cluster analysis on the 13 sampling locations and the annual averages of the parameters specified in table for the years 1984-2008 we were able to point out the alignment of the similar sampling locations for each year. The sampling locations basically grouped up according to the two constructional phases, but there were interesting exceptions (Fig. 4).

- Between 1997 and 1998 sampling location 202i disconnects from Phase II and connects to the cluster group, covering Phase I. This is the only location that changes its orientation and keeps it for the rest of the time analyzed. This occurred because of the constantly high water level of the area. Because of that the reeds died out and the surroundings of sampling location 202i became similar to Phase I, an open and eutrophic water space became dominant and the area of the "classic" wetland decreased along with the system's efficiency (Hatvani et al., 2011).

- Sampling locations Kb9 and 205 form separate groups, the former in 1997 and 1999, and the latter in 1996 and 1997. Previous because it is located in the Cassette, this constituted a separate waterspace with almost no water flow and high water retention time, so it is highly eutrophic, while with the latter (Kb205), because of the Combined Belt Canal and drainage pipes of fishing lakes from Somogy County joining the system here.

If one uses cluster analysis the results always have to be verified. Discriminant analysis results pointed out that 100.0% of the original grouped cases proved to be correctly classified.

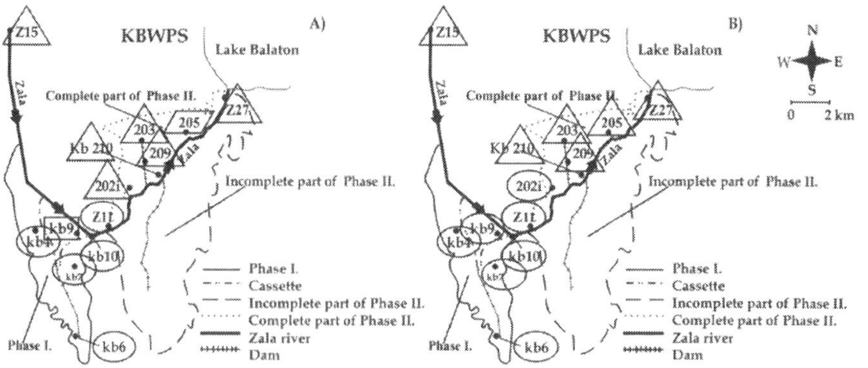

Fig. 4. Cluster groups of the KBWPS in 1997 (A) and control years 2007 & 2008 (B) (based on Hatvani et al., 2011)

Using cluster and discriminant analyses it was possible to point out key problems in the water level management of the system (Hatvani et al., 2011). For better interpretation the visualization of the spatial cluster results on a map of some sort in every possible case is suggested.

As discussed previously, Wilks' lambda distribution determines the parameters that effect the conformation of the previously discussed cluster groups the most. A Wilks' λ quotient was assigned to every parameter for every year and then clustered. Grouping the eleven quotients, the parameters were separated to three different groups according to how much they determine the original spatial cluster groups (Fig. 5).

Group 1 contains: chlorophyll-a, dissolved P, suspended solids and pH, CO_3^{2-}, HCO_3^-, dissolved O_2. These parameters had the smallest Wilks' λ quotients (average: 0.32), so they affected the conformation of the cluster groups the most and were responsible for the separation of the cluster group covering the eutrophic pond.

Group 2 contains the parameters which also stand in close relation to the eutrophication processes: Ca^{2+}, COD^C, total N, and total P, their average quotient was 0.55.

In fine the parameters in Group 3 (NH_4-N, NO_2-N , NO_3-N , Mg^{2+}, SO_4^{2-}, Cl^-, Fe^{2+}, Mn^{2+}, Na^+, K^+) generated an average quotient of 0.69 during the Wilks' λ distribution, so they affect the orientation of the spatial cluster groups the least and play a great role in the separation of the cluster group covering the Wetland (Phase II).

With these three methods used together a global picture was obtained regarding the similarities of the KBWPS' sampling locations and the parameters that drive these similarities. After gaining knowledge concerning the connection between the sampling locations and the parameters behind them, the next step would be to familiarize oneself with the processes evolving in the different areas of the water protection system.

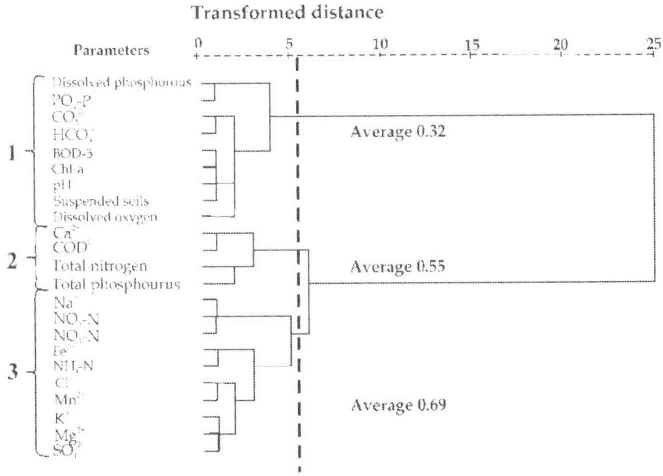

Fig. 5. Dendrogram of Wilks' lambda quotients

3.1.3.2 Stochastic connections

From the correlation matrix it became clear that at Z15 only the phosphorus forms (TP, SRP, dissolved P) in case of Z11 and Z27 besides the phosphorus forms the Na^+ and Cl^- ions have correlation coefficients (in absolute value) higher than 0.71. Meanwhile, at all three sampling locations the number of weak linear connections (values of correlation coefficients is between ±0.2) is 137-120-136 respectively.

In summary it can be said that in the eutrophic area of the KBWPS (Phase I) there are fewer weak linear connections than in the riverine or wetland areas. However, regarding the whole system, the linear connection between the 22 parameters cannot be considered generally significant. This is reflected in the following PCA results.

3.1.3.3 Principal component analysis

The PCA results presented concern the three cardinal, and one peculiar sampling locations of the KBWPS (Z15, Z11, Z27 and Kb9) and their surroundings. Their summer datasets were analyzed for the time interval 1984-2008.

The aim was to find the parameters that determine the processes evolving -mainly- in Phase I of the Kis-Balaton Water Protection System (KBWPS).

The biggest problem which had to be faced in the course of the PCA was that the datasets were time series, where data follow each other and are therefore not independent‡ because

‡The spatial or temporal distance when consecutive data turn independent can be measured for e.g. with variogram analysis (Kovács et al., 2011),

they are too close to each other in time, and the PCA cannot handle this kind of input[§]. To solve this problem only three summer months (May-July-September) were analyzed with two-month gaps in between. This is the growing season when primal production is most intensive (Wetzel, 2001).

Regarding the results, only the first factor can be considered as significant. It explains 25 to 30% of the data's variance, while the second PC only explained 15-19%, and so was discarded. The low explanatory value may originate from the parameters' weak linear connection, reflected in the correlation matrixes discussed in section 3.1.3.2.

If the constitution of the first component is analyzed in more detail it can be said that at:

- **Z15** (inlet, the River Zala) where the parameters TP, Dissolved P and SRP are present in the first PC. It corresponds what is already known about the River Zala, that there is technically no planktonic life in its water, and bentic eutrophication is not dominant in its sediment. These parameters originate from diffuse loads, thus their concentration only depends on the runoff of the River Zala. The highest concentrations were observable at peak flooding times.
- **Kb9** (Cassette) the parameters Chl-a and TP are present in the first PC. These are the main parameters that the OECD (Vollenweider & Kerekes, 1982) uses to classify the trophic conditions. Their presence is no surprise, because at Kb9 the water is still, with almost no water flow, and long hydraulic residence time. Bottom-up[**] processes dominate at this sampling location and its processes can be described with the Vollenweider model (Vollenweider, 1976).
- **Z11** (interface, representing Phase I) the parameters TP, Chl-a, TN Ca^{2+}, Suspended solids and Cl^- are present in the first PC, indicating Phase I's eutrophic and algae dominant waterbody. Again, the Vollenweider model can be used to describe this environment. The calcium ion indicates biogenic carbonate precipitation which is a dominant process in certain locations of Phase I.
- **Z27** (outlet, describing Phase II) the parameters K^+, Na^+, Cl^- and NO_3^- are present in the first PC, where decomposition processes are dominant. The reason for NO_3^- being one of the dominant parameters is that nitrification is the main process in this section of the system indicating aerobic conditions.

In summary it can be stated that PCA is a universal tool for determining the dominant processes of a certain sampling location's area or a whole system. The facts that were perceptible to the naked eye, such as the eutrophic conditions of the Cassette are now written down in numbers and can be subject of further scientific studies.

3.1.4 Conclusions regarding the results obtained from the KBWPS

The development and the functioning of the KBWPS is a good testing ground for new habitat remediation techniques and a great example of the presentation of the application and the use of the methods described above. The result of each method gave extra information to scientists and confirmed their previous suspicions. For example the constant water level deteriorates the wetland vegetation (Pomogyi et al., 1996), therefore decreases

---

[§] Instead of PCA dynamic factor analysis was developed to handle temporal dependence

[**] Bottom-up control: ecological scenario in which the abundance or biomass of organisms is mainly determined by a lack of resources and mortality owing to starvation (Pernthaler, 2005)

the efficiency of the system and that this process reached a peak in 1997 and 1998. It was always known that the water need of different vegetation types is different; the key result was that the irreversible change at 202i happened in those particular years. This was just one example highlighted regarding the mitigation wetland.

In the next section (3.2) the same methods were applied to the largest tributary of Europe's second longest river the Danube.

3.2 River Tisza

The River Tisza collects the waters of the Carpathian Basin's eastern region. According to Lászlóffy Woldemár (1982), its watershed area is 157,186 km^2. Less than one third of this is located in Hungary (Fig. 6).

Fig. 6. Watershed of the River Tisza (Based on Istvánovics et al., 2010)

From its spring in the Maramureşului Mountains to its confluence with the Danube, it stretches for 966 km across the Ukraine, Romania, Slovakia, Hungary and Serbia (Sakan et al., 2007). The Hungarian section of the Tisza from border to border is 594.5 km long. Its average runoff is 25.4 billion m^3 per year (Pécsi, 1969).

Despite the fact that in the last one and a half centuries numerous anthropogenic activities have influenced this area, in comparison to Europe's other large rivers it is still considered to have one of the most natural river valleys in Europe (Zsuga & Szabó, 2005). It is for this reason that it is in our common interest to protect it. If we only take Hungary into account, then approximately 400 settlements and 1,500,000 inhabitants' lives depend on its runoff and water quality.

3.2.1 Sampling locations and examined parameters on the River Tisza

Many surface waters are monitored as part of the National Sampling Network. In case of the Tisza, data from the first five Hungarian sampling locations were analyzed (258.7 river km) (Fig. 7). The exact specifications of these monitoring locations can be found in Hungarian Standard No. MSZ12749:1993.

The River Tisza reaches the Hungarian border at Tiszabecs. The next sampling location (SL) is at Záhony. There are two tributaries of considerable size between these two locations, the Szamos and the Kraszna. The next SL can be found at Balsa, just before the mouth of the Tokaj and Bodrog rivers. The Tiszalök SL is located just after the Eastern Trunk Sewer and the Tiszalök Water Barrage System. The last SL analyzed is at Polgár downstream from the River Sajó's mouth.

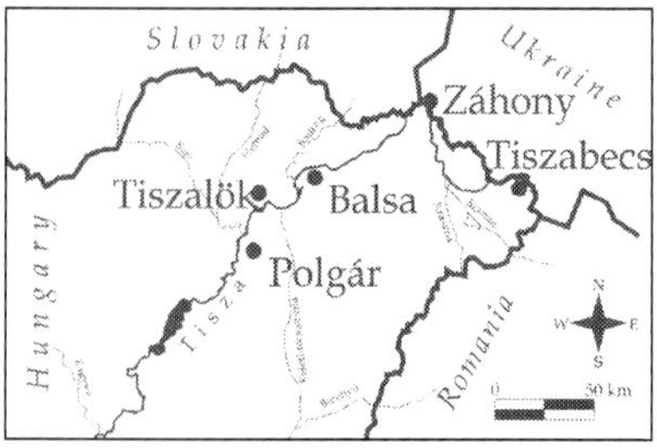

Fig. 7. River Tisza's first five sampling locations in Hungary

During the research, a 31-year-long dataset (1974-2004) was analyzed, consisting of 300,000 data. From 1970 only one sample a week was taken at Tiszabecs and Polgár, according to COMECON[††]'s specifications (T. Nagy et al., 2004). At Balsa only one sample was taken per month. At Záhony 26 samples were taken every year. In 1994 Hungarian Standard No. MSZ 12749:1993 came into force. As a result, since 1993 26 samples have been taken every year uniformly.

The parameters used were as follows: Runoff (m^3 s^{-1}), pH, Conductivity (μS cm^{-1}), M-alkalinity (mval l^{-1}), Oxygen saturation (%), Dissolved O_2, BOD-5, COD^C, Ca^{2+}, Mg^{2+}, Na^+, K^+, Total hardness, Carbonate hardness, Cl^-, SO_4^{2-}, HCO_3^-, NH_4-nitrogen, NO_2-nitrogen, NO_3-nitrogen, Mineral nitrogen[‡‡] (mg l^{-1}), PO_4-phosphorus, Chlorophyll-a (Chl-a) (μg l^{-1}).

3.2.2 Results of the analyses conducted on the River Tisza

During our work many temporal approaches can be employed, meaning the whole year can be examined or the dataset can be separated to seasons. In the former case the whole year's processes, while in the latter obviously the seasonal changes can be followed. To shed light on the seasonal changes the winter and summer data were analyzed separately. Summer was considered to last from June to October, while winter from November to March.

3.2.2.1 Cluster- and discriminant analyses and Wilks' lambda statistics

Regarding the River Tisza's data series, cluster analysis was conducted on averages formed from the parameters. This approach gives a longer perspective on the data. One average was formed at each sampling location from each parameter's total dataset.

The cluster analysis resulted in three groups. As a verification tool, discriminant analysis was applied.

[††]Council for Mutual Economic Assistance founded by the Soviet Union 1949.

[‡‡]Mineral-N is the summary of the NH_4-nitrogen, NO_2-nitrogen, NO_3-nitrogen.

The main question was whether these groups (formed from the averages) present at all and whether they are discoverable if the data is analysed in more detail (not in averages). The answer is yes.

By using discriminant analysis on the discrete data, the cluster groups formed from the averages were realized for example 94.8% in summer and 89% in winter. This means that the clustering from the averages is correct and representative.

To stay with our examples, Table 1 shows the Wilks' lambda coefficient in summer and winter for each parameter.

	Wilks' lambda coefficients	
Parameter	Summer	Winter
SO_4^{2-}	0.41	0.53
Na^+	0.74	0.87
Mg^{2+}	0.75	0.78
pH-lab	0.96	0.99

Table 1. Parameters' Wilks' lambda coefficients in summer and winter.

It is clear from Table 1 that the sulphate ion is the most determining in both seasons. Parallel to the Wilks' lambda distribution, it is suggested that the parameters' variability be analyzed, because it gives a much wider picture of certain parameters. In Fig. 8, three parameters are presented on box-and-whiskers plots. One with a small (sulphate, Fig. 8/A) one with a medium (calcium, Fig. 8/B) and one with a high (pH, Fig. 8/C) Wilks' lambda coefficient.

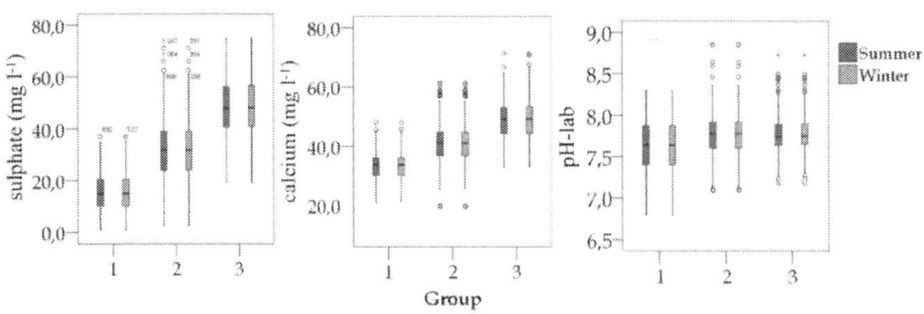

Fig. 8. Sulphate ion (A), calcium ion (B), pH (C) parameters' boxplots in increasing order of their Wilks' lambda quotients.

It is clear that the sulphate is the most variable, calcium is less so, and pH (which was the most influential regarding the Wilks' lambda distribution) is the least variable parameter.

3.2.2.2 Stochastic connections

The stochastic connections were analyzed using correlation analysis. The connection between the parameters was analyzed through different approaches. First the whole year, then the different seasons (winter, summer) were taken into account.

According to each approach (whole year, winter/summer), the number of strong correlations increases downstream. At Tiszabecs, the number of strong correlations is only seven (Table 2), at Tiszalök this number reaches 36. This can be explained by the flow conditions. In the area of the water barrage system, the water-flow slows down, and (according to the spiral model[§§]) so does physical transport; suspended solids are deposited, the water becomes more transparent and light limitation decreases (Padisák, 2005). This gives an opportunity for organisms to compose the nutrients into their systems faster and more efficiently. Usually the River Tisza is autotrophic during the summer months, but tributary input may considerably exceed net autochtonous production (Istvánovics et al., 2010).

	Number of linear connections [pcs.]		
	Annual	Winter	Summer
Tiszabecs	7	5	7
Záhony	14	17	9
Balsa	17	18	10
Tiszalök(wbs)	37	37	14
Polgár	36	41	18

Table 2. The number of strong linear connections ($|r| \geq 0.7$) at each sampling location by temporal distribution

During the summer months there were fewer strong linear connections than in the winter months. If the results from the whole year are compared to the summer and winter ones, it can be stated that the correlation matrix obtained from the winter data more closely resembles the annual correlation matrix than does the summer one.

As can be seen in Table 2, in the Tiszalök area the number of correlating parameters suddenly rises in both winter and summer.

Summarizing the correlation analysis, it can be stated that the number of correlations increases downstream. The annual, winter and summer results are different in the case of the number of correlations and in the case of the parameters which correlate as well. During the summer there are fewer linear connections, but these few are between the parameters in relation to the organic processes.

As previously stated, if the results for the whole year are compared to the summer and winter ones, it can be seen that the correlation matrix obtained from the winter data more closely resembles the annual correlation matrix than does the summer one. So, if just the whole year had been analyzed, vital differences between the summer and winter would have been lost.

After analyzing the connections between the SL based on the sampled parameters the next step is to take a closer look at the processes evolving in the river.

[§§]The riverine spiral model describes recycling of nutrients together with physical transport downstream (Padisák, 2005).

3.2.2.3 Principal component analysis

To answer the question which background processes determine the water quality and processes of the River Tisza, PCA was applied to the summer, winter and whole year data.

Before the PCA was conducted, the number of parameters had to be decreased, either because the parameter was not sampled during certain time periods or because it was unsystematically sampled over the whole investigated time period. Or, in other cases, the parameter itself contained information concerning other parameters (e.g. specific conductance). There are other cases when the dataset has to be decreased, more examples can be seen in section 3.3.2.3.

In terms of their importance, only the parameters with a factor score (in absolute value) higher than 0.7 were taken into account in the first and second principal components (PC). The summarized results of the PCA can be found in Table 3.

Season / Sampling location	Summer		Winter		Annual	
	First PC	Second PC	First PC	Second PC	First PC	Second PC
Tiszabecs	N-forms	Major ions	N-forms	*None*	Major ions	N-forms
Záhony	N-forms	Major ions	Major ions	N-forms	Major ions	N-forms
Balsa	N-forms	Major ions	Major ions	N-forms	Major ions	N-forms
Tiszalök	Major ions	N-forms	Major ions	N-forms	Major ions	N-forms
Polgár	Major ions	N-forms	Major ions	N-forms	Major ions	N-forms

Table 3. Summarized results of the PCA, None: There were no factor scores ≥ 0.7

Regarding the results, it can be said that the first two components explain approximately 50% of the data's total variance, independent of their spatial and temporal distribution.

Regarding the summer results, at Tiszabecs, Záhony and Balsa mostly the N-forms can be found in the first PC. In the second PC, the ions responsible for halobity (Mg^{2+}, Na^{+}, K^{+}, Cl^{-}) take place. Between Balsa and Tiszalök the background processes show a peculiar change: the scale tilts from the organic components towards the inorganic ones. At Tiszalök in the summer (according to the first PC), the major ions (e.g. Ca^{2+}, Mg^{2+}, Na^{+}, K^{+}, Cl^{-}) play a determining role; the Polgár SL shows the same pattern. The fact that at Tiszabecs, Záhony and Balsa the N-forms are the most determining in the first PC leads us to the assumption that biological processes such as saprobity and trophic conditions are responsible for the background processes. Since there was no direct relationship observed between nutrient levels and phytoplankton biomass (Istvánovics et al., 2010) other factors are responsible for changes in the N-forms. In contrast, the results from Tiszalök and Polgár show a change in the determining processes. After Tiszalök, the inorganic processes (e.g. aggregation, dissolution) take the place of the N-forms in the first PC.

From the perspective of the winter results, the first PC's explanatory power varies between 20% and 40%. At each SL except for Tiszabecs the ions determine the background processes. In the second factor, the N-forms are dominant.

Regarding the whole year's PCA results it can be stated that the annual conditions resemble the winter ones to a high degree (just as in the case of the correlation results). In the first PC (explanatory power: 21-39%) the ions take on the determining role, while in the second one the N-forms are dominant.

Regarding the temporal distribution, it is clear that during the winter inorganic processes are dominant in determining the Tisza's water quality and these results represent the annual conditions much more than the summer results do.

As in the case of the correlation analyses, and of the PC as well, results which are not temporally divided are not satisfactory, but this can only be confirmed if the summer and winter data are analyzed separately, as it was done in this case.

3.2.3 Conclusions regarding the result obtained from the River Tisza

It is clear that the methodology applied in the case of the KBWPS can successfully be applied to a river as well. The advantage of the methodology used is that it delimits sub-systems not based on geography but mathematics. It was known from the literature, that three large hydro-geomorphic sections can be distinguished along the river (Istvánovics et al., 2010). The meandering upstream and downstream sections are separated by the impact by the two reservoirs. The PCA for example clearly separated the Hungarian section of the River Tisza to two sub areas (Tiszalök SL (water barrage system)in the middle), where Tiszabecs and Balsa SLs belong to the upstream and Tiszalök and Polgár SLs to the reservoir section. This section however causes discontinuity in the environmental gradient along the river. The separation is the strongest in summer when autotrophic processes become driving forces in the reservoir. After presenting the two surface waters the final case study, the analysis of SE Hungary's groundwater will be discussed.

3.3 Groundwater

3.3.1 Sampling locations, and examined parameters

Drinking water has high arsenic content in south-eastern Hungary. Within the framework of the pilot project (Sustainable management and treatment of arsenic bearing groundwater in Southern Hungary (SUMANAS) –LIFE05 ENV/H/000418), the formation of the arsenic (geological in origin) and the development of its decontamination were examined. The geological part of the study was prepared by Smaragd GSH Ltd. Appropriate water chemistry analysis results made the utilization of multi-variate data analysis methods possible.

Subsurface water resources of the Pleistocene aquifer in the south-eastern part of the Great Hungarian Plain contain more or less arsenic. SE Hungary is a subsiding back arc basin that was loaded with river sediments (aleurit, sand and clay) in the Pleistocene. One of the two rivers, the Tisza, has a metamorphic, volcanic and re-accumulated sedimentary catchment area, while the Maros River mainly derives its sediments from volcanic territory (Nádor et al., 2007). Arsenic has been transported by fluvial fine-grained sediments (fraction: <2 µm), adsorbed on the surfactant amorphous iron-oxyhydroxides (Varsányi & Ó. Kovács, 2006), clay minerals and organic materials.

The primary transportation, accumulation and desorption of the arsenic from the sediments into the water is determined by absorbents (amorphous iron-hydroxides and the surface of the organic matter, clay minerals) (Lin & Puls, 2000; Varsányi & Ó. Kovács, 2006), physicochemical conditions (Redox conditions, pH), and changes in the groundwater's parameters (recharge, flow regime). These agents display great variability over time and space, causing divergent arsenic concentrations in the groundwater.

Groundwater sampling was carried out at 202 groundwater monitoring wells (like those mentioned in section 2.1.1) All the wells were located in SE Hungary and the bordering area of Romania). Most of these were supply wells plus a few monitoring wells situated at an average elevation of 93 m above Baltic sea level. Groundwater temperature varies according different depth intervals. The average water temperature is 19.6 °C, the lowest is 12.1 °C, while the highest is 81 °C.

In the course of water chemistry analysis, the form of the separate As formulas is the most important step, because As (III) is 60 times more toxic to the human body than As (V). For this reason determining the quantity of the different arsenic formulas present in the groundwater is a crucial point in this study. The method for measuring As (V) and As (III) separately was developed during the project by Bálint Analitika Ltd. In: *Körös Valley District Environment and Water Directorate, 2008.*

The parameters analysed can be found in Table 4, excluding Na^+ (mg l^{-1}) and conductivity (μS cm^{-1}). One of the biggest problems concerning the database was when the concentration of the analysed parameter was beneath the detection limit. This occurred in many cases regarding a few parameters (cadmium, mercury, lead etc.), so these were simply left out of the calculations.

In the case of a few other parameters (ammonium, nitrite, nitrate, sulphate), concentrations both above and below detection limit were observable. On occasions when a parameter is beneath the detection limit, a common practice is that the values are supplemented with for example half of the detection limit (other solutions are mentioned in section 2.1.2). If it is done in this way so the examination of stochastic relationship (PCA) may generate huge errors. In the case of cluster analysis, however, supplementing data (beneath detection limit) with values close to zero may only lead to minor errors, which can be accepted. In summary, it must be stated that keeping in mind which method can handle the values under the detection limit and which cannot is a key question. In some cases using these values will lead to huge errors (e.g. PCA) in other cases the results can still be considered to be correct (CA).

3.3.2 Results of the analyses conducted on the study area

3.3.2.1 Stochastic connections

The correlation matrix shows whether there is a linear relationship between the measured parameters or not. The example is as follows.

The Correlation matrix shed light on a strong linear connection between conductivity (measured on site) and Na^+ and HCO_3^-. If conductivity and Na^+ content are plotted on a diagram (Fig. 9), besides the linear relationship one may easily recognize the different character of the connection in cases of different concentration ranges. Thus, the application of diverse regression functions should be practical if the aim is to determine the relationship to this resolution. The above-mentioned case is depicted in Fig. 9/A. This graph is split into three parts (Fig. 9/ B, C & D). The groups of sampling points presented in these figures are the results of the subsequent cluster analyses. They represent different geographical regions.

The figure series draws attention to the fact that great differences exist between the presented groups that can be interpreted as the result of different hydrogeological conditions (Fig. 9/ B, C, D). Regarding the correlation relationships, space does not allow us to enter into further details in the present paper.

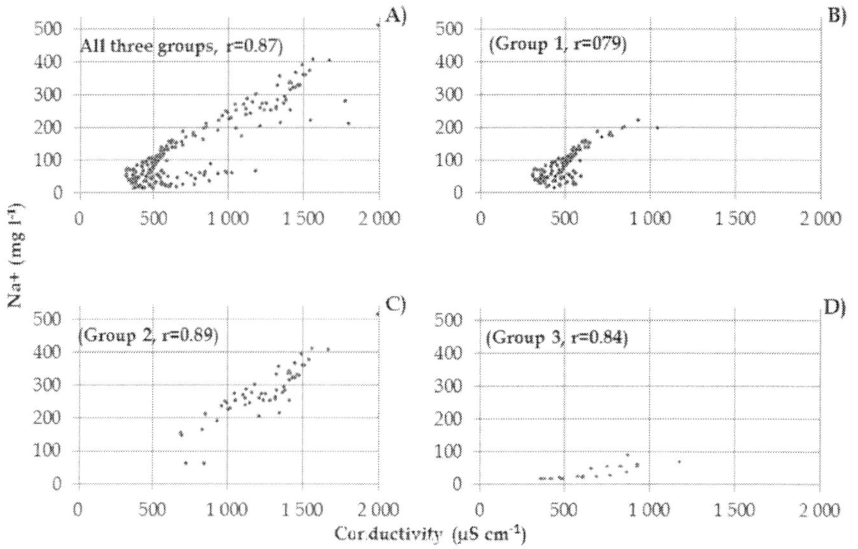

Fig. 9. x-y scatterplot showing the relationship between Na^+ and conductivity for all the groups (A), group 1 (B), group 2 (C), and group 3 (D)

3.3.2.2 Cluster-, discriminant analyses and Wilks' lambda statistics

In the next section the cluster results are presented, one containing the "problematic[***]" values and one conducted only on the subsurface waters' anions and cations used for facies determination.

During cluster analysis, the first step was to examine the datasets containing the parameters with values lower than the level of detection, and then to replace non detectable values with half of the detection limit (as mentioned above); as a result, the amount of data available for the cluster analysis increased. Two out of three strongly correlating parameters were removed (conductivity and Na^+) in order to reduce their collective effect. This is the reason why these cannot be found in the following analyses (e.g. Table 4).

Four groups were determined and placed on a map, three of which separated explicitly (Fig. 10). Different groups are marked with different colours. The fourth group contains outliers (e.g. shallow monitoring wells and a deep thermal well), thus there is as yet no explanation for its constitution. For the purposes of better interpretation, the three groups were given names after the geographical region they are located in and the fourth one Outlier. From

[***]In section 3.3.2.2 the term „problematic" refers to parameters with few datum under the detection limit. If their numbers is high the dataset's variance is low.

now on these will be referred to using the following names. The **first group** is the **Maros group** in the area of Maros alluvial fan and Makó graben. The **second group** is the **Körös group** in the area of the Körös basin. The **third group's** wells are geologically and geographically situated in the Maros alluvial fan, however it was given the name **Arad group** after the city around which they are located. As mentioned before, the clustering was repeated with a dataset without the "problematic" values. This time only anions and cations used for subsurface waters' facies determination were considered to be valid. Results were again visualized on a map (Fig. 11).

In the next paragraph the similarities and differences will be discussed between the results of the two clusterings (Fig. 10 & 11). The main difference is that the wells in the **second group (Körös)** in Fig. 10 after the second clustering (Fig. 11) with the decreased database (containing parameters: Ca^{2+}, Mg^{2+}, Na^{+}, K^{+}, HCO_3^{-}, Cl^{-} and SO_4^{2-}.) connect to the **first (Maros)** and the **fourth (Outlier) groups'** wells. Nevertheless a few wells kept their original relationship and stayed in the **second group (Körös)**.

Using the database without the "problematic" parameters, the data was plotted on a Piper diagram (Fig. 12) (the colour codes of the groups were retained). It is easy to see that the spatially separated groups are no longer present. This is no surprise because the application of the two methods has different intentions.

To verify the grouping, discriminant analysis was used, which pointed out that 94.6% of the original grouped cases were correctly classified using the grouping based on the data including the "problematic" parameters. Then the contents of the actual groups were changed according to the software's (SPSS) suggestions. Finally as a result 100% of the original grouped cases were proved to be correctly classified with the cross-validation resulting in a value of 96%.

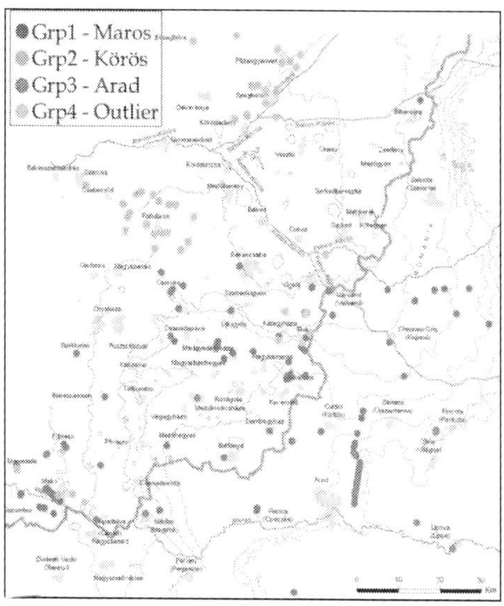

Fig. 10. Cluster results obtained from the database including the "problematic" parameters, where the same colour indicates the same group.

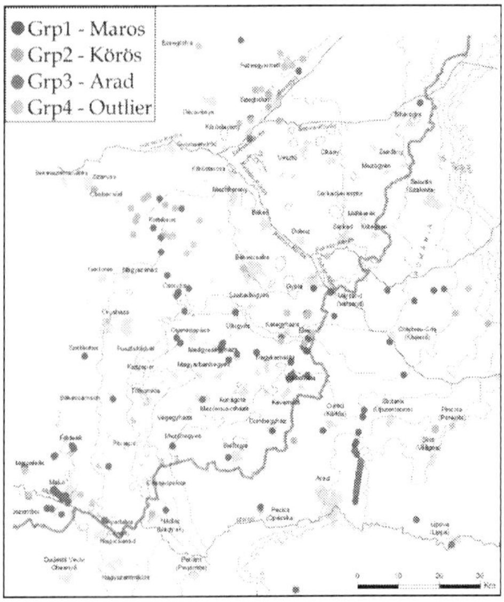

Fig. 11. Cluster results obtained from the database excluding the "problematic" parameters. The same color indicates the same group.

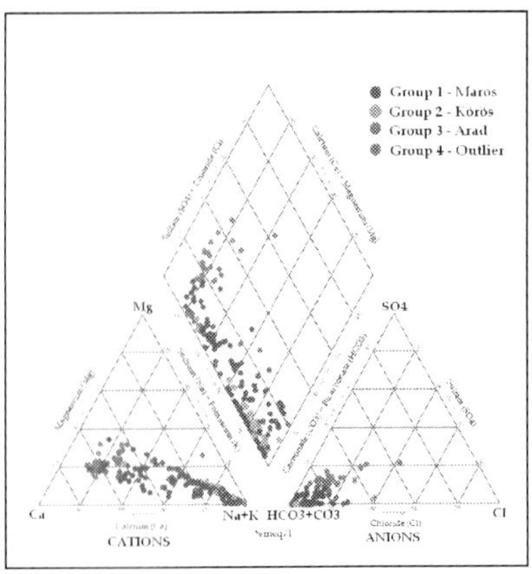

Fig. 12. Piper diagram obtained from the database excluding the "problematic" parameters

The result of the discrimination analysis can be visualized on Fig. 13 on the surface stretched between the first two discriminating planes (function 1 & function 2). Separation of the different groups can be easily observed. The wells in the **Outlier group** separated significantly from all the other groups.

In order to demonstrate the effect of the individual parameters on the formation of the cluster groups, a Wilks' lambda quotient was calculated for each parameter (Table 4).

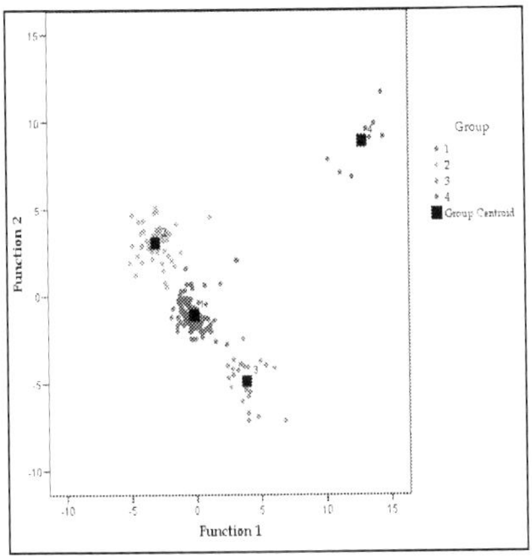

Fig. 13. Visualized results of the cross-validated discriminant results with the "problematic" parameters included.

Parameter	Measurement unit	Wilks' λ coefficients
NO_3^-	mg l^{-1}	0.249
HCO_3^-	mg l^{-1}	0.362
Fe_{total}	mg l^{-1}	0.404
SO_4^{2-}	mg l^{-1}	0.544
NH_4^+	mg l^{-1}	0.608
pH		0.610
Ca^{2+}	mg l^{-1}	0.690
As_{tot}	µg l^{-1}	0.723
Mg^{2+}	mg l^{-1}	0.748
F^-	mg l^{-1}	0.759
K^+	mg l^{-1}	0.797
Cl^-	mg l^{-1}	0.808
As (III)	µg l^{-1}	0.817
As (V)	µg l^{-1}	0.818
Dissolved O_2	mg l^{-1}	0.841
Br^-	µg l^{-1}	0.852
NO_2^-	mg l^{-1}	0.880
Water temperature	°C	0.889
Total organic carbon	mg l^{-1}	0.908
Mn^{3+}	mg l^{-1}	0.935

Table 4. Wilks' lambda quotients of each parameter in increasing order

As expected the most influencing parameter was the organic NO_3^- . Among the inorganic chemical components, the cluster grouping was notably influenced by anions such as: HCO_3^-, $SO_4^{2-}$and cations Mg^{2+}, Ca^{2+} and pH and Fe_{total} as well.

At this point enough results had been obtained to be able to see clearly why the two cluster results (Figs 10 & 11) show a resemblance to each other. The reason for this is that the main parameters that are necessary for determining hydrogeological facieses play an important part in forming the groups (based on the Wilks' lambda quotients), while the origin of the differences is the other influencing parameters. Spatial separation is much more obvious in the latter case. It is suggested to present the statistics of the different groups' parameters on box-and-whiskers diagrams. As an example HCO_3-and As (V) are presented (Figs 14 & 15). Based on these results, the individual groups can be described.

The most important aim of the introduced area's hydrogeological investigation was to determine the distribution of arsenic in the groundwater. The groundwater's different As (V) concentrations can be seen in Fig. 15. It is obvious that the arsenic accumulated mostly in the waters of the deep and a few shallow wells (**Outlier group**), and the least in the **Arad group's** wells. In the **Arad group** arsenic was traceable in 5 out of 33 wells. Two of the five sampled wells fall to the area of the Maros alluvial fan. The different forms of arsenic show great variance in the groundwater of the other areas but the arsenic content of groundwater is clearly higher in the area of Körös basin.

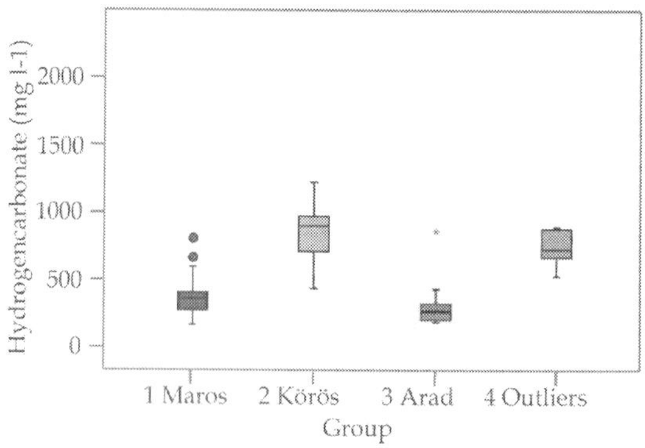

Fig. 14. Staticstics of hydrogencarbonate for each group

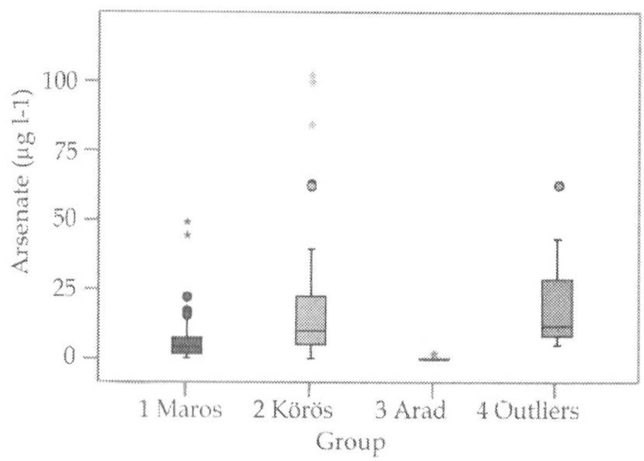

Fig. 15. Statistics of Arsenic (V) for each group

3.3.2.3 Principal component analysis

According to the literature, two out of the three **groups (Arad and Maros)** are located in different parts of the same flow system, while the third represents another flow direction. It is worth examining the dominant processes taking place in the different groups. However, a new problem had to be faced: a few parameters' variance was critically small in the different groups. In the cases of many hydrogeologically important parameters, their standard deviation is too small, so they had to be discarded. Examples included: sulphate in the **Körös group**, or As (III) and As (V) in **Arad group**, but the same thing can be said regarding the total iron content in the Arad area.

The most important results of the PCA - regarding the parameters and groups - are presented in Table 5. Absolute values of factor loadings higher than 0.71 are indicated in bold and **red.** Regarding the PCA result it can be said that no consistent conclusion can be drawn without discrepancy, either globally, or regarding the different groups and parameters.

	Group 1 (Maros)			Group 2 (Körös)			Group 3 (Arad)		
	PC1	PC2	PC3	PC1	PC2	PC3	PC1	PC2	PC3
Water temperature	-0.721	0.267	-0.374	0.714	-0.277	0.061	0.117	0.860	-0.289
Conductivity	0.287	0.940	0.112	0.644	0.737	-0.136	0.981	-0.112	0.011
pH	-0.748	-0.061	0.173	0.365	-0.673	0.339	-0.272	0.489	0.481
Dissolved O_2	-0.184	-0.277	0.420	-0.395	0.325	-0.378	-0.656	-0.423	0.410
Na^+	-0.498	0.850	0.076	0.891	0.393	-0.155	0.888	0.260	-0.345
K^+	0.796	0.132	-0.165	0.289	0.585	0.575	0.090	-0.572	-0.593
Ca^{2+}	0.928	-0.263	-0.021	-0.791	0.504	0.160	0.795	-0.488	0.129
Mg^{2+}	0.902	0.026	-0.114	-0.729	0.602	-0.141	0.898	-0.083	0.322
Mn^{3+}	0.872	0.077	0.150	-0.713	0.319	0.391	0.812	0.259	0.310
Cl^-	0.159	0.329	0.842	0.388	0.520	0.469	0.799	0.038	-0.226
HCO_3^-	0.278	0.885	-0.240	0.594	0.741	-0.230	0.829	-0.112	0.385

Table 5. The factor scores of the parameters which suited the conditions of the PCA in each cluster group

3.3.3 Conclusions regarding the results obtained from the groundwater system's analysis of SE Hungary and the bordering Romanian area

The explicit separation of the groundwater characteristics in the different parts of the Great Hungarian Plain has been a well-known fact for a long time (Rónai, 1985). Based on the dissolved cations, the individual water types are related to the three extended hydrogeological units (Duna-Tisza interfluve South-Tiszántúl, Maros alluvial fan, Körös basin) (Rónai 1985; Varsányi & Ó. Kovács, 2006). Our investigations confirm this. Furthermore, the investigations expand what is already known with the result from the Romanian area. Based on the results of the applied multi-variate data analysis methods, the groundwater sampled in the Makó graben originating from the Duna-Tisza interfluve does not separate from the characteristic water type of the Maros alluvial fan.

Based on hydraulic modelling, results obtained from the upper 600 m (screen depths of drinking water supply wells) and water age data, the Maros alluvial fan consists of one uniform gravitational flow system. Towards Romania hydraulic heads gradually increase.

Based on the literature and results obtained, it seems that the **Arad group** is located at the beginning of the regional flow system, while its middle part and discharge area is situated in the Hungarian part of the Maros alluvial fan. Along the flow path in the Maros alluvial fan, depending on the quantity of bounded cations in the clay minerals, the ions with one and two valences may interchange. This results in the systematic change of the dissolved cations' concentration in the direction of the flow. In the cases of the $Ca(HCO_3)_2$ and $Mg(HCO_3)_2$ water types, the concentration of cations with two valences decreases, while the concentration of Na^+ increases in the direction of the flow (Varsányi, 2001).

The Körös basin, (bordering the Maros alluvial fan), is an individual hydrogeological system that is - based on the water ages and high Na^+content of the groundwater - situated at the end of a NE-SW oriented gravitational flow system. Besides the gravitational flow, the area can be characterized with slow up flow originating from sediment compaction.

These statements do not contradict the results explained above; nevertheless, our results show that differences in the three areas' groundwater chemistry are affected not only by gravitational flow systems. It is important to mention that data analysis methods may provide significant extra information during the exploration of a certain area's hydrogeological conditions, but separating different flow systems and flow regimes based only on data analysis is not possible.

The results of the PCA may reveal background processes taking place in a gravitational flow system, like cation change processes, or the role of Na^+, which has an important place in both **Körös** and **Arad group**. In the case of the **Arad group** this fact contradicts with the group's location in the flow system. Regarding other parameters (for example high chloride and sulphate concentrations), anthropogenic contamination can be in the background or regarding river's ablation area geological origin is feasible as well.

In order to determine the origin of the contaminants further investigations are needed. The place of the **Körös group** in the flow system does not contradict the high factor score of the Na^+ , however this high value in comparison to its factor score in the Maros alluvial fan implies a background process, the albite-montmorillonite reaction in the sediments of the basin at depths of 60-500 meters (Varsányi, 2001).

4. Summary

In our chapter we introduced a few methods known for decades in earth sciences and geology (Davis, 2003). We developed an order of application, which seemed beneficial during our work. The results obtained were of great use in studies when a hypothesis needed verification or discarding. When choosing the cases studies our direct aim was to present data sets with problems commonly faced by scientist through the three water environments: Kis-Balaton Water Protection System, River Tisza and a groundwater system of SW Hungary and SW Romania. We hope this chapter will be of use for every scientist who has to work with water quality data.

5. Acknowledgements

We the authors would like to thank Paul Thatcher for his work on our English versions, Antal Füst D.Sc. for his helpful advices, and say thanks for the stimulating discussions with our colleagues and students at the Department of Physical and Applied Geology of Eötvös Loránd University.

6. References

Afifi, A., Clark, V. A., May, S. & Raton B. (2004). Computer-Aided Multivariate Analysis (4th ed.). Chapman & Hall/CRC, p. 489. ISBN 1-58488-308-1, USA

Davis, J. (2003). *Statistics and data analysis in geology*, Wiley, p. 656, ISBN 978-0-471-17275-8, USA

Dryden, I., Márkus, L., Taylor, C. & Kovács, J. (2005). Non-Stationary spatio-temporal analysis of karst water levels, *Journal of the Royal Statistical Society*, Series C-Applied Statistics, Vol.54., No.3., pp. 673-690, ISSN 1467-9876

Füst, A. (1997). *Geostatistics* (in Hungarian: *Geostatisztika*). Eötvös Kiadó, p. 427, Budapest Hungary

Füstös, L., Meszéna, Gy. & Simonné, M. N. (1986). *The methods of multi-variate data analysis* (in Hungarian: *A sokváltozós adatelemzés statisztikai módszerei)*, Akadémiai Kiadó, ISBN 963-05-3859-8, Budapest Hungary

Geiger, J. (2007). *Geomathematics (in Hungarian: Geomatematika)*. Jate Press, p.116, Szeged Hungary

Hans, G. (2005) *Uni-variate Time Series in Geosciences*, Springer-Verlag, p. 718, ISBN 3540238107, New York USA

Hatvani, I. G., Kovács, J., Kovácsné Székely, I., Jakusch, P. & Korponai, J. (2011). Analysis of long term water quality changes in the Kis-Balaton Water Protection System with time series-, cluster analysis and Wilks' lambda distribution. *Ecol. Eng.*,Vol. 37, No.4, pp. 629-635, ISSN 0925-8574

Helsel, D. R. & Hirsch, R. M. (2002). *Statistical Methods in Water Resources Techniques of Water Resources Investigations*, Book 4, chapter A3. U.S. Geological Survey. 522 p.

Isvánovics, V., Honti M. & Vörös L. (2010). Phytoplankton dynamics in relation to connectivity, flow dynamics and resource availability—the case of a large, lowland river, the Hungarian Tisza. *Hydrobiologia*. Vol. 637, No. 1. pp. 121-141, ISSN 0018-8158

Ketskeméty, L. & Izsó, L. (2005). *Introduction into the SPSS system* (in Hungarian: *Bevezetés az SPSS programrendszerbe)*, ELTE Eötvös Kiadó, ISBN 963-463-823-6, Budapest Hungary

Körös Valley District Environment and Water Directorate [KÖRKÖVIZIG], Makó Térségi Vízmű, Smaragd-GSH Kft., Bálint Analitika, Selor eeig, Geological Survey of Finland, Regia Autonomia Apa - Canal Arad (2008): LIFE 05 ENV/H/000418: Sustainable management and treatment of arsenic bearing groundwater in Southern Hungary (SUMANAS)

Korponai, J., Braun, M., Buczko, K., Gyulai, I., Forro, L., Nedli, J. & Papp, I. (2010). Transition from shallow lake to a wetland: a multi-proxy case study in Zalavari Pond, Lake Balaton, Hungary. *Hydrobiologia* Vol. 641, No. 1. pp. 225-244, ISSN 0018-8158

Kovács, B. & Szanyi, J. (2005). *Hydrodinamic and transport modeling in MODFLOW environment* (in Hungarian: *Hidrodinamikai és transzportmodellezés Processing MODFLOW környezetben*) II., Miskolci Egyetem - Szegedi Tudományegyetem - GÁMAGEO, p. 160, ISBN 963 661 637 X

Kovács, J. & Kovácsné Székely, I. (2006). The problems in interpreting the sample: theory and practice (in Hungarian: A minta értelmezési problémái: elmélet és gyakorlat). *Földtani Közlöny*, Vol. 136, pp. 139–146, ISSN 0015-542X

Kovács, J. (2007). The application of modern geomathematical methods in practical hydrogeology, Ph. D. dissertation p. 122 (manuscript)

Kovács, J., Czauner, B., Kovácsné Székely, I., Borsodi, A. & Reskóné Nagy, M. (2008). The description of the changes in the different water quality areas of Lake Balaton and their time series' patterns (in Hungarian: A Balaton eltérő vízminőséggel rendelkező térségeinek változásai és a mért vízminőségi idősorok mintázatai, 1985-2004 között). *Hidrológiai Közlöny*, Vol. 88,No. 6, pp. 172-174, ISSN 0018-1323

Kovács, J., Hatvani, I. G., Kovácsné Székely, I., Tanos, P. & Korponai, J. (2011). Key question of sampling frequency estimation during system calibration, on the example of the Kis-Balaton Water Protection System's data series. *Georgikon for Agriculture* Vol. 14, No. 1. pp. 53-68 ISSN 0239-1260

Kovács, J., Hatvani, I. G., Korponai, J. & Kovácsné Székely, I., (2010). Morlet wavelet and autocorrelaiton analysis of long term data series of the Kis-Balaton Water Protection System (KBWPS). *Ecol. Eng.*, Vol. 36, No.10, pp. 1469-1477, ISSN 0925-8574

Kovács, J., Márkus, L. & Halupka, G. (2004). Dynamic Factor Analysis for Quantifying Aquifer Vulnerability. *Acta Geol. Hung*, Vol. 47, No. 1, pp. 1–17, ISSN 1588-2594

Lászlóffy, W. (1982). *The Tisza River. Water development and management in its river basin* (in Hungarian: *A Tisza, vízi munkálatok és vízgazdálkodás a tiszai vízrendszerben*) Akadémiai Kiadó, p.610, Budapest Hungary.

Lin, Z. & Puls, R. W. (2000). Adsorption, desportion and oxidation of arsenic affected by clay minerals and aging process. *Environmental Geology*, Vol. 39 , No. 7 , pp 753-759, ISSN 1866-6299

Lotz, Gy. (1988). The Kis-Balaton Water Protection System (in Hungarian: A Kis-Balaton Vízvédelmi Rendszer). *HidrológiaiTájékoztató*, Vol. 28, No.2

Márkus, L., Berke, O., Kovács, J. & Urfer, W. (1999). Spatial Prediction of the Intensity of Latent Effects Governing Hydrogeological Phenomena. *Environmetrics*, Vol. 10, No. 5., pp. 633-654, ISSN 1180-4009

Nádor, A., Thamó-Bozsó, E., Magyari, Á. & Babinszki, E. (2007). Fluvial responses to tectonics and climate change during the Late Weichselian in the eastern part of the Pannonian Basin (Hungary). *Sedimentary Geology*, Vol. 202 , No. 1-2 , pp. 174–192, ISSN 0037-0738

Nguyen, H. L., Leermakers, M., Kurunczi, S., Bozo, L. & Baeyens, W.,(2005). Mercury distribution and speciation in Lake Balaton, Hungary. *Science of the Total Environment*, Vol. 340, No.1-3, pp. 231-246, ISSN 0048-9697

Norušis, M. J. (1993). SPSS for Windows Professional Statistics Release 6.0. SPSS Inc. p. 385. ISBN 0-13-178831-0, USA

Padisák J. (2005). *General limonology* (in Hungarian: *Általános limnológia)*. ELTE Eötvös Kiadó, p. 310 ISBN 963-463-721-3, Budapest Hugary

Padisák, J. & Reynolds, C. S. (2003). Shallow lakes: the absolute, the relative, the functional and the pragmatic, *Hydrobiologia, Vol.506-509 , No.1-3* , 506-509, 1-11, ISSN 0018-8158

Pécsi, M. (Eds.) (1969) *The River Tisza in the Great Hungarian Plane* (in Hungarian: *A tiszai Alföld)*. Akadémiai Kiadó, p. 381, Budapest Hungary.

Pernthaler, J. (2005). Predation on prokaryotes in the water column and its ecological implications. *Nature Reviews Microbiology*, Vol. 3, pp. 537–546 ISSN 1740-1526.

Pomogyi, P., Szeglet, P. & Csato, E. (1996). The change in the reed community of the KBWPS, Feneki pond, based on vegetation-mapping results (In Hungarian). In: Pomogyi, P. (Ed.), 2. Kis-Balaton Conference: Summary of the Research Result on the KBWPS from Between 1991–1995. University of Pannonia Georgikon, Faculty of Agriculture, Keszthely, pp. 206–207.

Pukelsheim, F. (1994). The Three Sigma Rule. *The American Statistician*, Vol. 48, No. 2 pp. 88-91, ISSN 00031305.

Ramsar Convention, "Convention on Wetlands of International Importance especially as Waterfowl Habitat. Ramsar (Iran), 2 February 1971. UN Treaty Series No. 14583.As amended by the Paris Protocol, 3 December 1982, and Regina Amendments, 28 May 1987."

Ritter, A., Munoz-Carpena, R. & Bosch, D.D. (2007) Agricultural land use and hydrology affect variability of shallow groundwater nitrate concentration in South Florida, *Hydrological Processes* Vol. 21, No. 18., pp. 2464-2473, ISSN: 1099-1085

Rónai, A. (1985): The Quaternary of the Great Hungarian Plain. Acta *Geol. Hung., Ser. Geol.*, Vol. 21, p. 446, ISSN 1588-2594, Budapest Hungary

Sakan, S., Gržetić, I. & Đorđevoć, D. (2007). Distribution and Fractionation of Heavy Metals in the Tisa (Tisza) River Sediments. *Environmental Science and Pollution Research*, Vol. 14, No. 4., pp. 229-236.

Shumway, R. H. & Davis, S. S. (2000) Time Series Analysis and Its Applications, Springer-Verlag, p. 549, ISBN 0-387-98950-1, New York USA

Stockburger, D. W. (2001). Multivariate Statistics: Concepts, Models, and Applications, *Missouri State University*, Available from <http://www.psychstat.missouristate.edu/multibook/mlt00.htm>

T. Nagy, M., Csépes, E., Aranyné Rózsavári, A., Bancsi, I., Kovács, P., Végvári, P. & Zsuga, K., (2004). A hosszú-távú adatsorok értékelésének korlátai. *Hidrológiai Közlöny*, Vol. 84. No. 5-6., pp. 162-165.

Tátrai, I., Mátyás, K., Korponai, J., Paulovits, G. & Pomogyi, P. (2000). The role of the Kis-Balaton Water Protection System in the control of water quality of Lake Balaton. *Ecol. Eng.*, Vol.16, No.16, pp. 73–78, ISSN 0952-8574

Varsányi, I. & Ó. Kovács, L. (2006). Arsenic, iron and organic matter in sediments and groundwater in the Pannonian Basin , Hungary, *Appled Geochemistry*, , Vol. 21, No. 6, pp. 949-963, ISSN 0883-2927

Varsányi, I. (2001). Groundwater of the southern Great Hungarian Plain: hydrogeochemical processes and hydrogeological conclusions. D.Sc. Thesis, Szeged, p. 126. (manuscript)

Vollenweider R. A. (1976). Advances in defining critical loading levels of phosphorus in lake eutrophication. *Mem. Ist. Ital. Idrobiol*, Vol. 33, No. 1. pp. 53-83.

Vollenweider, R. A. & Kerekes, J. (1982). Eutrophication of waters. Monitoring, assessment and control. OECD Cooperative programme on monitoring of inland waters (Eutrophication control), Environment Directorate, OECD, Paris, 154 p.

Voudouris, K., Panagopoulous & A., Koumantakis, J. (2000). Multivariate Statistical Analysis in the Assessment of Hydrochemistry of the Northern Korinthia Prefecture Alluvial Aquifer System (Peloponnese, Greece). *Natural Resources Research*, Vol. 9, No. 2. Pp.135-146. ISSN 1573-8981

Voudouris, K.S., Lambarkis, N.J., Papatheothorou & G., Daskalaki, P. (1997). An Application of Factor Analysis for the Study of the Hydrogeologycal Conditions in Pilo-Pleistocene Aquifers of NW Achaia (NW Peloponnesus, Greece). *Mathematical Geology*, Vol. 29. No. 1., pp. 43-59. ISSN 0882-8121

Wetzel, R. G. (2001). *Limnology, Lake and River Ecosystems*. 3rd edition. Academic Press, Elseveir Science, p.1006 ISBN 0-12-744760-1, California USA

Wilkinson, D.J. (2006). Stochastic Modelling for Systems Biology. Chapman & Hall/CRC, p. 254. ISBN 978-158488-540-5, USA

Zsuga, K. & Szabó, A. (2005). The ecological problems of the Tisza's Hungarian cathchment area, and its environmental problems (in Hungarian: A Tisza hazai vízgyűjtőterületének ökológiai állapota, környezetvédelmi problémái.) *Hidrológiai Közlöny*, Vol. 85. No. 6., pp. 168-170, ISSN 0018-1323

4

Detecting and Estimating Trends of Water Quality Parameters

Janina Mozejko
West Pomeranian University of Technology, Szczecin
Poland

1. Introduction

Water quality variables frequently exhibit variability in time. This variability may be cyclical with the seasons, steadily (a trend), abruptly (a step-change) or some other established variation over time. It may affect the mean, median, variance, autocorrelation or almost any other aspect of the data. Detection of temporal trends is one of the most important objectives of environmental monitoring. Trend analysis indicates whether pollution concentrations are increasing or decreasing over time. In addition, an estimate of the trend's magnitude can help to determine whether a statistically significant trend is of practical concern. Noise in the observed data adds uncertainty to detect possibly trends and means that statistical methods are necessary to have accurate detection. The detection and estimation of trends is also complicated by problems associated with characteristics of pollution data. These characteristics are a possible presence of seasonality, skewness, serial correlation, non-normal data, "less-than" (censored) values, outliers and missing values (Hirsch et al., 1982; Hirsch & Slack, 1984), therefore special statistical tests have been developed to deal with these possibilities.

Many different statistical approaches are currently available for detecting and estimating trends that may be present in water quality variables of interest. These range from simple correlation and regression analyses, time-series analyses, and methods based on non-parametric statistics. Many are complex and too advanced for a basic water quality monitoring programme. Also useful for the detection of trends there are a variety of graphical techniques. For a detailed description of such methods readers may refer, for example, to texts such as Hirsch et al., 1982; Lettenmaier et al., 1982; Hirsch & Slack, 1984; van Belle & Hughes, 1984; Helsel, 1987; Lettenmaier, 1988; Hirsch et al., 1991; McLeod et al., 1991; Esterby, 1993, 1996; Reckhow et al., 1993; Kundzewicz & Robson (eds), 2000; Hess et al., 2001; Kundzewicz & Robson, 2004; Hesel & Frans, 2006; Visser et al., 2009. Several trend tests have also been described in statistical textbooks (e.g. Gilbert, 1987; Helsel & Hirsch, 1992; Chatfield, 2003; McBride, 2005). Readers who are interested in more advanced methods can refer to recently published book containing the practical application of modern statistical approaches to the analysis of trends in real environmental studies (Chandler & Scott, 2011).

This chapter is focused on presenting commonly used, basic methods for detecting monotonic increasing or decreasing trends in water quality variables, which may be useful for routine analysis of trends in environmental monitoring. It provides an overview of some of the main statistical concepts and terminology required for the statistical testing of trend and summarizes and examines some of the major issues and choices involved in detecting and estimating the magnitude of temporal trends in measures of water quality. Presented methods are rather simple and can be easily implemented using various commercial or open-source statistical software packages. As an example the trend analysis of some water quality parameters of the lower Odra River at the monitoring site in Police was performed.

2. Statistical methods for testing and estimating trends

Trend analysis can be defined as the use of an empirical approach to quantify and explain changes in a system over a period of time (Chandler & Scott, 2011). The purpose of trend testing is to determine if the values of a random variable generally increase (or decrease) over some period of time in statistical terms (Helsel & Hirsch, 1992).

Generally parametric (distribution-dependent) or non-parametric (distribution-free) statistical tests can be used to decide whether there is a statistically significant trend. A test is said to be parametric if the change evaluated by the test can be specified in terms of one or more parameters. Linear regression is an example of a parametric test. Parametric testing procedures are widely used in classical statistics. In parametric testing, it is necessary to assume an underlying distribution for the data (often the normal distribution), and to make assumptions that data observations are independent of one another. If parametric techniques are to be used, it may be necessary to (a) transform data so that its distribution is nearly normal and (b) restrict analyses to annual series, for which independence assumptions are acceptable, rather than using the more detailed monthly, daily or hourly data (Helsel & Hirsch, 1992). Statistical tests designed for normal distributions are very sensitive to outliers and difficult to apply to water-quality records with large numbers of "less-than" values (Schertz et al., 1991). In non-parametric and distribution-free methods, fewer assumptions about the data need to be made (Kundzewicz & Robson, 2004). With such methods it is not necessary to assume a distribution. However, many of these methods still rely on assumptions of independence. They are robust towards missing values, values reported as "below the detection limit" and seasonal effects (Gilbert, 1987). Statistics based on the ranks of observations are one example of such statistics and these play a central role in many non-parametric approaches.

The success of a statistical trend analysis largely depends on selecting the right statistical tools considering various aspects of the available data: whether the data are normally distributed or can be described by an alternative distribution function, whether the data contain seasonality, whether the trend is monotonic or abrupt, and whether the trends are expected to be univariate or multivariate (Visser et al., 2009). The main stages of trend analysis are (Kundzewicz & Robson, 2000)

- Obtaining and preparing a suitable dataset
- Exploratory analysis of the data
- Application of statistical tests
- Interpretation of the results

2.1 Obtaining a suitable dataset

There are many important aspects that may need to be considered when obtaining and preparing data. These include (Kundzewicz & Robson, 2000)

- Data should be quality controlled before commencing an analysis of change (e.g. via outlier detection).
- Data series should be as long as possible. Short data series can be strongly affected by seasonal variability which can give misleading results. For investigation of quality change, a minimum of 5 years of monthly data for monotonic trend analysis is suggested (Lettenmaier et al., 1982)
- Missing values and gaps in a data series make analysis harder and raise questions of data quality. Gaps in the data may have been filled, and the assumptions used in filling the data will invariably affect the results (e.g. infill techniques by use of historical values under assumption of stationarity, or interpolation based on values of neighbouring points in time and space). Many of the methods presented in this chapter can still be applied to incomplete data series provided that the gaps are not too extensive and that they occur randomly.
- Frequency of data. Hourly, daily, monthly and annual data series are commonplace. In a few cases, the data may be irregular. Very frequent data contains more information but can also be harder to analyse both computationally and because more restrictive assumptions must be made.
- Use of summary measures. It is often appropriate to analyze time series that have been derived from the raw data. For example, it may be sensible to calculate annual means or medians, or to derive annual maximum values.
- Use of transformation. Water quality data is often highly skewed and non-normal. In such cases, data analysis can sometimes be assisted if the data is first transformed.

2.2 Exploratory data analysis

Examination of the time series is very important before performing and interpreting the statistical calculations. Exploratory data analysis (EDA) is a very powerful graphical technique that is a key component of any data analysis (Kundzewicz & Robson, 2004; Chandler & Scott, 2011). EDA is an advanced visual examination of the data. It involves using graphs to explore, understand and present data, and is an essential component of any statistical analysis. The first use of EDA is usually to examine the raw data in order to identify such features as data problems (outliers, gaps in the record, etc.); temporal patterns (e.g. trend or step-change, seasonality); and regional and spatial patterns. Exploratory data analysis also plays an important role in checking out test assumptions such as independence, or statistical distribution of data values. Finally, EDA is invaluable when it comes to interpreting and presenting the results of a statistical analysis, e.g. for examining residuals, trend gradients and significance levels. A well-conducted EDA is such a powerful tool that it can sometimes eliminate the need for a formal statistical analysis. Alongside EDA, statistical tests become a way of confirming whether an observed pattern is significant. (Kundzewicz & Robson, 2004).

Common types of graph that can be useful for water quality data series include histograms, normal probability and quantile plots, time series plots, correlograms, scatter plots and smoothing curves. There are also other plot types, such as box plots or cumulative sum (CUSUM) charts. Use of these graphs should allow identification of the most important features of the data. In particular, they enable visual assessment of any trend or step-change – and can indicate how great trend is relative to overall variation.

Time series plots. The obvious starting point in a trend analysis is usually a plot of the observations against time. Time series plots are designed to illustrate trends with respect to time, together with any seasonality effects. When plotting time series graphs, the data should generally be displayed as either (i) individual points connected with lines if there are up to about 100 values, (ii) connected lines, if there are many values, or (iii) unconnected points if the data is irregular. If there are missing values in an otherwise regular series then the line should be broken at these points. It is sometimes necessary to plot lines through irregular data when there are very many values. To produce a time series plot that best displays the features of the data it may be necessary to plot the data on more than one scale or to transform the data. For water quality data the most common transformation is to take logarithms of the data. If the data series is very long, the display may be improved by spreading the data series over several plots, or by plotting summary statistics (Kundzewicz & Robson, 2000). Interpretation of the time series plot is often aided by adding a smoothing curve (and sometimes a regression line) to follow the general trend in the data. Smoothing methods include methods such running means, or locally weighted regression (lowess) (Cleveland & Devlin, 1988; Cleveland et. al., 1988). Care must be taken that the degree of smoothing is appropriate for the data, e.g. if seasonality is present the smoothing window must cover a number of years. When data from several sites or variables are available, it can be informative to examine the series together, e.g. presenting data for several sites within a region on a single page.

Boxplots (box and whisker plots). Time series plots are conceptually simple, but they do not fulfill all of the previously described aims of an exploratory analysis. The boxplot is an alternative graphical technique that remedies some of these deficiencies (Chandler & Scott, 2011). Boxplots are designed to facilitate comparisons between the distributions of observations falling in different groups - for example different months of the year, different years or different spatial locations. For each group of observations, the 'box' extends from the first to the third quartile of the data; the height of the box thus represents the interquartile range (IQR). The median is indicated in each box by a horizontal line. The whiskers extend in each direction to the most extreme data point that is at most 1.5 IQRs away from the box; observations more extreme than this are marked individually. The idea is that the whiskers represent the main body of each distribution and observations falling beyond the whiskers are possible outliers. Symmetric data would case the median to lie in the middle of the rectangle and the lengths of the upper and lower whiskers would be about the same. In summary, by allowing a visual comparison of entire distributions, boxplots provide a powerful supplement to time series plots in an exploratory analysis. They are particularly useful for assessing changes in the variability and shape of distributions and for identifying possible outliers.

Cumulative sum (CUSUM) charts are also effective graphical tool. The CUSUM curves (representing at a defined time cumulative concentration value in function of cumulative time) indicate existing trends and their type; continuous or discontinuous. A sudden change in the direction of the CUSUM indicates a sudden shift in the average. A period where the

CUSUM chart follows a relatively straight path indicates a period where the average does not change. Cumulative methods are easy and rapid techniques used for quality control and climatology. They have been developed by Letenmaier, Guilbot et al. and later on by Chassande for their application in the visualization of water quality data evolution (Letenmaier, 1976; Guilbot et al., 1986; Chassande, 1989, as cited in Cun & Vilagines, 1997).

Histograms, normal probability and quantile plots. These plots can be used to examine the distribution of raw data or residuals (Kundzewicz & Robson, 2004). Histograms show the general shape of the distribution. Quantile plots simply plot the data values against their rank, or against the equivalent quantiles from a reference distribution e.g. the normal. If the quantile plot gives a straight line then the data can be assumed to come from the required distribution. If it deviates significantly from a straight line then this indicates departure from the assumed distribution and indicates which part of the data deviates from this distribution. The normal probability plot is a graphical technique for assessing whether or not a data set is approximately normally distributed. The data are plotted against a theoretical normal distribution in such a way that the points should form an approximate straight line. Departures from this straight line indicate departures from normality.

Autocorrelation plots. Most tests for detecting change assume independence of the sample values. Independence means that knowing the current value of a variable provides no information about what the next value will be. This clearly does not hold for time series data, which are usually correlated due to being observed frequently or seasonally. One way to quantify the extent of the correlation (dependence) is to calculate the autocorrelation function ACF (see e.g. Kendall & Ord, 1990). Autocorrelation is a measure of the correlation of a variable with itself, but with the time shifted. For example, a lag 1 autocorrelation for a daily series is the correlation between the series and the same series but moved 1 day. The lag 2 autocorrelation is the correlation with a time difference of 2. The autocorrelation plot shows the correlations at a series of lags. If autocorrelation is present at one or more lags then the data is not independent. To aid interpretation of a correlogram, it is common to add horizontal lines showing the magnitude of coefficients that should be considered 'significantly' different from zero. These lines define approximate 95% confidence limits for individual coefficients, under the assumption that the observations are an uncorrelated sequence of values drawn from probability distributions with a common mean and variance (Chatfield, 2003).

For further details of exploratory data analysis the user should refer to Kundzewicz & Robson, 2000. Excellent presentations of the general principles of EDA can also be found in Cleveland, 1994 and Tufte, 1983.

2.3 Basics of statistical testing for trend

The main stages in statistical testing for trend are (Kundzewicz&Robson (eds), 2000):

- Decide what type of series/variable to test depending on the issues of interest (e.g. monthly averages, annual maxima, deseasonalized data, etc.).
- Decide what types of change are of interest (gradual trend or step-change).
- Check out data assumptions (e.g. use exploratory data analysis, or a formal test).
- Select a statistical test (more than one is good practice).
- Evaluate significance levels.
- Investigate and interpret results.

In order to carry out a statistical test, it is necessary to define the null and alternative hypotheses; these are statements that describe what the test is investigating (Helsel & Hirsch, 1992). For example, to test for trend in the mean of a series the null hypothesis (H_0) would be that there is no change in the mean of a series, and the alternative hypothesis (H_1) would be that the mean is either increasing or decreasing over time. In carrying out a statistical test, one starts by assuming that the null hypothesis is true, and then checks whether the observed data are consistent with this hypothesis. The null hypothesis is rejected if the data are not consistent. To compare between the null and alternative hypotheses, a test statistic is selected and then its significance is evaluated, based on the available evidence. The test statistic is simply a numerical value that is calculated from the data series that is being tested. A simple example of a test statistic is the linear regression coefficient; this can be used to test for a trend in the mean. If there is no trend (the null hypothesis) then the regression coefficient should have a value near to zero. If there is a large trend in the mean (the alternative hypothesis), then the value of the regression coefficient would be very different from zero, being positive for increasing trend and negative for decreasing trend.

Two types of errors can result from a hypothesis test.

- **Type I error**. A Type I error occurs when the researcher rejects a null hypothesis when it is true. The probability of committing a Type I error is called the **significance level**. This probability is also called **alpha**, and is often denoted by α.
- **Type II error**. A Type II error occurs when the researcher fails to reject a null hypothesis that is false. The probability of committing a Type II error is called **Beta**, and is often denoted by β. The probability of not committing a Type II error is called the **Power** of the test.

The significance level measures whether the test statistic is very different from the range of values that would typically occur under the null hypothesis. It is the probability that a test erroneously detects trend when none is present. Popular levels of significance are 10% (0.1), 5% (0.05), 1% (0.01). Choosing level of significance is an arbitrary task, but for many applications, a level of 5% is chosen. Thus a 5% significance level would be interpreted as strong evidence against the null hypothesis—with a 1 in 20 chance of that conclusion being wrong.

The analysis plan includes decision rules for rejecting the null hypothesis. In practice, statisticians describe these decision rules in two ways - with reference to a P-value or with reference to a region of acceptance.

- P-value. The strength of evidence in support of a null hypothesis is measured by the **P-value**. Suppose the test statistic is equal to S. The P-value is the probability of observing a test statistic as extreme as S, assuming the null hypothesis is true. If the P-value is less than the significance level, we reject the null hypothesis.
- Region of acceptance. The **region of acceptance** is a range of values. If the test statistic falls within the region of acceptance, the null hypothesis is not rejected. The region of acceptance is defined so that the chance of making a Type I error is equal to the significance level. The set of values outside the region of acceptance is called the **region of rejection**. If the test statistic falls within the region of rejection, the null hypothesis is rejected. In such cases, we say that the hypothesis has been rejected at the α level of significance.

These approaches are equivalent. Some statistics texts use the P-value approach; others use the region of acceptance approach.

A test of a statistical hypothesis, where the region of rejection is on only one side of the sampling distribution, is called a **one-tailed test**. For example, suppose the null hypothesis states that the mean is less than or equal to 10. The alternative hypothesis would be that the mean is greater than 10. The region of rejection would consist of a range of numbers located on the right side of sampling distribution; that is, a set of numbers greater than 10. A test of a statistical hypothesis, where the region of rejection is on both sides of the sampling distribution, is called a **two-tailed test**. For example, suppose the null hypothesis states that the mean is equal to 10. The alternative hypothesis would be that the mean is less than 10 or greater than 10. The region of rejection would consist of a range of numbers located on both sides of sampling distribution; that is, the region of rejection would consist partly of numbers that were less than 10 and partly of numbers that were greater than 10 (Stat Trek).

When interpreting test results it is necessary to remember that no statistical test is perfect, even if all test assumptions are met. If more than one test has been applied to the data, interpretation of results can be complex. The presence of a single significant test result may only be weak evidence of change- even if this test is highly significant. If more tests are significant then this provides stronger evidence of change, unless they are very similar, in which case multiple significance is not an extra proof of change. It is important to examine the test results alongside graphs of the data, and with as much historical knowledge about the data as possible (Kundzewicz & Robson (eds), 2000).

2.4 Selecting an appropriate test

There are many approaches that can be used to detect trends in water quality data. In deciding which approach to take it is necessary to be aware of which test procedures are valid (i.e. the data meets the required test assumptions) and which procedures are most useful (likely to correctly find change when it is present). Type of used statistic (parametric or non-parametric) depends on data characteristics

- Distribution (normal, skewed, symmetric, heavy tailed)
- Outliers (wild values that can't be shown to be measurement error)
- Cycles (seasonal, weekly, tidal, diurnal)
- Missing values (a few isolated values or large gaps)
- Censored data (less-than values, historical floods)
- Serial Correlation

The principal method for assumption-checking is to use visual techniques, such as are described in Section 2.2. They include

- Histograms, normal probability plots and boxplots – to examine distribution and identifying possible outliers.
- Time series plots - to spot time dependent patterns or possibly changes in variance
- Autocorrelation plots.

Assumption checking may need to be carried out both prior to and after application of tests. For example, if a trend is detected, then the trend should be estimated and removed from the data, and the residuals checked for autocorrelation and for constancy of distribution. Use of visual methods for assumption checking will usually be sufficient, however formal tests described in many standard statistical books are also available for checking some assumptions: tests for normality of data (e.g. Kolmogorov-Smirnov and the Shapiro-Wilk's test of normality) and tests for data independence (e.g. Bartlett's test) If the assumptions made in a statistical test are not fulfilled by the data, then test results can be meaningless, in the sense that the estimates of significance level would be grossly incorrect.

General applicability of tests is listed bellow.

- If data are normally distributed, independent and non-seasonal either parametric or nonparametric tests should be suitable.
- Traditional parametric test statistics based on normal distribution could not have been used if data is not normally distributed, is seasonal, contains extreme values, contains values less than a detection limit, or irregular sampling frequency.
- When the assumptions required by parametric statistical tests cannot be met, any of the distribution-free tests are suitable, because they are not particularly sensitive to missing data or outliers, and requires no assumption of normality

2.5 Some commonly used tests and test statistics

This section presents a number of standard tests for detection of temporal trends. The tests are described in their standard or basic form.

2.5.1 Linear regression

If plots of data versus time suggest a simple linear increase or decrease over time, a linear regression of the water quality variable Y against time T may be fit to the data (Gilbert, 1987). The test statistic for linear regression is the regression coefficient (slope). This is one of the simplest and most common tests for trend and, in its basic form, assumes that data are roughly normally distributed (i.e., symmetric and unimodal). It can be misleading if seasonal cycles are present, the data are not normally distributed, and/or the data are serially correlated. In general, the normality assumption is less important than the independence assumption, and the proposed procedures give reasonable estimates of the trend even if the normality assumption is violated but the independence assumption holds (PMFine/Reports, 1991). Linear regression may be able to use with transformed data (Helsel & Hirsch, 1991).

The model is formulated as follows

$$Y = a + bT + \varepsilon_T \tag{1}$$

where a is the intercept, b is the slope and ε_T are random errors. T represents the year with the initial year taken as year 1. The errors are assumed to be independent and identically distributed. This assumption is more likely to be satisfied if yearly average is used as the response variable.

The regression coefficients can be estimated using the method of least squares (OLS). Hence, the estimates of the slope and intercept are given by (Montgomery et al., 2001).

$$b = \frac{\sum (T - \overline{T})(Y - \overline{Y})}{\sum (T - \overline{T})^2} \tag{2}$$

$$a = \overline{Y} - b\overline{T} \tag{3}$$

where $\overline{Y}$ is the mean of Y, $\overline{T}$ is the mean of T

The standard error of the slope is

$$SE(b) = \sqrt{\frac{\sum (Y - a - bT)^2}{(n-2)\sum (T - \overline{T})^2}} \tag{4}$$

The OLS estimations of the slope b and the standard error of the slope can be obtained using statistical computing software.

A t test may be used to test that the true slope is not different from zero.

$$t_b = \frac{b}{SE(b)} \tag{5}$$

This test statistic follows a Student distribution with df = n-2 degrees of freedom under the null hypothesis. We conclude that linear trend exist if P-value of t_b statistic is less than α. A positive (negative) value of b indicates an upward (downward) linear trend and b is the magnitude of that trend. 100(1-alfa)% two-sided confidence interval for b is b ± t(1-α, df=n-2)SE(b). The strength of the trend can also be reported as a percent change over time by dividing the slope of the regression line by the mean value of the water quality parameter of interest over time.

To determine how much (or how little) of the change in the water quality parameter of interest is correlated with time, or stated another way, to determine the percentage of the variability in the water quality parameter of interest that is explained by the variability in time, we look at the coefficient of determination. The coefficient of determination (denoted by R^2) is interpreted as the proportion of the variance in the dependent variable that is predictable from the independent variable. $R^2 = 0.10$ means that 10 percent of the variance in *Y* is predictable from *X*; $R^2 = 0.87$ means that 87 percent is predictable; and so on.

After the model is fit, test for autocorrelation in the residuals using the Durbin-Watson statistic can be performed (Montgomery et al., 2001). The Durbin-Watson test is a test for first-order serial correlation in the residuals of a time series regression. The test statistic is

$$d = \frac{\sum_{i=1}^{n} (e_{i+1} - e_i)^2}{\sum_{i=1}^{n} e_i^2} \tag{6}$$

where $e_i = y_i - \hat{y}_i$ and y_i and $\hat{y}_i$ are, respectively, the observed and predicted values of the response variable for individual i. Value of *d* always lies between 0 and 4 and *d* becomes smaller as the serial correlations increase. The residuals are autocorrelated if P value for the d statistic is less than a selected level of significance.

The simple regression model of Equation (1) can be modified to account for seasonality in the observations and/or information from other covariates (Helsel & Hirsch, 1992). If autocorrelation seems to be present, then the regression model can still be used. However, it should be fitted using a method that is more appropriate than ordinary least-squares (see e.g. Zetterqvist, 1991). Some statistical packages provide one or more of these methods as options.

2.5.2 Spearman's rank correlation coefficient

This is a rank-based test for correlation between two variables that can be used to test for a correlation between time and the data series (Gauthier, 2001). It can be used for trend detection but cannot be used for trend estimation.

Spearman's rank correlation coefficient or Spearman's rho, often denoted by the Greek letter ρ (rho) or as r_s, is a non-parametric measure of statistical dependence between two variables. It assesses how well the relationship between two variables can be described using a monotonic function. The sign of the Spearman correlation indicates the direction of association between X (the independent variable) and Y (the dependent variable). If Y tends to increase when X increases, the Spearman correlation coefficient is positive. If Y tends to decrease when X increases, the Spearman correlation coefficient is negative. A Spearman correlation of zero indicates that there is no tendency for Y to either increase or decrease when X increases. The Spearman correlation increases in magnitude as X and Y become closer to being perfect monotone functions of each other. When X and Y are perfectly monotonically related, the Spearman correlation coefficient becomes +1 or -1.

The idea behind the rank correlation coefficient is simple. Each variable is ranked separately from lowest to highest (e.q. 1, 2, 3, etc.) and the difference between ranks of each data pair is calculated. Tied values (equal-valued) are assigned a rank equal to the average of their positions in the ascending order of the values.

The Spearman correlation coefficient is defined as the Pearson correlation coefficient between the ranked variables RX and RY

$$r_s = \frac{S_{RX,RY}}{\sqrt{S_{RX}S_{RY}}} \tag{7}$$

where

$$S_{RX,RY} = \sum_{i=1}^{n}(RX_i - \overline{RX})(RY_i - \overline{RY}) \tag{8}$$

$$S_{RX} = \sum_{i=1}^{n}(RX_i - \overline{RX})^2 \tag{9}$$

$$S_{RY} = \sum_{i=1}^{n}(RY_i - \overline{RY})^2 \tag{10}$$

$RX_i, RY_i, \overline{RX}, \overline{RY}$ refer to ranks and n is the number of data pairs

When there are no tied ranks, then there is simpler expression that may be used to obtain the Spearman correlation coefficient

$$r_s = \frac{1 - 6\sum_{i=1}^{n} d_i^2}{n^3 - n} \tag{11}$$

where d_i is the difference between ranks.

For small values of n, the significance level of the r_s test statistic can be looked up in special tables (Gilbert, 1987; Helsel & Hirsch, 1992). For samples with more than 20 values a t statistic can be calculated using equation

$$t = r_s \sqrt{\frac{n-2}{1-r_s^2}} \tag{12}$$

which is distributed approximately as Student's t distribution with $n - 2$ degrees of freedom (df). Hence if P-value associated with that t statistic is less than α, we reject the null hypothesis and conclude that there is a trend in the data. If r_s is positive, we conclude that there is an increasing trend, and if it is negative, we conclude that there is a decreasing trend.

Spearman correlation coefficient can be used to test for detecting trends in time series data with no seasonal effects. If seasonal cycles are present in the data, that test can be used after removing seasonal variation, or by examining data collected in the same month over several years (Gauthier, 2001).

2.5.3 Non-seasonal Mann-Kendall test

The Mann-Kendall test is a nonparametric trend test which has the same power as the Spearman's rho test in detecting monotonic trends (Yue et al., 2002). It is appropriate for data that do not display seasonal variation, or for seasonally corrected data, with negligible autocorrelation.

The non- seasonal Mann-Kendall test (M-K) is applicable in cases when the data values Y_i of a time series can be assumed to obey the model

$$Y_i = f(T_i) + \varepsilon_i \tag{13}$$

where $f(T_i)$ is a continuous monotonic increasing or decreasing function of time and the residuals ε_i can be assumed to be from the same distribution with zero mean. It is therefore assumed that the variance of the distribution is constant in time.

The M-K test is based on the statistic S (Gilbert, 1987). When only one datum per time period is taken, each pair of observed values Yi, Yj ($i > j$) of the random variable is inspected to find out whether $Yi > Yj$ or $Yi < Yj$. Let the number of the former type of pairs be P, and the number of the latter type of pairs be M. Then S is defined as

$$S = P - M \tag{14}$$

If n is 10 or less, the absolute value of S is compared directly to the theoretical distribution of S derived by Mann and Kendall (Gilbert, 1987). Then Ho is rejected in favor of H_1 if the probability value corresponding to the absolute value of S is less than the a priori specified α significance level of the test. A positive (negative) value of S indicates an upward (downward) trend. For time series time series with 10 or more data points the normal approximation is used. The test procedure is to first compute S using Eq.14 as described before. Then compute the variance of S by the following equations

$$Var(S) = \begin{cases} \dfrac{\left\{ n(n-1)(2n+5) - \sum_{j=1}^{p} t_j(t_j-1)(2t_j+5) \right\}}{18} & \text{if ties} \\ \dfrac{\{n(n-1)(2n+5)\}}{18} & \text{no ties} \end{cases} \tag{15}$$

where n is the number of data, p is the number of tied groups in the data set and t_j is the number of data points in the jth tied group.

Then S and Var(S) are used to compute the test statistic Z as follows

$$Z = \begin{cases} \dfrac{S-1}{Var(S)^{1/2}} & \text{if } S > 0 \\ 0 & \text{if } S = 0 \\ \dfrac{S+1}{Var(S)^{1/2}} & \text{if } S < 0 \end{cases} \tag{16}$$

There is a correction for ties (±1 added to the S) when $y_i = y_j$ (Salas, 1993, as cited in (Gilbert, 1987).

The standardized test statistic Z is approximately normally distributed. A positive (negative) value of Z indicates an upward (downward) trend. To test for the either upward or downward trend (a two-tailed test) at the α level of significance, Ho is rejected if $|Z| > Z_{(1-\alpha/2)}$. If the alternative hypothesis is for an upward trend (a one-tailed test), Ho is rejected if $Z > Z_{(1-\alpha)}$. We reject Ho in favor of the alternative hypothesis of a downward trend if Z is negative and $|Z| > Z_{(1-\alpha)}$. Using P- value calculated for Z, Ho is rejected if $P < \alpha$.

The Kendall's correlation coefficient, a measure of the strength of the correlation, can be calculated as (Kendall, 1975)

$$\tau = \frac{S}{D} \tag{17}$$

where

$$D = \begin{cases} \sqrt{\left\{ \dfrac{n(n-1)}{2} - \sum_{j=1}^{p} t_j(t_j-1) \right\}} \sqrt{\left(\dfrac{n(n-1)}{2} \right)} & \text{if ties} \\ \dfrac{n(n-1)}{2} & \text{no ties} \end{cases} \tag{18}$$

It attains values from the interval (-1, +1), where the sign indicates the slope and the absolute values indicates the strength of the relationship.

When there are multiple observations per time period, there are two ways to proceed (Gilbert, 1987). First, we could apply the Mann-Kendall test to the medians or means calculated for each time period. An alternative approach is to consider the $n_i \geq 1$ observations at time i (or time period i) as ties in the time index. For this latter case the statistic S is still computed by Eq.14, where n is now the sum of the n_j, that is, the total number of observations. The differences between data obtained at the same time are given the score 0 no matter what the data may be, since they are tied in the time index. When there are multiple observations per time period, the variance of S is computed by the following equations, which account for ties in the time indcx

$$Var(S) = \frac{1}{18}\left[n(n-1)(2n+5) - \sum_{j=1}^{p} t_j(t_j-1)(2t_j+5) + \sum_{q=1}^{h} u_q(u_q-1)(2u_q+5) \right] + \\ + \frac{\sum_{j=1}^{p} t_j(t_j-1)(t_j-2)\sum_{q=1}^{h} u_q(u_q-1)(u_q-2)}{9n(n-1)(n-2)} + \frac{\sum_{j=1}^{p} t_j(t_j-1)\sum_{q=1}^{h} u_q(u_q-1)}{2n(n-1)} \quad (19)$$

where p and t_j are defined following (Eq.15), h is the number of time periods that contain multiple data, and u_q is the number of multiple data in the qth time period. Equation 19 reduces to Equation 15 when there is one observation per time.

To estimate the magnitude of an existing trend (as change per unit time) the Sen's nonparametric method may be used (Sen, 1968). Sen's Slope estimator is a nonparametric alternative for estimating a slope. This approach involves computing slopes for all the pairs of time points Q_i and then using the median of these slopes as an estimate of the overall slope Q. As such, it is insensitive to outliers and can handle a moderate number of values below the detection limit and missing values.

$$Q_i = \frac{Y_j - Y_i}{T_j - T_i} \quad (20)$$

where Y_j and Y_i are data values a times (or during time period) j and i, respectively, and where j>i; N is the number of data pairs for which j>i. If there is only one datum in each time period, then N= n(n-1)/2, where n is the number of time periods. If there are multiple observations in one or more time periods then N < n(n-1)/2, where n is now the total number of observations, not time periods, since Eq.20 cannot be computed with two data from the same time period, that is, when j=i. The median of N slope estimates is obtained in the usual way. That is, the N values of Q are ranked from smallest to largest and we compute

$$Q = Q_{(N+1)/2} \text{ if N is odd}$$

or

$$Q = \tfrac{1}{2}(Q_{N/2} + Q_{(N+1)/2}) \text{ if N is even} \quad (21)$$

Trend slope Q is a measure of monotonic change during the selected study period. The trend slopes represent the median rate of change in constituent concentrations or values for the selected period. They assist the user in comparing the magnitudes of trends that represent the same period for stations in a study. A 100(1-alfa)% two- sided confidence interval about the true slope may be obtained by the nonparametric technique (Sen, 1968). This procedure based on the normal distribution is valid for n as small as 10 unless there are many ties. At first we compute

$$C_{\alpha} = Z_{1-\alpha/2}\sqrt{Var(S)} \tag{22}$$

where Var(S) is computed from Eq. 15 or 19. The latter equation is used if there are multiple observations per time period. Next $M_1=(N-C_{\alpha})/2$ and $M_2=(N+C_{\alpha})/2$ are computed. The lower and upper limits of the confidence interval Q_{min} and Q_{max} are the M_1th largest and (M_2+1)th largest of the N ordered slope estimates Q_i, respectively. If M_1 is not a whole number the lower limit is interpolated. Correspondingly, if M_2 is not a whole number the upper limit is interpolated (Gilbert, 1987).

This method is very useful in cases where the trend can be assumed to be linear (Salmi et all., 2002). This means that $f(T_i)$ in Equation 13 is equal to

$$f(T_i) = QT_i + B \tag{23}$$

where Q is the slope and B is a constant.

To obtain an estimate of B in Equation 23 the n values of differences $Y_i - QT_i$ are calculated. The median of these values gives an estimate of B. The estimates for the constant B of the 100(1-alfa)% confidence intervals are calculated by a similar procedure.

2.5.4 Regional non-seasonal Mann-Kendall test

When data are collected at several stations within a region, there may be interest in making a general statement about trends. This statement about the presence or absence of monotonic trends will be meaningful if the trends at all the stations are in the same directions - that is, all upward or all downward. Time plots of the data at each station, preferably on the same graph, may indicate when general statements are possible. In many situations an objective testing method will be needed to help make the decision. The following procedure was originally proposed by van Belle and Hughes to test for homogeneity of trends between seasons (van Belle & Hughes, 1984).

To test for homogeneity of trend at multiple stations, chi-square statistic, χ^2_{homog} can be used

$$\chi^2_{homog} = \chi^2_{total} - \chi^2_{trend} = \sum_{j=1}^{M} Z_j^2 - M\bar{Z}^2 \tag{24}$$

where Z_j is the Mann-Kendall Z statistic for jth station and M is the number of stations.

$$\bar{Z} = \frac{1}{M}\sum_{j=1}^{M} Zj \tag{25}$$

χ^2_{homog} has a chi-square distribution with M-1 degrees of freedom (df).

To test for trend homogeneity between stations, we can compare the P-value calculated for this chi-square statistic to the significance level. If $P < \alpha$ we reject the Ho of homogenous station trends (trends have significantly different directions at different stations). In that case no regional - wide statements should be made about trend direction. However, a Mann-Kendall test for trend at each station may be used. If $P > \alpha$ then the statistic $\chi^2_{trend} = \overline{Z}^2$ is referred to the chi- square distribution with 1 df to test that the (common) trend direction is significantly different from zero. The trend is significant if P value associated with χ^2_{trend} is less then α.

2.5.5 Seasonal Kendall test

If seasonal cycles are present in the data, tests for trend after removing these cycles from data or are not affected by them should be used. Hirsch, Slack, and Smith proposed the seasonal Kendall (SK) test when seasonality is present (Hirsch et al., 1982). This test is suitable for seasonal data with a moderate level of autocorrelation and may by used even though there are missing, tied or ND values. Furthermore, the validity of the test does not depend on the data being normally distributed. The SK test is a generalization of the Mann-Kendal test. It was proposed for use with 12 seasons (months), but it can be adapted to apply to non-monthly seasonal data (e.g. quarters of the year, weeks). The test consists of computing the Mann-Kendall test statistic S_i and its variance, $Var(S_i)$, separately for each month (season) with data collected over years. These seasonal statistics are then summed, and a Z statistic is computed.

$$S_K = \sum_{i=1}^{K} S_i \tag{26}$$

$$Var(S_K) = \sum_{i=1}^{K} S_i \tag{27}$$

$$Z = \begin{cases} \dfrac{S_K - 1}{Var(S_K)^{1/2}} & \text{if } S_K > 0 \\ 0 & \text{if } S_K = 0 \\ \dfrac{S_K + 1}{Var(S_K)^{1/2}} & \text{if } S_K < 0 \end{cases} \tag{28}$$

K is the number of seasons. The ±1 added to the S_K in Eq.28 is a correction factor. This correction is not necessary if there are ten or more data for each season (ni >=10).

To test the null hypothesis, Ho, of no trend versus the alternative hypothesis, H_1, of either an upward or downward trend (a two-tailed test), Ho is rejected if $|Z| > Z_{(1-\alpha/2)}$. If the alternative hypothesis is for an upward trend (a one-tailed test), Ho is rejected if $Z > Z_{(1-\alpha)}$. We reject Ho in favor of the alternative hypothesis of a downward trend if Z is negative and $|Z| > Z_{(1-\alpha)}$. Using P- value calculated for Z, Ho is rejected if P < α. To test for homogeneity of trend direction in different seasons at a given station, the procedure developed by van Belle and Hughes can be used (van Belle & Hughes, 1984). This latter test is important, since if the trend is upward in one season and downward in another, the seasonal Kendall test and slope estimator will be misleading. The procedure is to compute

$$\chi^2_{homog} = \chi^2_{total} - \chi^2_{trend} = \sum_{i=1}^{K} Z_j^2 - K\bar{Z}^2 \tag{29}$$

$$\bar{Z} = \frac{1}{K}\sum_{i=1}^{K} Z_i \tag{30}$$

where Z_i is the Mann-Kendall Z statistic for ith season and χ^2_{homog} has a chi-square distribution with K-1 degrees of freedom (df).

To test for trend homogeneity between seasons, we can compare the P-value calculated for the χ^2_{homog} statistic to the significance level. If P<α we reject the Ho of homogenous seasonal trends over time (trends have significantly different directions in different seasons). In that case the seasonal Kendall test and slope are not meaningful, and it is best to compute the Mann-Kendall test and Sen's slope estimator for each individual seasons. If P > α then the statistic $\chi^2_{trend} = K\bar{Z}^2$ is referred to the chi-square distribution with 1 df to test for a common trend in all seasons. This trend is significant if P value for χ^2_{trend} statistic is less than α.

The seasonal Kendall slope estimator is a generalization of Sen's estimator of slope. It is computed as the median of all slopes between data pairs within the same season. No cross-season slopes contribute to the overall estimate of the Seasonal Kendall trend slope. This slope is the median rate of change of *Y* over time. A 100(1-alfa)% confidence interval about the slope is obtained in the same manner as in Section 2.5.3.

An assumption with the seasonal Kendall test is that the statistics for the different seasons are independent. When this is not the case an adjustment for serial correlation can be made when calculating Var(S) (Hirsch & Slack, 1984).

2.5.6 Regional seasonal Kendall test

When data are collected at several stations special procedures developed by van Belle and Hughes for M > 1 stations can be used (van Belle & Hughes, 1984). These procedures allow one to test for homogeneity of trend direction at different stations when seasonality is present. The first step is to compute the Mann-Kendall statistic for each season at each station by Eq.14. Next, the appropriate chi-square statistics should be calculated.

The chi-square statistics to test for common trend at different stations when seasonality is present are collected in Table 1.

Chi-square statistic	df	Test	Decision
$\chi^2_{total} = \sum_{i=1}^{K}\sum_{m=1}^{M} Z^2_{im}$	KM		
$\chi^2_{trend} = KM\bar{Z}^2$	1	Test for significant common trend	Trend is significant if $P < \alpha$
$\chi^2_{homog} = \chi^2_{total} - \chi^2_{trend}$	KM-1	Test for homogeneity of trend direction at different stations in different seasons	Trend is homogenous if $P > \alpha$
$\chi^2_{season} = M\sum_{i=1}^{K} \bar{Z}^2_{iM} - \chi^2_{trend}$	K-1	Test for seasonal heterogeneity	Nonsignificant if $P>\alpha$
$\chi^2_{station} = K\sum_{m=1}^{M} \bar{Z}^2_{kM} - \chi^2_{trend}$	M-1	Test for station heterogeneity	Nonsignificant if $P>\alpha$
$\chi^2_{station-season} = \chi^2_{homog} - \chi^2_{station} - \chi^2_{season}$	(M-1) (K-1)	Test for interaction between stations and seasons	Nonsignificant if $P>\alpha$

Table 1. Chi-square statistics to test for common trend at different stations

K is the number of seasons and M is the number of stations and

$$\bar{Z}_{iM} = \frac{1}{M}\sum_{m=1}^{M} Z_{im}$$ - mean over M station for the ith season, i=1, 2, …, K

$$\bar{Z}_{Km} = \frac{1}{K}\sum_{i=1}^{K} Z_{im}$$ - mean over K seasons for the mth station, m=1, 2, …, M

$$\bar{Z} = \frac{1}{KM}\sum_{i=1}^{K}\sum_{m=1}^{M} Z_{im}$$ - grand mean over all KM stations and seasons

If both $\chi^2_{station}$ and χ^2_{season} are significant or if $\chi^2_{station-season}$ is significant, then the χ^2_{trend} test should not be done. The only meaningful trend tests in that case those for individual station-seasons.

3. Example – Trend analysis of water quality parameters of Odra River

The following paragraphs present the results of applying the statistical methodology outlined in the previous section. As an example trend analysis of water quality parameters of the lower Odra River at the monitoring site in Police was performed. Police is a small city located about 10 km north of Szczecin. Data used in the study were collected monthly by the National Inspection Board for Environmental Protection in Szczecin between 1991 and 2007. pH, dissolved oxygen (DO), biological and chemical oxygen demand (BOD and COD) were investigated in detail. The trend analysis was carried out for annual means and the original (monthly) observations. All statistical analyses were completed using Statistica 9 and Microsoft Excel computer programs at the 0.1 significance level.

3.1 Trend analysis of annual means

Times series and box plots of annual means of the selected variables are given in Fig. 1 and 2. The boxes show the interquartile interval and the black line in the middle is the median. The front whisker goes from Q1 to the smallest non-outlier in the data set, and the back whisker goes from Q3 to the largest non-outlier; observations more extreme than this are marked individually with a circle (outliers), and the extreme data with an asterisk. For symmetric data the median would lie in the middle of the box, and the lengths of the upper and lower whiskers would be about the same.

A visual inspection of the time series plots (Fig.1) indicates that a downward trend in BOD and COD may exist. The boxplots (Fig.2) show that the data depart from a normal distribution and there are no outliers and extreme values for the variables. Thus, nonparametric tests for trend are likely to be more powerful than conventional parametric techniques in the analysis of data. Time series graphs present no seasonality during the 1991-2007 period which allow to use tests for trend without any seasonal modification.

Firstly, non-parametric Spearman and Mann-Kendall tests for trend were applied and the magnitudes of statistically significant trends were estimated with the corresponding 90% confidence intervals. The trend slopes were also expressed as a percent of the mean water quality concentration by dividing the slope (in original units per year) by the mean and multiplying by 100 (relative trend slopes). Table 2 presents the Spearman and Mann-Kendall test statistics, slope estimates and the conclusions of the tests. The Spearman's rho test provides results almost identical to those obtained for the Mann-Kendall test.

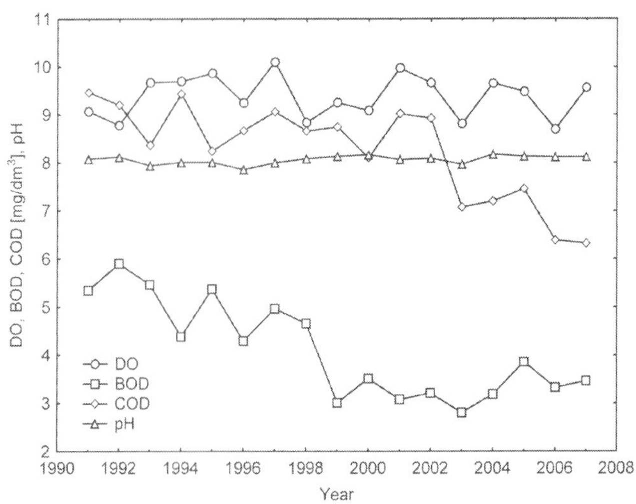

Fig. 1. Times series plots of annual means

Next temporal trends in data were also examined through linear regression analysis. The validity of the regression models are checked by the Durbin-Watson statistic (D-W), which is used for testing the serial correlation of error terms. The results showed insignificant autocorrelation structure of residuals (P-values > 0.1). Normal plots of residuals from linear models (Fig. 3) indicated deviation from normality only at the very extreme ends of the data range, allowing accepting regression analysis as an appropriate tool for detecting trends. Trend analysis results are given in Table 3.

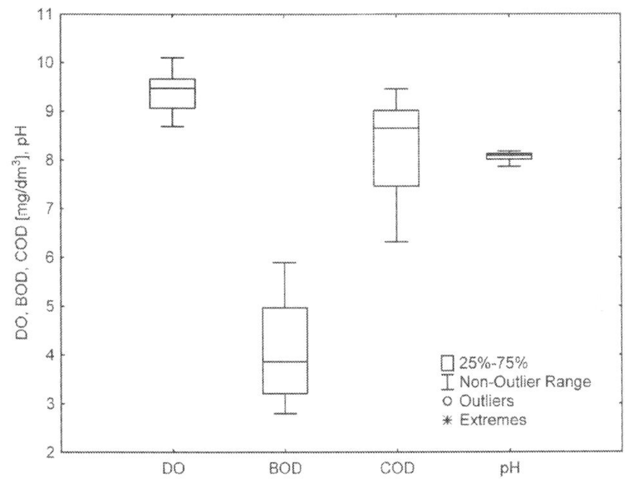

Fig. 2. Box plots of annual means

	Spearman test			Mann-Kendall test			Decision	Trend slopes			
	r_S	t	P value	Tau	Z	P value		Q	Q_{min} (90%)	Q_{max} (90%)	Q %
DO	-0.096	-0.37	0.715	-0.059	-0.33	0.742	No trend				
BZT	-0.748	-4.36	0.001	-0.529	-2.97	0.003	( ) trend	-0.164	-0.213	-0.107	-3.99
COD	-0.765	-4.60	0.000	-0.588	-3.29	0.001	(-) trend	-0.159	-0.223	-0.090	-1.98
pH	0.443	1.90	0.077	0.320	1.79	0.073	(+) trend	0.008	0.002	0.015	0.10

Table 2. Nonparametric tests results

When serial correlation is negligible, regression analysis can be a useful tool for detecting trend for normally and moderately skewed distributions. The t-test for regression coefficient is almost as powerful as the non-parametric Spearman or Mann-Kendall test. Therefore, these tests can be used interchangeably in practical applications, with identical results in most cases, as was shown on the presented examples.

According to the least-squares method and both non-parametric tests for trend, significant, negative trends in annual mean values of biological (BOD) and chemical oxygen demand (COD) at the significance level of 0.1 were detected. Over the selected study period of 1991-

2007, the estimated decrease in BOD and COD are about -0.16 mgO_2/dm^3 per year. pH shows a significant increase, but the yearly changes are very low (about 0.1% per year). It should be noted that pH data does not have a significant trend at the significance level of 0.05, but showing significant upward trend at α= 0.10. Trend in DO is no statistically significant.

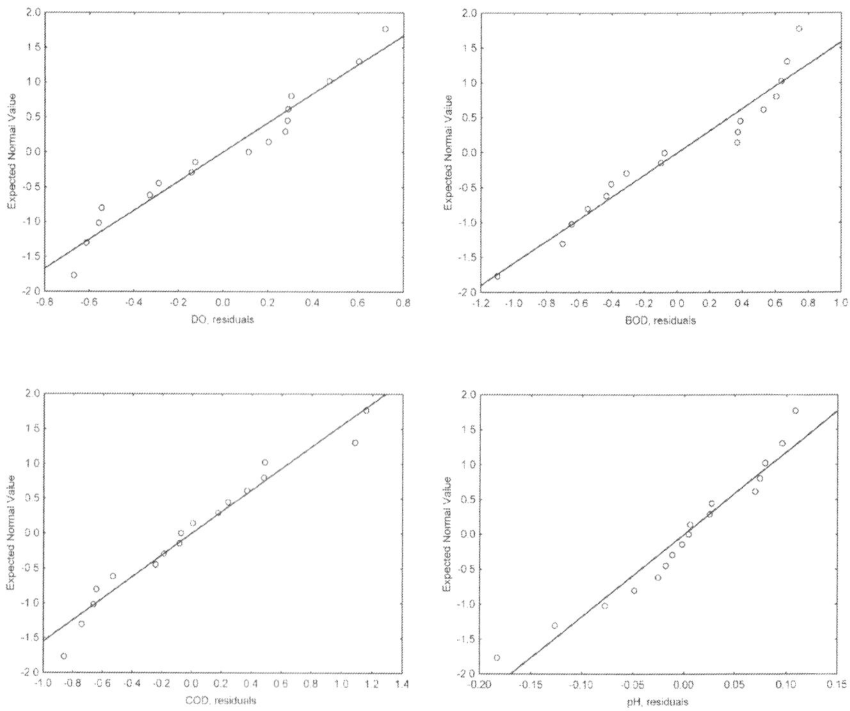

Fig. 3. Normal plots of residuals from linear models.

	slope b	t	P value	Decision	D-W statistic	P value	b_{min} (90%)	b_{max} (90%)	b%
DO	-0.003	-0.11	0.9150	No trend	2.39	0.128			
BZT	-0.165	-5.58	0.0001	Decreasing trend	1.57	0.114	-0.217	-0.113	-4.02
COD	-0.162	-5.29	0.0001	Decreasing trend	1.59	0.123	-0.216	-0.108	-2.01
pH	0.008	1.80	0.0928	Incrasing trend	1.66	0.154	0.0006	0.015	0.09

Table 3. Linear regression tests results

3.2 Trend analysis of monthly data

The data were first plotted to provide a general overview of possible trends as well as seasonal variations in the data material. Figure 4 shows time series plots of the monthly data. It can be seen that all variables are characterized by strong annual fluctuations.

Figure 5 depicts boxplots for individual years. The boxplots show that the data depart from a normal distribution not only in skewness, but also by the number of outliers and the extreme values which motivates the choice of the non-parametric Seasonal Mann-Kendall test for the following trend analysis which is robust against non-normal data and the presence of seasonality and outliers in the data.

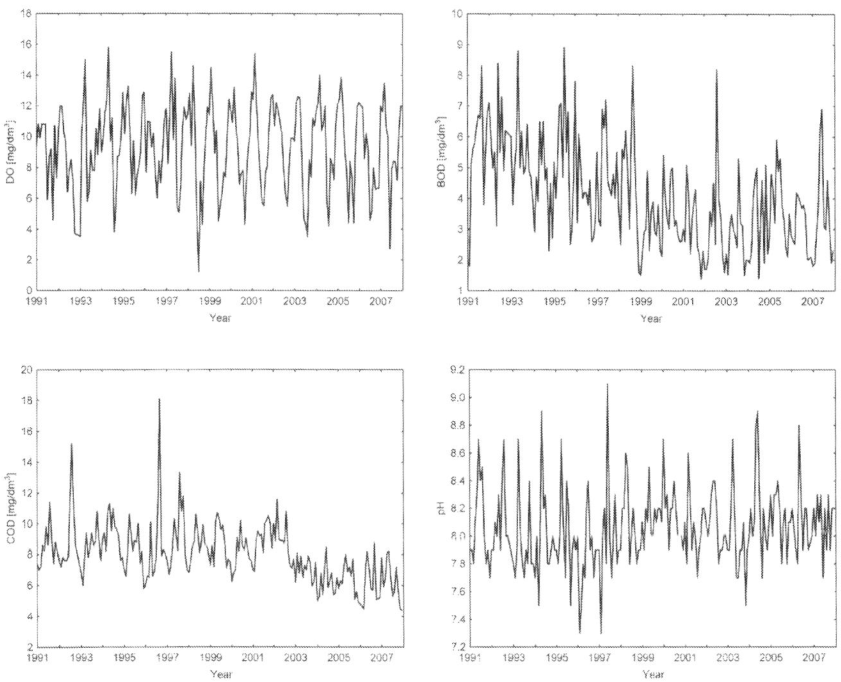

Fig. 4. Time series plots of selected variables

To calculate trends, a combination of statistical tests was used. First, the Mann-Kendall statistics for trend over years were computed for each "season". In this study, each of the 12 months of the year was defined as a season. Then the chi-square statistic measuring the overall change in data regardless of direction was computed. This statistic was partitioned into a chi-square due to homogeneity of trend among months and one due to trend. Each chi-square statistic was separately tested to determine significance. For data with a statistically significant monotonic trend the rate of change over time was estimated using a seasonal Sen Slope estimator, expressed as a change in units per year. It was computed as the median of all slopes between data pairs within the same season. The significance level of 0.1 was selected for all the trend tests.

Results of the seasonal Kendall trend tests and the Kendall slope estimators are given in tables 5-6. The results indicate significant, negative trends in biological (BOD) and chemical oxygen demand (COD), indicating improved water quality of Odra River. The estimated relative decrease in BOD based on the seasonal Sen Slope (about 4% per year) is twice larger then the relative decrease in COD (about 2% per year). No statistically significant trend in DO was detected. These trend results confirm the main conclusions from previous section.

pH shows trend but in differing directions in different seasons of the year. According to the M-K test (Table 4), the increasing trends are significant in January, March, September and December; however, the decreasing trend is significant in July and August. The increasing trends in winter months seem more significant than the decreasing trends in summer months. This heterogeneity of trend patterns results in a finding of "no consistent trend in one direction across all seasons" because values are increasing in some seasons while decreasing in others. This seasonal Kendall test result showed that only analyzing yearly means of pH values would not utilize all information in the data. Through the use of monthly values, the monthly variations are kept and the loss of information may be prevented.

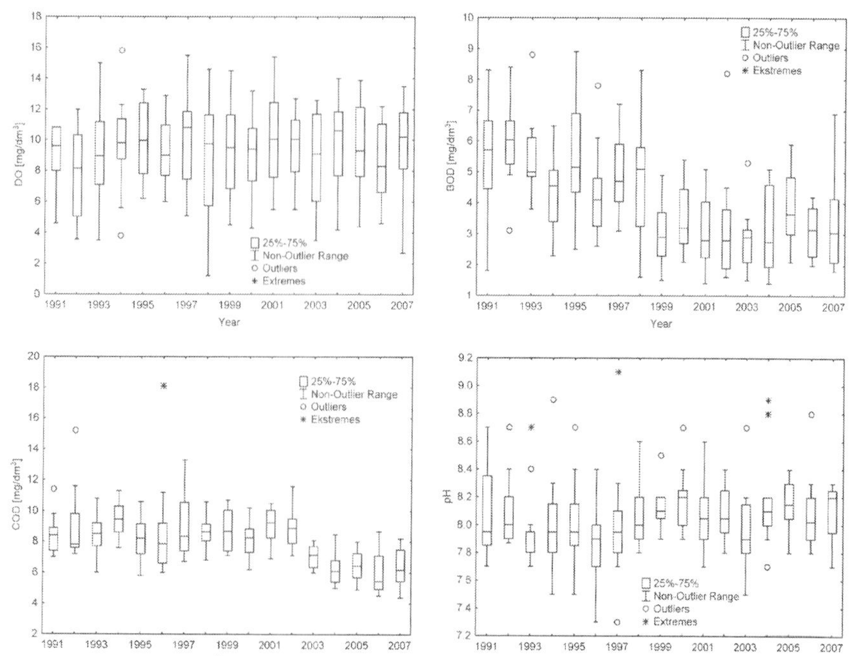

Fig. 5. Box plots for individual years

DO				BOD			
Month	Z	Tau	P value	Month	Z	Tau	P value
1	1.64	0.2922	0.1017	1	-2.11	-0.3764	0.0350
2	1.36	0.2521	0.1732	2	-1.58	-0.2929	0.1136
3	1.94	0.3469	0.0520	3	-2.28	-0.4061	0.0229
4	-1.12	-0.1993	0.2643	4	-2.27	-0.4059	0.0230
5	-0.46	-0.0818	0.6468	5	-1.29	-0.2305	0.1966
6	-0.33	-0.0593	0.7399	6	-1.51	-0.2687	0.1323
7	-1.49	-0.2667	0.1352	7	-2.88	-0.5147	0.0039
8	-0.96	-0.1710	0.3380	8	-2.51	-0.4478	0.0121
9	0.54	0.0967	0.5882	9	-2.19	-0.3911	0.0285
10	-0.63	-0.1124	0.5290	10	-1.70	-0.3026	0.0900
11	0.60	0.1160	0.5468	11	-2.78	-0.5359	0.0054
12	0.73	0.1345	0.4676	12	-2.62	-0.4852	0.0088
COD				pH			
Month	Z	Tau	P value	Month	Z	Tau	P value
1	-1.67	-0.2985	0.0944	1	3.12	0.5573	0.0018
2	-0.59	-0.1088	0.5567	2	1.60	0.2968	0.1088
3	-1.85	-0.3309	0.0638	3	2.24	0.4004	0.0249
4	-2.06	-0.3676	0.0394	4	-0.22	-0.0393	0.8257
5	-0.67	-0.1203	0.5002	5	1.06	0.1887	0.2903
6	-1.79	-0.3197	0.0733	6	0.17	0.0306	0.8641
7	-3.49	-0.6222	0.0005	7	-1.78	-0.3185	0.0744
8	-3.10	-0.5535	0.0019	8	-1.78	-0.3185	0.0744
9	-2.69	-0.4797	0.0072	9	2.43	0.4337	0.0151
10	-2.82	-0.5037	0.0048	10	-0.40	-0.0713	0.6894
11	-3.08	-0.5933	0.0020	11	1.45	0.2793	0.1467
12	-2.20	-0.4068	0.0279	12	3.56	0.6591	0.0004

Table 4. Mann-Kendall tests results for individual months

	$\bar{Z}$	χ^2_{total}	Test for trend heterogeneity			Test for significance of trend			Q	Q%
			χ^2_{homog}	p	Decision	χ^2_{trend}	p	Decision		
DO	0.15	14.61	14.33	0.2153	Non significant	0.28	0.5984	Non significant		
BOD	-2.14	58.14	3.01	0.9906	Non significant	55.13	1.1E-13	Decreasing trend	-0.16	-3.90
COD	-2.17	65.77	9.37	0.5878	Non significant	56.40	5.9E-14	Decreasing trend	-0.16	-2.03
pH	0.95	45.76	34.84	0.0003	Significant	10.93	0.0009	No consistent overall trend		

Table 5. Seasonal Kendall test results and slopes of the significant overall trends

4. Conclusion

The water quality of water is a subject of ongoing concern. Monitoring is the main tool used to determine the current status of water. An important goal of water quality monitoring is also trend analysis, driven by the desire to know whether the increased effort in waste and land-use management is having a beneficial effect.

This chapter has outlined the key components required for a study of change in water quality data, embracing stages such as preparing a suitable data set, exploratory analysis, application of adequate statistical tests and interpretation of results. The basic statistical tests and estimators described here, along with the use of exploratory data analysis procedures (including some of the types of graphics shown in this chapter), can be of great use in providing insights about water quality trends at a given site and about water quality trends over entire regions.

There is no unique solution to detect trends in water quality across widely differing catchments and monitoring systems. The choice of the method for trend detection and estimation should firstly be made on the basis of the specific goals of the study (only trend detection or also extrapolation), the available resources, and the system under study.

In general, distribution-free methods are recommended because they allow minimal assumptions to be made about the data and are therefore particularly suited to water quality series, which are often neither normally distributed nor independent. These methods are also well suited for analyzing datasets that have outliers, missing or tied data. Another advantage of the non-parametric procedures over parametric alternatives is in many cases, their relative simplicity. One disadvantage of these approaches is the relatively low power (i.e. a low probability of detecting a trend) in cases where the assumptions for a corresponding parametric test are reasonable. Another disadvantage for these procedures is that the some non-parametric tests (such as the Spearman's rho) may only be able to determine whether a statistically significant trend exists (trend detection) and cannot determine the size of the trend. The non-parametric techniques are particularly convenient to use in investigations of multiple data sets. However, in an analysis of an individual record, parametric methods, including use of transformations, can be very suitable. Their use requires careful checking of model fit and residuals. They are often more informative than the non-parametric procedures in more complex applications (Hirsch et al., 1991; Reckhow et al.,1993; Visser, A. et al., 2009).

This chapter showed that statistical methods are useful tools in water quality assessment provides useful information on the possibility of change tendency of the variables in the future. They are believed to assist decision makers in water quality evaluation and also determining priorities in management practices.

5. References

Chandler, R.E, & Scott, E.M. (2011). *Statistical Methods for Trend Detection and Analysis in the Environmental Sciences*, John Wiley & Sons, Ltd, ISBN 978-0-470-01543-8

Chatfield, C. (2003). *The Analysis of Time Series – An Introduction*, 6th edition. Chapman &Hall/CRC Press, ISBN-13: 978-1584883173, Boca Raton, Florida

Cleveland,W.S. & Devlin, S.J. (1988). Locally-weighted regression: an approach to regression analysis by local fitting. *Journal of the American Statistical Association*, Vol. 83, pp. 596-610.

Cleveland, W. S., Devlin,S. J. & Grosse, E. (1988). Regression by local fitting: methods, properties and computational algorithms. *Journal of Econometrics*, Vol. 37, pp. 87-114

Cleveland,W. S. (1994). *The Elements of Graphing Data*. Hobart Press, ISBN-13: 9780412742705, New Jersey, USA

Cun, C. & Vilagines, R. (1997). Time series analysis on chlorides, nitrates, ammonium and dissolved oxygen concentrations in the Seine river near Paris. *Science of the Total Environment*, Vol. 208, pp. 59-69, ISSN 0048-9697

Esterby, S.R. (1993). Trend analysis methods for environmental data. *Environmetrics*, Vol.4, pp. 459–481, ISSN 1099-095X

Esterby, S.R. (1996). Review of methods for the detection and estimation of trends with emphasis on water quality applications. *Hydrological Processes*, Vol.10, pp. 127-149, ISSN 1099-1085

Gauthier, T.D. (2001). Detecting trends using Spearman's rank correlation coefficient. *Environmental Forensics*, Vol.2, pp. 359-362, ISSN 1527-5930

Gilbert, R.O. (1987). *Statistical Methods for Environmental Pollution Monitoring*. Van Nostrand Reinhold, ISBN 0-471-28878-0, New York

Helsel, D.R. (1987). Advantages of nonparametric procedures for analysis of water quality data. *Journal of Hydrological Sciences*, Vol. 32, No. 2, pp. 179-189

Helsel, D.R. & Frans, L.M. (2006). Regional Kendal test for trend. *Environmental Science & Technology*, Vol. 40, No. 13, pp. 4066- 4073, ISSN 1547-6537

Helsel, D.R. & Hirsch, R.M. (1992). *Statistical Methods in Water Resources*, Elsevier, ISBN 0-444-81463-9, Amsterdam

Hess, A., Iyer, H. & Malm, W. (2001). Linear trend analysis: a comparison of methods. *Atmospheric Environment*, Vol. 35, pp. 5211–5222, ISSN 1352-2310

Hirsch, R.M., Alexander, R.B.& Smith, R.A. (1991). Selection of methods for the detection and estimation of trends in water quality. *Water Resources Res.*, Vol.27, pp.803-814

Hirsch, R.M. & Slack, J.R. (1984). A nonparametric trend test for seasonal data with serial dependence. *Water Resources Research*, Vol. 20, No.1, pp. 727-732

Hirsch, R.M., Slack, J.R. & Smith, R.A. (1982). Techniques of trend analysis for monthly water quality analysis. *Water Resource Research*, Vol.18, No.1, pp. 107–121

Kendall, M.G. (1975). *Rank correlation methods* (4th ed.), Charles Griffin, London

Kendall, M. & Ord, J.K. (1990). *Time Series*. 3rd edition, Edward Arnold, London

Kundzewicz, Z. W. & Robson, A. J. (eds) (2000). Detecting Trend and Other Changes in Hydrological Data. *World Climate Programme – Water, World Climate Programme Data and Monitoring*, WCDMP-45, WMO/TD no.1013. World Meteorological Organization, Geneva, Switzerland

Kundzewicz, Z. W. & Robson, A.J. (2004). Change detection in river flow records - a review of the methodology. *Hydrological Sciences Journal*, Vol. 49, No. 1, pp. 7-19, ISSN 2150-3435

Lettenmaier, D.P., Conquest, L.L. & Hughes J.P. (1982). *Routine Streams and Rivers Water Quality Trend Monitoring Review*. C.W. Harris Hydraulics Laboratory, Technical Report No. 75, Seattle, WA.

Lettenmaier, D.P. (1988). Multivariate nonparametric tests for trend in water quality. *Water Resources Bulletin,* Vol.24, pp. 505-512

McBride, G.B. (2005). *Using statistical methods for the water quality management: Issues, Problems and Solutions,* John Wiley & Sons, ISBN 0-471-47016-3, Hoboken, NJ. USA

McLeod, A.I. et al. (1991). Trend analysis methodology for water quality time series. *Environmetrics,* Vol.2, pp. 169–200, ISSN 1099-095X

Montgomery, D. C., Peck, E. A. & Vining, G. G. (2001). *Introduction to Linear Regression Analysis,* 3rd Edition, John Wiley & Sons, New York

PMFine/Reports (1991), Appendix. Statistical trend detection and analysis methods, Available from http://capita.wustl.edu/PMFine/Reports/TrendDetect/APXARAOP.PDF

Reckhow, K.H., Kepford, K. & Hicks W.W. (1993). *Statistical methods for the analysis of lake water quality trends.* U.S. Environmental Protection Agency Report 841-R-93-003

Salmi, T., M, Määttä, A, Anttila, P., Ruoho-Airola, T. & Amnell, T. (2002). *Detecting trends of annual values of atmospheric pollutants by the Mann–Kendall test and Sen's slope estimates - the Excel template application MAKESENS.* Publications on Air Quality, Finish Meteorological Institute, Helsinki

Sen, P.K. (1968). Estimates of the regression coefficient based on Kendall's Tau. *Journal of American Statistical Association,* Vol. 63, pp. 1379-1389

Schertz, T. L., Alexander, R. B. & Ohe, D. J. (1991). *The computer program estimate trend (estrend), a system for the detection of trends in water-quality data.* U.S. Geological Survey, Water-Resources Investigations Report 91-4040, Reston, Virginia

Stat Trek - Free Online Tutorials, Available from http://stattrek.com/Lesson5/HypothesisTesting.aspx?Tutorial=AP

Tufte, E. R. (1983). *The Visual Display of Quantitative Information.* Graphics Press, Cheshire, Connecticut, USA

Van Belle, V. G. & Hughes, J.P. (1984). Nonparametric tests for trends in water quality. *Water Resources Research,* Vol. 20, No. 1, pp. 127-136

Yue, S., Pilon, P.& Caradias, G. (2002). Power of the Mann-Kendall and Spearman's rho tests for detecting monotonic trends in hydrological series. *Journal of Hydrology,* Vol. 259, pp. 254-271, ISSN 0022-1694

Visser, A. et al. (2009). Comparison of methods for the detection and extrapolation of trends in groundwater quality. *Journal of Environmental Monitoring,* Vol.11, pp. 2030–2043, ISSN 1464-0333

Zetterqvist, L. (1991). Statistical estimation and interpretation of trends in water quality time series. *Water Resources Research,* Vol. 27, No. 8, pp. 1637–1648

5

Combining Statistical Methodologies in Water Quality Monitoring in a Hydrological Basin – Space and Time Approaches

Marco Costa[1] and A. Manuela Gonçalves[2]
[1]*CMAF and Escola Superior de Tecnologia e Gestão de Águeda, Universidade de Aveiro,*
[2]*CMAT and Departamento de Matemática e Aplicações, Universidade do Minho, Portugal*

1. Introduction

As water is a precious asset as well as a potential inducer of riches, water quality monitoring networks are important tools in the management and assessment of surface water quality and they could be improved by means of accurate forecasts of surface water variables.

The administration of hydrologic resources has been deserving a special prominence in the context of domestic and international politics in order to solve the complexity and uncertainty of the problems associated with a worldwide and local scale of sustainable administration (environmental, social, and economical) of natural water resources. Directive 2000/60/CE (UE 2000) of the European Parliament and Council (Water Framework Directive - WFD) establishing a framework for the community action in the field of water policy was incorporated into the domestic legislation in 2005 by Law nr 58/2005 and by Decree-Law nr 77/2006. According to this Directive, each member-state has to project, improve and recover all surface waters in order to achieve a good qualitative and quantitative status of all water bodies by 2015 (Machado et al., 2010). The relevance and interest that these subjects have been raising, specially in the Portuguese community, originated an entire group of state strategies and a new legal framework (INAG, 2008a, 2008b) derived from the WFD.

The river basin, which is the primordial unity of water resources planning and management, is usually submitted to pressures and changes due to human activities. Each hydrological basin is unique because it is a system that comprises orographic properties, ecological status, natural and anthropogenic factors, and where a network of water monitoring sites and a network of meteorological stations are integrated, thus allowing a correct monitoring of water quality.

At a river basin scale there is a need to establish a methodology for systematic data monitoring, characterization of surface water quality and a correct analysis of collected data (Vega et al., 1998). Surface water quality monitoring has as its main objective the characterization of water resources, as well as the monitoring of its space-time evolution in order to achieve an appropriate administration.

Water quality monitoring is an area encompassing a large set of disciplines. Statistical methodologies have been applied and developed with a particular emphasis on the last decades. Usually, data sets of environmental issues, namely of water quality measurements, have a significant complexity because they may have different properties simultaneously, and this implies a high level of tight, multi-disciplinary connection between water management technical bodies and instruments of analysis for decision-making (Vieira, 2003).

Multivariate statistical analyses have become widely applied in water quality assessment and sources apportionment of water over the last years (Wunderlin et al., 2001; Simeonov et al., 2003; Shrestha & Kazama, 2007). In several works, multivariate statistical analyses are applied to sets of water quality variables, usually comprised of quantitative analytical data. If the goal is to investigate the evaluation of water quality temporal or spatial variations as in Helena et al. (2000) or the natural and anthropogenic origin of contaminants in surface or ground water, as in Ato et al. (2010), the most suitable and applied approach is the principal components analysis (Liu et al., 2003; Lischeid, 2009; Varol & Sen, 2009). In some practical studies, there is data from a group of sample sites, usually from water monitoring sites, and in these cases it is useful to compare sampled data by means of several statistical methodologies: for instance, parametric and non-parametric correlation analysis tests (Elhatip et al., 2008).

Water quality data may present diverse dimensions of analysis, as is the case of spatial and time dimensions and multivariate and univariate dimensions. These dimensions can be separately analysed by means of suitable methodologies. However, if one dimension, space or time, is neglected in the analysis of the other, it can limit results and may even misrepresent conclusions. Also, multivariate and univariate dimensions require different approaches and statistical techniques. This aspect is decisive in water quality monitoring in a river basin featuring a set of water monitoring sites located in a main river and in its main adjacent streams.

In this work are discussed some statistical approaches that combine multivariate statistical techniques and time series analysis in order to describe and model spatial patterns and temporal evolution by observing hydrological series of water quality variables recorded in time and space. These approaches are illustrated with a data set collected in the River Ave hydrological basin located in the Northwest region of Portugal.

2. Study area and data description

Statistical methodologies will be illustrated based on a rather extended data set relative to the River Ave basin in Northwest Portugal (Figure 1) and consist mainly of monthly measurements of physicochemical and microbiological variables in a network of water quality monitoring sites.

In the last thirty years, the River Ave hydrological basin has been subjected, with the exception of its upstream areas, to a growing rhythm of untreated effluent discharges from industrial activities, namely from the textile sector strongly implanted in this region. This whole situation is instrumental for the water quality deterioration, resulting in inappropriate water for several uses—human consumption, industrial use, recreational uses, fishing and irrigation—, thus posing a serious danger for public health (Oliveira et al., 2005).

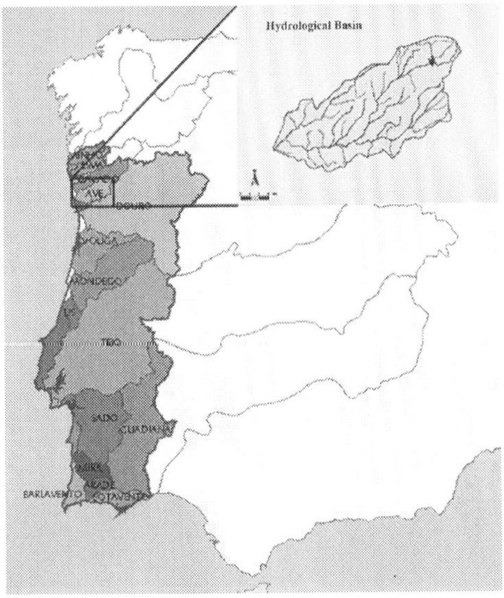

Fig. 1. River Ave hydrological basin.

The River Ave hydrological basin has an approximate area of 1390 km2, from its source in Serra da Cabreira to its mouth in Vila do Conde; the main river length is 101 km and the average flow at the mouth is 40 m3/s. In the Northwest region of Portugal summer is dry and winter is mild with plenty of rain. So, the highest levels of precipitation take place between October and March: this represents 75 percent of the yearly precipitation. The main adjacent streams of River Ave are the River Este (flowing from the North) and the Rivers Selho and Vizela (from the South).

The River Ave differs from other Portuguese rivers not only because of its high pollution levels, but also because of the large spatial and temporal variability of pollutants concentration.

Since 1988, and as part of a national plan, the Central Administration—through the Northern Regional Directory for the Environment and Natural Resources—and the Institute of Water periodically (monthly) monitored the quality of surface water along the River Ave and its main adjacent streams by means of 20 monitoring sites: twelve of them are located in the River Ave's mainstream: Cantelães (CANT), Garfe (GAR), Taipas (TAI), Pevidém (PEV), Pedome (PED), Riba d'Ave (RAV), Caniços (CAN), Portos (POR), Santo Tirso (STI), Ponte Trofa (PTR), Ponte Velha do Ave (PVA) and Formariz (FOR); Ponte Brandão (PBR) in the terminal segment of the adjacent stream River Selho; five in the adjacent stream River Vizela: Ferro (FER), Golães (GOL), Vizela Santo Adrião (VSA), Caldas de Vizela (CVI); and Louro (LOU), Balazar (BA) and Ponte Junqueira (PJU) in the adjacent stream River Este, (See Figure 2).

The data set comprises 11 quality variables: 10 physicochemical and a microbiological one (although there were more than 23 water quality variables available).

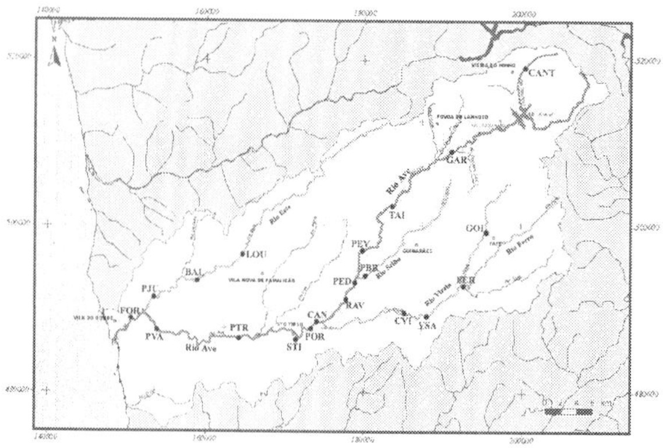

Fig. 2. Spatial distribution of the water quality monitoring sites of River Ave's hydrological basin.

Physicochemical Variables	Measurement Units
pH	Sorensen scale
WT - Water Temperature	°C (Celsius degrees)
COND - Conductivity	$\mu S / cm$
TSS - Total Suspended Solids	mg / l
DO - Dissolved Oxygen	mgO_2 / l
OD - Oxygen Demand	mgO_2 / l
COD - Chemical Oxygen Demand	mgO_2 / l
BOD_5 - 5-day Biological Oxygen Demand	mgO_2 / l
NH_4-N - Ammonical Nitrogen	$mgNH_4 / l$
NO_3-N - Nitrate-Nitrogen	$mgNO_3 / l$
Microbiological Variables	
FC - Faecal Coliforms	$n^{\circ} / 100ml$

Table 1. Water quality variables and measurement units.

According to the National Department for Pollution Control, these are relevant variables for the evaluation of surface water quality of rivers subjected to industrial effluent discharges. Table 1 summarizes these water quality variables and their measurement units. An exploratory analysis of all 11 quality variables allowed a general diagnosis of surface water quality of this hydrological basin during the period under observation. The samples presented extreme values (too high or too low) which were not excluded because they can denote serious situations from an environmental point of view. These values were confirmed as far as possible. Taking into consideration that modelling and forecasting procedures will be performed, the data set is divided into two parts: one for modelling proposes (until September 2004) and another part for the forecast and assessment stage (until October 2006).

3. Multivariate statistical analysis

In this section, different methodologies will be applied in the field of Multivariate Statistics—Clusters Analysis (CA) and Principal Component Analysis (PCA)—with the aim of evaluating and interpreting the space-time variations of a large and complex amount of data on surface water quality of any given hydrological basin. These methodologies have allowed the identification of homogeneous regions (i.e., groups of monitoring sites with similar characteristics in terms of quality variables) and the obtention of classification patterns that enabled us to generate hypotheses about the revealed structure involving modifying phenomena and thus better understand the mechanisms responsible for the surface water quality of the River Ave's basin.

The application of CA and PCA has achieved a meaningful classification of river water samples and has allowed the identification and assessment of spatial/temporal sources of variation affecting river water quality. In particular, cluster analysis (CA) allowed to reduce the large number of monitoring sites into a small number of homogeneous groups.

3.1 Cluster analysis

A first step consists of establishing a strategy to deal with a large data set collected in a group of water monitoring sites: for instance, twenty sites as in the present study case. CA is a group of multivariate techniques whose primary purpose is to assemble objects based on their characteristics (see Kaufman et al., 1990; Gordon, 1999, and Everitt et al., 2001). Hierarchical agglomerative clustering is the most common approach, providing intuitive similarity relationships between any given sample and the entire data set.

The strategy pursued in this paper follows a similar approach to Simeonov et al. (2003), Costa & Gonçalves (2011) and Gonçalves & Alpuim (2011), with the aim of geographically classifying homogenous groups of water quality monitoring sites based on water quality variables assessed by means of cluster analysis.

These approaches present different procedures in order to achieve the classification objectives. The first two papers consider clustering procedures that are implemented with a dissimilarity measure based on Euclidian distance and the last paper is performed with a dissimilarity measure based on Kullback information obtained in the state-space modelling process. It will be developed a discussion of its advantages and difficulties.

3.1.1 Cluster analysis of a global set of variables

In this study, hierarchical agglomerative CA was performed on the normalized data set (the 11 variables mentioned before). For the hierarchical agglomerative CA procedure purposes it will be considered the measure of dissimilarity proposed in Gonçalves & Alpuim (2011). The main problem is that for all locations and variables there are not observations for all months under study.

Therefore, let us consider x_{ikt} the value of the quality variable k, measured at location i, in time t. Let P_t be the set of all quality variables measured at the same time t, in sites i and j. The Euclidean distance between locations i and j at time t is given by the expression

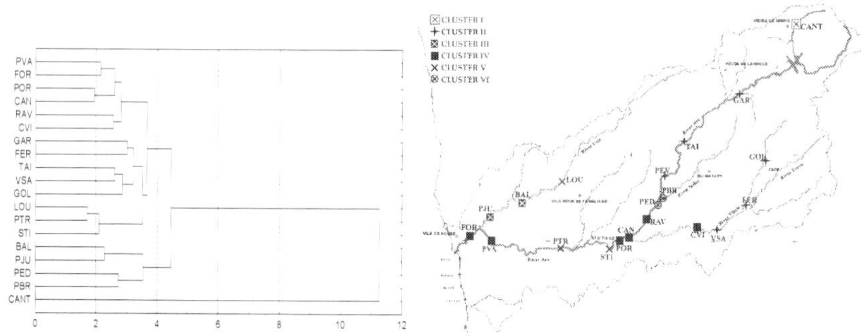

Fig. 3. Dendrogram of the 11 quality variables without PEV and the spatial representation of River Ave's clusters.

$$dist_{ij}(t) = \left[\sum_{k \in P_t} (x_{ikt} - x_{jkt})^2 \right]^{1/2} . \tag{1}$$

This dissimilarity measure corresponds to the average of this distance over all months t, where there is at least one quality variable with measurements in the two sites, that is

$$d_{ij} = \frac{1}{\# M_{ij}} \sum_{t \in M_{ij}} \left[\sum_{k \in P_t} (x_{ikt} - x_{jkt})^2 \right]^{1/2} \text{, with } i, j = 1, \ldots 20 \tag{2}$$

where M_{ij} is the set of all months with at least one variable measured in both sites i e j. The monitoring sites Pevidém (PEV) and Ponte Trofa (PTR) are the only ones lacking simultaneous measurements for any given quality variable. In order to avoid this problem, in a first phase PEV was dropped from the construction process of the clusters since it was the location with the smallest number of observations. Later, PEV was included in the closest neighbouring cluster. In order to ascertain the dissimilarity between PEV and the cluster including PTR, the latter was not inputted into the calculations. The construction used three different hierarchical cluster methods—Ward's, complete linkage and the unweighted pair-group average approach—because in this case they rendered well-defined clusters that were according to the reality of this particular river basin. As the results from the three methods are similar, the final result of the obtained groups was discussed according to the complete linkage method: in this method, the distance between the groups is defined as the distance between the most distant pair of objects, one from each group. According to the dendrogram analysis, we decided to form the clusters at a cut distance of d=3.4, thus obtaining six well-differenced clusters. The dendrogram of the monitoring sites obtained by means of the complete linkage method and the clusters geographical representation are shown in Figure 3. The resulting dendrogram has a cophenetic correlation coefficient of 0.85 (correlation coefficient between the original dissimilarity matrix and the "cophenetic matrix"), which validates the clustering procedure. Taking into consideration the quality variables averages within each cluster, they are classified into five categories according to their pollution levels established by the National Department for Pollution Control (NDPC): "Without Pollution (WP)", "Moderately Polluted (MP)", "Polluted (P)", "Very Polluted (VP)" and "Extremely Polluted (EP)". Each cluster is classified based on the NDPC criteria which are determinated according to the worst value of a given variable observed in the cluster. The resulting classifications of the six clusters confirm the previous knowledge about effluent discharge according to the economic activities located along the basin. Also, the effect of these discharges on water quality varies according to natural and geographical/economical reasons.

Cluster I (consisting of just one monitoring site-CANT) may be characterized as Without Pollution and corresponds to the source of River Ave. Then there is a set of locations which can be defined as Moderately Polluted (Cluster II, composed by GAR, TAI, PEV, GOL, FER and VSA), including 6 sites in both adjacent streams Este and Vizela situated upstream the Rivers Ave and Vizela. These stations receive pollution mostly from domestic wastewater and from agricultural and manure discharges.

Cluster III, classified as Polluted 1 (P1), is composed by BAL and PJU located in River Este, where the quantity of nitrate-nitrogen has been relatively high (the River Este tends to present the largest problems in relation to this quality variable).

In Cluster IV, six of the monitoring sites (RAV, CAN, POR, PVA, FOR, and CVI) are situated in the River Ave and only one, CVI, at the most downstream site of River Vizela. In fact, the polluted area (Cluster IV) corresponds to the segment of the River Ave that goes from around the station RAV down to its mouth. This is a densely populated region, with high industrial productivity, and here the River Ave receives similarly polluted waters (Polluted 2 (P2)) from its adjacent rivers. In Cluster V, with three monitoring sites, LOU (in river Este, downstream of the Municipality of Braga), STI and PTR (located near the most polluted area of the Municipality of Ponte Trofa and Santo Tirso), there is a growing urban population and a high concentration of industrial activity. Cluster V was classified as Very Polluted.

Finally, the most polluted cluster, Cluster VI (Extremely Polluted), consists of two monitoring sites, PBR and PED, located near the mouth of the Selho tributary and represents a highly polluted area. These monitoring sites receive pollution from domestic wastewater and industrial effluents located in city areas.

3.1.2 Cluster analysis for DO

As in Section 4, the analysis will focus on the modelling of Dissolved Oxygen (DO) concentration in water (measured in mg/l) because it is one of the most important variables in the evaluation of river water quality and because of its continuity in measurement at all selected water quality monitoring sites under analysis. This methodology intends to classify the water quality monitoring sites into spatial homogeneous groups based on the DO concentration, a variable considered relevant to characterize water quality. Furthermore, this type of analysis allows reducing the number of models in the modelling process. However, in this case the aim is to perform a clustering procedure for a univariate dimension, i.e., for a single water quality variable. Thus, the DO clustering procedure will be performed in an attempt to refine a methodology based on Kullback information measures that are obtained in the state space modelling process, as applied in Costa & Gonçalves (2011).

In order to identify homogeneous groups of water monitoring sites based on similarities in the temporal dynamics, we selected the modelling data sets by means of state space models and Kullback information measure, adapting here the methodology adopted in Bengtsson & Cavanaugh (2008). As the DO concentration showed much diversity regarding tendency and seasonality components in the River Ave and in its main adjacent rivers, we wanted to identify homogenous clusters of water monitoring sites considering the magnitude of DO concentration and then adopted a simple univariate state space model (SSM) for each location that considered DO in its true magnitude. By using a discrepancy measure suggested in Bengtsson & Cavanaugh (2008), we obtained a discrepancy matrix that allowed us to identify homogenous groups by applying clustering techniques.

In this modelling process are considered data series from 16 water monitoring sites (CANT, GAR, TAI, PBR, RAV, CAN, POR, STI, PTR, PVA, FOR, PJU, GOL, FER, VSA and CVI) between 1988 and 2006, because in the remaining four monitoring sites (PEV, PED, LOU, and BAL) the data is so scarce that it difficulted time modelling.

Briefly, it followed the main steps of the clustering proceeding. The discrepancy measure suggested in Bengtsson & Cavanaugh (2008) assumes that the variable $Y_{i,t}$ observed in location i at time t is modelled by a state space model as

$$Y_{i,t} = X_{i,t} + e_{i,t}, \tag{3}$$

$$X_{i,t} - \mu_i = \phi_i(X_{i,t-1} - \mu_i) + \varepsilon_{i,t}. \tag{4}$$

where (3) is the measurement equation and (4) is the transition or state equation. As usually, errors $e_{i,t}$ and $\varepsilon_{i,t}$ are uncorrelated white noises.

The unknown parameters $\Theta_i = \{\mu_i, \phi_i, \sigma_\varepsilon^2, \sigma_e^2\}$ in each location i are estimated by means of Gaussian maximum likelihood performed by EM algorithm (for more details, see subsection 4.1.3). Figure 4 reproduces parameters estimates for the 16 water monitoring sites.

The pseudo-distance between two monitoring sites i and j suggested by Bengtsson & Cavanaugh (2008), and defined as a form of the J-divergence (Kullback, 1978), accounts for the different lengths of each series of data sets by averaging over time

$$\overline{J}^X(Y_i,\Theta_i;Y_j,\Theta_j) = N_i^{-1} d^X(Y_i,\Theta_i;\Theta_j) + N_j^{-1} d^X(Y_j,\Theta_j;\Theta_i). \tag{5}$$

By employing output from the EM algorithm, including the maximum likelihood estimates, the sample $\overline{J}^X$ - divergence is given by

$$\overline{J}^X(Y_i,\hat{\Theta}_i;Y_j,\hat{\Theta}_j) = \frac{1}{2N_j\hat{\sigma}_{\varepsilon_i}^2}(S_{11}^{(j)} - 2\hat{\phi}_i S_{10}^{(j)} + \hat{\phi}_i^2 S_{00}^{(j)}) + \frac{1}{2N_i\hat{\sigma}_{\varepsilon_j}^2}(S_{11}^{(i)} - 2\hat{\phi}_j S_{10}^{(i)} + \hat{\phi}_j^2 S_{00}^{(i)}) - 1 \tag{6}$$

where smoothing quantities $S_{11}^{(k)}$, $S_{10}^{(k)}$, $S_{00}^{(k)}$ and parameters estimates $\hat{\Theta}_k$ are computed based on the model (Y_k, Θ_k). For more details, see Costa & Gonçalves (2011).

By using the parameters estimates of Figure 4 and the partial results of EM algorithm, the calculation of sample values $\overline{J}^X(Y_i,\hat{\Theta}_i;Y_j,\hat{\Theta}_j)$, $i,j = 1,\ldots 16$ allowed to obtain a matrix of pseudo-distances. The discrepancy matrix was subjected to Ward's, single linkage and complete linkage clustering procedures (Everitt et al., 2001). Because these three methods produced similar results, we only discussed the results obtaind through Ward's method. As shown in the dendrogram in Figure 5, the identified clusters are comprised by sites: Cluster I (CANT, TAI, GOL, FER, VSA); Cluster II (GAR, PJU, CVI); Cluster III (RAV, PVA, FOR); and Cluster IV (PBR, CAN, POR, STI, PTR).

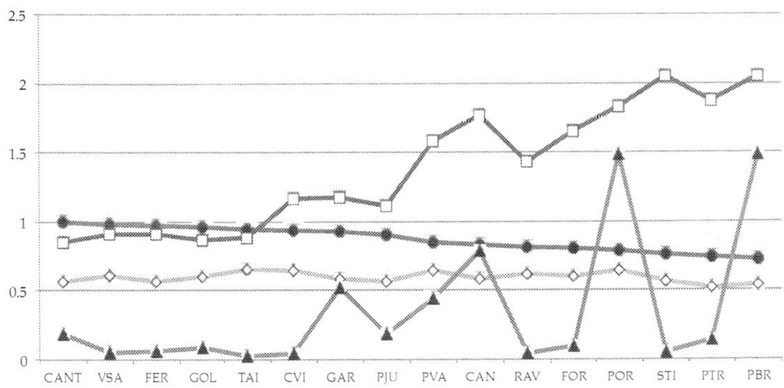

Fig. 4. Graphical representation of the parameters estimates to the 16 water monitoring sites (● - $\hat{\mu}\times 10^{-1}$, ◇ - $\hat{\phi}$, ▲ - $\hat{\sigma}_e$ □ - $\hat{\sigma}_\varepsilon$).

Considering the estimates of the processes mean obtained in the estimation procedure, it was clear that the clustering procedure performed a classification of the monitoring sites into a possible water quality scale in what concerns the annual mean DO concentration. In fact, the estimates of the processes mean in Cluster I monitoring sites presented the highest values obtained from DO concentration: the five monitoring sites of Cluster I presented the best water quality annual indicators, while the worst indicators are observed in Cluster IV monitoring sites. On the one hand, this methodology allowed classifying the water monitoring sites in four categories, considering the annual mean DO concentration: from best water quality (Cluster I) to worst water quality (Cluster IV). On the other hand, clustering procedure performs a discrimination of water monitoring sites based on state noise variance. Indeed, Cluster I corresponds to locations with the lowest state noise variances; Cluster II has state noise variances greater than Cluster I, and so on. Since the discrepancy measure tends to compare state densities $f(X_i\,|\,\Theta_i)$ and $f(X_j\,|\,\Theta_j)$, it is natural that clustering procedure depicts some patterns on the parameters of these distributions.

It is interesting to note that Cluster IV has water monitoring sites located downstream the confluences of Rivers Selho and Vizela, i.e., where the River Ave receives highly polluted waters from these adjacent streams. The water monitoring sites located in this middle stretch of the River Ave are much more polluted, probably because they are close to densely populated areas with high industrial production units.

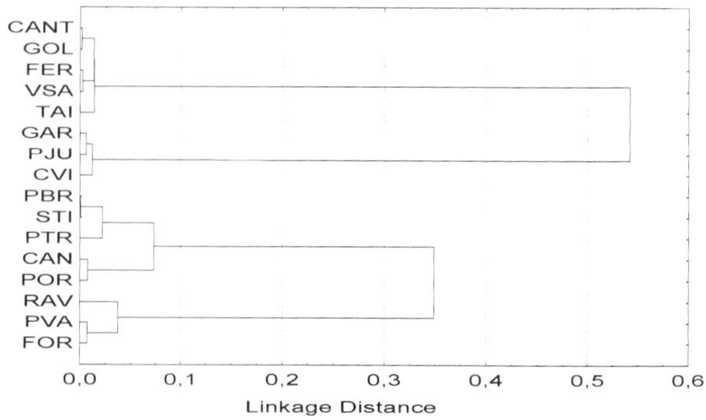

Fig. 5. Dendrogram showing the clustering of monitoring sites according to DO characteristics based on Ward's method.

3.2 Principal Component Analysis

The Principal Component Analysis (PCA) is designed to transform the original variables into new, uncorrelated variables called the principal components (PCs), which are linear combinations of the original variables (see Barnett, 1981, and Johnson et al., 1992). PCA allows us to explain and evaluate the correlation structure between observed variables in water quality sampling stations and to identify relevant factors. The PCA technique is separately applied to the homogeneous groups of water monitoring sites (six clusters), as obtained in the first clustering procedure, by taking into account all 11 water quality variables.

The Kaiser-Meyer-Olkin (KMO) statistics and Bartlett's test were performed in order to examine the data suitability for PCA. High values (close to 1) generally indicate that the principal component analysis may be useful, as is the case in this study (KMO=0.85). Also, the significance levels for Bartlett's test under 0.05 in this study indicate that there are significant relationships among variables.

Spearman rank-order correlations were used to study the correlation structure between variables in order to account for non-normal distribution of water quality variables.

PCA was separately performed on the raw data sets (11 variables) for the six different regions (clusters) WP, MP, P1, P2, VP and EP, as delineated by CA techniques, in order to compare the compositional pattern among the analysed water monitoring sites and to identify the factors influencing each one. To further reduce the contribution of variables with minor significance, the PCs were subjected to varimax rotation (raw) generating varifactors (VFs). The PCA of the six data sets yielded four PCs for the WP and MP monitoring sites, three PCs for the P1, P2, VP and EP monitoring sites with eigenvalues >1, explaining 69.86, 63.52, 69.91, 69.31, 65.66 and 73.12 of the total variance in the respective water quality data sets. An eigenvalue gives a measure of the factor significance: the factors with the highest eigenvalues are the most significant. Eigenvalues of 1.0 or greater are considered significant. Corresponding VFs, variable loadings and explaining variance are obtained for six clusters. In this chapter we will only present the results for Cluster VI (EP), in Table 2, and then their respective interpretation.

Variables	Varifactor 1	Varifactor 2	Varifactor 3
BOD_5	0.235	**0.778**	0.222
COD	0.251	**0.864**	0.063
TSS	-0.303	**0.751**	-0.260
DO	**-0.885**	-0.262	-0.093
OD	0.279	**0.905**	0.117
NH_4-N	**0.806**	0.086	0.092
NO_3-N	**-0.776**	-0.171	0.184
FC	**0.379**	0.273	0.301
COND	0.560	**0.588**	0.290
WT	**0.885**	0.019	-0.079
pH	-0.121	0.064	**0.913**
Eigenvalue	4.803	2.133	1.108
% Variance explained	43.663	19.387	10.070
Cumulative % variance	43.663	63.050	73.119

Table 2. Loadings of experimental variables (11) on the first three rotated PCs for EP sites data set (Bold values indicate the most important loadings).

Concerning the data set pertaining to EP sites, and based on the group of information resulting from the water quality analysis of the monitoring sites PBR (in River Selho) and PED (in River Ave), among the three varifactors kept in the application of ACP, Varifactor 1 explains 43.66 percent of the total variance, has strong negative loadings on the variables DO and nitrate-nitrogen and also strong positive loadings on ammonical nitrogen and temperature variables. This varifactor contains the variables most related to pollution of anthropogenic origin.

The same happens with Varifactor 2, which explains 19.38 percent of the total variance and presents strong positive loadings on BOD_5, COD, TSS, OD and COND. This organic factor can be interpreted as representing influences from point sources such as discharges from domestic wastewater and industrial effluents, but it cannot be interpreted only in terms of organic pollution, because it is also participated by the conductivity (mineral composition of the water).

Varifactor 3 explains 10.07 percent of the total variance, with strong positive loadings on pH, and presents the influence of this variable in the chemical processes in extremely polluted waters, from the nitrification processes of the nitrogen to the deposition of heavy metals. It is known that this basin's water should present low pH values (a natural characteristic of its own geomorphology). It is known that the River Selho is heavily polluted by the urban and industrial sewage of Guimarães and Pevidém, and so this varifactor reflects these modifier phenomena responsible for this serious environmental situation.

The obtained latent multifactors, with hydrochemical meaning, indicate that the responsible variables for the variation of the basin's water quality are mainly related with effluent discharges of anthropogenic origin (agricultural domestic and industrial origin) along the River Ave and its tributary streams. Only in areas Without Pollution or Moderately Polluted do latent factors represent the variability inherent to the natural climatic seasonality and the variability associated to the basin's geomorphological characteristics, both of which naturally influence the hydrochemistry of the rivers' surface water.

Although the principal component analysis did not result in a significant data reduction—by explaining the correlation among a set of variables in terms of a small number of principal components without losing much information—, nonetheless it helped to extract and identify the factors/sources responsible for variations in the rivers' water quality at six different sampling sites, and it also allowed to assess associations among variables, since they indicate the participation of individual physicochemical and microbiological variables in several influence factors. Varifactors obtained from component analysis indicate that the quality variables responsible for water quality variations are mainly related to discharge and temperature (natural origin), nutrients and organic pollution in relatively less polluted areas, pollution by organic matter and nutrients from anthropogenic sources (mainly as discharges of industrial and municipal wastewater), and manure affecting the quality and hydrochemistry of river water in highly polluted areas in the basin. For instance, in the less polluted group it is identified the most significant varifactor with positive weights in faecal coliforms (FC), water temperature (WT), and high negative weight in DO concentration. This means that high water temperatures associated with water contamination by organic matter cause an increase in the lack of DO in water. By contrast, in a group classified as "Very Polluted", the first varifactor (with 45.51 percent of the total variance) incorporates a significant set of organic-type variables, conductivity (COND) and pH variables. Besides representing an organic factor, it indicates anaerobic fermentation, hydrolysis of materials and the presence of mineral products (inorganic). This shows the strong influence of anthropogenic pollution in the much polluted clusters.

4. Statistical modelling of DO concentration

In this section, a temporal statistical analysis is exposed to illustrate the potential of some statistical approaches which, when combined, can be useful in understanding the evolution of water quality within a watershed. The starting point is based on the study case presented in Costa & Gonçalves (2011) that is developed and discussed here. Although there are more water quality variables available, DO concentration was selected due to its continuity in measurement at all selected water quality monitoring sites and to its importance in the evaluation of river water quality. The DO concentration analysis measures the amount of gaseous oxygen (O_2) dissolved in an aqueous solution. Oxygen gets into water by diffusion from the surrounding air by aeration and as a waste product from photosynthesis.

The DO in water is one of the most important quality variables to assess the degree of pollution existent in the surface waters of a river's hydrological basin. Low values indicate bad water quality. Oxygen levels also can be reduced through over-fertilization of water plants by run-off from farm fields containing phosphates and nitrates. DO concentration can be affected by organic pollution, which is the most common type of pollution in the River Ave's basin, and consequently, a frequent problem is a deficit of DO. This could be aggravated by the existence of a sequence of small dams in the River Ave and in its main adjacent rivers, which limit the oxygen's transfer by aeration. DO concentration is also affected by temperature: indeed, if water is too warm and there are too many bacteria or aquatic animal in the area, they may overpopulate, using DO in great amounts.

The evolution of DO concentration in the River Ave's basin is evaluated considering the clustering results presented in subsection 3.1.2. Thus, the modelling procedure will be performed for each cluster.

4.1 Prediction models

In order to satisfy prediction purposes, prediction models as linear regression models and state-space models are considered for each cluster. These approaches allow discussing the mode of incorporating seasonality and trend components in prediction models because state-space models integrate a dynamic structure that incorporates time dependence sometimes presented in environmental data.

Usually, linear models are preferred in respect to more complex models for being a primary tool in the context of environmental problems. Linear models are simple, have good statistical properties and are very robust statistical methods, which makes them a very attractive framework to describe the quality variables under study. However, the standard linear regression model does not include a possible time dependence of data. If errors of the linear models are correlated, the standard deviations of the coefficients given by linear model are not corrected, which may lead to a wrong decision given by the *t*-test. It is possible to overcome these difficulties considering that errors follow an AR(p) Gaussian stationary process. For more details, see Alpuim & El-Shaarawi (2008). Alternatively, in this work it is considered a linear state-space model whose formulation as presented can be interpreted as a calibration model of seasonal coefficients. As it will be shown, this formulation will allow making some useful interpretations.

4.1.1 Linear regression approach

Standard linear regression models are fitted to DO concentration data. In this case study, the authors consider that the observations at different locations in the cluster are treated as independent observations referenced in time, because there is no measure of space continuity.

Within each cluster it is considered a variable observation $Y_{j,t}^{(i)}$, where i represents the cluster, $i=1,2,3,4$, j represents the monitoring site running along all sites in the cluster i, $j=1,2,...,k_i$ and $t=1,2,..,n_j^{(i)}$ stands for the month. In order to contemplate both trend and seasonal components, the model in cluster i includes two additive components besides the error, that is,

$$Y_{j,t}^{(i)} = T_t^{(i)} + S_t^{(i)} + e_{j,t}^{(i)}. \tag{7}$$

After a graphical inspection, the trend is considered a simple linear function of time $T_t^{(i)} = \alpha^{(i)} + \beta^{(i)} t$. The seasonal component $S_t^{(i)}$ is a periodic function taking 12 different values $\lambda_s^{(i)}$ with $s=1,...,12$ associated with each month of the year, that is,

$$S_t^{(i)} = \begin{cases} 1, & \text{if date } t \text{ corresponds to month } s \\ -1, & \text{if date } t \text{ corresponds to month } 12. \\ 0, & \text{otherwise} \end{cases} \tag{8}$$

	Cluster I	Cluster II	Cluster III	Cluster IV
Intercept	9.197	9.867	8.634	7.711
Trend	0.004	-0.009	-0.003	-
January	0.960	1.382	2.045	2.540
February	1.117	1.410	2.122	2.457
March	0.602	0.577	0.983	1.323
April	-	0.784	0.708	1.027
May	-	-	-	-
June	-0.440	-1.093	-1.025	-1.352
July	-1.160	-1.557	-3.143	-3.391
August	-1.426	-2.037	-1.755	-1.501
September	-0.980	-0.912	-2.441	-3.240
October	-0.520	-0.546	-0.780	-1.143
November	0.720	0.711	1.183	1.118
December	1.125	1.281	2.103	2.161
$\hat{\sigma}^{2(i)}$	0.854	1.019	1.198	1.766
R^2	0.50	0.58	0.69	0.57

Table 3. Results for the linear models adjustment to the four clusters.

Note that the seasonal component is described with 12 dummy variables. In order for these parameters to be estimable, when the model has a constant term it is considered the restriction $\sum_{s=1}^{12} \lambda_s^{(i)} = 0$ with $\lambda_{12}^{(i)} = -\sum_{s=1}^{11} \lambda_s^{(i)}$. The choice of the twelfth month to be written as a linear combination of others is arbitrary. The model includes a stochastic term, $e_{j,t}^{(i)}$, which is taken as a sequence of uncorrelated zero mean random variables, with constant variance $\sigma^{2(i)}$.

In order to ensure the optimality properties of Ordinary Least Squares (OLS) method as the power of the t and F tests performed, a check of residuals shows that there were no significant violations of the normality. In order to obtain the final models, one for each cluster, it is used a backwards elimination procedure to select the significant variables. Thus, at each step, the regressor with the largest p-value for its t-statistics was removed, until all regressors were significant at level 0.05.

Table 3 presents the final results of the linear models' fitting to the four clusters. Briefly, as expected, the models adjustments show a seasonal pattern with lower values of DO concentration in the warmer months as compared to fall and winter months. Three clusters present weak trends: Cluster I shows a positive trend and Cluster II and III show a weak decreasing trend. The most polluted cluster does not present a significant trend, i.e., it has a stable behaviour, which may be justified if taken the highest error variance estimate into consideration.

4.1.2 State-space approach

State-space models are very versatile and can assume several state-space representations, namely by including linear models. For instance, if regression parameters $\boldsymbol{\beta}$ of the usual linear model $\mathbf{Y}=\boldsymbol{\beta}\mathbf{X}$ may vary over time, i.e., $\boldsymbol{\beta}_t$, the model may be treated as a dynamic model, thus admitting a state-space representation.

In order to identify possible changes over time, namely in trends, it is formulated a linear state-space model that takes into account only a seasonal component, since trends in linear models—even when trend is statistically significant—is very weak.

Thus, for cluster i it is considered the state-space model

$$\mathbf{Y}_t^{(i)}=\mathbf{S}_t^{(i)}\beta_t^{(i)}+\mathbf{e}_t^{(i)} \tag{9}$$

$$\beta_t^{(i)}=1+\phi^{(i)}(\beta_{t-1}^{(i)}-1)+\varepsilon_t^{(i)}. \tag{10}$$

The observation equation (9) relates the $k_i\times 1$ vector of observed DO concentration in the k_i monitoring sites of cluster i at month t, $\mathbf{Y}_t^{(i)}$, with the seasonal coefficient (mean value) of month t of cluster i, $s_t^{(i)}$, by means of a product of a stochastic calibration factor $\beta_t^{(i)}$ by $\mathbf{S}_t^{(i)}=\mathbf{1}_{(n_i)}s_t^{(i)}$ and adding to a zero mean vector error $\mathbf{e}_t^{(i)}$. Note that $\mathbf{1}_{(n_i)}$ represents a single matrix column $n_i\times 1$ with all elements equal to the unity.

It is assumed that the state, or in this case the calibration factor $\beta_t^{(i)}$, follows a stationary autoregressive process of order 1, AR(1), i.e., $|\phi^{(i)}|<1$, with unitary mean, expressed by (10). Both errors $\mathbf{e}_t^{(i)}$ and $\varepsilon_t^{(i)}$ are uncorrelated white noises with constant variances $E[\mathbf{e}_t^{(i)}\mathbf{e}_t'^{(i)}]=\sigma_e^{2,(i)}\mathbf{I}_{k_i}$ and $E[\varepsilon_t^{2,(i)}]=\sigma_\varepsilon^{2,(i)}$.

The state-space model (9) - (10) associated to the Kalman filter allows predicting calibration factor values at each month taking into account observed data up to that month (filtered predictions) and, if useful, one-step forecasts. An additional advantage of this approach consists of the filtered prediction of the calibration factor $\beta_t^{(i)}$ as a measure of the discrepancy between the observed value of the quality variable in a given month and the excepted value (the seasonality mean value) for this month.

For instance, if a calibration factor is predicted at 1.05 it means that this month's DO concentration value exceeded the seasonal coefficient of 5 percent. Thus, calibration factor predictions allow analysing the temporal dynamic of DO concentration by taking into account the expected structure translated by seasonal coefficients.

For each cluster it is necessary to estimate a set of parameters $\Theta^{(i)}=\{\phi^{(i)},\sigma_e^{2,(i)},\sigma_\varepsilon^{2,(i)}\}$. If the errors are normally distributed, the conditional log-likelihood of the random $(\mathbf{Y}_1^{(i)},\mathbf{Y}_2^{(i)},\dots,\mathbf{Y}_n^{(i)})$ can be written as

$$\log L(\Theta^{(i)};\mathbf{Y}_1^{(i)},\mathbf{Y}_2^{(i)},\dots,\mathbf{Y}_{n_i}^{(i)})=-\frac{n_i}{2}\log(2\pi)-\frac{1}{2}\sum_{t=1}^{n_i}\log\left(\left|\boldsymbol{\Omega}_t^{(i)}\right|\right)-\frac{1}{2}\sum_{t=1}^{n_i}\boldsymbol{\eta}_t'^{(i)}\boldsymbol{\Omega}_t^{-1,(i)}\boldsymbol{\eta}_t^{(i)} \tag{11}$$

where $\Omega_t^{(i)}=\mathbf{S}_t^{(i)}p_{t|t-1}\mathbf{S}_t'^{(i)}$ and $\eta_t^{(i)}=\mathbf{Y}_t^{(i)}-\hat{\mathbf{Y}}_{t|t-1}^{(i)}$ is the innovation (note that $p_{t|t-1}$ and $\hat{\mathbf{Y}}_{t|t-1}^{(i)}$ are obtained by Kalman filter algorithm: see the Annex).

Thus, it is possible to obtain the maximum likelihood estimates by maximizing the conditional log-likelihood in order to obtain the unknown parameters $\Theta^{(i)}$, for each cluster

$i=1,2,3,4$, by using numerical algorithms, namely the EM or the Newton-Raphson algorithms. Figure 6 reproduces estimates obtained in Costa & Gonçalves (2011). It is quite clear that the adopted approach reveals the differences between clusters in the context of DO concentration. As expected, the standard deviations estimated from the calibration factors are lower than the standard deviations resulting from the observation of the equations errors. Indeed, the state-space model separates two sources of variability reflected in two errors: observation and state equations. Thus, it is reasonable that calibration factors have less variability while the observation errors incorporate measurement errors and, most importantly, sporadic and unpredictable events as illegal pollutants discharges, among others. However, Cluster IV stands out from others because it presents the largest variability in error equation as much as in state equation error. So, Cluster IV presents the worst monthly averages of DO concentration but it also shows the highest variability.

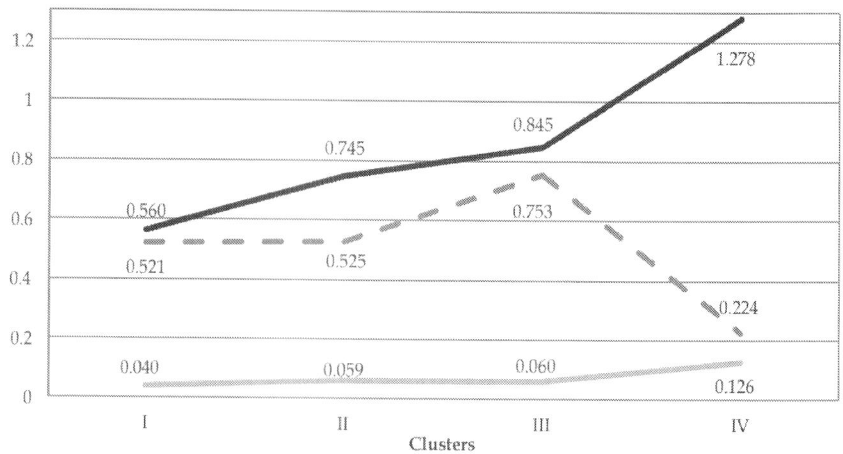

Fig. 6. Graphical representation of the four clusters estimated values via Gaussian maximum likelihood (- - $\hat{\phi}$, —— - $\hat{\sigma}_{\varepsilon}$, —— - $\hat{\sigma}_{e}$).

4.1.3 Model's adjustment discussion

As mentioned before, the modelling process focuses on a subset of available data. The remaining data (from October 2004 to October 2006) was used to assess forecast performance of models. Costa & Gonçalves (2011) shows that the state-space approach provides the least Root of the Mean Square Errors of forecasts (RMSE). Indeed, the RMSE obtained in state-space models was RMSE=0.846, while in the linear regression models the result was RMSE=0.961. Thus, the state-space approach improved the forecast accuracy in the sense of the mean square error of forecasts. Besides providing better forecasts of DO concentration, the state-space approach allows analysing temporal evolution in a dynamic way by observing calibration factor predictions. Other potentialities of the state-space models will be explored in the next section.

4.2 Online water quality monitoring by using Kalman filter predictions

The forecast accuracy is not always the most important characteristic of a model. Indeed, in water quality monitoring it is useful to analyse temporal evolution in order to identify possible factors or changes. This analysis can be performed in two ways:

- trends analysis by looking into the historical data;
- dynamic or online procedure.

The first approach can be achieved largely through linear regression models by analysing trends parameters and it allows diagnosing any global tendency (linear or other, as exponential, polynomial, etc.) that can be statistically significant.

Thus, the next discussion focuses on the dynamic monitoring procedure based on the state-space approach (associated to the Kalman filter algorithm). Indeed, the state-space approach adds a dynamic component to the usual linear models that can be useful in the water quality monitoring procedure. Considering that the DO concentration at each month can be regarded as an updated value of the expected value for a given month s_t (seasonal coefficient), that update is done by a multiplicative factor β_t by the observation equation

$$\mathbf{Y}_t = \mathbf{S}_t \beta_t + \mathbf{e}_t , \tag{12}$$

and so the online prediction of β_t for each month t by Kalman filter algorithm indicates how much any given monthly DO concentration diverges from the its monthly average. In order to achieve this, it is necessary to compute filtered values of β_t (in this case for all clusters) for all months.

Given the probability distribution of predictor $\hat{\beta}_{t|t}$,

$$\hat{\beta}_{t|t} \sim N\left(\beta_t, p_{t|t}\right), \tag{13}$$

the 95% point-wise prediction intervals are computed by the boundaries

$$\hat{\beta}_{t|t} \pm 1.96\sqrt{p_{t|t}} , \tag{14}$$

where $\hat{\beta}_{t|t}$ is the filtered value of β_t when Y_t is observed and $p_{t|t}$ is the estimated MSE of $\hat{\beta}_{t|t}$. If the DO concentration measurement shows no statistical change in comparison to the monthly mean, then, according to the statistical inference theory, the unity must be in the point-wise prediction interval corresponding to that month, that is,

$$1 \in \left]\hat{\beta}_{t|t} - 1.96\sqrt{p_{t|t}}, \hat{\beta}_{t|t} + 1.96\sqrt{p_{t|t}}\right[\tag{15}$$

considering a significant level of 5%.

Figure 7 represents the 95% point-wise predictions intervals for the four clusters. This procedure allows revealing the monthly measurements that are statistically different from the global seasonal component, because the value is higher or lower than the expected.

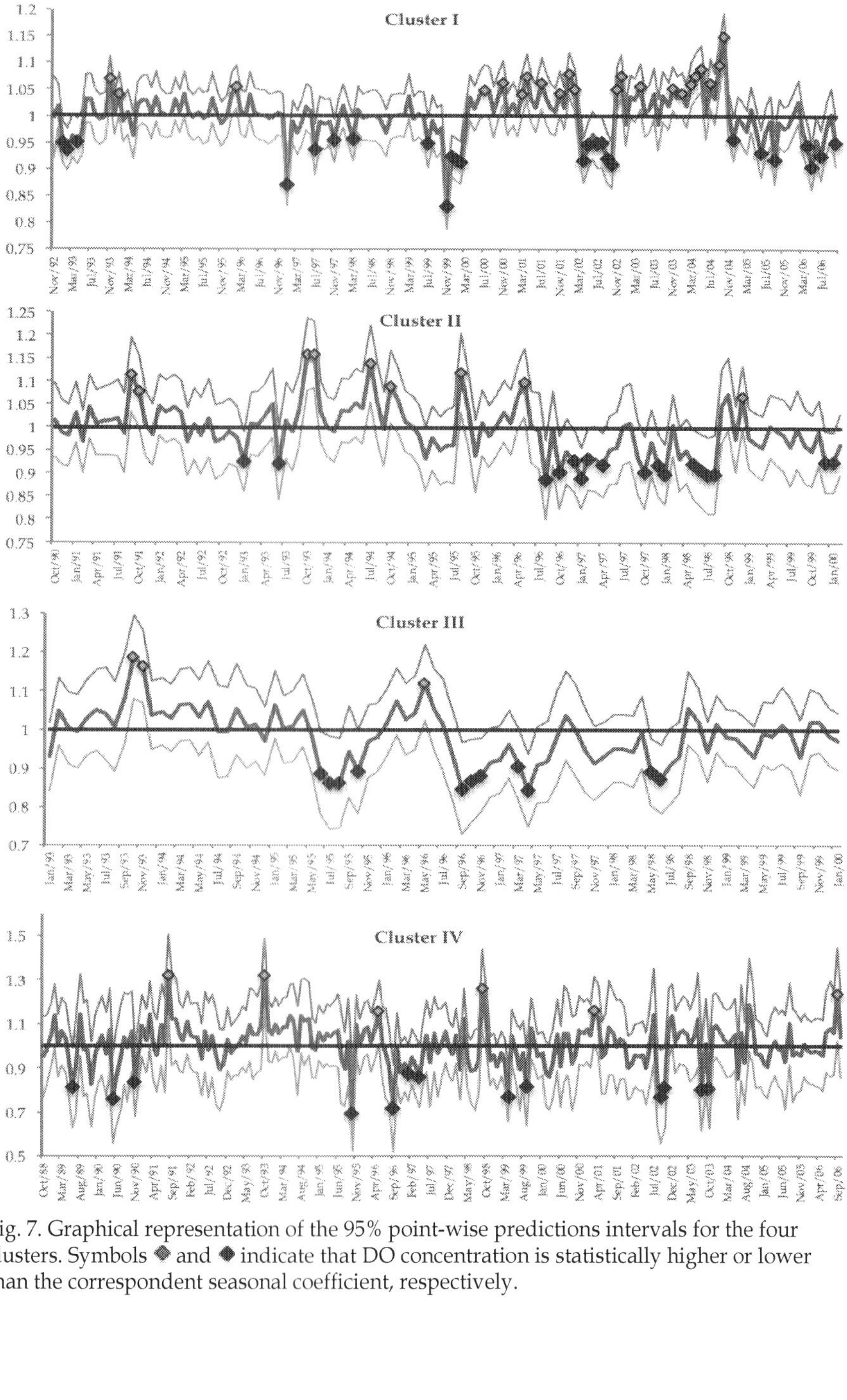

Fig. 7. Graphical representation of the 95% point-wise predictions intervals for the four clusters. Symbols ◆ and ◆ indicate that DO concentration is statistically higher or lower than the correspondent seasonal coefficient, respectively.

Cluster I results show that there were two periods of high water quality besides the expectation, namely from July 2000 to February 2002, and from November 2002 to October 2004. Since December 2004 some measurements indicate water quality deterioration in comparison to the expected results. For Cluster II, during a significant number of months from August 1996 to August 1998 the DO concentration measurements show lower values than the expected, while up to this time there were some sporadic months with a significant statistical difference in relation to seasonal coefficient. The graphical representation in the case of Cluster III presents two short time periods with lower relative values of DO concentration, from June 1995 to October 1995, and from September 1996 to April 1997. Cluster IV (the most polluted one) does not present evident periods of water quality improvement or deterioration and sporadically shows statistical unexpected values. Table 4 summarizes the percentage of months with DO concentration values that are statistically different from the expected. Clusters I, II and III show higher percentages than Cluster IV. This result is consistent with the linear regression results, in which these clusters present significant linear trends. In order to analyse the four clusters global behaviour, Figure 8 shows a simultaneous graphical representation of the months that were identified in the previous procedure. This representation shows that there are periods with a consistent behaviour in the generality of the clusters: for instance, from October 1993 to January 1994 were identified higher values of DO concentration, or from September 1996 to June 1998, where the DO concentration was lower than the expected in the whole area of the river basin.

Cluster	-	+	Total
I	15%	13%	28%
II	15%	8%	23%
III	13%	4%	16%
IV	6%	3%	9%

Table 4. Percentages of months during which DO concentration values are statistically different from the seasonal coefficients per cluster. Symbols "+" and "-"indicate whether the difference is by excess or defect, respectively.

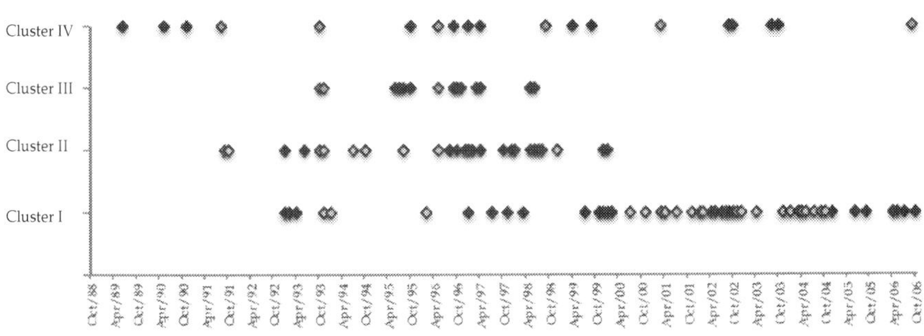

Fig. 8. Graphical representation of the most significant values for the four clusters over the same time axis. Symbols ◈ and ◆ indicate that DO concentration is statistically higher or lower than the correspondent seasonal coefficient, respectively.

5. Conclusion

The several statistics techniques applied in this work allow an integrate analysis of some environmental data dimensions, particularly of water quality data. The combination of multivariate statistical methodologies (as, for instance, CA and ACP procedures) with the temporal dimensions (as in linear and state-space approaches) has shown to be very useful in order to obtain global and more accurate results. For instance, hierarchical cluster analysis grouped 20 monitoring sites into six clusters of similar water quality characteristics and, based on the obtained information; it is possible to design a future, optimal spatial sampling strategy which could reduce the number of sampling monitoring sites and associated costs. The results of CA confirm the expected behaviour of the temporal/spatial dynamics of pollutants concentration (along the river and its main streams) and agree with those produced by the performed classification, thus allowing to reduce the large number of monitoring sites into a small number of homogeneous groups and yields an important data reduction.

An important conclusion from the CA procedure is the possibility of obtaining groups that can be classified according to their pollution level, as established from a set of criteria, and taking into account spatial and time dimensions.

The ACP analysis indicates that clusters have distinct factors/sources responsible for variations in River Ave's water quality and it helps to identify environmental, social and industrial aspects which influence water quality variations. The varifactor analysis shows very clearly that the industrial activity location has an impact on water quality.

Linear models and state-space models showed to be complementary in accordance to the proposed objectives. Linear models are useful when it is needed to identify global trends. State-space models have proven to be more accurate when the main objective is to obtain an accurate forecast of DO concentration. In addition, the state-space approach allows doing an online monitoring procedure to detect DO concentration values that are statistically unexpected. On the other hand, the state-space formulation presented in this work performs the measurement in percentage variation from the observed value of the seasonal coefficient.

The statistical modelling procedure was applied to a set of water monitoring sites grouped in homogeneous clusters. However, the modelling methodology can be applied to a single time series of any given quantitative water quality variable in a single location. This combination of statistical methodologies can be applied to other environmental issues, because statistical techniques are very versatile.

6. Annex

Briefly, the Kalman filter is an iterative algorithm that produces an estimator of the state vector $\mathbf{X}_t$ at each time t, which is given by the orthogonal projection of the state vector onto the observed variables up to that time. Considering the general formulation of a state-space model

$$\mathbf{Y}_t = \mathbf{H}_t \mathbf{X}_t + \mathbf{e}_t \cdot \tag{16}$$

$$\mathbf{X}_t = \mathbf{\Phi} \mathbf{X}_{t-1} + \boldsymbol{\varepsilon}_t \cdot \tag{17}$$

Let $\hat{\mathbf{X}}_{t|t-1}$ represent the estimator of $\mathbf{X}_t$ based on the information up to time $t-1$, that is, based on $\mathbf{Y}_1$, $\mathbf{Y}_2$, ..., $\mathbf{Y}_{t-1}$, and let $\mathbf{P}_{t|t-1}$ be its mean squared error (MSE) matrix. As the orthogonal projection is a linear estimator, the predictor for the next variable, $\mathbf{Y}_t$, is given by

$$\hat{\mathbf{Y}}_{t|t-1} = \mathbf{H}_t \hat{\mathbf{X}}_{t|t-1} \cdot \tag{18}$$

When, for time t, $\mathbf{Y}_t$ is available, the prediction error or innovation, $\boldsymbol{\eta}_t = \mathbf{Y}_t - \hat{\mathbf{Y}}_{t|t-1}$, is used to update the estimate of $\mathbf{X}_t$ through the equation

$$\hat{\mathbf{X}}_{t|t} = \hat{\mathbf{X}}_{t|t-1} + \mathbf{K}_t \eta_t , \tag{19}$$

where $\mathbf{K}_t$ is called the Kalman gain matrix and is given by $\mathbf{K}_t = \mathbf{P}_{t|t-1}\mathbf{H}_t'\left(\mathbf{H}_t\mathbf{P}_{t|t-1}\mathbf{H}_t' + \Sigma_{\mathbf{e}}\right)^{-1}$. Furthermore, the MSE of the updated estimator $\hat{\mathbf{X}}_{t|t}$ verifies the relationship $\mathbf{P}_{t|t} = \mathbf{P}_{t|t-1} - \mathbf{K}_t\mathbf{H}_t\mathbf{P}_{t|t-1}$. In turn, for time t, the forecast for the state vector $\mathbf{X}_{t+1}$ is given by the equation $\hat{\mathbf{X}}_{t+1|t} = \mathbf{\Phi}\hat{\mathbf{X}}_{t|t}$ and its MSE matrix is $\mathbf{P}_{t+1|t} = \mathbf{\Phi}\mathbf{P}_{t|t}\mathbf{\Phi}' + \Sigma_{\varepsilon}$.

This recursive process needs initial values for the state vector, $\mathbf{X}_{1|0}$, and for its MSE, $\mathbf{P}_{1|0}$, that will be later seen in more detail. As usual, the orthogonal projection corresponds to the best linear unbiased predictor. When the disturbances $\mathbf{e}_t$ and $\boldsymbol{\varepsilon}_t$ are normally distributed, the state vector and the observed variables are also normal. Therefore, in this case the orthogonal projection is also the conditional mean value and the Kalman filter is optimal.

7. Acknowledgments

This research was partially financed by FEDER Funds through "*Programa Operacional Factores de Competitividade - COMPETE*" and by Portuguese Funds through FCT - "*Fundação para a Ciência e a Tecnologia*", within the Project Est-C/MAT/UI0013/2011. A. Manuela Gonçalves was partially financed by FEDER Funds through "Programa Operacional Factores de Competitividade - COMPETE" and by Portuguese Funds through FCT - "Fundação para a Ciência e a Tecnologia", within the Project Est-C/MAT/UI0013/2011. Marco Costa was partially supported by Fundação para a Ciência e a Tecnologia, PEst OE/MAT/UI0209/2011.

8. References

Alpuim, T. & El-Shaarawi, A. (2008). On the efficiency of regression analysis with AR(p) errors. *Journal of Applied Statistics*, Vol.35, No.7, pp. 717-737

Ato, A., Samuel, O., Óscar, Y. & Moi, P. (2010). Mining and heavy metal pollution: assessment of aquatic environments in Tarkwa (Ghana) using multivariate statistical analysis. *Journal of Environmental Statistics*, Vol.1, No.4, 1-13

Barnett, V. (1981). *Interpreting Multivariate Data*, John Wiley & Sons, Sheffield

Bengtsson, T. & Cavanaugh, J.E. (2008). State-space discrimination and clustering of atmospheric time series data based on Kullback information measures. *Environmetrics*, Vol.19, No.2, pp. 103-121

Costa, M. & Gonçalves, A.M. (2011). Clustering and forecasting of dissolved oxygen concentration on a river basin. *Stochastic Environmental Research and Risk Assessment*, Vol.25, No.2, pp. 151-163

Elhatip, H., Hinis, M. & Gülbahar, N. (2008). Evaluation of the water quality at Tahtali dam watershed in Izmir-Turkey by means of statistical methodology. *Stochastic Environmental Research and Risk Assessment*, Vol.22, No.3, pp. 391-400

EU (2000). Directiva 2000/60/CE do Parlamento Europeu e do Conselho, de 23 de Outubro de 2000 (DQA, 2000), Estabelece um Quadro de Acção Comunitária no Domínio da Política da Água. *Jornal Oficial das Comunidades Europeias*, L327:1-72

Everitt, B., Dunn, G. & Leese, M. (2001). *Cluster Analysis*, 4th Edition, Arnold, London

Gonçalves, A. & Alpuim, T. (2011). Water Quality Monitoring using Cluster Analysis and Linear Models. *Environmetrics*, DOI:10.1002/env.1112

Gordon, A. (1999). *Classification*, Chapman and Hall/CRC, London

Helena, B., Pardo, R., Vega, M., Barrado, E., Fernandez, J. & Fernandez, L. (2000). Temporal evolution of groundwater composition in an alluvial aquifer (Pisuerga river, Spain) by principal component analysis. *Water Research*, Vol.34, No.3, pp. 807-816

INAG, I.P. (2008a). Tipologia de Rios em Portugal Continental no âmbito da implementação da Directiva Quadro da Água. I-Caracterização Abiótica, Ministério do Ambiente, do Ordenamento do Território e do Desenvolvimento Regional, Instituto da Água, I.P.

INAG, I.P. (2008b). Manual para a avaliação biológica da qualidade da água em sistemas fluviais segundo a Directiva Quadro da Água-Protocolo de amostragem e análise para os macroinvertebrados bentónicos. Ministério do Ambiente, do Ordenamento do Território e do Desenvolvimento Regional, Instituto da Água, I.P.

Johnson, R. & Wichern, D. (1992). *Applied Multivariate Statistical Analysis*, 3rd Edition, Prentice-Hall

Kaufman, L. & Rousseeuw, P. (1990). *Finding Groups in Data: An Introduction to Cluster Analysis*, John Wiley & Sons, New York

Lischeid, J. (2009). Non-linear visualization and analysis of large water quality data sets: a model-free basis for efficient monitoring and risk assessment. *Stochastic Environmental Research and Risk Assessment*, Vol.23, No.7, pp. 977-990

Liu, C.W., Lin, K.H. & Kuo, Y.M. (2003). Application of factor analysis in the assessment of ground-water quality in a blackfoot disease area in Taiwan. *The Science of the Total Environment*, Vol.313, No.1-3, pp.77-89

Machado, A., Silva, M. & Valentim, H. (2010). Contribute for the evaluation of water bodies status in Northern Region. *Revista Recursos Hídricos*, Vol.31, No.1, pp. 57-63

Oliveira, R.E.S., Lima, M.M.C.L. & Vieira, J.M.P., (2005). An Indicator System for Surface Water Quality in River Basins. *In The Fourth Inter-Celtic Colloquium on Hydrology and Management of Water Resources*, Universidade do Minho, Guimarães, Portugal

Shrestha, S. & Kazama, F. (2007). Assessment of surface water quality using multivariate statistical techniques: A case study of the Fuji river basin, Japan. *Environmental Modelling & Software*, Vol.22, No. 4, 464-475

Simeonov, V., Stratis, J., Samara, C., Zachariadis, G., Voutsa, D., Anthemidis, A., Sofoniou, M. & Kouimtzis, Th. (2003). Assessment of the surface water quality in Northern Greece. *Water Research*, Vol.37, No.17, 4119-4124

Varol, M. & Sen, B. (2009). Assessment of surface water quality using multivariate statistical techniques: a case study of Behrimaz Stream, Turkey. *Environmental Monitoring and Assessment*, Vol.159, No.1-4, pp. 543-553

Vega, M., Pardo, R., Barrado, E. & Debán, L. (1998). Assessment of seasonal and polluting effects on the quality of river water by exploratory data analysis. *Water Research*, Vol.32, No.12, 3581-3592

Vieira, J.M.P. (2003). Water management in national water plan challenges. *Revista Engenharia Civil*, Vol.16, pp.5-12

Wurderlin, D.A., Diaz, M.P., Ame, M.V., Pesce, S.F., Hued, A.C. & Bistoni, M.A. (2001). Pattern recognition techniques for the evaluation of spatial and temporal variations in water quality. A case study: Suquia river basin (Cordoba-Argentina). *Water Research*, Vol.35, No.12, pp. 2881-2894

6

Statistical Tools for Analyzing Water Quality Data

Liya Fu[1] and You-Gan Wang[2*]

[1]*School of Science, Xian Jiaotong University, China and Centre for Applications in Natural Resource Mathematics (CARM), School of Mathematics and Physics, the University of Queensland*

[2]*Centre for Applications in Natural Resource Mathematics (CARM), School of Mathematics and Physics, The University of Queensland Australia*

1. Introduction

Water quality data are often collected at different sites over time to improve water quality management. Water quality data usually exhibit the following characteristics: non-normal distribution, presence of outliers, missing values, values below detection limits (censored), and serial dependence. It is essential to apply appropriate statistical methodology when analyzing water quality data to draw valid conclusions and hence provide useful advice in water management. In this chapter, we will provide and demonstrate various statistical tools for analyzing such water quality data, and will also introduce how to use a statistical software R to analyze water quality data by various statistical methods. A dataset collected from the Susquehanna River Basin will be used to demonstrate various statistical methods provided in this chapter. The dataset can be downloaded from website *http://www.srbc.net/programs/CBP/nutrientprogram.htm*.

2. Graphical analysis of water quality data

Graphs provide visual summaries of data, quickly and clearly describe important information contained in the data, and provide insight for the analyst into the data under scrutiny. Graphs will help to determine if more complicated modeling is necessary. In this section, three particularly useful graphical methods are presented: boxplots, scatter plots, and Q-Q plots. R codes for plotting graphs in the following subsections will be given in detail.

2.1 Boxplots

A boxplot is a very useful and convenient tool to provide summaries of a dataset and is often used in exploratory data analysis. A boxplot usually presents a dataset through five numbers: extreme values (minimum and maximum values), median (50th percentile), 25th percentile, and 75th percentile. It also indicates the degree of dispersion, the degree of skew, and unusual values of the data (outliers). Furthermore, boxplots can display differences

**Address for correspondence*: Centre for Applications in Natural Resource Mathematics (CARM), School of Mathematics and Physics, the University of Queensland, St Lucia, QLD 4072, Australia between different populations without making any assumptions of the underlying statistical distribution. Boxplots of concentrations of total phosphorus (mg/L) at four stations from the Susquehanna River Basin from 2005 to 2010 are constructed (Fig. 1). R codes for constructing Fig. 1 are as follows:

```
> yl<- "Concentrations of total phosphorus (mg/L)"
> boxplot(TP ~ Station, ylab = yl, data = dat, boxwex = 0.5, outline = TRUE)
```

In Fig. 1, it can be seen that the four stations have nearly identical median values, and outliers could be present at all four stations. The distributions are right skewed. Further details on construction of a boxplot can be found in McGill et al. (1978), and Tukey (1977). More details for plotting boxplots are available in Adler (2009), Crawley (2007), and Venables & Ripley (2002).

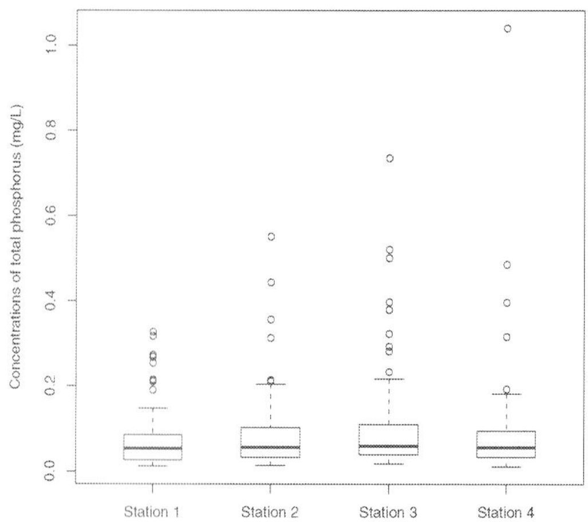

Fig. 1. Boxplots for total phosphorus at four stations at the Susquehanna River Basin

2.2 Scatter plots

A scatter plot is a very useful summary of a set of bivariate data (two variables), usually drawn before obtaining a linear correlation coefficient or fitting a regression line. It can be used to detect whether the relationships between two variables are linear or curved, and aids the interpretation of the correlation coefficient or a regression model. Fig. 2 is a scatter plot of the concentration of total phosphorus (mg/L) versus instantaneous flow (feet^3/s) in log scale at Station 1.

```
> xl<- "Instantaneous flow on log scale in cubic feet per second"
> plot(log(UNA$Flow), UNA$TP, xlab = xl, ylab = yl)
> points(log(UNA$Flow)[39], UNA$TP[39], col=2, pch =16)
> points(log(UNA$Flow)[18], UNA$TP[18], col=2, pch =16)
> points(log(UNA$Flow)[50], UNA$TP[50], col=2, pch =16)
```

We can generate a scatter plot using the data from two stations and can also use the function

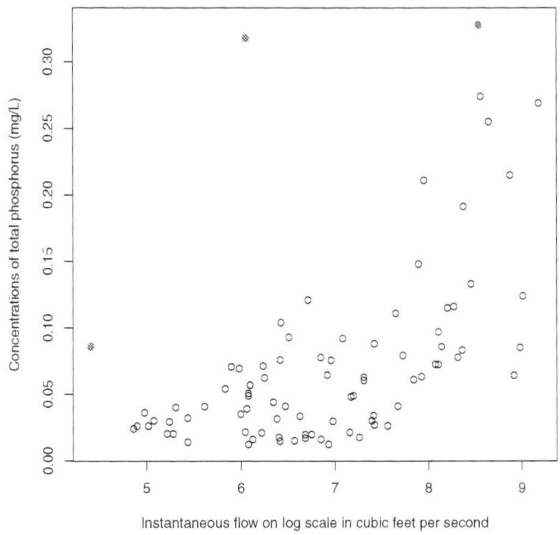

Fig. 2. Scatter plot of total phosphorus and instantaneous flow at Station 1, three possible outliers are in red

xyplot (Venables & Ripley, 2002) to split the data into different panels based on station (Fig. 3 (a), (b)).

```
> plot(log(UNA$Flow), UNA$TP, xlab = xl, ylab = yl, type = "p", pch = 1, col = 1)
> points(log(CKL$Flow), CKL$TP, pch = 3, col = 2)
> legend(4.5, 0.3, c("Station 1", "Station 2"), pch = c(1, 3), col=c(1, 2))
> library(lattice)
> xyplot(TP ~ log(Flow) | Station, data = dat, xlab = xl, ylab = yl, col = 1)
```

Concentration varies with natural log of instantaneous flow, as illustrated using a scatter plot. A linear regression model could be used to fit the data in Fig. 2, but true changes in slope are difficult to detect from only a scatter plot. Various methods have been developed to construct a central line to detect variation of slope locally in response to the data themselves, such as the locally weighted scatter plot smoothing (LOWESS) method (Fig. 4) (Cleveland et al., 1992).

```
> plot(log(UNA$Flow), UNA$TP, xlab = xl, ylab = yl)
> lines(lowess((UNA$TP) ~ log(UNA$Flow)), col=2)
```

2.3 Q-Q plots

A Q-Q plot presents the quantiles of a dataset against the quantiles of another dataset (Chambers et al., 1983; Gnanadesikan & Wilk, 1968). It can be used to determine whether two datasets come from populations with the same distribution. The greater the departure from the reference line, the greater the evidence to conclude that these two datasets come from populations with different distributions. If their distributions are identical, the Q-Q plot follows a straight line. Q-Q plots can be applied to compare the distribution of a sample

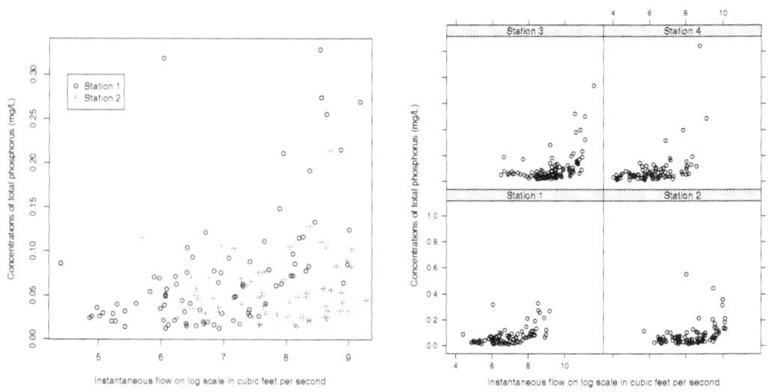

(a) two stations (b) four stations

Fig. 3. Scatter plots of total phosphorus and instantaneous flow

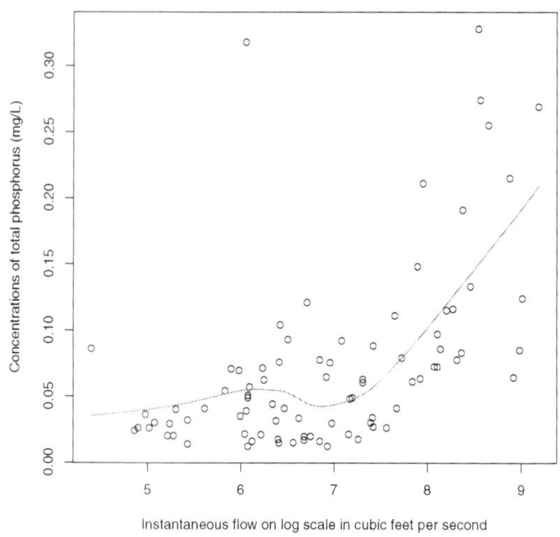

Fig. 4. The data of Fig. 2 fitted with the locally weighted scatter plot smoothing method

to a theoretical distribution (often a normal distribution). Therefore, Q-Q plots provide a very efficient way to tell how a sample distribution deviates from an expected distribution. The advantages of Q-Q plots are that (a) the sample sizes of two datasets do not need to be equal; (b) many distributional aspects can be simultaneously tested, such as shifts in location and scale and changes in symmetry; (c) the presence of outliers can also be detected. The functions *qqnorm* and *qqplot* can be used to construct a Q-Q plot (Adler, 2009; Crawley, 2007; Venables & Ripley, 2002). Fig. 5 (a) and (b) are two Q-Q plots to test whether the distributions of total phosphorus concentrations and the values of on the natural log scale at Station 1 are normal, respectively.

```
> qqnorm(UNA$TP); qqline(UNA$TP, col = 2)
> qqnorm(log(UNA$TP)); qqline(log(UNA$TP), col = 2)
```

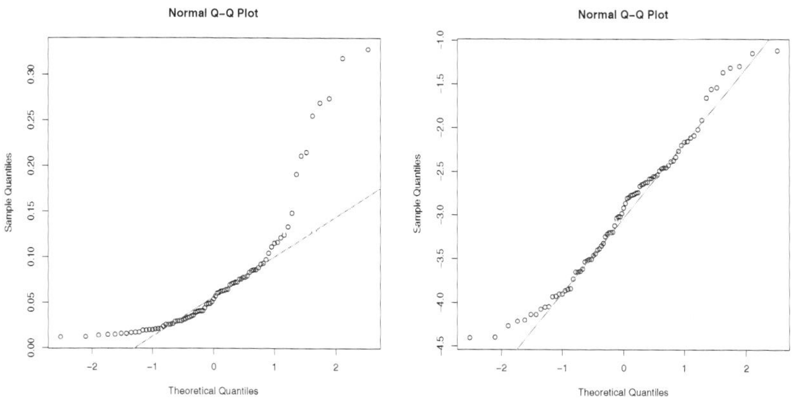

(a) (b)

Fig. 5. A Q-Q plot of total phosphorus concentrations at Station 1 versus the standard normal distribution

Fig. 5 (a) indicates that the distribution of total phosphorus concentrations is skewed to the right. Fig. 5 (b) shows an S-shape, but there is not sufficient evidence to prove that the distribution of total phosphorus on the natural log scale is non-normal. Fig. 6 is a Q-Q plot comparing whether two sample datasets are from populations with a common distribution. Note that there are also a few outliers appearing (possible outliers are in red). Otherwise, the plot suggests that the two samples have the same distribution.

```
> qq <-qqplot(UNA$TP, CKL$TP, plot.it = TRUE, xlab = "Concentrations of total phosphorus
at Station 1", ylab ="Concentrations of total phosphorus at Station 2")
> points(qq$x[85],qq$y[85], pch=16, col=2)
> points(qq$x[84],qq$y[84], pch=16, col=2)
> points(qq$x[83],qq$y[83], pch=16, col=2)
```

3. Water quality index

Sometimes it is difficult to assess water quality from a large number of water quality parameters. Traditional methods to evaluate water quality are based on the comparison of experimentally determined parameter values with an existing local normative, which does not provide a global summary on the spatial and temporal trends in the overall water quality (Debels et al., 2005; Kannel et al., 2007). To integrate complex water quality data and provide a simple and understandable tool for informing managers and decision-makers about the overall water quality status, various water quality indices (WQI) have been developed, which can be used to give a global vision on the spatial and temporal changes of the water quality. An early water quality index was proposed by Horton (1965), and then developed by Brown et al. (1970), Dojlido et al. (1994), McClelland (1974), and Pesce & Wunderlin (2000).

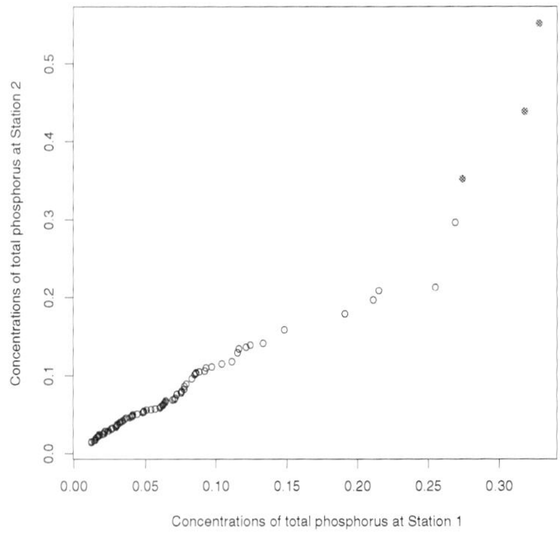

Fig. 6. A Q-Q plot comparing distributions of total phosphorus at Station 1 and Station 2

Water quality indices have been employed frequently in the public domain to assess water quality, such as the US National Sanitation Foundation Water Quality Index (Brown et al., 1970), the Canadian Water Quality index (CCME, 2001), the British Columbia Water Quality Index (Zandbergen & Hall, 1998) and the Oregon Water Quality Index (Cude, 2001). Main steps to derive a water quality index are as follows: select the most important water quality parameters (such as dissolved oxygen, total phosphorus, temperature); transform the parameters into a common scale; assign parameter weights; and aggregate scores to a single score. In this section, various water quality indices, such as those based on the weighted/unweighted arithmetic/geometric/harmonic mean functions, will be presented and compared. Their uses and limitations will be also discussed.

3.1 Weighted water quality indices

Water quality indices are usually obtained by assigning a suitable weight to each water quality parameter index and averaging them using some type of average functions. In this subsection, we consider three different weighted water quality indices. The water quality index proposed by Pesce & Wunderlin (2000) is:

$$WQI_{SA} = k \sum_{i=1}^{n} \omega_i S_i, \tag{1}$$

where n is the number of the water quality parameters, S_i is the score of the ith parameter, and ω_i is the relative weight given to S_i satisfying $\sum_{i=1}^{n} \omega_i = 1$. k is a subjective constant representing the visual impression of river contamination. The value of k ranges from 0.25 (for highly contaminated water) to 1 (for water without contamination). WQI_{SA} tends to overestimate the pollution due to the use of a subjective constant, which is not correlated with the measured parameters (Kannel et al., 2007).

Let $k = 1$ in Equation (1). In general, we have the objective water quality index originally proposed by Horton (1965), hereafter called the weighted arithmetic water quality index. It has been used by many researchers (Brown et al., 1970; Prati et al., 1971; Sanchez et al., 2007):

$$WQI_{WA} = \sum_{i=1}^{n} \omega_i S_i. \tag{2}$$

The third water quality index is based on the weighted geometric mean function (Brown et al., 1970; McClelland, 1974), which is always smaller than WQI_{WA} if all values of S_i are positive:

$$WQI_{WG} = \prod_{i=1}^{n} S_i^{\omega_i}. \tag{3}$$

The above weighted water quality indices indicate that each water quality parameter may have different weights based on the importance of the water quality situation. This characteristic could be desirable when water quality indices are specific to the protection of aquatic life. However, when sensitivity to changes in each water quality parameter is more desirable than sensitivity to the most heavily weighted water quality parameter, such weighting could be unnecessary (Cude, 2001; Gupta et al., 2003; Landwehr & Deininger, 1976). Some unweighted water quality indices were therefore explored (Cude, 2001; Dojlido et al., 1994; Landwehr & Deininger, 1976) and are now introduced in the following subsection.

3.2 Unweighted water quality indices

In this subsection, we introduce three unweighted water quality indices. The first two are arithmetic/geometric water quality indices proposed by Landwehr & Deininger (1976),

$$WQI_A = 1/n \sum_{i=1}^{n} S_i, \tag{4}$$

$$WQI_G = (\prod_{i=1}^{n} S_i)^{1/n}, \tag{5}$$

which is a special case of (2) and (3) with $\omega_i = 1/n$ for any i, respectively. As with the relationship between WQI_{WG} and WQI_{WA}, WQI_G is always lower than WQI_A. The third is the harmonic square water quality index,

$$WQI_H = \sqrt{\frac{n}{\sum_{i=1}^{n} \frac{1}{S_i^2}}}, \tag{6}$$

which has been suggested as an improvement over both WQI_{WA} and WQI_{WG} (Cude, 2001; Dojlido et al., 1994). Compared to WQI_{WA} and WQI_{WG}, WQI_H is the most sensitive to changes in single water quality parameter (Cude, 2001).

3.3 Harkins' water quality index

An objective water quality index was proposed by Harkins (1974), which is based on Kendall's nonparametric multivariate ranking procedure.

$$WQI_{HR} = \sum_{i=1}^{n} \frac{(R_i - R_{ic})^2}{var(R_i)}, \tag{7}$$

where

$$var(R_i) = \frac{1}{12M}[(M^3 - M) - \sum_{j=1}^{k_i}(t_{ij}^3 - t_{ij})],$$

R_i and R_{ic} correspond to the rank and control values of the ith water quality parameter, respectively. M is the number of water quality parameters plus the number of control values, t_{ij} is the number of elements involved in the jth tie encountered when ordering the measured values of the ith water quality parameter, and k_i is the total number of ties encountered in ranking the measurements of the ith parameter. Landwehr & Deininger (1976) and Gupta et al. (2003) compared WQI_{HR} with water quality indices WQI_{WA}, WQI_{WG}, WQI_G, and WQI_A. Their results indicated that these five indices are correlated well with the opinions of experts, and although the five indices showed significant correlation with each other, WQI_{HR} was the lowest of the five. Therefore, they suggested adopting any of the four indices except WQI_{HR}.

4. Methods for handling data below detection limits

One feature of water quality measurement is that some data will fall above or below the detection limit, and therefore not be captured, because of limitations of the measurement procedures or the analytical tools used in the laboratories. Data below a detection limit are also referred as left-censored data. There could also be multiple detection limits involved if an instrument is upgraded during the project period or data are combined from multiple laboratories. Even data below the detection limits are still of considerable importance because of the potential health hazard. The data below the detection limits complicate the analysis of the water quality data. Various strategies have been developed to analyze the data that fall below detection limits (Fu & Wang, 2011; Helsel, 1990; Shumway et al., 2002). In the following subsections, simple substitution methods, parametric methods, and nonparametric methods will be introduced.

4.1 Simple deletion/substitution methods

Simple deletion/substitution methods delete/replace the measurements below detection limits (DL) with fixed values, such as zero, 1/2DL, $1/\sqrt{2}$DL or DL (Helsel, 1990; Hornung & Reed, 1990). Hornung & Reed (1990) proposed using $1/\sqrt{2}DL$ when the data are not highly skewed and 1/2DL substitution otherwise. Hewett & Ganser (2007) found that 1/2DL and $1/\sqrt{2}$DL perform well when the sample size is less than 20 and the percent censored is less than 45 percent. It is easy and convenient to use the substitution methods. However, all tend to be biased and cause a loss of information (El-Shaarawi & Esterby, 1992; Helsel & Cohn, 1988; Lubin et al., 2004). When the results strongly depend on the values being substituted, particularly for data with multiple detection limits (Shumway et al., 2002), the substitution methods are not generally suitable. In particular, when there is a high proportion of data below detection limits, results for standard errors are also far less desirable, and the biased standard errors may further distort the inference (Helsel, 1990; 1992; Shumway et al., 2002).

4.2 Parametric methods

Assume that the distribution of measurements is known, such as normal or lognormal. The data below the detection limits can be filled using values randomly selected from the distribution or replaced with their conditional expected values (conditional on being less than the detection limits) (Helsel, 1990). Suppose that there are n detected measurements $(y_1, \ldots, y_n)$ and m measurements below the detection limits $(c_1, \ldots, c_m)$. The likelihood function is

$$L(\theta) = \prod_{i=1}^{n} f_\theta(y_i) \prod_{j=1}^{m} F_\theta(c_j), \tag{8}$$

where θ is a vector of parameter, $f(\cdot)$ is the probability density function of y, and $F(\cdot)$ is the cumulative density function (c.d.f.) of y. The parameter estimates of θ and summary statistics can be obtained by the maximum likelihood method (ML) (Cohen, 1976; Cohn, 1988; Helsel, 1992). Results based on a lognormal distribution assumption by the maximum likelihood method can be easily obtained using statistic software R (NADA package) (Lee & Helsel, 2005). If the distributional assumption is appropriate and the sample size is large, the maximum likelihood method is the most efficient (Cohn, 1988; Helsel, 1992; Hewett & Ganser, 2007). To incorporate the covariate effects when analyzing the water quality data that fall below detection limits, the following regression models can be considered.

4.2.1 Tobit regression

Tobit regression model (Tobin, 1958) has been widely used to analyze censored data. The model can be written as

$$log(y_i^*) = \beta_0 + \beta_1 x_i + \epsilon_i, \tag{9}$$

where y^* is a latent variable and $y_i = y_i^*$ if $y_i^* > c_i$ and $y_i = c_i$ otherwise. Random error term ϵ_i follows a normal distribution $N(0, \sigma^2)$. The likelihood function (8) can be written as

$$L(\beta_0, \beta_1) = \prod_i \left[\frac{1}{\sigma}\phi\left(\frac{log(y_i) - \beta_0 - \beta_1 x_i}{\sigma}\right)\right]^{\delta_i} \prod_i \left[\Phi\left(\frac{log(c_i) - \beta_0 - \beta_1 x_i}{\sigma}\right)\right]^{1-\delta_i},$$

where $\delta_i = 1$ if $y_i^* > c_i$ and $\delta_i = 0$ otherwise. The maximum likelihood estimates (MLE) of parameters can be obtained from the function *survreg* (survival package) and *vglm* (VGAM package) in R if the detection limit is a single number. An example to obtain MLE of parameters in a Tobit regression model $log(\text{DP}) = \beta_0 + \beta_1 log(\text{Flow}) + \epsilon$ is given,

```
> library(survival)
> fit<- survreg(Surv(log(DP), DP>=0.01, type = 'left') ~ log(Flow), data = UNA, dist =
'gaussian')
> summary(fit)
```

For multiple detection limits, the estimates can be derived by a Newton-Raphson algorithm. The Wald type test or the likelihood ratio test can be applied to test the group difference or covariate effects (by testing $\beta = 0$). Tobit regression is also applicable when both the measurements of the response and covariate variable are with detection limits (Helsel, 1992). When the distribution is known and the error terms are homeostatic, the estimate derived by the maximum likelihood method is optimal (Helsel, 2005b).

4.2.2 Logistic regression

Let $\tilde{y} = 1$ if the response is above a detection limit c and $\tilde{y} = 0$ otherwise. Assume that the probability of $\tilde{y} = 1$ is p, then $p = p(y > c)$. A binary logistic regression modeled as a linear function of covariate x is given by

$$\log(\frac{p}{1-p}) = \alpha_0 + \alpha_1 x.$$

The likelihood function is

$$L(\alpha_0, \alpha_1) = \prod_i p_i^{\tilde{y}_i}(1-p_i)^{1-\tilde{y}_i},$$

where $p_i = exp(\alpha_0 + \alpha_1 x_i)/[1 + exp(\alpha_0 + \alpha_1 x_i)]$. The maximum likelihood estimates of parameters can be obtained from *glm* function in R. The significance of the covariate effect can be tested using the likelihood ratio statistic (Helsel, 1992). For multiple detection limits, the ordered logistic regression can be used. More details can be seen in Helsel (1992). An example to obtain parameter estimates from a logistic regression $\log(p/(1-p)) = \alpha_0 + \alpha_1 log(\text{Flow})$ is given as follows, where $p = p(\text{DP} > 0.01)$.

```
> UNA$ DPd<- 1-(UNA$ DPrem ==“<”)
> logitfit<- glm(DPd ~ log(Flow), data = UNA, family = binomial(“logit”))
> summary(logitfit)
```

Parametric methods generally perform well for summary statistics when the dataset is large and the underlying distribution can be approximated by a known distribution. Specification of the underlying distribution of a dataset may be difficult in practical problems. The ML method does not work well when the distributional assumption is invalid or the sample size is small (<20) (Gleit, 1985; Helsel, 2005b; Helsel & Cohn, 1988). Furthermore, the ML method is sensitive to outliers, which usually exist in water quality data. An implementation of fully parametric methods is a robust and efficient semi-parametric regression method on order statistics (ROS) and will be introduced in the following subsection.

4.2.3 ROS method

The ROS method was provided by Helsel & Cohn (1988), which is a simple imputation method that fills in data below detection limits based on a probability plot of detections (Helsel & Cohn, 1988; Lee & Helsel, 2005; Shumway et al., 2002). It can be used to obtain summary statistics, plot modeled distributions, and predict values based on the modeled distributions (Fig. 7). The ROS method has been evaluated as one of the most reliable approaches for estimating summary statistics of data with multiple detection limits (Shumway et al., 2002). Lee & Helsel (2005) developed software implementation that performs the ROS method, and it is a part of the NADA library in statistical software R. R codes for Fig. 7 are as follows:

```
> library(NADA)
> UNA<- UNA[!is.na(UNA$OP), ]
> UNA$CenOP<- UNA$OPrem == “<”
> rosop<- cenros(UNA$OP, UNA$CenOP, forwardT =“log”, reverseT = “exp”)
> plot(rosop, plot.censored = TRUE)
> lines(rosop, col = 2)
```

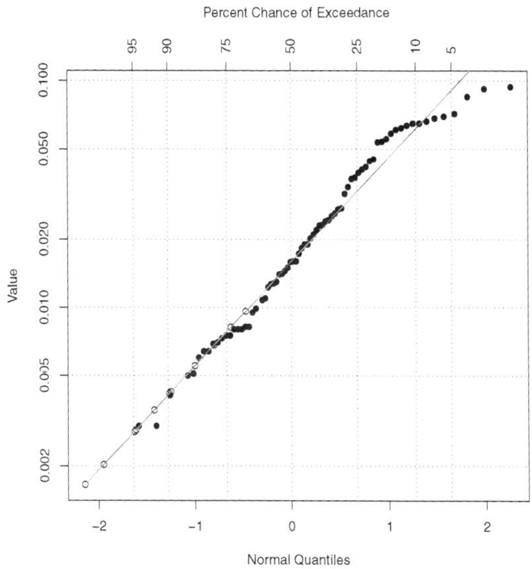

Fig. 7. A normal Q-Q plot for a ROS model. Solid circles are detected data. Open circles are modeled undetected values.

4.3 Nonparametric methods

Parametric and semi-parametric methods are based on the assumption of the underlying distribution of the data. Nonparametric methods provide an alternative that does not require specifying a distribution and filling in the data below detection limits. The nonparametric methods are generally used to analyze the right censored data. Left censored data can be converted into right-censored data by flipping the data over the largest observed value. Lee & Helsel (2007) provided software tools for direct analysis of data with multiple detection limits (left-censored data) by nonparametric modeling and hypothesis testing.

4.3.1 Kaplan-Meier

The Kaplan-Meier (K-M) method is the standard method for computing descriptive statistics of data that fall below detection limits (Helsel, 2005; Lee & Helsel, 2007). K-M method has been widely used in survival analysis, where it is employed with right-censored time-to-failure data. The K-M method can estimate the percentiles or c.d.f. for a dataset, and can test hypotheses. It can describe and compare the shapes of different datasets (Figs. 8 (a) and (b)).

```
> KM<- cenfit(UNA$OP, UNA$CenOP)
> plot(KM)
> dat2<- dat2[!is.na(dat2$OP), ]
> dat2$CenOP<- dat2$OPrem == "<"
> g2<- cenfit(dat2$OP, dat2$CenOP,dat2$Station)
> plot(g2,lty = c(1 : 3), col=c(1, 2, 4))
> legend(0.002, 0.8, c("Station 1","Station 2","Station 4"), lty = c(1:3), col=c(1, 2, 4))
```

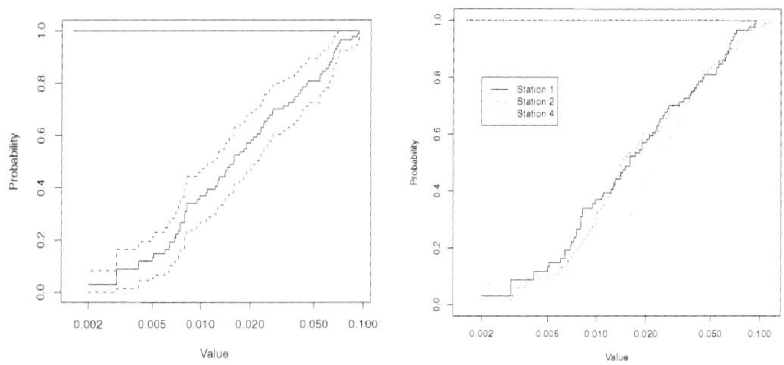

(a) Dashed lines are 95% confidence limits

(b) Empirical c.d.f at three stations

Fig. 8. Empirical cumulative distribution functions for datasets with multiple detection limits

Zhang et al. (2009) developed a nonparametric estimation procedure, and under a fixed detection limit and some mild conditions, they established the theoretical equivalence of three nonparametric test statistics: the Wilcoxon rank sum, the Gehan, and the Peto-Peto tests. Their simulation studies indicated that nonparametric methods work well for a range of small sizes and censoring rates (Zhang et al., 2009). For hypothesis testing with multiple detection limits, one robust method is to censor all data at the highest detection limit and then perform an appropriate nonparametric test (Helsel, 1992). This can result in a loss of information, however, the accelerated failure time (AFT) model can integrate the Gehan and logrank tests, incorporate covariate effects, and compare the differences between two/multiple data groups with multiple detection limits (Jin et al., 2006; Wei, 1992; Zhang et al., 2009).

4.3.2 AFT model

Assume that $\{Y_i, i = 1, \ldots, N\}$, $\{C_i, i = 1, \ldots, N\}$ and $\{X_i, i = 1, \ldots, N\}$ are measurements, detection limits and $p \times 1$ covariate vector, respectively. Let $\Delta_i = 1$ if Y_i is below the detection limit C_i and $\Delta_i = 0$ otherwise. Let $\tilde{Z}_i = \min\{-\log(Y_i), -\log(C_i)\}$; therefore $(\tilde{Z}_i, \Delta_i, X_i)$ are the observations. The accelerated failure time model is

$$Z_i = X_i'\beta + \epsilon_i,$$

where $Z_i = \log(Y_i)$, β is an unknown regression parameter vector, and ϵ_i is the error term. Suppose that $\{\epsilon_i, i = 1, \ldots, N\}$ are independent and identically distributed and their underlying distribution is unknown.

Estimation and inference of the regression parameters are based on the estimating functions given by

$$U(\beta) = N^{-2} \sum_{i=1}^{N} \Delta_i \omega(e_i) \left\{ X_i - \frac{\sum_{j=1}^{N} X_j I(e_i \geq e_j)}{\sum_{j=1}^{N} I(e_i \geq e_j)} \right\},$$

where $\omega(e_i)$ is a weight function and $e_i = \log(Y_i) - X_i'\beta_t$, where β_t is the true value of β. Let $\omega(e_i) = 1$ and $\omega(e_i) = \sum_{j=1}^{N} I(e_i \geq e_j)$; $U(\beta)$ correspond to the log-rank and Gehan statistics, respectively. The estimating functions $U(\beta)$ are step functions and discontinuous in the regression parameters, which makes it difficult to find consistent estimators and their asymptotic variance and covariance matrices. Much progress has been made to overcome these difficulties (Brown & Wang, 2006; Heller, 2007; Jin et al., 2003; Lee et al., 1993), and the function *lss* (lss package) can be used to obtain various statistics from an AFT model.

```
> library(lss)
> UNA$status<- 1-(UNA$OPrem=="<")
> aftfit<- lss(cbind(log(OP), status)~ log(Flow), data=UNA, gehanonly=FALSE, cov=TRUE)
> print(aftfit)
```

Jin et al. (2006b) extended marginal accelerated failure time models to multivariate censored data. Their method, which is based on an independence "working" model, may ignore the within-site correlations in obtaining parameter estimates, while taking account of the correlation in calculating the standard errors. More efficient estimators with similar computational complexity were developed for multivariate censored data analysis, when measurements from the same site exhibit strong temporal correlations (Fu & Wang, 2011).

5. Trend detection

In recent years, concentrations of various water quality parameters have been collected. Tests for trends specific to various water quality parameters have been of keen interest in environmental science (Helsel, 1992). A number of methods have been proposed to detect and assess changes in water quality. In this section, a variety of approaches will be introduced and their strengths and weaknesses investigated. The exogenous variable effects and serial dependence will be considered when testing water quality trends.

5.1 Parametric methods

Under the normality of residuals and constant variance assumptions, simple/multiple linear regressions are preferable for detecting trends of water quality.

5.1.1 Simple linear regression

Let Y be the random variable of interest in a trend test, such as concentrations of water quality parameters. T denotes time (often expressed in years). If Y is linear over time T, the linear simple regression of Y on T is a test for trend.

$$Y = \beta_0 + \beta_1 T + \epsilon, \tag{10}$$

where β_1 is the rate of change in Y. The null hypothesis for testing the trend of y can be stated as a test for $\beta_1 = 0$. The Wald type statistic (t-statistic) can be used. If the null hypothesis is rejected, it indicates that there is a linear trend in Y over time. If Y is not linear over time T, some transformation of Y, such as a log transformation, may be necessary. An example using a linear regression to detect the trend of total phosphorus concentrations at Station 1 is presented in Fig. 9. The results indicate that the trend of total phosphorus is not significant.

5.1.2 Multiple regression

Most concentrations of water quality parameters have strong seasonal patterns (see Fig. 10). They are influenced by the changes in biological activity, both natural and managed

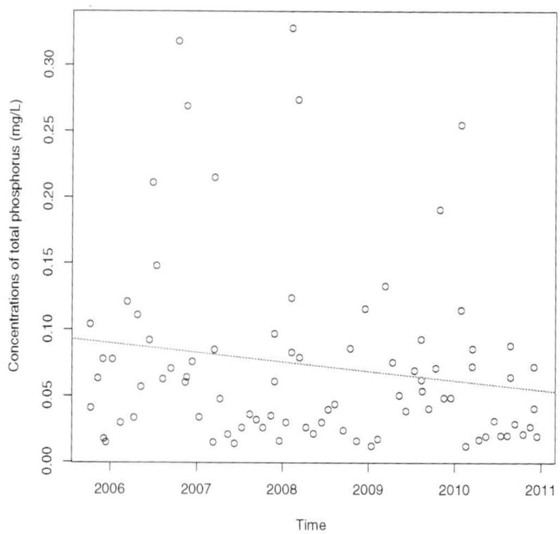

Fig. 9. Linear regression trend line for total phosphorus concentrations
Regression: $C = 0.09 - 0.007\text{Time}$, and $p = 0.14$

activities such as agriculture (Helsel, 1992; Hirsch et al., 1991). Therefore, it is important to consider seasonal effects when evaluating changes in water quality data. In parametric procedures, multiple regression with periodic functions can be used to describe seasonal variation. Consider the following simple case,

$$Y = \beta_0 + \beta_1 T + \beta_2 \cos(2\pi T) + \beta_3 \sin(2\pi T) + \epsilon, \tag{11}$$

where T is expressed in years and β_1 indicates the change rate of Y. Terms $\sin(2\pi T)$ and $\cos(2\pi T)$ capture the annual cycle and account for seasonality. Residuals ϵ must follow a normal distribution (or approximately normal). The trend test can be constructed by testing $\beta_1 = 0$.

If residuals still show a seasonal pattern (see Fig. 11), additional periodic functions should be included in model (11) to remove the seasonal variation. A general multiple linear regression is given by

$$Y = \beta_0 + \beta_1 T + \sum_{k=0}^{K} [\beta_{2k+1} \cos(2\pi k T) + \beta_{2k+2} \sin(2\pi k T)] + \epsilon. \tag{12}$$

The cases of $K = 0$ and $K = 1$ correspond to model (10) and (11). If $K = 2$, a period of 1/2 year is then also included in model (12). Fig. 11 shows that the residuals of the linear regression in Subsection 5.1.1 represent a seasonal pattern, therefore periodic functions should be included in the model.

When Y or some transformation of Y is linear with time T, and residuals follow a normal distribution with a constant variance, the parametric regression is optimal. However, the

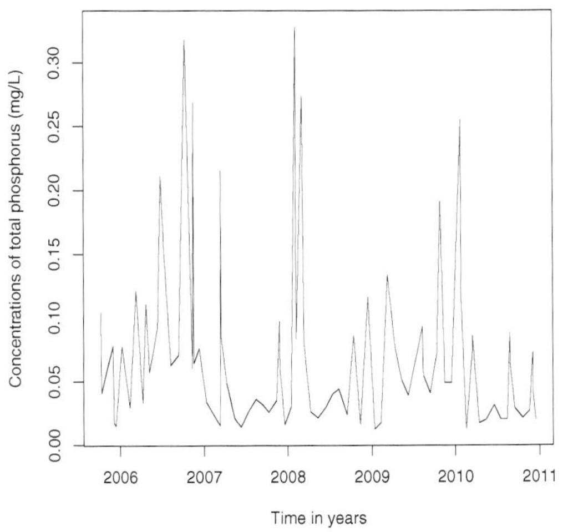

Fig. 10. Time series plot of total phosphorus concentration at Station 1

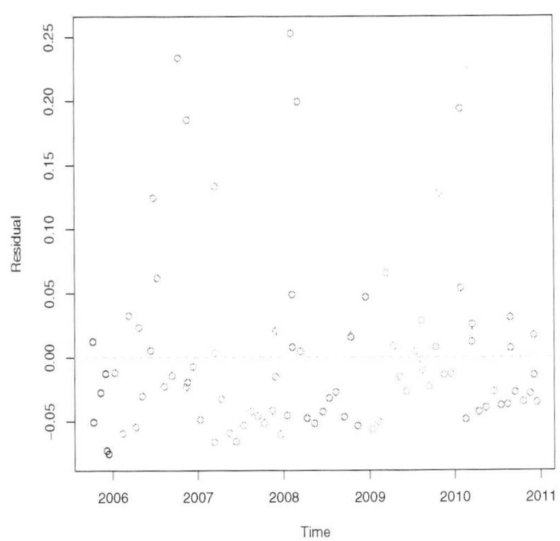

Fig. 11. Residuals of the linear regression mentioned in above subsection versus times in year.

distribution of water quality data is usually highly skewed, in particular, data related to discharge, as well as biological indicators (biomass, chlorophyll) (Helsel, 1992; Hirsch & Slack,

1984). The test that depends on the normality assumption may be inappropriate. The following subsection introduces several nonparametric methods that do not require the normality assumption.

5.2 Nonparametric methods

Water quality data usually have the following characteristics: nonnormal data, missing values, values below detection limits, and serial dependence. The nonparametric methods are robust and can handle these problems easily.

5.2.1 Mann-Kendall test

Mann (1945) and Kendall (1975) proposed a nonparametric test for randomness against trend. According to Mann (1945), the null hypothesis H_0 states that $(x_1, \ldots, x_n)$ are a sample of n independent and identically distributed random variables. The alternative hypothesis $H_1$of a two-sided test is that the distributions of x_k and x_j are not identical for all $k, j \leq n$, and $k \neq j$. The test statistic S is defined as

$$S = \sum_{k=1}^{n-1} \sum_{j=k+1}^{n} \operatorname{sgn}(x_j - x_k),$$

where

$$\operatorname{sgn}(\theta) = \begin{cases} 1 & \text{if } \theta > 0 \\ 0 & \text{if } \theta = 0 \\ -1 & \text{if } \theta < 0. \end{cases}$$

Under the null hypothesis, Mann (1945) and Kendall (1975) obtained the mean and variance of S.

$$E(S) = 0,$$

$$\operatorname{var}(S) = [n(n-1)(2n-5) - \sum_t t(t-1)(2t-5)]/18,$$

where t is the extent of any given tie (number of xs involved in a given tie) and $\sum_t$ denotes the summation over all ties.

Both Mann (1945) and Kendall (1975) derived the exact distribution of S for $n \leq 10$; proved that the distribution of S is normal as $n \to \infty$; and further showed that even for $n = 10$, the normal approximate is excellent if one calculates the standard normal variate Z by

$$Z = \begin{cases} \frac{S-1}{\{\operatorname{var}(S)\}^{1/2}} & \text{if } S > 0 \\ 0 & \text{if } S = 0 \\ \frac{S+1}{\{\operatorname{var}(S)\}^{1/2}} & \text{if } S < 0. \end{cases}$$

Hence, in a two-sided test for trend, the H_0 should be rejected if $|Z| \geq z_{\alpha/2}$, where $\Phi(z_{\alpha/2}) = 1 - \alpha/2$, $\Phi(\cdot)$ is the standard normal c.d.f. and α is the significance level for the test. A position value of S indicates an "upward" trend, and a negative value of S presents a "downward" trend. For an example from Station 21 at Susquehanna River basin (Fig. 10), the statistic

$S = -90$, the var$(S) = 1096.67$ under the null hypothesis and the p value is 0.0072, which indicates a downward trend in the concentration of total phosphorus at Station 21.

```
> library(Kendall)
> TP<- ts(CONY$TP, frequency=1, start=1990)
> mk<- MannKendall(TP)
> summary(mk)
```

The seasonality is a common phenomenon, which indicates that the distributions differ for different times of year. The Mann-Kendall test therefore is sensitive to seasonality. Hirsch et al. (1982) developed a modified Mann-Kendall test to detect the trend of data with seasonality.

5.2.2 The seasonal Kendall test

Hirsch et al. (1982) presented a modified Mann-Kendall test that detects trends in time series with seasonal variation and called as a seasonal Kendall test. Let $X = (X_1, X_2, \cdots, X_m)$ and $X_i = (x_{i1}, x_{i2}, \cdots, x_{in_i})$, where X is the entire sample consisting of m subsamples X_i, and m is the number of seasons. Each subsample X_i contains n_i annual values. The null hypothesis H_0 is that X is a sample of independent random variables (x_{ij}), and that X_i is a subsample of independent and identically distributed random variables for $i = 1, \cdots, m$. The alternative hypothesis H_1 is that the subsample is not distributed identically. The test statistic proposed by Hirsch et al. (1982) is given as follows. For month i,

$$S_i = \sum_{k=1}^{n_i-1} \sum_{j=k+1}^{n_i} \operatorname{sgn}(x_{ij} - x_{ik}). \tag{13}$$

Under the null hypothesis, S_i is a Mann-Kendall test statistic and

$$E(S_i) = 0,$$

$$\operatorname{var}(S_i) = [n_i(n_i-1)(2n_i-5) - \sum_{t_i} t_i(t_i-1)(2t_i-5)]/18.$$

The distribution of S_i is normal as $n_i \to \infty$ (t_i is the extension of a given tie in month i). Define the seasonal Kendall statistic

$$S^* = \sum_{i=1}^{m} S_i, \tag{14}$$

and its expectation

$$E(S^*) = \sum_{i=1}^{m} E(S_i) = 0,$$

and variance

$$\operatorname{var}(S^*) = \sum_{i=1}^{m} \operatorname{var}(S_i) + \sum_{i=1}^{m} \sum_{j \neq i}^{m} \operatorname{cov}(S_i, S_j). \tag{15}$$

Under the null hypothesis, S_i and S_j $(j \neq i)$ are independent, therefore

$$\text{var}(S^*) = \sum_{i=1}^{m}[n_i(n_i-1)(2n_i-5) - \sum_{t_i} t_i(t_i-1)(2t_i-5)]/18.$$

The standard normal variate Z^* is defined as

$$Z^* = \begin{cases} \frac{S^*-1}{\{\text{var}(S^*)\}^{1/2}} & \text{if } S^* > 0 \\ 0 & \text{if } S^* = 0 \\ \frac{S^*+1}{\{\text{var}(S^*)\}^{1/2}} & \text{if } S^* < 0. \end{cases}$$

The approximation is adequate for $n_i = 3$ and $m = 12$ for all i (Hirsch et al., 1982). For the example from Station 21 at Susquehanna River basin (Fig. 10), the statistic $S^* = -360$, the $\text{var}(S) = 10779.67$ under the null hypothesis, and the p value is 0.0005, which indicates a downward trend in the concentration of total phosphorus at Station 21.

```
> library(Kendall)
> TPS<- ts(c(t(RCON[,-1])), frequency = 12, start = c(1990, 1))
> smk<- SeasonalMannKendall(TPS)
> summary(smk)
```

A limitation of the seasonal Kendall test is one observation per month. If there are multiple observations in each of the months, Hirsch et al. (1982) suggested using the medians of the multiple observations in the seasonal Kendall test. Another limitation is that the seasonal Kendall test is not robust against serial dependence. When serial dependence exists, $\text{cov}(S_i, S_j)$ in Equation (15) does not equal zero. Hirsch & Slack (1984) provided a modification of the seasonal Kendall test which is robust against serial dependence, except when the data have very strong long-term persistence or when the sample sizes are small. More details can be found in Hirsch & Slack (1984) and Letternmatier (1988). In addition to detecting the trend, the magnitude of such a trend may also be desirable. In model (10), an estimate of β_1 can be used to estimate the trend. For a seasonal Kendall test, calculate $d_{ijk} = (X_{ij} - X_{ik})/(j-k)$ for all pairs (X_{ik}, X_{ij}) and $(k < j)$. Hirsch et al. (1982) proposed using the median of d_{ijk} as an estimator of the slope, which is robust against extreme values.

5.2.3 Sen's T test

Farrel (1980) proposed an aligned-rank test for detecting trends, which is distribution free and not affected by seasonal fluctuations (Van Belle & Hughes, 1984; Yu et al., 1993). Let x_{ij} be the measurement in the ith month of the jth year at a sampling station, and $i = 1, \ldots m$, $j = 1, \ldots, n$. Let R_{ij} be the rank of $(x_{ij} - x_{i+})$ among the mn values of differences, where $x_{i+} = \sum_{j=1}^{n} x_{ij}$. If ties occur, the average of the ranks is taken as the rank of each tie. The statistic is

$$T = \left\{ \frac{12m^2}{n(n+1)\sum_{i,j}(R_{ij} - R_{i+})^2} \right\}^{1/2} \left\{ \sum_{i=1}^{n} (i - \frac{n+1}{2})(R_{+j} - \frac{nm+1}{2}) \right\},$$

where $R_{i+} = \sum_j R_{ij}/n$ and $R_{+j} = \sum_i R_{ij}/m$. Under the null hypothesis of no trend, the distribution of T tends to the standard normal distribution. Simulation results indicated that the normal approximation for the statistic T was reasonable even for a small sample (Van Belle & Hughes, 1984).

The three nonparametric methods for detecting trends mentioned above have practically the same power at a statistical significance level of 0.05 (Yu et al., 1993). It is worth noting that there may exist water quality parameters which exhibit strong evidence of a download trend in some months and then exhibit strong evidence of an upload trend (step trend) (Helsel, 1992; Hirsch et al., 1991). The methods described above all assume a single trend across all seasons, provide a summary statistic for the entire record (monotonic trend), and do not indicate when there are trends in opposing directions in different months. Van Belle & Hughes (1984) developed a statistic for testing homogeneity of trends. The statistic is

$$\chi^2_{homogeneous} = \chi^2_{total} - \chi^2_{trend} = \sum_{i=1}^{m} Z_i^2 - m\bar{Z}^2,$$

where $Z_i - S_i/\sqrt{var(S_i)}$, S_i is the Mann-Kendall statistic in Equation (13), and $\bar{Z} = \sum_{i=1}^{m} Z_i/m$. Under the null hypothesis that the seasons are homogeneous with respect to trend, $\chi^2_{homogeneous}$ approximates the chi-square distribution with $m - 1$ degree of freedom. If $\chi^2_{homogeneous}$ exceeds the critical value, it indicates that there are different trends among different seasons. In that case, the three nonparametric methods are not meaningful, and the Mann-Kendall statistic can be used to test the trend for each individual season.

5.3 Adjusting covariate effects on trend tests

Several variables (X) other than time trend usually have considerable influence on water quality parameters (Y) (see Fig. 12). These variables are natural and random phenomena such as rainfall, temperature, and stream flow. To detect the trend of water quality parameters with time (T), these variable effects on water quality parameters need to be removed. The removal process includes modeling and explaining variable effects with regression methods and the LOWESS method (Helsel, 1992).

```
> xyplot(log(TP)~ log(Flow) | Year, col.line = 2, type=c("p", "r"), data = UNA, xlab = "Log values of flow", ylab = "Log values of total phosphorus concentrations")
```

5.3.1 Parametric methods

Consider a linear regression of Y versus time T and one or more covariates X,

$$Y = \beta_0 + \beta_1 T + \beta_2 X + \epsilon.$$

For the trend test, the null hypothesis is $\beta_1 = 0$. The t statistic can be used for the trend test. This model simultaneously explains the covariate effect and detects the trend with time. If the covariate changes with time, the following regression can be considered.

$$Y = \beta_0 + \beta_1 T + \beta_2 X + \beta_3 T * X + \epsilon,$$

where $T * X$ is the interactive term. For regression models, the relationship (linear function) between Y and X must be checked. Residuals should have no outliers and a constant covariance. The functions in R for testing these assumptions can be found in Subsection 5.4. According to the previous analysis, we use the following multiple linear regression to detect the trend of total phosphorus at Station 1. This model adjusts the flow effect and also captures the annual cycle.

$$\begin{aligned} ln(C) &= 2.25 - 0.74\text{Time} - 0.43\sin(2\pi\text{Time}) + 0.22\cos(2\pi\text{Time}) \\ &\quad -1.81 ln(\text{Flow}) + 0.096\text{Time} * ln(\text{Flow}) + 0.15 ln(\text{Flow})^2 \end{aligned}$$

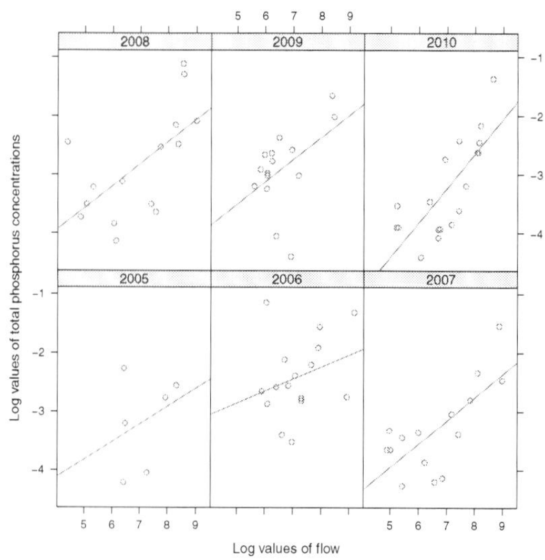

Fig. 12. Log values of total phosphorus concentration (C) versus log values of flow (F)

The results indicate an annual cycle and flow effect exist ($p<0.05$). After adjusting exogenous variables effects, the concentration of total phosphorus significantly decreases ($p = 0.02$).

5.3.2 Semi/nonparametric methods

Hirsch & Slack (1984) provided an adjusted seasonal Kendall test and proposed the following mixture procedure to test trends: (a) Use a regression model $Y = f(\beta X) + \epsilon$ to find the relationship between the concentration and covariates, where $f(\cdot)$ is a certain function of covariate X; (b) If there exists a significant relationship, compute the adjusted concentration $Y_{ik} - \hat{Y}_{ik}$, where $\hat{Y}_{ik} = f(\hat{\beta}X)$ is the estimated concentration of Y_{ik}; (c) Then apply the seasonal Kendall test for trend and slope estimator to the time series of $Y_{ik} - \hat{Y}_{ik}$.

The nonparametric LOWESS method (Cleveland, 1979; Helsel, 1992) can be used to remove a covariate effect without previously assuming the form of the relationship between Y and X. It is solely determined by the dataset and therefore it is robust to the distribution of the data pattern. The function *lowess* in the statistical software R can be used to obtain the fitted values $\hat{Y}$ of Y. The seasonal Kendall statistic (14) is calculated from $Y - \hat{Y}$ and T data pairs.

The parametric regression method can simultaneously check the covariate effect and detect the trend. When the linearity and normality assumptions are met, the parametric regression method outperforms for detecting and estimating the magnitude of trends. Otherwise, the LOWESS method is a desirable alternative. To examine the trend of a water quality parameter, the covariate and seasonal effects need to be removed. According to the real datasets, choose a reasonable statistical approach to test for trends. Various methods for trend tests are given in Table 1. For water quality data with detection limits, parametric Tobit regression (9) can be used. When a fixed detection limit exists, all the data below the fixed detection limit can be considered to be tied with each other. The nonparametric procedures such as Mann-Kendall, and the seasonal Kendall statistics can be used directly. If multiple detection limits exist, censor the data at the highest detection limit and then use an appropriate method to test the trend. Some information is certainly lost by making this change.

	No exogenous covariate (X) effects	
	No seasonality	Seasonality
Parametric	Regression of Y on T	Regression of Y on T and Seasonal terms
Nonparametric	M-K test	S-K test
Mixed		S-K test on residuals from regression of Y on X
	Exogenous covariate (X) effects exist	
	No seasonality	Seasonality
Parametric	Regression of Y on (T, X)	Regression of Y on (T, X, S)
Nonparametric	M-K of residuals from regression of Y on (T, X)	S-K of residuals from lowess of Y on X
Mixed	M-K of residuals from regression of Y on (T, X)	S-K of residuals from regression of Y on (T, X, S)

Table 1. Classification of various types of tests for monotonic trend. M-K indicates Mann-Kendall test, S-K indicates Seasonal Kendall test, and S denotes seasonal terms.

5.4 Computational implementation for linear regression models using R

In this subsection, we will show how to use the statistical software R to fit, evaluate and modify a linear regression model. More details can be seen in Adler (2009), Crawley (2007), and Venables & Ripley (2002).

A linear regression model is one of the most classic and popular methods in statistical practice. It is a very important tool for the statistical analysis of water quality data. It assumes that there is a linear relationship between a response variable (continuous) and some covariate variables. To fit a linear regression model to a dataset, the primary function is *lm*. We begin with the dataset mentioned in Section 2 to show how to fit a linear model in R. R codes for fitting the dataset are as follows.

```
> con.lm <- lm(log(TP) ~ log(Flow) + pH, data = UNA)
```

To print a simple display of the fitted information, use the *print* function:

```
> print(con.lm)
```

To obtain conventional regression analysis results, use the *summary* function:

```
> summary(con.lm)
```

To extract the regression coefficients, use the *coef* or *coefficients* function:

```
> coef(con.lm)
```

To obtain the variance-covariance matrix for the model fitted above, use the *vcov* or *Var* function:

```
> vcov(con.lm)
```

To calculate the confidence intervals for the coefficients in the fitted model, use the *confint* function:

```
>confint(con.lm, level = 0.95)
```

To get the residuals, use the *resid* or *residuals* function:

```
> resid(con.lm)
```

To obtain the fitted values, use the *fitted* or *fitted.values* function:

```
> fitted(con.lm)
```

To return the deviance of the fitted model, use the *deviance* function:

```
> deviance(con.lm)
```

To refit the model, it is better to use the *update* function, which can save some typing. For example, a slightly different model is used to fit the data above, which considers an extra covariate "Temp" besides "Flow" and "pH".

```
> con.lm2<-update(con.lm, . ~ . + Temp)
```

To compare models con.lm and con.lm2 which are used to fit the same dataset, use the *anova* function:

```
> anova(con.lm, con.lm2)
```

The main arguments to the function *lm* are

```
> lm(formula, data, weights, subset, na.action),
```

where *formula* is the model formula that specifies the form of the model to fit; *data* is an optional data frame containing the variables in the model; *weights* is a positive numeric vector containing weights to be used in the fitting process; *subset* is an optional vector specifying a subset of observations to be used in the fitting process; and *na.action* is a function which indicates how to handle missing values contained in the data.

The least-squares method performs well when the following key assumptions are satisfied: (1) There is a linear relationship between any pair of covariate variables (linearity); (2) The error terms are normally distributed (normality) with a constant variance (homoscedasticity); (3) The error terms are not correlated with each other (autocorrelation). However, because these assumptions may not be met in water quality data, linear regression is therefore not always appropriate. The test functions can be used to check these assumptions in R. The function *ncv.test* in the *car* package can be used to test the homoscedasticity. The function *durbin.watson* (*car* package) is used to test autocorrelation in a linear regression model. Diagnostic plots can also provide checks for homoscedasticity, normality, and influential observations (see Fig. 13), which can be obtained using the function *plot(con.lm)*.

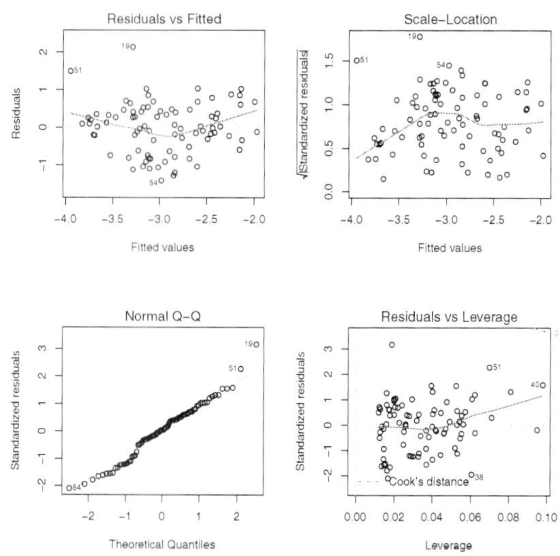

Fig. 13. Diagnostic plots for a linear regression model

6. Conclusions

Statistical methods are important in water quality analysis because much of what is known about water quality comes from numerical datasets. In this chapter, various statistical methods for analyzing water quality data have been introduced. Three typical graphs, boxplots, Q-Q plots, and scatter plots, which contain appropriate summarized information about datasets, are used to provide insight for analysts into datasets. A variety of classic water quality indices are applied to give a global assessment of water quality. Weighted water quality indices are relatively subjective; unweighted water quality indices and Harkins' water quality index are more objective. Other more advanced methods can be found in Raican et al. (2011) and Qian et al. (2007). To handle water quality data with detection limits, simple substitution methods as well as parametric and nonparametric approaches are investigated. Substitution methods are simple but possibly biased. Nonparametric methods which do not require the distributional assumption are robust and efficient (Helsel, 2005). Several popular methods, such as Mann-Kendall, the seasonal Kendall test, and multiple regression methods, are provided to detect and assess changes of various water quality parameters (Helsel, 1992). Meanwhile, nonlinear trends, serial dependence, covariate effects, and irregular measurement patterns need to be considered (Abaurrea et al., 2011; Morton & Henderson, 2008). Computational implementation using R for linear regression models is introduced. Examples using a real dataset are given to illustrate some very useful R functions.

7. Acknowledgments

Dr. Liya Fu is a postdoc fellow whose research was supported by the Centre for Applications in Natural Resource Mathematics (CARM), School of Mathematics and Physics, the University of Queensland, Australia.

8. References

Abaurrea, J., Asín, J., Cebrián, C. C. & García-Vera, M. A. (2011). Trend analysis of water quality series based on regression models with correlated errors, *Journal of Hydrology*, Vol. 400, 341–352.

Adler, J. (2009). *R in a nutshell*, O'Reilly Germany.

Brown, R. M., McClelland, N. I., Deininger, R. A. & Tozer, R. G. (1970). A water quality index: Do we dare? *Water and Sewage Works*, 117, 339–343.

Brown, B. M. & Wang, Y-G. (2006). Induced smoothing for rank regression with censored survival times, *Statistics in Medicine*, 26, 828–836.

Canadian Council of Ministers of the Environment (CCME). (2001). Canadian water quality guidelines for the protection of aquatic life, CCME water quality Index 1.0, Technical Report. In: Canadian environmental quality guidelines, 1999. Winnipeg: Canadian Council of Ministers of the Environment. *http://www.ccme.ca/assets/pdf/wqi_techrprtfctsht_e.pdf*.

Cleveland, W. S. (1979). Robust locally weighted regression and smoothing scatterplots, *Journal of the American Statistical Association*, 74, 829–836.

Cleveland, W. S., Grosse, E. & Shyu, W. M. (1992). Local regression models, Chapter 8 of *Statistical Models in S* eds Chambers and Hastie, Wadsworth & Brooks/Cole.

Chambers, J. M., Cleveland, W. S., Kleiner, B. & Tukey, P. A. (1983). *Graphical methods for data analysis*, PWA-Kent Publishing Co., Boston.

Cohen, A. C. (1976). Progressively censored sampling in the three parameter log-normal distribution, *Technometrics*, Vol. 18, No. 1, 99–103.

Cohn, T. A. (1988). Adjusted maximum likelihood estimation of the moments of lognormal populations from type I censored samples, *U.S. Geological Survey Open File Report*, 88–350,

Crawley, M. J. (2007). *The R book*, John Wiley & Sons Inc.

Cude, C. G. (2001). Oregon water quality index: a tool for evaluating water quality management effectiveness, *Journal of the American Water Resources Association*, 37, 125–137.

Debels, P. and Figueroa, R. and Urrutia, R. and Barra, R. and Niell, X. (2005). Evaluation of water quality in the Chillán river (Central Chile) using physiochemical parameters and a modified water quality index, *Environmental Monitoring and Assessment*, Vol. 110, No. 1301–322.

Dojlido, J. R., Raniszewski, J. & Woyciechowska, J. (1994). Water quality index applied to rivers in the Vistula River Basion in Poland, *Environmental Monitoring and Assessment*, Vol. 33, 33–42.

El-Shaarawi, A. H. & Esterby, S. R. (1992). Replacement of censored observations by a constant: an evaluation, *Water Research*, Vol. 26, Nol. 6, 835–844.

Farrel, R. (1980). Methods for classifying changes in environmental conditions, Technical Report VRF-EPA7. 4-FR80-1, Vector Research Inc., Ann Arbor, Michigan.

Fu, L. & Wang, Y-G. (2011). Nonparametric rank regression for analyzing water quality concentration data with multiple detection limits, *Environmental Science and Technology*, Vol. 45, No. 4, 1481–1489.

Gleit, A. (1985). Estimation for small normal data sets with detection limits, *Environmental Science and Technology*, 19, 1201–1206.

Gupta, A. K., Gupta, S. K. & Patil, R. S. (2003). A comparison of water quality indices for coastal water, *Journal of Environmental Science and Health*, A38, 2711–2725.

Gnanadesikan, R. & Wilk, M. B. (1968). Probability plotting methods for the analysis of data, *Biometrika*, Vol. 55, 1–17.

Harkins, R. D. (1974). An objective water quality index, *Journal of Water Pollution Control Federation*, Vol. 46, No. 3, 500–591.

Heller, G. (2007). Smoothed rank regression with censored data, *Journal of the American Statistical Association*, Vol. 102, No. 478, 552–559.

Helsel, D. R. (1990). Less than obvious-statistical treatment data below the detection limit, *Environmental Science and Technology*, Vol. 24, No. 12, 1766–1744.

Helsel, D. R. (2005). *Nondetects and data analysis: statistics for censored environmental data*, Wiley: New York.

Helsel, D. R. (2005b). More than obvious: better methods for interpreting nondetect data, *Environmental Science and Technology*, 15, 419–423.

Helsel, D. R., Cohn, T. (1988) Estimation of descriptive statistics for multiply censored water quality data, *Water Resources Research*, 24, 1997–2004.

Helsel, D. R. & Hirsch, R. M. (1992). *Statistical methods in water resources*, Elsevier Amsterdam, The Netherlands.

Hewett, P. & Ganser, G. H. (2007). A comparison of several methods for analyzing censored data, *Applied Occuputional and Environmental Hygiene*, Vol. 51, No. 7, 611–632.

Hirsch, R. M., Alexander, R. B. & Smith, R. A. (1991). Selection of methods for the detection and estimation of trends in water quality, *Water Resources Research*, Vol. 27, No. 5, 803–813.

Hirsch, R. M., Slack, J. R. & Smith, R. A. (1982). Techniques of trend analysis for monthly water quality data, *Water Resources Research*, Vol. 18, 107–121.

Hirsch, R. M. & Slack, J. R. (1984). A nonparametric trend test for seasonal data with serial dependence, *Water Resources Research*, Vol. 20, No. 6, 727–732.

Hornung, R. W. & Reed, L. D. (1990). Estimation of average concentration in the presence of nondetectable values, *Applied Occupational and Environmental Hygiene*, 5, 46–51.

Horton, R. K. (1965). An index-number system for rating water quality. *Journal of Water Pollution Control Federation*, Vol. 37, No. 3, 300–306.

Jin, Z., Lin, D. Y. & Wei, L. J. (2003). Rank-based inference for the accelerated failure time model, *Biometrika*, 90, 341–353.

Jin, Z., Lin, D. Y. & Ying, Z. (2006). On least-squares regression with censored data, *Biometrika*, 93, 147–161.

Jin, Z., Lin, D. Y. & Ying, Z. (2006b). Rank regression analysis of multivariate failure time data based on marginal linear models, *Scandinavian Journal of Statistics*, 33, 1–23.

Kannel, P. R., Lee, S., Kanel, S. R. & Khan, S. P. (2007). Application of water quality indices and dissolved oxygen as indicators for river water classification and urban impact assessment, *Environmental Monitoring Assessment*, Vol. 132, 93–110.

Kendall, M. G. (1975). *Rank correlation methods*, Charles Griffin, London.

Landwehr, J. M. & Deininger, R. A. (1976). A comparison of several water quality indices, *Journal of Water Pollution Control Federation*, Vol. 48, No. 5, 954–958.

Lee, L. & Helsel, D. R. (2005). Statistical analysis of water-quality data containing multiple detection limits: S-language software for regression on order statistics, *Computer & Geosciences*, 31, 1241–1248.

Lee, L. & Helsel, D. R. (2007). Statistical analysis of water-quality data containing multiple detection limits II: S-language software for nonparametric distribution modeling and hypothesis testing, *Computer and Geosciences*, 33, 696–704.

Lee, E. W., Wei, L. J. & Ying, Z. (1993). Linear regression analysis for highly stratified failure time data, *Journal of the American Statistical Association*, 88, 557–565.

Lettenmaier, D. P. (1988). Multivariate nonparametric tests for trend in water quality, *American Water Resources Association*, Vol. 24, No. 3, 505–512.

Lubin, J. H., Colt, J. S., Camann, D., Davis, S., Cerhan, J. R., Severson, R. K., Bernstein, L. & Hartge, P. (2004). Epidemiologic evaluation of measurement data in the presence of detection limits, *Environmental Health Perspectives*, 112, 1691–1696.

Mann, H. B. (1945). Non-parametric tests against trend, *Econometrica*, 13, 245–259.

Alternative for handling detection limits data in impact assessments, *Ground Water Monitor Remediation*, 4, 42–44.

McClelland, N. I. (1974). Water quality index application in the Kansas river basin. U.S. Environmental Protection Agency Region 7, Kansas City, Missouri.

McGill, R., Tukey, J. W. & Larsen, W. A. (1978). Variations of box plots, *The American Statistician*, Vol. 32, No.1, 12–16.

Morton, R. & Henderson, B. L. (2008). Estimation of nonlinear trends in water quality: An improved approach using generalized additive models, *Water Resourse Research*, Vol. 44, W07420, doi:10.1029/2007WR006191.

Prati, L., Pavenello, R. & Pesarin, F. (1971). Assessment of surface water quality by single index of pollution, *Water Research*, 5, 741–751.

Pesce, S. F. & Wunderlin, D. A. (2000). Use of water quality indices to verify the impact of Cordoba city (Argentina) on Suquýa river, *Water Research*, 34, 2915–2926.

Qian, Y., Migliaccion, K. W., Wang, Y. S. & Li, Y. C. (2007). Surface water quality evaluation using multivariate methods and a new water quality index in the Indian River Lagoon, Florida. *Water Recourses Research*, 43, 1–10.

Raican, S. M., Wang, Y-G., Harch, B. (2011). Water quality assessments for reservoirs using spatio-temporal data from balanced/unbalanced monitoring designs. *submitted*

Sanchez, E., Colmenarejo, M. F., Vicente, J., Rubio, A., Garcia, M. G., Travieso, L. & Borja, R. (2007). Use of the water quality index and dissolved oxygen deficit as simple indicators of watersheds pollution, *Ecological Indicators*, 7, 315–328.

Shumway, R. H., Azari, R. S. & Kayhanian, M. (2002). Statistical approaches to estimating mean water quality concentrations with detection limits, *Environmental Science and Technology*, 36, 3345–3353.

Tobin, J. (1958). Estimation of relationships for limited dependent variables, *Econometrica*, Vol. 26, No. 1, 24–36.

Tukey, J. W. (1977). *Exploratory data analysis*, Addison-Wesley Pub., Reading, MA.

Van Belle, G. & Hughes, J. P. (1984). Nonparametric tests for trend in water quality, *Water Resources Research*, Vol. 20, No. 1, 127–136.

Venables, W. N. & Ripley, B. D. (2002). *Modern applied statistics with S*, Springer.

Wei, L. J. (1992). The accelerated failure time model: a useful alternative to the Cox regression model in survival analysis, *Statistics in Medicine*, Vol. 11, 1871–1879.

Yu, Y.-S., Zou, S. & Whittemore, D. (1993). Non-parametric trend analysis of water quality data of rivers in Kansas, *Journal of Hydrology*, Vol. 150, 61–80.

Zhang, D. H., Fan, C. P., Zhang, J. & Zhang, C-H. (2009). Nonparametric methods for measurements below detection limits, *Statistics in Medicine*, 28, 700–715.

Zandbergen, P. A. & Hall, K. J. (1998). Analysis of the British Columbia water quality index for watershed managers: a case study of two small watershelds. *Water Quality Research Journal of Canada*, 33, 519–549.

7

An Innovative Nitrate Pollution Index and Multivariate Statistical Investigations of Groundwater Chemical Quality of Umm Rijam Aquifer (B4), North Yarmouk River Basin, Jordan

Mutewekil M. Obeidat[1], Muheeb Awawdeh[2], Fahmi Abu Al-Rub[3] and Ahmad Al-Ajlouni[4]

[1]Jordan University of Science and Technology, Faculty of Science and Arts, Irbid,
[2]Yarmouk University, Faculty of Science, Irbid,
[3]Jordan University of Science and Technology, Faculty of Engineering, Irbid,
[4]Jordan University of Science and Technology, Faculty of Science and Arts, Irbid, Jordan

1. Introduction

The chemical quality of the groundwater is modified by several factors, such as interaction with solid phases, residence time of groundwater, seepage of polluted river water, mixing of groundwater with pockets of saline water and anthropogenic impacts (Stallord and Edmond, 1983; Dethier, 1988; Faure, 1998; Umar et. al., 2006; Giridharan et al., 2008). Recently, there has been a tendency for groundwater quality deterioration, which has been caused by human activities (Dragon, 2008). The quality of water is of vital importance for humans, since it is directly linked with human welfare (Gupta et al., 2008). Poor water quality adversely affects the plant growth and human health (WHO, 1984, Hem, 1991; Karanth, 1997). Globally, nitrate is among the most common groundwater contaminants (Rajmohan and Elango, 2005). Potential sources of nitrate in groundwater include: fertilizers, septic tank effluent, municipal sewage, animal feedlots, decaying vegetation, and atmospheric deposition (Spalding and Exner, 1993; Wilhelm et al., 1996). In addition to the presence or absence of potential sources, field characteristics such as soil conditions, recharge rates, and depth to groundwater ultimately dictate an aquifer's vulnerability to nitrate contamination (Enwright and Hudak, 2009). Water quality index (WQI) is defined as a technique of rating that provides the composite influence of individual water quality parameters on the overall quality of water for human consumption (Vasanthavigar et al., 2010). It is an important parameter for demarcating groundwater quality and its suitability for drinking purposes (Mishra and Patel, 2001; Naik and Purohit, 2001; Avvannavar and Shrihari, 2008). There is a wide range of water quality indices that have been developed and used to classify water quality, which can be categorized based on the used variables (Terrado et al. 2010). Based on the considered variables, three classes of WQIs can be recognized:

1. Physiochemical variable WQIs such as those developed by Canadian Council of Ministers of the Environment (2001).

It is worth mentioning that the types of these WQIs depend on the used variables and the ultimate use of the index.

2. Biological WQIs such as those developed by Armitage et al. (1983). Biological water quality can be assessed based on different types of aquatic organisms (Terrado et al. 2010).
3. Hydro-morphological indices

Hydro-morphological indices take into consideration some characteristics of the fluvial ecosystem that do not relate to water quality directly (Terrado et al. 2010).

Many difficulties in data presentation, handling and interpretation could arise if the number of the data becomes large. Therefore, multivariate statistical techniques have become a powerful tool to handle and reduce large volume of water quality data. Statistical techniques, particularly multivariate statistics such as factor analysis are widely used tools for the identification of groundwater contamination (Grande et al., 1996). Many hydrochemical studies worldwide have shown that multivariate statistics significantly help classify and identify different factors controlling groundwater quality (Cloutier et al., 2008; Farnham, 2003; Belkhiri et al., 2010; Prasanna et al., 2010). Multivariate statistics help identify spatial and temporal variations in water quality and sources of contamination (natural and anthropogenic) by analyzing similarities/dissimilarities among the sampled sites (Andrade et al., 2008). Cluster analysis is carried out to reveal specific links between sampling points, while factor analysis/principal component analysis is used to identify the ecological aspects of pollutants on environmental systems (Ganfopadhyay et al., 2001; Kim et al 2009). Factor analysis enables both the classification of groups of data set and hydrochemical facies investigation and also the interpretation of their origin (Dalton and Upchurch 1978; Lawrence and Upchurch 1983; Dragon 2008). Jordan is an arid to semi-arid country with a land area of 89,206 km^2 that has suffered deficits in water resources since the 1960s. It ranks as one of the world's four most water-stressed countries. Moreover, Jordan is facing the problem of water resources contamination with different types of pollutants. The main objectives of the present study are (1) to assess the Umm Rijam (B4) groundwater pollution by nitrate using an innovative nitrate pollution index, and (2) to assess the groundwater quality of the B4 using multivariate statistical methods, namely cluster analysis (CA) and factor analysis (FA).

2. Study area

The Yarmouk River originates on the south-eastern slopes of Mount Hermon in Syria. The main trunk of the Yarmouk forms the present boundary between Syria and Jordan for 40 km before it becomes the border between Jordan and Israel. Yarmouk River Basin (YRB) is shared between Jordan and Syria. Only 1,424 km^2 of the basin total area (7,242 km^2) lie within the borders of Jordan (Fig. 1). The Yarmouk River average flow of 500 mcm/yr provides almost half of the surface water resources of the Jordan River. The climate is semi-arid with annual rainfall ranging from about 133 mm in the east to about 486 mm in the west. Geologically, Wadi Shallala Chalk Formation (B5) of early Middle-early Late Eocene age and Umm Rijam Chert Limestone Formation (B4) of Paleocene age overly the study area. It is composed mainly of limestone, chalk, and chert and reaches a maximum thickness of more than 200 m. The B5/B4 is the uppermost aquifer in the northern part the YRB (Fig. 2), where groundwater is stored under phreatic conditions. The B5/B4 is moderately jointed and fractured, and it is only slightly karstified. Many springs emerging from the B4 aquifer show chemical or bacteriological contamination (Chilton 2006). The subsurface geology consists of the Upper Cretaceous rock formations. From a hydrogeological point of view there are two important formations: Wadi Sir Limestone formation (A7) of Turonian age and Amman Silicified Limestone Formation (B2) of Campanian age. The two formations are hydraulically interconnected and considered as one aquifer, and they form the middle aquifer system. The B2/A7 consists of massive limestone, chert, dolomite, dolomitic limestone with intercalations of marl and chalk. The upper aquifer system (B5/B4) is separated from the middle aquifer system by the marly Muwwaqqar aquiclude (B3).

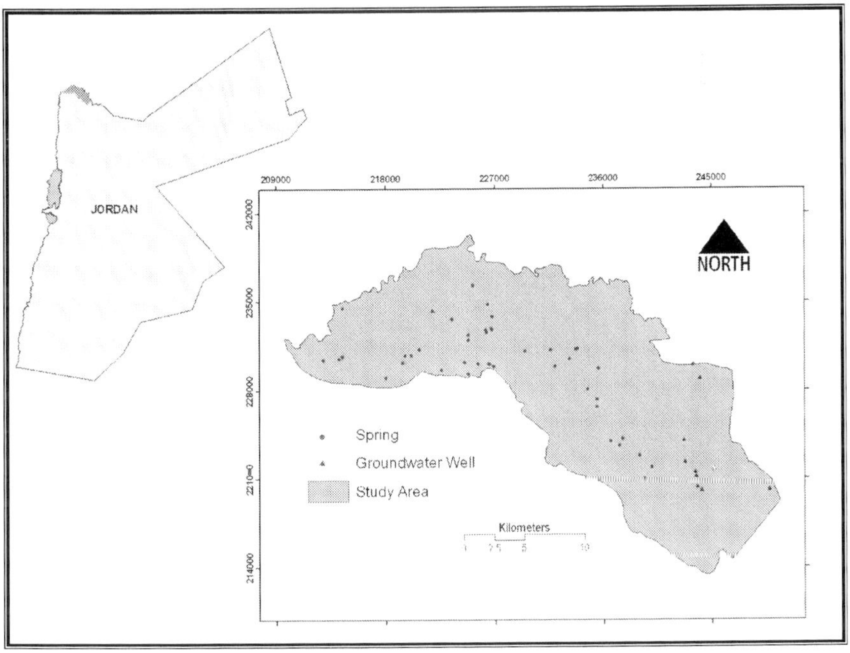

Fig. 1. Location map of the study area

3. Material and methods

3.1 Sampling and analytical procedures

Altogether 881 samples were used in the present study. The samples were taken from 12 wells and 43 springs representing the upper aquifer system (B4). The data involve those samples taken by the authors in 2007 and those retrieved from the open files of Water Authority of Jordan (WAJ). The data covers the period 1969-2009. The methods described by

(APHA, 1998) were followed during field and laboratory work. Electrical conductivity (EC), temperature and pH were measured in situ using portable meters. Prior to sample collection, well purging was performed for those wells which were not being pumped at the time of sampling. The goal was to ensure that the water sample truly represents the properties and conditions of the subsurface environment. Water was pumped from the well until the temperature, EC and pH became constant. The collected samples were investigated for the following hydrochemical parameters: calcium (Ca^{2+}), magnesium (Mg^{2+}), sodium (Na^{+}), potassium (K^{+}), chloride (Cl^{-}), bicarbonate (HCO_3^{-}), sulfate (SO_4^{2-}), nitrate (NO_3^{-}), total hardness (TH), and total dissolved solids (TDS). Na^{+} and K^{+} were determined by using flame photometer. Ca^{+2}, Mg^{2+}, HCO_3^{-} and Cl^{-} were analyzed by volumetric titration method. SO_4^{2-} and NO_3^{-} were analyzed spectrophotometrically.

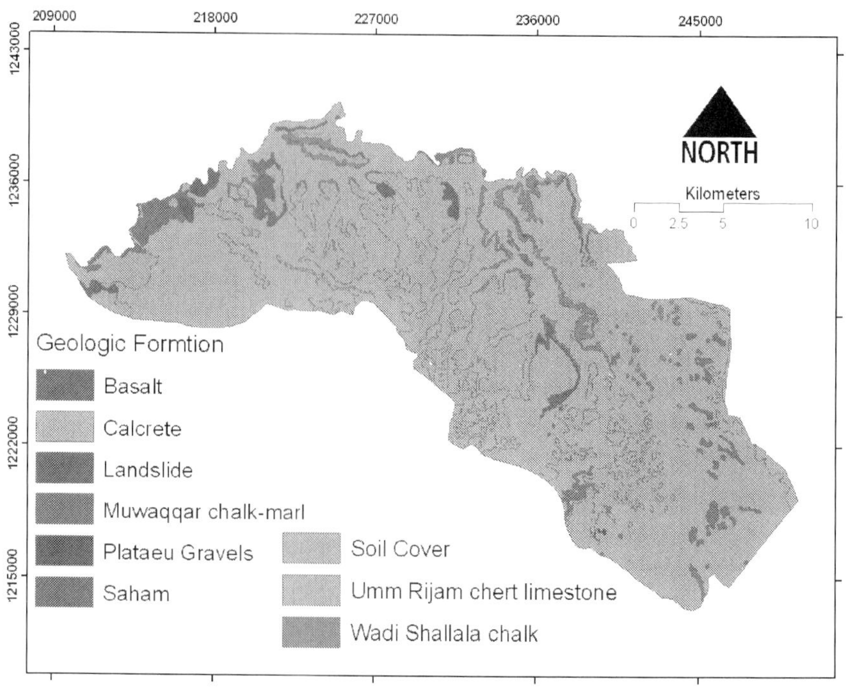

Fig. 2. Geological map of the study area

The total dissolved solids (TDS) content was calculated by using equation 1 (APHA, 1998), and the total hardness (TH) was calculated according to equation 2 (Todd, 1980)

$$TDS(mg/l) = Ca + Mg + Na + K + 0.5 * HCO_3 + Cl + SO_4 + NO_3 \quad (1)$$

$$\text{TH as } CaCO_3 \text{ (mg/l)} = 2.5 * Ca + 4.1 * Mg \quad (2)$$

Only samples with error in the cations-anions balance of ≤ 5% were used in this study. The error was calculated by using equation 3 (Edmond et al. 1995):

$$\text{Error\%} = \left|\frac{\sum \text{cations} - \sum \text{anions}}{\sum \text{cations} + \sum \text{anions}}\right| * 100 \tag{3}$$

3.2 Nitrate pollution index (NPI)

For the purpose of this study, a single-parameter water quality index called nitrate pollution index (NPI) is developed following two steps:

First step. The selection and justification of the pollution indicator namely the nitrate ion (NO_3^-)

Nitrate is the most frequently human-introduced pollutant into groundwater, and nitrate contamination of groundwater has become a common problem worldwide (Wakida et al. 2005). Potential sources of nitrate in groundwater resources include nonpoint sources such as intensive agricultural activities and unsewered sanitation in densely populated area, and point sources such as irrigation of land by sewage effluents (McLay et al., 2001). Because of its weak coordination to positive species and high solubility in water, nitrate is not strongly adsorbed to soil colloids and is highly mobile within the soil liquid phase. In strong oxidizing groundwater, nitrate the stable form of nitrogen, and it moves with no or littie transformation and little or no retardation (Freeze and Cherry, 1979). Chronic exposure to high nitrate concentration in drinking water has been linked to adverse health effects on humans, such as colon and rectum cancers, methemoglobinemia in infants, and non-Hodgkin's lymphoma (Ward et al., 1996; Knobeloch et al., 2000; De Roos et al., 2003). Groundwater with nitrate concentration exceeding the threshold of 20 mg/L as NO_3^- is considered contaminated due to human activities (referred to as human affected value, HAV) (Spalding and Exner, 1993). The maximum acceptable nitrate concentration for drinking water is 50 mg/L as NO_3^- (WHO, 1993). The main goal of the proposed NPI is to assess the extent, to which the groundwater in the area under consideration is polluted with nitrate due to human activities. The HAV of 20 mg/L is considered.

Second step. Calculation of the NPI

The NPI for a given site (well/spring) was calculated by using the following relation:

$$NPI = \frac{C_s - HAV}{HAV} \tag{4}$$

Where

Cs is the analytical concentration of nitrate in the sample

HAV is the threshold value of anthropogenic source (human affected value) taken as 20 mg/L

Then, the water quality was classified into five types based on the NPI values: clean (unpolluted), light pollution, moderate pollution, significant pollution, very significant pollution with the NPI value: <0, 0-1, 1-2, 2-3, > 3, respectively.

3.3 Cluster analysis (CA) and factor analysis (FA)

According to Liu et al. (2008), factor analysis and cluster analysis yield the common and similar relationships among hydrochemical variables by revealing multivariate patterns that may help classify the original data. Before analysis, the data used in the present study was subjected to standardization so as to increase the influence of variables whose variance is small, to reduce the impact of variables whose variance is large, to eliminate the influence of different units of measurement, and to render the data dimensionless (Dillon and Goldstein, 1984). It was done by a preparation of a correlation matrix of the data from which initial factor solutions were extracted by the principal component analytical method (Olobaniyi and Owoyemi 2006). Factor extraction was done with a minimum acceptable eigenvalue of one. The varimatrix rotation method was used to extract the final factors. Finally, factor's score was computed. Cluster analysis groups cases into classes based on similarities within a class and dissimilarities among different classes (Belkhiri et al., 2010). In the present study, the k-means cluster analysis was applied. This procedure attempts to identify relatively homogeneous groups of cases based on selected characteristics, using an algorithm that can handle large numbers of cases. The distance between clusters centers is calculated. Distances are computed using simple Euclidean distance. Factor and cluster analyses were performed using SPSS 13.0 for Windows.

4. Results and discussion

4.1 General hydrochemistry

The descriptive statistics of the hydrochemical parameters used in the study are presented in table 1. The pH value ranges between 6.8 and 8.5 with an average of 7.65. The groundwater can be described as neutral to slightly alkaline. The electrical conductivity is in the range 295-4220 µS/cm with an average of 791 µS/cm. Calcium concentration ranges between 26.45 and 281.72 mg/L with an average of 80.04 mg/L. Magnesium concentration is in the range 0-166.47 mg/L with an average of 37.61 mg/L. Sodium is in the range 2.3-565.8 mg/L

Parameter	Average	Minimum	Maximum	St.d.
pH	7.65	6.8	8.5	0.29
EC (µS/cm)	790.83	295	4220	480.42
Ca^{2+} (mg/L)	80.04	26.45	281.72	26.45
Mg^{2+} (mg/L)	37.61	0	166.47	24.15
Na^{+} (mg/L)	52.74	2.3	565.8	60.16
K^{+} (mg/L)	7.00	0	163	12.82
HCO_3^-(mg/L)	241.99	91.5	431.27	59.19
Cl^- (mg/L)	88.45	8.52	927.26	102.09
SO_4^{2-}(mg/L)	32.35	0	863.04	48.90
NO_3^- (mg/L)	45.75	0	794.81	66.87
TDS (mg/L)	593.12	221.25	3165	98.90
TH mg/l as $CaCO_3$	345.40	73.65	1075.31	131.62
CA-I	0.314	-0.025	0.565	
CA-II	0.076	-0.004	0.285	

Table 1. Univariate statistics of the hydrochemical parameters

with an average of 52.74 mg/L. Potassium concentration ranges between 0-163 mg/l with an average of 7 mg/L. Bicarbonate concentration is in the range 91.5-431.27 mg/L with an average of 242 mg/L. Chloride concentration is in the range of 8.52-927.26 mg/L with an average of 88.45 mg/L. Sulfate concentration ranges between 0-863.04 mg/l with an average of 32.35 mg/L.

The TDS content ranges between 221.25-3165 mg/L with an average of 593.12 mg/l. As shown in Fig. 3a, about 87% of the samples can be classified as freshwater based on the classification of Davis and Dewiest (1967). Total hardness is in the range 73.65-1075.31 mg/L with an average of 345.40 mg/L. Only 1.51% of the samples can be classified as soft-moderately hard water (Fig. 3b).

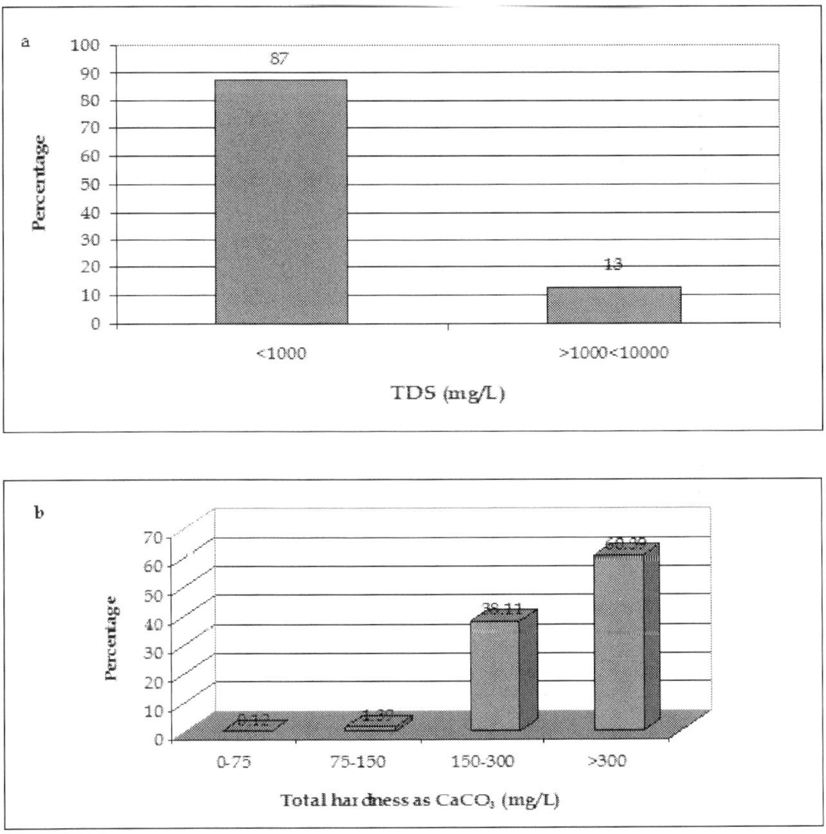

Fig. 3. Groundwater classification based on the: (a) TDS content (mg/L), and (b) total hardness as $CaCO_3$ (mg/L)

By using the trilinear (Piper) diagram, the groundwater can be categorized into the following groups (Fig. 4):

a. Normal earth alkaline water with prevailing bicarbonate
b. Earth alkaline water with increased portion of alkalies with prevailing bicarbonate
c. Earth alkaline water with increased portion of alkalies with bicarbonate and sulfate

Schoeller (1967) indicated that the ion exchange between the groundwater and its host environment during residence or travel can be understood by studying the chloro-alkaline indices, i.e. CA-I [(Cl^- - Na^+ + K^+) / Cl^-], and CA-II [(Cl^- – Na^+ + K^+) / (SO_4^{2-} + HCO_3^- + CO_3^{2-} + NO_3^-)]. Na^+ and K^+ ions in water are exchanged with Mg^{2+} and Ca^{2+} ions; positive value of the indices indicates base-exchange reaction whereas negative value indicates chloro-alkaline disequilibrium (Gupta et al., 2008). This reaction is known as cation–anion exchange reaction. During this process the host rocks are the primary sources of dissolved solids in the water. Schoeller indices reveal that more than 96% of the B4 groundwater has positive values indicating a base-exchange reaction (table 1).

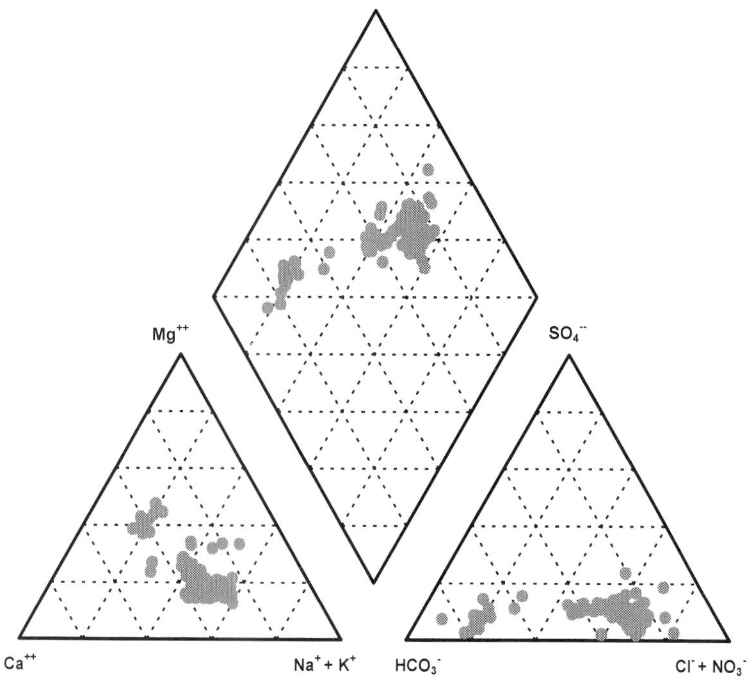

Fig. 4. Trilinear presentation of the groundwater

4.2 Spatial and temporal variation

The temporal variations of some hydrochemical parameters of the groundwater are presented in Fig. 5. For well AD1251, there was a slight increase in all hydrochemical parameters from February 1985 until August 1987, but since October 1988 until September 2002, there was a dramatic increase. Since September 2003, there was a dramatic decrease in all hydrochemical parameters. This is can be attributed to the fact that Al Ramtha wastewater treatment plant was established in 1988. At the beginning, the plant worked as stabilization ponds, but in 2003, it was changed to activated sludge. Well AD1251 is located just a few kilometers downstream of the plant. The spatial distribution of the EC and nitrate concentration is presented in Figs. 6 and 7, respectively. Three zones of the electrical conductivity can be observed: (1) east, where the electrical conductivity reaches 2668 µS/cm, (2) northwest, where the electrical conductivity ranges between 1520 and 2268 µS/cm, and (3) central part of the study area, where the electrical conductivity is less than 662 µS/cm.

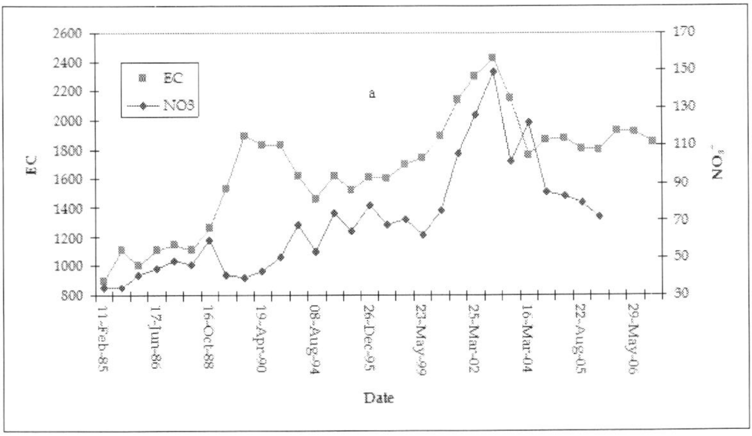

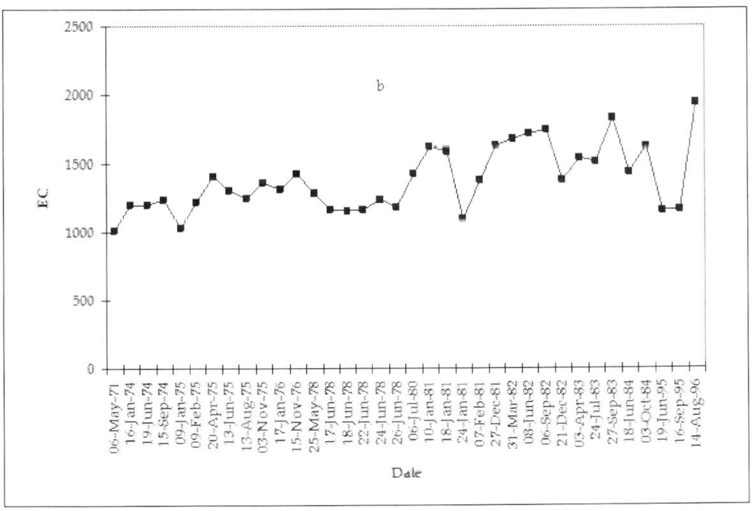

Fig. 5. Temporal variation of the EC (µS/cm) and NO_3^- (mg/L) for well AD1251 (a), and EC (µS/cm) for well AD1213 (b).

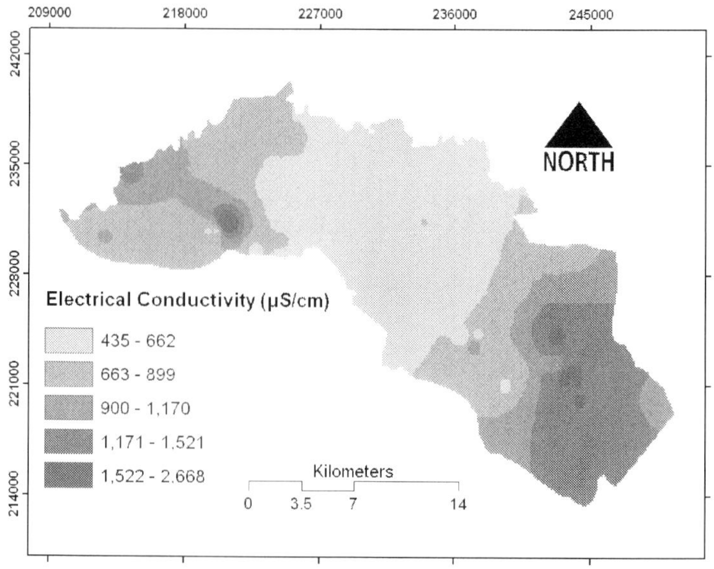

Fig. 6. Spatial distribution of the average of the EC (µS/cm)

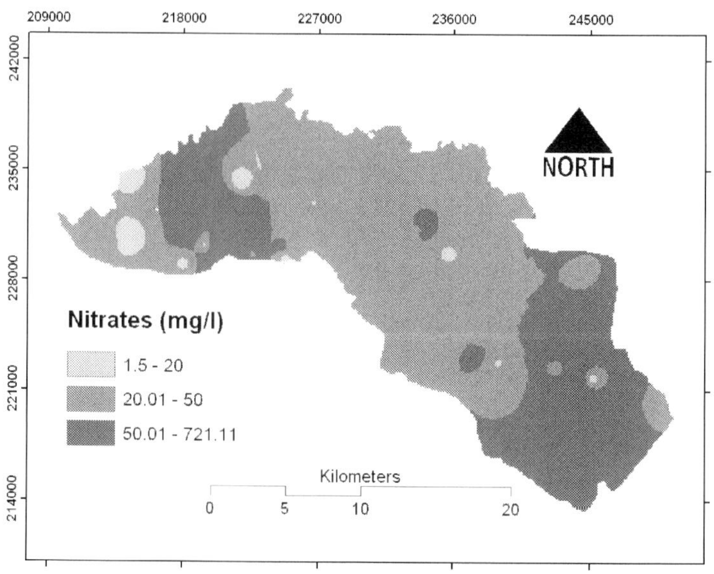

Fig. 7. Spatial distribution of the average nitrate concentration (mg/L as NO_3^-)

Nitrate concentration higher than 50 mg/L is found in the eastern and northwest parts of the study area, and concentration in the range of 20-50 mg/L is found in the central part of the study area. The spatial variation of the hydrochemical parameters reflects the hydrogeological situation and land use practices in the study area. Generally, high concentrations of all parameters were observed where the depth to the water table is less than 40 m, and where the land use is urban.

4.3 Nitrate concentration and the NPI

Nitrate concentration in the study area is in the range 1-794.81 mg/L with an average of 45.75 mg/L. The nitrate concentration was grouped into one of three classes (Fig. 8a).

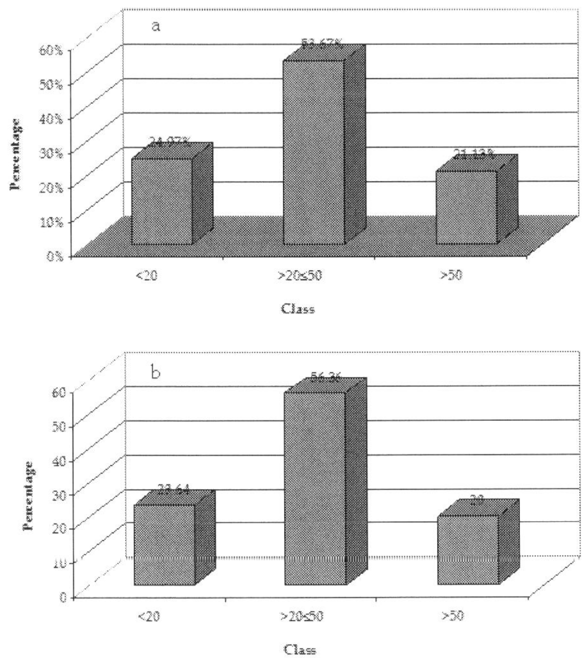

Fig. 8. Frequency distribution of the nitrate concentration (mg/L): (a) total samples, and (b) sampled sites (wells/springs)

i) low (<20 mg/L), ii) medium (≥ 20 mg/L to <50 mg/L), and iii) high (≥50 mg/L). Nitrate concentrations in the high class exceed the recommendations for drinking water set by WHO (1993). The medium class involves samples with nitrate concentrations high enough to indicate the influence of human activities (Spalding and Exner, 1993). The low class involves samples with a low risk for humans or the environment. About 75% of the samples has nitrate concentration exceeding the threshold value of anthropogenic source or the human affected value (HAV). More than 56% of the sampled sites has nitrate concentration of more than 20 mg/L and less than 50 mg/L, and 20% of the sampled sites has nitrate concentration of more than 50 mg/L (Fig. 8b). Based on the NPI values, five water classes were identified: clean, light pollution, moderate pollution, significant pollution, and very significant pollution, corresponding to NPI values of <0, 0-1, 1-2, 2-3, and >3, respectively. The sites (wells and springs) belonging to the classes significant and very significant pollution have nitrate concentration exceeding the maximum permissible limit of drinking water quality standards given by the WHO (1993). The spatial distribution of the NPI is presented in Fig. 9. Two major zones of very significant pollution is revealed: northwest and southeast of the study area. In the study area, there are two major potential sources of nitrate in groundwater: untreated/treated domestic wastewater and agricultural fertilizers. This can be indicated by the high content of E. Coli and total coliform recorded in wells and spring such as AD1251, AD1296, AD1046, AD1050, AD3027, AD1105, AD3057, AD3015, AD0536, AD1281, AD1280, AD0630, AD0600, AD0580, AD0560, AD1296, and AD0741. Moreover, a significant correlation coefficient exists between nitrate and potassium concentrations (table 2).

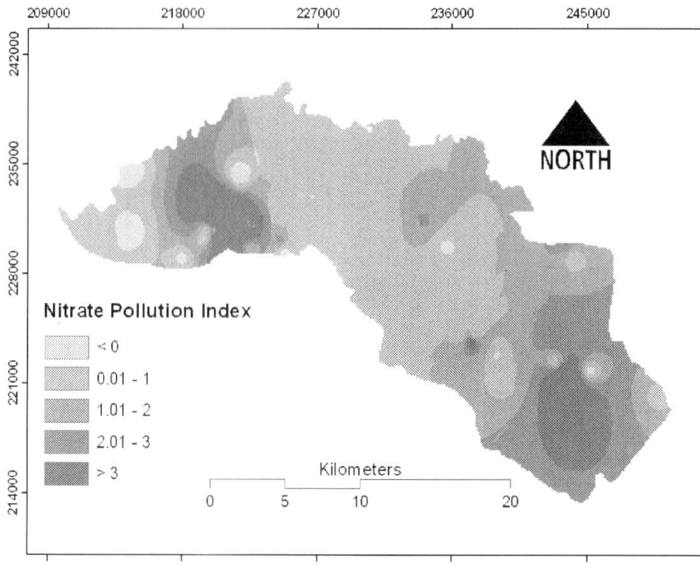

Fig. 9. Spatial distribution of the NPI

Parameter	pH	HCO_3	Ca	Cl	EC	Mg	K	Na	SO_4	NO_3
pH	1.000	-0.419	-0.382	-0.219	-0.253	-0.204	-0.074	-0.212	-0.099	-0.147
EC	-0.253	0.269	0.631	0.917	1.000	0.147	0.525	0.900	0.624	0.638
HCO_3^-	-0.419	1.000	0.584	0.126	0.269	0.350	0.099	0.141	0.123	0.046
Ca^{2+}	-0.382	0.584	1.000	0.487	0.631	0.192	0.623	0.446	0.344	0.648
Cl^+	-0.219	0.126	0.487	1.000	0.917	0.103	0.375	0.967	0.582	0.541
Mg^{2+}	-0.204	0.350	0.192	0.103	0.147	1.000	0.062	0.097	0.265	0.011
K^+	-0.074	0.099	0.623	0.375	0.525	0.062	1.000	0.321	0.246	0.823
Na^+	-0.212	0.141	0.446	0.967	0.900	0.097	0.321	1.000	0.593	0.503
SO_4^{2-}	-0.099	0.123	0.344	0.582	0.624	0.265	0.246	0.593	1.000	0.213
NO_3^-	-0.147	0.046	0.648	0.541	0.638	0.011	0.823	0.503	0.213	1.000

Table 2. Bivariate statistics of the hydrochemical parameters

4.4 Multivariate analysis

4.4.1 Cluster analysis

Cluster analysis is a classification that places objects into more or less homogeneous groups in a manner so that the relation between groups is revealed. The results of k-means clustering of the B4 groundwater are compiled in table 3. Three clusters were identified with distinct cluster centers. **Cluster 1** has the highest ionic concentration and comprises only 0.3% of the total samples. This cluster is highly nitrated with a nitrate concentration of 355 mg/L. **Cluster 2** has total dissolved solids content intermediate to cluster 1 and 3, and it involves 81% of the samples. The groundwater belongs to this cluster can be classified as freshwater. This cluster has a nitrate concentration of 30 mg/L, which exceeds the threshold value of anthropogenic source. **Cluster 3** comprises 18.6% of the total samples. The water belongs to this group can be classified as brackish water, and it contains high nitrate concentration (110 mg/L) exceeding the WHO (1993) standards for drinking water. The three clusters are shown in Fig. (10a) as Schoeller diagram on a semilogarithmic paper, with the data plot on parallel lines. The data which plot on parallel lines reflect dilution of a saline water type with fresh water (Mazor, 1991). The total dissolved content (TDS) and chloride concentration were plotted using a scatter diagram as shown in Fig. (10b). Three distinct groups can be clearly seen. Such pattern indicates mixing and progressive change in concentrations.

Parameter	Cluster		
	1	2	3
pH	8	8	8
EC	3,442	609	1,634
Ca^{2+}	146.9189	74.5637	104.7756
Mg^{2+}	37.4560	35.7162	40.7068
Na^{+}	127.58	29.11	168.16
K^{+}	69.5	4.7	13.9
HCO_3^-	239.61	240.27	251.34
Cl^-	278.297	49.441	289.685
SO_4^{2-}	95.60	19.36	87.00
NO_3^-	355	30	110

Table 3. Results of the k-means cluster analysis

4.4.2 Factor analysis

Table 4 shows the eigenvalues (>1) of the three extracted factors, their percentage of variance, and cumulative percentage of variance of hydrochemical parameters of the B4 groundwater. It is found that the three factors account for 76.78% of the total variance. Table 5 shows the loadings for the varimatrix-rotated factor matrix in the three factor model. According to Liu et al. (2003), the terms strong, moderate, and weak, which are applied to factor loadings, refer to absolute loading values of >0.75, 0.75-0.5 and 0.5-0.3, respectively.

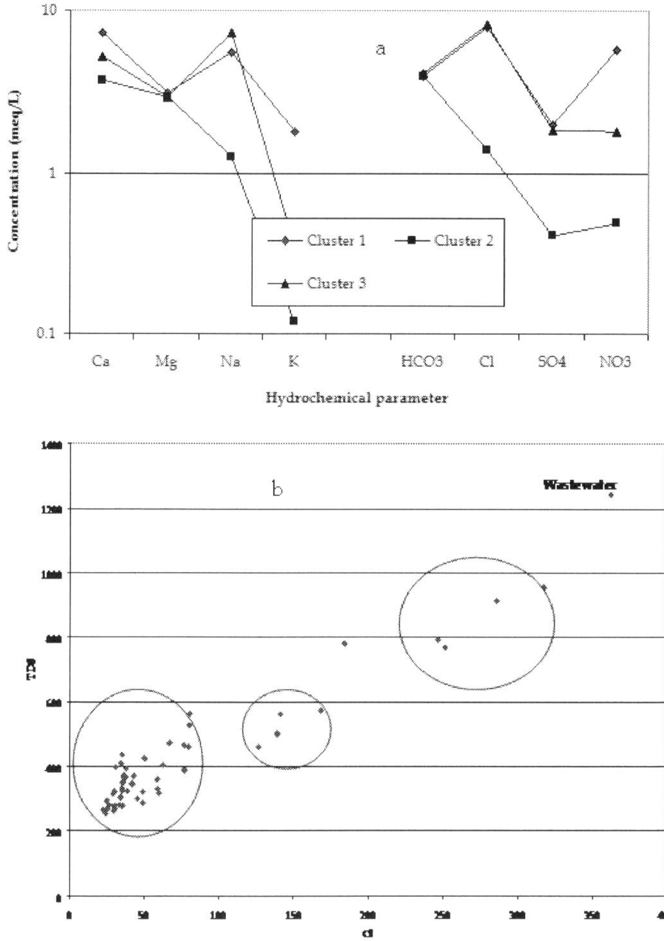

Fig. 10. Schoeller diagram presentation of the three clusters (a), and composition diagram of the chloride versus TDS (b)

Factor 1, which accounts for 47.05% of the total variance, has strong positive loadings on electrical conductivity, chloride, sodium and sulfate. This factor is termed as the "**salinity factor**" in reference to salt halite (Cloutier et al. 2008), indicating salinization of the groundwater due to mixing freshwater with saline water introduced into the aquifer by overpumping or infiltration of treated/untreated wastewater. **Factor 2** explains 16.27% of the total variance, and has strong positive loadings on nitrate and potassium, and moderate loading on calcium, and weak loadings on the electrical conductivity and chloride. This factor is termed as "**pollution factor**", indicating anthropogenic impacts from agricultural and domestic wastewater. **Factor 3** accounts 13.46% of the total variance, and has strong positive loading on bicarbonate, moderate positive loadings on calcium and magnesium, and moderate negative loading on the pH. The presence of bicarbonate, calcium, and magnesium reflects signatures of natural water recharge and rock-water interaction (Prasanna et al. 2010). This factor can be termed as the "**hardness factor**", since calcium and magnesium are generally used to calculate the hardness. The factors' scores were mapped to assess the processes affecting the groundwater quality in the study area (Fig.11). According to Olobaniyi and Owoyemi (2006), a factor score larger than +1 indicates intense influence by the process, whereas a very negative factor score (< -1) indicates that the area is unaffected by the process. Near zero factor score suggests moderate influence. Groundwater in the east and northwest parts of the study area is more saline than groundwater in the central part. Groundwater with highest hardness is found in the northwest part of the study area. Contamination of groundwater by nitrate together with potassium is prevailing in the west part of the study area.

Factor	Eigenvalue	% of variance	Cumulative variance %
1	4.705	47.050	47.050
2	1.627	16.271	63.322
3	1.346	13.460	76.781

Table 4. Eigenvalue (>1), percentage of variance, and percentage of cumulative variance of the three factors

Parameter	Factor		
	1	2	3
pH	-0.093	-0.125	**-0.678**
EC	**0.835**	0.454	0.185
Ca^{2+}	0.270	**0.705**	**0.526**
Mg^{2+}	0.163	-0.142	**0.629**
Na^{+}	**0.921**	0.249	0.073
K^{+}	0.163	**0.900**	0.012
HCO_3^-	0.009	0.141	**0.859**
Cl^-	**0.908**	0.305	0.069
SO_4^{2-}	**0.784**	-0.002	0.163
NO_3^-	0.304	**0.896**	-0.030

Table 5. Loadings of the varimatrix-rotated factors

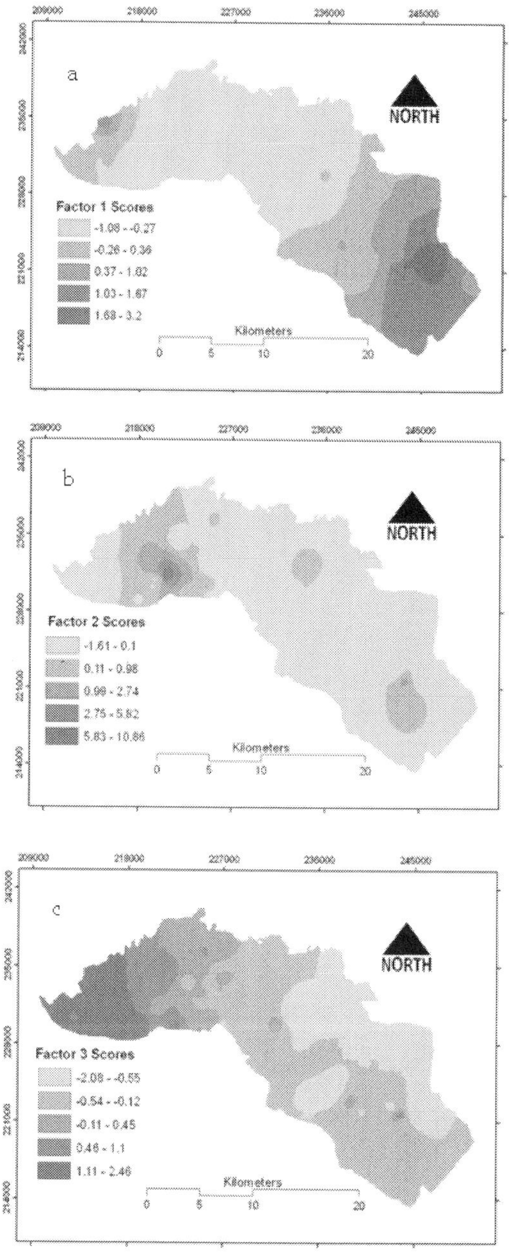

Fig. 11. Map of the three factors' scores

5. Summary and conclusions

This study has revealed that the phreatic groundwater in the northern Yarmouk basin is contaminated by nitrate to a high and variable degree. It is found that nitrate concentration (as mg/L NO_3^-) ranges from 1 and 795 mg/L, with an average of about 46 mg/L. More than 76% of the sampled sites has nitrate concentration in excess of the human affected value (20 mg/L), and about 20% of the sampled sites has nitrate concentration higher than the WHO standards of drinking water quality (> 50 mg/L). The highest nitrate concentrations were found for those wells/springs where the depth to groundwater is less than 30 m below ground surface. This study has developed a new water quality index named nitrate pollution index (NPI). It was generated using a simple mathematical equation which relates the analytical nitrate concentration to the threshold value of anthropogenic source of nitrate (HAV). The sampled sites were classified by using the NPI into the following classes: clean, light pollution, moderate pollution, significant pollution, and very significant pollution, corresponding to NPI values of <0, 0-1, 1-2, 2-3, and >3, respectively. The wells/springs which fall in the significant and very significant pollution classes have nitrate concentration exceeding the WHO standards. Multivariate statistical techniques (CA and FA) were applied on the hydrochemical parameters of the groundwater in the study area. CA resulted in three clusters which can be differentiated by the total ionic concentration and the nitrate concentration. FA resulted in a three-factor model which accounts for 76.78% of the groundwater chemistry variation. Factor 1 "salinity factor" is responsible for 47.05%of the total variance and has strong positive loadings on electrical conductivity, chloride, sodium and sulfate. Factor 2 "pollution factor" accounts for 16.27% of the total variance and has strong positive loadings on nitrate and potassium. Factor 3 "hardness factor" accounts for 13.46% of the total variance and has strong positive loading on bicarbonate and moderate loadings on calcium and magnesium. To deduce the processes that control the groundwater chemistry, factors' scores were mapped. There are two agents that control the groundwater quality in the study area: human activities and natural processes, as it is pointed by the factor analysis. Human activities which led to modification on the water quality include agricultural activities mainly crop cultivation and application of fertilizers. This is well indicated by the very significant correlation (r = 0.82) between nitrate and potassium. Treated/untreated wastewater is another major source of groundwater pollution, which is indicated by the presence of bacteriological contamination in the wells surrounding Ar Ramtha wastewater treatment plant. Total coliform content up to 1600 MPN/100 ml, and E. coli content of 500 MPN/100 ml were recorded. Rock/soil-water interaction plays an important role in the modification of the groundwater chemistry, especially water hardness.

6. References

Andrade, E. M., Palacio, H.A.Q., Souza, I.H., Leao, R.A.O., & Guerreiro, M.J. (2008). Land use effects in groundwater composition of an alluvial aquifer (Trussu River, Brazil) by multivariate techniques. *Environmental Research*, Vol.106, pp. 170-177, ISSN 0013-9351

APHA, (1998). Standard Methods for the Examination of Water and Wastewater. 20th Edn., American Public Health Association, ISBN 0875530478, Washington.

Armitage, P.D., Moss D., Wright J.F. & Furse M.T. (1983). The performance of a new biological water quality score system based on macroinvertebrates over a wide range of unpolluted running water sites. *Water Research*, Vol.17, pp. 333-347, ISSN 0043-1354

Avvannavar, S. M., & Shrihari, S. (2008). Evaluation of water quality index for drinking purposes for river Netravathi, Mangalore, South India. *Environmental Monitoring and Assessment*, Vol.143, pp. 279–290, ISSN 0167-6369.

Belkhiri, L., Boudoukh, A., Mouni, L., & Baouz, T. (2010). Application of multivariate statistical methods and inverse geochemical modeling for characterization of groundwater-A case study: Ain Azel plain (Algeria). *Geoderma*, Vol.159, pp. 390-398, ISSN 0016-7061.

Canadian Council of Ministers of the Environment (CCME, 2001), in: Canadian Environmental Quality Guidelines, CCME, Winnipeg, Canada.

Chilton, J. (2006). Assessment of Aquifer pollution vulnerability and susceptibility to the impacts of abstraction. Protecting ground- water for health: managing the quality of drinking-water sources. In: Schmoll O et al (eds) WHO drinking-water quality series. IWA, ISBN 1843390795, London.

Cloutier, V., Lefebvre, R., Therrien, R., & Savard, M.M. (2008). Multivariate statistical analysis of geochemical data as indicative of the hydrogeochemical evolution of groundwater in a sedimentary rock aquifer system. *Journal of Hydrology*, Vol.353, pp. 294-313, ISSN 0022-1694.

Dalton, M.G., & Upchurch, S.B. (1978). Interpretation of hydrochemical facies by factor analysis. *Ground Water*, Vol.16, No.4, pp. 228-233, ISSN 1745-6584.

Davis, S.N. and Dewiest, R.J.M. (1967). Hydrogeology. John Wiley and Sons Inc., ISBN 0894646389, New York.

De Roos, A.J., Ward, M.H., Lynch, C.F., & Cantor K.P. (2003). Nitrate in public water supplies and risk of colon and rectum cancers. *Epidemiology*, Vol.14, No.6, pp. 640-649, ISSN 1044-3983.

Dethier, D. P. (1988). A hydrochemical model for stream chemistry, cascade range, Washington U.S.A. *Earth Surface Processesand LandForms*, Vol.13, pp. 321–333, ISSN 1096-9837.

Dillon, W.R., & Goldstein, M. 1984. Multivariate analysis methods and applications, 3ed. John Wiley and Sons, ISBN 0-471-08317-8, New York.

Dragon, K. (2008). The Influence of Anthropogenic Contamination on the Groundwater Chemistry of a Semi-confined Aquifer (The Wielkopolska Buried Valley Aquifer, Poland). *Water Resources Management*, Vol.22, pp. 343-355, ISSN 0920-4741.

Edmond, J.M., Palwer, M.R., Measures, C.F., Grant, B., & Stallard, R.F. (1995). The fluvial geochemistry and denudation rate of the Guayana Shield in Venezuela. *Geochimica Coscochimica Acta*, Vol.59, pp. 3301-3323, ISSN 0016-7037.

Enwright, N., and Hudak, P.F. (2009). Spatial distribution of nitrate and related factors in the High Plains Aquifer, Texas. *Environmental Geology*, Vol.58, pp. 1541-1548, ISSN 1866-6280.

Faure, G. (1998). Principles and applications of geochemistry, 2nd ed., Prentice-Hall, ISBN 0023364505, Englewood Cliffs, NJ.

Farnham, I.M., Johannesson, K.H., Singh, A.K., Hodge, V.F., & Stetzenbach, K.J. (2003). Factor analytical approaches for evaluating groundwater trace element chemistry data. *Analytica Chimica Acta*, Vol.490, 123-138, ISSN 0003-2670.

Freeze, R.A., & Cherry, J.A. (1979). Groundwater. Prentice Hall Inc., ISBN 0-13-365312-9, Englewood Cliffs, New Jersey.

Ganfopadhyay, S., Gupta, A., & Nachabe, M.H., (2001). Evaluation of ground water monitoring network by principal component analysis. *Ground Water*, Vol.39, No.2, pp. 181-191, ISSN 1745-6584.

Giridharan, L., Venugopal, T., & Jayaprakas, M. (2008). Evaluation of the seasonal variation on the geochemical parameters and quality assessment of the groundwater in the proximity of River Cooum, Chennai, India. *Environmental Monitoring Assessment*, Vol.143, pp. 161-178, ISSN 0167-6369.

Grande, J.A., Gonzalez, A., Beltaran, R., & Sanchez-Rodas, D. (1996). Application of Factor Analysis to the study of contamination in the aquifer system of Ayamonte-Huelva (Spain). *Ground Water*, Vol.34, No.1, pp. 155-161, ISSN 1745-6584.

Gupta, S., Mahato, A., Roy, P., Datta J.K., & Saha, R.N. (2008). Geochemistry of groundwater, Burdwan District, West Bengal, India. *Environmental Geology*, Vol.53, pp. 1271-1282, ISSN 1866-6280.

Hem, J.D. (1991). Study and interpretation of the chemical characteristics of natural waters, Book 2254 (3rd edn). Jodhpur: Scientific Publishers.

Karanth, K.R. (1997). Groundwater assessment, development and management. Tata McGraw-Hill, ISBN 9780074517123, New Delhi.

Kim, K.H., Yun, S.T., Choi, B.Y., Chae, G.T., Joo, Y., Kim, K., & Kim, H.S. (2009). Hydrochemical and multivariate statistical interpretations of spatial controls of nitrate concentrations in a shallow alluvial aquifer around oxbow lakes (Osong area, central Korea). *Journal of Contaminant Hydrology*, Vol.107, pp. 114-127, ISSN 0169-7722.

Knobeloch, L., Salna, B., Hogan, A., Postle, J., & Anderson, H. (2000). Blue babies and nitrate-contaminated well water. *Environmental Health Perspectives*, Vol.108, No.7, pp. 675-678. ISSN 0091-6765.

Lawrence, A.R., & Upchurch, S.B. (1983). Identification of recharge areas using geochemical factor analysis. *Ground Water*, Vol.20, No.6, pp. 680-687, ISSN 1745-6584.

Liu, C-W, Jang, C-S, Chen,C-P, Lin, C-N, & Lou, K-L. (2008). Characterization of groundwater quality in Kinmen Island using multivariate analysis and geochemical modelling. *Hydrological Processes*, Vol.22, pp. 376-383, ISSN 1099-1085.

Liu, C-W, Lin, K-H, & Kuo, Y-M. (2003). Application of factor analysis in the assessment of groundwater quality in a blackfoot disease area in Taiwan. *The Science of the Total Environment*, Vol.313, pp. 77-89, ISSN 0048-9697

Mazor, E. (1991). Chemical and Isotopic Groundwater Hydrology: The Applied Approach, second ed., Marcel Dekker Inc. ISBN 0-8247-4704-6, New York.

McLay, C.D.A., Dragten, R., Sparling, G., & Selvarajah, N. (2001). Predicting groundwater nitrate concentrations in a region of mixed agricultural land use: a comparison of three approaches. *Environmental Pollution*,Vol.115, pp. 191-204, ISSN 0269-7491.

Mishra, P. C., & Patel, R. K. (2001). Study of the pollution load in the drinking water of Rairangpur, a small tribal dominated town of North Orissa. *Indian Journal of Environment and Ecoplanning*, Vol. 5, No.2, pp. 293-298, ISSN 0972-1215.

Naik, S., & Purohit, K. M. (2001). Studies on water quality of river Brahmani in Sundargarh district, Orissa. *Indian Journal of Environment and Ecoplanning*, Vol.5, No.2, pp. 397-402, ISSN 0972-1215.

Olobaniyi, S.B., & Owoyemi, F.B. (2006). Characterization by factor analysis of the chemical facies of groundwater in the deltaic plain sands aquifer of Warri, Western Niger delta, Nigeria. *African Journal of Science and Technology. Science and Engineering Series* 7 (1), pp. 73-81.

Prasanna, M.V., Chidambaram, S., & Srinivasamoorthy, K. (2010). Statistical analysis of the hydrogeochemical evolution of groundwater in hard and sedimentary aquifers system of Gadilam river basin, South India. *Journal of King Saud University (Science)*, Vol.22, pp. 133-145, ISSN 1018-3647.

Rajmohan, N., & Elango, L. (2005). Nutrient chemistry of groundwater in an intensively irrigated region of Southern India. *Environmental Geology*, Vol.47, pp. 820–830, ISSN 1866-6280.

Schoeller, H. (1967). Geochemistry of groundwater. An international guide for research and practice. UNESCO, chap 15, pp 1–18.

Spalding, R.F., & Exner, M.E. (1993). Occurrence of nitrate in groundwater- a review. *Journal of Environmental Quality*, Vol.22, pp. 392-402, ISSN 0047-2425.

Stallord, R. F., & Edmond, J. M. (1983). Geochemistry of Amazon River: the influence of the geology and weathering environment on the dissolved load. *Journal of Geophysical Research*, Vol.88, pp. 9671–9688, ISSN 0148–0227.

Terrado, M., Borrell, E., de Campos, S., Damia, B., & Tauler, R. (2010). Surface-water-quality indices for the analysis of data generated by automated sampling networks. *Trends in Analytical Chemistry*, Vol.29, No.1, pp. 40-52, ISSN 0165-9936.

Todd, D.K. 1980. Ground water hydrology.Wiley. ISBN 047187616X, New York

Umar, R., Muqtada, M., Khan, A., & Absar, A. (2006). Groundwater hydrochemistry of a sugarcane cultivation belt in parts of Muzaffarnagar district, Uttar Pradesh, India. *Environmental Geology*, Vol.49, pp. 999–1008, ISSN 1866-6280.

Vasanthavigar M, Srinivasamoorthy K, Vijayaragavan K, Rajiv Ganthi R, Chidambaram S, Anandhan P, Manivannan R, & Vasudevan S. (2010). Application of water quality index for groundwater quality assessment: Thirumanimuttar sub-basin, Tamilnadu, India. *Environmental Monitoring and Assessment*, Vol.171, pp. 595-609, ISSN 0167-6369.

Wakida, T.F., & Lerner, D.N. (2005). Non-agricultural sources of groundwater nitrate: a review and case study. *Water Research*, Vol.39, pp. 3-16, ISSN 0043-1354.

Ward, M.H., Mark, S.D., Cantor, K.P., Weisenburger, D.D., Correa-Villasenor, A., & Zahm, S. H. (1996). Drinking water nitrate and the risk of non-Hodgkins lymphoma. *Epidemiology*, Vol.7, No.5, pp. 465-471, ISSN 1044-3983.

WHO (World Health Organization), 1993. Guidelines for drinking water quality, vol. 1, Recommendations (2nd edn.). WHO, Geneva.

WHO. 1984. Guidelines for drinking water quality-II. Environmental Health Criteria 5. WHO, Geneva.

Wilhelm, S.R., Schiff, S.L., & Robertson, W.D. (1996). Biogeochemical evolution of domestic waste water in septic systems 2. Application of conceptual model in sandy aquifers. *Ground Water*, Vol.34, pp. 853-864, ISSN 1745-6584.

8

Monitoring and Modelling of Water Quality

Katarzyna Samborska, Rafal Ulanczyk and Katarzyna Korszun
Institute for Ecology of Industrial Areas
Poland

1. Introduction

Monitoring of water quality in the European Union is mainly regulated by the Water Framework Directive (Water Framework Directive [WFD], 2000) and other Directives, e.g.:

- Urban Waste Water Treatment Directive (Urban Waste Water Treatment Directive [UWWT], 1991),
- Drinking Water Directive (Drinking Water Directive [DWD], 1998),
- Bathing Waters Directive (Bathing Waters Directive [BWD], 1976),
- Shellfish Waters Directive (Shellfish Water Directive [SWD], 2006),
- Nitrates Directive (Nitrates Directive [ND], 1991),
- Directive on Integrated Pollution Prevention and Control (Directive on Integrated Pollution Prevention and Control [IPPC], 2008),

These Directives were established for particular water types or problems. The European legislation introduced and implemented the river basin approach to water resources management and monitoring of water quality. By means of the latter one, good water quality status should be provided. The Water Framework Directive divides the monitoring of water quality into three types: Surveillance (to establish baseline quality status and assess long term changes), Operational (to provide extra data on water bodies at risk or failing to meet the environmental objectives of the WFD) and Investigative (to determine the causes of such failure where they are unknown).

In Poland (similarly to other European countries) the European legislation on the monitoring of water quality is implemented by federal law (for example by Water Law Act) and appropriate executive regulations. These acts concern, among others: the classification of surface and groundwater regarding with its quality, the methodology and interpretation of monitoring results, requirements for the quality of potable water. Monitoring of water quality in Poland is conducted for surface water, groundwater and rainwater. The aim of the surface water monitoring is, first of all, to determine a current physico-chemical status of water and consequently to find solutions for improving it. Finally, the results enable also to design the plans for preventing water pollution. The emphasis is put on the elimination of pollutions originated from agriculture and urban areas. In addition, monitoring should provide the reliable forecasting of further changes.

The monitoring of surface water is carried out for both rivers and lakes. It is worth mentioning that Poland is involved in international systems of the surface water monitoring.

These systems are destined for monitoring of the quality of rivers which cross the boundaries of the country (transboundary monitoring). There are also systems which are addressed to the continuous observations of the surface water quality regarding to its acidification and euthropication (ICP Waters). The monitoring of potable water needs a special type of water examinations, which should be carried out with the high accuracy of measuring and should encompass as many parameters as possible. This type of monitoring should be also coupled with modelling tools, which have to alarm the potential users in case of water quality deterioration and make the prognosis for the future usage based on different scenarios of land development.

Monitoring of groundwater quality in Poland is conducted by the Polish Hydrogeological Survey, within the basins called groundwater bodies. The monitoring is arranged in the three types: diagnostic, operational and research. Since 1991, results of examinations have been collected in the database MONBADA. Each year, the report describing the status of the groundwater quality is made on the basis of physico-chemical examinations. However, the interpretation of the chemical conditions within groundwater bodies conducted with the annual frequency might be not sufficient for the accurate evaluation of the processes responsible for the changes. Data obtained during monitoring works are available and could be used in the scientific projects though there is no possibility to work on current results on-line.

There are many modelling tools which can be fed with data derived from monitoring. Hence, measurements performed in a frame of monitoring should not by only gathered in databases and analysed sometimes several weeks or months after the sampling but rather simultaneously transformed into real signals, which can be used immediately by the management unit. Mentioned modelling tools (presented in section 7) are designed for simulation of processes occurring in atmospheric water, surface water and groundwater or for all of them together. These tools are designed for scenarios analyses or for real-time simulations. This paper presents also examples of use of coupled monitoring and modelling systems for surface and ground water (section 8).

Summarising, it can be said that creation of the real-time monitoring systems is the key issue for the water quality improvement. Moreover, coupling of real-time monitoring tools with the real-time modelling is the most suitable solution for making prognosis on the basis of monitoring results.

2. Legislation acts connected to monitoring of water quality

The Water Framework Directive (WFD) introduces a strategic legislative framework to couple the various concerns of water policy and provide a more holistic and integrated approach to water management in entire river basins (Naddeo et al., 2005). The objectives of the Water Framework Directive are to protect water of high quality (high status water), prevent potential deterioration and to restore degraded surface and groundwater to good status by 2015. The key aim of the WFD is to prepare integrated systems of water resources management functioning within hydrological catchment (river basins), hence often crossing administrative boundaries (Teodosiu el al., 2003).

Article 8 of the Water Framework Directive sets out the requirements for the monitoring of surface water status, groundwater status and protected areas:

"Monitoring programmes are required to establish a coherent and comprehensive overview of water status within each river basin district."

The programmes had to be operational at the latest by 22 December 2006, and must be in accordance with the requirements of Annex V of the WFD. Annex V indicates that monitoring information from surface water is required for, among others, (WFD, 2000b):

- The classification of status;
- Supplementing and validating risk assessment procedure (Appendix II);
- The efficient and effective design of future monitoring programmes;
- The assessment of long-term changes in natural conditions and those resulting from human impact;
- Estimating pollutant loads transferred across international boundaries or discharging into seas;
- Assessing changes in status of those bodies identified as being at risk;
- Ascertaining causes of water bodies failing to achieve environmental objectives where the reason for failure has not been identified;
- Ascertaining the magnitude and impacts of accidental pollution;
- Use in the intercalibration exercise;
- Assessing compliance with the standards and objectives of Protected Areas; and,
- Quantifying reference conditions (where they exist) for surface water bodies.

Annex V also indicates that monitoring information from groundwater is required for:

- Providing a reliable assessment of quantitative status of all groundwater bodies or groups of bodies
- Estimating the direction and rate of flow in groundwater bodies that cross Member States boundaries;
- Supplementing and validating the impact assessment procedure and use in the assessment of long term trends;
- Establishing the presence of significant upwards trends in the concentrations of pollutants;
- Assessing the reversal of such trends in the concentration of pollutants, and;
- Establishing the chemical status of all groundwater bodies or groups of bodies determined to be at risk

The monitoring of drinking water is established by The Drinking Water Directive (DWD, 1998). The legislative act sets out quality standards for drinking water quality at the tap (microbiological, chemical and organoleptic parameters). In addition, it constitutes the general obligation of cleanliness and wholesomeness of drinking water. The Directive obliges each Member State to conduct regular monitoring of drinking water quality and as a result of that to inform all consumers about the water quality. Hence, the main aim of the Drinking Water Directive is to protect the health of the consumers in the European Union by providing them with the best quality of tap water. To achieve this goal DWD established standards for the 48 most common substances - parameters, which can be found in potable water. These microbiological and chemical substances must be examined in drinking water regularly (Hecq et al., 2006).

The Urban Wastewater Treatment Directive (UWWT, 1991) and the Integrated Pollution Prevention and Control Directive (IPPC, 1996) are mainly dealing with urban and industrial wastewater. The UWWT Directive is related to collection, treatment and discharge of urban wastewater. It also aims to protect the environment from the negative influence of the disposal/discharge of insufficiently treated urban wastewater. The directive encompasses four activities related to wastewater management: planning, regulation, monitoring, and information and reporting. Monitoring requires ensuring (Quevauviller et al., 2006):

- appropriate monitoring capacity of parameters to be monitored;
- accurate analysis of samples by using standard methods;
- timely frequency of monitoring for: monitoring of discharges from urban wastewater treatment plants; and monitoring of waters receiving those discharges.

The Nitrates Directive's main objective is to protect water quality across Europe by eliminating the jeopardy of nitrates pollution and byagricultural practices. All Member States have to examine water with regard to nitrate concentrations and trophic state. Hence, good monitoring procedures and networks are crucial in executing the acts of the Directive. There are currently 31 000 groundwater sampling sites in the EU, and 27 000 surface water stations. There are differences between countries in the designing and using their monitoring networks. This may be due to the fact that there are no specific official guidelines and/or protocols in the European Union (Fraters et al., 2003). On the other hand, the existing (in a draft form) guidelines for monitoring under the Nitrates Directive outline the monitoring of both agriculture (nutrient balances, changes in land usage and manure storage capacity) and water quality (effects of nitrate input to surface water and groundwater).

EC Shellfish Waters Directive (SWD, 2006) aims to protect shellfish populations. Hence it specifies the way how shellfish water should be monitored throughout the year. The frequency of sampling depends on the importance of the parameter being measured. The parameters encompass physico-chemical indicators including toxic organic and metal contaminants.

In Poland, the obligatory examinations of surface water and evaluation of its quality is enclosed in "Water Law" act (Water Law Act, 2010). In turn, the operating range, procedures and criteria of the quality's evaluation might be found in several regulations concerning: habitat requirements for fishwater, extinguishing of water vulnerable for nitrates pollution, standards for drinking water derived from surface resources, classification of surface water status and types and procedures of surface water monitoring. The aim of the surface water monitoring is the establishment of basis for further activities which should improve the status of water. These actions include also the prevention from pollution, especially the elimination of the eutrophication process. The holistic assessment of the surface water condition is the main tool of water resources management within the river basins, which stand for the cardinal units in water policy.

The legal basis of groundwater monitoring constitutes two acts: "Environmental Protection" (Environmental Protection Act, 2001) and "Water Law" acts. Similarly to surface water monitoring, the detailed description of procedures and performance might be found in a set of regulations concerning: extinguishing of water vulnerable for nitrates pollution, criteria and methods of the groundwater status evaluation and forms and procedures for carrying out monitoring of surface and groundwater bodies. The monitoring actions are directed towards the creation of an information base including chemical condition of groundwater within outlined bodies. The continuous control of groundwater quality aims at further planning of both clean-up activities and a protection from pollution, which leads to achieve good groundwater status before the date set by the European Commission.

3. Monitoring of surface water quality in Poland

The "Water Law" act introduced a division of the state area into water basins and water regions; initially, there were two main areas of water basins - the Vistula and the Oder river basins. Amendment to the Water Law act from 2005 introduced eight new water basins, which replaced the initial two (Inspectorate of Environmental Protection, 2010).

In case of examinations and evaluations of surface water quality based on monitoring, years 2010-2012 are the first period of a six-year-long project of water management within outlined water bodies. During this time, monitoring of surface water quality is performed in three types of activities: surveillance/diagnostic, operational and investigative/research. Fist mentioned aims to provide general assessment of water quality for each catchment and sub-catchment within the whole river basin. Information gathered during this type of research enables the determination of long-term changes in natural conditions. Second one, operational monitoring is carried out within those homogenous water bodies which, during the diagnostic phase, were described as endangered of failure to achieve the good status. This monitoring aims at evaluation of changes following the implementation of repair programmes. The last type - research monitoring - aims at observation and describing of unrecognised threats of pollution within surface water bodies.

In case of surface water monitoring, there are ten types of sampling/monitoring points: diagnostic (MD), operational (MO), operational for water vulnerable to nitrates pollution (MORO), operational for water exposed to eutrophication (MOEU), operational for fishwater (MORY), operational for surface water abstracted for public water supply (MOPI), operational for water qualified as recreational (MORE), operational for water bodies in boundaries of which there are protected ecosystems strictly dependent on quality of water (MONA), operational related to the execution of the international agreements (MOIN), research (MB).

The range and the frequency of monitoring depend on the type of monitoring points. For the diagnostic points, the frequency of measurements is from 1 to 8 times a year for biological factors and from 1 to 12 times a year for physic-chemical parameters. For the operational monitoring points the range of measured parameters depends on the type of pressure put on the environment within water bodies. In case of units in which effluents/discharges hazardous substances, especially priority hazardous substances were observed, or in monitoring points where the amount of these substances exceeded permissible limits, the monitoring is performed annually. For other operational monitoring points the frequency of examinations is once in 3 years for fishwater and once in a year in case of surface water used for public water supply.

The key role in the monitoring system play points located in estuaries of big rivers and those flowing directly to the Baltic Sea. In these places, the frequency of sampling is not less than 12 times a year and the range of parameters being examined include heavy metals, biogenic substances and indicators describing the oxygen conditions.

Each year the evaluation of surface water status is made for water bodies which were included in the diagnostic monitoring. In year 2013, after finishing the entire diagnostic programme, the summary specification of the status within surface water bodies will be prepared. On the basis of that, using the extrapolation method, the rest of the water bodies which are not included in the diagnostic monitoring will be assessed. The scheme of evaluation of surface water status is presented on figure below (Fig. 1).

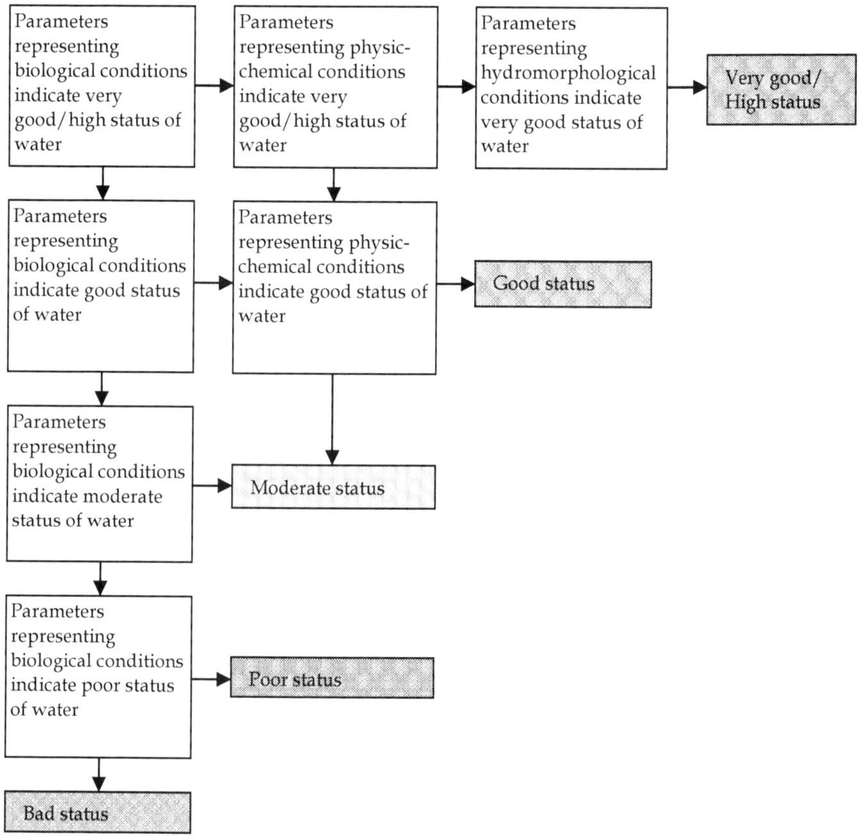

Fig. 1. Block diagram of the evaluation of surface water status, according to the Water Framework Directive (after: Loga and Sawicka, 2009)

The evaluation of the chemical status of water included in the operational monitoring depends on the type of ongoing programmes. In any case, the classification of ecological status is made on the basis of the limited number of factors depending on the potential threats of pollution and pressure put on the water environment by anthropogenic discharges. Hence, the evaluation of surface water status is fraught with uncertainties and rather informs about the success or failure in the implementation of repair programmes than about real conditions of the environment.

The State Environmental Monitoring of surface water in Poland is led by the Voivodship Inspectorates for Environmental Protection, and coordinated by the Main Inspectorate for Environmental Protection. The detailed plans of sampling campaigns, carried out in a frame of surface water monitoring, might be found in Monitoring of Environment Programmes, which are available on the website of Voivodship Inspectorate for Environmental Protection. In 2009, the number of monitoring points was 1616 and they were localised within 1328 homogeneous surface water bodies.

4. Monitoring of groundwater quality in Poland

On a national scale, the institution responsible for monitoring of groundwater is Polish Hydrogeological Survey (PSH). Their tasks are performed by the Polish Geological Institute. These tasks include (Sadurski 2005):

- performing hydrogeological measurements and observations;
- collecting and processing the data concerning groundwater condition and resources available,
- performing current analyses and assessments of hydrogeological situation;
- compiling and forwarding to the appropriate administration authorities, the forecasts of the changes in both resources and quality of groundwater, and about threats of pollution;
- compiling and forwarding to the public administration warnings about dangerous phenomena occurring in recharge zones and groundwater intakes.

Examinations of the groundwater quality are carried out within groundwater bodies defined as "determined volume of groundwater occurring within an aquifer or aquifers". This term refers to the functioning in Polish hydrogeology term "free water occurring in the saturation zone" (Pazdro 1977).

The delineation of groundwater bodies was made taking into consideration various factors, such as circulation conditions (i.e. locations of recharge and drainage zones), administrative and hydrological boundaries (catchments, basins), lithology and stratigraphy of water-bearing rocks. Groundwater bodies are complex structures, they might consist of few water horizons.

Measurements of the ground water quality are carried out within the national observational-research network. This network was designed to evaluate the groundwater status and to determine trends in chemical changes. Monitoring actions are performed for both the qualitative and quantitative assessment of the groundwater. The latter one is embraced by diagnostic, operational and research monitoring. The density of observation points reflects the complexity of geological and hydrogeological conditions and natural and anthropogenic pressures put on the water quality. Additional attention in monitoring is paid to areas around national boundaries and to protected areas. Hence, the final location of points depends on the formulated aim of monitoring, and the network comprises of (Fig. 2):

- Hydrogeological monitoring stations;
- Points of groundwater state observations;
- Points of groundwater quality monitoring;
- Observation piezometers;
- Built-up wells.

Fig. 2. Types of monitoring points: A) Piezometer, B) Spring, C) Well. (Sources of photos: Glubiak-Witwicka et al., 2003)

Diagnostic monitoring aim is to determine anthropogenic impacts on the groundwater quality as well as the long-term trends in chemical status changes. Accordingly, the changes might be induced by natural processes and human activities. The results of this type monitoring actions help to design the further operational monitoring system. Parameters examined during the diagnostic phase of groundwater monitoring are: pH, TOC (total organic carbon), conductivity, temperature, DO (dissolved oxygen), ammonia, arsenic, nitrates, barium, boron, chloride, chromium, zinc, fluoride, phosphates, aluminium, cadmium, magnesium, manganese, copper, nickel, lead, potassium, sulphate, sodium, calcium, carbohydrates, iron, organic substances: AOX – adsorbed organochlorine compounds. The range of parameters might be extended by substances expected to be present in groundwater regarding the potential pollution sources.

Operational monitoring should provide data necessary to reach the proper level of certainty in classifying groundwater endangered by not reaching the good quality status. The second major aim is to identify upwards trends in the chemical composition which might indicate the pollution, especially human-induced. The range of measured parameters includes: temperature, conductivity, pH, DO, ammonia, nitrites, nitrates, chloride, sulphate, phosphates, bicarbonates, sodium, potassium, calcium, magnesium, manganese and iron. In addition, if some parameters measured in the diagnostic phase of monitoring, classified the groundwater status as poor or bad, they are also included into operational observations.

Research monitoring aims at extending the recognition achieved after two previous phases of monitoring. This type of researches leads to determine the reasons, sizes and influences of the accidental pollution. This monitoring might be performed for those groundwater bodies which might not achieve the good quality status by the fixed date, and which are not subjected to the operational actions (Kazimierski & Pilchowska-Kazimierska, 2006).

There are 5 classes of groundwater quality:

- I class - water of very good quality - the chemical composition results only from natural conditions and represents natural geochemical background. The measured values of parameters do not reflect any human activities,

- II class - water of good quality - values of some parameters are increased, what is caused only by natural factors. Hence, the chemical status of water is the result of geogenic processes and human impact is negligible,
- III class - water of satisfactory quality. Values of several parameters are increased mainly as a result of natural processes. However, changes in chemical composition might reflect a feeble human impact,
- IV class - water of non-satisfactory quality. Values of some parameters are significantly increased as a result of natural or human-induced processes,
- V class - water of bad quality. Chemical composition of this type of water confirms the predominant human impact.

Classes from I to III stand for good status of groundwater, whereas classes IV and V - bad status. The status of groundwater bodies is presented on the maps in a following way: area filled with green colour - good status, area filled with red colour - bad status and black points - observed trends in chemical composition of groundwater (example on Fig. 3).

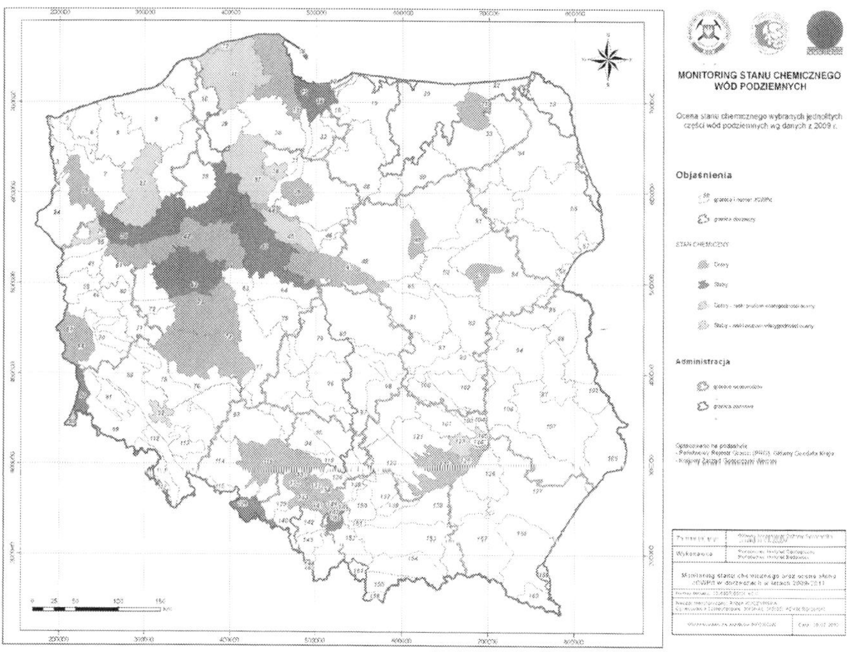

Fig. 3. The classification of groundwater status in year 2009: green colour stands for the good status of groundwater, red for bad status, light green and orange represent a high uncertainty of obtained results (source: http://www.gios.gov.pl)

The frequency of chemical examinations depends on both type of monitoring and type of aquifer. In case of unconfined aquifer the frequency is one time in 3 years (diagnostic monitoring) or twice a year (operational monitoring). On the other hand, in case of deeper aquifers, which are characterised by piezometric surface, the frequency is once in 6 years (diagnostic monitoring) or once a year (operational monitoring).

Results obtained during the monitoring are widespread via different publications, such as Quarterly Bulletin of Groundwater. Results are also gathered in databases, e.g.:

- SOH (Hydrogeological Observations System). This database collects information about water table level fluctuations (every Monday) and chemical condition (every year).
- MONBADA (MONitoring DataBAse) is the database of National Environmental Monitoring. It was created in 1991. In MONBADA data regarding the chemical status of groundwater and its evaluation are gathered. On the basis of collected results annual reports are prepared. These reports are published in series "Library of Environmental Monitoring" and on the National Monitoring web site (http://www.gios.gov.pl/wodypod).

5. Monitoring of precipitation in Poland

Qualitative monitoring of precipitation in terms of its chemical composition and occurrence of the acidic atmospheric deposition has been largely studied in different sites and locations in Central and Western Europe, North America and East Asia, during the last years (Kulshrestha et al., 2009; Li et al., 2007; Menz & Seip, 2004; Tsitouridou & Anatolaki, 2007; Wang et al., 2000).

In Poland, the National Chemistry Monitoring of Precipitation and Assessment of Pollutants Deposition were established in 1998 as a subsystem of National Environmental Monitoring. Monitoring tests in the full annual cycle were carried out for the first time in 1999. The Institute of Meteorology and Water Management in Wroclaw carried substantive supervision over implementation of these tasks. The purpose of monitoring of the precipitation chemistry and of the deposition of pollutants is to define, at a national scale, spatial and temporal decomposition of pollutants entering from the wet precipitation to the ground. National network of measurement - monitoring consists of: 25 research stations (to ensure the representativeness of precipitation chemistry measurements) and 162 rainfall stations (which characterise the average rainfall for the Poland area). At all 25 research stations the rainwater is collected continuously and analysed on a monthly basis. At the time of sampling, the amount and type of precipitation are measured. Monthly precipitation samples are analysed for the concentration of acidic compounds, nutrients and metals (including heavy metals). The monitoring includes: chlorides, sulphates, nitrites and nitrates, ammonia nitrogen, total nitrogen, total phosphorus, potassium, sodium, calcium, magnesium, zinc, copper, iron, lead, cadmium, nickel, chromium and manganese. Measurements include also pH and electrical conductivity of the precipitation. Basing on the results from all monitoring stations (25 research and 162 rainfall stations) maps of wet deposition of monitored substances are prepared.

The atmospheric precipitation collectors are characterised by a different degree of automation; the most sophisticated can record meteorological data and also some characteristics of precipitation (rate, electrolytic conductivity, pH) (Chief Inspectorate of Environmental Protection, National Chemistry Monitoring of Precipitation and Assessment of Pollutants Deposition, http://www.gios.gov.pl/chemizm/index.html.) The simplest collectors for total precipitation sampled for chemical analysis are glass, metal, and plastic containers. The containers are equipped with glass, steel, or polyethylene funnels. Figure 4 presents the scheme of the simplest collector (Bijsman et al., 1991).

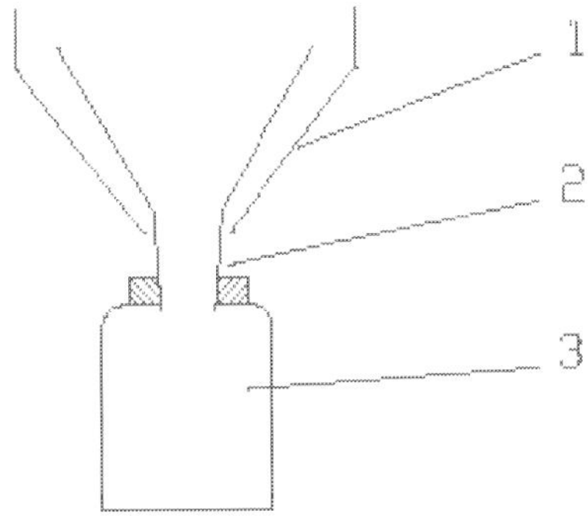

Fig. 4. Scheme of the simplest collector for total precipitation samples: 1 - outer funnel casing, 2 - funnel, 3 - bottle.

The more complicated version has an automatic cover and a humidity sensor; in this type of system dry precipitation is excluded from a sample (Fig. 5) (Ligocki et al., 1985).

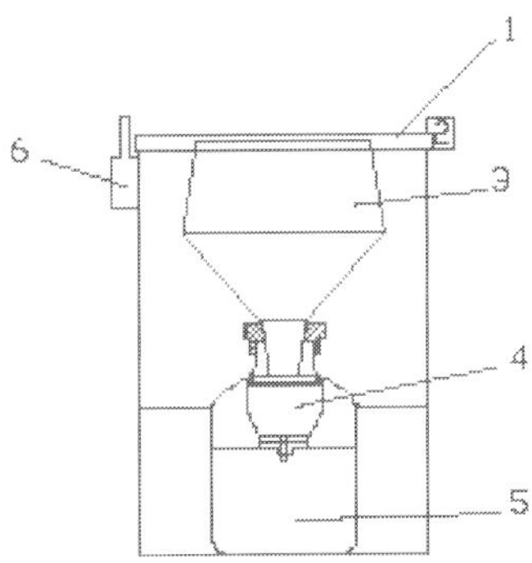

Fig. 5. Scheme of automatic collector for sampling the atmospheric precipitation: 1 - cover, 2 - electric motor, 3 – polyethylene funnel, 4 - filtration funnel, 5 - polyethylene bottle for collecting filtrate, 6 - humidity sensor.

Samplers should be placed ca. 1.5 m above the ground, in an open area, at best, overgrown with grass. The detailed conditions for collector location are given in the Polish standard (PN-ISO 5667-8), which is a translation, without any changes of the international standard ISO 5667-8:1993 (ISO, 1993).

Collection of atmospheric precipitation has to satisfy a number of conditions, especially with respect to a collection system. In compliance with recommendations of the European Monitoring Environmental Program (EMEP) the collectors for wet precipitation and total precipitation should be characterised by:

- the sampler should not be too large or bulky, because this will obstruct the air flow around the sampler; the diameter of the collector must be large enough to provide samples large enough for chemical analysis - a diameter of 20 cm is sufficient;

- collector must be made of a material, which does not alter the chemical composition of the sample, and shall give a reliable measure of the amount of precipitation on a daily basis;
- an appropriate container capacity to avoid any loss of a sample even when precipitation is very intensive.

6. Water quality modelling

Mathematical model is described in different ways. The encyclopaedia of life support system described a model as an approximate description of a class of real-world objects and phenomena expressed by mathematical symbolisms (Agoshkov, 2002). Concise Oxford Dictionary (1990) described a model as a simplified form of a system that assists calculations and predictions of a condition of a system in a given situation. The United States Environmental Protection Agency USEPA (EPA, 2009) described water quality models as tools for simulating the movement of precipitation and pollutants from the ground surface through pipe and channel networks, storage treatment units and finally to receiving waters (Kannel et al., 2011).

Model predictions might be used in addition to or instead of monitoring data for several reasons:

- Modelling might be feasible in some situations where monitoring is not.
- Integrated monitoring and modelling systems could provide better information than one or the other alone for the same total cost.
- For example, regression analyses that correlate pollutant concentration with some more easily measurable factor (such as streamflow) could be used to extend monitoring data. Models can also be used in a Bayesian framework to determine preliminary probability distributions of impairment that can help direct monitoring efforts and reduce the quantity of monitoring data needed for making decisions.
- Modelling can be used to assess (predict) future water quality situations resulting from different management strategies. For example, assessing the improvement in water quality after a new wastewater treatment plant is built, or the effect of increased industrial growth and effluent discharges (UNESCO, 2005).

Water quality modelling for decision-making occurs at a disciplinary divide between science and management. Workers in science and management operate in fields that traditionally have different objectives, priorities and expectations. These differences can create barriers to the effective use of scientific models by watershed managers (McNamara, 2004).

Water quality models include both mathematical expressions and expert scientific judgement. They include process-based (mechanistic) models and data-based (statistical) models. The models should be characterised by following features:

- link management options to meaningful response variables (such as pollutant sources and water quality standard parameters),
- incorporate the entire chain from stressors to responses,
- be consistent with scientific theory,
- have reported the uncertainty,

- compatible with the quantity and quality of available data (The use of complex mechanistic models for water quality prediction in situations with little useful water quality data does not compensate for that lack of data. Model complexity can give the impression of credibility, but this is usually misleading).

It is often preferable to begin with simple models and then, over time, add additional complexity as justified by the collection and analysis of additional data. This strategy makes efficient use of resources. It targets the effort toward information and models that will reduce the uncertainty as the analysis proceeds (UNESCO, 2005).

Water quality models can be applied to many different types of water system, including atmospheric water, groundwater, wetlands, streams, rivers, lakes, reservoirs, estuaries, coastal waters and oceans.

The fact that most, if not all, water quality models cannot accurately predict what actually happens does not detract from their value. Even relatively simple models can help managers understand the real world prototype and estimate at least the relative, if not actual, change in water quality associated with given changes in the inputs resulting from management policies or practices (UNESCO, 2005).

7. Tools for water quality modelling – Examples and purposes of use

Models simulating water quality are used for three main purposes:

- Planning remediation of degraded areas or protection of water resources, soils, health and ecosystems.
  In this case models calculate results of different scenarios. This type of models answers "what if" question and finds the solution which is best according to a specified objective and satisfying physical, technical, legal and other constraints (Wagner, 1992).
  Groundwater quality models can analyse possibilities for removing a plume of polluted water or control its flow away from wells, streams, lakes. For such purpose models can simulate scenarios including: inserting curtain walls, removing polluted soil, in situ (bio)chemical remediation or installing pumping/recharge wells (Wagner, 1992). For surface waters this type of models is used for development of long-time management scenarios for river basins or lakes and for assessment of the results of planned land use changes and investments (Kardaetz et al., 2008; Thebault & Qotbi, 1999; Yoshimura et al., 2009). Models of this type for atmospheric water quality are used widely to predict the impact of pollutants emissions on the eutrophication and acidification processes. For all types of water (including water distribution networks and wastewater collecting systems) models can be used also for planning better and cost-effective monitoring and controlling systems.
- Simulating ongoing processes using real-time modelling systems. Real-time refers to a state where data referring to a system is analysed and updated at the rate at which it is received (i.e. at the rate at which the system operates). Real-time modelling (also referred to as online modelling) refers to the process of employing numerical models to make predictions about current or near future system states and outputs based on newly received (and forecasted) data. Real-time modelling is employed in a range of environmental fields, including meteorology, hydrology, and in urban water systems (UWS), typically for one of the following purposes (Hutton et al., 2010):

- To provide warnings of future events,
- To inform management of future system states and potential anomalies,
- To explore a range of possible control strategies such that the control solution that optimises some function is implemented (typically a system property such as: combined sewer overflows discharge, operational cost, surface water intakes management or groundwater in-situ remediation processes).
- Analysing past events. This type of models is used for example when there is a need for information about:
- Influence of historical human activities or natural processes (as climate changes) on the present water quality status (Olsson et al., 2009).
- Trends in long-term processes occurring in the water environment (e.g. euthropication) (Heiskary & Swain, 2002).

Surface water quality models include processes to simulate (Deltares, undated; Hipsey et al., 2009):

- dissolved oxygen, organic matter (detritus),
- inorganic and organic nutrients (nitrogen, phosphorus, silicon),
- chlorophyll, phytoplankton species, microphytobenthos species, macrophyte species, vegetation species, zooplankton, zoobenthos species, fish species and other higher biology,
- carbon dioxide, methane,
- sulphates, sulphides,
- organic micropollutants (including European Water Framework Directive priority pollutants),
- heavy metals, iron,
- bacterial contaminants,
- salinity, chlorides,
- suspended sediment,
- dissolved conservative substances,
- pH, pe, light regime, temperature.

Groundwater quality models can simulate among others:

- Groundwater ages,
- Species transport,
- Propagation of mineral dissolution and precipitation fronts,
- Mineral buffering during acid mine drainage,
- Cation exchange,
- NAPL dissolution,
- BTEX degradation and corresponding geochemical changes,
- Temperature-dependent pyrite oxidation during deep well injections,
- Natural attenuation of ammoniacal liquor: Phenol degradation, nitrification and ion exchange of ammonium,
- Seawater intrusion and mixing,
- Fixed-pressure gas-phase equilibria,
- Isotope mole balance,
- any kinetically controlled reactions.

Atmospheric water quality models are usually parts of tools simulating air quality. In those models atmospheric water quality is simulated mostly as an element of

- aerosols formation,
- wet deposition.

In following tables are presented examples of modelling tools used for simulation of:

- atmospheric water quality (Table 1),
- surface water quality (Table 2),
- ground water quality (Table 3),
- surface-groundwater quality (Table 4).

Model name	**Short description**	**Availability / Producer**	**Source**
FRAME (Fine Resolution Atmospheric Multi-pollutant Exchange)	FRAME is a Lagrangian atmospheric transport model with high spatial (5x5km) and horizontal resolution (33 layers). Model can be used to assess the long-term annual mean deposition of reduced and oxidised nitrogen and sulphur. FRAME was developed initially to focus in particular on transport and deposition of reduced nitrogen and was named the Fine Resolution AMmonia Exchange model. Recent developments in the treatment of sulphur and oxidised nitrogen mean that it may now be considered as a robust multi-chemical species tool.	http://www.frame.ceh.ac.uk// Center for Ecology & Hydrology	(Anthony et al., 2003)
RAQM (regional air quality model)	RAQM is a three dimensional Eulerian model, which is constructed on a spherical and terrain-following coordinate system. A mass conservation equation in model includes a series of physical and chemical processes (advection, diffusion, multiphase chemistry, dry deposition, cloud mixing and wet scavenging). The continuity equation is solved by a time-splitting technique, which allows different time steps to be used for various processes for computational efficiency and for easy incorporating or updating of new modules. RAQM, which includes 37 species and 91 reactions employs a modified QSSA (quasi-steady-state assumptions) technique with a predictorecorrector algorithm to solve the chemical equations.		(Han, 2006; Han et al, 2006)

REM-CALGRID (Regional Eulerian Model - California Grid Model)	REM-CALGRID (RCG) is a chemical transport model development designed to fulfil the requirements of the ambient air quality framework directive 96/62/EC of the European Commission. It is an Eulerian grid model of medium complexity that can be used on the regional and the urban scale for short-term and long-term simulations of oxidant and aerosol formation. Intended field of application is the modelling of pollutants (photo-oxidants, aerosols) on different scales from Europe-wide domains down to urban domains.	http://www.geo.fu_berlin.de/met/ag/trumf/RCG/index.html / Freie Universität Berlin	(Stern, 2009)
RAINS (Regional Air Pollution Information and Simulation)	The Regional Air Pollution Information and Simulation (RAINS) model developed by the International Institute for Applied Systems Analysis (IIASA) combines information on economic and energy development, emission control potentials and costs, atmospheric dispersion characteristics and environmental sensitivities towards air pollution. The model addresses threats to human health posed by fine particulates and ground-level ozone as well as risk of ecosystems damage from acidification, excess nitrogen deposition (eutrophication) and exposure to elevated ambient levels of ozone. These air pollution related problems are considered in a multi-pollutant context quantifying the contributions of sulphur dioxide (SO2), nitrogen oxides (NOx), ammonia (NH3), non-methane volatile organic compounds (VOC), and primary emissions of fine (PM2.5) and coarse (PM10-PM2.5) particles. The RAINS model also includes estimates of emissions of relevant greenhouse gases such as carbon dioxide (CO2) and nitrous oxide (N2O). Work is progressing to include methane (CH4) as another direct greenhouse gas as well as carbon monoxide (CO) and black carbon (BC) into the model framework.	http://www.iiasa.ac.at/rains/documentation.html?sb=11 / International Institute for Applied Systems Analysis	(Amann et al., 2004)

Table 1. Models for simulation of atmospheric water quality

Model name	Short description	Availability / Producer	Source
SPARROW - SPAtially Referenced Regressions On Watershed attributes	SPARROW (SPAtially Referenced Regressions On Watershed attributes) is a watershed modeling technique for relating water-quality measurements made at a network of monitoring stations to attributes of the watersheds containing the stations. The core of the model consists of a nonlinear regression equation, describing the non-conservative transport of contaminants from point and diffuse sources. The model predicts contaminant flux, concentration, and yield in streams and has been used to evaluate alternative hypotheses about the important contaminant sources and watershed properties that control transport over large spatial scales.	http://water.usgs.gov/nawqa/sparrow/sparrow-mod.html / US Geological Survey (free)	(Shwarz et al., 2006)
SMS (Surface Water Modeling System)	The Surface Water Modeling System (SMS) is environment for one-, two, and three-dimensional hydrodynamic modelling. A pre- and post-processor for surface water modelling and design, SMS includes 2D finite element, 2D finite difference, 3D finite element modelling tools. Supported models include RMA2, RMA4, ADCIRC, CGWAVE, STWAVE, BOUSS2D, CMS Flow, and CMS Wave models. An interface has also been developed for facilitating the use of the FHWA commissioned analysis package FESWMS. The TUFLOW numerical model with flood analysis, wave analysis, and hurricane analysis is supported. SMS also includes a generic model interface, which can be used to support models which have not been officially incorporated into the system. Primary applications of the models include calculation of water surface elevations and flow velocities for shallow water flow problems, for both steady-state or dynamic conditions. Additional applications include the modelling of contaminant migration, salinity intrusion, sediment transport (scour and deposition), wave energy dispersion, wave properties (directions, magnitudes and amplitudes) and others.	http://www.aquaveo.com/sms / AQUAVEO (commercially available)	(AQUAVEO, 2011)

QUAL2K (River and Stream Water Quality Model)	QUAL2K (or Q2K) is a river and stream water quality model that is intended to represent a modernised version of the QUAL2E (or Q2E) model. Q2K is similar to Q2E in the following respects: • The channel is well-mixed vertically and laterally. • The system can consist of a mainstem river with branched tributaries. • Non-uniform, steady flow is simulated. • The heat budget and temperature are simulated as a function of meteorology on a diel time scale. • All water quality variables are simulated on a diel time scale. • Point and non-point loads and withdrawals are simulated. The QUAL2K model simulate among others: • slowly oxidizing form of BOD (slow CBOD) and a rapidly oxidizing form (fast CBOD). • Sediment-water fluxes of dissolved oxygen and nutrients as a function of settling particulate organic matter, reactions within the sediments, and the concentrations of soluble forms in the overlying waters. • attached bottom algae with variable stoichiometry. • light extinction as a function of algae, detritus and inorganic solids. • alkalinity and total inorganic carbon. • generic pathogen as a function of temperature, light, and settling. • hydraulics of weirs as well as the effect of weirs and waterfalls on gas transfer.	http://www.epa.gov/athens/wwqtsc/html/qual2k.html / *Environmental Protection Agency* (free)	(Chapra et al., 2007)
SWAT (Soil and Water Assessment Tool)	SWAT predicts the effect of management decisions on water, sediment, nutrient and pesticide yields with reasonable accuracy on large, ungaged river basins. Model includes: • elements of water balance: surface runoff, return flow, percolation, ET, transmission losses, pond and reservoir storage, crop growth and irrigation, groundwater flow, reach routing, water transfer, • nutrient and pesticide loading, • nutrients, pesticide, bacteria, conservative metals and sediment transport routing, • simulation of water quality parameters: algae, CBOD, dissolved oxygen, • plant growth cycle, • set of land management practices (e.g. planting, harvesting, tillage, fertiliser/pesticide application, street cleaning)	http://swatmodel.tamu.edu/ / Texas A&M University (free)	(Neitsch et al., 2004, Neitsch et al., 2005)

HSPF (Hydrological Simulation Program-Fortran)	HSPF simulates for extended periods of time the hydrologic, and associated water quality, processes on pervious and impervious land surfaces and in streams and well-mixed impoundments. HSPF uses continuous rainfall and other meteorological records to compute streamflow hydrographs and pollutographs. HSPF simulates interception soil moisture, surface runoff, interflow, base flow, snowpack depth and water content, snowmelt, evapotranspiration, ground-water recharge, dissolved oxygen, biochemical oxygen demand (BOD), temperature, pesticides, conservatives, fecal coliforms, sediment detachment and transport, sediment routing by particle size, channel routing, reservoir routing, constituent routing, pH, ammonia, nitrite-nitrate, organic nitrogen, orthophosphate, organic phosphorus, phytoplankton, and zooplankton. Program can simulate one or many pervious or impervious unit areas discharging to one or many river reaches or reservoirs.	http://water.usgs.gov/software/HSPF/ / US Geological Survey (free)	(Bicknell et al., 1997)
CAEDYM (Computational Aquatic Ecosystem Dynamics Model)	The Computational Aquatic Ecosystem Dynamics Model (CAEDYM) is an aquatic ecological model that may be run independently or coupled with hydrodynamic models DYRESM or ELCOM. CAEDYM consists of a series of mathematical equations representing the major biogeochemical processes influencing water quality. At its most basic, CAEDYM is a set of library subroutines that contain process descriptions for primary production, secondary production, nutrient and metal cycling, and oxygen dynamics and the movement of sediment.	http://www.cwr.uwa.edu.au/software1/models1.php?mdid=3 / Center for Water Research (commercially available)	(Hipsey et al., 2009; Hipsey, 2009)

WASP (Water Quality Analysis Simulation Program)	WASP is a generalised framework for modelling contaminant fate and transport in surface waters. Its flexible, compartmental approach allows it to address problems in one, two, or three dimensions. It is designed to allow easy substitution of user-written routines into the program structure. WASP has been used to answer questions regarding biochemical oxygen demand, dissolved oxygen dynamics, nutrients and eutrophication, bacterial contamination, and organic chemical and heavy metal contamination. The current WASP (v 7.41 - 2011) software package includes the scientific modules TOXI, EUTRO, and DYNHYD. TOXI models the transport and transformation of chemicals, EUTRO simulates dissolved oxygen and eutrophication processes, and DYNHYD is a hydrodynamic model used for prediction water flow and volume.	http://www.epa.gov/athens/wwqtsc/html/wasp.html / *Environmental Protection Agency* (free)	(EPA, 2006)
AQUATOX	AQUATOX is a simulation model for aquatic systems. AQUATOX predicts the fate of various pollutants, such as nutrients and organic chemicals, and their effects on the ecosystem, including fish, invertebrates, and aquatic plants. This model is a valuable tool for ecologists, biologists, water quality modellers, and anyone involved in performing ecological risk assessments for aquatic ecosystems.	http://water.epa.gov/scitech/datait/models/aquatox/index.cfm / *Environmental Protection Agency* (free)	(Park, Clough, 2009)

Table 2. Models for simulation of surface water quality

Model name	Short description	Availability / Producer	Source
biomoc - A multispecies solute-transport model with biodegradation	BIOMOC is a two-dimensional model that can simulate the transport and biotransformation of multiple reacting solutes. The program is general and flexible, allowing for any combination of biodegradation processes. A number of expressions for biological transformation rates have been included as options in the code. These include single, multiple, and minimum Monod kinetics and competitive, noncompetitive, and Haldane inhibition. The kinetic parameters can be formulated to simulate zero-order or first-order approximations of biodegradation rates. The growth and decay of several microbial populations performing the transformations is also accounted for. The microbial growth can be disabled, limited by biomass inhibition, or limited by the availability of a specified nutrient.	http://water.usgs.gov/software/BIOMOC/ / US Geological Survey (free)	(Essaid, Bekins, 1997)
PHREEQC - Computer Program for Speciation, Batch-Reaction, One-Dimensional Transport, and Inverse Geochemical Calculations	PHREEQC version 2 is a computer program written in the C programming language that is designed to perform a wide variety of low-temperature aqueous geochemical calculations. PHREEQC is based on an ion-association aqueous model and has capabilities for (1) speciation and saturation-index calculations; (2) batch-reaction and one-dimensional (1D) transport calculations involving reversible reactions, which include aqueous, mineral, gas, solid-solution, surface-complexation, and ion-exchange equilibria, and irreversible reactions, which include specified mole transfers of reactants, kinetically controlled reactions, mixing of solutions, and temperature changes; and (3) inverse modeling, which finds sets of mineral and gas mole transfers that account for differences in composition between waters, within specified compositional uncertainty limits.	http://wwwbrr.cr.usgs.gov/projects/GWC_coupled/phreeqc/ / US Geological Survey (free)	(Parkhurst, Appelo, 1999)

PHAST - A Computer Program for Simulating Groundwater Flow, Solute Transport, and Multicomponent Geochemical Reactions	The computer program PHAST (*PH* REEQC *A* nd H *ST* 3D) simulates multicomponent, reactive solute transport in three-dimensional saturated groundwater flow systems. PHAST is a versatile groundwater flow and solute-transport simulator with capabilities to model a wide range of equilibrium and kinetic geochemical reactions. The flow and transport calculations are based on a modified version of HST3D that is restricted to constant fluid density and constant temperature. The geochemical reactions are simulated with the geochemical model PHREEQC, which is embedded in PHAST. Major enhancements in PHAST Version 2 allow spatial data to be defined in a combination of map and grid coordinate systems, independent of a specific model grid (without node-by-node input). At run time, aquifer properties are interpolated from the spatial data to the model grid; regridding requires only redefinition of the grid without modification of the spatial data.	http://wwwbrr.cr.usgs.gov/projects/GWC_coupled/phast/ / US Geological Survey (free)	(Parkhurst et al., 2010)
HST3D - Heat and Solute Transport in Three-Dimensional Ground-Water Flow Systems	The Heat- and Solute-Transport Program (HST3D) simulates ground-water flow and associated heat and solute transport in three dimensions. Over the years since the release of Versions 1.0 through 1.4, various additions, modifications, and corrections have been made to the original simulator. Major changes included in Version 2 are (1) a revised data-input file with all spatial information described by coordinate location; (2) a new iterative solver for the matrix equations based on a generalised conjugate-gradient method; (3) an evapotranspiration boundary condition; (4) a division of the simulator output into many files; (5) a new set of output files designed for use by post-processing programs for graphical visualization and for flow totalization; (6) a pre-processor for evaluating dimensioning requirements; and (7) a post-processor for totalizing boundary flow rates and cumulative amounts. The post-processing program for graphical visualization must be supplied by the user. Version 2 of the simulator has been verified using five test problems selected from the published literature. One involves heat transport, four involve solute transport, and all have variable-density fluids.	http://wwwbrr.cr.usgs.gov/projects/GW_Solute/hst/ / US Geological Survey (free)	(Kenneth, Kipp, 1997)

SUTRA (Saturated-Unsaturated Transport)	SUTRA (Saturated-Unsaturated Transport) is a computer program that simulates fluid movement and the transport of either energy or dissolved substances in a subsurface environment. The original version of SUTRA was released in 1984. The version SUTRA 2.2 (2010) is an upgrade that adds to the previous version the ability to specify time-dependent sources and boundary conditions (without programming) and to output information pertaining to source and boundary condition nodes in a convenient format. The code employs a 2D or 3D finite-element and finite-difference method to approximate the governing equations that describe the two interdependent processes that are simulated: 1) Fluid-density-dependent saturated or unsaturated ground-water flow; and either 2) (a) transport of a solute in the ground water, in which the solute may be subject to: equilibrium adsorption on the porous matrix, and both first-order and zero-order production or decay; or (b) transport of thermal energy in the ground water and solid matrix of the aquifer.	http://water.usgs.gov/nrp/gwsoftware/sutra/sutra.html / US Geological Survey (Free)	(Voss, Provost, 2010)

GMS - Groundwater Modeling System	GMS provides tools for every phase of a groundwater simulation including site characterization, model development, calibration, post-processing, and visualization. GMS supports both finite-difference and finite-element models in 2D and 3D including MODFLOW 2000, MODPATH, MT3DMS/RT3D, SEAM3D, ART3D, UTCHEM, FEMWATER, PEST, UCODE, MODAEM and SEEP2D. GMS can be used for simulation of:
	• Solute Transport
	  • Simple analytical transport modelling with ART3D
	  • Simple 3D transport with MT3D, MODPATH, or FEMWATER
	  • Reactive 3D transport with RT3D or SEAM3D
	  • Multi-phase reactive transport with UTCHEM
	• Unsaturated Zone Flow and Transport
	  • Fully 3D unsaturated/saturated flow and transport modeling with FEMWATER or UTCHEM
		http://www.aquaveo.com/gms / AQUAVEO (commercially available)	(AQUAVEO, 2011b)

PMWIN (Processing MODFLOW)	Processing MODFLOW (PMWIN) is a integrated groundwater modeling system including: • 3D finite-difference groundwater model MODFLOW. Supports multiple versions, including MODFLOW-96, MODFLOW-2000, and MODFLOW-2005.MODFLOW-88, MODFLOW-96, and MODFLOW 2000. • Solute transport models MT3DMS (v5.2), MT3D, RT3D (v2.5), and MT3D99. • Multicomponent reactive transport model PHT3D (v2.0) that incorporates MT3DMS (v5.2) for the simulation of three-dimensional advective-dispersive multi-component transport and the geochemical model PHREEQC-2 (v2.14) for the quantification of reactive processes. • Variable-density groundwater flow and multi-species solute and heat transport model SEAWAT (v4.0) . • Nonlinear parameter estimation package PEST (v12.0). Supports advanced features such as regularization, truncated singular value decomposition (SVD), and powerful SVD-Assisted Parameter Estimation. • Particle-tracking Model PMPATH which uses a semi-analytical particle-tracking scheme similar to MODPATH to calculate the groundwater paths and travel times. Through the interactive graphical modelling environment of PMPATH, the user can place particles and perform particle tracking with just a few mouse clicks.	http://www.pmwin.net/pmwin5.htm / Wen-Hsing Chiang (free)	(Chiang, Kinzelbach, 2001)
MARS (Multiphase Areal Remediation Simulator)	MARS (Multiphase Areal Remediation Simulator) can be used to model recovery and migration of light nonaqueous phase liquids in unconfined heterogeneous, anisotropic aquifers. MARS writes input flow files for the BIOF&T model which simulates multispecies dissolved phase transport in heterogeneous, anisotropic, fractured media, or unfractured granular porous media. It models flow of water and light nonaqueous phase liquid (LNAPL) and aqueous phase transport of up to five species in ground water with multiple pumping and/or injection wells. MARS is a finite-element model that allows accurate representation of highly-irregular material and physical boundaries in a heterogeneous and anisotropic media.	http://www.scisoftware.com /environmental_software/product_info.php? products_id=46 / (commercially available)	(SSG, 2011)

VisualMODFLOW	VisualMODFLOW simulate groundwater and surface water interactions, and changes to groundwater chemistry. This modelling system includes: • MODFLOW-2000, 2005 the standard for 3D groundwater flow • SEAWAT v.4, for 3D variable-density groundwater and heat flow • MODPATH, the standard package for particle tracking • MT3DMS, the multi-species contaminant transport engine • PHT3D, multi-component geochemical reactive transport • MT3D991, for multi-species reactive transport • RT3D, for reactive transport simulations • Zone Budget, for sub-regional water budget calculations • LAK Package, for surface water/groundwater interactions • Multi-Note Well (MNW) package for well simulations • ETS1 Package, for evapotranspiration modeling • WinPEST, automated calibration and predictive analysis • VMOD 3D-Explorer, 3D visualization and animations • SAMG Solver1, for convergence of complex models • GMG, the USGS Geometric Multi-Grid solver • MGO, for pumping well optimization	http://www.swstechnology.com/groundwater-software/groundwater-modeling / Schlumberger (commercially available)	(Schlumberger, 2010)
BIOF&T (Biodegradation, Flow and Transport)	BIOF&T models biodegradation, flow and transport in the saturated and unsaturated zones in two or three dimensions in heterogeneous, anisotropic porous media or fractured media. Model simulates convection, dispersion, diffusion, adsorption, desorption, and microbial processes based on oxygen-limited, anaerobic, first-order, or Monod-type biodegradation kinetics as well as anaerobic or first-order sequential degradation involving multiple daughter species.	http://www.scisoftware.com/environmental_software/product_info.php?products_id=33 / (commercially available)	(SSG, 2011)

HYDROGEOCHEM	HYDROGEOCHEM is a numerical model of fluid flow, thermal transport, hydrologic transport, and biogeochemical kinetic/equilibrium reactions in saturated/unsaturated media. It iteratively solves three-dimensional fluid flow, heat transfer, and reactive biogeochemical transport equations. The Richards equation is solved for fluid flow, the thermal transport equation is solved for heat transfer, and the advection-dispersion- reactive transport equations are solved for all mobile component equations and kinetic-variable equations. HYDROGEOCHEM is designed for generic applications to reactive transport problems controlled by both kinetic and equilibrium reactions in subsurface media.	http://www.scisoftware.com/environmental_software/product_info.php?products_id=44 / (commercially available)	(ORNL, 2004)

Table 3. Models for simulation of groundwater quality

Model name	**Short description**	**Availability / Producer**	**Source**
WATEQ4F	WATEQ4F is a chemical speciation code for natural waters. It uses field measurements of temperature, pH, Eh, dissolved oxygen and alkalinity, and the chemical analysis of a water sample as input and calculates the distribution of aqueous species, ion activities, and mineral saturation indices that indicate the tendency of a water to dissolve or precipitate a set of minerals. The model assumes homogeneous aqueous phase equilibria, except for redox species. Equilibrium with respect to mineral solubilities is not assumed. The program results are used primarily to examine the tendency of a water to reach mineral solubility equilibria as a constraint on interpreting the chemistry of natural waters.	http://wwwbrr.cr.usgs.gov/projects/GWC_chemtherm/software.htm / US Geological Survey (free)	(Ball, Nordstrom, 2001)

MODHMS	MODHMS is a physically based, spatially distributed, integrated surface/subsurface modeling framework hydrologic system. MODHMS extends MODFLOW-SURFACT subsurface modeling code to include overland and channel flow and transport. Model addresses complex water resources management issues including: • Integrated water resource assessment • Groundwater availability & safe-yield analysis • Conjunctive surface water-groundwater use • Stream flow restoration • Flood prediction and mitigation • Agricultural irrigation management • Cleanup of industrial contaminants • Watershed-scale analysis of point source and nonpoint source pollution • Fluvial hydraulic analysis	http://www.hglsoftware.com/Modhms.cfm / HGL (commercially available)	(HGL, 2006)

Table 4. Models for simulation of surface and groundwater quality

8. Examples of use of integrated systems for water quality monitoring and modelling

In this section two examples of integrated real-time monitoring and modelling systems are presented. Both systems are up-to-date (are being still developed under ongoing projects). Both are also aimed at providing managers with additional information about the environment, information needed for efficient operation. First of mentioned systems is the system for feedback-driven in-situ remediation developed in the UPSOIL project (Sustainable Soil Upgrading by Developing Cost effective, Biogeochemical Remediation Approaches) (UPSOIL, 2011). The second is system developed in the ZiZOZap project, aimed at supporting the management and protection of dammed reservoir (ZiZOZap, 2011).

Monitoring-modelling system for in-situ chemical oxidation

One of the UPSOIL project test sites (used here as an example) was located in an industrial area at the port in Belgium. Until 2001 a warehouse for non-hazardous (no soil polluting) goods was located on this site. Since 2002 the storage of hazardous products (acetone, isopropyl alcohol, methyl ethyl ketone, vinyl acetate, acrylates, ethyl acetate, fatty acids, acetic acid) was started on the site. Contaminants present at the site included: heavy metals, mineral oil, polycyclic aromatic hydrocarbons, volatile organic compounds, benzene, toluene, xylene, extractable organic halogen, oils and grasses, trichloroethylene and total organic carbon.

The remediation process was based on the optimal interaction between the injected remedial agents and the target contaminants. The first version of the injection system (MIP-IN) was a combination of the commercially available Geoprobe MIP system for in-situ detection of volatile contaminants and a system for in situ injection of remediation liquids into the soil. The system, for 12 months was under iterative "modification - evaluation" process, incorporating suggestions and technical requirements defined by the stakeholders involved within project. The whole "contaminant detection - injection cycle" encompasses the following scheme:

- Volatile contaminants enter the MIP device and are detected over 0.3 m vertical soil column.
- App. 25 samples from each 0.3 m depth interval are transported by the carrier gas to the field Gas Chromatograph (GC) which is giving a very high data resolution of the contaminant spreading.
- The field GC signal from the 0.3 m interval and correlating depth interval is shown on the interface. The MIP signal response is logged automatically.
- When contamination is detected the operator has to define a balanced volume of solution to be injected at the corresponding 0.3 m interval. The volume to be injected is entered on the laptop and the program automatically gives a signal to inject this volume when the injection part of the probe is at the same depth as the corresponding MIP signal.
- Under pressure the injection pump delivers at a defined flow rate the volume of solution defined for a given depth interval. Pressure, flow rate and injected volume are logged automatically.
- The solution is entering the soil under pressure where it is spread due to the geological settings.

Real-time monitoring system used in this test site included 14 monitoring wells, which were equipped with sensors logging values of pH, redox, temperature and conductivity with 15-minutes intervals (Fig. 6). This system was established before the injections. It operated for four months after the injection and during the field work it was a complement to the measurements done with the gas chromatograph coupled with a system for in-situ injection.

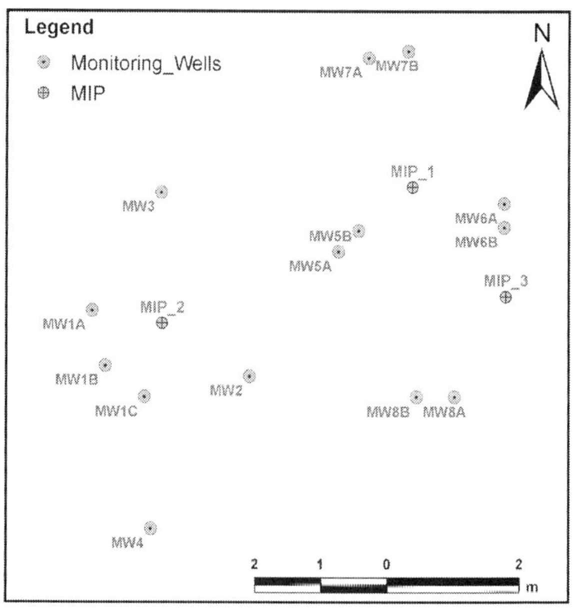

Fig. 6. Location of the 3 MIP-IN injection points and 14 monitoring points

Apart from real-time monitoring system, six sampling campaigns were done and key physical and chemical parameters were measured in each monitoring well. Samples were taken 17 weeks before injection (campaign T0b), 1 week before injection (T0a), 1 week after injection (T1), 3 weeks after injection (T2), 8 weeks after injection (T3) and 16 weeks after injection (T4).

During this remediation test a mathematical transport model was developed to enable optimal designing of the injection process as well as the re-arranging the whole procedure after the first step of the remediation (e.g. add new wells, change the volume of oxidants). Within the UPSOIL project relatively simple and operational software was developed for preliminary site specific design and further optimisation of the injection. In addition, during and after the injection field work this software serves additionally as a tool gathering all the field data including the real-time logged parameters (like pH, temperature, oxidation-reduction potential, conductivity) and the chemical data measured periodically during the sampling campaigns. The model consists of the Excel-Python module for calculations of physical parameters of the injection, and PhreeqC based module for hydro-geochemical simulations. The outcome of the first part of the model (physical module) includes:

- suggested radius of influence (ROI),
- total bulk treatment volume,
- total volume of reactant to be delivered,
- maximum volume of oxidant required at each injection interval,
- duration of injected oxidant effectiveness,
- effective velocity of the groundwater,
- injection time to reach the suggested ROI,
- suggested number of injection wells in the treatment area,
- suggested injection depth intervals to be used,
- suggested concentration of oxidant to be used (at averaged amounts of pollutants + natural demand).

The physical module of the model can be used to define the optimal injection strategy before moving into the field. Later, during the injection field work additional site specific data can be gradually inserted into the model in order to observe the soil absorption capacity (actual volume of liquid that can be injected at a given depth interval at a reasonable flow and with no "surface break through") and radius of influence at a given injection volume and flow/pressure rate.

In combination with the physical module (Excel-Python application) a 1D transport and chemical reactions module was elaborated for the oxidation processes (permanganate as an oxidant) using the PhreeqC code. This model requires input data including: parameters of the liquid to be injected, the baseline geochemistry (situation before injection) and the concentrations levels and type of contamination in the soil. The objective of the transport and chemical model is to simulate the chemical response to the injection and predict when (if required) it would be optimal to carry out next injection campaign. The prediction is gradually more solid when field monitoring data after injection are put into the model. Thanks to the continuously updated database of the developed tool it is possible to get actual comparisons of results derived from loggers, chemical monitoring and numerical modelling.

During the field works, at 3 points (Fig. 6) 435 kg NaMnO4 in 5.2 m^3 of water was injected at a depth between 2 and 7 m below ground level. Successful reaching of targeted interval by oxidant was observed via:

- visual detection of the purple colour in monitoring wells,
- significant decrease of the oxidation-reduction potential
- increase in electric conductivity.

Results of the remediation are shown also as an example of benzene concentrations before and after the injections (Fig. 7).

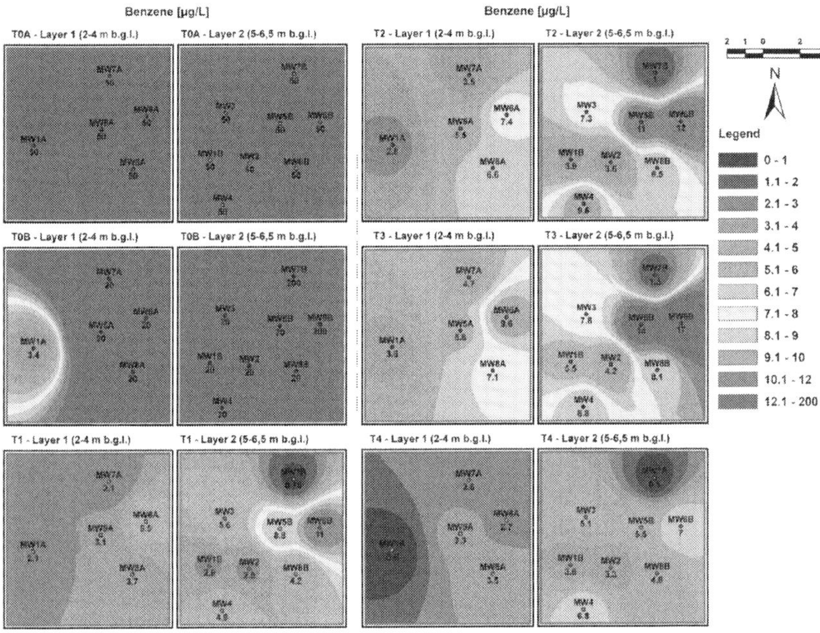

Fig. 7. Concentration distribution map for benzene and six sampling campaigns

Monitoring-modelling system for the dam reservoir management and protection

The ZiZOZap project is a part of the efforts to develop numerical multidimensional models of water reservoirs, which are undertaken in leading world countries. These models facilitate management of reservoir for their ability to predict the effects of the decisions undertaken. The purpose is to enhance the environmental safety and public heath of local population exploiting water resources. The object of the project study, Goczalkowice Water Reservoir, is a crucial water resource for 3,5 millions of inhabitants of Upper Silesia region (southern Poland). The Reservoir is also a unique natural area - a habitat for number of valuable species of birds, mammals, fish, small invertebrates and plants. Accommodation of environmental values and human needs in the area of The Reservoir is one of major tasks of the project.

The model which will be obtained as a result of the project will support the process of decision making and management of the Reservoir. Moreover, the model after necessary adaptations should be applicable for other water reservoirs (IPIS, 2011).

For the case study reservoir a monitoring and analyses of various environmental elements are carried out by interdisciplinary groups of experts. Data on hydrological, hydrogeological and physico-chemical conditions of water and bottom sediments as well as parameterised ecological and hygienic indicators referring to fauna, flora and the natural environment around the reservoir are collected in integrated databases. Project partners try to identify the key issues related to optimal management of the dam reservoir and build an information system together with a database. Based on the obtained research results and water management scenarios a numeric model of the reservoir is being developed. This will allow continuous assessment of the quality and functional state of the reservoir as well as stimulation and forecast of its changes. It is expected that in practice the system of reservoir models will enable to predict qualitative and quantitative changes in water resources which will affect water treatment and minimisation of costs, as well as forecasting water fertility and changes in the reservoir or the surrounding ecosystems (IETU, 2010).

As a part of created modelling system, a model for reservoir hydrodynamics and ecosystems was developed using ELCOM and CAEDYM software. ELCOM (Estuary, Lake and Coastal Ocean Model) is a three-dimensional hydrodynamics model for lakes and reservoirs, and is used to predict the variation of water temperature and salinity in space and time. The model forms the three-dimensional hydrodynamics driver to the CAEDYM water quality model (Hodges & Dallimore, 2009). The CAEDYM (Computational Aquatic Ecosystem Dynamics Model) is an aquatic ecological model that consists of a series of mathematical equations representing the major biogeochemical processes influencing water quality. At its most basic, CAEDYM is a set of library subroutines that contain process descriptions for primary production, secondary production, nutrient and metal cycling, and oxygen dynamics and the movement of sediment (Hipsey et al., 2009; Hipsey, 2009). Variables simulated by CAEDYM model include: light, inorganic particles, sediments and resuspension, dissolved oxygen, carbon, nitrogen, phosphorus and silica, phytoplankton dynamics, bacteria, zooplankton, higher biology, pathogens and microbial indicator organisms, geochemistry and metals, sediment diagenesis (Hipsey et al., 2009).

Within the ZiZOZap project the integrated ELCOM-CAEDYM model was coupled with mentioned above databases, which are used for storing the periodically collected and real-time measured data. Monitoring database includes among others:

- Real-time measured parameters: inflow (main river, pumping stations), outflows, water intake, water conductivity, pH, water temperature (at 1m depth intervals), dissolved oxygen, chlorides, nitrates, chlorophyll, redox potential, turbidity, air temperature, wind speed and direction, air humidity, air pressure, precipitation, solar radiation.
- Periodic measurements: complex water chemistry, physical parameters, pathogens and microbial composition, plankton, benthic organisms.

Sample output from the ELCOM model (water velocity at different weather and inflow conditions) is presented on Figure 8. Example of output from the CAEDYM model (dissolver oxygen, temperature) is shown on the Figure 9.

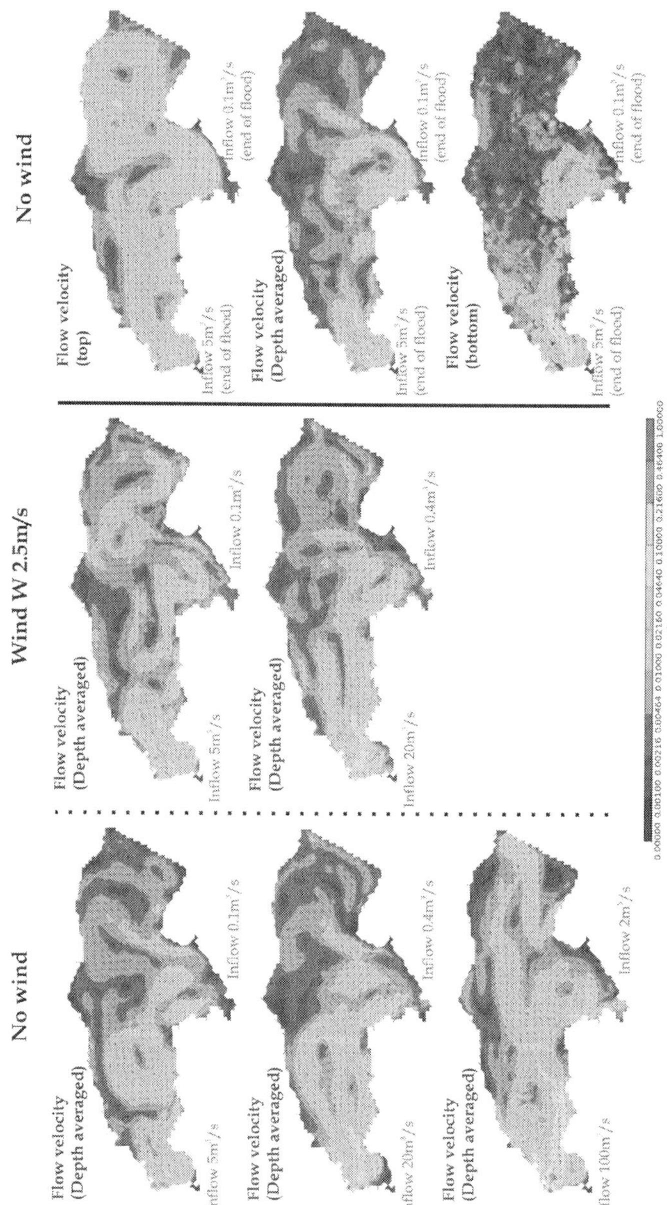

Fig. 8. Sample output from the ELCOM model for Goczalkowice Reservoir (velocities distribution)

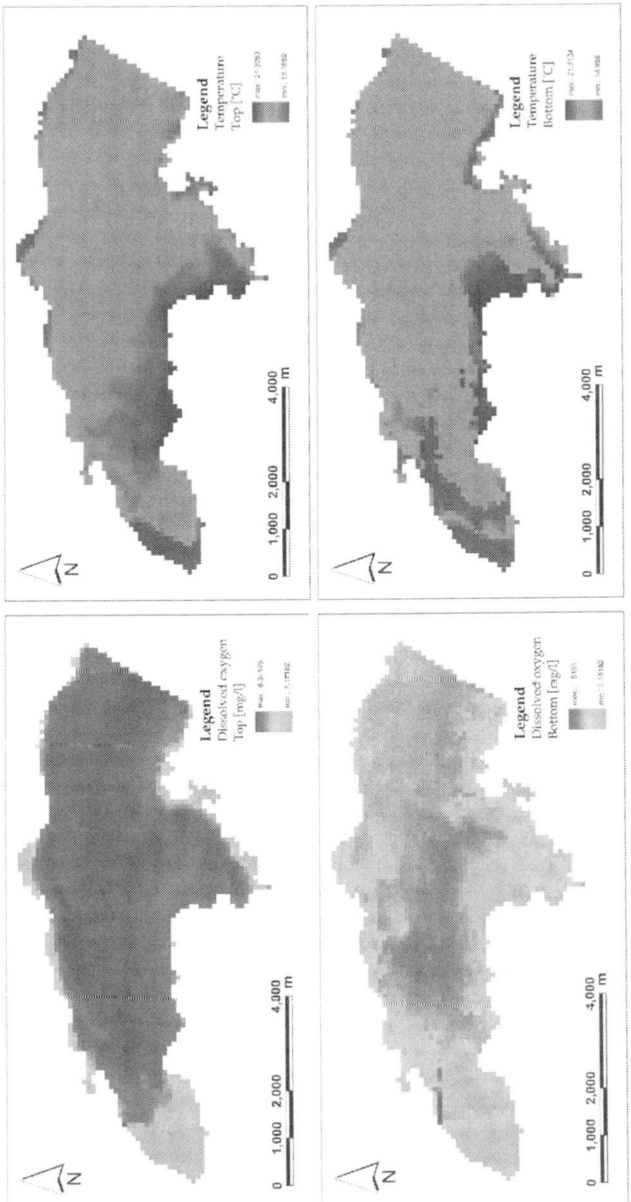

Fig. 9. Sample output from the CAEDYM model for Goczalkowice Reservoir (dissolved oxygen, temperature)

9. Conclusions

General conclusions of this chapter are as follow:

- Monitoring of the water quality (for atmospheric water, surface water and groundwater) should be designed in a way:
    - Ensuring spatial and temporal comparisons of the results - It is possible only if the monitoring is integrated in time and space, and if the structure of the monitoring system is uniform in all monitoring points.
    - Addressing local and regional problems.
- Results of the monitoring should be stored in a form allowing their immediate use (e.g. as online databases).
- If there are no resources or technical/environmental possibilities for monitoring of needed water quality parameters, appropriate models should be used. Models can be coupled for this purpose with monitoring systems including simple parameters, directly related to the water quality.
- Each short and long-term activities potentially affecting the water quality should be preceded by detailed analyses of different scenarios and their impacts. For such tasks a set of widely used modelling tools is easily available.

10. References

Agoshkov, V. I. (2002). Mathematical Models of Life Support Systems. In V. I. Agoshkov (Ed.), Knowledge for Sustainable Development, An Insight into the Encyclopaedia of Life Support Systems. Vol. 1 UNESCO/EOLSS. pp 335-281

Aman, M., Cofala, J., Heyes, C., Klimont, Z., Mechler, R., Posch, M., Schöpp, W. (2004). The RAINS model. Documentation of the model approach prepared for the RAINS peer review 2004.

AQUAVEO (2011). Aquaveo, LLC, The Surface-water Modeling Solution, http://www.aquaveo.com/sms

AQUAVEO (2011). Aquaveo, LLC, The New Groundwater Modeling System, http://www.aquaveo.com/gms

Ball, J.W., Nordstrom, D.K. (2001). USER'S MANUAL FOR WATEQ4F, WITH REVISED THERMODYNAMIC DATA BASE AND TEST CASES FOR CALCULATING SPECIATION OF MAJOR, TRACE, AND REDOX ELEMENTS IN NATURAL WATERS, U.S. Geological Survey, Open-File Report 91-183

Bathing Waters Directive {BWD] (1976). Directive 76/160/EEC. (The Bathing Waters Directive). Official Journal, Vol. OJ L 32 5.2.1976, pp. 1-7 with the amending act Regulation (EC) No 1137/2008, OJ L 311, 21.11.2008.

Bicknell, B.R., Imhoff, J.C., Kittle, J.L., Jr., Donigian, A.S., Jr., & Johanson, R.C. (1997). Hydrological Simulation Program--Fortran, User's manual for version 11. U.S. Environmental Protection Agency, National Exposure Research Laboratory, Athens, Ga., EPA/600/R-97/080, 755 p.

Bujsman, E., Jonker, P. J., Asman, A. H., Ridder, T.B. (1991). Chemical composition of precipitation collected on a weathership on the North Atlantic. Atmos. Environ., 25 A, 1991, 873

Chapra, S.C., Pelletier, G.J. & Tao, H. (2007). QUAL2K: A Modeling Framework for Simulating River and Stream Water Quality, Version 2.07: Documentation and Users Manual. Civil and Environmental Engineering Dept., Tufts University, Medford, MA., Steven.Chapra@tufts.edu

Chiang, W.-H. & Kinzelbach, W. (2001). 3D-Groundwater. Modeling with. PMWIN. (with CD-ROM). 2001, Springer Verlag

Common Implementation Strategy for Water Framework Directive [WFD] (2000b). Policy summary to Guidance Document No. 7. Monitoring under the Water Framework Directive. Produced by Working Group 2.7 - Monitoring, pp. 1-20.

Deltares (undated). Water Quality Modelling: Accurate, quantitative analysis and prediction, www.deltares.nl

Directive on Integrated Pollution Prevention and Control [IPPC] (2008). Directive 2008/1/EC. (The Directive on Integrated Pollution Prevention and Control). Official Journal, Vol. OJ L 24 29.1.2008, pp. 1-22.

Dore, A., Heywood, E., Vieno, M., Fournier, N., Smith, R., Weston, K., Hall, J., Sutton, M. (2003). Modelling the Transport and Deposition of Sulphur and Reduced and Oxidised Nitrogen in the UK. Status Report to DEFRA, as a contribution to Long Range Transport of Pollutants in the UK, Centre for Ecology and Hydrology

Drinking Water Directive [DWD] (1998). Directive 98/83/EC (1998). (The EU Drinking Water Directive). Official Journal, Vol. OJ L 330 of 05.12.1998, pp. 1-23.

Environmental Protection Act (2001). Journal of Laws No. 62 item. 627, pp. 1-165.

EPA (2006). Water Quality Analysis Simulation Program (WASP) Version 7.1, Watershed and Water Quality Modeling, Technical Support Center, US EPA, Office of Research Development. National Exposure Research Laboratory, Ecosystems Research Division, Athens, GA, Release Notes: 2/7/2006, http://www.epa.gov/athens/wwqtsc/html/wasp.html

EPA (2009). Water quality models and tools. In United States Environmental Protection Agency (USEPA). http://www. epa.gov/waterscience/models/

Essaid, H.I., Bekins, B.A. (1997). BIOMOC, A Multispecies Solute-Transport Model with Biodegradation. U.S. Geological Sorvey. Water-Resources Investigations Report 97-4022

Fraters, B., Kovar, K., Willems, W.J., Stockmarr, J. and Grant, R. (2003). Monitoring effectiveness of the EU Nitrates Directive Action Programmes Results of the international MonNO3 workshop in the Netherlands, 11-12 June 2003, pp. 1-290.

Glubiak-Witwicka, E., Starozenko, Z., Paszek, L. (2003). Groundwater. Pp. 1 12 (http://bip.katowice.pios.gov.pl/wody_podziemne.htm)

Han, Z. (2006). A regional air quality model: Evaluation and simulation of O3 and relevant gaseous species in East Asia during spring 2001. Environmental Modelling & Software. 22 (2007) 1328-1336

Han, Z., Ueda, H., Sakurai, T. (2006). Model study on acidifying wet deposition in East Asia during wintertime, Atmospheric Environment, 40, 2006, 2360–2373

HGL (2006). MODHMS: a comprehensive MODFLOW based hydrologic modeling system, Version 3.0, Code documentation and user's guide. HydroGeoLogic Inc., Herndon, VA, 2006.

Hipsey, M.R., Antenucci, J.P., & Hamilton, D. (2009). Computational Aquatic Ecosystem Dynamics Model: CAEDYM v3.2 Science Manual (DRAFT). Centre for Water Research, University of Western Australia, December 15, 2009

Hipsey, M.R. (2009). Computational Aquatic Ecosystem Dynamics Model: CAEDYM v3.2 User Guide (DRAFT). Centre for Water Research, University of Western Australia, December 15, 2009

Hecq, P., Hulsmann, A., Hauchman, F.S., McLain, J.L. & Schmitz, F. (2006). Drinking water regulations. In: Analytical Methods for Drinking Water Edited by P. Quevauviller and K. C. Thompson. John Wiley &Sons, pp. 1-38.

Heiskary, S. & Swain, E. (2002). Water Quality Reconstruction from Fossil Diatoms: Applications for Trend Assessment, Model Verification, and Development of Nutrient Criteria for Lakes in Minnesota, USA. Environmental Outcomes Division, Minnesota Pollution Control Agency

Hodges, B. & Dallimore, C. (2009). Estuary, Lake and Coastal Ocean Model: ELCOM v2.2 Science Manual. Centre for Water Research, University of Western Australia, December 15, 2009

Hutton, C.J., Vamvakeridou-Lyroudia, L.S., Kapelan, Z., & Savic, D.A. (2010). Real-time modelling and Data Assimilation techniques for improving the accuracy of model predictions: Scientific report. PREPARED 2011.010. http://www.prepared-fp7.eu/

IETU (2010). Integrated Support System for Management and Protection of Dam Reservoir. Institute for Ecology of Industrial Areas. IETU News, No.3, (August 2010),

Inspectorate of Environmental Protection (2010). The state of river water cleanliness based on the national monitoring results performed in years 2007-2009. Library of the Environmental Montoring. Warsaw. p. 209. (in Polish)

IPIS (2011). Integrated System Supporting Management and Protection of Water Reservoir (ZiZOZap). Institute of Environmental Engineering of the Polish Academy of Sciences,
http://www.ipis.zabrze.pl/index.php?option=com_content&view=article&id=269%3Aintegrated-system-supporting-management-and-protection-of-water-reservoir-zizozap&catid=71%3Aprojekty-zakladow&lang=pl

ISO (1993). 5667-8:1993, Water quality - Sampling - Part 8: Guidance on the sampling of wet deposition

Kannel, P.R., Kanel, S.R., Lee, S., Lee, Y.-S., Gan, T.Y. (2011). A Review of Public Domain Water Quality Models for Simulating Dissolved Oxygen in Rivers and Streams. Environ Model Assess (2011) 16:183–204

Kardaetz, S.,, Strube, T., Brüggemann, R., Nützmann, G. (2008). Ecological scenarios analyzed and evaluated by a shallow lake model. Journal of Environmental Management, Volume 88, Issue 1 (July 2008), pp. 120-135

Kazimierski, B., Pilichowska-Kazimierska, E. (2006). Monitoring i ocena stanu wód podziemnych (The monitoring and estimate of the underground water condition in Poland), Monitoring Środowiska Przyrodniczego, nr 7, s. 9-19, Kieleckie Towarzystwo Naukowe, Kielce. (in Polish).

Kenneth, L. Kipp, Jr. (1997). Guide to the Revised Heat and Solute Transport Simulator: HST3D -- Version 2. U.S. Geological Survey, Water-Resources Investigations Report 97-4157, 1997, 149 p.

Kulshrestha, U.C., Reddy, L.A., Satyanarayana, J., Kulshrestha, M.J. (2009). Real-time wet scavenging of major chemical constituents of aerosols and role of rain intensity in Indian region. Atmospheric Environment, 43, 2009, 5123–5127

Li, Ch., Kang, S., Zhang, Q., Kaspari, S. (2007). Major ionic composition of precipitation in the Nam Co region, Central Tibetan Plateau. Atmospheric Research, 85, 2007, 351–360

Ligocki, M. P., Leuenberger, C., Pankow, J. (1985). Trace organic compounds in rain. II. Gas scavenging of neutral organic compoundsm. Atmos. Environ., 19, 1985, 1609-1619

Loga, M & Sawicka, K. (2009). Guidebook. Statistical methods enable the assessment of water status within homogenous water bodies which are not included in the direct observations. P. 103. (in Polish)

McNamara, L. (2004). Water quality modelling for decision-making: the drinkingwater watersheds of Sydney, Australia, Proceedings of the Frontis Workshop From Landscape Research to Landscape Planning: Aspects of Integration, Education and Application Wageningen. The Netherlands 1-6 June 2004

Menz, F.C., Seip, H.M. (2004). Acid rain in Europe and the United States: an update. Environmental Science and Policy, 7, 2004, 253-265

Naddeo, V., Zarra, T., Belgiorno, V. (2005). European procedures to river quality assessment. Global NEST Journal, Vol 7, No 3, pp 306-312.

Neitsch, S. L., Arnold, J. G., Kiniry, J. R., Srinivasan, R., Williams, J. R. (2004). Soil and Water Assessment Tool input/output file documentation. Grassland, Soil and Water Research Laboratory, Blackland Research Centre, Texas

Neitsch, S. L., Arnold, J. G., Kiniry, J. R., Williams, J. R. (2005). Soil and Water Assessment Tool theoretical documentation. Grassland, Soil and Water Research Laboratory, Blackland Research Centre, Texas

Nitrates Directive [ND] (1991). Directive 91/676/EEC. (The Nitrates Directive). Official Journal, Vol. OJ L 375, 31.12.1991, pp. 1-10.

Olsson, O., Ikramova, M., Bauer, M., Froebrich, J. (2009). Applicability of Adapted Reservoir Operation forWater Stress Mitigation Under Dry Year Conditions. Water Resour Manage, (2010) 24:277–297

ORNL (2004). HYDROGEOCHEM 5.0: A Three-Dimensional Model of Coupled Fluid Flow, Thermal Transport, and HYDROGEOCHEMical Transport through Variably Saturated Conditions: Version 5.0. Oakridge national laboratory,

Park, R.A., Clough, J.S. (2009). Aquatox (RELEASE 3) modeling environmental fate and ecological effects in aquatic ecosystems, volume 2: Technical documentation. U.S. Environmental Protection Agency, EPA-823-R-09-004

Parkhurst, D.L., Appelo, C.A.J. (1999). User's Guide to PHREEQC (Version 2)--A Computer Program for Speciation, Batch-Reaction, One-Dimensional Transport, and Inverse Geochemical Calculations. U.S. Geological Survey, Water-Resources Investigations Report 99-4259

Parkhurst, D.L., Kipp, K.L. & Charlton, S.R. (2010). PHAST Version 2—A program for simulating groundwater flow, solute transport, and multicomponent geochemical reactions. U.S. Geological Survey Techniques and Methods 6–A35, 235 p.

Pazdro, Z. (1977). Hydrogeology. Wyd. Geol. Warszawa, (in Polish).

Quevauviller, P., Thomas, O., van der Beken, A. (eds.) (2006). Wastewater quality monitoring and treatment. Water Quality Measurements Series. John Wiley & Sons, p. 397.

Sadurski, A. (2005). Transboundary groundwater bodies as a task of the Polish State Hydrogeological Survey. Polish Geological Institute Special Papers 18, 79-85.

Schlumberger (2010). Visual MODFLOW Premium - Three-dimensional groundwater flow, heat, and contaminant transport modeling software, http://www.swstechnology.com/pdfs/brochures/Visual_MODFLOW.pdf

Schwarz, G.E., Hoos, A.B., Alexander, R.B. & Smith, R.A. (2006). The SPARROW Surface Water-Quality Model—Theory, Applications and User Documentation: U.S. Geological Survey, Techniques and Methods 6–B3, 248 p. and CD -ROM

Shellfish Water Directive [SWD] (2006). Directive 2006/113/EC. (The Shellfish Water Directive). Official Journal, Vol. OJ L 376/14 of 27.12.2006, pp. 1-7.

SSG (2011). Scientific Software Group, http://www.scisoftware.com/environmental_software/product_info.php?products_id=46

Stern, R. (2009). Das chemische Transportmodell REM-CALGRID, Modellbeschreibung, Freie Universität Berlin, Institut für Meteorologie Troposphärische Umweltforschung

Teodosiu, C., Barjovenau, G., Teleman, D. (2003). Sustainable water resources management. 1. River Basin management and the EC Water Framework Directive. Environmental Engineering and Management Journal, December 2003, Vol.2, No.4, 377-394.

Thebault, J.M., Qotbi, A. (1999). A model of phytoplankton development in the Lot River (France).: Simulations of scenarios. Water Research, Volume 33, Issue 4, (March 1999), pp. 1065-1079

Tsitouridou, R., Anatolaki, Ch.(2007). On the wet and dry deposition of ionic species in the vicinity of coal-fired power plants, northwestern Greece. Atmospheric Research, 83, 2007, 93–105

UNESCO (2005). Water resources systems planning and management - ISBN 92-3-103998-9 - © UNESCO 2005

UPSOIL (2011). Sustainable Soil Upgrading by Developing Cost effective, Biogeochemical Remediation Approaches, http://www.upsoil.eu/

Urban Waste Water Treatment Directive [UWWT] (1991). Directive 91/271/EEC (1991). (The EU Urban Waste Water Treatment Directive). Official Journal, OJ L 135/40, pp. 1 - 13.

Voss, C. I. & Provost, A.M. (2002). (Version of September 22, 2010). SUTRA, A model for saturated-unsaturated variable-density ground-water flow with solute or energy transport. U.S. Geological Survey Water-Resources Investigations Report 02-4231, 291 p.

Yoshimura, C., Zhou, M., Kiem A.S., Fukami, K., Prasantha, H.H.A., Ishidaira, H., Takeuchi, K (2009). 2020s scenario analysis of nutrient load in the Mekong River Basin using a distributed hydrological model. Science of The Total Environment, VoL.407, Issue 20 (1 October 2009), PP. 5356-5366

Wagner, J.M., Shamir, U. & Nemati, H.R. (1992). Groundwater Quality Management Under Uncertainity: Stochastic Programming Approaches and the Value of Information. Water Resources Research, Vol.20, No.5, (May 1992), pp. 1233-1246

Wang, T.J., Jin, L.S., Li, Z.K., Lam, K.S. (2000). A modeling study on acid rain and recommended emission control strategies in China. Atmospheric Environment, 34, 2000, 4467-4477

Water Framework Directive [WFD] (2000). Directive 2000/60/EC (2000) (The EU Water Framework Directive). Official Journal, Vol. OJ L 327, pp. 1-72.

Water Law Act (2001). Journal of Laws No. 115, item. 1229, pp. 1-80.

Exploring Potentially Hazardous Areas for Water Quality Using Dynamic Factor Analysis

József Kovács[1], László Márkus[2], József Szalai[3], Márton Barcza[6],
György Bernáth[1], Ilona Kovácsné Székely[4] and Gábor Halupka[5]
[1]*Eötvös Loránd University, Department of Physical and Applied Geology,*
[2]*Eötvös Loránd University, Department of Probability Theory and Statistics,*
[3]*VITUKI Environmental and Water Menagement Research Institute Non-Profit Ltd.,*
[4]*Budapest Business School, Institute of Methodology,*
[5]*Repét Ltd., Environmental Consulting Office,*
[6]*University of Szeged, Department of Mineralogy, Geochemistry and Petrology,*
Hungary

1. Introduction

Just like rivers, lakes and seas, subsurface drinking water resources are not safe from human impact and pollution. For sustainable development it is imperative to specify the areas where and the ways how these resources can be best protected. The aim of this study is to present a mathematical method for analysing spatially and temporally dependent data that is capable of assisting groundwater protection, by specifying potentially dangerous areas. In order to demonstrate the power of the proposed method the results obtained in three entirely different geological and hydrogeological environments are presented as case studies.

Natural and anthropogenic activities influence the fluctuation of groundwater level. Clearly, the main influencing factor is the precipitation recharging the subsurface water. Artificial influencing factors are the communal water withdrawals. If canals and rivers exist in the study area, they also control the natural random fluctuations of the groundwater table, they either feed or drain the shallow-groundwater depending on their water level. Considering the above-mentioned facts a mathematical model is formed on the basis of probability theory. Groundwater level data provided by monitoring wells are considered as time dependent random quantity, so hydrographs of each well are regarded as realizations of stochastic processes. Processes belonging to individual wells are not separate phenomena, but the occurrence of the same global natural phenomenon under different local conditions. That is why it is natural to treat these processes as the components of one multidimensional course, and as a matter of fact these components are probabilistically interdependent. It has to be emphasized that this interdependence is related to a spatial structure, and then the data is interpreted as the realization of a one-dimensional but *space-time* dependent stochastic process. We do not wish, however, to analyse the dynamics of this spatiotemporal process to a full extent. In our approach we suppose that the observed processes at different locations are governed by the same essential impacts, such as the mentioned recharge from precipitation, rivers and water withdrawal. It is the intensity of these impacts that depends on locality, creating the main aspect of the spatial dependence. So the goal is to identify these impacts, and to determine their spatially dependent intensities. In order to decompose the hydrograph time series into linear combination of the influencing effects - called factors - dynamic factor analysis has been applied. Dynamic factor analysis (DFA) produces factor time series and their loadings as an output; then the series must be identified with those processes of Nature that influence or drive the analysed phenomenon. The obtained loadings represent factor-intensities that provide essential information on the geological environment, improving the chance of correct decisions when environmental issues are on the agenda.

It has to be mentioned at this point that the whole factor analysis is scale invariant, and therefore, not absolute but only relative intensities can be inferred. These, however, are quite sufficient for our purposes, since we only wish to compare locations or subareas as less or more vulnerable to contamination. DFA readily provides the value of factor loadings belonging to a fixed factor and a fixed hydrograph that is to a fixed monitoring location. Since in our model the intensity is the sole source of dependence on location, the spatial aspect of interdependence can be revealed by the further analysis of the loadings. Consequently, if we wish to investigate a location where *no observation* is available, it can be done by spatially predicting the factor loadings and combining them with the factors; ultimately the hydrograph can be predicted. However, from hydrogeological point of view factor loadings and corresponding intensities are far more informative than the predicted hydrograph itself. At certain locations where no data are available, the loadings provides information about the expected intensity of infiltration, water-migration upward or downward, leakage effect or depression. Of course the uncertainty of the prediction has to be accounted for.

In the given applications the factors correspond to the infiltration, the water withdrawal and the river influence when relevant, hence the intensities appear to be connected to aquifer vulnerability. Intensive water withdrawal increases the danger of contamination of an aquifer, since overpumping may establish contact with a distant, already polluted storage. On the other hand high intensity infiltration increases vulnerability by helping local entry of any surface contaminant into the aquifer. Intense river impact may increase the danger of infiltrating contaminants carried by the river. Hence, the loadings, the measures of intensity can be regarded as *important quantitative markers of vulnerability* of the aquifer.

2. Dynamic factor analysis

Often the statistician encounters measurements of a very complex but well-observable time-dependent random phenomenon that is induced by only a few basic but unobservable (latent) effects of relatively simple dynamic structure. The behaviour of the measured phenomenon can be understood much better when these common driving forces are identified. In the conventional setup when independent observations are at hand factor analysis provides the variables representing the latent effects. As Anderson (1963) alerted, this technique can be misleading when applied to multidimensional time series with delayed interdependence among its components. Lagged interdependence invalidates the results of conventional factor analysis that has been elaborated for independent observations. Correct identification of the governing effects became possible with the invention of *Dynamic Factor Analysis (DFA)*, that is capable to take into account the dynamic structure of both observations and factor time series.

The term dynamic factor analysis goes back to the pioneering work of Geweke (1977). Shortly after different concepts and procedures had become known as dynamic factor analysis, all of them being generalisations of this or that properties of the conventional factor model. For example Picci & Pinzoni (1986/1, 1986/2), Van Schuppen & Van Putten (1985) generalise the known property that observations are conditionally independent given the factors. Others, like Deistler & Scherrer (1991) decompose in every frequency the spectral density matrix function as sum of a diagonal and a singular matrix. So, by constructing factor spectral densities, they produce realisations of the factor time series on the base of the observed realisations. Gourieroux et al. (1995), Gourieroux & Monfort (1997), consider factor representations for Markov-processes. These as well as other early contributions to the literature on dynamic factor models like Sargent & Sims (1977), Engle & Watson (1981), Watson & Engle (1983), Connor & Korajczyk (1993) consider time series mostly with limited panel dimensions. Factor models for spatio-temporal processes became also a focus of interest, see e.g. Mardia et al. (1998) and Dryden et al. (2005).

The increasing availability of high-dimensional data sets has intensified the quest for computationally efficient estimation methods, leading to a renewed interest in dynamic factor analysis. DFA consolidated to state space representations of structural time series models. The new wave of literature was headed by Forni et al., (2000), Stock & Watson (2002) and Bai (2003). These methods are typically applied to high dimensional panels of time series. Exact maximum likelihood methods such as proposed in Watson & Engle (1983) have traditionally been dismissed as too computationally intensive. However Jungbacker & Koopman (2008) present new results that allow the application of exact maximum likelihood methods to large panels. Examples of recent papers employing likelihood-based methods for the analysis of dynamic factor models are Doz & Reichlin (2006), and Reis & Watson (2007). Alhtough the majority of publications for DFA is concerned with the problems of economics, environmental applications are also widespread (Márkus et al., 1999; Kovács et al., 2004; Ritter et al., 2007), a few of them assesses water quality (Kovács et al., 2002; Kovács et al., 2004; Muñoz-Carpena et al., 2005; Ritter et al., 2006).

The model, we briefly describe in the present paper, does not differ much of the ones based on state space representation, but the algorithmic solution lays on different concept. The idea, as we use, originates in Bánkövi et al. (1979), though it also relates to Box & Tiao (1977) on canonical transformations of time series vectors. While prescribing an autoregressive structure to the factor time series our factor analysis model minimises a cost function, which is a linear combination of the conditional variance of the prediction error and the state estimation error. The problem of finding the optimal factors leads to a minimisation problem on Stiefel manifolds much in the focus of recent investigations in operational research (see e.g. Rapcsák, 2002). The theoretical solutions are very complex and difficult to give, the explicit solutions hardly go as far as 10 dimensions, falling far short to our case. So instead, this optimisation problem can be solved by an iterative method, which relies on the maximization algorithm of sums of heterogeneous quadratic forms developed in Bolla et al. (1998).

Let's consider the usual static factor model equation

$$Y = \mathbf{A} \cdot F + \varepsilon \qquad (1)$$

expressing that the observations Y are described by linear combinations of several latent factors F plus a random uncorrelated noise ε. Usually the number of observed variables N is significantly higher than that of the factors, M. The crucial difference when dynamic factor models are considered is that both observations and factors are empirical time series instead of independent samples/measurements of the variables, as is in ordinary models. To complete the model the dynamic structure of the factors has to be specified. The linear transformation expressed through the **A** matrix, however, should not depend on time. Supposing the observed N-dimensional time series

$$Y(t) = (Y_1(t),\dots,Y_N(t))', \quad 0 \le t \le T.$$

to be weakly stationary apart from a possible linear trend, and emphasising time dependence rewrite (1) as

$$Y(t) = \mathbf{A} \cdot F(t) + \varepsilon(t) \qquad (2)$$

with the **N** x **M** matrix **A**, the factor time series vector

$$F(t) = (F_1(t),\dots,F_M(t))', \quad 0 \le t \le T,$$

of M *uncorrelated, stationary* time series, and the N-dimensional Gaussian white noise

$$\varepsilon(t) = (\varepsilon_1(t),\dots,\varepsilon_N(t))', \quad 0 \le t \le T.$$

We are aimed at finding optimal, in a certain sense, *estimations* of the factors:

$$\hat{F}(t) = (\hat{F}_1(t),\dots,\hat{F}_M(t))'$$

The estimation of our model should focus on the following three natural requirements:

i. The estimation of the factors should be a *time-independent* homogeneous linear transformation of the observations.

$$\hat{F}(t) = \mathbf{B} \cdot Y(t) \qquad (3)$$

ii. The factor time series components $F_j(t)$ should be linearly well predictable from their past. It is certainly fulfilled supposing them to be *autoregressive* processes of order L_j with a constant included in the autoregression:

$$F_j(t) = c_{j,0} + \sum_{k=1}^{L_j} c_{j,k} \cdot F_j(t-k) + \delta_j(t) \qquad (4)$$

where the components of the Gaussian white noise $\delta(t) = (\delta_1(t),\dots,\delta_M(t))'$ are independent from each other and from $\varepsilon(t)$.

iii. Again, time-independent linear transformation of the factors should provide a "good" estimation - called factor-estimator - of $\mathrm{Y}(t)$, as expressed by equation

$$\hat{Y}(t) = \mathbf{D} \cdot \hat{F}(t) \tag{5}$$

where the **D** matrix is in fact the estimation of **A**.

The choice of autoregression in (ii) justified not only by its simple dynamic structure, but also by the fact that hydrographs of monitoring wells can reliably be modelled by autoregressive processes. In the process of model estimation it is supposed that the structure prescribed for the *unobservable factors* is inherited to its *estimations*. Were the components of $\left(\hat{F}_1(t),\ldots,\hat{F}_M(t)\right)$ observable, their *best forecast* $\widetilde{F}_j(t)$ could be obtained as

$$\widetilde{F}_j(t) = c_{j,0} + \sum_{k=1}^{L_j} c_{j,k} \cdot \hat{F}_j(t-k) \tag{6}$$

In order to relate $\widetilde{F}_j(t)$ to $F_j(t)$ we call $\widetilde{F}_j(t)$ the *empirical best forecast* of $F_j(t)$. In other words it is just the plug in of the predicted factors into the best forecast of the autoregression. As the true values are not known, the coefficients $c_{j,k}$ have to be estimated. The optimality of this forecast, guaranteed for a truly autoregressive process with known coefficients only, cannot in general be preserved for the plug in. Keeping this in mind we will use (6) for the forecast of the estimator $\hat{F}(t)$ given its past. Since the observations and thus the predictions of the factors can be computed for all $\mathrm{t}, 0 \leq \mathrm{t} \leq \mathrm{T}$, it is possible to compare the forecast with the estimator itself, and by centring, get an unbiased estimation $\hat{\delta}(t)$ of the noise $\delta_j(t)$ in (4) as

$$\hat{\delta}_j(t) = \widetilde{F}_j(t) - \hat{F}_j(t) - \left[\overline{\widetilde{F}_j - \hat{F}_j}\right]$$

(For any $X(t)$ $\overline{X}$ denotes the average

$$\frac{1}{T+1}\sum_{t=0}^{T} x(t))$$

The squared sum $\varepsilon^{(d)}$ of $\hat{\delta}_j(t)$ is called the estimated *dynamic error*:

$$\varepsilon^{(d)} = \sum_{j=1}^{M}\sum_{t=L_j}^{T} \hat{\delta}_j(t)^2$$

Similarly, $\hat{Y}(t)$, the factor-estimator of observations in (5), opens the way to estimate $\varepsilon(t)$, the noise in (2) by taking the centered difference as

$$\hat{\varepsilon}_i(t) = Y_i(t) - \hat{Y}_i(t) - \left[\overline{Y_i - \hat{Y}_i}\right],$$

the squared sum $\varepsilon^{(s)}$ of which is called the predicted *static error:*

$$\varepsilon^{(s)} = \sum_{i=1}^{N}\sum_{t=0}^{T} \hat{\varepsilon}_i(t)^2$$

If the importance of this or that observation is to be emphasized, or the precision of the forecast of this or that factor is of major concern then *weights* can be introduced in the definition of both dynamic and static errors, to achieve this end. To fulfil the requirement given in (iii), the estimation of the model is regarded to be "good" if the sum of the estimated static and dynamic errors is minimal. This means the minimization of the following functional:

$$\Psi(T) = \varepsilon^{(s)} + \varepsilon^{(d)} = \sum_{i=1}^{N}\sum_{t=0}^{T}\hat{\varepsilon}_i(t)^2 + \sum_{j=1}^{M}\sum_{t=0}^{T}\hat{\delta}_j(t)^2 \tag{7}$$

on the constraints,

$$\mathrm{var}\left(\hat{\mathrm{F}}\right) = I_M \tag{8}$$

stemming from the uncorrelatedness of the factors (I_M denotes the M×M unit matrix).

The real statistical difficulty lays in the estimation of the model parameters, that is the matrices **B**, **C**, **D**. Remark, that **C** is the matrix of $c_{j,k}$-s from (4) endowed with zeros when necessary.

The usual ML methodology results in very complicated computations, and even though one can determine the density function, but it seems rather hopeless to find its place of global maximum. In state space models the EM algorithm provides a way to tackle the estimation problem. Alternatively, Markov Chain Monte Carlo (MCMC) estimation may also prove to be viable.

Our approach originates in Bánkövi et al. (1979), where instead of finding a direct optimal solution to (7) and (8) an iterative approximation by a criss-cross algorithm is suggested. It can be developed further by using the optimisation procedure of heterogeneous quadratic forms as described in Bolla et al. (1998), where the analysis of the optimisation can also be found. For a detailed description see Ziermann and Michaletzky (1995) or Márkus et al. (1999).

Introducing a new appropriate orthonormal system $\{\mathbf{e}_j\}_{j=1,\dots,M}$ as in Márkus et al.(1999), the functional, rewritten as

$$\Psi = \sum_{j=1}^{M} \mathbf{e}_j^T \mathbf{Q}_j \mathbf{e}_j$$

with the $\mathbf{Q}_j$ matrices computable from the observations and the **C** matrix, has exactly the same structure - that is the sum of heterogeneous quadratic forms - as the one treated in Bolla et al. (1998). By applying Lagrange's multipliers it is easy to obtain a necessary condition for the existence of stationary point, namely that equation

$$[\mathbf{Q}_1\mathbf{e}_1,\dots,\mathbf{Q}_M\mathbf{e}_M] = [\mathbf{e}_1,\dots,\mathbf{e}_M]\mathbf{S}$$

must hold with **S** being a symmetric M×M matrix. Specifically the vectors $\mathbf{Q}_j\mathbf{e}_j$ are included in the subspace spanned by $\mathbf{e}_1,\dots,\mathbf{e}_M$.

For the place of global maximum **S**≥0 holds. Introducing the notations $[\mathbf{e}_1,\dots,\mathbf{e}_M] = \mathbf{E}$, $[\mathbf{Q}_1\mathbf{e}_1,\dots,\mathbf{Q}_M\mathbf{e}_M] = \mathbf{Q}(\mathbf{E})$ and writing the condition formally, we have

$$\mathbf{Q}(\mathbf{E})=\mathbf{ES}, \qquad \mathbf{E}^T\mathbf{E}=\mathbf{I}_M, \mathbf{S}\geq 0,$$

which is nothing, but the polar decomposition of the matrix **Q**(**E**). A one to one correspondence of the polar, and the singular value decompositions is established in Bolla et al. (1998) and a criterion for **S**≥0 i.e. **S** being positive semidefinit is also given there. So, the actual computations rely on the singular value decomposition instead of the polar one. It seems that this is the computationally most demanding step of the algorithm, and sometimes it does not converge fast enough.

Summarising, the algorithm is as follows. For a given $\mathbf{E}_1 = \mathbf{E} = [\mathbf{e}_1,...,\mathbf{e}_M]$ vector system the next one $\mathbf{E}_2 = \mathbf{H} = [\mathbf{h}_1,...,\mathbf{h}_M]$ is defined from the polar decomposition of **Q**(**E**) by

$$\mathbf{Q}(\mathbf{E})=\mathbf{HS}, \qquad \mathbf{H}^T\mathbf{H}=\mathbf{I}_M, \qquad \mathbf{S}\geq 0,$$

It can be shown - as is in Bolla et al. (1998) - that the algorithm increases the value of the Ψ functional, and the E_1, E_2,... matrices are getting closer and closer to each other, and so the accumulation points of the algorithm are their stationary points as well. The set of accumulation points is a connected one. If any point obtained from the algorithm falls "near" to the place of global maximum then the algorithm will be *convergent*, and it will converge to this *place of global maximum*. However the algorithm is not necessarily globally convergent, because in general there may exist fixpoints of it different from the place of global maximum. Every stationary point of the functional is fixpoint of it at the same time.

After the detailed presentation of the mathematical method we turn to the case studies of hydrogeological applications. Although all of the presented applications study groundwater levels, the data are registered in three different parts of Hungary (see Fig. 1), and therefore the geological and hydrogeological environment is entirely different.

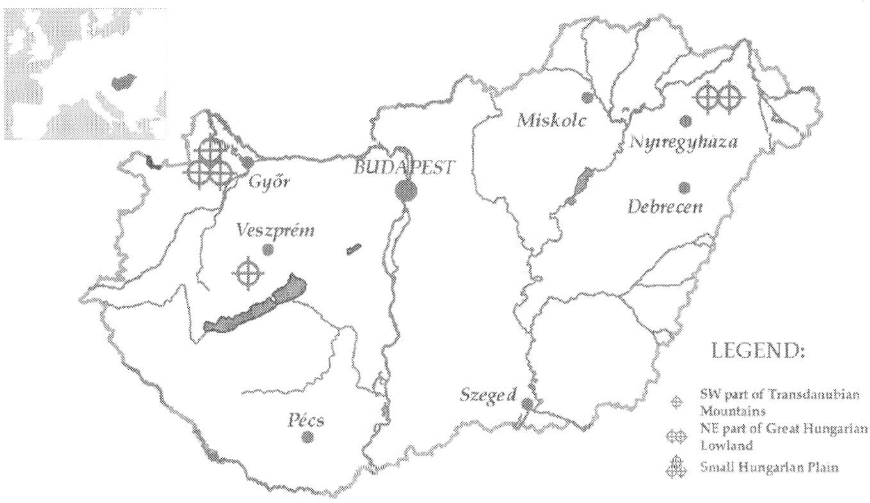

Fig. 1. The three different areas of Hungary, subject to case studies

3. Application of dynamic factor analysis in karstic environment

In Western Hungary the mining industry withdrew a large amount of water from the ground in the period of 1950-1990, when intense bauxite and coal mining characterised the Transdanubian region. In the south-western side of the Transdanubian Mountains (Bakony, Keszthelyi Mountains and Balaton-Highland) especially large quantity of water was pumped out from the main Triassic karst aquifer, the water conducting geological layer consisting mainly of Triassic dolomite and limestone. Both Bakony and Keszthelyi Mountains are built up mainly from this Triassic dolomite and limestone. In the Balaton-Highland a more various geological structure can be found, the Triassic aquifer consists mainly of middle and lower Triassic formations divided by aquitard layers. The Bakony and the Balaton-Highland are separated by a deep rift (Lesence-valley) from the Keszthelyi Mountains. This rift is filled with impermeable formations, and those hold back or slow down considerable part of the water flow. A reverse fault separates the Balaton Highland from the Bakony, and this creates an impermeable obstacle for flows between them (Fig. 2).

At its peak the withdrawal continuously exceeded the infiltrated water supply, and as a consequence water capacity of some springs declined to the half or one third and some others dried out completely. As the deterioration of the environment became more and more apparent, a monitoring system has been set up to observe karstwater levels. The system was extended over time and observations became more frequent. For our analysis it provided the *hydrographs*, the water levels registered in karstwater monitoring wells in the south-western side of the Transdanubian Mountains (Bakony, Keszthelyi Mountains and Balaton-Highland).

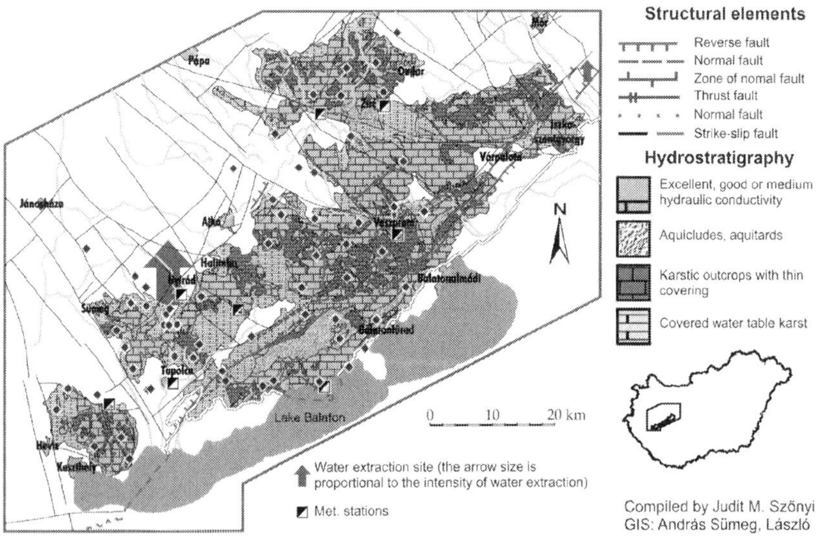

Fig. 2. Structural elements, hydrostratigraphy, water extraction and karstwater monitoring wells in the Transdanubian Mountanis (compiled by Mádlné, Szőnyi J. , GIS: L. Füle, 1998)

A hydrograph of a (monitoring) well reflects, on the one hand, all characteristics of the route of water from the surface to the observation point, and on the other hand, any kind of change of geology along this route. So, information on geological conditions can be obtained *after* water has reached the aquifer, moreover, the basis of this information-providing is just the water that completed its migration. This allows for the characterisation of the aquifer "bypassing" the direct study of uncertain geological parameters, since every influencing effect of the geological setting is already left behind by the water: the *final result* of roaming of water is reflected in the fluctuating level in the monitoring well. Consequently, ours is an "upside down" approach: the influence of geological parameters can indirectly be inferred by studying the hydrographs.

The individual hydrographs can be regarded as temporal measurements or samples from the above mentioned aggregated phenomenon, the fluctuating karstwater level. Only 64 monitoring sites provided sufficiently regular readings and even in the period of 1970-1990. The scattered areal distribution of the selected 64 monitoring wells can be seen on Fig. 2. The sporadic timing of observations forced us to study hydrogeological time series of yearly averages, derived from the non time-equidistant raw data. Observations after 1990, though available, are not taken into consideration in this study, since the social and economic changes taking place at that time in Hungary triggered sharp decline in mining and mounting concern towards environmental damages. This led in particular to the closure of the mines at the main water withdrawal site. Since recovery processes of nature are usually different of human interference, new changes were induced in the karstwater reservoir system. Therefore, it seems to be reasonable to exclude this time period from the present investigations.

We applied dynamic factor analysis (as described above) to the data in order to decompose the observed empirical time series (the hydrographs) into linear combinations of several underlying factors. Since there was no reason to emphasise the behaviour of any particular monitoring well, or factor, we used equal unit weights in the static and dynamic error terms.

On the different possibilities to determine the number of factors as well as the order of autoregressions we refer the reader to Márkus & Kovács (2003). In our case three factors proved to be sufficient to describe the behaviour of hydrographs by satisfactory precision, and among the concurrent models the AR(1) - AR(1) - AR(1) proved to be the easiest and most natural to be interpreted. The three effects, that is the corresponding three dynamic factors explain about 92% of the overall variance. It has to be noted though that generalization of this customary characterization of ordinary factor analysis for the dynamic case is not quite straightforward. We were able to identify the first two factors only, and conjecture that the third is describing a small territorial variance in infiltration.

The immediate aim of the application of dynamic factor analysis was to *identify the main underlying effects* shaping the hydrographs, and give spatial prediction of the intensity of these effects. The interpretation of the obtained factors reiterate the well known fact that infiltration and water withdrawal (where relevant) are the two main effects shaping the hydrographs. The more valuable gain from the analysis is that the loadings of the dynamic factors serve as measures of the intensity of the effects at a given location and thus endowing a certain area with a comprehensive hydrogeological characteristic.

Remark here, that throughout the considered period monitoring wells near the main mines registered about 100 meters drop in karstwater level. Even in wells further away from the main site decline could exceed 20 meters. The decline was not uniform, its pace changed with time. In the area of study there was one main pumping centre at Nyírád, near several closely located bauxite mines, and some other, much smaller water withdrawal sites near coal and bauxite mines, furthermore, water has also been withdrawn for community use (Fig. 2). The withdrawn amount of water at these latter sites combined, was of considerably smaller magnitude than that of the main site at Nyírád. For this reason it seems to be sufficient to use the data of the dominating Nyírád withdrawal in the analysis, noting that local withdrawals can still play an important role regionally, but the use of these data does not improve significantly the identification of the factor.

However, the water withdrawal data cannot be used without transformations for comparisons with the decline in water levels. On the one hand, water withdrawal is partly balanced by the decrease in water capacity of springs, and that has to be compensated for in comparisons. On the other hand the amount, dynamically supplied to the aquifer, has to be subtracted, because only the remaining amount lowers the static water reserve of the aquifer. The detailes of the actual computations can be found in Márkus et. al., 1999, Kovács, 2007.

The result is surprisingly spectacular. In Fig. 3 we compare the graph of the differences of the first factor (blue line) with the sum of the registered water withdrawal and the spring capacities (red and green lines). In order to present these data on the same graph one has to bring them to the same scale; the simplest way - as it is done in Fig 3 - is to standardise them. The red and green lines differ in the amount of withdrawn water, the red includes the Nyírád withdrawal only, whereas the green takes into account all the registered withdrawals in the area.

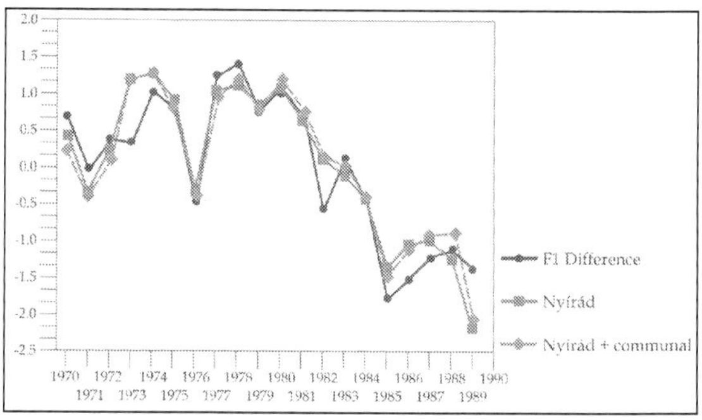

Fig. 3. Comparison of the differenced 1st factor (blue) with the rescaled water withdrawal at Nyírád (red) and Nyírád plus communal (green) withdrawals

Second dynamic factor represent the overall tendencies in infiltration throughout the whole investigated area. In order to prove it, infiltration has to be computed by conventional methods.

Infiltration is a very complex process, because at a given location the amount of infiltrating water depends on geomorphology, water conductivity of superficial rocks, air temperature, amount, duration and physical state of precipitation, and also vegetation and many other factors. Therefore, the effect of infiltration varies throughout the mountains. Nevertheless, the area is not so large that the main tendencies in the yearly amount of infiltrated water would differ too heavily among locations. This circumstance allows us to identify only one factor (the second) with infiltration and obtain comparable weights (intensities) of the infiltration effect in every monitoring well.

Different empirical methods exist for the computation of the infiltration. When used in the analysis, infiltrations are computed by methods developed or refined on the basis of observations in Hungary, rather then elsewhere, (cf. e.g. Primault (1963), Dracos (1980), Burman and Pochop (1994), Jeannin & Grasso (1995) and references therein), because karstic infiltration strongly depends on local characteristics. For comparisons we used four methods, called by the names of their creators as Böcker, Kessler, Maucha and Morton.

The method of Kessler (1954) is based on several decades long observations in the Mecsek Mountains. Böcker's method (1974) was evaluated at several different locations. Maucha (1990) computes infiltration on the basis of observations in the northern part of Hungary, where the karst knowingly has different characteristics (e.g. water conductivity) than the one we study. Using Morton's evapotranspiration model (Morton, 1983) Csepregi (1995) developed a way to calculate infiltration on monthly basis. As it was pointed out in Márkus et al. (1999), the different methods produce pretty different results - especially so in temporal dynamics - correlating only 0.5-0.7 among themselves. All the methods mentioned, compute infiltrations on the basis of precipitation data, and do not depend on location, or local geological structure. On the contrary to this, the factor representing infiltration was computed from the hydrograph - that is from the groundwater level data.

To identify the second factor with the infiltration, it has to be compared to the computed conventional ones. We got highly significant correlations (around 0.8) among the factor and the conventionally computed infiltrations, for details see Márkus et al., 1999; Kovács, 2007. The corresponding graph (Fig. 4) emphasizes the fit further.

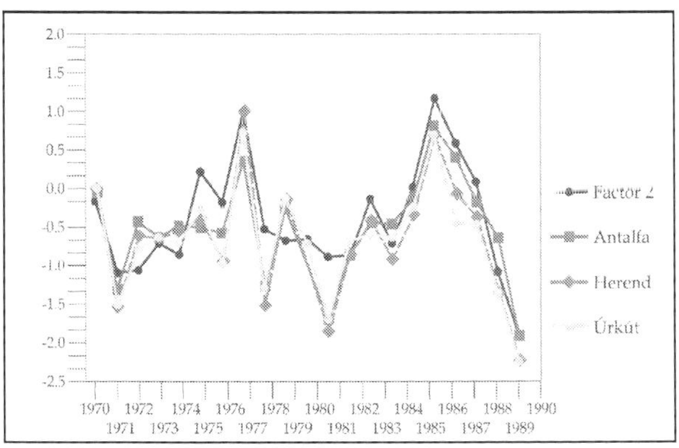

Fig. 4. Graphs of factor 2 and standardised Kessler's infiltration computed from the data of three different meteorological stations

The dynamic factor loadings reflect the proportion of the underlying effects in influencing the fluctuation of the hydrographs. When these loadings indicate a dominant role among the underlying effects either for withdrawal or infiltration then important conclusions can be drawn for the aquifer vulnerability or sensitivity for contamination.

We would like to present a relationship between the above-mentioned factors and vulnerability, in two steps. First we describe a connection between the factors and geological parameters influencing hydrographs. Afterwards, factors are considered as influencing effects in vulnerability of an aquifer by means of a different scale of groundwater flow regimes.

To reach this goal it is essential to define vulnerability. The following few approaches can help to see essence of the concept despite the lack of a widely accepted definition.

* "*Vulnerability is an intrinsic property of a groundwater system. It depends both on the geological and hydrogeological characteristics of an area, and on the sensitivity of groundwater system to human and natural impacts.*" *(COST Action 65 Final Report, 1995)* * "*Vulnerability as a system-approached term means the ability of an environmental system and it surroundings to compensate human activities.*" *(Mádl-Szőnyi & Füle, 1998)* * "*Intrinsic vulnerability is the term used to define the vulnerability of groundwater to contaminants generated by human activities. It takes account of the inherent geological, hydrological and hydrogeological characteristics of an area, but is independent of the nature of the contaminants.*" *(Daly et al., 2001)*

Briefly: *sensitivity* expresses a chance of contamination of an aquifer, whereas *vulnerability* is like a summary of conditions (human and/or other factors), that can activate the risk of any geological settings into real danger. Nevertheless it is important to take into consideration that sensitivity refers to an aquifer as rock-type, while vulnerability concerns a complete groundwater providing system, as Alföldi (1994) pointed out. Our opinion is that the presented results below refer to vulnerability, since water withdrawal is a kind of human activity and infiltration reflects the existing climatical, and other meteorological conditions. So neither of them is part of intrinsic attributes of an aquifer.

Mathematical study of the time series have disclosed two factors which account for more than 90% of fluctuations of well-hydrographs. One of them is *water withdrawal,* the other one is *infiltration.* How to connect these factors to geological parameters and what kind of parameters are these?

1. Water withdrawal. It is a kind of anthropogenic activity, non-geological parameter. Nevertheless it provides indirectly information about hydraulic conductivity, karstic network, and material of the aquifer.
2. Infiltration. It is itself an important characteristic of an aquifer, on the other hand it has relationships with precipitation, features of cover material (soil, as well), and it is in inverse ratio to runoff.

If the referred parameters are considered, it can be seen that all of them are key parameters in one or more vulnerability-assessment methods. To show this table 1. summarizes 6 very remarkable methods, in point of view of the used parameters.

It can be claimed that a close connection exists among the identified factors and the geological key parameters of vulnerability-assessment methods. Consequently factors (and so the mathematical method, as well) should have an indirect relationship to vulnerability, itself. The next point assesses the question, how.

Parameter / Method	1	2	3	4	5	6
Infiltration, recharge and/or its type	*	*	*	*	*	*
Precipitation, climate						*
Epikarst (thickness, quality)			*			
Deepness of groundwater level (thickness of unsaturated zone)	*	*				
Run-off				*		*
Hydraulic conductivity	*	*		*	*	
Karstic network			*			*
Material of unsaturated zone (thickness, quality - [no epikarst specifically])	*	*		*	*	*
Topography	*	*				
Protective cover material, soil (thickness, quality)	*	*	*	*		*
Material of the aquifer (thickness, quality)	*	*				

Table 1. Parameters of vulnerability assessment methods. The numbers refer to the following vulnerability-assessment methods: 1.) *DRASTIC* (Aller et al., 1985), 2.) *SINTACS* (Civita & Regibus, 1995), 3.) *EPIK* (Doerfliger, 1996), 4.) *PI* (Goldscheider et al., 2000), *Irish method* (Daly & Drew, 1998), 6.) *European Approach* (Daly et al., 2001)

One of the above mentioned hydrograph-determining factors is *water withdrawal*. Its effect is indirect. If an aquifer is under over-pressure, it means that it is protected against the entry of contaminant. The effect of extensive withdrawal decreases the resistance, so entry of pollutant may occur by either vertical (local) or horizontal (regional) recharge - as Mádl-Szőnyi pointed out (1997). Consequently, vulnerability of the aquifer increases in both local and regional scales. However, pressure-fluctuation spreads throughout an area faster than water or contaminant, decreasing of overpressure by water withdrawal in large, closed or quasi-closed aquifers does not mean immediately increasing vulnerability (Alföldi, 1994). However our opinion is that it is essential to take into consideration role of *time*. Dynamic factor analysis used data of 21 years, a period that seems long enough to assume that the process of decreasing pressure definitely has negative effect on vulnerability. When a distant aquifer is contaminated and the available time is long enough for spreading of the contaminant a direct connection can be established between extensive water withdrawal and increasing regional vulnerability. On the other hand increasing water withdrawal and local vulnerability can only be connected indirectly, because in a period when there is no significant precipitation to induce remarkable infiltration the medium carrying any soluble contaminant is missing. So pollution will not immediately follow the decrease or disappearance of overpressure protection of the aquifer.

The other determining factor is *infiltration*. In areas of high intensity infiltration a contaminant (general, soluble type is assumed) can enter in great quantities into the aquifer, directly causing high local vulnerability. At the same time infiltration can generate an increase in regional vulnerability when the investigated area has a regional recharge function, too (ignoring here dilution, degradation of the contaminant). If the investigated area is situated in a non-regional recharge area, fluctuations in hydrographs refer to local vulnerability.

Summarising, it can be stated that the factor of *water withdrawal* has direct connection with regional, but indirect connection with local vulnerability, while factor of *infiltration* directly refers to local and indirectly to regional vulnerability, depending on which hydraulic regime does the pilot area belong to. Observing, that in withdrawal-dominated wells infiltration-generated fluctuations are covered up by the effect of water-withdrawal, *regional vulnerability can be described by the factor of water-withdrawal.* When infiltration-dominated wells are in non-regional recharge areas, *local vulnerability can be characterized by the factor of infiltration.* Both statements concern the whole aquifer as the subject of vulnerability assessment.

As it could be read above, we have revealed a connection between vulnerability and factors by dynamic factor analysis, approaching both from basic geological parameters used in vulnerability assessments and from groundwater flow regimes in different scales. So a new, exact description of vulnerability of an aquifer can arise from this mathematical method.

4. A case study in an alluvial environment

The next case study concerns shallow-groundwater in alluvial environment, where the influence of surface waters is negligible. The study presents the results of dynamic factor analysis on the shallow-groundwater level time series of monitoring wells located in the Nyírség and Hajdúság area, which is a somewhat higher elevated plain in the northeastern part of the Great Hungarian Plain. The studied area lays between and around two major cities of Hungary, Debrecen and Nyíregyháza (with a population of 200,000 and 120,000 people), and it is surrounded by other plains of lower altitude. The hydrology of the studied area is quite complex. Most of the higher elevated Nyírség is a recharge area, while in the surrounding lower elevated areas features of discharge indicated for example by the saline lakes and soil.

The Nyírség and its vicinity was part of the Pannonian Sea that separated from the Central-Paratethys about 5.4 million years ago, and was slowly filled by clastic sediments. Later, due to the Pliocene erosion and tectonic events during Pliocene/Pleistocene considerable parts of the area lost sedimentation. As a result of vertical plate movements some basin areas, including the studied one had subsided 150-700 meters while the surrounding areas ascended forming the Carpathian Mountains. During Pleistocene, those basins were filled by coarse sediments (gravel, coarse sand) from northeast and northwest (Urbancsek, 1965).

The layers of alluvial sediments wedge out suddenly, they have lenticular appearance, and toward south both frequency and thickness of their occurrence increases.

During middle Pleistocene, tectonical activity lessened, so the amount of sediment and its grain size were decreasing, clay intercalations became more frequent. During late Pleistocene, tectonical activity intensified again, and average grain size of the alluvium increased. During Holocene, the typical sediment of the study area was sand.

The Pliocene sediment sequence contains thermal water with high salt and gas contents, so it is not suitable for drinking water. The lower Pleistocene sequence is the most favourable drinking water aquifer from both qualitative and quantitative point of view; it contains water suitable for communal use.

Middle Pleistocene sediments are poorer aquifers, though the drinking water supply of the largest city of the area Debrecen is supplied partially from the sand intercalations of this sequence. The hydraulic conductivity of the upper Pleistocene sediments is lower than that of lower Pleistocene. Shallow aquifers should also be mentioned though those are not thick enough for serving considerable demands (Fig. 5, Virág et al., 2005).

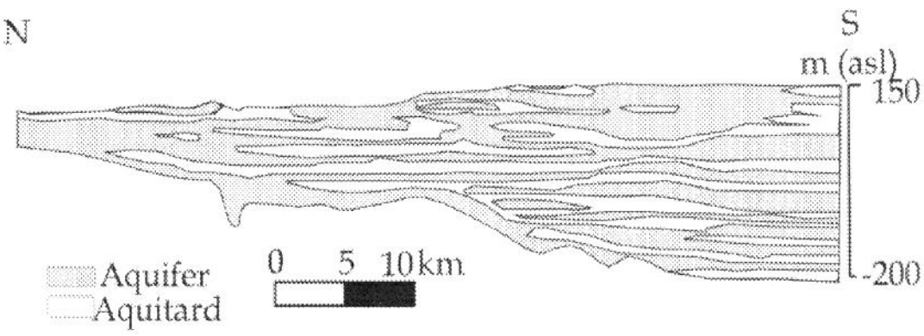

Fig. 5. Geological cross-section of Pleistocene sequence in the studied area (based onVirág et al., 2005)

Strata of the studied area suddenly wedge out. These have lenticular appearance. As a hydrogeological consequence there can be significant communication between the aquifers that is very complex in both levels and directions (Marton & Szanyi, 1997a, 1997b). Regional and local flow systems of the area were explored in earlier studies. Water age measurements data proved that the Nyírség region is a recharge area and the recharge intensity may reach 8 to 16 mm/year. On the other hand the margins of the area are the discharge zones of the flow system (Fig. 6 & 7) (Deák, 1979; Marton, 1981; Deák et al., 1987; Marton & Szanyi, 2000; Székely, 2003, 2006).

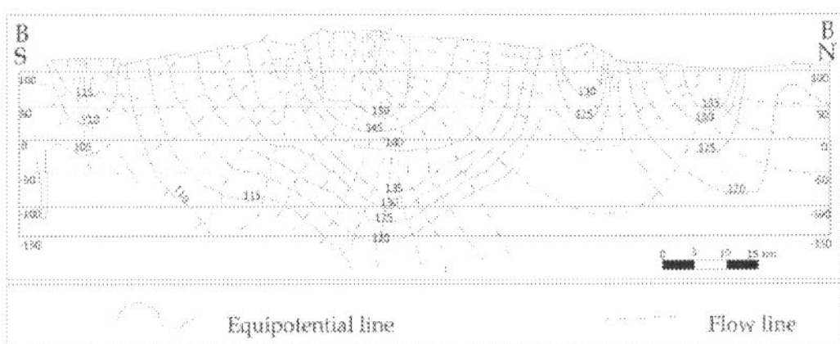

Fig. 6. Hydrological cross section (north to south) of Nyírség (Marton, 1981)

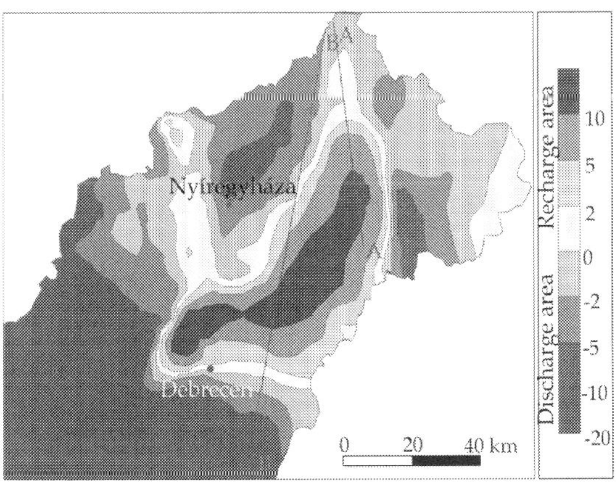

Fig. 7. Calculated piezometric head difference (m) between the layers of upper and lower Pleistocene, and the location of cross sections based on Székely, 2003

In the vicinity of larger cities, significant depressions are formed due to the withdrawal of subsurface water. Piezometric levels of the lower Pleistocene main aquifer declined 30 meters at Debrecen and 7 meters at Nyíregyháza. In the surroundings of Debrecen the compaction due to water withdrawal resulted in a 62 cm descent of the surface (Szanyi, 2004). According to other studies, descent of the shallow groundwater level can reach 1 meter at certain locations, e.g. at Debrecen. At other, smaller settlements (with a population of 20,000-30,000 people) the depressions are shallower, but still of substantial extent. A future equilibrium state expected to commence at a 4% decrease in the groundwater supply. The decreasing water table causes decreasing evapotranspiration and increasing infiltration from surface water flow (Table 2) (Székely, 2003, 2006).

Replenishment	Percentage
From small surface flows	30.75
From rivers	5.87
Decrease of Evapotranspiration	59.57
Decline of stored supply	3.81

Table 2. Replenishment forms related to the 209 m^3/day of production plus the 146,000 m^3/day marginal outflow of 2020 (after Székely, 2003)

Based on hydrogeological models the piezometric water levels predicted to rise due to decline of water withdrawal in the early 1990's. However, due to the increasing number of individual, private and often illegal well drilling and the connected water withdrawal, shallow groundwater level may decline even further (Székely, 2003, 2006).

Hydrographs of 215 monitoring wells were at our disposal for the analysis. The period of observation varied in time. Observation of some well started in 1933 and lasts up to today. The measurement frequencies were variable as well, changing from monthly, to daily. To correct for this situation and make the results comparable with other explanatory data we had to consider only the annual mean water level as our input data.

In the period of 1986-2000 122 hydrographs proved to be sufficiently homogeneous in order to apply dynamic factor analysis to them. Spatial distribution of the wells was uneven, the north part of the study area had less wells.

To be able to identify the factor time series it is necessary to have the data of the water balance. These are the time series of water withdrawal, precipitation and temperature. Though water withdrawal data is not registered with sufficient accuracy for the whole area, fortunately the communal withdrawal for the largest city at Debrecen Water Works has proper representative power (Fig. 8). So, we used it for the comparison. The necessary precipitation data were provided by 20 gauge stations and 4 other stations served us with daily temperature data.

So, we were in the position to apply dynamic factor analysis and analyse its results. The estimated factor time series were correlated to the mentioned elements of water balance, such as precipitation, evapotranspiration and withdrawal.

The first factor time series identifies the most influential background effect. The estimation of reference evapotranspiration (ET) was computed by the Hargreaves method, and then the actual evapotranspiration was calculated. The estimation is based on the daily maximum and minimum temperature and the solar radiation values (Allen et al., 1998). This method is preferred by a number of authors (Diodato & Geccarelli, 2006; Magliulo, 2003) and is widely regarded as a reliable one. The correlation coefficient between time series of the first factor and evapotranspiration is 0.73, which indicates that the most important effect is related to the evapotranspiration. To obtain a clearer relationship it is worth considering the subprocesses of the evapotranspiration as well.

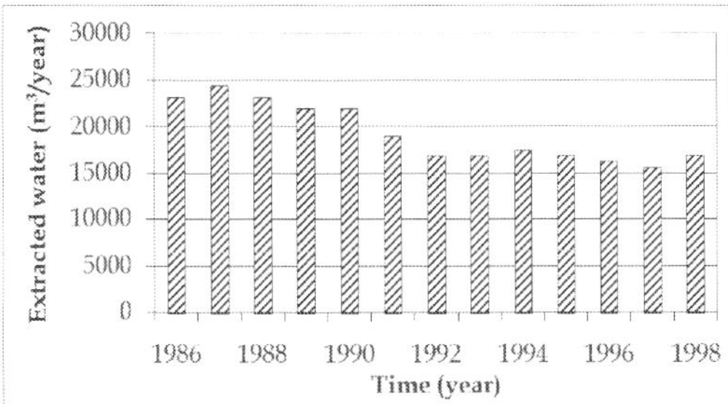

Fig. 8. Water consumption (in thousand m^3) from confined aquifers at Debrecen between 1951 and 1998 (Marton & Szanyi, 2000)

We employed the soil water characteristic curve to determine the evaporation of the shallow-groundwater (Kovács, 1981). The evaporation was computed for the 122 observation wells, and then the mean values were used to describe the study area. Fig. 9 displays the time series of evaporation and of factor one. The 0.96 correlation between the two time series indicates a very strong relation. Based on the results the first factor is identified as the evaporation of shallow-groundwater.

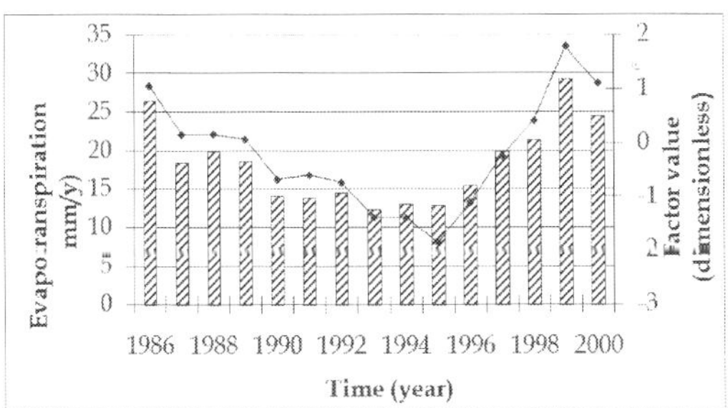

Fig. 9. The evaporation (columns) and the first factor (line) between 1986 and 2000

The other important element of the water-balance is the infiltrated part of precipitation. We compared this time series to the second factor. Because the studied area's climatic conditions, the majority of winter precipitation reaches the water table, while during vegetation it is absorbed by plants. The start and end of the vegetation period could shift by one or two months. On the long run, however, the sum of precipitation between December and March represents best the yearly infiltration. The water supply infiltrating to the shallow groundwater was calculated from the precipitation data of 20 gauge stations. It is more useful to consider the relation of each observation year to the long time average. For this purpose we created the time series of deviation from that average. The created deviation time series has 0.79 correlation with the second factor time series. Fig. 10 also underpins the strong relationship.

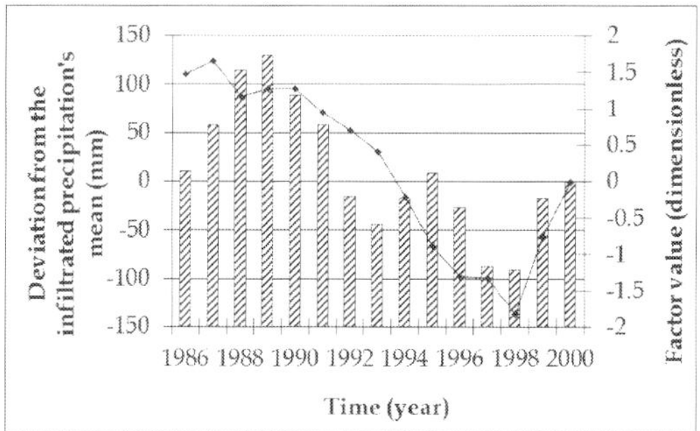

Fig. 10. Deviation from the mean of infiltrated precipitation (columns) and the second factor (line) between 1986 and 2000

Aside of infiltration, another downward moving effect exsits in water balance; it is the leakage effect, due to water withdrawal from deeper (100-150 m deep) layers. Several authors (Juhász, 1987; Marton & Szanyi, 2000) draw attention to its significance. For this reason we estimated the strength of the relationship between the shallow-groundwater level and the communal water withdrawal. Since leakage is a long-term process, the influence of the withdrawal will appear in water table with some delay; hence an appropriate time lag has to be taken into account. So, the relationship should be represented by cross-correlation. Table 3 shows the obtained cross-correlations. The dependence between the shallow-groundwater level and the withdrawal increases up to two-years lag, then it starts to decrease. We found that the second factor has strong relation to water withdrawal that reaches its maximum in a two-year lag at 0.89 correlation.

Time lag (year)	0	1	2	3	4	5
Groundwater level	0.63	0.68	0.69	0.58	0.39	0.35
2nd factor	0.83	0.88	0.89	0.85	0.76	0.60

Table 3. Cross correlation coefficients of water withdrawal with lagged shallow-groundwater level (first row) and the 2nd factor (second row)

As it was mentioned earlier, not only the factors themselves but also their loadings contain valuable information as they represent the intensity of a given effect in creating the actually observed time series, i.e. the hydrographs. The DFA-provided values of factor loadings belong to a fixed monitoring location. As we wish to extend the analysis to locations where no observation is available, the factor loadings should be spatially predicted by standard geostatistical methods, such as kriging etc. To this end the empirical semivariograms has to be computed first from the values of the factor loadings given at each observation points. In case of first factor the obtained semivariogram has two thresholds, and the corresponding ranges are 10 and 25 km-s. This allows us to conclude that the loadings possess a genuine spatial structure, that has a finer and a coarser scale, meaning that the effects appear at two different scale. Evaporation depends on the depth of water table, and is affected by two impacts. One of them is the local decline of the water level caused by water withdrawal from the deeper strata: this may correspond to the finer scale. The other impact may be related to topography, since the surface of shallow-groundwater follows the surface morphology. Lower values on the map of factor loadings appear in the vicinity of larger cities where the water level declined, indeed, by the extensive water withdrawal of the cities, thus decreasing evaporation (Fig. 11). Higher values mark the recharge areas and surroundings of rivers.

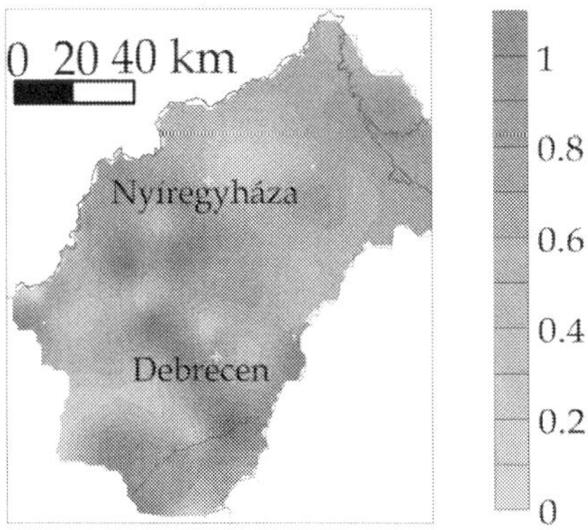

Fig. 11. Map of intensities (loadings) for the first factor

The loadings of the second factor (Fig. 12) were analysed in a completely similar way. In order to characterise the spatial structure of the loadings the empirical semivariogram was computed and fitted with a theoretical one. This time, too, we got a two-threshold semivariogram and the corresponding ranges were 12 and 35 kilometers. In our opinion the appearance of two ranges is the consequence of the existence of two processes that create the second factor.

The increase of the communication of water among strata is a local impact, hence creates a smaller range, while the infiltration from precipitation is influenced by the variability of the uppermost layers near to the surface, and that is reflected in the appearance of the larger range. The smaller range coincides with the one obtained for the first factor, and that is quite understandable in view of the generating process, the local decline of shallow-groundwater level.

The monitoring sites are also marked on the map of the loadings of the second factor, and notably their distribution is quite uneven. In subareas where observation locations are dense we could get an accurate information on the intensity of downward migration of water. On elevated ridges, that can be found in recharge areas, factor loadings are larger; here the effect of infiltrating precipitation is stronger hence more detectable in the hydrograph. The map displays higher loadings in the northeastern edge of the study area, however this does not reflect the reality; it is caused by an interpolation error that is the consequence of the lack of observation sites.

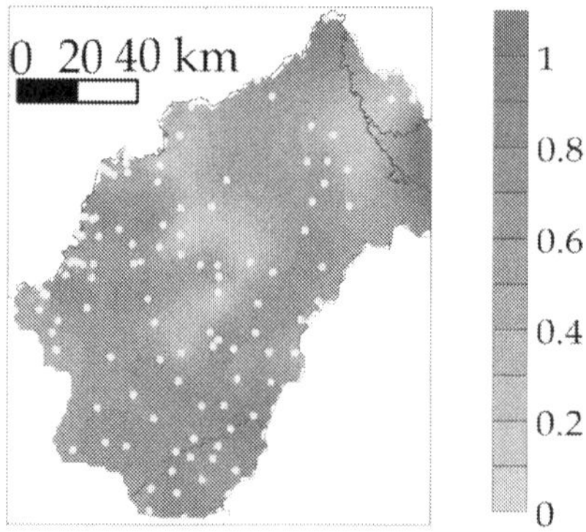

Fig. 12. Map of intensities (loadings) for the second factor (Yellow points indicate the monitoring wells.)

5. Study of a river influenced area

Our third area of case studies located on the southern part of the West Pannonian Basin, so called Kisalföld (Small Hungarian Lowland) at the NW state border of Hungary (Fig. 13).

The West Pannonian Basin is the second largest alluvial plain of the Carpathian-basin. It is located among the Transdanubian Central Range, the Alps and the Western Carpathians forming an irregular square depression (Bulla, 1962). Its basin was formed after periods of sedimentary deposition and depression surrounded by transversal and edge faults. The basin bottom at Northwest is formed from Variscian altered rock on top of which Miocene and Pannon-sea sediment was deposited. In the middle and Southeast regions the bottom is covered by thick Mesozoic limestone and dolomite beds on top of which Neogene marine sediment was deposited. After the Pannonian inland sea recessed the area was first occupied by the Pannonian Lake and in the Upper Pliocene fluviocarstic lake system (Bulla, 1962).

After the disappearance of the lake system in the beginning of the Pleistocene, the Danube and other rivers arriving from the Western-Carpathians passed through the area in a southern direction. In the Lower Pleistocene as a result of tectonics the area subsided, on which the Danube was able to form an extensive alluvial cone (Pécsi, 1959). At the Southwest part of the West Pannonian Basin the Rába and the smaller tributaries (arriving from the Alps) created the alluvial cone sequence. In the second part of the Pleistocene at the middle of the basin another subsidence was observable resulting in an alluvial cone with a diameter of approximately 200 km (Fig. 13) formed by the Danube, arriving from North-west - across the Gate of Devin (Porta Hungarica) - at that time. This subsidence (which was here and there still observable in the Holocene) defined the actual hydrographical picture of the Danube and its branches (Mosoni-Duna, Maly Dunaj) and the tributaries. The thickness of the Pleistocene and Holocene sediments together is 200-250 m (Don et al., 1993; Góczán, 1999; Neppel et al., 1999).

The top capping stratum of the alluvial cone was formed from the deposition of the fine grained, muddy river load in the flood plain originating from the Danube's floods and the branches. The fine grained Holocene top capping is 0-5 m thick and shows diverse regional distribution. In the valleys of the Danube's tributaries it consists of Holocene alluvial sediment and in case of the Hanság region it consists of the lacustrine and wetland sediments. The NW part of the Small Hungarian Lowland (Kisalföld) is covered by coherent Pleistocene sand and pebbles that reaches the Rába in South-east direction in a narrow line. In the middle regions spots with alluvial and eolic sand make the surface more diverse.

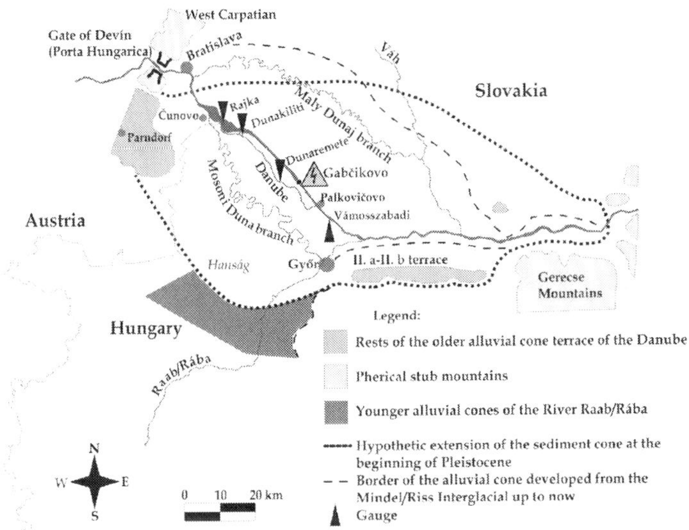

Fig. 13. Older and younger alluvial cones of the Danube on the West Pannonian Basin (based on Pécsi, 1959), with the names of the main geographical units occurring in the text

The West Pannonian Basin is not only a geographic but a hydrogeological unit as well because of the characteristics of the alluvial cones. From south the sandy-pebbly sediments of Rába valley connect to the alluvial cone of the Danube in Hungary (Erdélyi, 1979; Góczán, 1999).

Central Europe's largest potable water supply is located at the Kisalföld in its Pleistocene porous aquifer alluvial cone. The water coming from the Danube and its branches influenced the shallow groundwater level of the alluvial cone. Based on the data obtained from the shallow groundwater monitoring wells set up in 1950's and later, it was possible to determine the main direction of the groundwater migration. The direction of the flow was confirmed by isotope-hydrological measurements and its speed became measurable as well (Szalai, 2008). Further, it was also possible to determine those specific areas where the major part of the shallow groundwater supply is provided by the Danube (Erdélyi, 1979; Major, 1976). This hydrological situation has changed in the early 1990's.

In October 1992, on the Slovakian territory at Čunovo a dam was constructed at the 1851+750 river km, therefore natural fluctuation of the Danube has changed. Since then majority of the Danube flow has been diverted to the insulated power-plant channel that rejoins to the original riverbed only at the 1811 river km, at Palkovičovo. Therefore, the original 2000 m3*s-1 mean discharge dropped drastically in the natural river network. The Mosoni-Duna branch had only a flow rate of 10-20 $m^3 s^{-1}$, while the original Danube bed had only a flow rate of 250-350 $m^3 s^{-1}$. Consequently, water level in the riverbeds dropped several meters, and by 1993, some of the Danube's branch rivers had dried out. As the subsurface aquifer of the area is in a close connection with the surface water network, the level of shallow-groundwater also dropped significantly. Since 1995, to improve the situation, Slovakia has increased the annual mean water flow to the original Danube bed up to 400 $m^3 s^{-1}$. Furthermore, Hungary has built a submerged weir at Dunakiliti with a spillover at 123 m a.s.l., so water level increased in the original bed backward 10 km from the weir. The uplifted water level supports the artificial recharge of shallow groundwater from the rehabilitated riverbed and across the built up channel-system. The water supply to Mosoni-Duna branch has been increased to 40 $m^3 s^{-1}$ that has resulted in a slight growth of discharge.

In our approach the observed processes at different locations are supposed to be governed by the same essential impacts, such as recharge from precipitation and rivers and water withdrawal, while intensity of these impacts depends on locality, creating the main aspect of the spatial dependence.

Our goal was to identify these impacts, and to determine their spatially dependent intensities. In order to do this, dynamic factor analysis has been applied to the data, registered from the Danube's diversion (1992) to 2009. A total of 93 shallow groundwater monitoring wells have been sampled. Annual averages were formed from the data of each well. For the purpose of identifying factor-time series precipitation and Danube water level data were used as well. During the analysis three factor-time series have been estimated by DFA.

The standardised first factor-time series corresponded to the combined deviance of precipitation during the period of January-April. These months are known to be the periods that determine the major part of the supply for the shallow ground water. The two data series showed strong correlation (r=0.79) (Fig. 14).

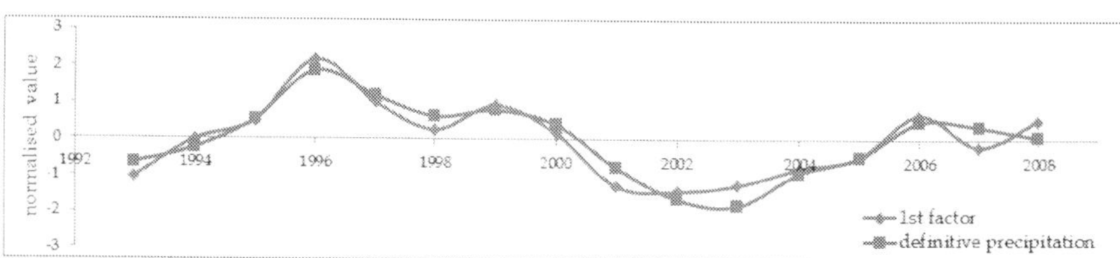

Fig. 14. Graphs of factor 1 and definitive precipitation

The isoline map prepared from the factor loadings contain information regarding the environmental impacts (Fig.15). Along the Danube the predicted factor loading values are small, because the dominant factors are the water levels of the river and the natural and manmade water supplementing channels. Higher values are observable North-west near to the state border. This particular area is covered by Pleistocene alluvial sand and pebbles, and their greater permeability increases infiltration.

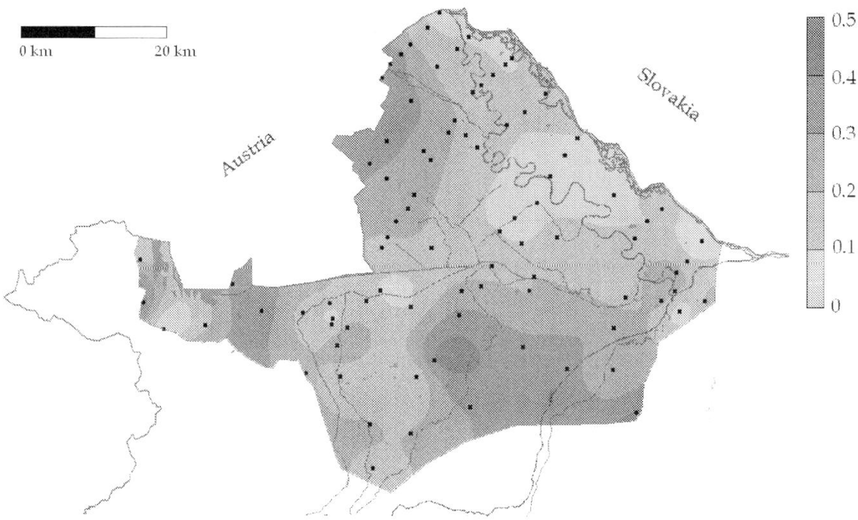

Fig. 15. Map of intensities (loadings) for the first factor

Second factor shows strong correlation (r=0.89) with the mean of registered water levels of Danube at two gauges, Dunaremete and Rajka (Fig. 16). Because of their location both water gauges are indirect indicators of river diversion and water level regulation. In the diagram the effect of the underwater weir installed in 1995 is remarkable. With the exception of floods this structure fixes the level of the upper water at a nearly constant value which is approximately 200-250 cm higher than what was measured immediately after the diversion. The increased water level with the recharging facilities make the gravitational recharge of underground water possible.

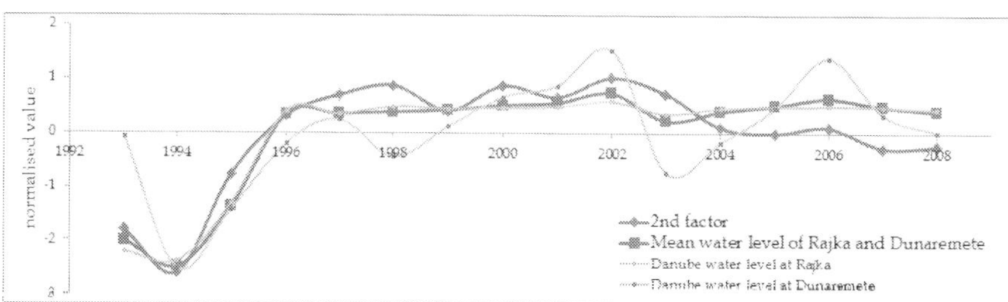

Fig. 16. Rescaled graphs of the factor 2 and water levels of Danube

Based on the map prepared from the loadings of the second factor (Fig. 17) the areas increasingly vulnerable to the diversion of Danube can be outlined. The highest values i.e. highest vulnerabilities were observed in the NW part of the area. In addition to manmade influence the reason may lay in the geological setting as well. The high factor loadings S and SE from this area are suspected to be indicating the effect of the gravitational recharge and the increased runoff of the Mosoni-Duna.

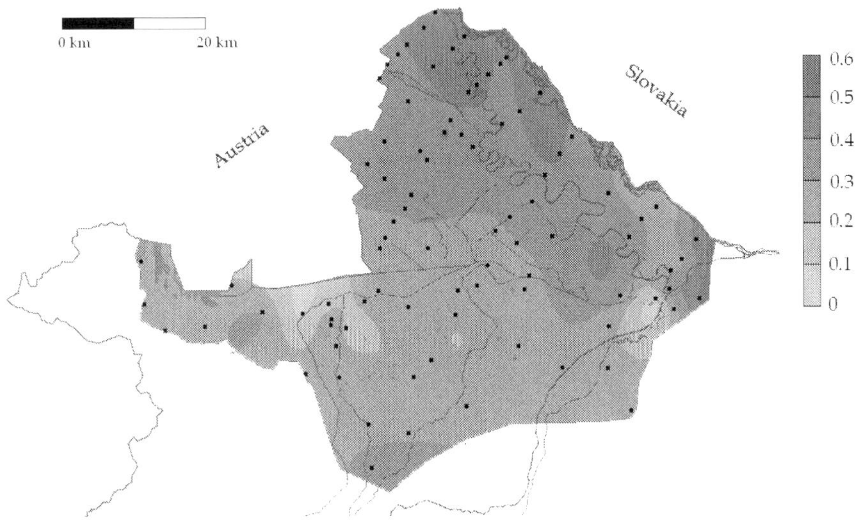

Fig. 17. Map of intensities (loadings) for the second factor

The third factor shows a relationship with Danube level values measured at Vámosszabadi water gauge (r=0.81) (Fig. 18). This gauge is located under the downstream channel of the hydroelectric power plant. By the installation of the plant hydraulic relations have radically changed. In addition to natural ones, considerable daily water level fluctuations appear according to the operation of the power plant. Pressure waves caused by water level fluctuations may be beneficial from the point of view of groundwater recharge, but their expansion is rather restrained by the fine-grained top capping stratum which is rather thick here (Fig. 19).

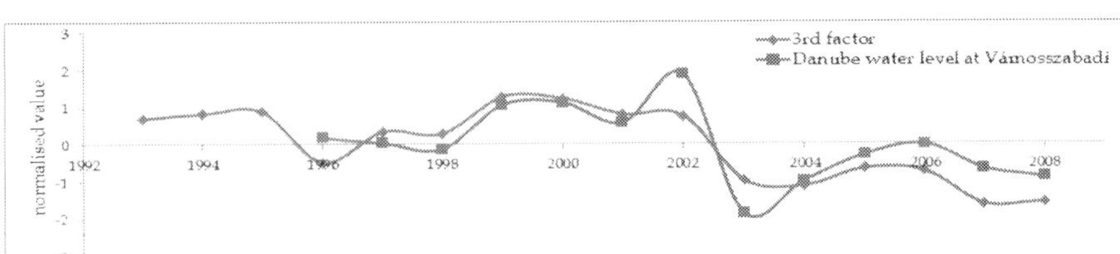

Fig. 18. Comparison of the third factor with the water level of the Danube at Vámosszabadi

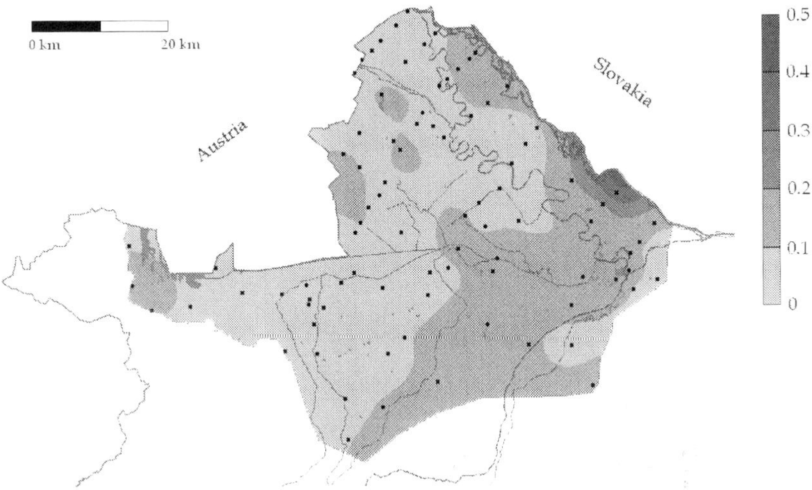

Fig. 19. Map of intensities (loadings) for the third factor

6. Summary

In the process of complex evaluation of water resources the quantitative and qualitative parameters cannot be considered separately. These parameters are influenced by several background effects that are variable in both time and space, hence it is dynamic factor analysis that makes their estimation mathematically possible. The identification of these effects and their intensity provide information on the area characteristics from either geological, hydrogeological or environmental point of view. The subject of the application of our proposed method is a well observable, measurable, physical parameter, the temporal fluctuation of water level at several monitoring sites of the area. Dynamic factor analysis decomposes the time series of water levels into the combination of factors that are then identified as the various occurrences of the precipitation, water withdrawal, and river effects. These effects primarily govern the water quantitatively but may also have an impact on the qualitative characteristics, therefore they might be informative and important markers on the water quality. That is why the factor loadings represent a quantitative measure of vulnerability of the studied areas and the corresponding maps constitute the main results of our works. As an extension of the present study, a joint analysis of the quantitative and qualitative characteristics can be considered in the sequel.

7. Acknowledgement

We the authors would like to thank all of our collegues from the Department of Phisycal and Applied Geology at the Eötvös Loránd University and for the help of Edit Borbás and Péter Tanos, and last but not least István Gábor Hatvani for reviewing our English version.

The European Union and the European Social Fund have provided financial support to the research under the grant agreement no. TÁMOP 4.2.1./B-09/KMR-2010-0003 and TÁMOP-4.2.1/B-091/1/KONV-2010-0005.

8. References

Alföldi, L. (1994). Ideas for the groundwater contamination's sensitivity questions (in Hungarian) *Hidrológiai Közlöny*, Vol.74, No.1, pp. 15-22, ISSN 0018-1823

R. G., Pereira, L. S.; Raes, D. & Smith, M., (1998). *Crop evapotranspiration: Guidelines for computing crop requirements.* Irrigation and Drainage Paper, No.56, FAO, Rome, Italy, pp. 300, ISBN 92-5-104219-5

Aller, L.; Bennet, T.; Lehr, J. H. & Petty, R. J. (1985). DRASTIC: A standardized system for evaluating groundwater pollution potential using hydrologic settings, U.S. EPA, Sobert S. Kerr Environmental Research Laboratory, Ada, OK, EPA/600/2-85/0108, p. 163, In: *A short overview on groundwater vulnerability assessment – basic statements for use in the framework of the COST 620 Action*, Gogu, R. & Dassargues, A. (1998).

Anderson, R. L. (1963). The use of factor analysis in the statistical analysis of multiple time series. *Psychometrika*, Vol.28, pp. 1-25, ISSN 0033-3123

Bai, J. (2003). Inference for factor models of large dimensions. *Econometrica 71*, pp. 135–172, ISSN 0012-9682

Bánkövi, Gy.; Veliczky, J. & Ziermann, M. (1979). Dynamic models for prediction of the development of national economies, In: *Models and Decision Making in National Economies*, Janssen, J. M. L.; Pau, L. F. & Straszak, A., (Eds.), pp. 257-267, ISBN 0-444-85280-8, Amsterdam, North-Holland

Böcker, T. (1974). *Determining infiltration in a karstic area using seasonal beoundary precipitaion method (in Hungarian)* VITUKI beszámoló, pp. 207-216, Budapest, Hungary

Bolla, M.; Michaletzky, Gy.; Tusnády, G. & Ziermann, M. (1998). Extremes of sum of heterogeneous quadratic forms. *Linear Algebra and Its Applications*, Vol.269, pp. 331-365, ISSN 0024-3795

Box, G. E. P. & Tiao, G. C. (1977). A canonical analysis of multiple time series. *Biometrika*, Vol.64, pp. 355-405, ISSN 0006-3444

Bulla, B. (1962). *Geography of Hungary* (in Hungarian), Tankönyvkiadó, Budapest, Hungary

Burman, R. & Pochop, L. O. (1994). Evaporation, Evapotranspiration and Climatic Data. *Developments in Atmospheric Science*, 22, Elsevier, p. 278, ISSN 0167-5117, Amsterdam, Holland

Civita, M. C. & De Regibus (1995). Determining aquifer vulnerability using different methods (in Italian: Sperimentazione de alcune metodologie per la valutazione della vulnerabilita degli aquiferi.) *Quaderni de Geologia Applicata*, 3, pp. 311-324, ISSN 1593-8433, Bologna, Italy

Connor, G. & Korajczyk, R. A. (1993). A test for the number of factors in an approximate factor model. *Journal of Finance* 48 (4), 1263–91, ISSN 0022-1082

COST Action 65 (1995). Hydrogeological aspects of groundwater protection in karstic areas, Final Report, ISBN 92-827-4682-8, Luxembourg

Csepregi, A. (1995). Investigation of groundwater recharge on the Hungarian Plain by precipitation and areal evapotranspiration data (in Hungarian), Technical Report of the Hungarian Water Researches Institute

Daly, D. & Drew, D. (1998). Irish methodologies for karst aquifer protection, Paper submitted for conference "Seventh Multidisciplinary Conference on Sinkholes and the Engineering and Environmental Impacts of Karst", Pennsylvania, USA

Daly, D.; Dassargues, A.; Drew, D.; Dunne, S.; Goldscheider, N.; Neale, S.; Popescu, Ch. & Zwahlen, F. (2001). Main concepts of the "European Approach" for (karst) groundwater vulnerability assessment and mapping, Draft for Working Group 1 of the COST 620 Action "Vulnerability and risk mapping for the protection of carbonate (karst) aquifers", p. 12

Deák, J. (1979). Environmental isotopes and water chemical studies for groundwater research in Hungary. *Isotope Hydrology 1978*, Vol.1, International Atomic Agency, Vienna, pp. 221-249, ISBN 92-0-040079-5

Deák, J.; Stute, M.; Rudolph, J. & Sonntag, C. (1987). Determination of the flow regime of quaternery and pliocene layers in Great Hungarian Plain (Hungary) by D, 18O, 14C and noble gas measurements. *Isotope Techniques in Water Research Development*, International Atomic Agency, Vienna, ISBN 92-0-040087-6.

Deistler, M. & Scherrer, W. (1991). Identification of linear systems from noisy data, In: *Modeling, Estimation and Control of Systems with Uncertainty* (Sopron, 1990) Progr. Systems Control Theory, 10. Birkhauser, Boston, pp. 91-112, ISBN 0817635807

Diodato, N. & Geccarelli, M. (2006). Computational uncertainty analysis of groundwater recharge in catchment. *Ecological Informatics*. Vol.1, pp. 377-389, ISSN 1574-9541

Doerfliger, N. (1996). Advances in karst groundwater protectin strategy using artificial tracer tests analysis and multiattribute vulnerability mapping (EPIK method) - Thesis presented to the Faculty of Sciences of the Neuchatel University for the obtention of the title Docteur ès sciences, University of Neuchatel (Switzerland), Faculty of Sciences, p. 308

Don, Gy.; Kaiser, M.; Marsi, I.; Scharek, P.; Síkhegyi, F.; Szeiler, R.; Tullner, T. & Zsámbok, I. (1993). Geological Setting of the Szigetköz - p. 63 +14 annexs. MÁFI Library, Budapest, Hungary (manuscript)

Doz, C.D. Giannone & Reichlin, L. (2006). *A quasi maximum likelihood approach for large approximate dynamic factor models*. European Central Bank, p. 37, ISSN 1725-2806, Frankfurt am Main, Germany

Dracos, Th. (1980). Hydrology, introduction for engineers (in German), Springer-Verlag, p. 194, ISBN 978-3540632924

Dryden, I.; Markus, L.; Taylor, C.; Kovacs, J. (2005). Non-Stationary spatio-temporal analysis of karst water levels, *Journal of the Royal Statistical Society, Series C Applied Statistics*, Vol. 54., No.3., pp. 673-690, ISSN 1467-9876

Engle, R. F. & Watson, M. W. (1981). A one-factor multivariate time series model of metropolitan wage rates. *Journal of the American Statistical Association* 76, pp. 774–81, ISSN 0162-1459

Erdélyi, M. (1979). Hydrogeology and Hydrodinamics of the West Pannonian Basin (In Hungarian). *Hidrológiai Közlöny*, Vol. 59, No.7, pp. 290-301, ISSN 0018-1823

Forni, M.; Hallin, M.; Lippi, M. & Reichlin, L. (2000). The generalized dynamic factor model: Identification and estimation. *The Review of Economics and Statistics* 82, pp. 540–54, ISSN 1530-9142

Geweke, J. F. (1977). The Dynamic factor analysis of economic time series models In: Latent Variables in Socio-economic Models, Aigner, D.J. & Goldberger, A.S., (Eds.), pp. 365-382, Amsterdam, North-Holland

Góczán, L. (1999). The Little Hungarian Plain (in Hungarian), In: *Pannon Encyclopaedia Magyarország földje* (Pannon Encyclopaedia The land that is Hungary - geology, geography and cartography of Hungary and its surroundings) Karátson, D., 302-305, Kertek 2000 Publishing House, ISBN 963-85792-3-4, Budapest, Hungary

Goldscheider, N.; Klute, M.; Sturm, S. & Hötzl, H. (2000). The PI method - a GIS-based approach to mapping groundwater vulnerability with special consideratin of karst aquifers, Z.angew. Geol., Vol.46, No3., pp. 157-166

Gourieroux, C. & Monfort, A. (1997). Time Series and Dynamic Models. *Cambridge University Press*, p. 665

Gourieroux, C.; Monfort, A. & Renault, E. (1995). Inference in Factor Models, In: *Advances in Econometrics and Quantitative Economics*, Maddala, G.; Phillips, P.C.B. & Srinivasan, T., (Eds.) pp. 311-353, ISBN 1-55786-382-2, Blackwell

Jeannin, P. Y. & Grasso, D. (1995). Estimation des infiltrations efficaces journali\`eres sur le bassin karstique de la Milandrine, (Ajoie, JU, Suisse). *Bulletin d'Hydrogéologie*, No.14., pp. 83-93, ISSN 1421-4946

Juhász, J. (1987). *Hydrogeology* (in Hungarian). Akadémiai Kiadó, pp. 466-548, 871, ISBN 963 05 4381 8, Budapest, Hungary

Jungbacker, B. & Koopman, S. J. (2008). Likelihood-based Analysis for Dynamic Factor Models, Tinbergen Institute Discussion Paper TI 2008-0007/4

Kessler, H. (1954). The percentage of infiltration and the permanently extractable water in karst-terrains (in Hungarian) *Vízügyi Közlemények (Journal of Hydraulic Engineering)*, 36, ISSN 0733-9429

Kovács, Gy. (1981). *Subterranean Hydrology*. Water Resources Publications, pp. 440-460. ISBN 978-0918334350 Colorado, Littleton, USA

Kovács, J.; Mákus, L.; Halupka, G. (2002) Measuring contamination-vulnerability of aquifers by dynamic factor analysis, 8th Annual Conference of the International Association for Mathematical Geology, *Terra Nostra*, Schriften der Alfred-Wegener-Stiftung, ISSN 0946-8978 Volume 2, pp.: 463 - 468 Berlin, Germany

Kovács, J. (2007). The application of modern geomathematical methods in practical hydrogeology (in hungarian, with english abstract), Ph. D. dissertation, p. 122 (manuscript)

Mádlné, Szőnyi, J. & Füle, L. (1998). Groundwater vulnerability assessment of the SW Trans-Danubian Central Range, Hungary. *Environmental Geology*, 35, pp. 9-18, ISSN 0943-0105

Mádlné, Szőnyi, J. (1997). Vunlerability analysis of aquifers on the example of the Transdanubian Central Range's karst aquifer (in Hungarian) *Földtani Közlöny*, Vol.127, No.1-2, pp. 19-83, ISSN 0015-542X

Magliulo, V.; d'Andria, R. & Rana, G. (2003). Use of the modified atmometer to estimate reference evapotranspiration in Mediterranean environments. *Agricultural Water Management*, Vol.63, No.1, (November 2003), pp. 1-14, ISSN 0378-3774

Major, P. (1976). Effects of the Dunakiliti hydropower station on the subsurface water body in Szigetköz, (in Hungarian) Vízgazdálkodási Tudományos Kutató Intézet (VITUKI report), VITUKI, Budapest

Mardia, K. V.; Goodall, C. R.; Redfern, E. J. &Alonso, F. J. (1998). The kriged Kalman filter. *Test*. Vol.7, pp. 217-285

Márkus, L. & Kovács J. (2003). On the Effects of Water Extraction of Mines in Karstic Areas. *Mathematische Geologie,* Vol.7, ISSN 1431-8016

Márkus, L.; Berke, O.; Kovács, J. & Urfer, W. (1999). Spatial Prediction of the Intensity of Latent Effects Governing Hydrogeological Phenomena. *Environmetrics,* Vol.10, pp. 633-654, ISSN 1180-4009

Marton, L. & Szanyi, J. (1997a). Geostatistical investigation of pleistocene sediments in Eastern-Hungary: 1. Mapping of the transmissivity (in Hungarian), *Journal of the Hungarian Hydrological Society,* Vol.77, No.5, pp. 233-240, ISSN 0018-1823

Marton, L. & Szanyi, J. (1997b). Geostatistical investigation of pleistocene sediments in Eastern-Hungary: 2. Mapping of the leakage (in Hungarian), *Journal of the Hungarian Hydrological Society,* Vol.77, No.5, pp. 241-248, ISSN 0018-1823

Marton, L. & Szanyi, J. (2000). Connection between the level of the shallow groundwater and the communal water extraction (in Hungarian), *Journal of the Hungarian Hydrological Society* Vol.80, No.1, pp. 3-13, ISSN 0018-1823

Marton, L. (1981). *Investigation of the groundwater in Nyírség using environmental isotopes (in Hungarian),* C. Sc. Thesis, Debrecen, p. 133.

Maucha, L. (1990). Calculation of karstic infiltration (in Hungarian). *Hidrológiai Közlöny (Journal of the Hungarian Hydrological Society),* Vol.70, pp. 153-161, ISSN 0018-1823

Morton, F. I. (1983). Operational estimates of areal evapotranspiration and their significance to the science and practice. *Journal of Hydrology,* Vol.66, pp. 1-76, ISSN 0022-1694

Munoz-Carpena, R.; Ritter, A.; Li, Y.C. (2005). Dynamic factor analysis of groundwater quality trends in an agricultural area adjacent to Everglades National Park, *Journal of Contaminant Hydrology,* Vol. 80., No. 1-2, pp. 49-70, ISSN 0169-7722

Neppel, F.; Somogyi, S.; Domokos, M. and the experts of the cooperating Danube Countries (1999). Paleogeography of the Danube and its cathment. A hydrographic monograph. Follow-up volume V. The riverbed conditions of the Danube. Part 2, Regional cooperation of the Danube Countries in the Framework of the International Hydrological Programme of UNESCO, Published by VITUKI, ISBN 963 511 128 2

Pécsi, M. (1959). A magyarországi Duna-völgy kialakulása és felszínalaktana (Genesis and surface morphology of the Hungarian Danube Valley, (in Hungarian). *Földrajzi Monográfiák,* Vol.3, Akadémiai Kiadó, Budapest, Hungary.

Picci, G. & Pinzoni, S. (1986/1). A new class of dynamic models for stationary time series, In: *Time Series and Linear Systems,* Bittanti, S.,(Ed.) Lecture Notes in Control and Information Sciences, Vol.86, pp. 69-114, ISBN 3-540-16903-2

Picci, G. & Pinzoni, S. (1986/2). Factor analysis models for stationary stochastic processes, In: *Analysis and Optimization of Systems (Antibes, 1986)* Lecture Notes in Control and Information Sciences Vol.83, pp. 411-424

Primault, B. (1963). Du calcul de l'evapotranspiration. *Arch. Meteorologie, Geophysik, Biologie,* Vol.12, pp. 124-150, ISSN 00039861, Zürich

Rapcsák, T. (2002). On minimization on Stiefel Manifolds. *European Journal of Operational Research,* Vol.143, pp. 365-376, ISSN: 0377-2217

Reis, R. & Watson, M. W. (2007). Relative Goods' Prices, Pure Inflation, and the Phillips Correlation (November 2007). NBER Working Paper Series, Vol. w13615, pp. -, 2007. Available from http://ssrn.com/abstract=1032868

Ritter, A. & Munoz-Carpena, R. (2006). Dynamic factor modeling of ground and surface water levels in an agricultural area adjacent to Everglades National Park, *Journal of Hydrology*, Vol. 317, No. 3-4, pp 340-354, ISSN 0022-1694

Ritter, A.; Munoz-Carpena, R.; Bosch, D. D. (2007). Agricultural land use and hydrology affect variability of shallow groundwater nitrate concentration in South Florida, *Hydrological Processes* Vol. 21, No. 18., pp. 2464-2473, ISSN 1099-1085

Sargent, T. J. & Sims, C. A. (1977). Business cycle modeling without pretending to have too much a priori economic theory. In C. A. S. et al. (Ed.), New methods in business cycle research. Minneapolis: Federal Reserve Bank of Minneapolis

Stock, J. H. & Watson, M. W. (2002). Forecasting using principal components from a large number of predictors. *Journal of the American Statistical Association* 97, pp. 1167–79, ISSN 0162-1459

Szalai, J. (2008). Recharging efforts of the subsurface water body in Szigetköz (in Hungarian) report, VITUKI, 2008

Szanyi, J. (2004). *Environmental effects of groundwater production from South-Nyírség's example (in Hungarian).* Ph.D. thesis, University of Szeged, Departement of Mineralogy, Geochemistry and Petrology, p. 81

Székely, F. (2003). *Developping of the hydrogeological model in Northeastern part of the Great Hungarian Plain (in Hungarian),* Report of the Environmental and Water Researh Institute, p. 61, Budapest

Székely, F. (2006). Results of hydrogeological modelling of porous sediments in the Northeastern part of the Great Hungarian Plain (in Hungarian). *Journal of the Hungarian Hydrological Society,* Vol.86, No.4, pp. 23-28, ISSN 0018-1823

Urbancsek, J. (1965). Structural geology of the Quarter in the Great Hungarian Plain (in Hungarian). *Journal of the Hungarian Hydrological Society,* Vol.47, No.3, pp. 111-124. ISSN 0018-1823

Van Schuppen, J. & Van Putten, C. (1985). Invariance properties of the conditional independence relation. *The Annals of Probability,* Vol.15, No.3, pp. 934-945, ISSN 1050-5164

Virág, M.; Szűcs, P. & Lakatos, A. (2005). Local hydrogeological models in the Nyírség area (in Hungarian), *Proceeding of the XXIII. Conference of the Hungarian Hydrological Society,* Nyíregyháza, July 2005

Watson, M. W. & Engle, R. F. (1983). Alternative algorithms for the estimation of dynamic factor, MIMIC and varying coefficient regression. *Journal of Econometrics* 23, pp. 385–400, ISSN 0304-4076

Ziermann, M. & Michaletzky, Gy. (1995). Factoranalysis of time series, Szigma, XXVI No. (3-4), pp. 77-91

10

Assessment of Groundwater Quality in Industrial Areas of Delhi, India by Indexing Method

Papiya Mandal[1]* and Sunil Kumar[2]

[1]*National Environmental Engineering Research Institute (NEERI), Delhi Zonal Laboratory, New Delhi,*

[2]*NEERI, Kolkata Zonal Laboratory, Kolkata, India*

1. Introduction

The GW contamination has become a grave problem due to rapid growth of population, expansion of irrigation activities, industrialization and high rate of urbanization in India. Over the last few decades, the land and water use patterns have changed drastically in India. GW is an important resource for drinking purpose which contains over 90% of the fresh water resources (Sabahi et al. 2009). The untreated industrial effluents discharged into the surface water sources cause severe GW pollution in the industrial belt (Mondal et al. 2005). People are becoming more aware of the complexity of the nature and the delicate balance that exist with the global ecosystem (Ahmet et al. 2006). The National Capital Territory (NCT) of Delhi, India is a water scare state with a deficit in the drinking water supply to its resident (Shekhar et al. 2009). The Yamuna flood plain is only potential fresh GW resources in Delhi. The exploration, exploitation and unscientific management of GW resources in the NCT of Delhi, India have posed a serious threat of reduction in quantity and deterioration of quality (Adhikary et al. 2009). In Delhi, GW level has gone down approximately 8 m per year in the last 10 years which is at the rate of about 0.8 m per year (CGWB 2006). Annual replenishable GW sources of NCT Delhi are about 297 million cubic meter while the GW draft is about 480 million cubic meter (Chatterjee et al. 2009). Thus out of nine districts, seven districts of NCT Delhi have been categorized as overexploited with regards to the dynamic GW resources (CGWB 2006).The city is forced to meet 50 % of its water requirement from GW (Kumar et al. 2006). Solid and liquid waste emanating from the industry are inevitability the by products of manufacturing process which contains toxic chemicals and trace metals. The open unlined drains and availability of dumping of toxic industrial waste in nearby recharge of GW areas, act as the source of GW pollution (Dutta et al. 1997).

The application of statistical approaches *viz.,* Pearson correlation coefficient, linear regression analysis and water quality index (WQI) offer a better understanding about the GW quality. WQI is a mathematical instrument used to transform large quantities of water

*Corresponding Author

quality data into a single number, which represents the water quality level (Saeedi et al. 2010). It is useful to identify the suitable location of GW resources in a particular region.WQI is generally calculated from the point of view of the suitability of GW for human consumption (Sikdar et al. 2007). It provides information to the citizens and policy makers. Moreover, GW quality assessment is important because of spreading water borne diseases. Several epidemiological studies advocated that about 80% of the diseases in the world are due to poor quality of drinking water. The present paper deals with the assessment of GW contamination, identification of suitable locations of GW resources and mitigation measures suggested for its sustainable management.

2. Materials and methods

2.1 Study area

Naraina industrial area (NIA) in Delhi covers approximately 2000 industries in two phases, out of which, phase I has approximately 700 industries. The total area covered by phase I is 114.5 acres. It has a latitude and longitude of 28 ^{0}38´´3.1´and 77^{0} 8´´0.1´, respectively. In Phase I, 400 industries are classified as green category and 300 industries are classified as orange category by Central Pollution Control Board (CPCB), Delhi according to pollution potential. The green industry includes garment, footwear, packaging, printing etc. The orange industry includes dye, pan masala, electroplating, service station etc. Piped water supply of Delhi Jal Board (DJB) at NIA Phase I is not enough, so people are to some extent dependent on GW. The wastewater collection system comprises of sewer and open brick masonry drain. Wastewater from both the phases I & II are discharged into the trunk sewer near Samta Ashram. The sampling locations of GW were identified through GIS map as depicted in (Fig. 1).

2.2 Sampling

The characterization of GW quality at every location is not always feasible in view of time and cost involvement for sample collection. So selectively ten numbers of GW samples were collected in clean polyethylene bottles from manually operated tube wells in the month of April 2009, covering entire NIA phase I. The bottles were washed thoroughly with dilute nitric acid and then rinsed with distilled water. Prior to sampling, sampling bottles were rinsed thoroughly with GW to be analyzed. For heavy metals analysis, separate glass bottles were used for sampling. Water samples were acidified with 1.5 ml nitric acid. Collected samples were protected from direct sunlight during transportation, refrigerated at 4 ^{0}C and analysed within 2 to 3 days. Glassware's used for analyses were rinsed thoroughly with nitric acid followed by double distilled water before use.

2.3 Methodology

Each samples were analyzed for 13 physico-chemical parameters, such as pH, turbidity, electrical conductivity (EC), total dissolved solids (TDS), total hardness (TH), sodium (Na^{+}), potassium (K^{+}), calcium (Ca^{++}), magnesium (Mg^{++}), chloride (Cl^{-}), fluoride (F^{-}), nitrate (NO_{3}^{-}) and sulphate ($SO4^{--}$). pH, turbidity, EC were measured by pH meter, turbidity meter and conductivity meter respectively. TDS was determined as the residue left after evaporation of filtered samples. The Whatman filter paper no.42 was used for filter the samples. TH, Ca^{++} and Mg^{++} were determined by EDTA titration method. Na^{+} and K^{+} were determined by flame photometric method. The amount of Cl^- present in GW was measured by silver nitrate method. The F^- concentration was measured by SPANDS spectrophotometric method. The phenol disulfonic acid method was employed for estimation of NO_3. SO_4 amount was derived from turbidimetric method. Trace elements namely iron (Fe), copper (Cu), lead (Pb), chromium (Cr) and cadmium (Cd) were analyzed using atomic absorption spectrophotometer (AAS). Analyses were carried out as per standard methods of APHA (20th edition, 1998). All units except pH, turbidity and EC were measured in mgl^{-1}. Turbidity was measured in NTU and EC in was measured μmohs/cm.

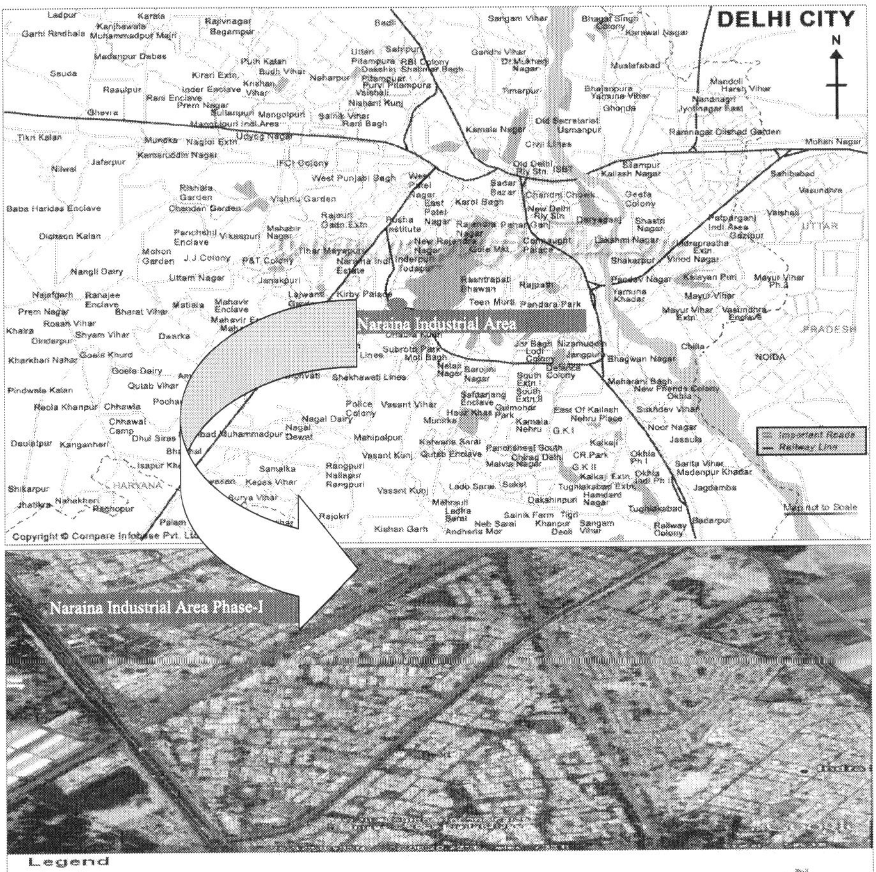

Fig. 1. Locations of GW Sampling

2.4 Statistical analysis

The Pearson's correlation coefficient and linear regression were performed between EC and chemical constituents. The significant correlation coefficient ($r \geq 0.7$) and linear regression with EC is summarized in Table 1.

Parameter	Parameters	Correlation Coefficient (r)	Regression	Regression Coefficient (R^2)
EC	TDS	0.995	0.5501 EC + 22.72	0.99
	TH	0.969	0.3537 EC -37.29	0.94
	Ca^{2+}	0.851	0.0679 EC + 25.82	0.72
	Mg^{2+}	0.868	0.0449 EC - 24.84	0.76
	Na^+	0.840	0.0589 EC + 12.68	0.71
	K^+	0.822	0.0027 EC -1.03	0.69
	Cl^-	0.928	0.2228 EC -125.01	0.86
	NO_3^-	0.882	0.0049 EC -1.74	0.91
	SO_4^{--}	0.80	0.0818 EC -33.26	0.64

Table 1. Significant correlations coefficient ($r \geq 0.7$) and linear regressions with EC

2.5 Characterization of GW Quality

The average concentration of physico-chemical parameters of GW samples and its percentage compliance with drinking water quality standard BIS (10500, 1991) are delineated in Table 2.

Parameters	Range		Average	Indian standard (desirable Limit)	Percent compliance
	Minimum	Maximum			
pH	7.7	8.2	7.7 ± 0.4	6.5-8.5	0
Turbidity	0.25	1.47	0.65 ± 0.4	5	100
EC	998	2180	1442 ± 402	–	–
TDS	560	1225	816 ± 222	500	
TH	290	730	473 ± 147	300	10
Na^+	47	140	98 ± 28	–	–
K^+	1.2	5.1	2.9 ± 1.3	–	–
Ca^{++}	75	160	124 ± 32	75	10
Mg^{++}	24	82	40 ± 21	30	50
Cl^-	68	370	196 ± 96	250	70
F^-	0.37	1.42	0.94 ± 0.5	1	50
NO_3^-	3	9	5 ± 2	45	100
SO_4^{--}	32	160	85 ± 41	200	100
Fe	0.01	0.09	0.04 ± 0.03	0.3	100
Cu	0.01	0.08	0.04 ± 0.02	0.05	70
Pb	0.01	0.08	0.03 ± 0.03	0.05	80
Cr	0.01	0.09	0.04 ± 0.03	0.05	80
Cd	0.01	0.07	0.04 ± 0.02	0.01	20

*All units except pH, Turbidity and EC are in mg/l, *Turbidity in NTU, *EC in μmohs/cm, ± Standard Deviation

Table 2. Comparison of GW quality with Indian Drinking Water Quality Standard (BIS 10500, 1991)

WQI is a useful tool used to obtain a comprehensive picture of GW quality and to identify the suitable location of GW resources in a particular region. For WQI analysis, 14 selected parameters in each sample were assigned a weight (w_i) according to its relative importance in the overall water quality for drinking purpose. Four numbers of selected parameters *e.g.*, NO_3^-, Pb, Cr and Cd were assigned maximum weight of 5 due to their major importance in water quality assignment. Cu is given minimum weight of 2 as it plays low significant role in the water quality assignment. The weight of other parameters varied from 2 to 5 depending on their significant importance in water quality determination. The relative weight of chemical parameters is shown in Table 3.

Chemical Parameters	Indian Standard (BIS 10500,1991)	Weight (w_i)	Relative Weight $W_i = \frac{w_i}{\sum_{i=1}^{n} w_i}$
TDS	500-2000	4	0.0784
pH	6.5-8.5	4	0.0784
TH	300-600	2	0.0392
Cl^-	250-1000	3	0.0588
SO_4^{--}	200-400	4	0.0784
NO_3^-	45-100	5	0.0980
F^-	1-1.5	4	0.0784
Ca^{++}	75-200	2	0.0392
Mg^{++}	30-100	2	0.0392
Fe	0.3-1.0	4	0.0784
Cu	0.05-1.5	2	0.0392
Pb	0.05	5	0.0980
Cr	0.05	5	0.0980
Cd	0.01	5	0.0980
		$\sum w_i = 51$	$\sum W_i = 1.000$

Table 3. Relative weight of chemical parameters

For each parameter, lower value indicates desirable level and higher level indicates permissible limit of BIS 10500, 1991.

In the second step, the relative weight (W_i) was calculated using following equation:

$$W_i = \frac{w_i}{\sum_{i=1}^{n} w_i} \quad (1)$$

Where W_i =relative weight, w_i =weight of each parameter and n = number of parameters. In the third step, the quality rating scale (qi) was calculated using equation 2:

$$q_i = (C_i / S_i) \times 100 \quad (2)$$

Where, q_i = quality rating, C_i = concentration of selected parameters in mg/L, and S_i = Indian drinking water standard (BIS 10500, 1991) of selected parameter in mg/l

$$SI_i = W_i \, q_i \quad (3)$$

Where SI_i = Sub index of *i*th parameter.

$$WQI = \sum SI_i \tag{4}$$

Where WQI= Water quality Index.

The WQI is categorized into five types from excellent water to water unsuitable for drinking. Water quality classification based on WQI value as presented in Table 4. The correlation coefficient matrix of chemical constituents is presented in Table 5.

WQI Value	Water Quality
<50	Excellent
50-100	Good Water
100-200	Poor water
200-300	Very poor water
>300	Water unsuitable for drinking

Table 4. Water quality classification based on WQI value

Parameters	TDS	TH	Na	K	Ca	Mg	Cl	F	NO_3	SO_4	Fe	Cu	Pb	Cr	Cd
TDS	1														
TH	.98**	1													
Na	.82**	.69*	1												
K	.81**	.78**	.77**	1											
Ca	.88**	.88**	.62	.60	1										
Mg	.86**	.89**	.62	.80**	.58	1									
Cl	.90**	.87**	.76*	.70*	.76*	.79**	1								
F	.07	.15	-.20	.04	-.10	.39	.02	1							
NO_3	.85**	.85**	.73*	.73*	.71*	.81**	.85**	.05	1						
SO_4	.78**	.77**	.65*	.79**	.50	.86**	.78**	.49	.76*	1					
Fe	.07	.54	.15	-.10	-.40	-.57	-.14	-.30	-.43	-.10	1				
Cu	-.43	.44	.29	.50	.23	.55	.16	.76*	.25	.62	-.37	1			
Pb	.14	.16	.21	.14	.39	.14	-.12	-.50	.19	.10	-.35	.38	1		
Cr	.68	.57	.66	.37	.61	.41	.58	.02	.24	.43	.074	.23	.24	1	
Cd	-.03	.07	-.18	.02	.16	-.04	-.34	.06	-.09	-.20	- .80*	.31	.57	-.60	1

* Correlations are significant at the 0.05 level (2 tailed) ** Correlations is significant at the 0.01 level (2 tailed)

Table 5. The correlation coefficient matrix of chemical constituents

3. Results and discussion

The average pH of samples was 7.7 ± 0.4 which is within the permissible limit of BIS, recommended as 6.5 to 8.5. Turbidity was much lower than the desirable limit of BIS. EC was observed to vary from 998 to 2180 µmohs/cm which was alarmingly high. EC had strong correlation with TDS, TH, Cl^- and NO_3^- as $R^2 \geq 0.9$. EC had moderate correlation with Ca^{++}, Mg^{++}, Na^+, K^+ and SO_4^{--} as $R^2 \geq 0.5$. The average concentrations of TDS and TH were found to fluctuate between 560 to 1225 mg/l and 290 to 730 mg/l, respectively as BIS recommended the safe limit of 500 and 300 mg/l, respectively. Concentration of water by evaporation and contamination of water due to industrial and municipal waste disposals might have caused a huge increase in the dissolved solids (Mondal et al. 2007). Sometimes high TDS is harmful for kidney and heart patient. It may cause laxative or constipation problem. The hardness is generally caused due to Ca and Mg ions present in the water samples. The high degree of hardness in the study area can definitely be attributed to the disposal of untreated/improperly treated sewage and industrial wastes (Shankar et al. 2008). The average concentrations of Ca^{++} and Mg^{++} were recorded 124 ± 32 and 40 ± 21 as BIS recommended 75 and 30 mg/l, respectively. The average Cl^- concentration of samples was 196 ± 96; indicated 70 % of samples were complying with BIS. In general, Cl^- is widely available in various forms in all type of rocks. The Cl^- is also considered to be conservative in GW of Delhi area (Datta et al. 1996). Sometimes soil porosity and permeability also plays a key role in increasing the Cl^- concentration in the GW. The average F^- concentration of samples were recorded 0.94 ± 0.5.In India, 25 million people in 8700 villages are consuming high fluoride water (Handa et al. 1998).The highest fluoride concentration in the present study area was 1.42 mg/litre. The average NO_3^- concentration of samples was 5 ± 2. NO_3^- concentration was very low which indicated that GW was not contaminated due to agricultural runoff. The average SO_4^{--} concentration of samples was 85 ± 41 which was found within the safe limit as prescribed by BIS. The average concentrations of anion and cation of GW sample are presented in Figs. 2 and 3.

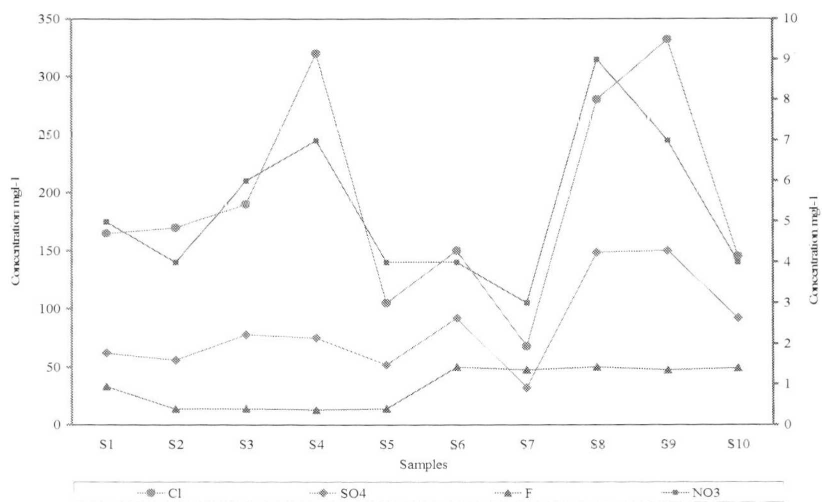

Fig. 2. Average concentrations of anion of GW sample

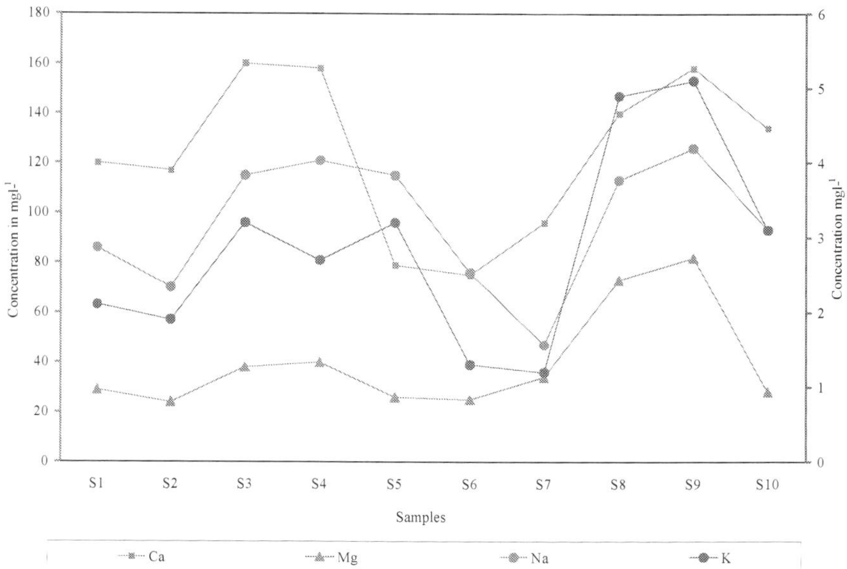

Fig. 3. Average concentrations of cation of GW sample

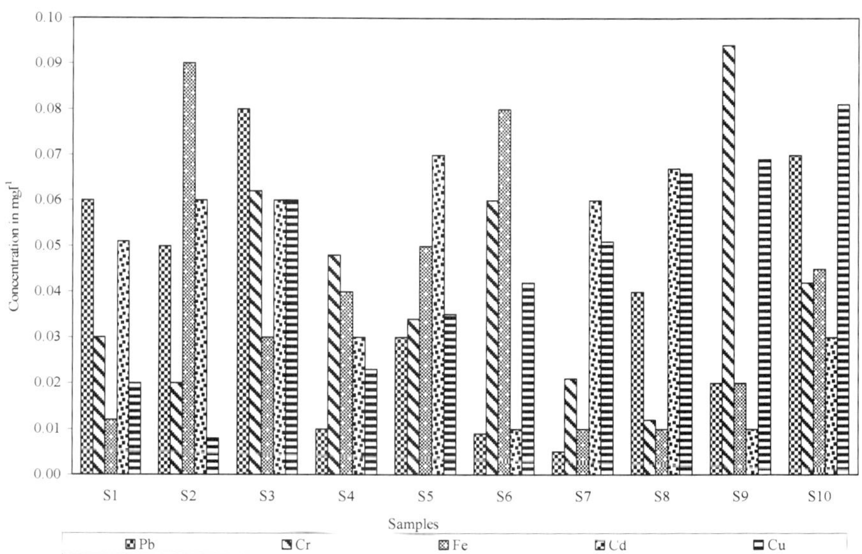

Fig. 4. Average concentrations of heavy metals of GW sample

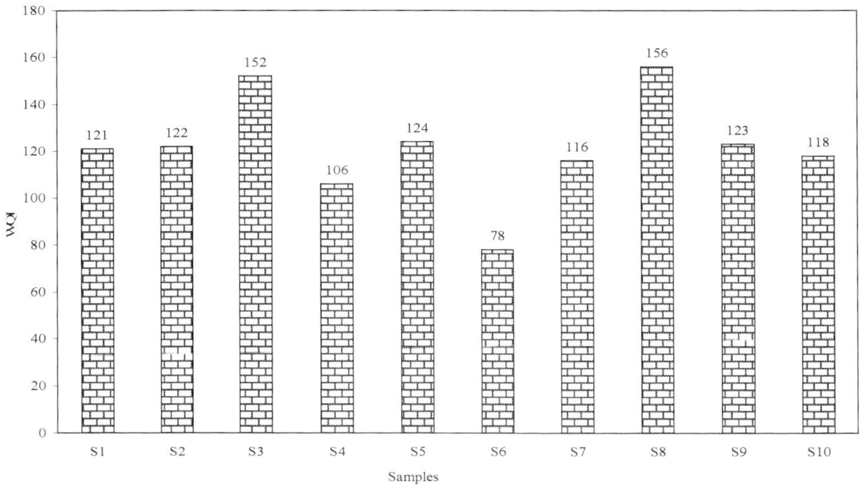

Fig. 5. WQI graph of GW Samples

Location wise concentration of trace metals is presented in Fig. 4. The average concentrations of Fe, Cu, Pb, Cr and Cd were 0.04 ± 0.03, 0.04 ± 0.02, 0.03 ± 0.03, 0.04 ± 0.03 and 0.04 ± 0.02 mg/litre, respectively.

Pb is generally present in GW samples due to plumbing accessories and industrial waste. At few locations, Cu, Cr, Fe and Cd concentrations could be high due to disposal of electroplating and dye industrial waste. Some significant correlations were observed between TDS, TH, Ca^{++}, Mg^{++}, Na^{+}, K^{+}, Cl^{-}, NO_3^{-} and SO_4^{--} is presented in Table 5. The WQI graph of GW samples is presented in Fig. 5.WQI values of GW samples varied from 78 to 156. Water quality with WQI value above 100 is considered as poor. The water is not fit for drinking purposes when WQI value is above 300. GW quality was most suitable only one particular location e.g., location of sixth number sample S6, where WQI value was 78. At other locations, WQI value varied from 106 to 156 during the study period. GW quality was poor at NIA, phase-I due to presence of more dissolved salt and trace elements. The qualitative analysis of study indicated though GW quality was poor at industrial areas of Delhi but more or less suitable for drinking purposes.

4. Conclusions

The assessment of GW quality of Naraina Industrial Area of Delhi by using the indexing method indicated that except one source, other sources of GW quality were poor due to presence of more amounts of dissolved salts and trace elements. Periodical assessment of GW quality and characterization of treated effluent of each orange type industry would provide an overall scenario about the sources of GW contamination. The augmentation of artificial GW recharge through rain water harvesting system and regulation of GW withdrawal might improve WQI value of GW samples. The GW quality is categorized as poor but overall safe for human consumption. Improvement of existing drainage system and scientific handling of industrial hazardous waste materials are the remedial techniques to be adopted for preventing and mitigating the GW contamination. The results of the study would be helpful in identifying the sources of GW contamination and open an avenue for better planning to achieve sustainable management of GW.

5. References

APHA (1998). American public health association, Standard Methods for Examination of waters and wastewaters, 20th Edition, Washington, DC, USA

Adhikary, P.P., Chandrasekharan, H., Chakraborty, D., Kamble, K. (2009). Assessment of groundwater pollution in West Delhi, India using geostatistical approach Environmental Monitoring and Assessment DOI 10.1007/s10661-009-1076-5

Ahmet, D., Fevzi, Y., Tuna, A.L., Nedim, O. (2006). Heavy metals in water, sediment and tissues of Leuciscus cephalus from a stream in southwestern Turkey Chemosphere 63, 1451-1458

CGWB (2006). Groundwater year book, NCT, Delhi, Report, Central Ground Water Board, Ministry of Water Resources, Government of India, New Delhi

Chatterjee, R., Gupta, B.K., Mohiddin, S.K., Singh, P.N., Shekhar, S., Purohit, R. (2009). Dynamic groundwater resources of National Capital Territory, Delhi: assessment, development and management options .Environmental Earth Sciences 59,669-686

Datta, P.S., Bhattacharya, S.K., Tyagi, S.K. (1996). Assessment of groundwater contamination from fertilizers in the Delhi area based on ^{18}O studies on recharge of phreatic aquifers and groundwater flow paths of mixing in the Delhi area. Journal of Hydrol 176, 25-36

Datta, P.S., Deb, D.L., Tyagi, S.K. (1997). Assessment of groundwater contamination from fertilizers in the Delhi area based on ^{18}O, NO_3 and K^+ composition. Journal of contaminated Hydrology 27,249-262

Handa, B.K. (1998). Fluoride occurrence in natural waters in India and its significance.Bhu-Jal News, 3, 31-37

Kumar, M., Ramanathan, A.l., Rao, M.S., Kumar, B. (2006). Identification and evaluation of hydrogeochemical processes in the groundwater environment of Delhi, India. Environ Geol 50,1025-1039

Mondal, N.C., Saxena, V.K. (2005). Assessment of groundwater pollution due to tannery industries in and around Dindigul, Tamilnadu, India. Environ Geol 48,149-157

Saeedi, M., Abessi, O., Sharifi, F. (2010). Development of groundwater quality index. Environmental Monitoring and Assessment 142,263-268

Sahu, P., Sikdar, P.K. (2007). Hydrochemical framework of the aquifer in and around East Kolkata Wetlands, West Bengal, India. Environmental Geology 10.1007/s00254-007-1034-x

Sabahi, E.A.l., Rahim, S.A., Zuhairi, W.Y.W., Nozaily, F.A.l., Alshaebi, F. (2009). The Characteristics of Leachate and Groundwater Pollution at Municipal Solid Waste Landfill of Ibb City, Yemen. American Journal of Environmental Sciences 5 (3),256-266

Shekhar, S., Prasad, R.K. (2009). The groundwater quality in the Yamuna flood plain of Delhi (India) and the management options. Hydrogeology Journal 17:1557-1560

Shankar, B.S., Balasubramanya, N., Reddy, M.T.M. (2008). Impact of industrialization on groundwater quality -a case study of Peenya industrial area, Bangalore, India. Environmental Monitoring and Assessment, 142,263-268.

Part 2

Water Quality Monitoring Studies

11

Sodium Levels in the Spring Water, Surface and Groundwater in Dalmatia (Southern Croatia)

Nives Štambuk-Giljanović
Institute of Public Health Split, Medical School, University of Split
Croatia

1. Introduction

A great number of water resources used for water supply occasionally have sodium concentrations exceeding 20 mgL-1 (Tuthill & Calabrese, 1975; EPA, 1996). Thus, in Massachusetts more than 50% of water from community collectors has more than 20 mgL-1 Na. In some areas the sodium concentration exceeds 100 mgL-1. In order to obtain data used for bringing national decisions on monitoring sodium in drinking water Tuthill & Calabrese (1975), from the University of Massachusetts in Amherst, analyzed blood pressure in students from two adjacent water intake areas where all environmental conditions, except the sodium concentration in drinking water, were similar. The students in the water intake area with 105 mgL-1 Na had significantly higher blood pressure than students in the area with a 5 mgL-1 Na concentration. This study points to the necessity of carrying out longitudinal investigations on sodium concentration together with an increased number of data on health.

In view of the influence sodium in blood has upon the increase in blood pressure, the author decided to classify water as hypotensive (< 10 mgL^{-1} Na), normotensive (from 11 to 20 mgL^{-1} Na) and hypertensive (>20 mgL^{-1} Na) according to the same principle of sodium levels in drinking water.

However, there are significant variations in sodium intake among different individuals as well as from one day to another for each individual (Bauber, 1967). In Western Europe and North America the present consumption of sodium chlorides is estimated to be 5-20 gday-1, with an average of approximately 10 gday-1 (i.e. 4g of sodium per day) (WHO, 1979). The minimum daily sodium demand is ca 50 mg in average adults (WHO, 1984). The average daily sodium demand for adults varies from 2-5 g (WHO, 1979). The allowed sodium levels from food and water have been estimated for various groups of adults and children (WHO, 1984).

Excessive sodium concentration can be easily detected through taste. In solutions at room temperature the taste threshold for sodium present in salts as sodium chloride and sodium sulfate is ca 130 to 140 mgL^{-1}. Generally, the taste is bad at a concentration of <200 mgL^{-1} (regardless whether it is a chloride or sulfate). However, sensitive individuals can detect a bad taste at concentrations varying from 175 to 185 mgL^{-1} (WHO, 1979). Therefore, there are built-in restrictions for allowed sodium levels in drinking water. The threshold for sodium taste in water can be changed and it depends on bonded anions, the solution temperature and personal habits for salt intake.

Sodium plays an important role in nutrition (Weast, 1983), contributes to electrolytes regulated by kidneys (Howard & Schrier, 1990), maintains water balance in the body, and affects muscle contraction and the production of adrenaline and amino acids. However, high concentrations of sodium can disrupt cell or blood chemistry. The excessive consummation of sodium may significantly contribute to many diseases, especially hypertension (Dahl & Love, 1957; Hoffman, 1988; Rutan et al., 1988). Sodium is also toxic to plants in high concentrations (Driscoll, 1986). Sodium concentrations above 70 mgL^{-1} are problematic for irrigation if water is absorbed by leaves (Bouwer, 1978). High concentrations of sodium applied to soils can reduce permeability and produce alkali conditions in which vegetation cannot grow. Elevated sodium levels can also cause problems for industrial uses - for example, foaming in steam boilers.

There is no maximum contamination level (MCL) for sodium, but its guidance level is 20 mg L^{-1} (EPA, 1996). Sodium-restricted diets can often limit one's total intake to less than 1000 mg day^{-1}. Assuming a water intake of 2 L mg day^{-1}, drinking water can contribute substantially to one's total sodium intake at concentrations above 100 mg L^{-1}. For patients on low sodium diets, drinking water may account for more that 60% of total sodium intake.

Considering the importance of sodium for human health the objectives of the investigations were 1) the systematic monitoring of the sodium levels in the spring water, surface and groundwater in Dalmatia and 2) identification of potential causes of the wide variability in this parameter.

2. Study area

Dalmatia belongs to karst regions with abundant rainfall. However, it is characterized by insufficient quantities of water since water sinks underground through fissures (Štambuk-Giljanović, 1999).

The groundwater in Dalmatia can be classified as fissure water which, according to hydrochemical and hygienic characteristics, resembles surface water. Groundwater becomes turbid and has to be treated before entering the water supply system. It is moderately hard with a fairly quick flow. Karst waters can be generally classified as calcium-hydrocarbonate water and differ according to the content of sulfates, chlorides, magnesium, sodium and dissolved CO_2 (Štambuk-Giljanović, 2006).

The majority of karst water resources have high quantities of chlorides resulting from the seawater intrusion, i.e. due to the direct mixing of seawater with fresh water or with groundwater flows. These water resources, in addition to increased chlorides content, have increased sulfates, sodium and magnesium contents.

3. Methods

Individual water samples in Dalmatia (29 taken from spring waters, 25 from surface water and six samples from groundwater) were taken monthly at 60 measurement stations (Figure 1) from January 1, 2010, to January 1, 2011.

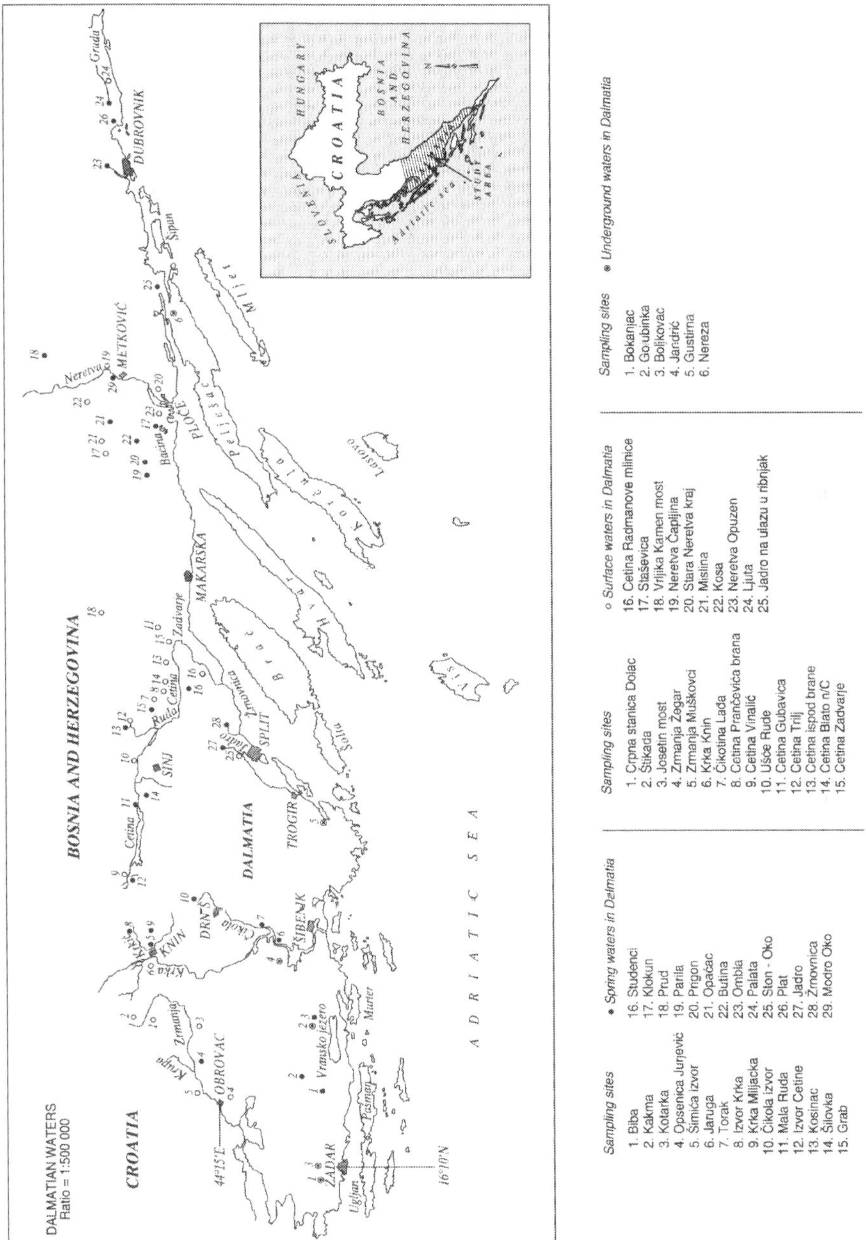

Fig. 1. Sampling sites of waters in Dalmatia

The sulfates concentrations were measured by employing the turbidimetric method, chlorides by the argentometric method and hardness by the complexometric method according to American standard methods (1995).

According to the concentrations of the mentioned quality parameters the following coefficients were computed: K_1 (Larson & Scold, 1958) (the ratio between the chlorides and sulfates sum and the carbonate hardness) and K_2 (the ratio between noncarbonate and carbonate hardness). Sodium was computed from the difference between K_1-K_2 and carbonate hardness according to the author's formula (Štambuk-Giljanović, 2003, 2004, & 2005).

$$Na\ (mgL^{-1}) = (K_1 - K_2) \cdot 2\ HCO_3^- \cdot M\ (Na) \qquad (1)$$

Wherein M (Na) = 23 $gmol^{-1}$.

According to coefficient K_1 values, the water types were classified into three groups (Štambuk-Giljanović, 2006):

a. non-corrosive water with K_1 lower than 0.2;
b. water with low corrosiveness, K_1 from 0.2 to 0.65;
c. very corrosive water with K_1 higher than 0.65.

The annual results of investigations of waters in Dalmatia were statistically analyzed. They were presented as arithmetic means and medians, since the average concentration values expressed by a median are better than those expressed by an arithmetic mean since the distribution of sodium concentrations is asymmetric. Standard deviations and variability coefficients were computed and the maximum and minimum sodium concentrations in the measurement stations were presented. Statistically analyzed data are presented in Tables 1-3. The tables also present the corrosiveness coefficient K_1 for all the measurement stations.

4. Results

Sodium concentrations in spring water, surface and groundwater in Dalmatia are presented in Tables 1-3 for 2010 at 60 measurement stations. Sodium concentrations expressed by a median are presented in Figures 2, 4 and 6 while the corrosiveness coefficient K_1 is presented in Figures 3, 5 and 7.

In the spring waters in Dalmatia (Table 1) the sodium concentrations expressed by a median (Figure 2) ranged from 1.8 (Ombla and Biba) to 17.6 mgL^{-1} (Klokun). The highest variations were noted at the Kotarka station (217%), and the lowest at Opsenica Jurjevići (32%). At the Šilovka station the sodium concentration was 21 mgL^{-1} with a significant variation (152%). That type of water is classified as hypertensive. According to corrosiveness coefficient K_1 (Figure 3) it is very corrosive water.

Maximum sodium concentrations in spring water varied from 4.6 (Opsenica Jurjevići) to 617 mgL^{-1} (Kotarka) while the minimum concentrations ranged from less than 0.1 (Biba, Kakma, Torak, Kosinac and Studenac) to 2.9 mgL^{-1} (Ston-Oko). The corrosiveness coefficient K_1 was from 0.1 in Parila water (Vrgorac area) to 0.88 in Kotarka (Zadar region). Out of 29 spring water resources analyzed, 83% can be classified as noncorrosive and 17% as water with a low degree of corrosiveness; 75.8% are classified as hypotensive, 20.7% as normotensive and 3.5% as hypertensive.

Sampling sites	Sampling sites number	Aritmetic mean concentration	Standard deviation	Median	Variability coefficient, %	Maximum concentration	Minimum concentration	Corrosion coefficient K_1
1. Biba	12	4.9	6.8	1.8	138.7	21.9	<0.1	0.18
2. Kakma	12	14.6	27.2	9.1	186	99	<0.1	0.20
3. Kotarka	12	80.9	176.1	11.9	217	617	0.5	0.88
4. Opsenica Jurjevići	12	2.5	0.8	2.2	32	4.6	1.5	0.19
5. Šimića izvor	12	8.1	7.2	7.3	88.8	21.8	0.1	0.17
6. Jaruga	12	13.5	12.2	9.6	90.3	36.4	2.0	0.16
7. Torak	12	8.6	8.6	6.3	100	27.7	<0.1	0.20
8. Izvor Krka	12	5	4.5	3.7	90	14.2	0.1	0.18
9. Krka Miljacka	12	13.8	10.3	13.2	74.6	32.7	0.9	0.17
10. Čikola izvor	12	2.9	3.3	1.9	113	11.7	0.4	0.16
11. Mala Ruda	12	3.2	2.7	2.9	84.3	9.8	0.3	0.20
12. Izvor Cetine	12	9.9	11.7	4.8	118	34.7	0.2	0.18
13. Kosinac	12	5.0	5.1	3.6	102	17.4	<0.1	0.19
14. Šilovka	12	35.7	54.6	21	152	199	0.2	0.66
15. Grab	12	3.2	2.4	2.6	75	9.5	0.2	0.11
16. Studenci	12	4.7	3.4	5.1	72.3	10	<0.1	0.10
17. Klokun	12	16.6	9.2	17.6	55.4	32.9	0.5	0.35
18. Prud	12	14.5	17.8	9.8	122	59.9	1.1	0.7
19. Parila	12	7.2	6.2	4.8	86	19.3	1.1	0.1
20. Prigon	12	13.5	18.3	7.1	135	55.8	0.5	0.5
21. Opačac	12	5.3	3.2	5.8	60.3	10.2	0.2	0.19
22. Butina	12	17.8	11.8	16.1	66.2	41.3	4.0	0.18
23. Ombla	12	2.4	1.5	1.8	86.2	6.1	0.9	0.20
24. Palata	12	5.5	3.7	5.1	67.2	14.3	0.0	0.19
25. Ston-Oko	12	10.8	4.7	12	43.5	19.9	2.9	0.25
26. Plat	12	3.3	2.4	1.8	72.7	8.4	0.8	0.21
27. Jadro	12	5.5	4.4	3.3	80	13.7	0.9	0.18
28. Žrnovnica	12	4.2	2.0	3.7	47.6	8.7	1.5	0.18
29. Modro Oko	12	13.1	14.3	9.4	109	52.1	0.5	0.68

Table 1. Sodium concentrations (mgL^{-1}) in Dalmatian spring waters (2010)

Sampling sites	Sampling sites number	Aritmetic mean concentration	Standard deviation	Median	Variability coefficient, %	Maximum concentration	Minimum concentration	Corrosion coefficient K_I
1. Crpna stanica Dolac	12	3.3	2.2	3.0	66.6	7.8	0.8	0.18
2. Štikada	12	2	1.3	1.6	65	4	0.2	0.16
3. Josetin Most	12	1.4	1.4	1.0	100	3.4	<0.1	0.17
4. Zrmanja Žegar	12	2.5	2.4	1.7	96	8.2	0.4	0.16
5. Zrmanja Muškovci	12	2.6	1.8	2.5	69.2	5.7	<0.1	0.19
6. Krka Knin	12	10.1	4.5	10.6	44.5	19.9	0.1	0.19
7. Čikotina Lađa	12	14.7	12.5	12.2	85	38.5	0.4	0.17
8. Cetina Prančevića brana	12	8.7	11.3	4.3	129	39.7	<0.1	0.16
9. Cetina Vinalić	12	17.2	18.1	9.6	105	63.9	1.1	0.10
10. Ušće Rude	12	4.5	3.4	3.5	75.5	9	0.3	0.10
11. Cetina Gubavica	12	9.3	8.3	7.5	89.2	24.8	0.4	0.12
12. Cetina Trilj	12	4.5	2.5	4.6	55.5	96	<0.1	0.17
13. Cetina ispod brane	12	17.2	21.9	6.3	127	74	2.0	0.19
14. Cetina Blato n/C	12	11.2	9.6	8.8	85.7	33.5	0.0	0.18
15. Cetina Zadvarje	12	13.7	6.8	14.4	49.6	23	1.6	0.19
16. Cetina Radmanove mlinice	12	22	15.3	22	69.5	46.9	0.9	0.17
17. Staševica	12	16.1	6.6	17.6	40	28	6.3	0.18
18. Vrljika Kamen most	12	2.6	2.1	2.6	80.7	6.4	<0.1	0.17
19. Neretva Čapljina	12	22.8	14.8	17.1	64.9	53.9	10.6	0.20
20. Stara Neretva kraj	12	177.9	217.2	95.3	122	733	4	2.1
21. Mislina	12	329.2	423.7	42.5	128	1144	1.9	3.2
22. Kosa	12	368	587	43.2	159	1793	3.0	2.6
23. Neretva Opuzen	12	642.6	722.7	502	112	2153	4.3	0.17
24. Ljuta	12	3.4	4.3	2.4	126	15.3	<0.1	0.19
25. Jadro na ulazu u ribnjak	12	4.1	3.9	3.1	95	12.3	<0.1	0.2

Table 2. Sodium concentrations (mgL^{-1}) in Dalmatian surface waters (2010)

In surface waters in Dalmatia (Table 2), sodium concentrations expressed by a median (Figure 4) varied from 1.0 (Josetin Most) to 502 mgL-1 (Neretva Opuzen). The sodium concentrations were highest at the Kosa station (159%) and lowest at Staševica (40%). The water at the Neretva Opuzen station is hypertensive and very corrosive (Table 2, Figure 5). Increased sodium concentration at that station results from the infiltration of seawater into the Neretva River due to level oscillations. The seawater influence is greatest from the Neretva estuary to Opuzen while it decreases towards Metković. After the Metković Bridge the seawater action is no longer felt. In other words, after the Metković Bridge the riverbed rises by 2-2.5 m which decreases the infiltration of seawater into the riverbed after the bridge. Maximum sodium concentrations in surface water were from 3.4 (Josetin Most) to 2153 mgL^{-1} at the Neretva-Opuzen station, while the minimum concentrations varied from less than 0.1 (Josetin Most, Cetina-Prančević Dam, Cetina-Trilj, Cetina-Blato on/C, Ljuta, Jadro at the fishpond entrance) to 10.6 mgL^{-1} (Neretva-Čapljina). The corrosiveness coefficient K_1 varied from 0.1 (Cetina-Vinalić and Ruda estuary) to 3.2 at Mislina so that this water can be classified as very corrosive water.

Sampling sites	Sampling sites number	Aritmetic mean concentration	Standard deviation	Median	Variability coefficient, %	Maximum concentration	Minimum concentration	Corrosion coefficient K_1
1. Bokanjac	12	19.5	16.1	14.1	82.5	53.2	3.1	0.84
2. Golubinka	12	14.1	10.9	11.1	77.3	34.7	3	1.8
3. Boljkovac	12	173	161	124.3	93	578	<0.1	1.6
4. Jandrić	12	23	28.4	15.3	123	111	5.6	0.8
5. Gustirna	12	25	23.1	14.4	92.4	71.2	5	0.6
6. Nereza	12	276	497.4	72.7	180	1610	2.5	0.7

Table 3. Sodium concentrations (mgL^{-1}) in Dalmatian underground waters (2010)

Among all analyzed 25 surface water resources according to K_1, 84% of them can be classified as noncorrosive water and 165% as very corrosive water; 64% of surface water can be classified as hypotensive, 20% as normotensive and 16% as hypertensive.

In the Dalmatian groundwater (Table 3) the sodium concentrations expressed by a median (Figure 6) varied from 11.1 mgL^{-1} at Golubinka to 124.3 mgL^{-1} at Boljkovac. The greatest variations were recorded at the Nereza station (Dubrovnik area), i.e. 180% while the lowest variations were observed at Golubinka (Zadar area), i.e. 77.3%. At the gauging station Boljkovac the sodium concentration expressed by a median was 124.3 mgL^{-1} while at Nereza it was 72.7 mgL^{-1} so that it can be classified as hypertensive water.

According to corrosiveness coefficient K_1 all analyzed groundwater resources are very corrosive (Figure 7). The maximum sodium concentrations in the groundwater ranged from 34.7 (Golubinka) to 1610 mgL^{-1} (Nereza) while the minimum concentrations varied from less than 0.1 (Boljkovac) to 5.6 mgL^{-1} (Jandrići). The analyzed groundwater can be classified as normotensive (67%) and hypertensive (33%).

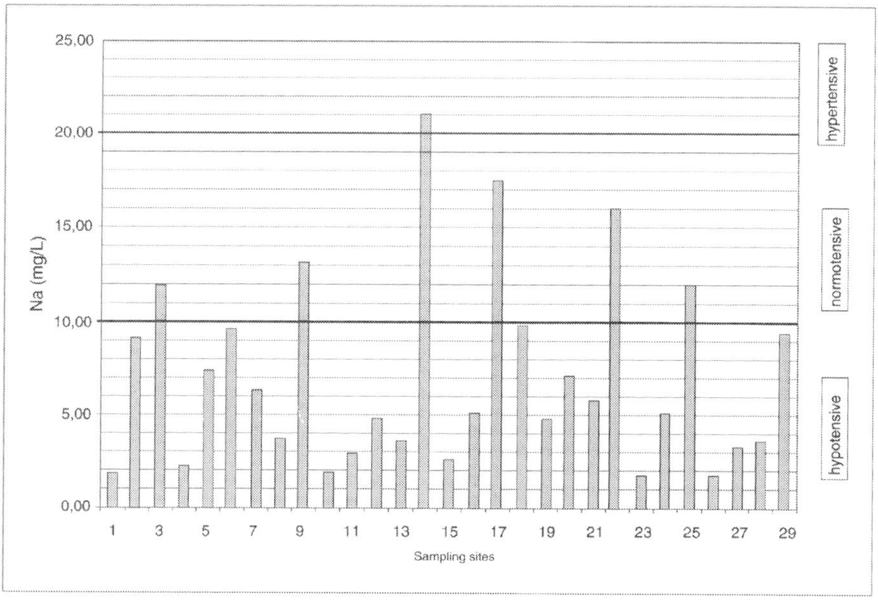

Fig. 2. Sodium concentrations (mgL^{-1}) expressed by the median in Dalmatian spring waters (2010)

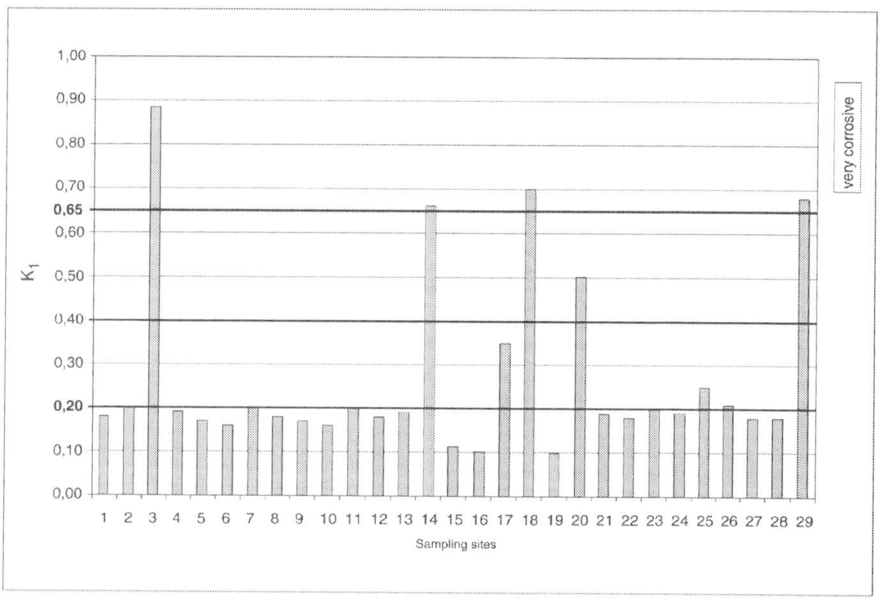

Fig. 3. Corrosion coefficient K_1 in Dalmatian spring waters (2010)

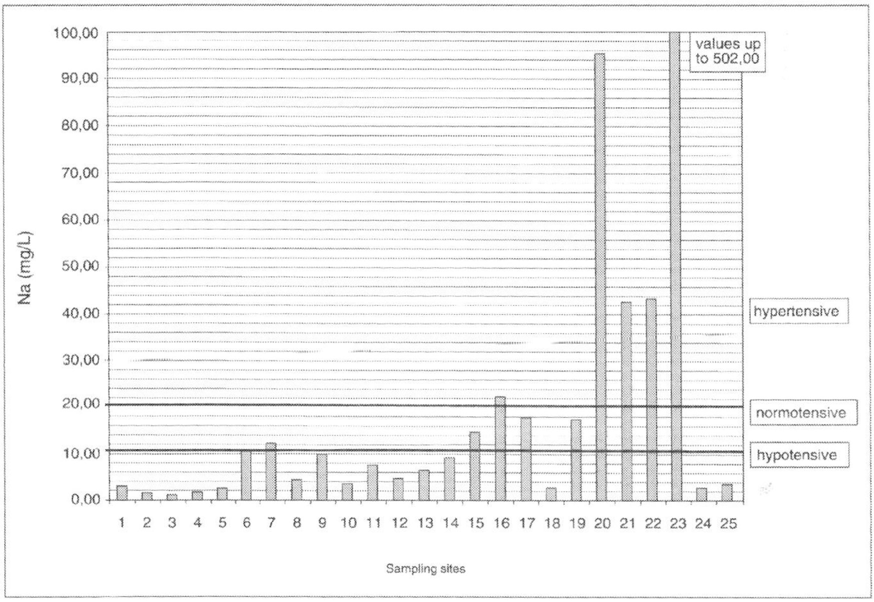

Fig. 4. Sodium concentrations (mgL-1) expressed by the median in Dalmatian surface waters (2010)

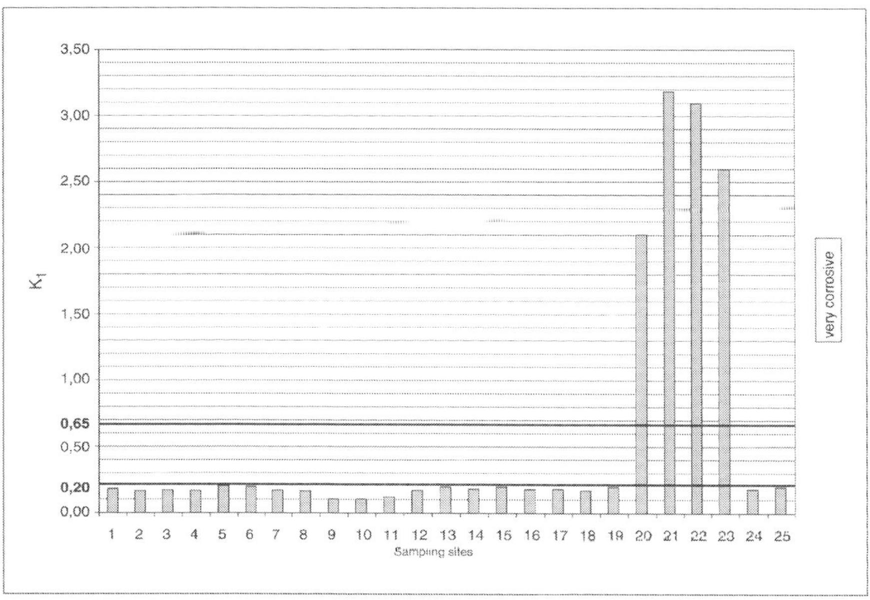

Fig. 5. Corrosion coefficient K_1 in the surface Dalmatian water (2010)

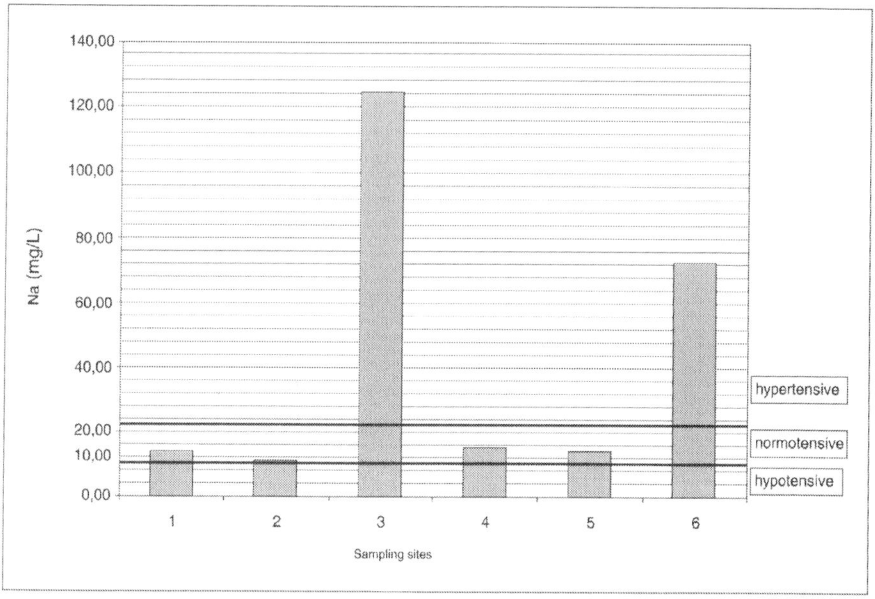

Fig. 6. Sodium concentrations (mgL^{-1}) expressed by the median in Dalmatian underground waters (2010)

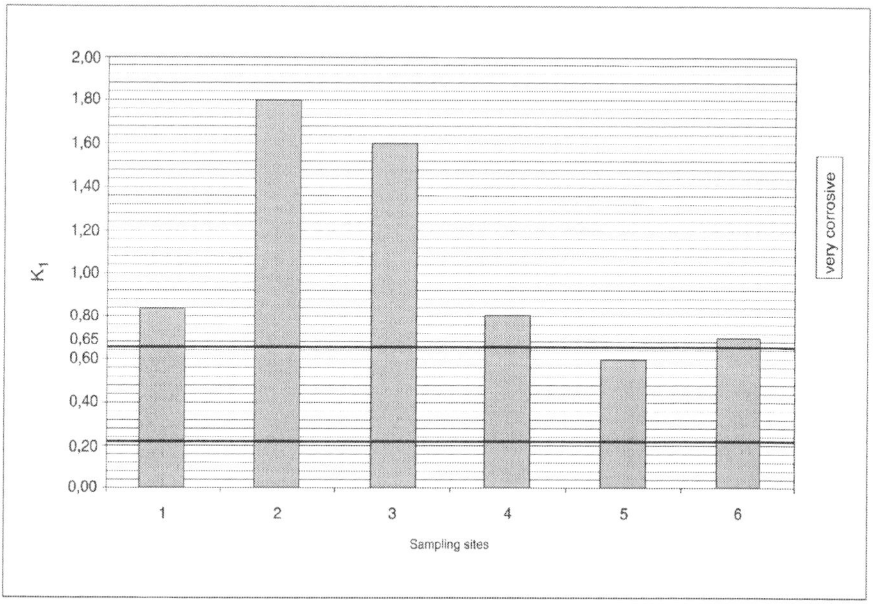

Fig. 7. Corrosion coefficient K_1 in the Dalmatian underground waters (2010)

5. Discussion

Dalmatia belongs to the karst region with surface and groundwater karst phenomena; the latter are more complex and result from the geological composition of karst. Thus, karst regions are characterized by permeable limestone with fissures caused by faults and folding. Rainfall water is lost from the surface by sinking underground through limestone fissures which expand due to water outflows and thus create a large storage of groundwater. The groundwater, stored in limestone, flows down to sea level; however, its flow is prevented by impermeable layers or rock layers of low permeability from different lithological formations. Dolomites and dolomite limestone are most important in stopping the groundwater outflow. Those rocks act as insulators preventing groundwater circulation. As barriers, they stop the groundwater flow and force it to appear on the surface in the form of concentrated springs; sometimes groundwater flows can act as siphons and appear in the sea as submarine springs. This barrier is often formed by the sea itself which slows down groundwater flows so that water appears as coastal springs. In the most abundant Dalmatian coastal springs groundwater comes to the surface, flowing from distant karst regions and flooded poljes (fields). It then sinks underground along faults in impermeable rocks or through ponors (swallow holes) along the mountains boundaries.

Closed karst poljes in Dalmatia are catchment areas in which the water is collected and then sinks underground, however, in some places it comes to the surface. (Štambuk-Giljanović, 2006). The majority of springs were used as bases for water intakes for city water supplies in Dalmatian coastal areas. Thus, the Bokanjac groundwater is used for the Zadar area water supply, the Kakma and Biba Springs supply Biograd and Benkovac, the Čikola Spring is used for the Drniš water supply, the Jaruga and Jandrić Springs for Šibenik, the Jadro Spring for Split and Trogir, the Klokun Spring for Ploče, the Ombla Spring for Dubrovnik and the Duboka Ljuta Spring for the water supply of Cavtat and Ćilipe. The Mala Ruda Spring is used for the new Sinj water collector while the Prud Spring supplies Pelješac and Korčula. The above mentioned large springs represent the only sources of water supply from by springs. The intakes for Dalmatian rivers are on the Cetina River near Gata and Zadvarje and on the Zrmanja River; soon there will be a new intake on the Krka River.

The water resources in Dalmatian karst have more bicarbonates than chlorides and sulfates since water frequently comes into contact with limestone and dolomites while layers of sulfate minerals are rare and less soluble and there are no chloride minerals. In many karst water resources there are significant chlorides concentrations which result from seawater either by direct mixing with freshwater or infiltrating into the groundwater flows. In these flows, apart from increased chlorides concentrations, the sodium and magnesium concentrations are also increased. Sulfates and chlorides in some karst water resources, unlike bicarbonates, vary significantly and depend upon groundwater levels, i.e. rainfall quantities.

Sodium concentrations in Dalmatian water resources expressed by a median during 2010, varied from 1.8 mgL^{-1} in spring water to 502 mgL^{-1} in surface water. According to a study carried out in Canada (Subramanian & Méranger, 1984) sodium concentrations expressed by a median ranged only from 0.3 to 242 mgL^{-1}.

According to the author's classification of drinking water as hypotensive (<10 mgL^{-1} Na), normotensive (from 11 to 20 mgL^{-1} Na) and hypertensive (>20 mgL^{-1} Na) by relating sodium concentration projectively with the arterial pressure and knowing that an increased concentration of sodium in blood increases blood pressure, the following can be concluded: among all analyzed 60 water resources in Dalmatia, 64% can be classified as hypotensive, 20% as normotensive and 16% as hypertensive which means that they exceed the recommended concentration of 20 mgL^{-1} which is characteristic both for hypotensive and normotensive drinking water.

According to the corrosiveness coefficient K_1 out of all analyzed 60 water samples in Dalmatia, 70% can be classified as noncorrosive water (K_1 lower than 0.2), 6.7% as water with low corrosiveness (K_1 from 0.2 to 0.65) and 23.3% can be classified as very corrosive water (K_1 higher than 0.65).

According to Freis (1973) the primary factor connecting sodium and hypertension is the out-cell volume which includes the plasma volume. Sodium is important for hypertension since it is the main factor in the formation of out-cell liquid. Excessive salt in food can be processed by a normal kidney without increasing the volume of out-cell liquid. However, if the kidney does not function properly this volume is increased and hypertension intensified even after a moderate sodium intake. Conversely, at that time sodium quantity should be limited in the diet until the out-cell liquid is reduced which will then significantly decrease blood pressure.

People suffering from hypertension should drink normotensive or even hypotensive water. Sodium concentration in drinking water should be longitudinally analyzed together with an increased number of health data. It is of special importance that data gathered in this way will facilitate the systematic monitoring of sodium in drinking water and ensure quick preventive actions for decreasing sodium concentrations, either by reverse osmosis or by substitution of drinking water sources. Sodium should be included in water quality monitoring since its concentrations can be important to doctors prescribing diets for their patients.

6. Conclusion

During the year 2010, sodium concentrations were monitored in spring water, surface and groundwater in Dalmatia, in order to study their influence upon human health, i.e. arterial hypertension and cardiovascular diseases.

Sodium concentrations expressed by a median in the spring water varied from 1.8 to 17.6 mgL^{-1} with variations from 32% to 217%. According to the corrosiveness coefficient K_1, 83% of the spring water can be classified as non-corrosive (K_1 lower than 0.2), 17% are waters with low corrosiveness (K_1 from 0.2 to 0.65). Out of 29 samples of analyzed spring water, 75.8% were classified as hypotensive (<10 mgL^{-1} Na), 20% as normotensive (from 11-20 mgL^{-1} Na) and 3.5% as hypertensive (>20 mgL^{-1} Na).

Sodium concentrations expressed by a median in surface water ranged from 1-502 mgL^{-1} with variations from 40% to 159%. According to corrosiveness coefficient K_1, 84% of surface water can be classified as non-corrosive and 16% as corrosive. Out of 25 samples of analyzed surface water 64% is classified as hypotensive, 20% as normotensive and 16% as hypertensive.

Sodium concentrations expressed by a median in groundwater ranged from 11.1-124.3 mgL^{-1} with a variation from 77.3% to 180%. According to corrosiveness coefficient K_1 all groundwater can be classified as very corrosive water. The analyzed groundwater was normotensive (67%) and hypertensive (33%). Of all the analyzed 60 samples in Dalmatia, during the year 2010, 70% of the water resources can be classified as non-corrosive, 6.7% as water with a low degree of corrosiveness and 23.3% as very corrosive water.

According to the research results obtained on 60 water samples, 64% can be classified as hypotensive, 20% as normotensive and 16% as hypertensive.

Sodium concentrations and/or water corrosiveness determined by chloride should be used to study the health of each group's drinking water with different sodium (and/or chlorides) concentrations. It can also be used to epidemiologically correlate thc pathological states such as arterial hypotension or hypertension with sodium intake from water; it should be also relevant to correlate other pathological states, such as stomach erosion due to water corrosiveness caused by chlorides.

7. References

APHA. (1995). *Standard Methods for the Examination of Water and Wastewater*, American Public Health Association, Washington, DC.

Bauber, T.R., Stamber, J., Stamber,R. & Pullman,T.N. (Eds). (1967). In: *Environmental factors in hypertension*. Grune and Stratton, New York, NY.

Bouwer, H.(1978). *Groundwater Hydrology*. Mc Graw-Hill, New York, NY.

Dahl, L.K. & Love, R.A. (1957). Etiological role of sodium chloride intake in essential hypertension in humans, *Journal of the American Medical Association*,Vol. 164,p. 397.

Driscoll, F.G. (1986). *Groundwater and Wells*, Johnson Division, St. Paul, MN.

EPA (U.S. Environmental Protection Agency). (1996). Drinking Water Regulations and Health Advisories, *Environmental Protection Agency*, Washington, DC.

Freis, E.D.(1973). *Modern management of hypertension*. Veterans' Administration, Washington, DC., p.2.

Hoffman, C.J.(1988)."Does the sodium level in drinking water affect blood pressure levels?", *Perspectives in Practice*,Vol. 88,No 11,pp. 1132 1135.

Howard R.L. & Schrier, R.W.(1990). A unifying hypothesis of sodium and water regulation in health and disease, *Hormone Research* ,Vol.34,pp. 118-123.

Larson, T.S. & Scold, R.W.(1958). Laboratory studies Relating Mineral Quality of Water to Corrosion of Steel and Cast Iron, *Corrosion* ,Vol.16,p. 285.

Rutan, G.H., Kuller, L.H., Neaton, J.D., Wentworth, D.N., McDonald, R.H. & Mc Fate-Smith, W.(1988). Mortality associated with diastolic hypertension and isolated systolic hypertension among men screened for the multiple risk factor intervention trial, *Circulation* ,Vol.77,No. 3,p. 504.

Subramanian, K.S. & Méranger, J.C.(1984). A survey for sodium, potassium, barium, arsenic, and selenium in Canadian drinking water supplies, *A Spectrosc*,Vol. 5,p. 34.

Štambuk-Giljanović, N.(1999). Water quality evaluation by index in Dalmatia. *Wat Res*, Vol. 33, No16, pp. 3423 3440.

Štambuk-Giljanović, N.(2003), Characteristics of water resources in Dalmatia according to established standards for drinking water, *J Water Supply Res Technol Aqua*, Vol.52, No.4, pp. 307-318.

Štambuk-Giljanović, N.(2004). Comparison of Dalmatian water evaluation indices, *Water Environ Res*,Vol.75,No.5, pp. 388-405.

Štambuk-Giljanović, N.(2005). The quality of water resources in Dalmatia, *Environ Monit Assess*, Vol. 104, pp. 235-268.

Štambuk-Giljanović, N.(2006). *Waters of Dalmatia*, ISBN 953-96304-7-9, Institute for Public Health Split, Split, Croatia.

Tuthill, R.W. & Calabrese, E.J.(1979). Elevates Sodium Levels in the Public Drinking Water as a contributor to Elevated Blood Pressure Levels in the Community, *Arch Environ Health*, Vol.34,pp. 197-202.

Weast, R.C.(1983). *CRC Handbook of Chemistry and Physics*, CRC Press, Boca Raton, FL.

World Health Organization (1979). Sodium, chlorides and conductivity drinking water. *EURO Reports and Studies No. 2.*, Regional office for Europe, Copenhagen, Danemark.

World Health Organization (1984). Health criteria and other supporting information, Ch.16, *Guidelines for drinking-water quality*. Vol. 2, Geneva, pp. 145-151.

12

Groundwater Quality Degradation in Obrenovac Municipality, Serbia

Nenad Zivkovic[1] et al.*
[1]*University of Belgrade, Faculty of Geography*
Serbia

1. Introduction

Water has no definition. Any interpretation, explanation, its formula or description, does not actually represent anything of what it really is. Although much is written about it and we think to know all, yet, it remains a great mystery to us. The water is like religion. Everywhere around us, does not impose, used as needed, usually mechanically, we are unable to recognize the salvation role in it. When we have enough water and turn it into a "servant" (means of use), we relate to it with negligence, just like we do with the faith in times of welfare. Only the lack of water and the need for it create a sense of ontological connection to us. And then we devote ourselves to prayer. And the greater the need for it becomes, the more zealous the prayer is.

In any well-organized society people's attitude toward water has a special status. If the need for quality water is greater, and its availability is lesser, awareness of the need to preserve and nurture this resource is also greater. During the medieval subsistence economy, the water quality problem did not exist and it was used in its natural state without any special treatment. By social division of labor, development of cities and urbanization, industrialization and intensive agricultural production, a man comes into conflict with the water. Although the conflict was inevitable, it seems that in many communities (societies) a good chance to keep it under control was omitted. So, from a natural ally, we have created ourselves an opponent to whom we are not worthy rivals, nor will we ever be. And all we needed to do was to show just a little respect.

In Serbia, we have long lived under the delusion that with 6000 m^3/s of water flowing off from our territory and 10 million inhabitants, which is 52 m^3/inhabitant/day, we fall into countries with the high-water supply. At the end of the last century, the intensification of climate oscillations brought frequent and long drought periods, which pointed to the weaknesses of our water management systems. Then we have remembered that a water crisis is threatening, because, in fact, only 8 % of all waters that flow off are the domicile

*Slavoljub Dragicevic[1], Ilija Brceski[2], Ratko Ristic[3], Ivan Novkovic[1], Slavoljub Jovanovic[1], Mrdjan Djokic[4] and Sava Simic[5]
[1]*University of Belgrade, Faculty of Geography, Serbia*
[2]*University of Belgrade, Faculty of Chemistry, Serbia*
[3]*University of Belgrade, Faculty of Forestry, Serbia*
[4]*University of Nis, Faculty of Science and Mathematics, Dep. of Geography, Serbia*
[5]*Institute for Nature Conservation of Serbia, Serbia*

waters (4 m^3/inhabitant/day). And with a shortage of water we have also found out that its quality is very bad, considering both surface and groundwater. In the meantime, the need for water has been growing, used resources have been diminishing, and our awareness of the water has not changed a bit since the eighties of the last century.

Although the groundwater participates with 70 % in water supply of the population, its importance is not nearly dedicated enough attention. All the advantages that this resource has over the surface water are often ignored, especially at the time of the river reservoirs construction. One of the indicators that proved poor treatment of the groundwater is considerably more modest monitoring that is carried out over it in comparison with the surface water. Perhaps because the last mentioned is in the sight of every one of us and because the impact of public opinion is extremely important and present in every society, especially ours. Thus, the control system of water quality, or monitoring in general, unfortunately, is a sensitive issue, a subject of finances in a great deal, and thus gets a political connotation as well. In cases of extreme phenomena of water degradation, as it is in the Borska River Basin (one of the largest copper mines in Europe), even the most terrible consequences that are visible, cannot change anything because this issue is under State jurisdiction. The situation is similar with the Topciderska River that flows through the urban core of Belgrade. Although it is known to be one of the most polluted in the country, as demonstrated by regular monitoring, little is changing in a positive way because in the background of the problem, bigger interests are at stake. A constant exceeding of the maximum permitted concentration of mercury and heavy metals, phosphorus and nitrogen compounds (Dragicevic et al., 2010) is occurring throughout the whole length of this river. In addition, the university experience indicates that students of natural sciences such as geography, biology, forestry, agriculture, come to study with a very superficial knowledge of the groundwater, accumulated in previous education. And this is also a sign that this issue is on the social margins.

Quality studies of well and spring waters in Serbia include only limited territory, insignificant compared to the total area. Continuous monitoring is performed by state institutions (Environment Agency and its services), and includes only the observation well stations, and the certain sources of the wells. It does not include the quality of well and spring waters in inhabited areas; the quality is usually monitored when it is already too late, only in cases of health problems of the population. Basic idea of this study is to show the quality of the groundwater on the territory that is under permanent anthropogenic pressure. The Municipality of Obrenovac was selected as the one of those that had a larger number of the springs and wells for the population water supply, and which were not part of the monitoring, more exactly the space with recorded environmental problems.

2. Research area

The Municipality of Obrenovac is an integral part of Belgrade region and is located in its western part. The two most striking morphological shapes of this area are the River Sava, which makes its northern border and the right tributary of the Sava, the Kolubara River, with its broad valley bottom. The Municipality consists of the town Obrenovac (24 000 inhabitants) and 28 villages with a total of 74 000 inhabitants. The town acquires particular attention in the seventies of the last century when two power plants (Nikola Tesla 1 and 2) were put into operation, which under the current conditions produce about half the electricity in Serbia. These two systems operate on the basis of the exploitation of lignite from the Kolubara basin that provides 70 % of coal in the country. The basin is located next to the Municipality of Obrenovac, its area is of 600 km^2, and the estimated reserves of lignite in it are 2.2 billion tons.

Fig. 1. Position of the Obrenovac Municipality in Serbia

Testing of the groundwater quality was conducted in the south-western, rural part of the Municipality which still does not posses water supply network and which is out of permanent monitoring. The central point of this area belongs to the plane of Dubrava, which is, compared to the surrounding terrain, raised about twenty meters and where the agricultural production is performed. This activity is highlighted as an important cause of degradation of the groundwater quality due to uncontrolled use of fertilizers and other agro-chemical substances. Next, even more important cause is very poor municipal sanitation. Namely, the construction of the septic tanks was carried out without any approval and necessary documentation, often in a primitive way without any insulation enabling contact of feces and groundwater. In addition, most households own livestock and that means collecting manure for fertilizing crops as well. We are convinced, however, that the disposal is not done in hygienic manner and that these plots are not secured, so smooth infiltration of the manure into the soil is carried out. Also, neither the collection nor disposal of municipal waste is regulated in this area which adds to the pollution. Unfortunately, despite the clear environmental degradation of Dubrava and its environment, it will not be easy to fight for a more favorable status. It is quite clear that, for example, each and most compelling criticism of the power plant does not force changes because its harmful effects require huge investments for rehabilitation. And in the time of general lack of money and in the knowledge that such power plants are the "engine" of the half of Serbia, it is not hard to see who will have an advantage in public. What influence they may have gives the fact that only 2 km from Dubrava there is the ashes disposal site where about 2.5 million tons of coal ash is disposed each year.

In addition to waters from the Dubrava area, this study has included the spring horizon, unique in Serbia, which drains the same groundwater zone at the foot of the scarp. The aforementioned spring series, represented by sixty hydrological springs is a rarity that needs state protection and in that sense, this research should be the driver of a broad campaign of its revitalization.

3. Previous studies

Environmental problems in Obrenovac Municipality are evident for many years. However, only in recent years the systematic observations are carried out, the data are collected in order to put monitoring in service of plan making to solve the problems (Dragicevic et al., 2008). Although pollution has been observed in all spheres, the existence of thermal power plants has caused water and soil to be of secondary importance, and that special attention is paid to the air. Therefore, since 2006 there has been a station for automatic measurement of air pollution in Obrenovac, which detect values of 5 parameters (NO, NO_2, NO_x, SO_2, PM_{10}). According to preliminary results of the environmental situation in Obrenovac (Municipality Press, 2010), it is stated that the concentration of the first four parameters has not seriously threatened the people's health, but the content of fly ash and particles with a diameter smaller than 10 μm (PM_{10}) are very high and there was several days period when it was constantly above the allowable limit. According to Nenadovic S. et al. (2010) the air pollution is significantly affected by weather conditions, where the situation is much worse in winter. Contamination of the air by nitrogen and sulfur compounds is very important for water as well, because their dissolution in the atmosphere create acid rains that are very harmful to soil and groundwater, and contributes to the mobility of heavy metals from soil to water. Wind flow in this area is such that Dubrava is the first targeted.

The only study to date, covering all aspects of environmental degradation in Obrenovac Municipality was published in 2007 titled: "The intersection of the current status of the presence of hazardous and harmful substances in the environment in the Municipality of Obrenovac" (CIP, 2007). Identification of pollutants of air, water and soil has been carried out and then by sampling from different locations, the composition of these areas has been analyzed. In the Dubrava area, 8 locations (wells) were selected for the groundwater quality analyses. Of these, 5 had the physical-chemical incorrectness (increased turbidity, ammonia, nitrates, nitrites, $KMnO_4$ consumption), while the bacteriological tests showed 6 incorrect samples (mainly the presence of coliform bacteria of fecal origin). The general conclusion was that environmental problems are most often a result of emission from power plants, the existing ash disposal sites, inadequate use of agro-technical products and poor sanitation utilities. This study did not analyze the spring water, nor was it anywhere referred to as groundwater resources.

The aim of this study was to check a much larger sample of well water condition of Dubrava, as well as the springs in its foothill. The first reason is that these are the only sources of water supply for about 6500 inhabitants of the related settlements, and the other is an attempt that, as part of water protection measures, the spring series gets the status of natural resource and restores the functions it had till the end of the eighties of the last century.

4. The origin and morphological characteristics of the terrain

Dubrava is the part of a large Vukodraz-Tamnava plane that binds to post-lacustrian phase of the Pannonian Basin. Loose sediments at the bottom and the rim of the Pannonian basin have a great distribution and great thickness, which indicates the long-term marine-lake period in the basin. The original lake phase of the basin was replaced by the marine phases, which lasted from the middle Miocene to middle Pliocene. The last marine phase of the basin was in the lower Pliocene, and after the middle Pliocene lake-continental phase in the southern rim of the basin resulted in the last lake phase, which lasted from the Upper Pliocene to Middle Pleistocene (Markovic, 1967). Disappearance of the lake and draining its bottom created the modern network of the flows unified in the Danube River system. The youngest flow has become the main stream (the Danube), and the older flows, the former tributaries of the sea and the lake (the Sava, the Kolubara) have become the direct and indirect tributaries of the Danube. Only with the advent of the Sava and its cutting into the central lake level, its tributaries have been activated which turn the former lake riparian landscape into fluvio-denudational shape.

Vukodraz-Tamnava plane is 140-115 m high and points to the more persistent erosive base, relatively tectonic state of rest and the favorable climatic conditions during its formation. Part of this area, which is important for this study (Dubrava), is a local divide, low tump between the Sava and the Kolubara in the length of 12 km, southwest-northeast direction extension and the surface of 97 km^2. The greater part of it, 65 km^2, slightly tilted to the south, southeast and east is intersected by shallow valleys that converge to the Tamnava and over a dozen sources drain in Ljubinic, Trstenica and Stubline. The smaller part, 32 km^2, is oriented to the north and ends with the Posavski scarp. It is one of the more outstanding forms of the lower Posavina. This scarp is both the right valley side of the Sava and the part of famous South-Pannonian scarp consisting of the Posavski and Podunavski scarp. At the rim of Dubrava, this scarp is twenty meters high, although in the west is higher, up to 38 m. Built by the River Sava while intersecting along the Sava trench, and the side moving toward the south taking off the soft lake sediments. Between the current Sava and the scarp there is a distance of ten kilometers, and in that zone a younger form exist, the Sava scarp, 7.5 m high, which follows the course of the Sava River through its entire length. It is the younger one, of the Holocene age, while the Posavski scarp is predominantly of the Virm age. There is a lot of evidence among them showing the building of the terrace plateau made by the Sava. These are its oxbow lakes which are nowadays still clearly observed in the relief (Velika Bara, Mala Bara, Jazmak).

A characteristic feature which distinguishes this area in the geological sense, and is important for the formation of groundwater, is almost exclusively represented by Neogene and Quaternary formations. They have a large surface and lie over the Paleozoic and Mesozoic rocks. These older ones are created at the bottom of the Pannonian Sea, and are mostly made of marl, sand and clay. The youngest sediments in the form of sand and gravel fill the lowest areas of the Sava and the Kolubara valley bottoms.

5. Lithological composition and hydro-geological properties

Hydro-geological conditions at Dubrava are showed best by the profile of the drill made in Vukicevica (130 m above the sea level) up to 373 m depth (Milojevic et al., 1975). The dominant layer is made of marly clay, thickness of 220 m and starting from 55 m of depth. Below it, clay, marl, sandstone and limestone take turns in different varieties, and usually mixed with each other. Above marly clay layer, there is a layer of gray sandy clay, about 20 m thick, whose upper limit is the lower limit of phreatic aquifers. Groundwater zone is composed of fine-grained quartz sand about 7 m thick, which on the given profile appeared at 26.6th meter. Yellow clayey sand, also water-bearing (3 m), gravelly clay (4 m) and yellow clay on the surface (19 m) are deposited over it.

The first water-bearing horizon in this zone is located in small-grained quartz sands at a depth of 23 to 33.5 m, and the absolute elevation of 97 to 107 m. Clayey sediments of limestone beginning at a depth of 273 m prevail below this series of sediments. Upper series of sand is formed of yellow clay which is pebbly in places. This clay cover has a wide distribution, so that it extends eastward from Vukicevica, through Trstenica to Stubline, then to the village of Grabovac, in the area of the village of Dren and elsewhere. In a series of thin sands there are intercalations of coal. This sand and gravel come to the surface of the terrain in the village of Dren, in the River Vukicevica itself, and then appear on a steep section of Drenovacka Voda at a level of 95 m, where a powerful spring also appears.

Formed aquifer is of freatic type. Upper clay aquifers are formations that are saturated when the water is practically water impermeable, so the aquifer is poorly fed with water from precipitation. Therefore, the amplitude fluctuations of the groundwater levels are low and usually amount to 1 m. During intense rainfall the terrain gets wet, and water is collected in the hollows and large swamps are created. Such phenomena are very common in the zone Grabovac-Vukicevica. Shallow surface water flows collect and drain the surface water, so that a little water is retained on the terrain and infiltrate the underground. In dry periods the clay is dried, cracks are formed which, in the first stage, could receive a certain amount of water and partly pass it further. Phreatic aquifer is of a large expanse. Depth to water level in some parts of the terrain is different, but typically ranges from 18 to 22 and even up to 25 meters.

Spring series under the Posavski scarp is probably unique according to numerousness in Serbia. As per our study, 60 springs have been noted, and it seems that there are even more of them (inaccessible, neglected springs...). They all feed from the same phreatic aquifers formed in Dubrava. The aquifer is tilted to the north and has no a great fall, and all drainages are at altitudes of 90 to 95 meters.

Southern rim of the area (Dubrava) is poorer in springs and they are tied to shallow valleys of the Josevica (one spring in Ljubinic) and the Trstenica (ten springs in Trstenica and Stubline), and four springs have been recorded on the eastern border in Stubline in the foothill of the scarp towards the alluvial plain of the Tamnava and the Kolubara. All of them belong to the same water-bearing horizon as the springs of Grabovac-Dren scarp, considering that the aquifer from the Dubrava peak profile is oriented to the south. Western rim of the Municipality belongs to the convergent area of the Vukicevica basin. Its underground is part of the same hydro-geological collector as well, and the spring series can be traced back from the spring bend of the river to the place where it breaks the scarp between Grabovac and Dren (16 springs).

6. Analysis of the well water

Well water at the site of Dubrava was studied in ten villages located on the area or around it. Sampling was conducted on two occasions and both times in the autumn of the years of 2008 and 2009. Selection of wells was such as to cover the largest possible area of Dubrava and discover possible directions of the pollutants movement. The organization of this research is reflected in carefully guided records of all wells. In that sense, the great number of information has been collected in order to provide all the relevant indicators of the current situation, but also to establish the proposals for possible remediation. In addition to basic data about the owner, a sketch of the position of wells in relation to housing and auxiliary facilities has been made. Each well has been photographed and the GPS coordinates have been recorded. The depth of the well, its diameter and the depth of water in it have been measured. Then the manner of how the water is drawn has been described, whether it is done mechanically or manually. The data on whether a well is dug or drilled, when it was built, when it was last cleaned, what is its coating and the condition of retaining wall are also important. The number of households served by each well has been recorded and also the total number of the well consummators. The distance from the septic tanks has been calculated and each facility that may affect water quality has been noted.

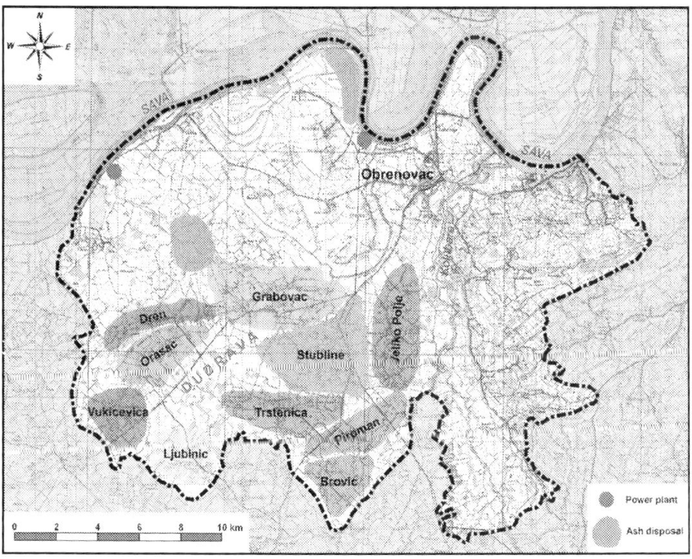

Fig. 2. The locations of wells in villages of Dubrava (Topographic Map 1:50000, VGI, 1970)

On the occasion of the on-site sampling, we have measured the parameters that can be easily changed by standing and during transport: conductivity, pH, temperature and oxygen content. Two samples were taken from each well, one of which was used for bacterial and the other for chemical analysis. Microbiological analysis for total mesophilic bacteria has been carried out using the method of membrane filtration and seeding on a suitable substratum for the total microorganisms. Determination of coliform bacteria and *E. coli* has been also done using the method of membrane filtration and seeding on specific substrata. Besides this, the analysis has included 22 parameters for which the maximum allowable concentration was regulated. These are: turbidity, color, odor, conductivity, pH, $KMnO_4$ consumption, ammonium ion, nitrates, nitrites, chlorides, sulfates, phosphates, calcium, magnesium, iron, anionic detergents and heavy metals (As, Hg, Pb, Cu, Cd , Cr). Analyses of all parameters are derived and aligned with national and international regulations (SZZZ, 1990; USEPA, 1983; Rump & Krist, 1988).

6.1 Physicochemical analysis

The total number of wells where the water quality has been tested is 351. The number of samples in the settlements is such that reflects their size (number of inhabitants) and spatial representation on Dubrava. Table 1 gives an overview of the results by settlements.

Settlement	Num. of samples	Acc. to one param.	Acc. to several param.	Defective samples		Correct samples	
				Num.	%	Num.	%
Piroman	27	2	13	15	56	12	44
Brovic	23	5	1	6	26	17	74
Trstenica	32	12	15	27	84	5	16
Ljubinic	22	9	7	16	73	6	27
Vukicevica	18	8	7	15	83	3	17
Stubline	89	45	37	82	92	7	8
V.Polje	28	8	3	11	39	17	61
Orasac	23	1	2	3	13	20	87
Dren	33	11	10	21	64	12	36
Grabovac	56	18	19	37	66	19	34
Sum	351	119	114	233	66	118	34

Table 1. Deviation from the MCL (maximum contaminant limit) according to physico-chemical parameters

Shown collectively, the correct (proper) 118 wells make only 1/3 of the total number. There is great disparity among settlements in terms of physicochemical correctness of water. The worst situation is in Stubline with only 8 % (7 of 89) satisfactory results, while in Orasac it is 87 % (20 of 23). Only three settlements have more than a half number of proper samples and it is a clear indication of poor water condition. In addition, half of the incorrect samples deviated in terms of quality by just one parameter, while the other half were invalid on multiple parameters. Such a relationship is more or less represented in the settlements as well.

Of all undesirable substances contained in water, the most common are nitrates. They were detected for 220 times, which means that in 63 % of the tested wells they exceed the MCL. Their presence is not of a local character, because they were found in all settlements. In addition to the spatial representation, what concerns is much greater presence than allowed. The mean value of nitrates among the improper samples is 182 mg/l, which is 3.6 times more than the MCL. In addition, 65 samples were with over 200 mg/l, and 8 with more than 500 mg/l.

	Turb.	E.C.	pH	NO_3^-	Ca	Mg	As	SO_4^{-2}	NH_4^+
NºD.S.	3	97	1	220	26	43	1	10	1
MCL	1	1000	6.8-8.5	50	200	50	0.01	250	1
Avg.	11	1195	5.2	182	224	64	0.04	366	1.16
Max.	15	1860	5.2	739	285	99	0.04	683	1.16

NºD.S. - number of defective samples, Turb. - turbidity (NTU), E.C. - electrolytic conducti-vity (μS/cm), Cations and Anions in mg/l, Avg. - average

Table 2. Analyses results of the parameters that overstepped the MCL

According to the number of samples that exceed the MCL, the next parameter is the electrolytic conductivity. There were 97 of such samples, that is, 28 %. Overruns have occurred in all settlements and usually in those wells where the nitrate excess was detected. Calcium and magnesium ions are present in 26; that is, in 43 wells in unallowable concentrations. This excess is not large, although it is evident that the calcium concentration in all wells is slightly increased in comparison to natural water and is usually above 100 mg /l. Increased presence of sulfates is noted in 10 wells, but it is important that all samples were detected in only two settlements on the eastern rim of Dubrava. Of all the other physicochemical parameters analyzed, the turbidity was problematic in three samples, and the pH, ammonium ion and the increased content of arsenic by one sample. Higher concentration of arsenic than allowed was detected in one well water in the village Brovic. It was found that it originated from the rodent control products and the well was immediately placed out of service, and the competent services were also informed about that.

6.2 Microbiological analysis

Microbiological analysis of well water in the area of Dubrava has showed that the situation is no better in comparison to the hydro-chemical analysis. There ware a total of 282 of the improper samples, which is 80 % of the total number of the wells. The vast majority of them (232) was contaminated with only one type of bacteria. How worrisome is the situation highlights the fact that in 7 of 10 settlements the number of incorrect samples is greater than 80 %, and that there was no proper sample in three settlements.

The most common of all are the aerobic mesophilic bacteria that existed in 279 wells, which is 79 % of the total number of the samples. The average number exceeding of these bacteria is very high and amounts to more than 4 times (431) compared to the number allowed in drinking water (100/1ml). In several settlements that number is even higher, and the worst situation is in Veliko Polje, where the average per improper well is 1060 (the maximum is 3500).

Total coliform bacteria are less common, in 53 wells with an average of 59 bacteria per contaminated well. This means that in those wells the average was 6 times larger than the allowed number set (regulated) for drinking water (10/100ml) - the maximum was 300.

In Orasac and Dren, under the Dubrava scarp, the average amounted to even more than a 100, while in two settlements in the southeast of the area these bacteria were not detected. The most dangerous of all parameters, *E. coli* was found in 15 wells. They are represented in the five settlements with an average of 21 bacteria per well.

Settlement	Num. of samples	Acc. to one param.	Acc. to several param.	Defective samples		Correct samples	
				Num.	%	Num.	%
Piroman	27	9	-	9	33	18	67
Brovic	23	6	-	6	26	17	74
Trstenica	32	17	1	18	56	14	44
Ljubinic	22	19	3	22	100	0	0
Vukicevica	18	15	1	16	89	2	11
Stubline	89	55	21	76	85	13	15
V.Polje	28	24	4	28	100	0	0
Orasac	23	15	4	19	83	4	17
Dren	33	29	3	32	97	1	3
Grabovac	56	43	13	56	100	0	0
Sum	351	232	50	282	80	69	20

Table 3. Deviation by the microbiological parameters

Typ.of bact.	Mark	S	T	V	P	B	Lj	V.P.	G	O	D	Sum
	N°wells	89	32	18	27	23	22	28	56	23	33	351
Colif.	Avg.	47	60	60	-	-	20	55	55	103	117	59
	N°D.S.	23	1	1	-	-	2	4	13	6	3	53
	%(D.S.)	26	3	6	-	-	9	14	23	26	9	15
Meso.	Avg.	272	159	221	201	220	342	1060	515	532	476	431
	N°D.S.	75	18	16	9	6	22	28	56	17	32	279
	%(D.S.)	84	56	89	33	26	100	100	100	74	97	79
E. coli	Avg.	18	-	-	-	-	10	10	32	20	-	21
	N°D.S.	5	-	-	-	-	1	2	5	2	-	15
	%(D.S.)	6	-	-	-	-	5	7	9	9	-	4

Letters in the first row are the first letters of the settlements, Colif. - Coliform bacteria, Meso. - Aerobic mesophilic bacteria, E. Coli - Escherichia coli, NoD.S. - number of defective samples

Table 4. Structure of bacteriological incorrect samples by settlements

7. Analysis of the spring water

Research of the springs on the territory of Dubrava was conducted in two phases. The first included their location, description and hydrological characteristics, and the other water sampling and determination of their hydro-chemical and microbiological characteristics. To achieve this previously needed to collect, systematize and analyze existing technical documentation to make the Cadastre of the springs. In that sense, the geo-referencing of topographical, geological and hydro-geological background was executed. This was the basis for further research and development of GIS.

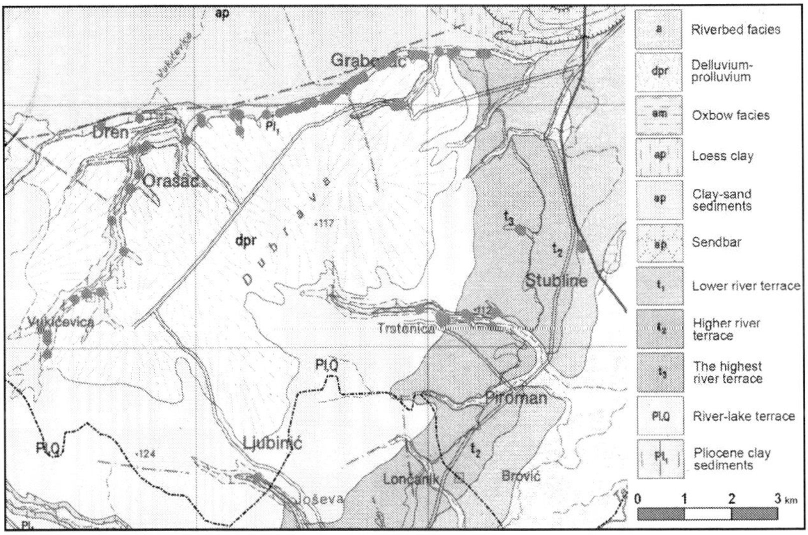

Fig. 3. The location of springs (red spots) (Geological Map 1:100000, Geozavod, 1976)

The first task in the field was to find the springs, and here we affectionately received the assistance of the settlement residents on Dubrava. The problem was that many of the springs were neglected and ingrown with weeds and it was impossible to locate them without the people who knew the terrain. After that, the baseline data about them have been collected, systematized in the "identity card" of each spring. These are the data on the settlement, the owner of the land on which the spring is located, brief description, geographical coordinates and altitude. Its steadiness during the year has been noted, and its purpose as well (water supply, livestock watering, irrigation). The description has also included the access to the spring, derived capping and its condition, the spring type (jet, diffuse). Their current yield has been measured; each spring has been photographed and located on geological map.

7.1 Biochemical analysis of the spring water

After processing all the physical indicators and assessments of their condition, out of the total of 88 springs, 30 were selected for which the biochemical analysis of water was done. Water sampling procedures and methods used for biochemical analysis fully meet the existing standards and are identical to those applied to the wells. Because of the special interests for the spring series under the Grabovac-Dren scarp, the largest number of samples was derived from Grabovac, 24 of them. Three samples are from Vukicevica, two from Stubline and one from Orasac. Hydro-chemical analysis has showed that all the indicators are good except the one for nitrate. This parameter deviates (varies) in 14 cases, which is about 50 % of the springs. Its mean value in the contaminated springs is 75 mg/l and it is 25 mg/l higher than the allowed value for drinking water. Regarding the bacteriological analysis, all the samples showed to be incorrect. The number of mesophilic bacteria everywhere exceeded the allowable number (100/1ml). Their average value was 303, which is three times more than the one usable for drinking, while the maximum reaching 450.

8. Discussion

Preceding analyses of well and spring waters on Dubrava have made the first step in the establishment of monitoring that will lead to groundwater conservation as one of the most valuable resources in this area. What have the analyses showed? First of all, it was known even before our study that the water is of a dubious quality, because the biochemical analyses were carried out occasionally and on demands of individuals. However, what has been unknown is the spatial definition of contamination, as well as the possibility of comparing the results of time-matched water sampling. Adhering to the strictly regulated values on drinking water quality standards, the groundwater situation is very bad.

Settlement	Num. of samples	Chem. param.	Bacter. param.	Defective samples		Correct samples	
				Num.	%	Num.	%
Piroman	27	15	9	19	70	8	30
Brovic	23	6	6	11	48	12	52
Trstenica	32	27	18	28	88	4	12
Ljubinic	22	16	22	22	100	0	0
Vukicevica	18	15	16	17	94	1	6
Stubline	89	82	76	88	99	1	1
V.Polje	28	11	28	28	100	0	0
Orasac	23	3	19	19	83	4	17
Dren	33	21	32	33	100	0	0
Grabovac	56	37	56	56	100	0	0
Sum wells	351	233	282	321	91	30	9
Springs	30	14	30	30	100	0	0
SUM	381	247	312	351	92	30	8

Table 5. Deviation from the MCL by biochemical parameters of the well water

From a total of 381 water samples, only 30 of them meet the required quality for drinking water. These are mostly the wells in the southeastern part of Dubrava, in Piroman and Brovic. These settlements are on the border of the studied area and under the stronger influence of the groundwater of the River Tamnava alluvium.

In other settlements, besides the slightly better situation in Orasac, the situation is such that clean water is a real rarity. According to relation of the physical-chemical and bacteriological indicators, it is clear that the well and spring water are of the same water horizon. However, comparing the water quality of the wells in Grabovac and the springs at the foothill of the scarp in the same settlement, there is a difference. Specifically, the average value of aerobic mesophilic bacteria in the wells is 515, while in the spring water it is 294. Also, the content of nitrates in the well water is on the average 166 mg/l, and amounts to 75 mg/l in the springs. Anyway, the horizontal distance from the spring to the nearest wells is about 200 m and about 2 km to the farthest analyzed wells. Similarities in the water quality between the wells in Grabovac and their differences in relation to the spring water have two meanings. The first is that there is a constant source of pollution on Dubrava that contaminates the groundwater to a level of the slightly higher amount than these parameters show in the springs themselves. And second, that almost all the wells are in very poor condition and that there is strong local infection of the water. Existing differences in the quality of well and spring waters are a sign that slowly trickling through the groundwater zone from the collector center (center of Dubrava) to the place of flowing out (springs), leads to the natural filtration of water and lowers the pollutants. This is also referred by the fact that the coliform bacteria were found in 13 wells in Grabovac, and *E. coli* in 5, which are indicators of fecal waters and are not present in spring waters.

8.1 Nitrate contamination

It is obvious that the water chemical improperness is mostly affected by increased nitrate content. They represent the final stage of a known process of nitrification, and are an important step in the cycling of nitrogen in the soil. Nitrates themselves are not particularly toxic; however, by transferring to nitrites in the organism (body) and a number of other conversions they can create serious health problems (e.g. methemoglobinemia). The normal daily intake of nitrates in the organism (body) in any way should be about 75 mg, and their recommended MCL in drinking water is 50 mg/l (SZZZ, 1990). Therefore, it is clear that the constant presence of nitrate as it exists in the groundwater of the investigated area, with the concentrations that exceed the allowed values, can harm the health of these people.

The origin of nitrates in the water, their transport and effects on humans are a subject that is most present in the studies of drinking water quality in recent decades (Keeney, 1989; Spalding & Exner, 1993; Kovar & Krasny, 1995; Goss, Barry et al., 1998; Foster et al., 2002; Visser, Dubus et al., 2009; Vaux, 2011). Groundwater nitrate contamination stemming from longstanding agricultural practices remains a worldwide and increasing problem with serious economic and health effects (USEPA, 1990). Of all agricultural contaminants, nitrate is the most widespread in exceeding national water standards. In agricultural areas of North America, for example, between 5-46 % of domestic wells in aquifers exceed the 45 mg/l NO_3 drinking water standard (Hamilton & Helsel, 1995; Goss, Barry et al., 1998, as cited in Wassenaar & Baisden, 2011).

How is the nitrate contamination formed on Dubrava? First of all, their presence in the surface and ground waters in Serbia is a normal occurrence, particularly in rural areas and local wells. However, this content can vary during the year, usually not exceeding 50 mg/l and only rarely reaches any excess values. In contrast, a spatially large and constant contamination has been established on Dubrava. There are several sources of pollution. They can be divided into dispersed and pointed (concentrated). The uncontrolled treatment of agricultural land by artificial fertilizers and manure, and the impact of thermal power plants as well, belong to the first group. Local (pointed) sources of pollutants are numerous and are related to the septic tanks, pit latrines, open sewer leakage, inappropriately stored animal manure, municipal waste, local landfill.

On the surface of Dubrava, out of 97 km^2, the agricultural land is represented at 95% of the area. All land is privately owned with the plots of an average size of about twenty acres. The most common crops are cereals (maize, wheat), clover and little less industrial plants. Much less land is under orchards and vineyards, until the vegetables are grown mostly in the housing estates. All these crops require additional nutrition by mineral and organic fertilizers. However, the agricultural production in Serbia is far away from the yield that was usually achieved in eighties of the last century. Starting from the nineties, the circumstances in which life took place were such that the society maintained only a basic existence. Investments in many areas ceased, so as in agriculture. Consumption of mineral fertilizers was around 1.45 million tons in 1985, and then fell to 220 000 tons in 2000, that is, 300 000 tons in the year of 2003. That is why the yields were also drastically reduced. The last ten years were not any better. The yield of wheat was reduced by 20% (from 4.5 to 3.7 t /ha), or sugar beet by 27% (from 47 to 37 t/ha). Mineral fertilizer consumption is nowadays about 36 kg per hectare, and in order to achieve the European yields, the consumption should be increased as many as eight times. All these findings support the preservation of groundwater quality on Dubrava, especially as agricultural production there has never had an intense character. Irrigation of large parcels has never been practiced and water supplies have been collected from precipitation. What is the actual impact of scientific farming methods (modern farming) today on the increased content of nitrates in groundwater is difficult to say, but this influence is certainly not negligible. Lack of funds for the purchase of nitrogen-based mineral fertilizers has been compensated by animal manure that the vast majority of the land owners in the Dubrava have. The fact that both sources of nitrate in groundwater are not present in quantities as before does not mean that their influence is less. In fact, there are studies that proved that the weather conditions, especially with frequent fluctuations in precipitation in semi-arid areas may encourage intake of nitrates into the groundwater (Lagerstedt, 1994). Precipitation in this part of Serbia is exactly of such characteristics with an average of 650 mm (Zivkovic & Dragicevic, 2003).

8.2 The impact of the thermal power plants on the water quality

Thermal power plants Nikola Tesla 1 and 2 which are in a direct proximity of Dubrava dually affect the quality of its waters. Both ways are in relation to atmospheric contamination; the first is pollution from the chimneys, and the other from the ashes dumping sites. The section "Previous studies" shows some effects that the thermal power plants have on the Obrenovac Municipality environment. In the absence of our own research of air pollution, we can only supplement the existing comment. Average emission values of the thermal power plants flue gases in the period of 1990-2002 were (in mg/m^3): SO_2 - 14 588 (allowed 650), NOx - 3479 (450), CO - 1363 (250) (CIP, 2007). This shows the extent to which the exceeding of the harmful gases allowable emissions was done in the long term period. The same publication shows the total emission of particles and gases in 2005 (in tons): particles 17 000, SO_2 - 124 000, NOx - 25 000, CO_2 - 20 million. Although the analyses carried out in the spring of 2007 showed that at some distance from the thermal power plants there was no excess of the harmful gases (apart from suspended particles), the conclusion was that the environment was seriously threatened (CIP, 2007). Here, we should bear in mind the long term period, as well as those weather conditions when wind and rain deposit huge amounts of harmful substances on the land cover. A consequence of lignite combustion leads to emissions of nitrogen oxides, sulfur and carbon, hydrocarbons, dust, soot, smoke and other suspended particles. "Acid rains" are formed under the influence of nitrogen oxides (N_2O, NO, NO_2), which have good solubility and corrosion in the presence of sulfur dioxide, gas without color and odor. This creates a permanent environmental degradation.

The ashes dumping site of the thermal power plant Nikola Tesla 2 is located in a direct proximity of Dubrava. Solid phase after the burning of coal is deposited here, and the quantity is from 12 to 14% of the primary mass of coal. Most of all elements, that are present in coal, remain in the solid phase, i.e., in the ash and smoke residue. Of particular concern is the significant proportion of heavy metals and radioactive elements. How all this affects the groundwater of Dubrava will remain a mystery for now, but one detail from the field has remained in our memory. According to narration of many local residents of Grabovac (nearest to the landfill), it happens that the wind from the north direction brings a huge cloud of ash and soot that deposit in the settlement. On that occasion, anyone who happens to be in his house immediately closes the doors and windows and removes all valuables from the yard. The cloud leaves, as they say, a layer of ash thick like a finger which cannot be removed long after.

8.3 Pointed sources of pollution

It may seem that these sources of pollution are not as dangerous as the first ones, but having in mind the microbiological analyses, it must be concluded that their role is very important. They all occur as a result of extremely poor communal sanitation. Lack of water supply network forces the residents to use local sources of drinking water (wells, springs), and the absence of a common system of the used water drains makes every household to solve the problem itself. Although there is the legislation on the facilities construction for the local water supply and drainage, it is not respected. Irregularities occur most often during the construction of septic tanks, pit latrines and manure disposal sites. First of all, the required distance of a septic tank to the well is not respected, so they are often a few meters apart. The next disadvantage is irregular retaining wall of the septic tanks and pit latrines where there is no isolation of fecal matters from the surrounding land. The same applies to the manure disposal sites, which are unsecured and subject to external influences long during the year. Because of all the above mentioned, the fecal waters penetrate the aquifers and make them constantly contaminated. This contamination is usually both, chemical and bacteriological.

Ammonia and organic nitrogen, mostly from the human urinary system, are commonly present in wastewater within septic tanks. Typically, almost all ammonia is converted into nitrate before leaving the septic-tank soil-absorption system drain field. Once nitrate passes below the zone of aerobic bacteria and the roots of plants, negligible attenuation occurs as it travels through the soil (Franks, 1972, as cited in Lowe et al., 2010). Once in ground water, nitrate becomes mobile and can persist in the environment for long periods of time. Areas having high densities of septic-tank systems risk elevated nitrate concentrations reaching unacceptable levels. In the early phases of ground-water quality degradation associated with septic-tank systems, nitrate is likely to be the only pollutant detected (Deese, 1986, as cited in Lowe et al., 2010). As the effluent from a septic-tank soil absorption system leaves the drain field and percolates into the underlying soil, it can have high concentrations of pathogens, such as viruses and bacteria. Organisms such as bacteria can be mechanically filtered by fine-grained soils and are typically removed after traveling a relatively short distance in the unsaturated zone. Pathogens can travel up to 40 feet (12 m) in the unsaturated zone in some soils (Franks, 1972, as cited in Lowe et al., 2010). Some viruses can survive up to 250 days (USEPA, 1987), which is the minimum ground-water time of travel for public water-supply wells or springs to be separated from potential biological contamination sources.

Therefore, neither by the depth of bacteria penetration into the soil, nor by the retention time in it, the groundwater on Dubrava can be immune to this contamination. If we count the longest path that the groundwater passes on Dubrava (from the hydro-geological divide to the spring horizon, about 3 km), it would need about 600 days (5 m/day). However, since the vast majority of the wells is located along the border areas and densely grouped, their interactions are very distinct. So, the eventual increased presence of bacteria in some part rapidly expands to all downstream wells. Apart from penetrating the soil, on Dubrava there is a problem of inadequate protection of the wells causing pollution directly to the groundwater. This type of contamination is usually bacteriological. It should be also noted that even in the settlements in the area of Dubrava there is no organized waste disposal site, so the illegal dumps of solid waste that have a local impact on the groundwater quality are very frequent.

8.4 Springs

All that leads to the contamination of the well waters on Dubrava is largely reflected in the spring waters as well. Concentrations of pollutants are not as drastic, but for nitrate and especially total aerobic mesophilic bacteria, they are beyond the allowable limits. If we followed the rules of drinking water properness, no spring would be usable. The survey conducted among owners of the parcels that these springs belong, has led to the data that out of the 88 tested, even 34 are used for drinking, and 43 also for watering livestock. And only a few of them use the basic disinfectants. Most of these springs were in operation in the seventies of the last century. What kind of a quality the water was, we would not know, but the springs were maintained and served as the pride to their owners. Many of them were tapped and equipped and, on weekends and holidays, "the whole village came down to the springs" (referring primarily to Grabovac and Dren). The current situation is such that most are hardly reachable, neglected and almost forgotten.

9. Conclusion

Before receiving the results of water quality on Dubrava, there have been some indications that the situation is not satisfactory. Therefore, at the same time with the field work, a wide action that will put the entire Dubrava under a certain degree of protection has been initiated. The implementation should involve, besides the local community, Obrenovac Municipality, also the professional services of the region of Belgrade administration, as well as those of the state level. It was concluded that the site of Dubrava is a unique combination of natural and anthropogenic values and with its environment is a challenge for everyone involved in environmental protection.

It has already been mentioned that the current spring series (from Grabovac to Dren), as part of the same water horizon, does not exist anywhere in Serbia in a similar form. Although the springs are of low yield (together with about 350 liters per minute, constant throughout the year and without oscillations), they are a real hydrological jewel. Besides the intention to try and help people on Dubrava when the water is in question, this research is also largely focused on the protection action of the springs and their revitalization.

Where should we start? In 58% of the examined (tested) local wells, the significant construction-technical and sanitary-hygienic deficiencies that affect the quality of drinking water have been observed. The sanitary protection zone has not been established in most of the wells, and there is no investment-technical documentation as well. So, the first step is training of the people who, by their carelessness, endanger all the residents of Dubrava, and then follows the action of the professional and inspection services in order to eliminate the local sources of contamination. This is primarily to check the condition and rehabilitation of the septic tanks, pit latrines and manure disposal sites that were improperly constructed. As for the wells themselves, it is sometimes enough to clear them from mechanical impurities and to hyper-chlorinate them. This also applies to the springs that were tapped. In the case of air pollution that originates from the thermal power plants, it can not be eliminated, but the regular control and filter installations replacement, as well as the prescribed rehabilitation of the ashes disposal sites, could be maintained at an acceptable level.

10. Acknowledgement

This paper was realized as a part of the project "Studying climate change and its influence on the environment: impacts, adaptation and mitigation" (43007) financed by the Ministry of Education and Science of the Republic of Serbia within the framework of integrated and interdisciplinary research for the period 2011-2014.

11. References

CIP (2007). *The intersection of the current status of the presence of hazardous and harmful substances in the environment of the Municipality of Obrenovac.* Belgrade. (in Serbian)

Deese, P.L. (1986). *An evaluation of septic leachate detection.* USEPA Project Summary EPA/600/52-86/052, 2 p.

Dragicevic, S.; Stepic, M. & Karic, M. (2008). *Natural potentials and degraded areas of Obrenovac municipality.* Jantar groupe, Belgrade. (in Serbian)

Dragicevic, S.; Nenadovic, S.; Jovanovic B.; Milanovic M.; Novkovic I.; Pavic, D & Ljesevic M. (2010). Degradation of Topciderska River water quality (Belgrade). *Carpathian Journal of Earth and Environmental Sciences*, October 2010, Vol. 5, No. 2, p. 177 - 184

Foster, S.; Hirata, R.; Gomes, D.; D'Elia, M. & Paris, M. (2002). *Groundwater Quality Protection: a Guide for Water Utilities,* Municipal Authorities and Environment Agencies. World Bank Publication. Washington, D.C., USA.

Geozavod (1976): Geological Map 1:100000, Institute of Geology, Belgrade.

Goss, M. J.; Barry, D. A. J. & Rudolph, D.L. (1998). Contamination in Ontario farmstead domestic wells and its association with agriculture: 1. Results from drinking water wells. *Journal of Contaminant Hydrology* 32(3-4): 267-293.

Keeney, D. R. (1989). *Sources of nitrate to groundwater.* In: Nitrogen management and ground water protection. ed R.F. Follett, pp. 23-33. New York, NY: Elsevier Science.

Kovar, K. & Krasny, J. (1995). *Groundwater Quality: Remediation and Protection.* IAHS Publication 225, IAHS Press, Wallingford, UK.

Lagerstedt, E.; Jacks, G. & Sefe, F. (1994). Nitrate in groundwater and N circulation in eastern Botswana. *Environmental Geology* 23:60–64

Lowe, M.; Wallace, J.; Sabbah W. & Kneedy, J. (2010). *Science-based land-use planning tools to help protect ground-water quality, Cedar valley, Iron County, Utah.* Special study 134, Utah geological survey. A division of Utah department of natural resources. ISBN 978-1-55791-836-9

Markovic, J. (1967). Relief of Macva, Sabacka Posavina and Pocerina. *A special issue of the Science Department of Geography BU*, Vol. I, Belgrade (in Serbian)

Milojevic, N.; Filipovic, B. & Dimitrijevic, N. (1975). *Hydrogeology of the city of Belgrade territory*. Obod, Cetinje. (in Serbian)

Municipality Press (2010). *Environmental situation in Obrenovac.* N° 1, june.

Nenadovic, S.; Matovic, Lj.; Milanovic, M.; Janicevic, S.; Grbovic, J.; Novakovic, I. & Ljesevic, M. (2010). Impacts of some meteorological parameters on the SO_2 concentrations in the City of Obrenovac, Serbia. *Journal of Serbian Chemical Society*, Belgrade, *75* (5) 703–715

Nolan, B. T. & Stoner, J. D. (2000). Nutrients in groundwaters of the conterminous United States 1992-1995. *Environmental Science & Technology* 34(7): 1156-1165.

Rump, H.H. & Krist, H. (1988). *Laboratory Manual for the Examination of Water, Waste Water and Soil*, VCH, Weinheim.

Spalding, R.F. & Exner, M.E. (1993). Occurrence of nitrate in groundwater: A review. *Journal of Environmental Quality* 22(3): 392-402.

SZZZ (1990). *Drinking water, Standard methods for the examination of hygiene*. Federal Institute for Health Protection, Belgrade (in Serbian)

USEPA (1983). *Methods for chemical analysis of water and wastes*, United States Environmental Protection Agency, Cincinnati, Ohio.

USEPA (1987). *Guidelines for delineation of wellhead protection areas*. United States Environmental Protection Agency, Washington, D.C.

USEPA (1990). *National Pesticide Survey: summary of results of EPA's national survey of pesticides in drinking water*. United States Environmental Protection Agency, Washington, D.C.

Vaux, H. (2011). Groundwater under stress: the importance of management. *Environmental Earth Sciences*. Volume 62, Number 1, 19-23

Visser, A.; Dubus, I. et al. (2009). Comparison of methods for the detection and extrapolation of trends in groundwater quality. *Journal of environmental monitoring : JEM* 11(11): 2030-2043.

VGI (1970): Topographic Map 1:50000, Belgrade.

Wassenaar, L. & Baisden, W.T. (2011). *Efficacy of BMPs to amend nitrate contamination in groundwater systems: a Canadian experience for comparison to New Zealand*. In: Currie, L.D.; Christensen, C.L. (eds) Adding to the knowledge base for the nutrient manager : 24th annual FLRC workshop, Massey University.

Zivkovic, N & Dragicevic, S. (2003). *The regime of precipitation in the region of Belgrade*. Faculty of Geography, Belgrade. ISBN 86-82657-31-7 (in Serbian)

13

Surface Water Quality Monitoring in Nigeria: Situational Analysis and Future Management Strategy

A.M. Taiwo[1], O.O. Olujimi[2], O. Bamgbose[3] and T.A. Arowolo[3]*
[1]*Division of Environmental Health Risk Management, University of Birmingham,*
[2]*Department of Chemistry, Cape Peninsula University of Technology,*
[3]*Department of Environmental Management and Toxicology,*
University of Agriculture, Abeokuta,
[1]*United Kingdom*
[2]*South Africa*
[3]*Nigeria*

1. Introduction

Water is a finite resource that is very essential for the human existence, agriculture, industry etc. Without any doubt, inadequate quantity and quality of water have serious impact on sustainable development. In developing countries, most of which have huge debt burdens, population explosion and moderate to rapid urbanization, people have little or no option but to accept water sources of doubtful quality, due to lack of better alternative sources or due to economic and technological constraints to treat the available water adequately before use (Calamari and Naeve, 1994; Aina and Adedipe, 1996). The scarcity of clean water and pollution of fresh water has therefore led to a situation in which one-fifth of the urban dwellers in developing countries and three quarters of their rural dwelling population do not have access to reasonably safe water supplies (Lloyd and Helmer, 1992).

Assessment of water is not only for suitability for human consumption but also in relation to its agricultural, industrial, recreational, commercial uses and its ability to sustain aquatic life. Water quality monitoring is therefore a fundamental tool in the management of freshwater resources. To underpin its importance, *World Health Organization* (WHO), *United Nations Environment Programme* (UNEP), *United Nations Educational, Scientific and Cultural Organization* (UNESCO) and *World Meteorological Organization* (WMO) launched in 1977, a water monitoring programme to collect detailed information on the quality of global ground and surface water.

Monitoring can be conducted for the following five major purposes:

- characterize waters and identify changes or trends in water quality over time;
- identify specific existing or emerging water quality problems;

*Corresponding Author

- gather information to design specific pollution prevention or remediation programs;
- determine whether program goals -- such as compliance with pollution regulations or implementation of effective pollution control actions -- are being met; and
- respond to emergencies, such as spills and floods.

Nigeria, the most populous country in Africa, has an area of 923,768 km^2, and located between Latitudes 4° 16' N and 13° 52' N, and Longitudes 2° 49' E and 14° 37 'E (see Figure 1). It is stretched across dimensions of 1,200 km from east to west and 1,000 km from north to south (http://ramsar.wetlands.org/portals/15/nigeria.pdf). The space occupied by inland water bodies in Nigeria is estimated at 900 km^2 (Ekiye and Zejiao, 2010) representing 0.1 % of the total land mass while the coastal area stretched up to 853 km (comprising inshore waters, coastal lagoons, estuaries and mangrove especially in the Niger Delta (http://www. unep. org/AbidjanConvention/docs/the%20status%20of%20the%20nigerian%20coastal%20zones%20version%202.pdf). The total area occupied by water in Nigeria is 13,000 km^2.

Nigeria has abundant water resources covering an enormous and diverse landscape, although they are unevenly distributed over the country (WHO/UNEP, 1997). The four major drainage systems in the Nigeria are:

- The Niger River Basin Drainage System with its major tributaries of Benue, Sokoto-Rima, Kaduna, Gongola, Katsina-Ala, Donga, Tarabe, Hawal and Anambara Rivers.
- The Lake Chad Inland Drainage System comprising the Kano, Hadejia, Jama'are Misau, Komadougou-Yobe, Yedoseram and Ebeji Rivers.
- The Atlantic Drainage System (east of the Niger) comprising the Cross, Imo, Qua Iboe and Kwa Rivers.
- The Atlantic Drainage System (west of Niger) made up of the Ogun, Oshun, Owena and Benin Rivers.

Apart from the Lake Chad Drainage System, the remaining three drainage systems terminate in the Atlantic Ocean with an extensive network of delta channels (Figure 2).

The natural quality of rivers and streams at any point reflects the quality of upstream contributions of surface run-off and groundwater discharge. The various water bodies in the country are used for various purposes such as fishing, transportation, sand mining, irrigation, recreation, abstraction for industrial and domestic purposes as well as electric generation.

Urbanization, intensive agriculture, recreation, and the manufacturing industry are affecting water quality throughout the world. Available literature on environmental monitoring of surface water indicated that streams and rivers in the country are showing increasing trend of water pollution due to increase population, industrialization and urbanization. Waste generations by the industries and households have continued to increase. These wastes are indiscriminately disposed-off into the water bodies. This has led to pollution of inland water bodies and coastal waters and subsequently increased water quality parameters such as heavy metals, nutrients and organic matter, soluble ions, oil and grease, and organic chemicals such as pesticides and poly-nuclear aromatic hydrocarbons (PAHs), (Esoka and Umaru, 2006; Adebayo et al., 2007; Jaji et al., 2007; Mashi and Alhassan, 2007; Solomon, 2009). Urgent attention is therefore necessary to mitigate water pollution problems in Nigeria through monitoring as well as enforcement of emission standards by industries (Ekiye and Zejiao, 2010).

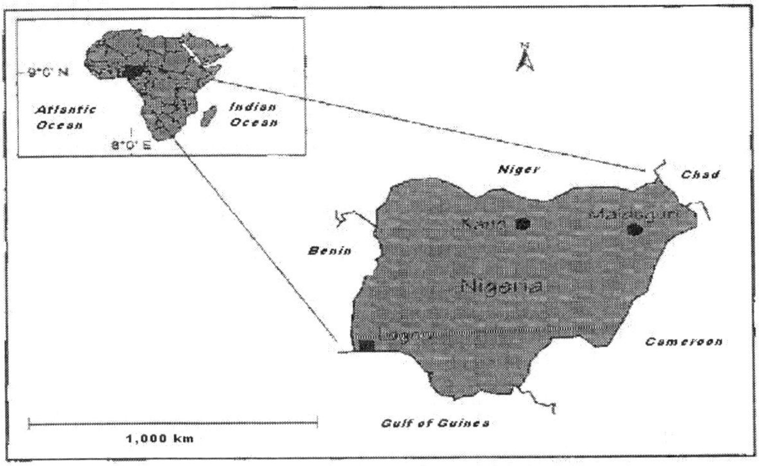

Fig. 1. Location map of Nigeria

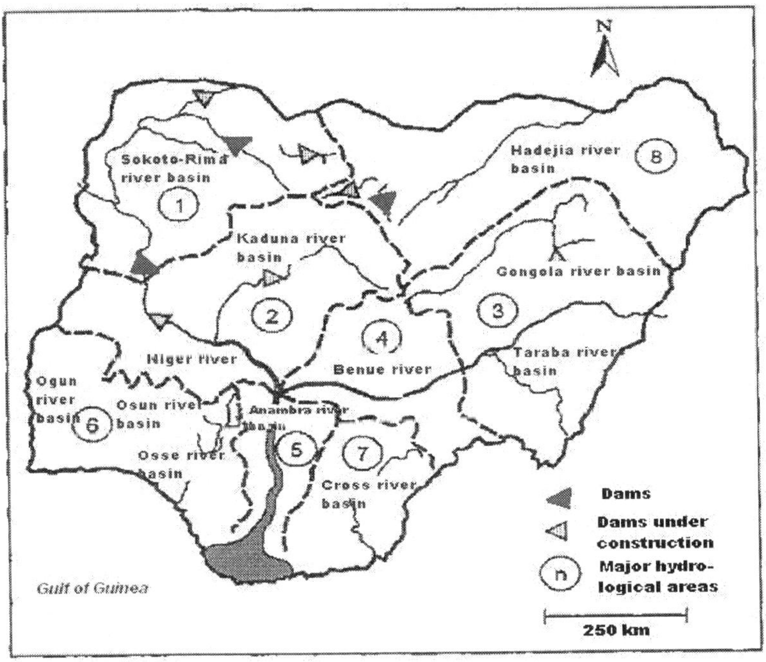

Fig. 2. Map of Nigeria showing major rivers and hydrological basins: 1 Niger North, 2 Niger Central, 3 Upper Benue, 4 Lower Benue, 5 Niger South, 6 Western Littoral, 7 Eastern Littoral, 8 Lake Chad (WHO/UNEP, 1997).

According to National Bureau of Statistics (2009), at least 27% Nigerians depended absolutely on streams, pond, river and rainwater for their drinking water source. Research has shown high prevalence of waterborne diseases such as cholera, diarrhoea, dysentery, hepatitis etc. among Nigerians (Oguntoke et al., 2009; Raji and Ibrahim, 2011). The need for water quality monitoring is paramount to safeguard the public health and also to protect the water resource in Nigeria. To emphasise the importance of environmental monitoring in Nigeria, the then Federal Environmental Protection Agency (FEPA) - now Federal Ministry of Environment and National Water Resources Institute (NWRI) organised a National Seminar on "Water Quality Monitoring and Environmental Status in Nigeria" between October 16 to 18, 1991. The proceeding of this National Seminar was published by FEPA in 1996 (Aina and Adedipe, 1996). The major objective of this paper is to assess the current situation of surface water quality monitoring in Nigeria two decades after the FEPA/NWRI National Seminar and provide future management strategy that may be adopted by the authorities. Certainly, information concerning the water quality of our rivers will provide a useful instrument for policy makers to formulate management strategy for control and abatement of water pollution.

2. Sources of water pollution in Nigeria

2.1 Industrial discharges and oil spills

Studies carried out in most cities in Nigeria had shown that industrial effluent is one of the main sources of surface water pollution in Nigeria (Ekiye and Zejiao, 2010). Industrial effluents when discharged directly into the rivers without prior treatment have capacity of increasing water quality parameters. Dada (1997) indicated that less than 10 % of industries in Nigeria treat their effluents before being discharged into the rivers. This has led to high load of inorganic metals such as Pb, Cr and Fe in most of water bodies (Ahmed and Tanko, 2000; Wakawa et al., 2008). Table 1 shows the results of physico-chemical parameters of effluents collected from some industries in Lagos, Kaduna and Port Harcourt. Most of the results were higher than the permissible limits set by FEPA (1991). The resultant effects of this will be on the receiving streams and rivers. The impacts could include water quality impairment, reduction in fish abundance and effect on water-usage for recreation, industrial and domestic purposes. High phosphate concentrations in these effluents could result into nutrient enrichment of the receiving water bodies thereby leading to ecological disaster. Metal pollution of Warri River by industrial discharges has been reported by Ayenimo et al. (2005). The River was monitored for heavy metals such as Fe, Cu, Ba, Pb, Cd, Cr, Ni and Co. Results showed elevated values of these metals at sampling point located near an industry. Correlation analysis of the metals also suggested common source. Other water quality parameters showed elevated values indicating pollution by the nearby industry.

The activities of the oil industries in the Niger-Delta region of Nigeria have impacted negatively on the surface water quality around the area. This has led to water scarcity, disruption of socio-economic activities and poor aesthetic quality of most of the water bodies polluted by the oil spills (Egborge, 1994). Most of the rivers around the Niger-Delta region of the country could not be abstracted for treatment for drinking purpose because of pollution by crude oil. The impact of oil activities in these areas had done much havoc to the environment of this region most especially on the water resources.

Effluent Parameters	Lagos (Adebayo et al., 2007)		Kaduna (Yusuf and Shonibare, 2004)*	Port Harcourt (Ajao and Anurigwo, 2002)*	FEPA (1991) standard
Temperature (°C)	27.6	30.3	30		<40
Color (CTU)			2400		
pH	7.6	4.8			6-9
Conductivity (mS cm^{-1})	761	1156.6			
Turbidity (NTU)	11.6	573.6			
Salinity (%)	0.1	0.5			
Alkalinity (mg L^{-1})	766.6	445			
Free Chlorine (mg L^{-1})			0.5		
Nitrate (mg L^{-1})			4.0	362	20
Ammonia (mg L^{-1})			1.0		
Phosphate (mg L^{-1})			1.0	836	5
Total Hardness (mg L^{-1})	1233.3	4083.3			
TSS (mg L^{-1})	320	833.3	400	3533	30
TDS (mg L^{-1})			1100		2000
Oil and Grease (mg L^{-1})	20	0.0	7.0	2343	10
BOD (mg L^{-1})	534	1352.3	300	4374	30
COD (mg L^{-1})	850	2253	1800		
H_2S (mg L^{-1})	17.2	130	0.6		

* Total effluent discharge from observed industries, TSS-total suspended solids, TDS-total dissolved solids, BOD-biochemical oxygen demand, COD-chemical oxygen demand, DO-dissolved oxygen (Data adapted from Ekiye and Zejiao, 2010).

Table 1. Effluent analysis of some industrial discharges in Nigeria

2.2 Municipal and agricultural wastes

Waste management is a major problem in most developing nations of the world including Nigeria (Taiwo, 2011). Indiscriminate disposal of municipal wastes remains a major threat to surface water pollution in Nigeria. In most cases, sewage and waste water from homes are routed into the rivers and streams. Jaji et al. (2007) found elevated water quality parameters in some sampling locations of Ogun River. These were partly attributed to the activities of abattoir located close to the River at a notable market in Abeokuta metropolis. The work of Arimoro et al. (2007) on the impact of sawmill activities on the water quality of River Benin reported high BOD and low DO values at the discharge point of the wastes into the River.

The impact of point source pollution from sewage treatment oxidation pond on a receiving stream was studied by Ogunfowokan et al. (2005). The researchers observed significant elevation of water indices such as pH, BOD, nitrate, phosphate and TSS. It is well known that oxygen depletion in water bodies could cause fish death while increase in BOD signifies high load of organic matter. Also, organic matter decomposition in surface water produced inorganic nutrients such as ammonia, nitrate and phosphorus with resultant effects of eutrophication and other serious ecological problems of such water body (Ogunfowokan et al., 2005). Taiwo (2010) has also observed high water quality parameters of a stream in Abeokuta due to direct discharge of poultry wastes into the stream. The use of pesticides and fertilizer for bumper food production is a well known policy of several Governments all over the world. However, agriculture remains the major source of nitrate and phosphate pollution of surface water. Nitrate in drinking water is detrimental to infant health due to the disease known as methemoglobineamia (Taiwo, 2010).

2.3 Urban run-off

Urbanization in most Nigerian cities has resulted in the concentration of large population in some areas living under poor sanitation conditions (Olade, 1987). This invariably has led to increased waste generations with heaps of waste everywhere. During rainfall, some of these wastes are washed into the poor drainage systems and subsequently, into nearby rivers (Taiwo et al. 2011). Lack of town planning principles and strategies in Nigeria's cities and towns had aggravated the risks of urban run-off with resultant effect on surface water. The poorly managed drainage system in the country had caused the surface water impairment due to erosions during rainfall. Rainfall runoff carries all sorts of pollutants from houses, industries, farmland and dumping sites. Research has shown that some of the water quality parameters of both ground and surface water often rise up during rainfall with high values of turbidity, solids and anionic species often been recorded (Jaji et al., 2007; Mustapha, 2008; Taiwo, 2010; Taiwo et al., 2011). The effect of urban run-off has been studied on the Epie Creek in the Niger Delta by Izonfuo and Bariweni (2001). The impact of human activities around the Creek was felt on the water body as low DO values were recorded during the wet season due to urban run-off.

Agricultural run-off of pesticides, plant and animals wastes is also a major contributing source of organic pollution to water bodies in Nigeria. The work of Mustapha (2008) had linked the periodic eutrophication of Oyun Reservior in Offa, Kwara state to run-off of phosphate fertilizers from nearby farms in addition to cow dungs washing from the watershed into the Reservoir. Water pollution through surface run-off has been reported in literatures with subsequent effects on nutrient enrichment, water quality impairment, marine lives spawning ground destruction and fish kill (Martin et al., 1998; Izonfuo and Bariweni, 2001).

3. Water quality of rivers in Nigeria

Figure 3 shows some important rivers in Nigeria while Table 2 shows the results of physico-chemical parameters of some of these rivers which have been monitored by several researchers in the country.

High TSS found in rivers in Nigeria has tendency of reducing the light penetration into the river leading to a reduced photosynthesis with consequent effects on both phytoplankton and zooplankton populations of the aquatic environment. A study by Ajibade (2004) has shown high TSS values in Asa River (Kwara state) while Osibanjo et al. (2011) has also reported similar findings for Rivers Ona and Alaro in Ibadan. Harrod and Theurer (2002) in a similar study reported that high TSS concentration could cause a reduced development and survival of salmonid eggs and larvae. Clogging of TSS on fish gills could also result into stress, reduced growth, suppressed-immune system leading to increased susceptibility to disease and osmotic dysfunction and death (Bilotta and Brazier, 2008). Elevated values of TSS are capable of shielding harmful organisms in drinking water (Taiwo et al., 2011). TSS could also act as a vector of nutrients such as phosphorus (Heathwaite, 1994), and toxic compounds such as pesticides and herbicides from the land surface to the water body (Kronvang et al., 2003) leading to proliferation of phytoplankton in rivers. In surface water, TSS could cause drift in invertebrate population (Bilotta and Brazier, 2008).

Another notable water quality characteristic of rivers in Nigeria is high turbidity. Turbidity values reported for most rivers in Nigeria were far greater than 5.0 NTU limit given by WHO (2008) (Table 2). Ajibade (2004), Adefemi et al. (2007) and Wakawa et al. (2008) have also reported elevated turbidity values in rivers in Nigeria. This could be linked to run-off effects as well as domestic and industrial discharges on the rivers.

Low BOD and COD values have been reported in New Calabar River and Kubanni River in Kaduna. However, a high BOD and COD values have been observed for Challawa River in Kano State with mean concentrations ranging between 10 to 30 mg L^{-1} and 170 to 260 mg L^{-1} respectively (Wakawa et al., 2008). Very low DO values (2.67-3.30 mg L^{-1}) were also observed in Challawa River. Pollution of the rivers was directly linked with the industrial effluent discharges. Osibanjo et al. (2011) also reported high COD values for the water samples from Rivers Ona and Alaro. The authors attributed these to leachate from dumpsites, agricultural and urban runoffs.

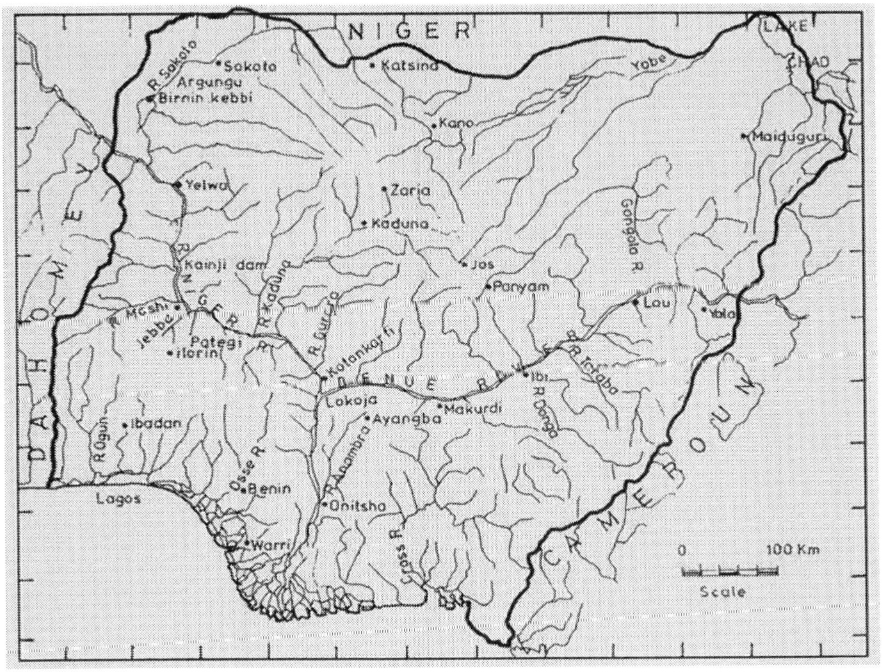

Fig. 3. Map showing rivers in Nigeria.

Nutrient enrichment is predominant in most rivers in Nigeria. The values of nitrates and phosphate depicted by authors in Table 2 were low. But reports by other authors on some rivers indicated high nutrient load of most rivers in Nigeria. For instance, Olajire and Imeokparia (2001) observed high concentrations of nitrate, ammonia and phosphate in Osun River, a consequence of human activities. Human activities observed along the study area include agricultural land-use, anthropogenic activities and industrialization. Farming operations around the area were said to have contributed immensely to elevated values of ammonia and phosphate (Olajire and Imeokparia, 2001). Edet and Worden (2009) assessed the seasonal and tidal effects on the physical parameters of river and groundwater in Calabar, Nigeria. The researchers reported a significant seasonal effect on temperature, redox potential (Eh), and DO in the river water. They also observed that a significant tidal influence existed on DO in both river-and groundwater. Comparison between groundwater and river water showed statistically significant difference in EC, TDS, Eh, DO, Na, Cl^- and NO_3^-. The significant differences in EC, TDS, Na and Cl^- were attributed to tidal flushing; the difference in Eh was due to geology of the area while NO_3^- was as a result of anthropogenic pollution. The study concluded that tidal flushing, anthropogenic effects and oxygen supply during recharge contribute to the shaping of water chemistry in the area.

	Ogun River (Jaji et al., 2007)				New Calabar River (Abu and Egenonu, 2008)		Kubanni River, Kaduna (Agbogu et al., 2006)		WHO, 2008
	Mean	SD	Mean	SD	Mean	SD	Mean	SD	
	Wet season		Dry season						
Temp (°C)	27.0	0.47	30.03	1.17	25.8	0.92	25.71	0.62	
pH	7.12	0.17	7.09	0.40	5.35	0.41	6.73	0.03	6.5-8.5
DO (mg L^{-1})	7.6	1.27	5.36	0.73	6.31	0.81			
BOD (mg L^{-1})					2.24	0.41	1.83	0.28	
COD (mg L^{-1})					6.4	2.72			
EC (μS cm^{-1})							224.08	9.38	
Salinity (%)					0.01	0.01			
SO_4^{2-} (mg L^{-1})					22.95	7.47			<250
Turbidity (NTU)	42.66	15.53	15.36	5.94			106.4	21.11	<5.0
TSS (mg L^{-1})							18.8	1.45	
TDS (mg L^{-1})	156.05	39.64	442.84	900.26	1.26	0.22	26.2	1.76	<600
Cl^- (mg L^{-1})	16.64	4.81	261.97	732.97	5.44	3.19	29.02	2.12	
NO_3^- (mg L^{-1})	3.80	5.27	3.64	2.86	0.32	0.12	1.86	0.09	11*, 50
PO_4^{3-} (mg L^{-1})	0.07	0.03	0.02	0.01	0.17	0.01	0.16	0.01	
Oil & Grease (mg L^{-1})	146.96	71.70	71.33	72.55	0.02	0.01			
Coliform (cfu 100 mL^{-1})	6180	3275.6	4645	6562.5			13000	4900	0.0
Pb (mg L^{-1})	0.02	0.02	0.04	0.03	0.14	0.04			0.01
Cd (mg L^{-1})	0.01	0.01	0.01	0.01	0.05	0.03			
Zn (mg L^{-1})	0.03	0.01	0.19	0.32					3-5
Fe (mg L^{-1})	3.28	0.82	1.86	0.49	0.24	0.09			0.3
Mn (mg L^{-1})	0.13	0.05	0.41	0.54					0.4
Cu (mg L^{-1})	0.03	0.02	0.12	0.19	0.07	0.03			2.0
Ni (mg L^{-1})					0.01	0.00			0.07
Hg (mg L^{-1})					0.01				0.06

SD-standard deviation (original data from the literatures have been subjected to simple descriptive statistics).

Table 2. Mean concentration of water quality parameters of some Rivers in Nigeria

Metal pollution of some rivers in Nigeria may be due to industrial discharges, corrosion of iron and steel materials in building, leachates from dumpsites and vehicles etc (Jaji et al. 2007). Fe concentrations in most Nigeria's rivers are usually greater than WHO standard of 0.3 mg L^{-1} in drinking water (Offiong and Edet, 1996). High Pb concentration in some surface water samples have been attributed to the use of leaded petrol in vehicles. Taiwo (2010) found elevated values of lead at a sampling site of Alakata stream in Abeokuta. High lead level in surface water may also be due to vehicular deposition. Dry deposition of particulate lead on water bodies is capable of increasing Pb level of surface water. Lead is a potential killer especially in children. In some villages in Gummi and Bukkuyum Local Government area of Zamfara state, more than 400 deaths were reported due to lead poisoning (Galadima et al., 2011). Heavy metals are potential threats to the environment and ecosystem due to their persistence and bio-accumulation in food chain.

The presence of oil and grease on the surface of water bodies in Nigeria is a sign of pollution which may have serious effect on the aquatic ecosystem of the nation. Oil and grease is toxic to aquatic organisms in general and also capable of reducing dissolved oxygen. Between 1997 and 2001 in the Niger Delta region of Nigeria, 2,097 oil spill incidents were recorded (Nwilo and Badejo, 2005). The activity of oil exploitation and refining has reduced the water quality of Nigeria's river in the Niger Delta region. The mangrove ecosystem along the coastal areas of the Niger Delta has been wantonly destroyed by oil spills (Nwilo and Badejo 2005). Besides the oil region of Niger Delta, which is synonymous with oil spill pollution, elevated levels of oil and grease have been reported by Osibanjo et al. (2011) at the downstream of Rivers Ona and Alaro in Ibadan, South-western Nigeria. The authors attributed these to urban run-off from auto repair workshop and petroleum depot.

The high values of coliform reported for some river water samples confirm faecal pollution from domestic sewage, dumping sites, abattoir activities etc. High coliform values are typical characteristics of many rivers in Nigeria. For instance, high population of faecal coliform counts has been reported in Ikpoba River, Edo state of Nigeria by Tatah and Ikenebomeh (1999). Similarly, Nwankwu (1992) also reported coliform values in the range of 3100-150, 000 cfu 100 mL^{-1} at Iddo area of the Lagos lagoon. The high population of these microbial pollutants was linked to contaminations from the dumpsites around the Lagoon. Pollution of surface water in Nigeria therefore calls for great attention. Some people see water body as medium for waste disposal. Feacal pollution of rivers in Nigeria signifies poor sanitation management as well as unhygienic manner of living among people, especially those living close to the riverine areas.

It has also been observed that pathogenic contamination of Nigeria's rivers comes from aquaculture practices involving fertilization of ponds with cow and poultry manures; and direct dumping of faecal matters into the rivers (Obasohan et al., 2010). Presence of microbial pollutants in drinking water could pose risks to public health of waterborne diseases such as cholera, diarrhoea, hepatitis, dysentery, poliomyelitis, typhoid etc.

Meinhardt (2006) has highlighted susceptible population to waterborne diseases to include pregnant women, children, immune-suppressed individuals, geriatric patients and patients with pre-existing clinical disorders and chronic diseases. Nearly 40,000 cases cholera epidemic were reported in 11 states in Nigeria from January–October 2010, out of which 1,500 people were confirmed dead (Adewara and Visser, 2011). Diarrhoea is the second largest cause of children's mortalities in Nigeria according to Water Initiative (2010). Nigerian children of age less than 5 years make up 17% of the total annual deaths of 1.8 millions being recorded globally due to poor sanitation (Water Initiative, 2010).

4. Organic pollutants monitoring in Nigeria freshwater systems

Unlike inorganic pollutants and pathogens monitoring in Nigeria water systems, organic pollutants monitoring had suffered major neglects in terms of qualitative and quantitative monitoring. Due to poor institutional funding by the Nigeria governments, studies on the levels of polynuclear aromatic hydrocarbons (PAHs) and other organic contaminants in the country are very scanty. Until recently, most studies on organic compounds dates back to the 1970's and 1980's when there was some equipment and little funding to conduct research. The poor funding had resulted in poor human resources base coupled with lack of instrumentation to monitor the level and distribution of these organic pollutants in the environment.

Despite these challenges, few studies had recently reported some levels of PAHs in the Nigeria water systems. PAHs have been listed as priority pollutants that need to be monitored in both industrial and domestic effluents and freshwater systems (Emoyan, 2009). The health impacts of PAHs include carcinogenic, cardiovascular, bone marrow and liver toxicity. Emoyan (2009) reported the concentration range of 0.2309 to 1.0468 mg L^{-1} for PAHs in surface water due to contamination from Kokori-oil field in the Niger-Delta. Fluoren was the dominant of the 16 PAHs priority pollutants investigated. The source of water contamination was identified to be mainly petrogenic. The reported concentrations were higher than the CCME (2008) and the Netherland MPC guidelines. In addition to PAHs, another important set of organic pollutants are pesticides. Pesticides are highly toxic, synthetic organic compounds that are widely used in agriculture and the industries (Adeboyejo et al., 2011; Adeyemi et al., 2011). Pesticides usage in Nigeria had soared in the past decades such that over 21 different types of organophosphates, organochlorine and carbamates insecticides have been introduced into the Nigeria market. Out of the broadly classified pesticides, organochlorine pesticides (OCPs) constitute the most widely used for agricultural purposes despite their ban in other countries (Bouman, 2004; Blaso et al., 2005; Ize-Iyamu et al., 2007; Adeyemi et al., 2011).

Organochlorine pesticides are known to be very stable, resistant to natural breakdown, lipo-soluble compounds that are capable of bioaccumulation and biomagnifications in the fatty parts of biological organisms (Adeyemi et al., 2011). The health effects of pesticides in man and animals include immune systems malfunction, endocrine disruption, breast cancer, irritation, dizziness, tremor, toxic and chronic convulsion (Ize-Iyamu et al., 2007; Ezemonye et al., 2009; Adeboyejo et al., 2011; Adeyemi et al., 2011). Like PAHs, studies on OCPs are very few. Pesticides unlike other organic compounds may enter ground water and surface water as run-off due to direct application to the soil surface or application to crops for destroying, preventing, repelling or mitigating against pests, animals, fungi and microorganisms (Ware and Whitacre, 2004; Ezemonye et al., 2009). Other sources of pesticides entrance into water bodies includes through point sources like sewage treatment plants, sewer overflows and poor management practices of farmers or from atmospheric deposition (Guruge and Tanabe, 2001; Mayon et al., 2006; Shukla et al., 2006).

In a recent study, Ezemonye and co-workers (2009) reported concentration ranges of diazinon (organophosphate insecticide) in water, sediment and fish samples from Warri River in Niger-Delta after extraction and quantification using high pressure liquid chromatography. The concentration range of 0.07 to 1.76 μg L^{-1}, 0.49 to 2.54 μg L^{-1} and 0.08 to 4.29 μg L^{-1} were reported for water, sediment and fish samples, respectively. Ize-Iyamu et al. (2007) investigated and reported seven OCPs namely lindane, aldrin, pp-DDE, pp-DDD, op-DDD, op-DDT and pp-DDT in water and two fish species (Tilapia Zilli and Claras lazera) from Ikoro, Ovia and Ogba rivers using gas chromatography - electron capture detector. Lindane and aldrin were the most dominant OCPs in all the samples analyzed ranging from 0.5712 to 0.8143 μg L^{-1} and 0.5031 to 0.3218 μg L^{-1}, respectively.

In a more recent study on Lagos Lagoon, Adeyemi et al. (2011) reported the levels of nine OCPs residue in water using solid-phase extraction followed by gas chromatography quantification. The concentration range for the nine OCPs are 0.006 to 0.950, ND to 0.067, ND to 0.123, 0.015 to 0.774, 0.015 to 0.996, 0.012 to 0.910, 0.005 to 0.477, 0.015 to 0.996 and 0.080 to 0.790 μg L^{-1} for chlordane, heptachlor, methoxychlor, hexachlorobenzene, endosulfan, dichelorophenyltrichloroethane, dichelorophenyldichloroethylene, dieldrin and aldrin, respectively. In another study on the Lagos lagoon, Adeboyejo et al. (2011) found the following range of OCPs in water samples: aldrin (ND - 658 ngl^{-1}), chlordane (218.0 - 702.0 ng L^{-1}), endrin (ND - 2551 ng L^{-1}), DDD (20.3 - 268 ng L^{-1}), pp-DDE (0 - 176 ngl^{-1}), α-HCH (ND - 2300 ngl^{-1}), γ-HCH (ND - 783 ngl^{-1}), β-HCH (28.9 - 518 ng L^{-1}), δ-HCH (0 - 498 ng L^{-1}), endosulfan (ND – 3726 ng L^{-1}) and heptachlor (ND – 1405 ng L^{-1},).

Also, some levels of OCPs have been reported in edible aquatic life in the Lagos lagoon. Adeboyejo et al. (2011) reported OCPs concentration ranges of ND to 30240, ND to 517, 26 to 5480, ND to 11430, ND to 37780, ND to 13260, ND to 13, 0 to 3510, ND to 11490, 0 to 55740, ND to 990, 38 to 421, 0 to 1030 and ND to 60.3 μg kg^{-1} for aldrin, dieldrin, chlordane, endrin, endosulfan, heptachlor, HCB, α-HCH, γ-HCH, β-HCH, δ-HCH, pp-DDD, pp-DDE and DDT, respectively.

Another group of organic pollutants of total neglect in the fresh and marine water systems are chemicals specifically classified as endocrine disruptors. According to Diamanti-Kandarakis et al. (2009), "an endocrine-disrupting substance is a compound, either natural or synthetic, which through environmental or inappropriate developmental exposures alters the hormonal and homeostatic systems that enable the organism to communicate with and respond to its environment." Endocrine disrupting chemicals (EDCs) consist of many natural and synthetic organic compounds, but are mostly manmade products such as alkylphenols, alkylphenols-ethonylates, and polychlorinated biphenyls (PCBs). Others include polychlorinated dibenzo-p-dioxins, polychlorinated dibenzofurans, organochlor pesticides, dichlorodiphenyl, dichloroethylene, nonylphenols, steroid hormones and phthalates (Arditsoglou and Voutsa, 2008; Hjelmborg et al., 2006; Mauricio et al., 2006). Of these groups of chemicals are Phthalates or phthalic acid esters (PAEs). They are diakyl or alkyl esters of 1,2 benzene carboxylic acid (Luks-Betlej et al., 2001; Alatriste-Mondragon et al., 2003; Adeniyi et al., 2008). They are formed when methanol, ethanol or other alcohols react with the carboxyl groups on the benzene ring of phthalic acids. The corresponding esters are formed with different alkyl chains, e.g., dimethyl phthalate (DMP), dibutyl phthalate (DBP) and di (2-ethylhexyl) phthalate (DEHP). Similar to other organic contaminants, EDCs are not been monitored in the freshwater systems to the best of our knowledge until recently when levels and the distribution pattern of phthalate esters were reported in some water samples from south western Nigeria.

Fatoki and Ogunfowokan (1993) reported PAEs concentrations in the river and tap water in the south western Nigeria. Water samples from the Lagos, Oyo, Osun, Ondo, and Delta states of Nigeria and the cities' water treatment plants were analyzed for the presence of phthalate esters. Extraction with $CHCl_2$ and liquid chromatography was used for compound separation. Several phthalate esters, dimethyi- (DMP), diethyl- (DEP), and di-n-butyl- (DBP) were found present at levels of 10 mg L^{-1} to 1472 mg L^{-1} in river water and 91 mg L^{-1} to 1219 mg L^{-1} in tap water from the cities' water treatment plants.

Ogunfowokan et al. (2006) investigated the levels and seasonal variations in the occurrence of PAEs in sewage oxidation pond at the Obafemi Awolowo University, Ile-Ife, Nigeria. The sewage lagoon and the receiving stream were grossly polluted as several phthalate ester plasticizers: dimethyl phthalate (DMP), diethyl phthalate (DEP), di-2-propyl heptyl phthalate (DPHP), dibutyl phthalate (DBP), diethyl hexyl phthalate (DEHP), di-octyl phthalate (DOP) and diisononyl phthalate (DINP) were found present at monthly mean levels of between 24.02 and 139.25 mg L^{-1} in the sewage treatment lagoon and 10.41 and 80.53 mg L^{-1} in the receiving stream. The results showed higher levels of phthalate esters in the sewage lagoon compared to the receiving stream. The sewage lagoon was identified as a pollution point source into the receiving stream. Levels of phthalates obtained from the receiving stream are much higher than the water criteria of 3 µg L^{-1} phthalates recommended by the United States Environmental Protection Agency (USEPA) for the protection of fish and other aquatic life in water and the Suggested No-Adverse Effect Levels (SNAEL) of 7.5-38.5 µg L^{-1} for drinking water. This should give cause for great environmental concern. Peoples' health downstream is at stake and so is the 'health' of the ecosystem.

In a more recent study on this group of compounds, Adeniyi et al. (2011) reported four phthalate esters— DMP, DEP, DBP and DEHP—in water, sediment, and some fish species using flame ionization gas chromatography. The samples were collected from the Ogun river catchments, Ketu, Lagos. The DMP was not detected in the water and fish samples but was detected in sediments collected from four of the six sampling sites. The concentration of DEP,DBP, and DEHP in the fish species ranged from 320.0–810.0, 380.0–1,080.0, and 40.0–150.0 µg kg^{-1} in *Tilapia* sp.; 310.0-860.0, 400.0–1,170.0, and 40.0–110.0 µg $^{-1}$ in *Chrysichthys* sp.; and 320.0-810.0, 400.0–3,970.0,and 30.0–300.0 µg kg^{-1} (DEHP) in *Synodontis* sp., respectively. The values of 1,480.0–1,755.0, 2,025.0–2,705.0, and 255.0–480.0 µg L^{-1}, were recorded respectively for DEP, DBP, and DEHP in the water samples, while the concentration ranges of phthalates in the sediment samples are ND to 850.0 µg kg^{-1} (DMP), 80.0–350.0 µg kg^{-1} (DEP), 190.0–1,420.0 µg kg^{-1} (DBP), and 20.0–820.0 µg kg^{-1} (DEHP).

5. Sediment pollution

Sediment acts as a sink for pollutants. Environmental contaminants such as hydrocarbons, heavy metals and pesticides have been known to have direct toxic effects when released into the aquatic environment (Forstner et al., 1998; Fleeger et al., 2003). There is a direct link between surface water and sediment contamination. Accumulated heavy metals or organic pollutants in sediment could be released back into the water with deleterious effects on human health. Adekola and Eleta (2007) have carried out studies on the heavy metal pollution of Asa River in Ilorin, Kwara state of Nigeria. The researchers reported metals in the concentration range of Mn (179.9-469.4 mg kg^{-1}), Fe (1998.4-4420.4 mg kg^{-1}) Cr (3.0-11.3 mg kg^{-1}), Zn (26.6-147.6 mg kg^{-1}) and Cu (1.9-13.3 mg kg^{-1}). Higher enrichment factors (EFs) were calculated for zinc, chromium, manganese and iron in the sediment samples. High values of EFs indicates anthropogenic source of contamination. Elevated levels of Fe, Ni and V have been observed in the sediments of Iko River, a gas flaring station in the Niger Delta areas of Nigeria (Benson and Etesin, 2008). Contamination of sediments and fauna in aquatic ecosystems of Niger Delta by a wide range of toxic and bio-accumulative metals and hydrocarbons have been reported severally by Ebong et al. (2006), Udosen and Benson (2006) and Benson et al. (2007). In another study on the Lagos Lagoon carried out by Adeboyejo et al. (2011), the levels of OCPs in sediment were investigated from three stations along the Lagoon. The concentration in sediment (μg kg^{-1}) ranged from ND to 2.96, 0 to 1.83, 0.62 to 43.54, ND to 15.6, ND to 32.13, ND to 7.15, ND to 1.3 and 0 to 2.42 for aldrin, dieldrin, chlordane, endrin, endosulfan, heptachlor, HCD and α-HCH, respectively. The findings showed that these chemicals tend to bind with river's sediment particles with resultant effects on water pollution.

6. Future management strategy for surface water quality monitoring in Nigeria

At present, the monitoring of surface and ground water in Nigeria is carried out mostly by individual researchers in the Universities, Research Institutes, Government Agencies and some other organisations. The monitoring is haphazard, short term and based on individual interest and the reagents and equipment available to the Scientist. The monitoring is not properly coordinated and quality assurance programme is not incorporated in most of the studies. Therefore, comprehensive data on the water quality of major rivers in the Nigeria is not available. Recalling that water quality monitoring is a scientifically designed system of long-term, standardized measurement, systematic observation, evaluation and reporting of water quality in order to define status and/or trends, the need to improve the monitoring of the country's surface water cannot be overemphasised.

According to Whitfield (1988), the goals for water quality monitoring should be directed towards expansive information needs, determination of compliance within objective and standard framework, assessment of environmental trend and effects, mass transport estimation, and performance of general surveillance. It is therefore necessary to set up monitoring goals, which must be carried out with an appropriate sampling plan, where collected data could be periodically reviewed (Whitfield, 1988). All these procedures are lacking in the context of water quality in Nigeria.

The Federal Ministry of Water Resources and Rural Development was established in 1984 to among other functions, safeguard the water resource of the nation through periodic monitoring. The Federal Environmental Protection Agency - FEPA (established in 1988) which later transformed to the Federal Ministry of Environment also had the mandate of monitoring the environment of the country. The functions of FEPA also included regulation of effluents discharge by industries and several other Institutions. Statutory power was given to the Agency to prosecute any offender. The essence of this is to protect the water resource of Nigeria from pollution.

Water quality degradation had been found most severe in Lagos, Rivers, Kano and Kaduna where most of the country's industries are located (Ekiye and Zejiao, 2010) with subsequent effects on public health and economic development (Ajibade, 2004, Adewolu et al., 2009). There is need for protection of water bodies in Nigeria as millions of the populace rely on it for daily water supply. The Federal Ministry of Environment must strengthen the present environmental laws such that the polluters of water bodies could be prosecuted. Industrial and agricultural sectors should also be compelled to treat their wastes before being discharged into the water bodies. Drastic measures must be taken by all authorities concern to minimize children morbidity and mortality due to poor sanitation and water quality problems.

There is an urgent need for the establishment of a National Water Monitoring Programme (NWMP) that will monitor the surface water of Nigeria. Quality Assurance Monitoring Plan (QAMP), which addresses quality control issues in detail, must be included in the monitoring programme. The water quality data that will be obtained from such a programme will be used to characterize waters, identify trends over time, identify emerging problems, determine whether pollution control programs are working, help direct pollution control efforts to where they are most needed, respond to emergencies such as floods and spills and also provide the regulatory community with baseline data needed to enforce environmental law.

In the USA, trained volunteers are monitoring the condition of their local streams, lakes, estuaries and wetlands. Environmental Pollution Agency encourages all citizens to learn about their water resources and supports volunteer monitoring because of its many benefits. Volunteer water monitors build community awareness of pollution problems, help identify and restore problem sites, become advocates for their watersheds and increase the amount of needed water quality information available on USA waters. A similar approach can be adopted in Nigeria. For example, a Directory of Environmental Monitoring Volunteers should be introduced using the National Youth Service Corps (NYSC) members ('Youth Corpers'). The young Science graduates that spend one year serving the country should be trained as Water Quality Monitoring Officers during their Orientation Programme. In addition, a special scheme to recruit young graduates as Water Quality Monitoring Officers should be introduced. The results of a comprehensive monitoring of our surface water on a weekly, monthly and quarterly basis will then be uploaded on a website. If the programme is adopted, quality data on our streams and rivers will be acquired.

In the light of new information and additional knowledge about the various environmental toxicants in the past two decades, there is an urgent need for a review of current environmental standards, regulations and laws in the country. The establishment of an Agency or a unit under the Federal Ministry of Environment that will enforce the laws and prosecute offenders cannot be over emphasised.

7. Conclusion

This paper has reviewed the water quality assessment and monitoring of surface water in Nigeria. The variation in water quality experienced in Nigeria reflects differences in land management and the physical environment. These differences occur both as a result of natural variability, societal development and pollutant inputs. In addition, water quality in the vicinity of urban areas is influenced by industrial and urban development. Understanding the condition of rivers and streams is critical if Nigeria is to develop effective plans to maintain, manage, and restore them. Water quality is affected in many ways. It changes with the seasons and geographic areas, even when there is no pollution present, therefore we must pay close attention to water quality by monitoring and testing. Like most other countries, the provision of quality water to the ever increasing population of Nigerians cannot be over emphasised. Without an adequate water supply, the millions of Nigerians will suffer, agriculture will be hampered, and the recreational industry will suffer. Therefore, a system that monitors the changing quality of the surface-water resources is highly necessary, and must be put in place by all Federal and State Agencies concerned. Water quality monitoring is therefore, pertinent for provision of data baseline that will be useful for policy makers and stakeholders to formulate policy that will favour protection and management of water resources. The issue of surface water pollution need to be addressed urgently by the Government to safeguard the public health.

8. References

Abu, G. O. and Egenonu, C. (2008). The current pollution status of the new Calabar River in the Niger Delta region of Southern Nigeria: A survey of antibiogram profiles of its bacterial isolates. *African Journal of Environmental Science and Technology,* Vol. 2, Issue 6, (June 2008), Pp. 134-14. ISSN 1996-0786.

Adebayo, O. T., Balogun, A. M. and Olubiyi, O. A. (2007). Chemical analysis of some industrial effluents that discharge into Lagos Lagoon, Nigeria. *Research Journal of Environmental Sciences,* Vol. 1, No. 4, pp. 196-199. ISSN 1819-3412.

Adeboyejo, O. A., Clarke, E.O. and Olarinmoye, M.O. (2011). Organochlorine Pesticide Residues in water, sediments, Fin and Shell-fish samples from Lagos Lagoon Complex, Nigeria. *Researcher,* Vol. 3, pp. 38-45. www.sciencepub.net/researcher/research0303/06_4630research0303

Adefemi, O. S., Asaolu, S. S. and Olaofe, O. (2007). Assessment of the physico-chemical status of water samples from major dams in Ekiti State, Nigeria. *Pakistan Journal of Nutrition,* Vol. 6, No. 6, pp. 657-659. ISSN 1680-5194.

Adekola, F. A. and Eleta, O. A. A. (2007). A study of heavy metal pollution of Asa River, Ilorin. Nigeria; trace metal monitoring and geochemistry. *Environmental Monitoring Assessment,* Vol. 125, No. 1-3 (October 2006), pp. 157–163. DOI: 10.1007/s10661-006-9248-z.

Adeniyi, A., Dayomi, M., Siebe, P. and Okedeyi, O. (2008). An assessment of the levels of phthalate esters and metals in the Muledane open dump, Thohoyandou, Limpopo Province, South Africa. *Chemistry Central Journal,* Vol. 2, No. 9, (May 2008). DOI: 10.1186/1752-153X-2-9.

Adeniyi, A. A., Okedeyi, O. O. , Yusuf, K. A. (2011). Flame ionization gas chromatographic determination of phthalate esters in water, surface sediments and fish species in the Ogun river catchments, Ketu, Lagos, Nigeria. *Environmental Monitoring and Assessment,* Vol. 172, No. 1-4, pp. 561-9. ISSN: 0167-6369.

Adewara, S. O. and Visser. M. (2011). *Use of anthropometric measures to analyze how sources of water and sanitation affect children's health in Nigeria.* Environment for Development. Discussion Paper Series, (April 2011). 23p. EfD DP11-02.

Adewolu, M. A., Akintola, S. L., Jimoh, A. A., Owodehinde, F. G., Whenu, O. O. and Fakoya, K. A. (2009). Environmental threats to the development of aquaculture in Lagos State, Nigeria. *European Journal of Scientific Research,* Vol. 34, No 3, pp. 337-347. ISSN 1450-216X.

Adeyemi, D., Anyakora, C., Ukpo, G. and Adedayo, A.and Darko, G. (2011). Evaluation of the levels of organochlorine pesticide residues in water samples of Lagos Lagoon using solid phase extraction method. *Journal of Environmental Chemistry and Ecotoxicology*, Vol. 3, No. 6, (June 2011), pp. 160-166. ISSN-2141-226X

Agbogu, V. N., Umoh, V. J., Okuofu, C. A., Ameh, J. B., Smith, S. I. (2006). Study of the bacteriological and physicochemical indicators of pollution of surface waters in Zaria, Nigeria. *African Journal of Biotechnology*, Vol. 5, No. 9, (May 2006), pp. 732-737. ISSN 1684-5315.

Ahmed, K. and Tanko A. I. (2000). Assessment of water quality changes for irrigation in the River Hadejia Catchment. *Journal of Arid Agriculture*, Vol. 10, pp 89-94.

Aina, E. O. A. and Adedipe, N. O. (Eds.) (1996). Water Quality Monitoring and Environmental Status in Nigeria. FEPA Monograph 6, FEPA, Abuja, Nigeria, p239.

Ajao, E. A. and Anurigwo, S. (2002). Land - based sources of pollution in the Niger Delta, Nigeria. *Ambio Journal of the Human Environment*, Vol. 31, No. 5, pp. 442 - 445. ISSN: 0044-7447.

Ajibade, L.T. (2004). Assessment of water quality near River Asa, Ilorin, Nigeria. *The Environmentalist*, Vol. 24, No 1, pp. 11-18. DOI:10.1023/B:ENVR.0000046342. 65791.07.

Alatriste-Mondragon, F., Iranpourb, R. and Ahringa, B. K. (2003). Toxicity of di-(2-ethylhexyl) phthalate on the anaerobic digestion of wastewater sludge. *Water Resources*, Vol. 37, No. 6, (March 2003), pp. 1260-1269. Doi:10.1016/S0043-1354(02)00387-1

Arditsoglou, A. and Voutsa, D. (2008). Determination of phenolic and steroid endocrine disrupting compounds in environmental matrices. *Environmental Science and Pollution Research*, Vol. 15, No. 3, pp. 228-236. DOI: 10.1065/espr2007.12.459

Arimoro, F. O., Ikomi, R. B. and Osalor, E. C. (2007). The impact of sawmill wood wastes on the water quality and fish communities of Benin River, Niger Delta Area, Nigeria. *International Journal of Science and Technology*. Vol. 2, No. 1, pp. 1-12. web.firat.edu.tr/fenbilimleri/Dergiler/TJST/arsiv/.../1_%20francis.pdf

Ayenimo, J. G., Adeeyinwo, C. E. and Amoo, I. A. (2005). Heavy metal pollutants in Warri River, Nigeria. *Kragujevac Journal of Science*, Vol. 27, pp. 43-50. ISSN 1450-9636

Benson, N. U., Essien, J. P., Williams, A. B., & Ebong, G. A. (2007). Petroleum hydrocarbon accumulation potential in shellfishes from littoral waters of the Bight of Bonny, Niger Delta (Nigeria). *Research Journal of Environmental Sciences*, Vol. 1, No. 1, pp. 11-19. ISSN 1819-3412.

Benson, N. U. and Etesin, U. M. (2008). Metal contamination of surface water, sediment and Tympanotonus fuscatus var. radula of Iko River and environmental impact due to Utapete gas flare station, Nigeria. *The Environmentalist*, Vol. 28, No. 3, (May 2008), pp. 195-202. DOI: 10.1007/s10669-007-9127-3.

Bilotta, G. S and Brazier, R. E. (2008). Review: Understanding the influence of suspended solids on water quality and aquatic biota. *Water Research*, Vol. 42, pp. 2849 - 2861. Doi:10.1016/j.watres.2008.03.018.

Blaso, C., Font, G. and Pico, Y. (2005). Analysis of pesticides in fruits by pressurized liquid extraction and liquid chromatography-ion trap-triple stage mass spectrometry. *Journal of Chromatograph A*, Vol. 1098, No 1-2, (December 2005), pp. 37-43. doi:10.1016/j.chroma.2005.08.037

Bouman, H. (2004). South Africa and the Stockholm Convention on Persistent Organic Pollutants. *South African Journal of Science*, Vol. 100, pp. 323-328.

Calamari, D. and Naeve, H. (Eds.) (1994). Review of pollution in the African aquatic environment. CIFA Technical Paper No. 25, FAO, Rome, 118 pp.

CCME, (2008). Canadian Soil Quality Guidelines for the Protection of Environmental and Human Health: Benzo [a] Pyrene (1997). *In: Canadian Environmental Quality Guidelines, 1999, Canadian Council of Ministers of the Environment*, Winnipeg.

Dada, A.Y. (1997). Evaluation of Kano State Environmental Management Programme-1995 on industrial pollution: Case study of Challawa and Sharada Industrial areas, Post-graduate Diploma Dissertation, Department of Geography, Bayero University, Kano.

Diamanti-Kandarakis E, Bourguignon,J- P., Giudice, L. C., Hauser, R., Prins, G. S., Soto, M., Zoeller, R. T. and Gore, A. C. (2009). Endocrine-Disrupting Chemicals: An Endocrine Society Scientific Statement. *Endocrine Reviews*, Vol. 30, No. 4, (June 2009), pp. 293-342. ISSN Online 1945-7197.

Ekiye, E. and Zejiao, L. (2010). Water quality monitoring in Nigeria; Case Study of Nigeria's industrial cities. *Journal of American Science*, Vol. 6, No. 4, pp. 22-28. ISSN: 1545-1003.

Ebong, G. A., Ita, B. N., Nyong, A. B. and Benson, N. U. (2006). Seasonal changes in the water quality of Qua Iboe River Estuary and its associated creeks in Ibeno, Nigeria. *Journal of Applied Sciences*, Vol. 9, No. 2, pp. 6469–6482. eISSN: 1812-5662

Edet, A. and Worden, R. H. (2009). Monitoring of the physical parameters and evaluation of the chemical composition of river and groundwater in Calabar (Southeastern Nigeria).*Environmental Monitoring and Assessment*, 157, Issues 1-4, 243-258. ISSN: 0167-6369

Egborge, A. B. M. (1994). *Water pollution in Nigeria - Biodiversity and Chemistry of Warri River. Ben Miller Books (Nig) Ltd.*

Emoyan, O. O. (2009). Quantification and distribution of Polynuclear Aromatic Hydrocarbons (PAHs) in surface waters in the vicinity of Kokori Oil Field, Nigeria. *Journal of Applied Science & Environmental Management*, Vol. 13, No. 4. Pp. 39 - 42. ISSN 1119-8362.

Esoka PA and Umaru JM. (2006). Industrial effluent and water pollution in Kakuri area, Kaduna South, Nigeria. *Journal of Industrial pollution and control*, Vol. 22, No 1, pp. 93-100. ISSN 0970-2083.

Ezemonye, L. I., Ikpesu, T. O. and Tongo, I. (2009). Distribution of endosulfan in water, sediment and fish from Warri river, Niger delta, Nigeria. *African Journal of Ecology*, 48, No 1, pp. 248–254. doi/10.1111/j.1365-2028.2009.

Fatoki, O. S. and Ogunfowokan, A. O. (1993). Determination of phthalate ester plasticizers in the aquatic environment of south western Nigeria. *Environment International*, 19: 619–623.

Federal Government of Nigeria. (1988). *Federal Environmental Protection Agency Decree 58, 1988*, Federal Ministry of Information and Culture, Lagos, Nigeria.

FEPA (1991) *Guidelines and Standards for Environmental Pollution Control in Nigeria.* National Environmental Standards-Parts 2 and 3, Government Press, Lagos, pp: 238

Forstner, U., Calmano, W., Hong, J., and Kersten, M. (1998). Effects of redox variations on metal speciation. *Waste Management and Research*, Vol. 4, pp. 95–104.

Fleeger, J.W., Carmana, K.R., & Nisbetb, R.M. (2003). Indirect effects of contaminants in aquatic ecosystem. *Science of the Total Environment*, Vol. 317, No. (1–3) (December 2003), pp. 207–233. Doi: 10.1016/S0048-9697(03)00141-4 . ISSN: 0048-9697.

Galadima, A., Garba, Z. N., Leke, L., Almustapha, M. N. and Adam, I. K. (2011). Domestic water pollution among local communities in Nigeria - causes and consequences. *European Journal of Scientific Research* , Vol.52, No.4, pp. 592-603. ISSN 1450-216X.

Guruge, K.S. and Tanabe, S. (2001). Contamination by persistent organochlorine and butylin compounds in West coast of Sri Lanka. *Marine Pollution Bulletin*, Vol. 42, (March 2001), pp. 179-186. Doi:10.1016/S0025-326x(00)00140-5. | Harrod, T. R. and Theurer, F. D. (2002). Sediment. *In: Haygarth, P.M., Jarvis, S.C. (Eds.), Agriculture, Hydrology and Water Quality*. CABI, Wallingford, p. 502.

Heathwaite, L. (1994). Eutrophication. *Geography Review*, Vol. 7, No. 4, pp. 31–37.

Hjelmborg, P. S., Ghisari, M. and Bonefeld-Jorgensen, E. C. (2006). SPE-HPLC purification of endocrine disrupting compounds from human serum for assessment of xenoestrogenic activity. *Analytical & Bioanalytical Chemistry*, Vol. 385, No 5, (July 2006) pp. 875-887. Doi: 10.1007/s00216-006-0463-9

Ize-Iyamu, O. K., Asia, I. O. and Egwakhide, P. A. (2007). Concentrations of residues from Organochlorine pesticide in water and fish from some rivers in Edo state, Nigeria. *International Journal of Physical Sciences*, Vol. 2, No. 9 (September 2007), pp. 237-241. ISSN 1992-1950.

Izonfuo, L. W. A. and Bariweni, A. P. (2001). The effect of urban runoff water and human activities on some physico- chemical parameters of the Epie Creek in the Niger Delta. *Journal of Applied Science and Environmental Management*, Vol. 5, No. 1, pp. 47-55. ISSN 1119-8362.

Jaji, M. O., Bamgbose, O., Odukoya, O. O. and Arowolo, T. A. (2007). Water quality assessment of Ogun River, South West Nigeria. *Environmental Monitoring Assessment*, Vol. 33, No 1-3, pp. 473-482. doi: 10.1007/s10661-006-9602-1. ISSN: 0167-6369

Kronvang, B., Laubel, A., Larsen, S. E., Friberg, N. (2003). Pesticides and heavy metals in Danish streambed sediment. *Hydrobiologia*, Vol. 494, No. (1–3), pp. 93–101. ISSN:1573-5117.

Lloyd, B. and Helmer, R. (1992). Surveillance of drinking water quality in rural areas. Longman Scientific and Technical, New York: Wiley, pp34-56.

Luks-Betlej, K., Popp, P., Janoszka, B. and Paschke, H. (2001). Solid-phase microextraction of phthalates from water. *Journal of Chromatography A*, Vol. 938, No. 1, (December 2011), pp. 93-101. Doi: 10.1016/S0021-9673(01)01363-2

Martin, J.C., Hoggart, C. and Matissa, A. (1998). Improvement priorities for sewage treatment in Latvian small and medium sized towns. *Water Science Technology*, Vol. 37, No. 8, pp. 137-144. ISSN Print: 0273-1223.

Mashi, S.A. and Alhassan, M. M. (2007). Effects of wastewater discharge on heavy metals pollution in Fadama soils in Kano City, Nigeria. *Journal of Biomedical and Environmental Sciences*. Vol. 20, No 1, pp. 70-77.

Mauricio, R., Diniz, M., Petrovic, M., Amaral, L., Peres, I., Barcelo, D. and Santana, F. (2006). A characterization of selected endocrine disruptor compounds in a Portuguese wastewater treatment plant. *Environmental Monitoring Assessment*, Vol. 118, No 1-3, pp. 75-87. Doi: 10.1007/s10661-006-0986-8. ISSN: 0167-6369

Mayon, N., Berrand, A., Lerroy, D., Malbrouck, C., Mandiki, S. N. M., Silvestere, F., Golffart, A., Thorme, J. and Kestemont, P. (2006). Multiscale approach of fish responses to different types of environmental contamination: A case study. *Science of the Total Environment*, Vol. 367, No 2-3, (August 2006), pp. 715-731. doi:10.1016/j.scitotenv.2006.03.005. ISSN: 0048-9697

Meinhardt, P. L. (2006). Recognizing waterborne disease and the health effects of water contamination: a review of the challenges facing the medical community in the United States. *Journal of Water and Health*, Vol. 4, suppl. 1, pp. 27-34. ISSN Print: 1477-8920.

Mustapha, M. K. (2008). Assessment of the water quality of Oyun Reservoir, Offa, Nigeria, using selected physico-chemical parameters. *Turkish Journal of Fisheries and Aquatic Sciences*, Vol. 8, pp. 309-319. www.trjfas.org/pdf/issue_8_2/309_319.

National Bureau of Statistics. 2009. Social Statistics in Nigeria. 399 p. http://www.nigerianstat.gov.ng/ext/latest_release/ssd09.pdf.Accessed: 10/10/2011.

Nwankwu, E. O. (1992). Microbial quality of Lagos lagoon water and important commercial fish species. *In: Annual Report, Nigeria Institute for Oceanography and Marine Research*, Lagos, pp. 37-38.

Nwilo, P.C. and Badejo, O. T. (2005). Oil spill problems and management in the Niger Delta. *International Oil Spill Conference*, (May 15-19, 2005), Miami, Florida, USA.

Obasohan, E. E., Agbonlahor, D. E. and Obano, E. E. (2010). Water pollution: A review of microbial quality and health concerns of water, sediment and fish in the aquatic African ecosystem. *Journal of Biotechnology*, Vol. 9, No. 4, (January 2010) pp. 423-427. ISSN 1684-5315.

Offiong, O. E. and Edet, A., (1996). *Environmental pollution: water, land, air and noise in Cross River State*. Technical Report submitted to Cross River Environmental Protection Agency, Calabar.

Ogunfowokan, A. O., Okoh, E. K., Adenuga, A. A. and Asubiojo, O. I. (2005). An assessment of the impact of point source pollution from a university sewage treatment oxidation pond on a receiving stream - a preliminary study. *Journal of Applied Sciences*, Vol. 5, No 1, pp. 36 - 43. Doi: 10.3923/jas.2005.36.43.

Ogunfowokan, A.O., Torto, N., Adenuga, A. and Okoh, E.K. (2006). Survey of levels of phthalate ester plasticizers in a sewage lagoon effluent and a receiving stream. *Environmental Monitoring and Assessment*, Vol. 118, pp. 457–480. ISSN: 0167-6369

Oguntoke, O., Aboderin, O. J, Bankole, A. M. (2009). Association of water-borne diseases morbidity pattern and water quality in parts of Ibadan City, Nigeria. *Tanzania Journal of Health Research*, Vol. 11, No. 4, 2009, pp. 189-195. ISSN: 1821-6404.

Olade, A. (1987). Heavy Metal Pollution and the need for monitoring: Illustrated for developing countries in West Africa. *In: Hutchinson, T.C. and Meema, K.M. (eds.) Lead, Mercury, Cadmium and Arsenic in the Environment*. John Wiley & Sons Ltd, USA. pp. 335-341.

Olajire, A. A. and Imeokparia, F. E. (2001). Water quality assessment of Osun River: Studies on inorganic nutrients. *Environmental Monitoring Assessment*, Vol. 69: 17-28. ISSN: 0167-6369

Osibanjo, O., Daso, A. P. and Gbadebo, A.M. (2011). The impact of industries on surface water quality of River Ona and River Alaro in Oluyole Industrial Estate, Ibadan, Nigeria. *African Journal of Biotechnology*, Vol. 10, No. 4, (January 2011), pp. 696-702. ISSN 1684–5315.

Raji, M. I. O. and Ibrahim Y. K. E. (2011). Prevalence of waterborne infections in Northwest Nigeria: A retrospective study. *Journal of Public Health and Epidemiology*, Vol. 3, No. 8, (August 2011) pp. 382-385. ISSN 2141-2316

Solomon, U.U. (2009). The state of solid waste management in Nigeria. *Waste Management Journal*, Vol. 29, No. 10, pp. 2787 - 2790.

Shukla, G., Kumar, A., Bhanti, M., Joseph, P.E. and Taneja, A. (2006). Organochlorine pesticide contamination of ground water in the city of Hyderabad. Environmental International, Vol. 32 (February 2006), pp. 244-247. doi:10.1016/j.envint.2005.08.027.

Taiwo, A. M. (2010). Environmental impact of poultry farm operations on Alakata Stream at Isolu in Abeokuta, Nigeria. Unpublished Master's Thesis. University of Agriculture, Abeokuta. 108 p.

Taiwo, A. M. (2011). Composting as a sustainable waste management technique in developing countries. *Journal of Environmental Science and Technology,* Vol. 4, No. 2, pp. 93-102. doi=jest.2011.93.102

Taiwo, A. M., Adeogun, A. O., Olatunde, K. A., Adegbite, K. I. (2011). Analysis of groundwater quality of hand-dug wells in peri-urban areas of Obantoko, Abeokuta, Nigeria for selected physico-chemical parameters. *Pacific Journal of Science and Technology,* Vol. 12, No. 1, (May 2011), pp. 527-534. www.akamaiuniversity.us/PJST12_1_527.pdf

Tatah, J. F. K. and Ikenebomeh, M. J. (1999). Influence of brewery effluent on heterotrophic counts and some physical parameters of Ikpoba Tropical River, Nigeria. *Nigerian Journal of Microbiology,* Vol. 13, pp. 55–58.

Udosen, E. D., & Benson, N. U. (2006). Spatio-temporal distribution of heavy metals sediments and surface water in Stubbs Creek, Nigeria. *Trends in Applied Sciences Research,* Vol. 1, No. 3, pp. 292–300.Doi: 10.3923/tasr.2006.292.300

Water Initiative. (2010). UN water global annual assessment of sanitation and drinking water (2010 report). Implemented by World Health Organization. Geneva: United Nations. http://www.who.int/water_sanitation_health/glaas/en/.Accessed January 2011

Wakawa, R. J ., Uzairu, A., Kagbu, J. A. and Balarabe, M. L. (2008). Impact assessment of effluent discharge on physico-chemical parameters and some heavy metal concentrations in surface water of River Challawa Kano, Nigeria. African *Journal of Pure and Applied Chemistry,* Vol. 2, No 9, (October 2008), pp. 100-106. ISSN 1996 – 0840.

Ware, G. W. and Whitacre, D. M. (2004). *The Pesticide Book,* 6th Ed. 496 pp. Meister Media Worldwide, Willoughby, Ohio. (ISBN 1892829-11-8).

Whitfield, P. H. (1988). Goals and data collection designs for water quality monitoring. *Water Resources Bulletin,* Vol. 24 No. 4, pp. 775-80. Doi: 10.1111/j.1752-1688.1988.tb00928.x. ISSN: 0043-1370

WHO. (2008). Guidelines for drinking water quality. 3rd Ed. Health criteria and supporting information, Geneva.

WHO/UNEP (1997). Water Pollution Control - A Guide to the Use of Water Quality Management Principles. In: *Helmer, R. and Hespanhol, I. (eds). United Nations Environment Programme, the Water Supply & Sanitation Collaborative Council and the World Health Organization.* ISBN 0 419 22910 8. http://www.who.int/water_sanitation_health/resourcesquality/wpccasestudy4.pdf.

Yusuf, R. and Shonibare, J. (2004). Characterization of textile industries' effluents in Kaduna, Nigeria and pollution implications. *Global Nest: the International Journal,* Vol. 6, No 3, pp. 212-221. ISSN 1790-7632.

14

Temporal Water Quality Assessment of Langat River from 1995-2006

Zalina Mohd Ali[1], Noor Akma Ibrahim[2], Kerrie Mengersen[3], Mahendran Shitan[2], Hafizan Juahir[4] and Faridatul Azna Ahmad Shahabuddin[1]

[1]*Faculty of Science and Technology, University Kebangsaan,*
[2]*Faculty of Science, University Putra,*
[3]*Faculty of Science and Technology, Queensland University of Technology,*
[4]*Faculty of Environmental Studies, University Putra,*
[1,2,4]*Malaysia*
[3]*Australia*

1. Introduction

Water quality is generally described according to biological, chemical and physical properties (Coke et al 2005). Based on these properties, the quality of water can be expressed via a numerical index (i.e. Water Quality Index, WQI) by combining measurements of selected water quality variables. The index is important in evaluating the water quality of different sources and in observing the changes in the water quality as a function of time and other influencing factors (Sarkar and Abbasi 2006). The time when samples are taken is one of the contributing factors that can influence the concentration of a particular water quality variable (Coke et al 2005). Thus, temporal assessment is a good indication in determining the presence or absence of trend and seasonality to which water quality is responding to changes in the catchment and time.

However, the assessment on the temporal effect of water quality sub indices variables and WQI are rarely carried out by the Malaysian Department of Environment and Malaysian Department of Irrigation and Drainage (DID). Several studies by past researchers investigated the water quality assessment in Langat River especially in spatial assessment such as evaluating the polluting effects from various land use pattern (Suki et al 1988), relationship between water quality and sewage discharge and location Lee et al (2006) and on spatial variations of water quality variables (Juahir et al 2010a). Most of the studies did not consider the temporal assessment in details. Therefore, the influence of time on selected water quality variables and water quality index of Langat River are studied by using box plot to examine the annual and quarterly pattern. Then, regression time series and decomposition analysis are carried out on normally distributed variables with no outliers at particular stations along the Langat River. Both methods are helpful in evaluating the changes with time and in determining the best fitted models of the selected variables.

2. Description of study area

2.1 Background

Pollution prevention improvement programme has been introduced by the Malaysian Department of Environment (DOE) from 2001 to improve the condition of polluted rivers in Malaysia. Langat River which is situated in the state of Selangor, Peninsular Malaysia with a total catchment area of approximately 1,815km^2 is chosen for this programme. The catchment area is shown in Fig. 1. Data used in the analysis for Langat River were collected from six monitoring stations as shown in Table 1 and Fig.1.

DOE station number	Station number	DOE Station Code	Distance from estuary (km)	Grid reference	Location
2814602	1	IL01	4.19	2°52.027′ 101°26.241′	Air Tawar Village
2815603	2	IL02	33.49	2°48.952′ 101°30.780′	Telok Datuk, near Banting town
2817641	3	IL03	63.43	2°51.311′ 101°40.882′	Bridge at Dengkil Village
2918606	4	IL04	81.14	2°57.835′ 101°47.030′	Near West Country Estate
2917642	5	IL05	86.94	2°59.533′ 101°47.219′	Kajang Bridge
3118647	6	IL07	113.99	3°09.953′ 101°50.926′	Bridge at Batu 18

Table 1. DOE sampling station at the study area

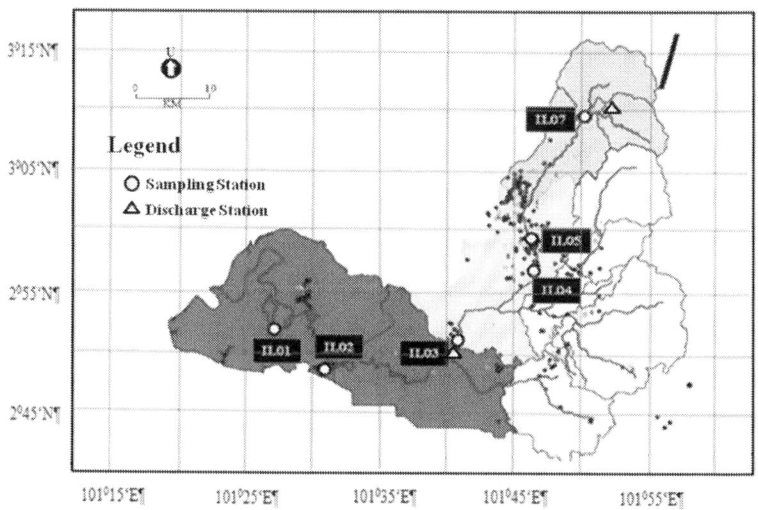

Fig. 1. Location of the selected sampling stations

2.2 Land use changes

The land use of Langat catchment consists of mainly agriculture, forest, urban areas (commercial and residential) and water bodies. There are 3 types of forest in Langat River catchment area such as dipterocarp, peat swamp and mangrove. Agriculture is the dominant land use, followed by forest, urban areas and water bodies of Langat catchment as presented by Table 2 and Fig. 2. However, Langat River as a tropical catchment area is experiencing rapid urbanization (Amini et al 2009) where the urban expansion occurred since 1981 (see Table 2 and Fig. 2). The gain in size of urbanised area was also reported by Jaafar et al (2009) and the urban development which occurred in the Langat River catchment was due to extensive land exchange from agriculture to urban-industrial-commercial use.

Year	Forest	Developed Area/Urban	Agriculture	Water bodies
1981	41.42	2.59	54.79	1.19
1988	30.32	9.61	57.97	2.10
1991	29.89	9.72	58.21	2.18
1996	27.47	10.56	60.22	1.75
2001	24.11	17.56	56.21	2.12

Table 2. Land use composition (in %) for Langat river basin (Source : Idrus et al 2003)

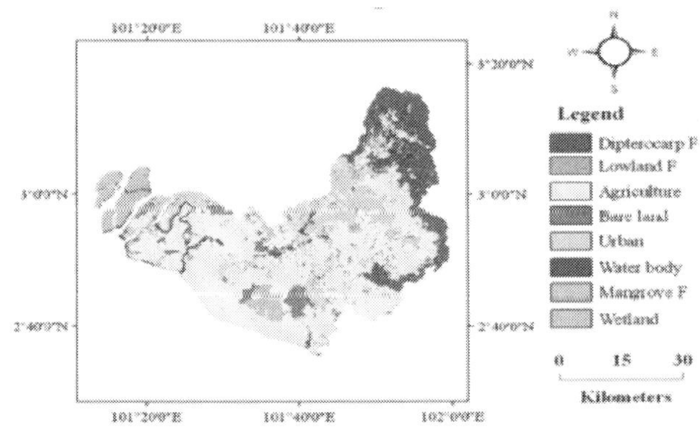

Fig. 2. GIS map of land uses in Langat River Basin for 2005 (Source : Mokhtar et al 2011)

2.3 Population change

Selangor is one of the most populated states in Malaysia with rapid growth from 1980 to 2010. Based on the Malaysian government census reports in 1980, 1991, 2000 and 2010, the recorded population of Selangor was 1.4 million in 1980, 2.3 million in 1990, close to 3.9 million in 2000 and rise to 5.4 million in 2010. In Selangor, a large portion of Langat River

Basin is located in three administrative districts of Selangor, namely Hulu Langat, Sepang and Kuala Langat as shown in Fig. 3. The rest of the basin covers part of Negeri Sembilan, Klang, Petaling and Federal Territory of Kuala Lumpur.

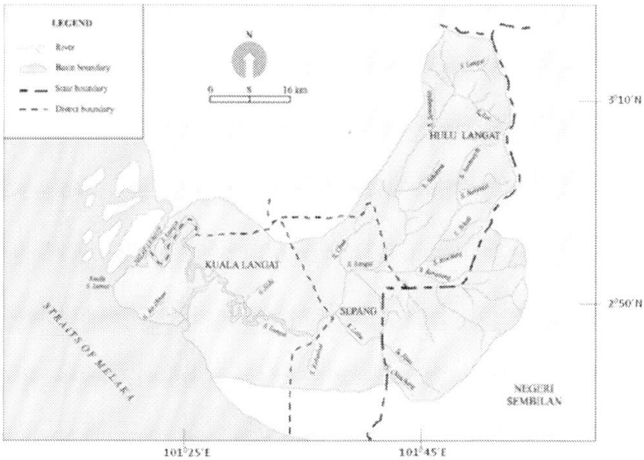

Fig. 3. Langat River Basin

Table 3 and Fig. 4 show the district population in the Langat River Basin for Kuala Langat, Sepang and Hulu Langat. The population of these three districts increased from 1980 to 2010 and the population increased was consistent with the urban development discussed above.

District	1980	1991	2000	2010
Kuala Langat	101578	130,090	192176	224648
Sepang	46025	54671	108640	211361
Hulu Langat	177877	413900	864451	1156585

Table 3. Number of population by districts along Langat River

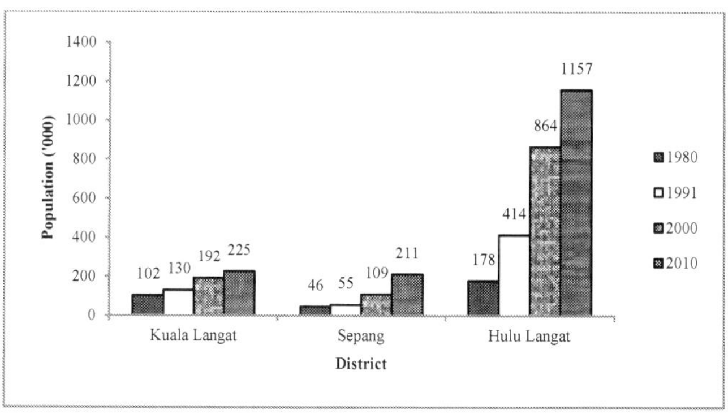

Fig. 4. Number of population along Langat River Basin, by district from 1980-2010.

2.4 Sources of pollutions

Langat River is one of the most important raw water resources for drinking water, recreation, industry, fishery and agriculture. The river flows from the highest peak of 1493 meter of Gunung Nuang across Langat Basin to Kuala Langat and land use activities along the river banks contribute to deterioration of river water quality (Charlie 2010). The sources of the Langat River pollution are identified as industrial discharge (58%), domestic sewage from treatment plants (28%), construction projects (12%) and pig farming (2%) (Khairuddin et al 2002). A study from Juahir et al (2010a) showed that major sources of surface water quality variations in Langat River come from industrial effluents, wastewater treatment plants, domestic and commercial areas.

The declining quality of the river water is caused by two main sources of water pollution, i.e. point sources and non-point sources. Point source (PS pollution) is single and identifiable source that discharge pollutants into the environment such as discharge from manufacturing and agro-based industries, sewage treatment plants and animal farms. On the other hand, non-point sources pollution (NPS pollution), also known as polluted runoff is pollution where sources cannot be traced to a single point. NPS is defined as pollution originating from diffused sources such as agricultural activities and surface runoffs which contributed by storm runoff, e.g. from rainfall, snowmelt, or irrigation over land surfaces into the drainage system (Sapari et al 2009). The increase of PS and NPS pollution loading such as discharges of surface runoff, domestic sewage, ship wastes and industrial discharges into coastal waters may resulted by rapid urbanization along the river. The NPS pollution is seen as the main contributor to the pollution load in Langat River compared to PS pollution (UPUM 2002). It is also obvious that the Langat River ecosystem is under stress from the discharge of effluents particularly domestic sewage (Lee et al 2006). Table 4 indicates some of the major source of pollution in the study area (Department Of Environment, DOE 2007).

Point Source (PS)	Non-Point Sources (NPS)
Industrial Estate	Agriculture
Landfills	Construction
Piggeries	Forest
Private Sewage Treatment Plants	Residential
Public Sewage Treatment Plants	Residential and Commercial
Restaurants	Urban
Sand Mining Activities	
Wet Markets	
Workshops	

Table 4. Sources of Pollution in Sub-catchment of Langat River

Due to rapid urbanization and changes from undeveloped to developed area, Langat River experienced changes of discharges and direct runoff volume. A study done by Juahir et al (2010b) has showed that there is a relationship between land use and discharge or flow rate and runoff in Langat River. The annual mean discharge and direct runoff of Langat River at two selected gauging stations, i.e. Dengkil Station (station number 2816441) and Lui Station (station number 2917401) are shown in Fig. 5 and Fig. 6. The Dengkil gauging station is located at the downstream of Langat River (2°51'20''N, 101°40'55''N) and Lui gauging station is located at the upstream of Langat River (3°10′25′′N, 101°52′ 20′′E). The increment trend of discharge and direct runoff at Dengkil Station compared to Lui Station is consistent with the increasing trend of urban development and the decrease of agriculture and forest areas within the region (Juahir et al 2010b).

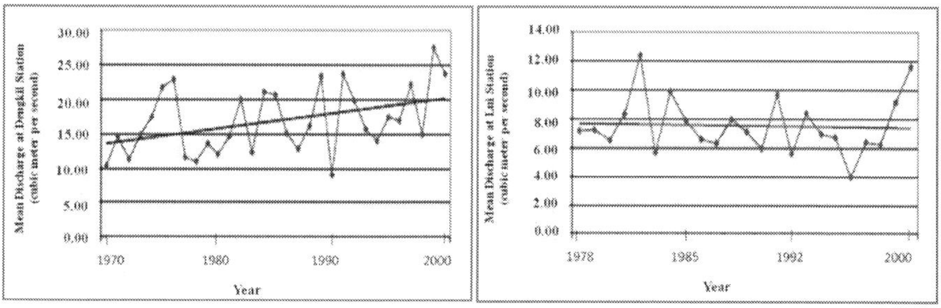

Fig. 5. Mean discharge at Dengkil Station and Lui Station (adapted from Juahir et al 2010)

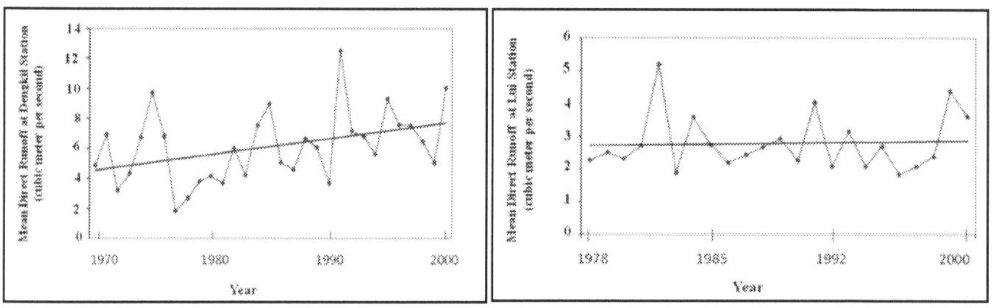

Fig. 6. Mean direct runoff at Dengkil Station and Lui Station (adapted from Juahir et al 2010)

3. Water quality data

Water quality data used in this study were obtained from the Malaysian Department of Environment (DOE). The data obtained however, were not collected at regular time intervals and to facilitate the analysis, quarterly data was used instead. Time series data from September 1995 until December 2007 for selected parameters and stations were used in the present study. Since quarterly data is used in the analysis, the data that represent the first quarter is taken from the last month of that quarter i.e. data from the month of March. Similarly, for the second quarter, data from the month of June will represent the data for that second quarter. However, if the last month of a quarter does not contain any data, then data from either the first or second month of that quarter will be considered . For example, for quarter 1, the data that represent quarter 1 will either be from the month of February or January, likewise for quarter 2 the data from the month of May or April will be taken to represent the second quarter data.

The six selected water quality variables used in this study are Suspended Solids (SS), Biochemical Oxygen Demand (BOD), Ammoniacal Nitrogen (AN), Chemical Oxygen Demand (COD), Dissolved Oxygen (DO) and pH. These variables were selected by the panel of experts as the variables that when calculated and use collectively will give some indication on the water quality level or water quality index of a river (DOE 1997). According to the best-fit relationship for each six parameters, the new variables of the 6 sub indices (SI) were determined and the overall trend for Langat River were obtained using the formula given by

$$WQI = 0.22\,SIDO + 0.19\,SIBOD + 0.16\,SICOD + 0.16\,SISS + 0.15\,SIAN + 0.12\,SIpH \quad (1)$$

where,

WQI = Water quality index; SIDO = Sub-index of Dissolved Oxygen; SIBOD = Sub-index of Biological Oxygen Demand; SICOD = Sub-index of Chemical Oxygen Demand; SIAN = Sub-index of Ammonical Nitrogen; SISS = Sub-index of Suspended Solid; SIpH = Sub-index of pH value.

Generally, WQI is a unitless number varies between 0 and 100. Measurements of each of these parameters are taken and compared to a classification table (see Table 5), where the water is identified as excellent, good, fair, poor or very poor (DID 2009).

Status	Polluted	Slightly Polluted	Clean
WQI value	0-59	60-80	81-100

Table 5. WQI categorisation scheme

4. Methods

4.1 Graphical analysis

Graphical analysis is very useful in data analysis and helps the researcher in seeing pattern, trends and other features not easily apparent using numerical summaries. Box-Whisker plot, normal probability plot and scatter plot were used to analyze data graphically in this study. Displaying data using graphs allow for more effective visualization and presentation of large data sets in a small space (Cooke et al 2005). By using graphical analysis, we can visualize any gaps in the data, relationship between variables and trends that might exist in the selected water quality data.

4.1.1 Box-whisker plot

Box-Whisker plot is a powerful exploratory data analysis tool. It is also called the five-number summary (Tukey, 1977). To plot, the given sub index data are ranked from smallest to largest value. Then, the five-number summary which include the smallest and largest values, the median (a measure of central tendency that is more robust and not sensitive to outlying values, indicates the 50th percentile), and the lower and upper hinges (i.e. 0.25 and 0.75 quartiles respectively) are obtained. This information is then represented by the Box-Whisker plot. The box represents the inter-quartile range and the whiskers are lines that extend from the box to the highest and lowest values, excluding outliers. The outliers are individual points with values beyond the highest and lowest limits and are plotted with asterisks. A line across the box indicates the median of the data. If the median lie in the middle of the box with upper and lower whiskers of similar length, it shows that data for any given year or quarter are symmetric. The lacks of symmetry suggest departure from normality. Apart from that, the plot can also give immediate visuals about the center, the spread, and the overall range of distribution. Additionally, the median confidence intervals can be plotted along the boxes in Minitab as shown in Fig. 8-Fig. 10 and are useful in offering a rough guide to determine the differences of medians. If the two boxes do not overlap, this offers evidence of a statistically significant difference between the medians (Mcgill et al 1978). In this study, the plots are used to evaluate several things, for example to ascertain how the sub indices and water quality index time series data are distributed, to demonstrate outlier and to evaluate normality of the data. It also can be used to visualize the median differences and to track the annually and quarterly changes to the water quality data.

4.1.2 Normal probability plot

Normal probability plot is a graphical method for testing normality. If our data follows the hypothesized normal distribution, then the plotted points fall approximately along a straight line. There are various test statistics for normality and Anderson-Darling statistic is a widely used test (Montgomery et al 2008). If the p-value is smaller than the critical value (usually 0.05), the underlying population is not normal.

4.1.3 Scatter plot

The sub indices water quality parameters with normal distribution will be selected based on the results from Box-Whisker plot and probability plot. The scatter plot of the selected parameters was constructed to see any trend and seasonal patterns. Scatter plot is used to show the relationship between dependent variable (sub indices water quality parameters) and independent variable (time). Each value of dependent variable is plotted against its corresponding time. If the sub indices for water quality parameters values tend to increase or decrease in a straight line fashion as the time increases, and if there is a scattering of the (time, sub index) points around the straight line, then it is reasonable to describe the relationship between the sub index and time by simple linear regression model. This could help to identify whether the differences in water quality variables are due to the actual trend or due to the changes in the water quality variables.

4.1.4 Analysis framework

Fig. 7 shows the analysis framework to be used in this study. The first step is to assess the temporal variations of the sub indices and water quality variables. The assessments were performed using Minitab 15.1. Then, exploratory assessment is carried out to determine the annual and quarterly changes of the selected water quality variables. From this analysis, if there seem to be some apparent trends or oscillations in the selected water quality variables, it will be evaluated. To assess the statistical significance of the changes and to examine the uncertainty about the possible trend and seasonality, regression and additive decomposition analysis were performed. To continue with these analyses, the variables considered are those without any outliers and are normally distributed. Models fitted for the selected variables at certain stations can be used to predict future values.

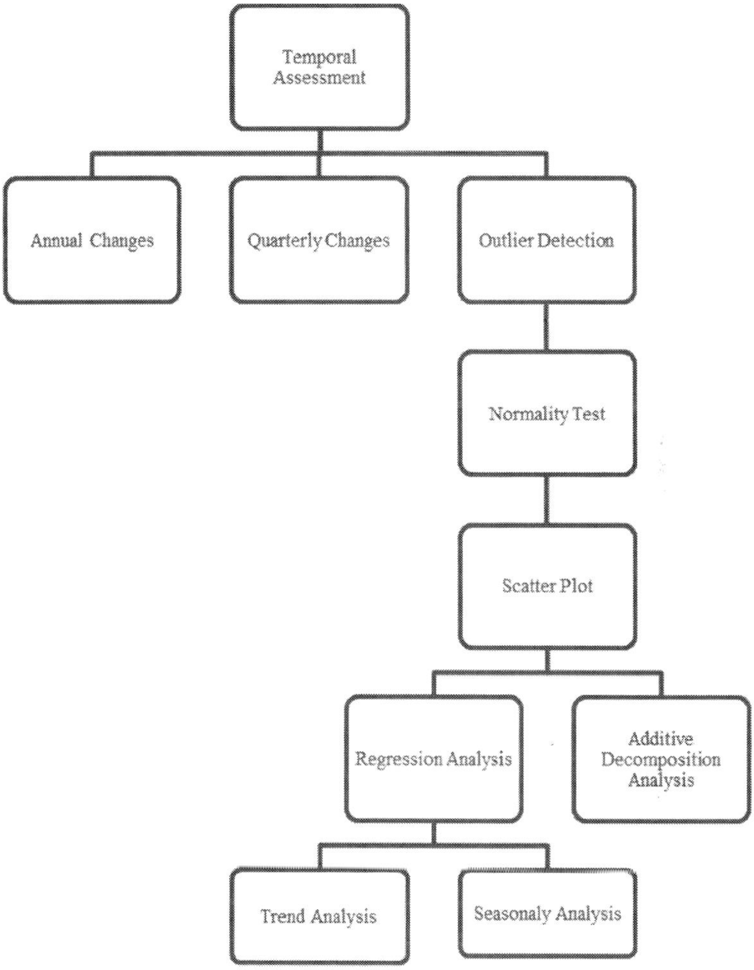

Fig. 7. Analysis Framework

4.2 Linear regression analysis

Linear regression analysis is an important parametric method to identify the monotonic trend in a time series. It is useful to describe the relationship between variables. The method is often performed to determine the slope of selected variables. In this study, the regression analysis was used to investigate and to model the relationship between the selected water quality variables versus time. The slope indicates the mean temporal change of the variables. Positive values of the slopes show increasing trends in the mean temporal change while negative values of the slopes indicate decreasing trends.

4.2.1 Trend analysis

The trend model

$$y_t = TR_t + e_t \quad (2)$$

$$\text{or} \quad y_t = \beta_0 + \beta_1 * time + e_t \quad (3)$$

where,

y_t = variable values, TR_t = trend and e_t = error term, in time period *t*.

This model explains that the time series y_t can be represented by an average level (denoted $\mu_t = \beta_0 + \beta_1 * time$) that changes over time according to the equation $TR_t = \mu_t$ and by the error term, e_t.

4.2.2 Quarterly analysis

The quarterly model

$$y_t = \beta_0 + \beta_1 * time + \beta_2 Q1 + \beta_3 Q2 + \beta_4 Q3 + \beta_5 Q4 + e_t \quad (3)$$

where,

$Q1$ = quarter 1, $Q2$ = quarter 2, $Q3$ = quarter 3 and $Q4$ = quarter 4.

4.3 Additive decomposition analysis

The additive model is useful when the seasonal variation is relatively constant over time. In the present study, the selected parameters were separated into linear trend, seasonal and error components.

$$y_t = TR_t + SN_t + CL_t + e_t \quad (4)$$

where,

TR_t = trend, SN_t = seasonal, CL_t = cyclical and e_t = error term in time period t.

Basic steps in decomposition method are:

1. Estimate the trend using centered moving average.
2. De-trend the series by subtracting the trend estimated in (1) from the original series.
3. Estimate seasonal factors. This entails an effect for each quarter by using the average of the de-trended values for a particular season. However, Minitab uses median rather than mean.
4. Final step is to determine the random component. The random component could be summarized using measures such as mean absolute percentage error (MAPE, %), mean absolute deviation (MAD, expressed in the same unit as the variables) and mean squared deviation (MSD). MAPE and MAD measure the accuracy of fitted time series values, $\hat{y}$ and the MSD is more sensitive to unusually large forecast errors (outliers). The formulas are as follows :

$$MAPE = \frac{\sum|(y_t - \hat{y}_t)/y_t|}{n} \times 100 \quad (5)$$

$$MAD = \frac{\sum|y_t - \hat{y}_t|}{n} \quad (6)$$

$$MSD = \frac{\sum|y_t - \hat{y}_t|^2}{n} \quad (7)$$

5. Results and discussion

5.1 Annual changes

Box plots provide a visual impression of the location and shape of the underlying distributions (Vega et al 1998). In this section, the presence or absence of trends over time was examined graphically through Fig. 8. From Fig. 8, SIDO and WQI showed a positive trend except for certain particular years for example in 1997 and 1999 for SIDO and 1997 for WQI where the medians were clearly lower than the median for the rest of the data. For SICOD and SIpH, the median were fairly stable without many variations for most of the years selected. However for SIBOD, the trend highly fluctuates from 1995 up to 1999 and after 2000 the fluctuation stabilised. The trend for SISS, however, showed a decreasing trend for three consecutive years from 1998 to 2000 and then in 2001 the trend increased. Similar decreasing pattern was observed again in 2004. The decreasing and increasing trend in SISS was also observed in SIAN but the decreasing trend started in 1995. The changes that occurred to the annual trend of water quality variables were influenced by many factors for example, certain hydrological events and/or developments in the river basin (Ravichandran 2003).

5.2 Quarterly changes

To plot the graphs in Fig.9, all quarterly data for each parameter and each station were combined from all selected years from 1995-2006. For example, quarterly data for SIDO for all selected years were combined for all stations i.e. Station 1 to Station 6. The distribution of each parameter and each quarter can then be examined as shown in Fig. 9. From Fig. 9, SIDO, SIBOD, SIAN, SISS and WQI showed that the quarterly median values did not differ much. The median values for SIDO in quarter 3 and quarter 4 were fairly equal. However, SISS showed a decreasing trend in quarter 4. The quarterly medians in SICOD and SIpH were generally similar for all quarters except in quarter 4 where SICOD showed a slight increase in the quarterly median.

5.3 Outliers detection

The box plots in Fig. 10 show that most of the variables depart from normality in their skewness. Many variables also have outliers and extreme values. To ease our analysis, variables with no outliers were selected i.e. SIDO at Station 1-5, SIBOD at station 5, SICOD at station 1 and 5, SIAN at station 2-5, SISS at station 2-3, SIPH at station 2 and WQI at station 1,4 and 5.

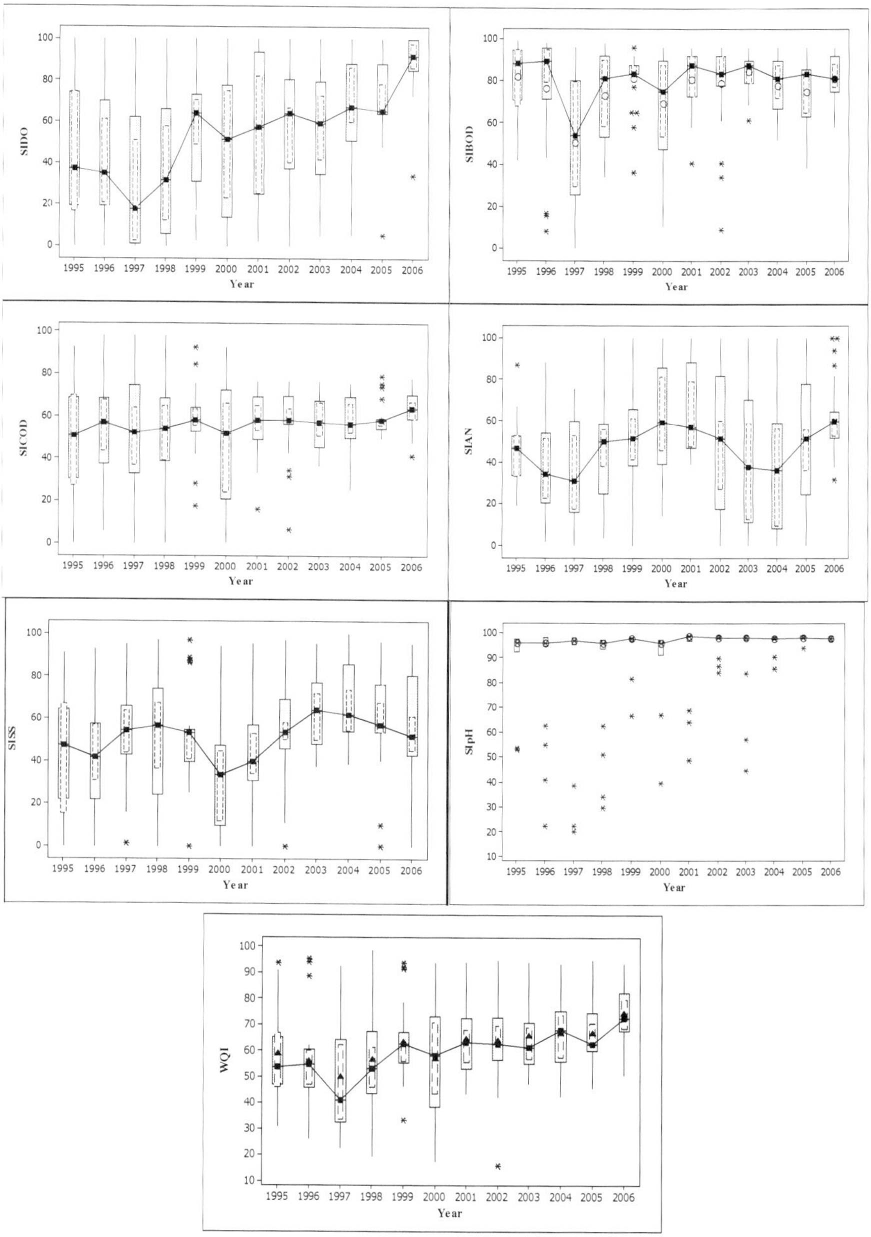

Fig. 8. Box plots of measured values of sub indices water quality parameters and water quality index in Langat River during 1995-2006

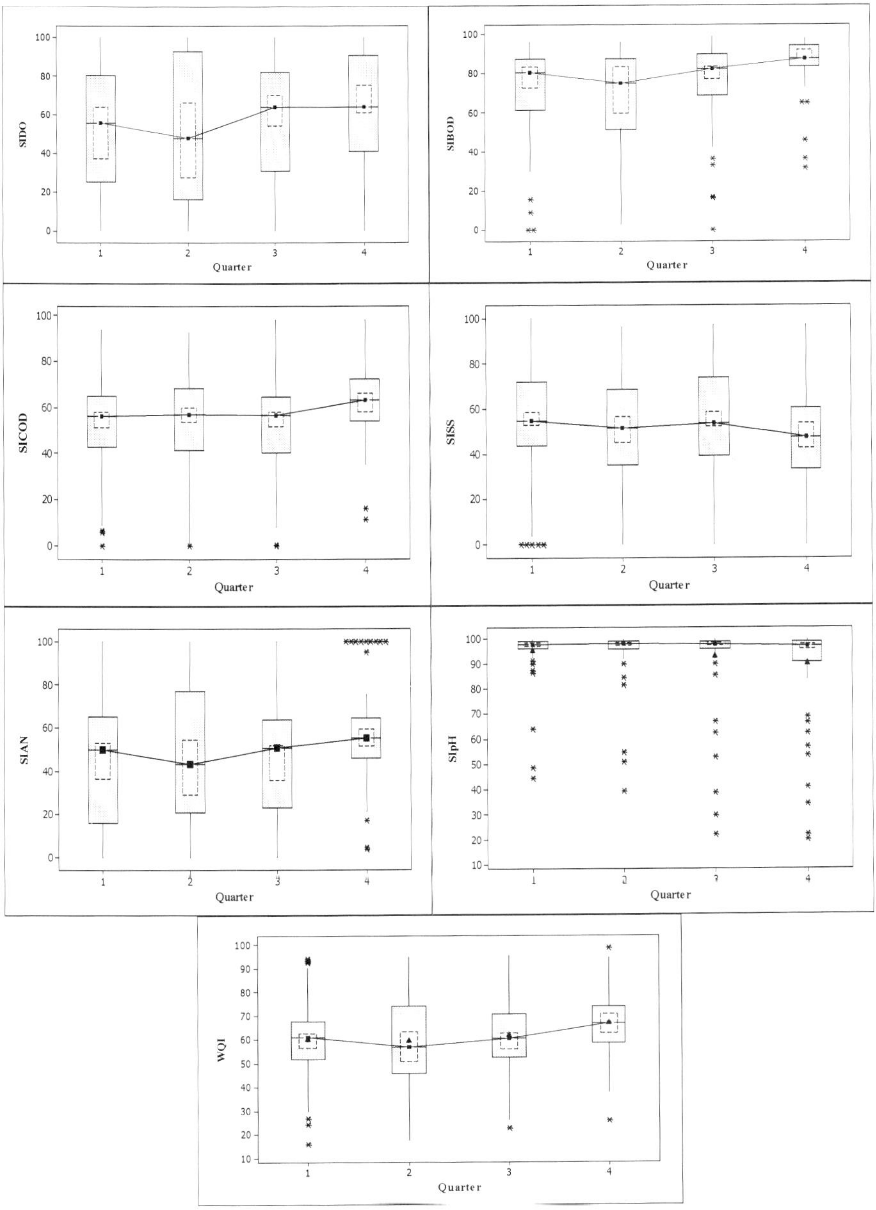

Fig. 9. Box plots of measured values of sub indices water quality parameters and water quality index in Langat River 1995-2006

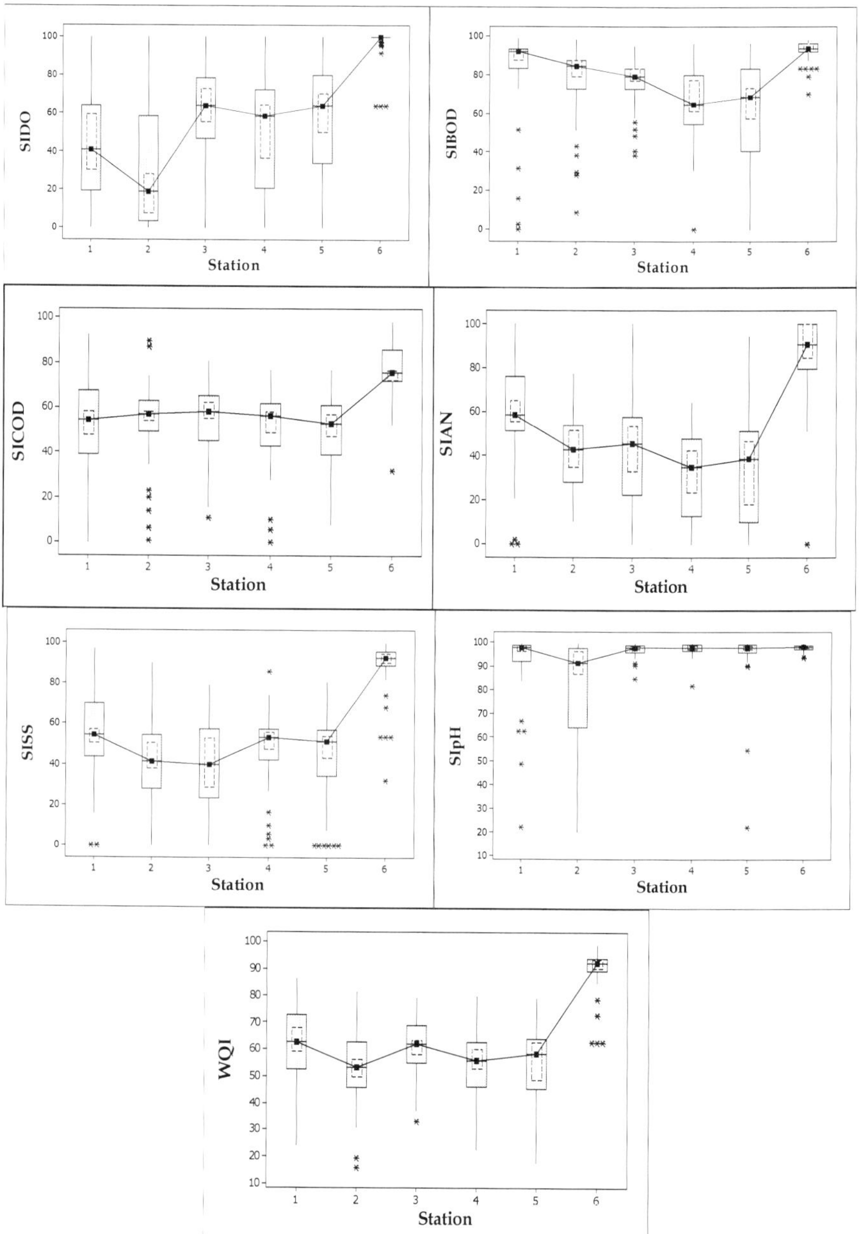

Fig. 10. Box plots of measured values of sub indices water quality parameters and water quality index at 6 main stations of the Langat River.

5.4 Normality test

Normality tests were applied to selected variables in Table 6 at particular sampling stations. From Fig. 11- Fig. 15, normal probability plots showed that SIDO at Station 2, 4 and 5, SIBOD,SICOD,SIAN at Station 5, SISS at Station 2, SICOD at Station 1 and SIAN at Station 4 were not normally distributed based on the Anderson-Darling test. Table 6 shows the list of the parameters with the Anderson-Darling test statistics and the corresponding *p*-values in the bracket.

Variables	Station 1	Station 2	Station 3	Station 4	Station 5	Station 6
SIDO	**0.456 (0.256)**	2.816 (<0.005)	**0.521 (0.175)**	0.887 (0.021)	1.117 (0.006)	
SIBOD					1.190 (<0.005)	
SICOD	0.840 (0.028)				1.119 (0.006)	
SISS		0.839 (0.028)	**0.576 (0.127)**			
SIAN		**0.745 (0.049)**	**0.540 (0.157)**	1.264 (<0.005)	1.156 (<0.005)	
SIpH						
WQI	**0.518 (0.179)**			**0.289 (0.600)**	**0.745 (0.049)**	

Table 6. Parameters selected with normality test results

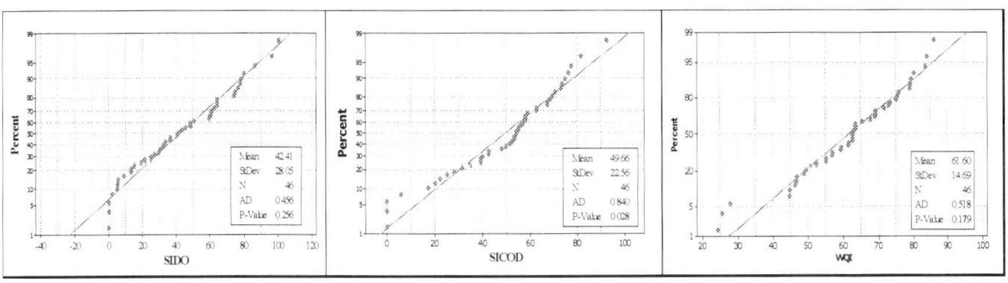

Fig. 11. Normal Probability Plot of SIDO, SICOD and WQI at Station 1

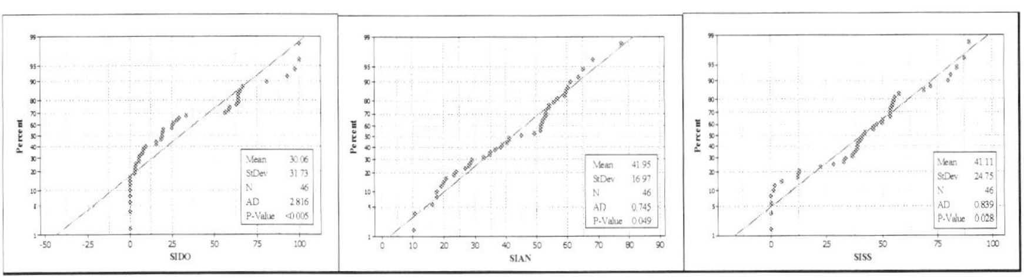

Fig. 12. Normal Probability Plot of SIDO, SIAN and SISS at Station 2

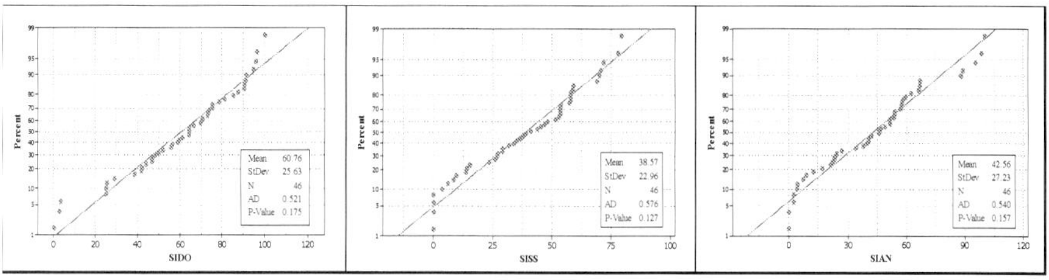

Fig. 13. Normal Probability Plot of SIDO, SISS and SIAN at Station 3

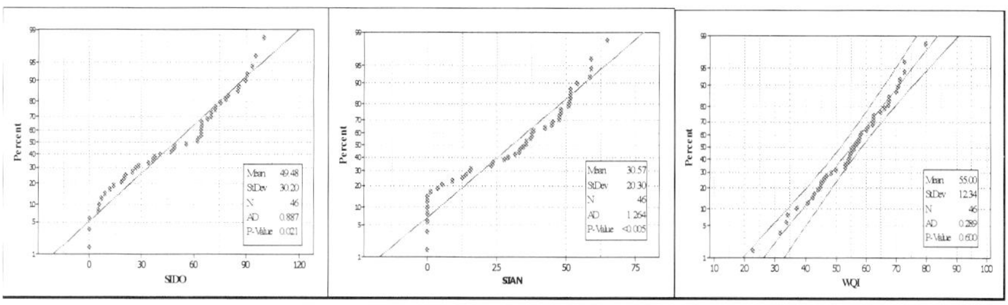

Fig. 14. Normal Probability Plot of SIDO, SIAN and WQI at Station 4

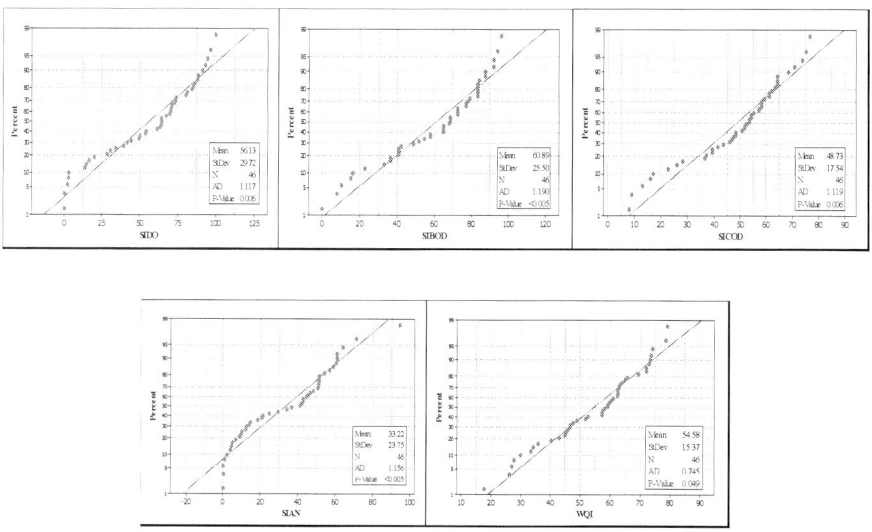

Fig. 15. Normal Probability Plot of SIDO, SIBOD, SICOD, SIAN and WQI at Station 5

5.5 Trend and seasonal analyses

Trend and seasonal analyses using scatter plot were performed for selected stations in Fig. 16 i.e. SIDO at Station 1 and 3, SIAN at Station 2 and 3, SISS at Station 3 and WQI at Station 1, 4 and 5. Scatter plots in Fig.16 showed positive trend patterns in SIDO and WQI as well as quarter-to-quarter variations; hence, the model should include both trend and seasonal variations. The magnitudes of the seasonal variations were fairly constant around the level of the series, so an additive model is appropriate. To check on the significance of trend and seasonality, regression analysis and decomposition analysis were carried out. The trend-only models in regression analysis exclude the seasonal variation for both variables. Notice that the R^2 values for the trend-only models in Table 7 are between 16.70 and 38.20. To improve the forecast accuracy, seasonal variations for both variables were taken into account. Four indicator variables which represent quarter 1 to quarter 4 were used to model the seasonal variations and to test whether the seasonality were statistically significant. In Minitab, the last indicator variable is removed because it is highly correlated with the first three indicator variables for both SIDO and WQI. The results in Table 8 show that the R^2 values for the trend-and-seasonal models were increased between 39.70 and 50.30. Even though the increase was not substantial, but it is acceptable for quarterly time series data.

Variables	Station	Regression Analysis			
		Trend Model	$R^2(\%)$	MSD	p-value
SIDO	1	$SIDO = 13.6 + 1.23 * time$ (0.005) (0.000)	34.50	527.00	0.000
	3	$SIDO = 33.03 + 1.18 * time$ (0.000) (0.000)	38.20	415.00	0.000
WQI	1	$WQI = 46.09 + 0.66 * time$ (0.000) (0.000)	36.40	140.40	0.000
	4	$WQI = 46.18 + 0.38 * time$ (0.000) (0.005)	16.70	129.80	0.000
	5	$WQI = 39.96 + 0.62 * time$ (0.000) (0.000)	29.50	170.30	0.000

Table 7. Results of regression trend-only models analysis for SIDO and WQI in selected stations

Result in Table 8 also shows that only the coefficients in WQI model at Station 5 were statistically different from zero for all variables (corresponding p-values in the bracket). Hence, the additive seasonal variations of each quarter can be interpreted relative to Quarter 4. From the model, the WQI in quarter 1 is, on average, 9.71 units lower than WQI in quarter 4. Similarly, quarter 2 and 3 are 18.8 and 13.7 units lower than quarter 4 respectively.

Variables	Station	Regression Analysis			
		Seasonal Model	R^2(%)	MSD	p-value
SIDO	1	$SIDO = 6.73 + 1.23 * time - 4.32 * Q1 + 11.5 * Q2 + 19.4 * Q3$ (0.433) (0.000) (0.635) (0.209) (0.034)	45.90	466.80	0.000
	3	$SIDO = 38.27 + 1.17 * time - 8.26 * Q1 - 6.71 * Q2 - 5.86 * Q3$ (0.000) (0.000) (0.348) (0.445) (0.495)	39.70	434.50	0.000
WQI	1	$WQI = 43.2 + 0.66 * time - 2.06 * Q1 + 5.25 * Q2 + 8.23 * Q3$ (0.000) (0.000) (0.669) (0.280) (0.087)	44.40	131.80	0.000
	4	$WQI = 54.41 + 0.38 * time - 5.98 * Q1 - 17.0 * Q2 - 10.4 * Q3$ (0.000) (0.001) (0.151) (0.000) (0.013)	42.50	96.04	0.000
	5	$WQI = 50.4 + 0.618 * time - 9.71 * Q1 - 18.8 * Q2 - 13.7 * Q3$ (0.000) (0.000) (0.047) (0.000) (0.005)	50.30	128.80	0.000

Table 8. Results of regression trend-and-seasonal models analysis for SIDO and WQI in selected stations

On the other hand, Fig. 17 showed the results from Minitab time series decomposition analysis of WQI showing the original data (labelled 'actual') along with the fitted line ("Trend") and the predicted values ("Fits") from the additive model which include both the trend and seasonal components. Details of the seasonal analysis were shown in Fig. 18 and Table 9. Estimates of the quarterly variation from the trend line for each season (seasonal indices) are shown in Fig. 18a with box plots of the actual differences shown in Fig. 18b. The percentage of variation by seasonal period was illustrated in Fig. 18c and model residuals by seasonal period in Fig. 18d. Unfortunately, many decomposition methods do not perform significance tests on seasonal indexes. The significance tests on seasonal indexes are important to believe that the seasonal variations exist. The indexes can be tested to confirm that they are statistically different from zero. Since decomposition methods do not perform such tests, the easiest way to test their significance is to create indicator variables as performed in the previous regression analysis. Additional details of the component analysis are shown in Fig. 19. Fig. 19a is the original time series. Fig. 19b is the plot of the time series with the trend removed. Fig. 19c is a plot of the time series with the seasonality removed (should see a trend pattern) and Fig. 19d is a residual plot of the detrended and seasonally adjusted data. The wave-like pattern in Fig. 19d suggests the constant variance over time in WQI at Station 5.

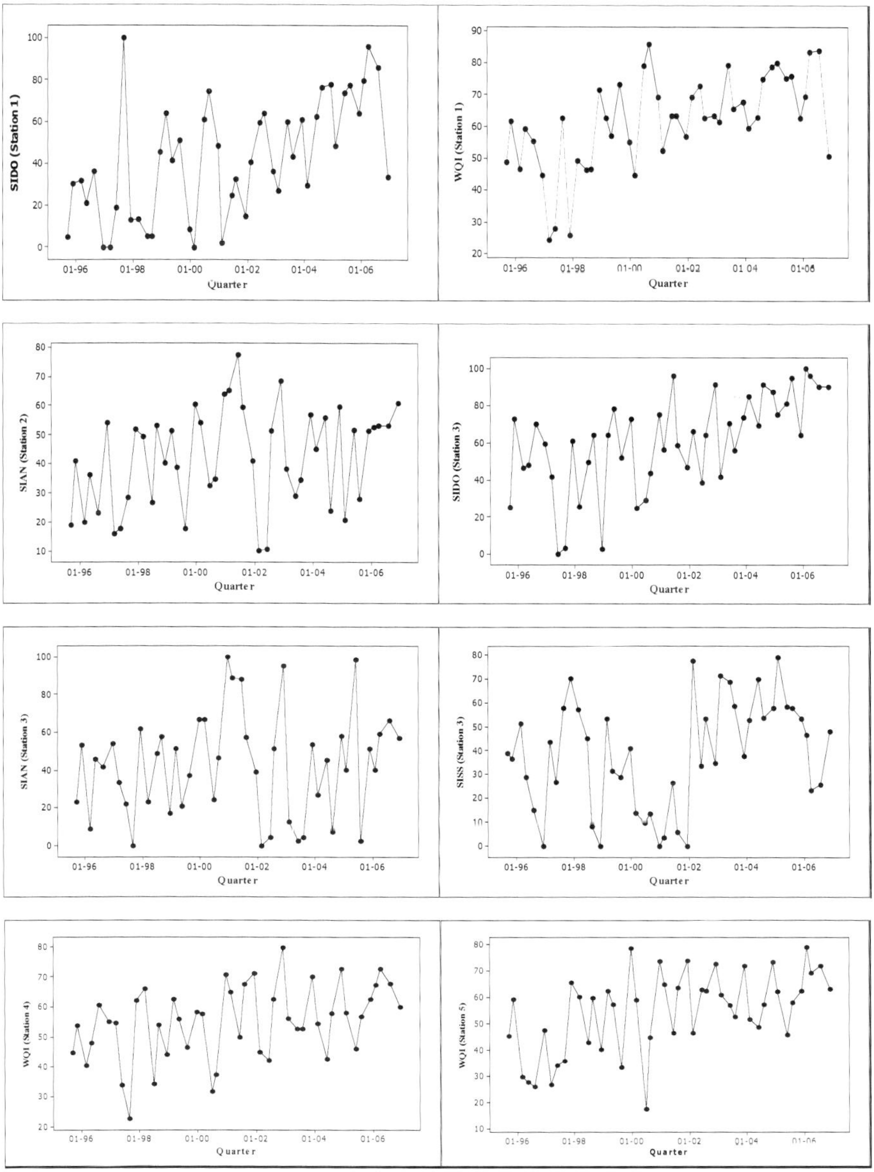

Fig. 16. Scatter plot of selected water quality variables in certain stations (in bracket)

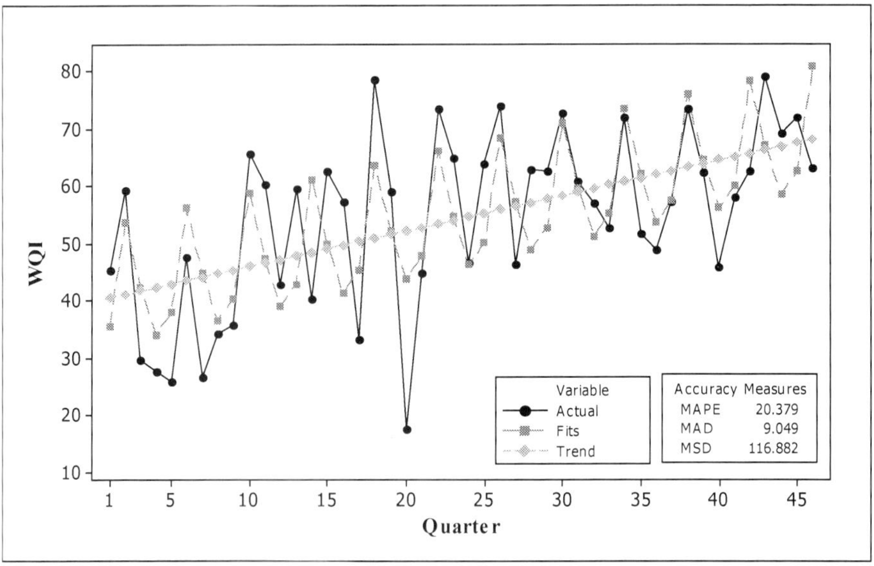

Fig. 17. Additive Time Series Decomposition Plot for WQI in Station 5

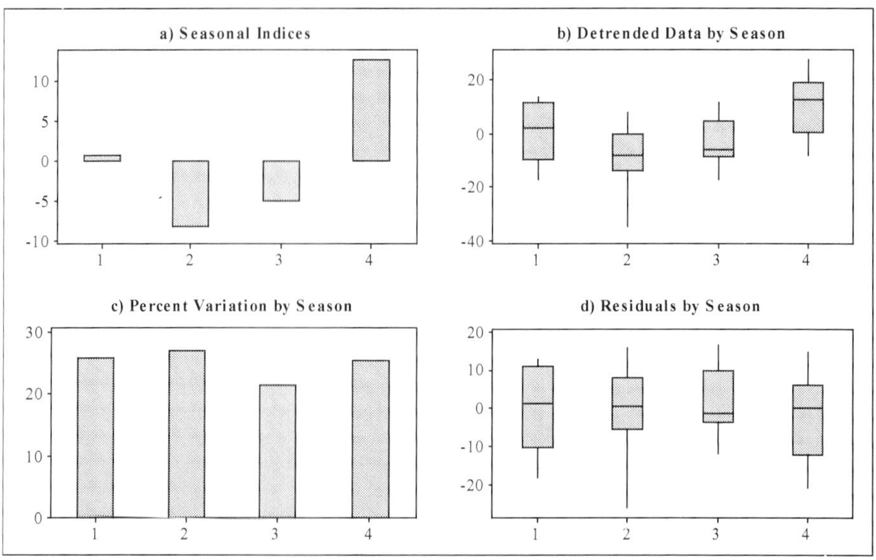

Fig. 18. Seasonal Analysis for WQI at Station 5

Variables	Station	Trend Model	Additive decomposition Models						
			Seasonal Indices				MAPE (%)	MAD	MSD
			Q1	Q2	Q3	Q4			
SIDO	1	$SIDO = 13.28 + 1.23 * time$	-14.30	5.84	16.98	-8.52	104.98	15.69	424.58
	3	$SIDO = 33.13 + 1.17 * time$	-7.00	-0.48	-0.50	7.97	96.93	15.64	393.25
WQI	1	$WQI = 46.08 + 0.658 * time$	-5.44	3.03	3.75	-1.35	16.56	8.20	118.90
	4	$WQI = 45.96 + 0.373 * time$	0.14	-12.09	-0.83	12.79	15.82	7.71	95.12
	5	$WQI = 39.96 + 0.615 * time$	0.70	-8.30	-5.03	12.63	20.38	9.05	116.88

Table 9. Results of additive decomposition model for SIDO and WQI in selected stations

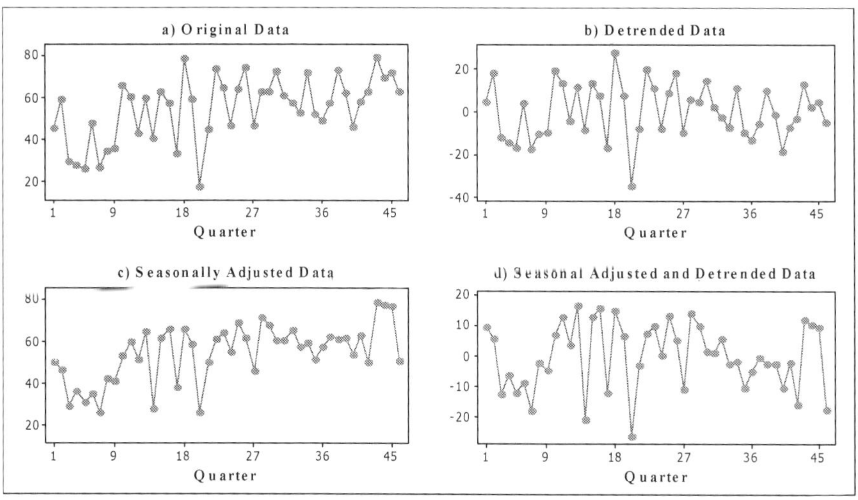

Fig. 19. Component Analysis of WQI in Station 5

5.6 Forecasting model evaluation

The measure of forecast accurracy were evaluated as part of model validation effort (Montgomery et al 2008). To evaluate the resulting model in Section 5.4, the accuracy measures of the model were determined as summarized in Table 9. There were three measures reported in this study, Mean Average Percentage Error (MAPE), Mean Average Median (MAD) and Mean Square Error (MSE). Focusing on the WQI model at Station 5, the accuracy measures were 20% for MAPE, 9.05 for MAD and 116.88 for MSE. Obviously, small variability in forecast errors is prefered in all forecasting models, but a larger forecast error (i.e. residual) or a relatively small one is very subjective. Therefore, the results of MAPE, MAD and MSE in this case can be reasonably accepted. Further, normality test on the distribution of forecast error was examined. From Fig. 20, the p-value is 0.211, so the hypothesis of normality for the forecast error could not be rejected at the 0.05 level. The forecasts that could adequately model all the structure in the data and the sequence of forecast errors would have no systematic or nonrandom pattern (Montgomery et. al 2008) as shown in Fig. 21. From Fig. 22, the sample autocorrelation function (ACF) shows that all spikes in the sample ACF at lower lags are inside the confidence interval limits. This suggest that there is no pattern in the forecast errors. Therefore, there is strong evidence to support the claim that the residuals are not correlated. Since quarterly data were used, the forecast values in Table 10 with MSD value is considered reasonable for this model.

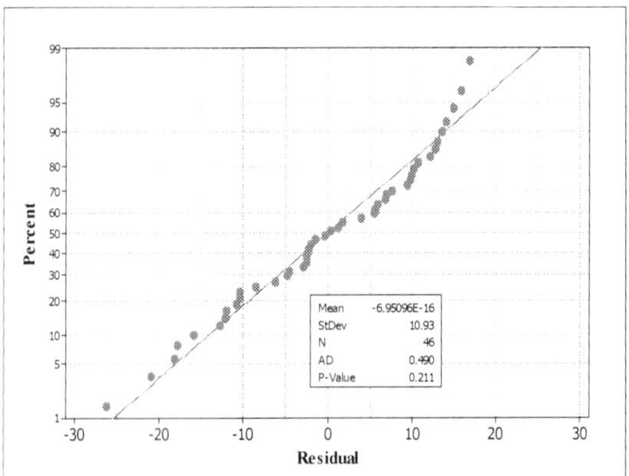

Fig. 20. Normal probability plot of residuals for additive model of WQI in Station 5

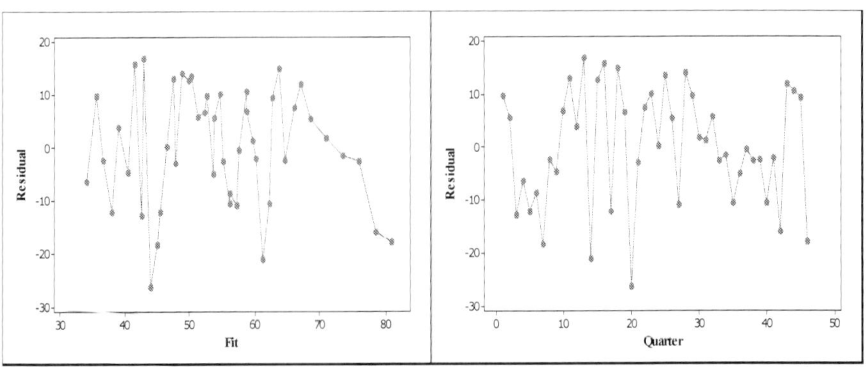

Fig. 21. Scater plot of WQI residual versus WQI fitted value and quarter

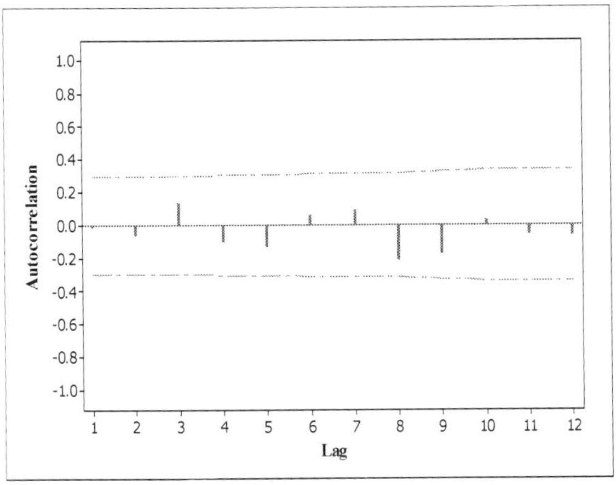

Fig. 22. Sample ACF of forecast error for WQI in Station 5

Variables	Station	Quarter	Forecast	2007	MSD
SIDO	1	1	78.25	61.50	597.91169
		2	90.62	77.50	
		3	66.35	76.86	
		4	61.80	67.68	
	3	1	88.77	91.75	609.25021
		2	89.92	76.99	
		3	99.55	80.96	
		4	85.75	95.11	
WQI	1	1	80.71	55.66	682.64048
		2	82.08	75.47	
		3	77.64	74.89	
		4	74.21	76.28	
	4	1	51.79	65.04	428.17771
		2	63.43	75.64	
		3	77.42	67.66	
		4	65.15	68.07	
	5	1	61.18	63.64	408.44976
		2	65.07	72.47	
		3	83.34	64.73	
		4	72.03	70.96	

Table 10. Forecast value for the selected water quality variables in 2007

6. Conclusion

In this study, various techniques were utilized to evaluate temporal variations in the surface water quality of Langat River. From box plot and median analysis, the annual changes of SIDO and WQI showed an oscillating trend in annual variation with noticeable increment. The annual variation for parameters SICOD, SIBOD, SIAN and SISS showed very little or no trend. There is no trend exhibited by SIpH and its quarterly values do not differ much. The quarterly analyses were examined and the results showed that the median of SIDO, SIBOD, SIAN, SIpH and WQI were lower in quarter 2 than the median of SISS in quarter 4. Significant trends in water quality were found in SIDO at Station 1 and 3 and WQI at Station 1, 4 and 5. However, the effects of quarters appear to be prominent only in WQI at Station 5. WQI is the most significant variable contributing to water quality variations for all quarters. Therefore, further analysis should be carry out to study the relationship between the location of the station (i.e. Station 5) and, sampling measurement time with all the variables that strongly influencing WQI such as urbanization, population density, water shortages and pollution (Cheng et al 2003). In addition, Sapari et al (2009) mentioned that urban NPS pollution has become a growing concern for most major towns in Malaysia due to the serious threat of pollution to river water quality in urban environment. Since this study focus only on the sub indices and index, it would be additionally informative if the same analyses are repeated to all available variables with longer data sets (i.e. monthly data in 1995-2006). Longer data sets could provide clearer indication of trend and seasonal pattern inherent in the time series data. Imputation method should also be considered to overcome the problem of unequal spacing in the measurements.

7. Acknowledgment

The authors would like to thank Malaysian Department of Environment for supplying the data on which this work was based.

8. References

Amini, A., Mohammad Ali, T., Ghazali, A.H. & Kim Huat, B. (2009). Adjustment Of Peak Streamflows Of A Tropical River For Urbanization. *American Journal Of Environmental Sciences*. 5 (3), 285-294.

Bowerman, B.L, O'Connell, R.T. & Koehler, A.B. (2005). *Forecasting, Time Series, And Regression*, ISBN 0-534-40877-6, Thomson Brooks/Cole, U.S.

Cheng, S. Chan, C.W. & Huang, G.H. 2003, An Intergrated Multi-Criteria Decision Analysis And Inexact Mixed Integer Linear Programming Approach For Solid Waste Management, *Engineering Applications Of Artificial Intelligence*, 543-554.

Cooke, S.E., S.M. Ahmed & N.D. Macalphine - Revised. (2005). *Introductory Guide To Surface Water Quality Monitoring In Agriculture*. Conservation And Development Branch, Alberta Agriculture, Food And Rural Development. Edmonton, Alberta.

Department Of Environment Malaysia - DOE (1997). *Malaysia Environmental Quality Reports*. Kuala Lumpur: Ministry Of Science, Technology And Environment.

Department Of Environment Malaysia - DOE (2007). *Malaysia Environmental Quality Reports*. Kuala Lumpur: Ministry Of Science, Technology And Environment.

Department Of Irrigation And Drainage - DID (2009). *Study On The River Water Quality Trends And Indexes In Peninsular Malaysia*. Putrajaya : Ministry Of Natural Resources And Environment.

Idrus, S., Shah, A.H.H. And Mohamed, A.F. (2003). Analysis Of Land Use And Land Cover Changes 1974-2001 In The Langat Basin, Malaysia Using Geographic Information System (GIS). In Mokhtar, M.B., Idrus, S. And Aziz, S. (Eds). Ecosystem Health Of The Langat Basin, Proceedings Of The 2003 Research Symposium On Ecosystem Of The Langat Basin, 209-225. LESTARI, UKM, Bangi.

Jaafar, O., S.A. Sharifah Mastura & Mohd Sood, A. (2009). Land Use and Deforestation Modelling of River Catchments in Klang Valley, Malaysia. *Sains Malaysiana*. 38(5), 655-644.

Juahir, H., M. Zain, S., Yusoff, M.K., Tengku Hanidza, T. I., Mohd Armi, A. S. Toriman, M.K & Mokhtar, M. (2010a). Spatial Water Quality Assessment Of Langat River Basin (Malaysia) Using Environmetric Techniques. *Environmental Monitoring And Assessment*. Vol. 173, No. 1-4, 625-641.

Juahir,H, M. Zain, S., Aris, A.Z., Yusof, M.K., Abu Samah, M.A. & Mokhtar, M. (2010b). Hydrological Trend Analysis Due To Land Use Changes At Langat River Basin. *EnvironmentAsia* 3 (Special Issue), 20-31.

Khairuddin M. I, Abd Malek A. 2002. Program Pencegahan Pencemaran Dan Peningkatan Kualiti Air Sungai Langat. Proceeding Simposium Penyelidikan Lembangan Langat 2001. Mazlin M., Shaharuddin I., Ahmad Fariz M. Abdul Hadi H. S. And Sarah A.A.G.A. (Eds). Lestari, UKM, Bangi. 183-189. In Lee, Y.H., Abdullah, M.P., Chai, S.Y., Mokhtar, M. & Ahmad, R. (2006). Development Of Possible Indicators For Sewage Pollution For The Assessment Of Langat River Ecosystem Health. *Malaysia Journal Of Analytical Sciences*. Vo. 10. No. 1. 15-26.

Lee, Y.H., Abdullah, M.P., Chai, S.Y., Mokhtar, M. & Ahmad, R. (2006). Development Of Possible Indicators For Sewage Pollution For The Assessment Of Langat River Ecosystem Health. *Malaysia Journal Of Analytical Sciences*. Vo. 10. No. 1. 15-26.

Mokhtar, M., Toriman, M.E. Abraham Hossain, M.A. Kok, W.T. (2011) Institutional Challenges For Integrated River Basin Management In Langat River Basin, Malaysia. *Water And Environment Journal*. 25. 495-503

Montgomery, D.C., Jennings, C.L. & Kulahci, M. (2008). Introduction To Time Series Analysis And Forecasting. Wiley Interscience. ISBN 978-0-471-65397-4.

Sapari, Nasiman And Adlan, M.N. (2009). *Non-Point Sources (NPS) Pollution Modelling And Water Quality Improvement Using Wetland.* In: International Engineering Convention, Danascus, Syria.

Ravichandran, S. (2003). Hydrological Influences On The Water Quality Trends In Tamiraparani Basin, South India. *Environmental Monitoring And Assessment*. Vol. 87. (October 2002). 293-309.

Mcgill, R., Tukey, J.W., & Larsen, W.A. (1978). "Variations Of Box Plots". *The American Statistician* Vol. 32, No. 1 (February 1978), 12-16.

Sarkar, C. Abbasi, S. A.(2006). QUALIDEX - A New Software For Generating Water Quality Indice. Environmental Monitoring And Assessment. Vol. 119, No. 1-3 (August 2006), 201-231.

Carlie, S.A. 2010. Assessment Of Water Quality Characteristics During Base And Storm Flow Events On Sungai Langat Basin. Master Thesis. Universiti Teknologi Malaysia.

Suki, A., Yusoff, M.K & Poe, M.K. 1988. Water Quality Profile Of Sg. Langat. *Pertanika*. Vol. 11, No. 2, 273-281.

Tukey, J.W. 1977. *Exploratory Data Analysis*. Addison-Wesley. ISBN 0-201-07616-0.

Universiti Malaya Consultancy Unit (UPUM) (2002). *Final Report Program Pencegahan Dan Peningkatan Kualiti Air Sungai Langat*. Kuala Lumpur.

Vega, M., Pardo, R., Barrado, E. & Deban, L. (1998). Assessment Of Seasonal And Polluting Effects On The Quality Of River Water By Exploratory Data Analysis. *Water Research*. Vol. 32. No. 12, 3581–3592.

15

Mining and Water Pollution

Hlanganani Tutu
Molecular Sciences Institute
School of Chemistry
University of the Witwatersrand, Johannesburg
South Africa

1. Introduction

Residual sulphide minerals, especially pyrite, in the tailings dumps are unstable when exposed to atmospheric oxygen and undergo oxidation resulting in the generation of acid mine drainage (AMD) and the subsequent release of heavy metals and metalloids (Jambor & Blowes, 1998). The problem of sulphide oxidation and the associated AMD, or more generally acid rock drainage (ARD), as well as the solution and precipitation processes of metals and minerals, has been a major focus of investigation for over half a century (Nordstrom et al., 1979; Morin & Hutt, 1997; Jambor & Blowes, 1998; Nordstrom & Alpers, 1999).

AMD is formed by a series of complex geochemical and microbial reactions. Notwithstanding, these reactions and processes are not yet fully understood and are still subject to ongoing research.

There are four sub-reactions involved in AMD (Singer & Stumm, 1970), the first reaction being the oxidation of sulphide to sulphate and ferrous ion:

$$FeS_2 + 7/2O_2 + H_2O \rightarrow Fe^{2+} + 2SO_4^{2-} + 2H^+ \quad (1)$$

Penetration of oxygenated water is believed to occur through expansion and contraction of pore-air due to diurnal and seasonal temperature changes. Therefore, erosion of the dumps or any other disturbance which would permit penetration of oxygenated water, for example reworking of the dump poses a potential pollution problem.

Reactions similar to that in (1) involving other sulphide minerals can lead to the dissolution of elements such as As, Cu, Cr, Ni, Pb, Co and Zn which can be leached into underlying aquifers by percolating rainwater. As groundwater exudes into surface water bodies, ferrous iron undergoes oxidation. This reaction is pH dependent and proceeds slowly under acidic conditions (pH≈2-3):

$$4Fe^{2+} + O_2 + 4H^+ \rightarrow 4Fe^{3+} + 2H_2O \quad (2)$$

This is the rate-determining step in the overall acid-generating sequence and is dependent on the presence of *Thiobacillus ferrooxidans* bacteria (which are acidophilic and chemolithotropic), oxygen and pH.

Neutralising reactions, with mineral phases in rock, often reduce the acidity resulting in a groundwater plume at slightly acidic pH but containing high concentrations of Fe^{2+} and SO_4^{2-}.

The ferric ion produced in reaction 2 precipitates as ferric hydroxide:

$$4Fe^{3+} + 12H_2O \rightarrow 4Fe\,(OH)_3\downarrow + 12H^+ \quad (3)$$

On losing a water molecule and undergoing structural re-arrangement, the iron hydroxide forms goethite (FeOOH), a common iron(III) oxide.

The fourth reaction involves the oxidation of additional pyrite by ferric ion, Fe^{3+}.

$$FeS_2 + 14Fe^{3+} + 8H_2O \rightarrow 15Fe^{2+} + 2SO_4^{2-} + 16H^+ \quad (4)$$

This is a self-sustaining reaction and as such renders tailings dumps as sources of AMD for a long time after mining operations have ceased (Nordstrom, 1979; Moses *et al.*, 1987; Ehrlich, 1996; Blowes *et al.*, 1998). This is prevalent in gold and coal mines which are considered to be the main long-term polluters (Pulles, 2003).

There has been less interest in the mineralogical and geochemical interactions taking place within the tailings facilities (Jambor, 1994), yet this is an essential aspect to understand the parameters controlling AMD formation and to develop effective prevention methods. To this end, this work aims at assessing this phenomenon and, based on the results, to draw conclusions about the potential impact this could have on the quality of receiving natural waters. The assessment is based on gold and coal mining case studies (Fig. 1). In the gold mining case, geochemical transformations in two tailings storage facilities (one undergoing reprocessing and the other a decommissioned active slimes dam) were assessed while in the coal mining case these processes were assessed in a decommissioned opencast operation.

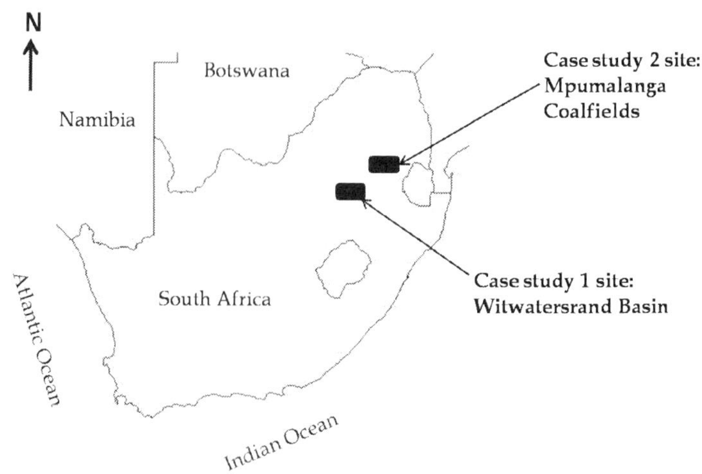

Fig. 1. Location of case study sites.

2. Case study 1: Water quality in gold mining operations in the Witwatersrand Basin, South Africa

The Central Rand goldfield of the Witwatersrand Basin, South Africa has been one of the most important gold mining areas in the world. It is, however, replete with gold mine tailings which have contributed significantly to water pollution as a result of AMD. Water quality is affected mainly by mine tailings and spillages, especially from active slimes dams, reprocessed tailings, as well as from footprints left behind after reprocessing.

Water pollution studies undertaken over the past decade have established good baseline data on the state of the water environment of the Central Rand. Water quality information from the Rand Water Board, Water Research Commission and Crown Gold Recovery shows that the drainage systems on the East and Central Rand have been affected by AMD and have high salt loads (Forstner & Wittmann, 1976; Marsden, 1986; Jones et al., 1988; Scott, 1995; Rosner & van Schalkwyk, 2000; Rosner et al., 2001; Davidson, 2003; Mphephu, 2004).

In this work, column leaching experiments are introduced to simulate (at a laboratory scale) the leaching process occurring in the tailings storage facilities (TSFs). They provide a conceptual approach to understanding the chemical weathering of the minerals in the tailings and the subsequent leaching of the constituents that pose a potential pollution threat to surface and ground water systems.

2.1 Sampling and analytical protocol

Tailings for column leaching were collected from a TSF undergoing reprocessing (Fig. 2).

Fig. 2. Monitoring method being used for preparing slimes tailings for reprocessing at Crown Mines.

The yellowish oxidised layers were scrapped off and the dark-greyish (unoxidised) tailings sampled. pH was measured by inserting a probe into the moist tailings. The material was homogenised and packed into columns (80 cm length x 4.5 cm width) in the laboratory. A small quantity was dried and used to determine the pore volume which is the volume of water required to fill up the pore spaces in the material. Pore volume is the amount or volume of water needed to replace water in a certain volume of saturated porous media. In this study, it is a measured volume of water needed to make slurry from dry fine tailings. The average pore volume obtained for each column was about 600 ml and as such twelve collections of 50 ml volumes made up one pore volume. The collections are reported as pore volume numbers which are fractions of 600 ml, for example, the first collection is 50 ml/600 ml = 0.0833; the second is 100 ml/600 ml = 0.167 and so on. Leaching was done at pH 3 and pH 7. pH 3 (obtained by acidification with sulphuric acid) was aimed at simulating the acidity of a plume leaching through tailings profiles whereas pH 7 simulated the pH of the water used in the spray jet technique (monitoring method) for creating a slurry of tailings during reprocessing. The columns were set up in duplicate such that T1 and T2 were used for leaching with the pH 7 solution while D1 and D2 were used for leaching with the pH 3 solution.

Samples were also obtained from an active slimes TSF that had just been decommissioned (sketch will be referred to later). An active slimes dam or TSF is one in which deposition, in the form of slurry, of tailings from a gold processing plant is occurring. When deposition of tailings is discontinued, usually after some years of operation, the slimes dam is said to be dormant. It is at this stage that reprocessing or remediation plans can be considered. Most of the processing on the Central Rand goldfield is of old dumps and is termed reprocessing as it is a way to salvage economically extractable gold in them. Sampling was done along the catena (at depths of 40-50 cm) from the edge of the TSF towards the central pond. The measurement of pH of the tailings material was done by inserting a probe into a 1:1 wt. ratio of solid:deionised water.

Pure sulphate salts were used as proxies for the commonly found salts in efflorescent crusts that usually occur in the mining area. These were dissolved in a solid:liquid ratio of 1 g: 100 mL in deionised water and paramaters such as pH and electrical conductivity (EC) assessed after stabilization.

Total analysis of elements in dried tailings material was conducted following digestion in a microwave system (Anton Paar, Switzerland). The digestion was carried out using hydrofluoric acid and aqua regia (a 3:1 mixture of hydrochloric acid and nitric acid). Elemental concentrations were determined by inductively coupled plasma spectroscopy (Spectro Genesis, Kleve, Germany).

Measurement of pH and EC was carried out with a portable kit (Multi Line F/Set 3, WTW, Germany), equipped with a pH electrode, an integrated temperature probe (SenTix 41), a standard conductivity cell (Tetra Con 375), and an oxidation-reduction potential probe (SenTix ORP). All reagents used were of analytical grade and certified reference materials were used for quality control.

2.2 Results and discussion

2.2.1 Leaching experiments

During sampling, a pH of 7.8 was recorded for the tailings. This slightly elevated pH in the slimes is attributed to the addition of lime during gold extraction. This is done to keep the cyanide in solution as it escapes at low pH posing a toxicity problem (Funke, 1990). The results for elemental concentrations in tailings following total digestion in the microwave are presented in Table 1. The results point to tailings as potential sources of leachable metals to the environment.

Element	Mg	Na	Ca	K	Al	Fe	Co	Ni	U	Zn	Pb	Cu	As
					mg kg^{-1}								
	1500	700	1600	8000	15000	16000	15	20	35	40	35	30	25

Table 1. Average concentrations (mg kg^{-1}) of selected elements in a composite sample from the TSF.

The results for leaching at pH 7 (Fig. 3a) showed an initial increase in pH to about 8 (roughly the pH of the tailings) and then a drop. This could be attributed to the effects of buffering by the tailings material. This effect was also apparent when pH 3 (Fig. 3b) was used for leaching. The leachates gave an initial pH above 4.5 which dropped gradually to 3 as leaching progressed up to two pore volumes. This trend of decreasing pH is likely to occur beyond two pore volumes.

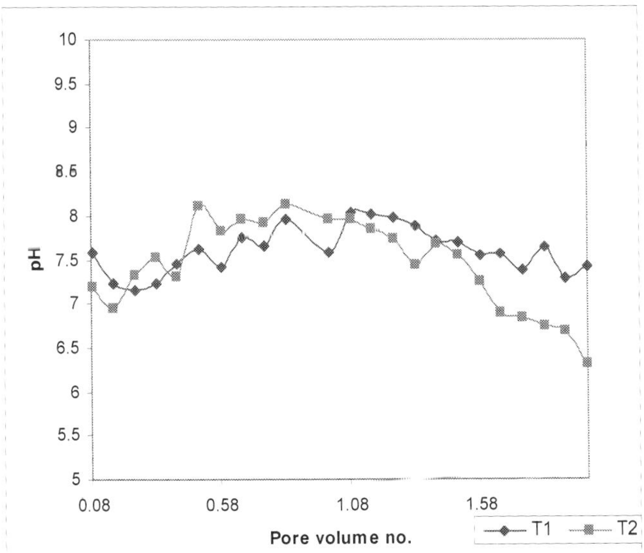

(a)

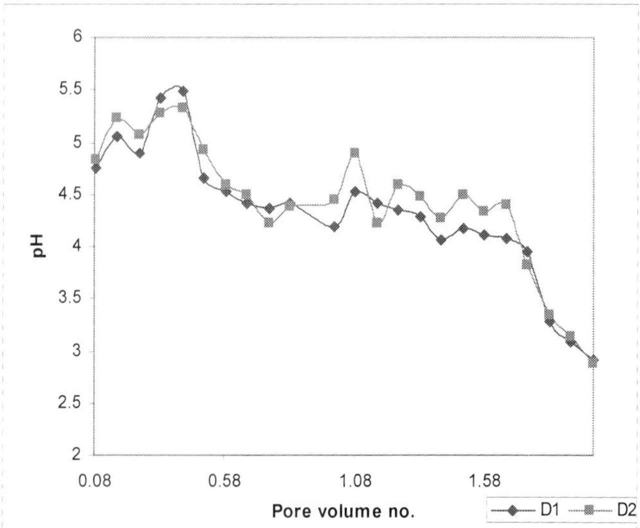

(b)

Fig. 3. pH versus leachate volume number for (a) pH 7 in columns T1 and T2 and (b) pH 3 in columns D1 and D2.

A plot of conductivity and pH versus the leachate volume number (Fig. 4) was done for pH 3. The results gave a correlation between the two parameters as expected, that is, the conductivity increased with a decrease in pH.

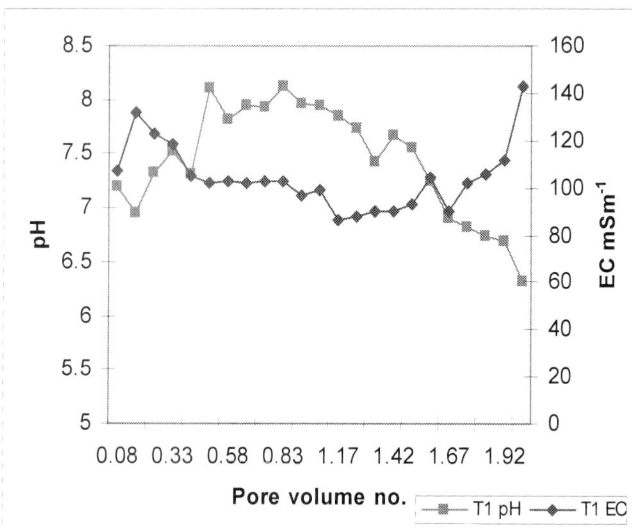

(a)

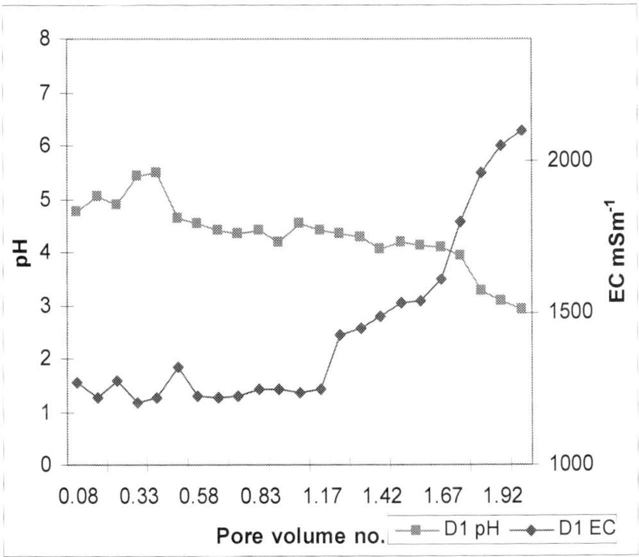

(b)

Fig. 4. pH and conductivity (EC) versus leachate volume number for (a) pH 7 in column T1 and (b) pH 3 in column D1.

The effect of pH on the leachability of metals is shown in Fig. 5.

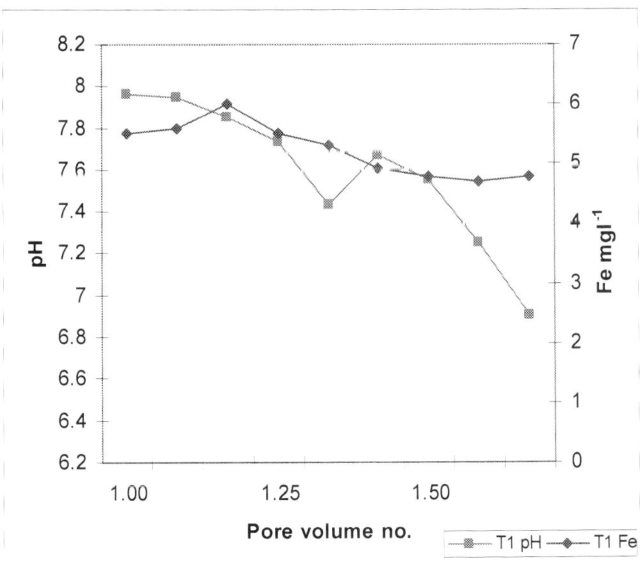

(a)

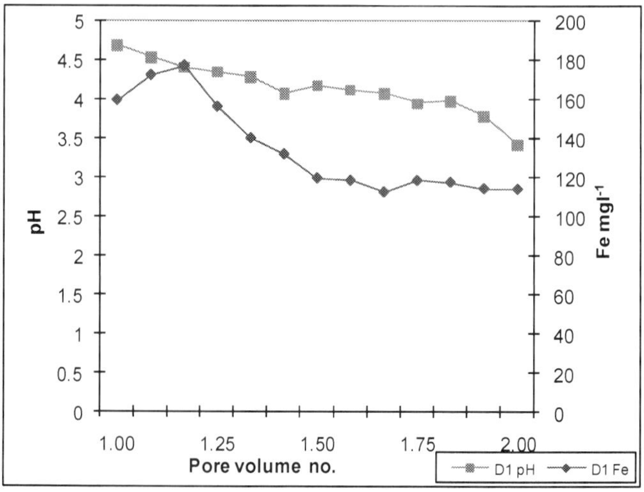

(b)

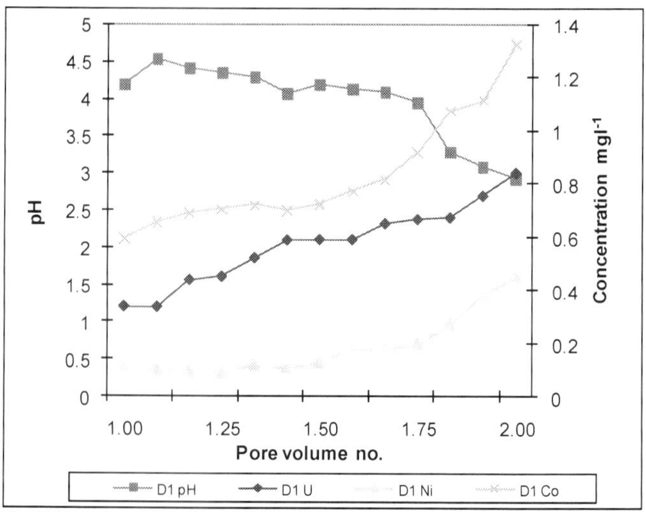

(c)

Fig. 5. pH and iron versus leachate volume number for (a) pH 7 and (b) pH 3; (c) pH, Co, U and Ni concentrations versus leachate volume number for pH 3.

The metal concentrations increase with a drop in pH. This is expected as the host ore in the material (sulphides) undergoes oxidation and releases the constituent metals. The iron concentrations, however, initially increased and then dropped (Fig. 5a and 5b). This drop could be attributed to the formation of hydrous ferric oxides. Once released, iron hydrolyses when it comes into contact with water (reactions 2 and 3 above). The resulting iron hydroxides have a low solubility and at a pH > 3 are quite stable in terms of reverse hydrolysis (Schwertmann, 1985). The low iron concentrations could be the reason for the non-balance, in terms of mass balance, between sulphates and iron (deriving from the oxidation of pyrite). However, no effect of co-precipitation was observed as the other metals had increasing concentrations in subsequent leachates. Further investigations would be required to assess this unusual phenomenon.

2.2.2 Water quality

The best representation of the water quality distribution in and around the gold mining areas in the East and Central Rand goldfields was given by Davidson (2003), Mphephu (2004) and Tutu et al. (2008). This work does not intend to re-invent the wheel in this regard, but rather to draw focus on some of the important contributing factors to variability in water chemistry in these areas as a result of interactions occurring in the tailings.

Firstly, tailings consist of a complex and dynamic heterogeneous physical-chemical matrix in which there are mineral↔water interactions. These interactions can lead to the dissolution of some minerals into the water column while precipitation and adsorption can lead to the formation of new minerals (Allen et al., 1993). This, in part, explains the leaching behaviour of the metals discussed above. It was observed that prolonged leaching of neutral to basic tailings eventually leads to acid generation, and consequently a release of most elements constituting the host ores. This is illustrated in Fig. 6. Over time, the outer layer of the tailings gets oxidised and forms an oxidation layer with an oxidation front defining the boundary with the unoxidised layer. The plume emerging from this is expected to be a result of the interaction between the acidic fluid exiting the oxidised zone and the host ore (and buffering constituents) of the unoxidised zone.

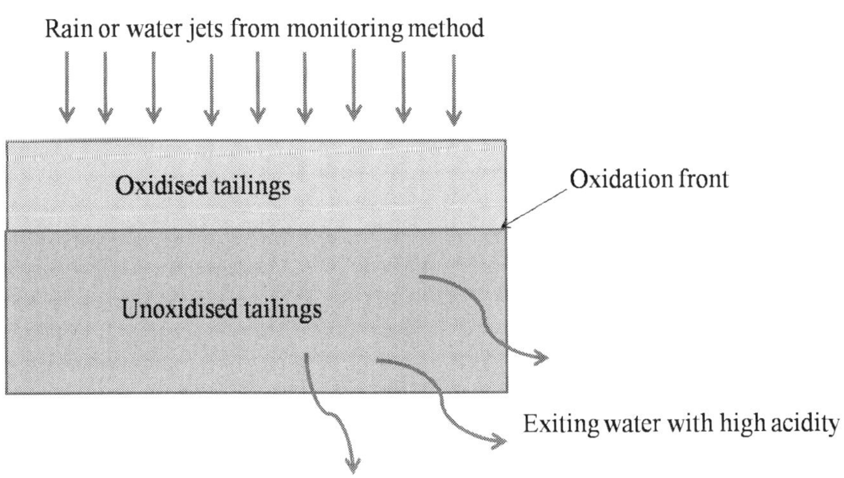

Fig. 6. A section through a TSF showing expected leaching outcomes for an AMD case.

An acidic plume will be expected to exit if the generated acidity in the oxidised zone exceeds the buffering capacity in both the oxidised and unoxidised zones ($m_{H+} > m_{HCO3^-}$). AMD depletes this buffering ability completely below a pH of about 4.2 by neutralizing carbonate and bicarbonate ions to form carbonic acid (H_2CO_3). This is depicted by the following chemical equations:

$$H^+ + CO_3^{2-} \rightarrow HCO_3^- \quad (5)$$

$$H^+ + HCO_3^- \rightarrow H_2CO_3 \quad (6)$$

The carbonic acid is unstable and readily breaks down into water and carbon dioxide:

$$H_2CO_3 \rightarrow H_2O + CO_2 \quad (7)$$

However, acidic water exiting may contain some cyanide species (typically strong-acid dissociable cyanide complexes) despite the acidic regime, resulting in a concomitant mixture of high-sulphate, high-metal and moderately high-cyanide water (Bakatula et al., 2008). The TSF itself usually consists of varying physical-chemical parameters within it. This is illustrated better in the case of a decommissioned active slimes dam (Fig. 7).

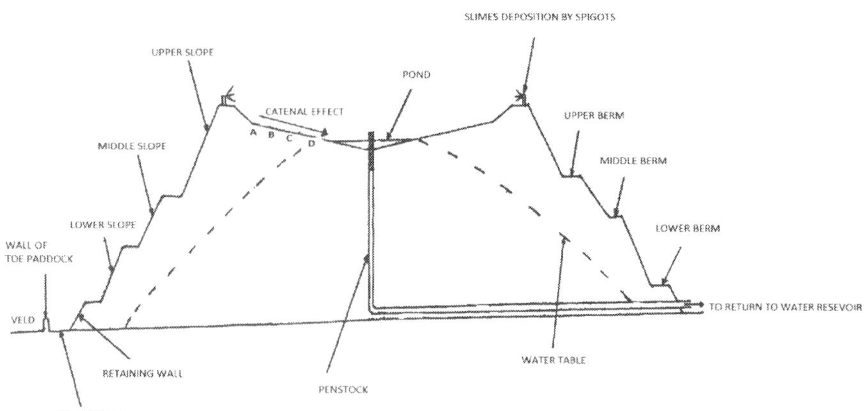

Fig. 7. General section of a TSF.

During operation, the material was discharged at the spigots from where it would spread out along the catena with coarse particles settling out first and the fines settling yonder, with some settling in the pond together with the water (Tutu, 2006). The material was basic at discharge (with pH values more than 9.5) as it was fresh from the processing plant where liming had been done.

The results for pH measurements on the TSF are presented in Table 2.

Level	Site	pH
Top	(A) 30 m in	7.60
Top	(B) 80 m in	7.87
Top	(C) 130 m in	7.95
Top	(D) 180 m in	8.63
Berm	Upper berm	3.65
Berm	Mid berm	2.54
Berm	Lower berm	2.70
Top	Pond	4.01

Table 2. pH measurements for selected areas within TSF.

The berms displayed low pH values. This could be attributed to their constituent coarse material being more susceptible to oxygen entrainment, thus oxidising quickly and releasing acidity. In some cases, such as in TSFs with vegetated slopes, the lower berms may have a neutral to basic pH and higher contents of organic matter due to accumulation of eroded material and debris. Along the catena, the pH is increasingly basic due to the basic slurry flowing down as well as the slow oxidation of the finer materials in that area. However, the pond water is acidic largely due to the fast oxidation of residual sulphides in the aqueous solution and surface flow from the acidic zones on the TSF edges (Tutu, 2006).

The results for metal concentrations along the catena are presented in Fig. 8.

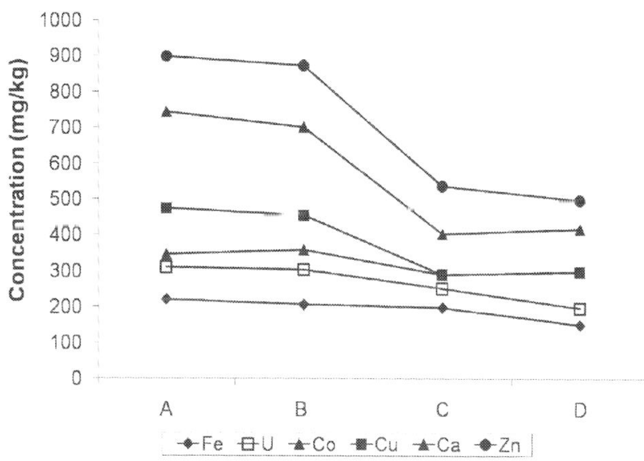

Fig. 8. Metal concentrations across the TSF catena from the edge (A) towards the central pond (D).

The metal concentrations decrease along the catena and this could be due to precipitation as the solid material is separated from the aqueous phase during material flow. The elevated pH of the slimes is above pH of hydrolysis of most of the metals. It should be noted that metal concentrations down the catena could also be determined by density segregation as the material settles, with heavier metals settling in coarser material near the edge and the lighter ones in finer material towards the pond (Tutu, 2006).

Secondly, in some cases where tailings footprints have been rehabilitated by liming, the alkalinity of the exiting water plume may exceed acidity, resulting in neutral to basic water. It is common practice in the Central Rand goldfield to plough the footprints followed by levelling of the surface and then application of lime to eradicate residual acidity in the soil. However, the water leaching through these facilities may still contain elevated concentrations of some elements (e.g. Ca, Mg, U, As, Cr, V) and sulphates in addition to cyanide species. Elements such as U, As, Cr, and V tend to form soluble carbonate complexes (Levinson, 1974; Langmuir, 1997; Zielinski et al., 1997; Tutu, 2006).

Thirdly, there can be added variability in water chemistry and quality as a result of efflorescent crusts. These salt crusts occur as a result of evaporation of shallow groundwater. They contain a variety of sulphate-based salts and fingerprint the local geochemistry of where they are found. The most common among the salts found in the Central Rand goldfield are: gypsum ($CaSO_4.2H_2O$), jarosite ($KFe(SO_4)_2(OH)_6$), melanterite ($FeSO_4.7H_2O$), copiapite ($(Fe, Mg)Fe_4(SO_4)_6(OH)_2.20H_2O$), goslarite ($ZnSO_4.7H_2O$), epsomite ($MgSO_4.7H_2O$) (Naicker et al., 2003; Tutu et al., 2008). They occur on the tailings, next to tailings ponds, along stream banks and on tailings footprints (Fig. 9). In some instances, particularly where a pH gradient or another factor such as incongruent precipitation is prevalent, conspicuous segregation of different salts occurs. Bands of brownish (Fe-containing), pinkish (Co and Mn-containing), bluish (Cu-containing), whitish (Ca and Mg containing) salts could be seen. These salts contain elevated concentrations of metals, metalloids, sulphates (Rosner et al., 2001; Rosner & van Schalkwyk, 2000; Naicker et al., 2003; Tutu et al., 2008) and cyanide species (Bakatula et al., 2008).

Fig. 9. Efflorescent crusts (from the top left in a clockwise direction): at a tailings footprint precipitates on the wall and base (floor); distinct white and yellowish precipitates along a flow path from the footprint; pinkish lumps of precipitates; and precipitates along a stream draining from a TSF (distinct Fe precipitates along the stream can be seen).

The contribution of individual salts to receiving water chemistry during dissolution is presented in Fig. 10 using the pure salt forms. In this case, the dissolution was done in de-

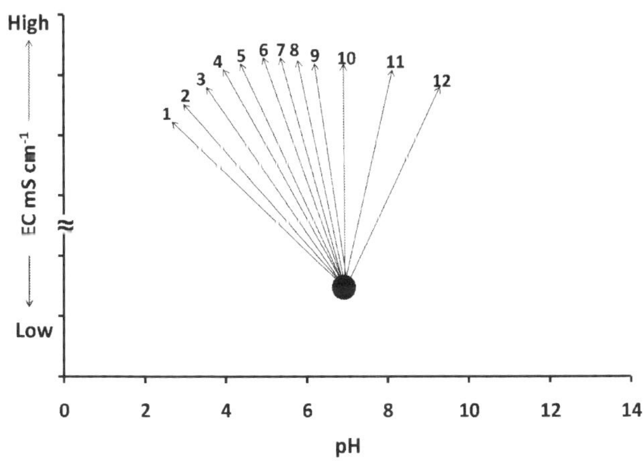

Fig. 10. Geochemical vectors showing expected changes in pH and EC of receiving water on dissolution of: (1) $Cr(SO_4)_3.15H_2O$; (2) $CuSO_4.5H_2O$; (3) $FeSO_4$; (4) $Al_2(SO_4)_3.16H_2O$; (5) $NiSO_4.6H_2O$; (6) $ZnSO_4.7H_2O$; (7) $(NH_4)_2SO_4$; (8) $MgSO_4$; (9) Na_2SO_4; (10) $MnSO_4.H_2O$; (11) $CaSO_4.2H_2O$; (12) K_2SO_4.

ionised water (EC = 0.018 mS cm^{-1} and pH = 7), represented by the central dot. The arrows show the direction of pH change and in all cases an increase in EC as the dissolution of the salts increases the ion concentrations.

Practically, the crusts are a mixture of the various salts and thus the resultant effect would be their combined effect, with the dominating salt in aqueous solution determining the overall effect on pH. It should be noted that by virtue of containing a mixture of minerals, the salts would likely undergo incongruent dissolution in water. Congruent dissolution occurs when all products are soluble, for example, the dissolution of pure gypsum. Incongruent dissolution is likely to contribute to episodic release of the respective elements in the salts as well as the distinct separation of the minerals when the salts re-precipitate.

3. Case study 2: Water quality at a decommissioned opencast coal operation in Mpumalanga Province, South Africa

Coal mining in South Africa is maturing and there are many decommissioning collieries in the major coalfields (Pinetown et al., 2007; Prevost, 2004). AMD is a major problem at several mines in the Witbank Coalfields in the Mpumalanga Province. In some cases, AMD has been found to originate from opencast workings which had been backfilled and has found its way to surrounding natural water courses (Geldenhuis, 1998). The effect of AMD from coal mines has led to serious ecological consequences and mining next to rivers puts the South African supply of fresh water at risk (McCarthy, 2008).

In this work, the behaviour of backfill spoil material in a decommissioned opencast coal mine was assessed with respect to acid-generating potential. The facility was partly rehabilitated by applying a top soil cover to the spoil material and vegetating with some grass. Once a year, lime is applied over the facility at a dosage of 700 kg per hectare as a control for acidity.

The quality of water exiting the facility was also assessed. The protocols followed are explained briefly in the following section and the results presented later.

3.1 Sampling and analytical protocol

The samples were collected from a decommissioned opencast mine operation (Fig. 11). For solids, top soils (labeled 1 e.g. A1) were collected at 40 cm depths while subsoils (labeled 2 e.g. A2) were collected at 1.5 m depths. The soils were collected into polypropylene containers. Both surface and groundwater (borehole water) samples were collected according to commonly accepted sampling protocols (Hermond and Fechner-Levy, 2000).

The solid samples were dried in an oven at 40°C and then pulverized. Total elemental concentrations were determined following microwave digestion (as in Case 1). Assessment of acid generation potential was conducted using the acid base accounting (ABA) technique.

Acid base accounting is a method that is widely used to evaluate the characteristics and behaviour of mineral ores in an effort to assess the potential of AMD generation (Usher et al., 2003). It provides values that help in assessing the acid-producing and acid-neutralising potential of sulphidic material (Skousen et al., 2002). It is applied to opencast coal mine spoil material and to predict the post-mining water quality at polluted sites (Skousen et al., 2002). The soil samples gathered at the site will allow for the estimation of the total overburden amounts. The recommended test procedures for ABA are as follows (Usher et al., 2003):

- Initial pH
- Final pH
- Acid potential (AP)
- Neutralisation potential (NP) and
- Net neutralisation potential (NNP)

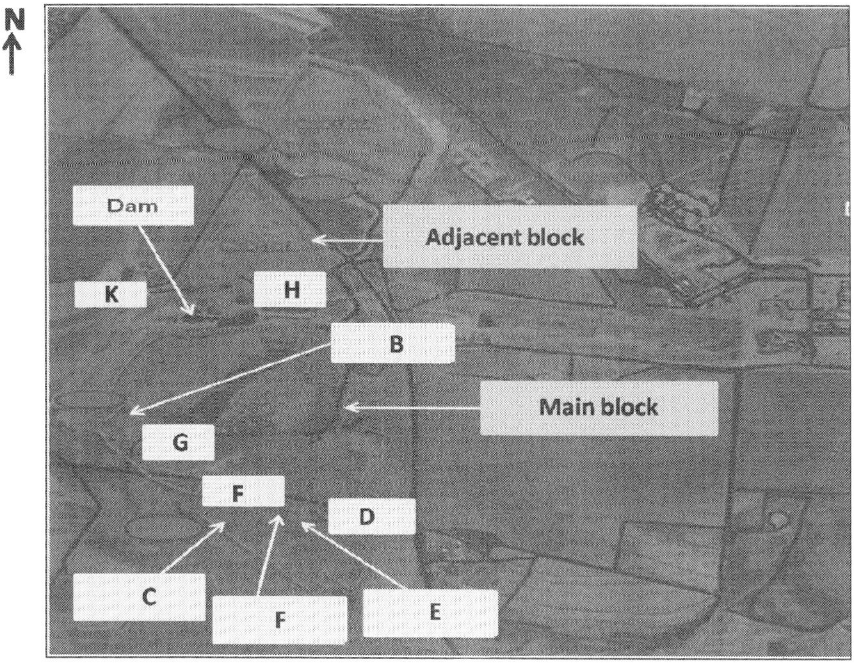

Fig. 11. Sketch of location of sampling points at the decommissioned opencast operation.

The initial pH is a measurement of the pH of the spoil material before oxidation. The recommended ratio of sample to deionised water is 1 g:10 mL, and the pH is measured after 24 hours. This recommendation allows for a more complete solubility reaction and a more realistic determination of the initial pH (Usher et al., 2003). The final pH is a measure of the pH of the spoil material after it has been completely oxidised. The NP is a measure of the amount of acid-neutralising compounds in the coal samples and overburden. It mostly measures carbonates and alkaline earth cations. The NP is calculated from the amount of acid neutralised by $CaCO_3$ in the sample. The AP is the maximum amount of sulphuric acid that can be produced from the oxidation of sulphur-containing minerals in the rocks or overburden material (Usher et al., 2003).

After the NP and AP are calculated, the total NNP is determined by subtracting the AP from the NP i.e. NPP = NP - AP. The NNP values are separated into categories which indicate either the acidity or alkalinity of the spoil material (Skousen et al., 2002). Conceptually, a sample with NNP < 0 is potentially acid-generating and a sample with NNP > 0 indicates more acid-neutralising potential (Skousen et al., 2002).

Field parameters and elemental concentrations were determined as in Case 1. Anions (Cl^- and SO_4^{2-}) were determined by ion chromatography (Metrohm 761 Compact Ion Chromatograph, Switzerland) while carbonates were determined by titrimetry (Clesceri et al., 1998). Water classification was conducted using Statistica ver 8 (Statsoft, USA).

3.2 Results and discussion

3.3.1 Analysis of solids

The analytical results of the total metal concentration in the soil samples of the mining area are shown in Table 3.

Sample	Description	Fe	Mn	Ni	Ca	Mg	S	Pb	Cu	Al	Cd	Cr
						mg kg^{-1}						
A1	Top soil	43689	375	541	540	1262	502	nd	54	23689	75	416
A2	Sub soil	48062	887	686	26841	18023	2616	2.91	48	21318	115	205
B1	Top soil	10019	158	890	297	582	135	0.96	38	30058	90	1029
B2	Sub soil	30517	295	842	14016	9543	604	1.99	49	26839	100	704
C1	Top soil	49138	483	889	28927	17050	2395	4.79	60	25096	121	512
C2	Sub soil	52908	453	775	7054	2097	2097	6.67	41	39085	119	251
D1	Top soil	52970	394	474	2142	1266	636	1.95	55	36417	72	483
D2	Sub soil	37880	227	801	499	1668	382	2.94	73	44063	116	565
E1	Top soil	37512	443	711	22951	12825	1061	0.96	59	26615	96	444
E2	Sub soil	53435	458	722	21851	13645	1145	4.77	48	24427	115	158
F1	Top soil	52207	792	811	32438	18810	3167	1.92	67	23417	109	480
F2	Sub soil	43288	323	406	936	1362	4767	nd	33	28988	76	131
G1	Top soil	44197	379	766	314	907	89	0.97	54	44004	84	868
G2	Sub soil	52069	532	745	28874	21655	1636	4.81	37	22137	118	135
H1	Top soil	24584	306	1360	572	790	26	3.92	44	26836	142	784
H2	Sub soil	75696	386	1011	648	1021	3554	0.96	66	21422	91	2402
J1	Top soil	55632	1959	857	26053	1567	274	3.92	63	9794	114	549
J2	Sub soil	77810	571	984	7143	2952	801	8.57	54	13905	116	952
K1	Top soil	78257	598	829	4212	486	5268	5.88	80	23898	101	988
K2	Sub soil	46223	599	1504	55268	32505	541	6.96	65	20676	165	1157

Table 3. Total metal concentrations in soil samples collected at the decommissioned opencast mine.

The results pointed to elevated concentrations of trace elements, particularly Fe, Mn, Cd, Cr. The concentration of Fe was notably high, exceeding 78000 mg kg^{-1} in sample K1. However, since this was a top soil sample, and its source not being the mining area, the source of Fe could not be ascertained. The sub-soils generally showed higher concentrations of sulphur compared to the top soils. A quick assessment of correlation revealed a strong correlation between iron and sulphur, an indication of the presence of pyrite (FeS_2) in the overburden material. However, there could be other Fe-sulphur compounds such as pyrrhotite ($Fe_{1-x}S$) and iron sulphate ($FeSO_4$).

The interaction between the top and sub-soil with respect to upward movement of pollutants e.g. due to capillary action was not ascertained in this study, but may not be ignored as the amount of evaporation exceeds precipitation in the study area. This is evidenced by the occurrence at certain spots of efflorescent salts similar to those described in Case 1. Other trace elements found include Pb and Cu which had concentrations of 8.57 and 80 mg kg^{-1}, respectively.

The results for initial and final pH of soil samples are presented in Fig. 12.

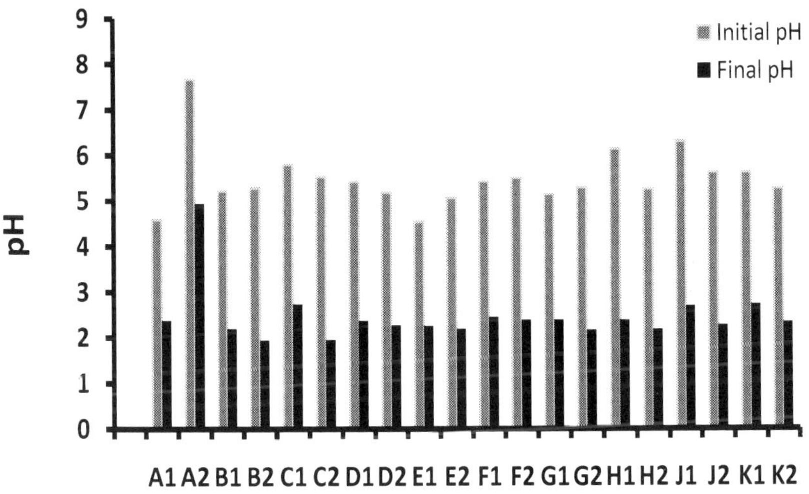

Fig. 12. The initial and final pH of soil samples before and after complete pyrite oxidation.

The initial pH values were in the range 4.54 to 6.29. As explained previously, the initial pH values indicate the immediate acidic or basic characteristic of the sample and it could also indicate if acid has already been generated. The results obtained point to acid generation as a result of the abundance of oxidisable pyrite in the area. An elevated initial pH of 7.66 was observed for sample A2 and this was attributed to the soil type that was a mixture of clay soil and ash. Coal ash is usually alkaline, but it is not clear how this came to be added to the spoil material at that point. The final pH of the samples showed that after complete oxidation, the pH of the soil would be less than 2.5 in most samples and acidic conditions would dominate. Sample B2 had a lower final pH of 1.94, suggesting that the total available carbonates could have already been neutralized or leached out leaving a lesser source of alkalinity to buffer the acid.

The results for NNP are presented in Fig. 13. Negative values of NNP were obtained for all samples in the area, with the lowest recorded value being -9.8 kg t^{-1} $CaCO_3$ for sample G2. The negative NNP values implied that the soil did not have enough $CaCO_3$ to neutralise the acidity generated in the soil hence there was a potential of acidic drainage in the area.

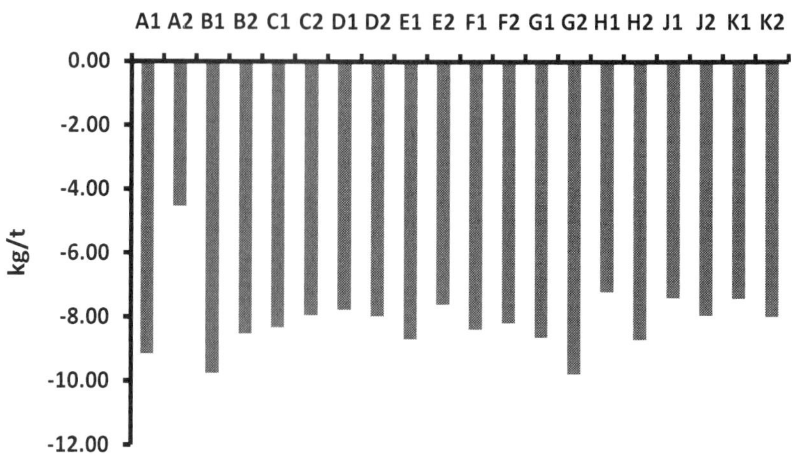

Fig. 13. Variation in the net neutralising potential (NNP) of soil samples.

3.3.2 Water analysis

The results for the water samples collected in the area are presented in Table 4. The results for the pH measurements showed that the pH of the borehole and surface water in the area ranged from 5.92 to 9.00.

Sample	Description	Water level	pH	Eh	EC	SO_4^{2-}	CO_3^{2-}	Cl^-	Na	K	Ca	Mg	Fe	Mn	Ni	Zn	Cu	Al	Co
			Field parameters			Major anions			Major cations				Trace elements						
		m		V	mS cm^{-1}	mg L^{-1}													
H	Borehole	2.3	7.26	0.6	0.4	108	18.0	17.3	12.5	8.9	18.9	12.1	25.8	1.2	nd	0.03	0.02	nd	0.02
F	Borehole	11.2	5.92	0.24	1.6	1146	16.5	17.2	18	13	285	110	24.5	16.7	0.15	0.2	0.01	0.34	0.3
A(g)	Borehole	0	7.11	0.16	0.3	72.9	66.0	17.9	37.7	15	50.6	15.3	7.3	0.6	nd	0.01	nd	0.02	nd
C	Borehole	5.1	7.12	0.34	0.9	435	19.5	14.2	37.3	20	47	84.4	0.04	0.5	nd	0.01	0.02	0.01	nd
K	Borehole	4.5	3.60	0.21	1.5	1170	45.0	7.1	36.8	41	25.6	327	0.1	0.4	nd	nd	nd	0.02	nd
G	Borehole	0.8	9.00	0.12	3.3	2352	30.0	28.4	53.8	34	406	454	0.05	1.5	nd	0.01	nd	0.02	nd
E	Borehole	14.5	6.65	0.15	0.8	327	82.5	10.6	18.7	7.7	99.2	56.5	0.17	1.3	nd	nd	0.02	nd	nd
A(s)	Stream		8.63	0.27	0.6	84.3	186	67.4	80.1	11	44.9	54.5	0.12	0.04	nd	nd	0.03	0.11	nd
J(s1)	Stream		6.66		1.4	708	123	31.9	68.6	13	134	179	1.07	2.5	0.02	0.01	0.02	0.01	nd
B(s2)	Stream		6.05	0.21	3.9	3522	9.0	17.4	34	59	610	577	216	46.2	0.64	0.06	0.01	nd	0.31
B(g)	Borehole	0	6.66	0.09	2.2	1794	10.5	7.1	24.2	22	352	314	197	18.9	0.2	0.1	0.02	nd	0.09
B(s1)	Stream		7.30	0.33	0.8	130	99.0	46.1	44.2	11	36.6	47.9	0.9	0.05	nd	nd	0.03	0.59	nd
Dam	Dam		8.82	0.26	0.7	188	114	24.8	59.9	8.2	60	36.9	0.05	0.02	nd	nd	0.03	0.22	nd
L	Borehole	0	7.16		0.1	2.0	37.5	7.1	12.7	3.2	6.5	5.1	0.3	0.05	nd	nd	0.03	0.18	nd
J(s2)	Stream		6.12		1.4	987	18.0	31.9	57.1	18	201	190	0.2	0.2	nd	nd	0.03	nd	nd
J(s3)	Stream		6.19		1.6	1188	19.5	28.4	69	29	261	237	0.08	0.5	nd	nd	0.03	0.04	nd

Table 4. Water analysis results for samples collected at the decommissioned opencast operation (nd – not detected; blank spaces are for parameters not analysed).

Within the former mining area (Main block), the surface water showed mostly neutral to alkaline pH while the borehole water showed moderately acidic to neutral pH. The pH variation showed that as the water flowed downward from the dam towards the decant point at B, there was a decrease in pH from alkaline to moderately acidic. It was observed that some borehole water within the Main block had alkaline pH values of between 8 and 9. Examples are boreholes at G and K. This was attributed to the depth of the borehole water from the surface. The lower the borehole water depth, the higher the pH because the borehole water was closer to the surface and its pH was increased by the presence of the lime that was applied on the top soil. The water depths of boreholes G and K were 0.8 m and 4.48 m, respectively. These depth values were contrasted to the values for boreholes E and F which had water depths of 14.5 m and 11.22 m and pH values of 6.66 and 5.92, respectively. This discrepancy could be explained by the lag in the movement of groundwater. The water percolating as rain water dissolves the lime on the top soil thus an elevated pH for shallow groundwater. Some of this water flows as surface water to recharge surface water bodies. When the percolating water reaches the deeper zones, however, the buffering capacity is depleted as more and more pyrite in the spoil material is oxidized and acid generated. It would be expected that once the top soil lime is exhausted, even the shallow groundwater will tend to be acidic. Liming is conducted every January, but previous results of the pH of the borehole water show that towards the end of the year, the pH decreased as the added carbonates got exhausted. This acid generation corroborates the results obtained for NNP. The dam in the area recorded an alkaline pH of 8.82. This would be expected as the dam is recharged mainly by surface flow and also dilution effects are important.

EC values were elevated in most of the cases, confirming what was noted previously about the possibility of having high dissolved ions despite the elevated pH. For instance, sample B was a soil sample collected next to a decant point. At this point, a borehole where the groundwater was emerging at the surface and analytical results showed moderately acidic water (pH 6.7), elevated concentration of the sulphates (1794 mg L^{-1}) and Fe (197 mg L^{-1}). A plot of log EC and pH for the water samples is presented in Fig. 14. The plot clearly shows two groups that can be identified by the direction of arrows. The one group is of samples showing a decrease in pH, largely samples that showed a tendency towards acid generation while the other group showed an increase in pH and constituted mainly of samples collected at borehole depths of less than 1 m and some surface water samples as well.

Water Classification

Trilinear plots (Piper diagrams) are used in hydrogeological studies as an effective graphical means of displaying data that contain three dominant components, each of which is typically expressed as a percentage of the total of the three. Analyses of ground water samples can be plotted on a trilinear plot to show relative percentages of cations (e.g., Na^+, K^+, Ca^{2+}, and Mg^{2+}) and anions (e.g. Cl^-, SO_4^{2-} and (HCO_3^- + CO_3^{2-})). Multiple samples can be plotted on the same trilinear plot, and trends, groupings, and mixing patterns can easily be observed. Once plotted, multiple samples can then be classified or grouped together. A piper diagram in Fig. 15 shows the classification of the borehole and surface waters at the site.

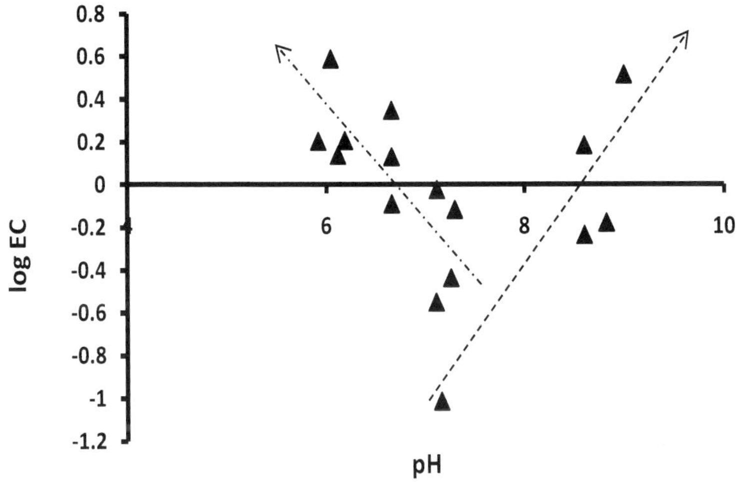

Fig. 14. EC-pH relationship of the borehole and surface water samples.

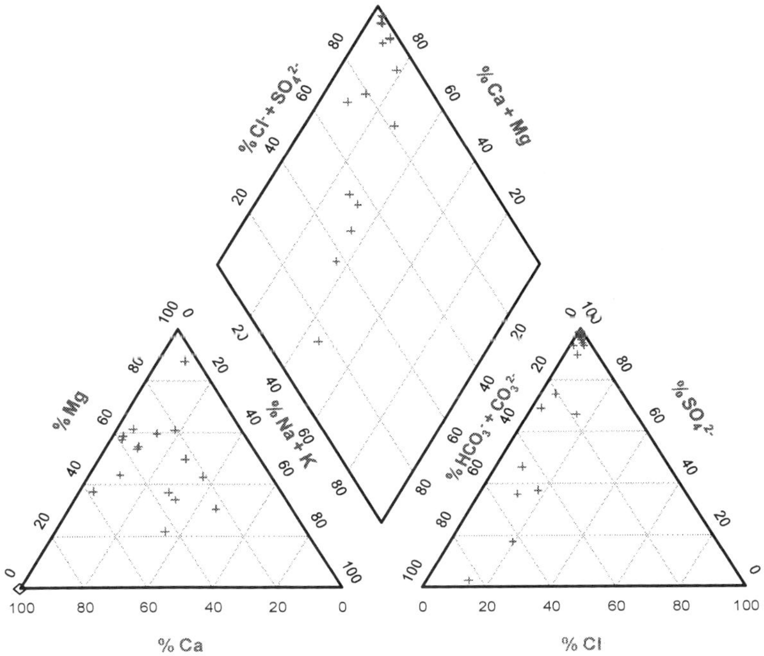

Fig. 15. Distribution of % ($Na^+ + K^+$), % Mg^{2+}, and % Ca^{2+} and % Cl^-, % SO_4^{2-}, % ($HCO_3^- + CO_3^{2-}$) depicting the classification of borehole and surface waters.

The cations Na^+, K^+, Ca^{2+} and Mg^{2+} were plotted on the left triangle and the anions Cl^-, SO_4^{2-} and (HCO_3^- + CO_3^{2-}) were plotted on the right triangle. Points on the anion and cation diagrams were projected upward to where they intersect on the diamond. The data plotted into two distinct groups: (1) the surface water samples plotted towards the increasing proportions of (bicarbonate + carbonate) anions and (2) the borehole samples plotted towards increasing proportions of the sulphate anion. Both surface and groundwater had high proportions of calcium and magnesium which could mainly be from liming. The grouped data indicated that the borehole waters had increased sulphate content, largely due to the onset of pyrite oxidation at greater depths. The water classification revealed the following facies: Ca-Mg sulphate type for the borehole water and Ca-Mg sulphate-bicarbonate for the surface waters.

4. Conclusion

The studies reported here have shown the potential of occurrence of acidic leachates from mining environments with a predominantly sulphidic (particularly pyritic) geological ore composition. Disturbance of old waste dumps by processes such as reprocessing has been shown to lead to an increased potential for release of pollutants. Column leaching was useful for simulating the real environmental scenario likely to occur in tailings impoundments. There was evidence of release of pollutants and their likely interaction with the soil matrix during percolation and downward leaching, and their subsequent ingress into groundwater systems. Water quality was shown to be influenced by plumes emerging from such leaching processes. The results pointed to loss of buffering capacity over prolonged leaching, leading to release of pollutants from the host ores.

Interventions to curtail AMD e.g. by liming have been shown to come short in some cases, particularly where there is no systematically worked out match between the sulphidic composition and the amount of lime to be applied as was demonstrated for the coal mining case. In this case, acid base accounting proved to be an excellent first-order tool and early warning system to determine whether mine waste has the potential to form acidic drainage. It underscored the fact that, when evaluating the extent or likelihood of AMD, it is important to know the amount of hydrogen ions remaining in solution after the buffering is complete.

While liming could be an important intervention for most divalent metals, the likelihood of some elements within the constituent ore, particularly the oxy-anions, to be leached to the receiving water is high. Thus, other interventions have to be considered alongside liming in attempting to curtail AMD and its attendant pollutants.

Variability of the composition of waste material, particularly due to prior metallurgical processes and disposal methods, has been shown to contribute significantly to the variability in the quality of impacted water. Thus, even within the same locality the water quality may differ significantly. Features such as efflorescent salts have also been assessed so as to understand their contribution to the overall quality of water. Their role as evaporation barriers has been shown to enable them to have a high capacity to store various pollutants contained in the shallow groundwater which are precipitated as the water exudes to the surface. As such, it has been possible to infer their impact on receiving water in the event of dissolution during reprocessing activities and flush rain showers.

Further work still needs to be conducted with respect to the detailed solid-liquid interactions occurring in the tailings. As these interactions are complex, geochemical modelling techniques could be applied to simulate the reaction of various components in the host ores with the pore solution as it flushes through the material. This could shed some light regarding aspects such as the observed reduction in Fe concentrations over the duration of leaching likely due to precipitation, a phenomenon that usually occurs with co-precipitation of other metals but which was not the case in this instance. For the decommissioned coal operation, some kinetic tests need to be performed on the samples to assess the rate of oxidation of pyrite in the area. This test is usually done on samples with a net neutralising potential value between -20 and 20 kg t^{-1} $CaCO_3$ as it is a grey range of uncertainty whereby the sample may or may not generate acidity. The test utilises humidity cells to simulate the weathering process in weekly cycles for a period of about 20 weeks. At the end of each week, the pH of the sample is measured and the weathered products are collected in the leach process.

Other interventions can be considered as rehabilitation strategies that could support the liming effort for the decommissioned coal opencast operation. These include use of reactive barriers constructed to cut across the general flow path of the shallow plume as it exits the block under study. This will enable the pollutants in the water to be trapped and thus clean up the water exiting into the natural streams outside the block. Reactive barriers can be made of clay, concrete or other cost-effective materials with a capacity to adsorb pollutants. Another option would be using a phytoremediation strategy. Phytoremediation is a technique that involves the use of metal hyper-accumulating plants to remove metals from contaminated soils. It is cost-effective and environmentally-friendly. It makes use of the capacity of plants to form phytochelatins in roots by releasing organic acids that dissolve metals, making them bioavailable and transporting them to shoots. Hyperaccumulators are conventionally defined as species capable of accumulating metals at levels 100-fold greater than those typically measured in common non-accumulator plants. This would be important than the grass cover currently being used considering the elevated concentrations observed in the spoil material. Hyperaccumulators can be used for phytomining which is a process whereby a metal-hyperaccumulating plant species is grown on polluted soils and the biomass harvested and burned to produce a bio-ore.

5. Acknowledgement

The author would like to acknowledge the following organizations for financial support towards this work: the National Research Foundation of South Africa, the Carnegie Corporation and the Friedel Sellschop Foundation. The contribution by the following people is also acknowledged: Prof. E. M. Cukrowska, Prof. T. S. McCarthy, Prof. L. K. Chimuka, I. M. Weiersbye, E. N. Bakatula, L. Maku, S. Sotyingwa, I. Khumalo, R. Repinga and B. P. Camden-Smith.

6. References

Allen, H. E., Perdue, E. M. and Brown, D. S. (1993). *Metals in groundwater*, Lewis Publishers, Chelsea, London, U. K.

Bakatula, E.N., Tutu, H., Cukrowska, E.M. and Weiersbye, I.M. (2008). Cyanide and cyanide complexes in gold-mine polluted land, *Proceedings of the Mine Closure Seminar*, ISBN 978-0-9804185-6-9, Johannesburg, South Africa, 14-17 October 2008

Blowes, D. W., Ptacek, C.J., Benner, S. G., Waybrant, K. R., Bain, J. G. (1998). Permeable reactive barriers for the treatment of mine tailings drainage water, *Proceedings of the International Conference and Workshop on Uranium Mining and Hydrogeology*, Freiberg, Germany, September, 1998, Vol.2, 113-119

Clesceri, S., Greenberg, E. and Rhodes, R. (Ed(s).). (1998). *Standards methods for the examination of water and wastewater*. (20th edition), American Public Health Association, Washington DC, USA

Davidson, C. (2003). Catchment diagnostic framework for the Klip River catchment, Vaal barrage, October 1998 - September 1999, *MSc Research Report* submitted to the University of the Witwatersrand, Johannesburg

Ehrlich, H. L. (1996). *Geomicrobiology*, Dekker, New York, 719

Forstner, U. and Wittmann, G. T. W. (1976). Metal accumulations in acidic waters from gold mines in South Africa. *Geoforum* 7, 44 - 49

Funke, J. W. (1990). The water requirements and pollution potential of South African gold and uranium mines, *Water Research Commission Report*, ISBN 0 947447 27X, Pretoria, South Africa

Geldenhuis, A.J., Bell, F.G., (1998). *Environmental Geology* 103, 345

Hermond, H.F., and E.J. Fechner-Levy. (2000). *Chemical fate and transport in the environment*, Academic Press, San Diego, USA.

Jambor, J. L. (1994). Mineralogy of sulfide-rich tailings and their oxidation products. In: Jambor, J. L. and Blowes, D. W. (eds.): *Short Course Handbook on Environmental Geochemistry of Sulfide Mine Waste*. Mineralogical Association of Canada, Nepean, Vol.22, 59-102

Jambor, J. L. and Blowes, D. W. (1998). Theory and applications of mineralogy in environmental studies of sulfide-bearing mine waste. In: Cabri, L. J. and Vaughan, D. J. (eds.): *Short Course Handbook on Ore and Environmental Mineralogy*. Mineralogical Association of Canada, Nepean, Vol. 27, 367-401

Jones, G. A., Brierley, S. E., Geldenhuis, S. J. J. and Howard, J. R. (1988). Research on the contribution of mine dumps to the mineral pollution load in the Vaal Barrage, *Report to the Water Research Commission*, WRC Report No. 136/1/89, Pretoria, South Africa

Langmuir, D. (1997). *Aqueous environmental geochemistry*. Prentice Hall, Upper Sadle River, N. J.

Levinson, A. A. (1974). *Introduction to exploration geochemistry*, 2nd edn, Photopress Inc., USA, 689-697

Marsden, D. D. (1986). The current limited impact of Witwatersrand gold-mine residues on water pollution in the Vaal River system. *Journal of the South African Institute of Mining and Metallurgy*, 86, 481-504

McCarthy, T.S. (2008). Coal mining plans put fresh water at risk, Sunday Independent newspaper, Accessed on 10 February 2008, Available from http://www.iol.co.za

Morin, A. K. and Hutt, N. M. (1997). *Environmental geochemistry of mine site drainage. Practical theory and case studies*, MDAG Publishing, Vancouver, Canada, 333

Moses, C. O., Nordstrom, D. K., Herman, J. S. and Mills, A. L. (1987). Aqueous pyrite oxidation by dissolved oxygen and by ferric iron. *Geochimica et Cosmocimica Acta*, Vol. 51, 1561-1571

Mphephu, N. F. (2004). Geotechnical environmental evaluation of mining impacts on the Central Rand. *PhD Thesis* submitted to the University of the Witwatersrand, Johannesburg

Naicker, K., Cukrowska, E. and McCarthy, T. S. (2003). Acid mine drainage arising from gold mining activities in Johannesburg, South Africa and environs, *Journal of Environmental Pollution* 122, 29-40

Nordstrom, D. K., Jenne, E. A. and Ball, J. W. (1979). Redox equilibria of iron in acid mine waters. In: Jenne, E.A. (Ed.): *Chemical modeling in aqueous systems*. Am. Chem. Soc. Symp. Washington, D.C., Series 93, 51-79

Nordstrom, D. K. and Alpers, C. N. (1999). Geochemistry of acid mine waste. In: Plumlee, G. S. and Logsdon, M. J. (Eds.), *Reviews in Economic Geology, The environmental geochemistry of ore deposits*. Part A: Processes, techniques, and health issues, Vol. 6A, 133-160

Pinetown, K.L., Ward, C.R., van der Westhuisen, W.A. (2007). *International Journal of Coal Geology* 70, 166

Prevost, X.M. (2004). South African reserves and the Minerals Act. Coaltrans South Africa, Cape Town, South Africa

Pulles, W. (2003). *Mining Weekly*, January 24-30, 2-3

Rosner, T. and van Schalkwyk, A. (2000). The environmental impact of gold mine tailings footprints in the Johannesburg region, South Africa, *Bull. Eng. Geol. Env.*, Vol. 122, 137-148

Rosner, T., Boer, R., Reyneke, R., Aucamp, P. and Vermaak, J. (2001). A preliminary assessment of pollution contained in the unsaturated and saturated zone beneath reclaimed gold-mine residue deposits, *Water Research Commission Report* No. 797/01/01, Pretoria, 210

Schwertmann, U. (1985). The effect of pedogenic environments on iron oxide minerals. In: *Advances in Soil Science*, Vol. 1, Springer-Verlag, New York, USA

Scott, R. (1995). Flooding of Central and East Rand goldmines: an investigation into controls over the inflow rate, water quality and the predicted impacts of flooded mines. *Water Research Commission Report* No. 224/1/72, Pretoria

Singer, P. C. and Stumm W. (1970). Acid mine drainage: rate determining step, *Science* 167, 1121-1123

Skousen, J., Simmons, J., McDonald, L.M., Ziemkiewicz, P. (2002). Acid-Base Accounting to Predict Post-Mining Drainage Quality on Surface Mines, *J. Environ. Qual.* 31, 2034-2044

Tutu, H. (2006). Determination and geochemical modelling of the dispersal of uranium in gold-mine polluted land in the Witwatersrand Basin. *Ph.D thesis* to the University of the Witwatersrand, Johannesburg

Tutu, H., T. S. McCarthy, E. M. Cukrowska. (2008). The chemical characteristics of acid mine drainage with particular reference to sources, distribution and remediation: the Witwatersrand Basin, South Africa, as a case study, *Applied Geochemistry* 23: 3666-3684

Usher, B.H., Cruywagen, L.M., de Necker, E. and F.D.I. Hodgson, F.D.I. (2003), On-site and Laboratory Investigations of Spoil in Opencast Collieries and the Development of Acid-Base Accounting Procedures, Water Research Commission, WRC Report No. 1055/1/03, ISBN 1-77005-053-1, Vol. 1., pp 3, 9, 11, 13, 25

Zielinski, R. A., AsherBolinder, S., Meier, A. L., Johnson, C. A. and Szabo, B. J. (1997). Natural or fertilizer-derived uranium in irrigation drainage: A case study in southeastern Colorado, U. S. A. *Applied Geochemistry* 12(1), 9-21

16

The Influence of Lignite Mining on Water Quality

Barbara Jachimko
University of Zielona Góra
Poland

1. Introduction

Mining is one of the oldest human activities in the world. Due to long history and changes in the economic situation there are many abandoned mine areas throughout the world. Mining is an industrial branch which is very destructive to the environment. It influences the environment in a number of ways, such as landscape conversion, changes in the water regime and a negative influence the quality of water. Waters polluted by mining wastes are called AMD (acid mine drainage). AMD, connected to lignite mining are extremely acidic and enriched with iron, sulphate and heavy metals. The pollution was caused by weathering products of different sulphides (mainly pyrite FeS_2), which exposed to water and oxidant weather and release protons, sulphate, ferrous and heavy metals into the water. It threatens to the biota and even completely destroys life for a long, difficult to estimate time. The are two ways of preventing such pollution: prevention of AMD formation (activity by source) and treatment AMD already formed. The aim of treatment techniques is to obtain a long-term effect after the treatment activity has been stopped. Until now no success has been achieved. A very interesting example of the influence of lignite mining on water quality is the "anthropogenic lake district" located in Mużakowski Łuk (Western Poland). It comprises about 100 reservoirs of different origin, size and chemical composition, which have never been reclaimed. Among them there are acidic reservoirs of the meromixic type, whose chemical properties have been monitored almost since the appearance of the youngest one, i.e. 1982. On the basis of almost thirty year long observations some interesting remarks, presented below have been formulated. The most important is the difference between the chemical composition of the upper and lower layer, especially the pH value. The chapter presents short review of the occurrence of post-mining areas, the chemistry of acid mine drainage and the description of the "anthropogenic lake district" as a case study of the influence of lignite mining on water quality.

2. Scale of the problem – Global problems with post-mining areas

World lignite deposits are mainly located in Australia, China, some EU countries, Russia, and the USA. China, the world's largest coal producer now accounts for nearly 40 per cent of the world total. Brown coal (lignite) is burnt in power stations for the production of electricity and is also used for other domestic and industrial purposes (steam coal). In 2007, in Europe, the biggest amounts of lignite were mined in Germany - $180*10^6$ t, Turkey $74*10^6$

t, Greece 66*10^6 t, Poland - 57*10^6 t, Czech Republic - 49*10^6 t, Romania - 75*10^6 , Bulgaria - 28*10^6 t. The are three main techniques of lignite mining: underground, open-cast and gasification (on the surface or underground). Nowadays, about 95 to 98% of lignite is mined by the open-cast method and a few percent by the underground one. The underground method was common from the 19th c. till the first half of 20th c. Then, it was discontinued due to low safety, efficiency and economical aspects. Many accidents and cases of excavation flooding occurred. The youngest mine technique is underground gasification in a temperature of ca. 1200°C. The method has not become popular due to numerous disadvantages (thermal conversion of the surrounding rocks, toxic gases and fumes, earth surface deformations, changes in the hydrological balance, groundwater pollution by minerals and benzene). Therefore, it can be used in uninhabited areas, such as deserts or sea shelves. (Kasztelewicz et al. 2008).

Europe has a long tradition of mining beginning with flint mines, then metals, such as copper, gold and silver, ending with coal. The Phoenicians mined metals in Spain in the Bronze Age. The first gold coins were produced in Bulgaria, Greece and Turkey around 700 BC. Techniques of underground mining and even dewatering were improved by The Romans. Some fields started to be abandoned in the middle of the nineteenth century and needed control of environmental impact and remediation (Wolkersdorfer & Bowell, 2005). In 2008 the 32 EU countries produced 10 per cent of the world's coal (all types) of which approximately half came from Germany and Poland. (Brown et al. 2010) - Table 1. During the last 100 years in Germany 10 lignite mining districts come into being (Niederrheinischer in Nordrhein-Westfalen, Thueringsch-Saechsischer, now Mitteldeutsches, Lausitzer, Niederhessischer, Braunschweig-Magdeburger, Oberpfaelzer, Westerwald, Rhein-Main, Ostdeutcher -after the war in Poland, Schlesischer - after the war in Poland).

In Poland there are five lignite mines: Adamów, Bełchatów, Konin, Turów, and Sieniawa, with a total production of 56*10^6 tonnes in 2010. In all of them lignite is produced by the open cast method, excluding "Sieniawa". "Sieniawa" located in Ziemia Lubuska, is one of the oldest brown coal fields in Poland. It was discovered in the middle of the 19th c. The exploitation started in 1873 in the mine called "Emiliensglück". The mine was destroyed during the Second Word War. Then it was activated in 1950 as the mine "Sieniawa". In 1983 it produced about 209*10^3 tonnes of coal. More then half of total production, ca. 33 mln Mg per year came from Bełchatów, 20% from Turów, 17% from Konin and 8,5% from Adamów.

3. Influence of mine activity on water resources

Mining must be preceded by seam dewatering. It changes the hydrological regime and upsets the water balance. Lowering the water table causes lack of water in domestic wells, it decreases soil humidity and river flow. After closure many pits are filled back with groundwater, which had been earlier polluted by discharge from mining wastes and overburden material. In Poland in 2002, about 361 million m^3 of water were pumped from open lignite mines, producing cones of depression of 1200 km^2 (Wilk, 2003). In the "Bełchatów" mine (Poland) the cone of depression was about 730 km^2. In Poland the drainage water is generally of good quality and not environmentally hazardous, with the exception of water from Turów with the salinity of over 5 g/l. Some of the serious AMD problems occur in Germany, in the Lausitz and around Leipzig. They come from the high AMD potential and deficit of water in the area. About one third of the total amount of about

500 open-cast pit lakes in Germany were acid with a pH below 3 (Fyson A., Nixdorf B., Kalin M., 2006). 168 pit lakes have been documented in the Lausitz and approximately half of them are extremely acidic (pH 2.4-3.4) (Table 1). Acidic mine lakes in the Lausitz contain huge amount iron (up to 800 mg/dm^3) and sulphate (up to 4636 mg/dm^3).

Parameter	unit	minimum	maximum
pH	-	2.4	3.4
Acidity	$mmol/dm^3$	0.1	26.6
Total Fe	mg/dm^3	0.2	800
Sulphate	mg/dm^3	460	4636
Total organic	mgC/dm^3	0.8	10.9
Total N	mg/dm^3	0.9	5.3
Total P	$\mu g/dm^3$	4	26

Table 1. Summary of some chemical parameters for 75 acidic mine lakes in the Lausitz region (Nixdorf et al. 2001).

4. Chemistry of AMD

4.1 Weathering of minerals accompanying coal and metal ores

Mining affects the natural cycles of some elements, mainly sulphur and iron. It changes the natural base acid balance by the increase of the amount of sulphides minerals exposed to water and atmosphere. Protons and other acid ions released due to accelerated dissolution should be neutralised by the reactions of carbonates and silicates weathering. Coal deposits often contain very small quantities of base materials, which inhibits neutralisation. Sulphur-bearing minerals, such as iron sulphides, sphalerite (ZnS), chalcopyrite ($CuFeS_2$), galena (PbS), arsenopyrite (FeAsS) often occur with coal and metal ores. The pyrite content in mine wastes was estimated at between 0,1 and 5% (weight) (Blodau, 2006). In Turów (Poland) the pyrite content was in the range of 0,3 to 0,5% (Kwiatkowski, 2006) and in Sieniawa below 0,8%. Iron-disulphides (FeS_2) exist in two main forms: pyrite and marcasite, which have the same chemical composition and different crystallography. Pyrite is the most stable and prevalent form.

Pyrite oxidation is the main process involved in acidity generation. Two oxidants are involved in pyrite oxidation : oxygen and ferric ion, Fe (III).

The reactions of pyrite weathering under the exposure to air and water are as follows:

$$FeS_2+{}^7/{}_2O_2+H_2O=Fe^{2+}+2SO_4^{2-}+2H^+ \quad [1]$$

$$Fe^{2+}+{}^1/{}_4O_2+H^+= Fe^{3+}+ {}^1/{}_2H_2O \quad [2]$$

$$Fe^{3+}+ 3H_2O=Fe(OH)_3+3H^+ \quad [3]$$

$$FeS_2+14Fe^{3+}+8H_2O=15Fe^{2+}+2SO_4^{2-}+16H^+ \quad [4]$$

According to the reactions [1] and [4] ferrous and sulphate ions are released at the rate of 1:2 (Lenk & Wisotzky, 2001). In the first stage sulphide oxidation one mole of oxidized pyrite gives two moles of protons and two moles of sulphides. In an anoxic milieu, ferrous ions are inert and can exist in high concentrations. The reaction [4] describes pyrite oxidation by ferric ions (Fe(III)). Ferric ions coming form pyrite weathering supply reactive Fe(III) and allow the process to proceed. The rate constant of the reaction [2] depends on pH, so oxidation of ferrous to ferric in a pH below 3 is extremely slow (Nordstrom, 1982). In the acidic range (pH from 2,5 to 3,5) ferrous oxidation is catalysed by microorganisms. In a neutral to alkaline milieu Fe(III) concentration is limited by the precipitation of hydrous ferric oxides and the importance of the reaction [2] decreases. The reaction of pyrite oxidation is surface controlled (Evangelou, 1995) and the nature of products depends on the origin and shape of pyrite.

The final AMD chemistry depends on the following factors:

- amounts of sulphides exposed to weathering factors,
- weathering reaction rate,
- secondary reactions, such as ferric minerals precipitation and redox reactions,
- availability of base minerals.

4.2 Secondary processes in the aspect of water acidity

After pyrite weathering products, such as protons, sulphates (VI) and ferrous ions are released into water, different secondary processes occur. The processes involve precipitation (mainly of iron, aluminium or calcium minerals), ion exchange, isomorphic substitution, redox reaction (often supported by microorganisms) and the transformation of solids. The processes influence water quality and on the basis of the impact on pH, they are classified as acidity or alkalinity generation. The most important process generating acidity is the precipitation different ferric compounds (Table 2). Their accumulation in acid reservoirs is an important part of the iron cycle in the environment (Cornell & Schwertmann, 2000).

Mineral	Formula	Reaction
Ferrihydrite	$Fe(OH)_3$	$Fe(OH)_3 + 3H^+ \Leftrightarrow Fe^{3+} + 3H_2O$
K-Jarosite	$KFe_3(SO_4)_2(OH)_6$	$K^+ + 3Fe^{3+} + 2SO_4^{2-} \Leftrightarrow KFe_3(SO_4)_2(OH)_6 + 6H^+$
H-jarosite	$HFe_3(SO_4)_2(OH)_6$	$3Fe^{3+} + 2SO_4^{2-} + 6H_2O \Leftrightarrow HFe_3(SO_4)_2(OH)_6 + 5H^+$
Schwertmannite	$Fe_8O_8(OH)_x(SO_4)_y$	$Fe_8O_8(OH)_x(SO_4)_y + (24\text{-}2y)H^+ \Leftrightarrow 8Fe^{3+} + ySO_4^{2-} + (24\text{-}2y+x)/H_2O$
Goethite	$FeOOH$	$FeOOH + 3H^+ \Leftrightarrow Fe^{3+} + 2H_2O$
melanterite	$FeSO_4{*}7H_2O$	$Fe^{3+} + SO_4^{2-} + 7H_2O \Leftrightarrow FeSO_4{*}7H_2O$
siderite	$FeCO_3$	$Fe^{3+} + CO_{2(g)} + H_2O \Leftrightarrow FeCO_3 + 2H^+$

Table 2. Influence of ferric minerals on acidity generation

Goethite (named after Johann Wolfgang von Goethe) is the most thermodynamically stable phase at a pH>2, but it occurs in minor quantities within precipitates, however, it is in common in bottom sludge. Optimum pH conditions for schwertmannite formation were found between 2.8 and 3.2 (Bigham, 1996). Jarosite forms at a pH<2.5 and ferrihydrite (basic formula $Fe_5O_8H{*}H_2O$) precipitates at a pH>4.5 (Blodau, 2006).

According to Langmuir (1997) the highest pH value, for the potassium concentration of 10^{-4} and sulphate of 10^{-2} mol/kg the pH value is 3.8. Ferric atoms involved in the crystalline structure of minerals are often replaced by other ions, such as the substitution of aluminium in goethite. The general formula of goethite incrusted by aluminium is $Fe_{1-x}Al_x)_{y/3}O_{1-y}(OH)_{1+y}$, where y may reach 0.25 (Wolska & Schwertman 1993).

Aluminium precipitates in the form of gibbsite and kaolinite and in the presence of high sulphate concentrations, in the form of different sulphates such as alunogen, alunite or jurbanite.

According to Nordstrom (1982), the stability for particular minerals, for the sulphate concentration of 10^{-4} mol/kg, ranges as follows:

- 0-3.3 pH - jurbanite AlSO4(OH)*5H2O
- 3.3-4.47 pH - alunite KAl3(SO4)2(OH)6
- pH>4.47 gibbsite Al(OH)3.

Schwertmannite is metastable to goethite and transforms in solid phase with protons release, according to the reaction:

$$Fe_8O_8(OH)_{5.5}(SO_4)_{1.25(s)}+2.5H_2O=8FeOOH_{(s)}+2.5H^++1.25SO_4^{2-} \quad [5]$$

Another mineral, which often precipitates in post-mine water is gypsum:

$$Ca^{2+}+SO_4^{2-}+2H_2O=CaSO_4*2H_2O \quad [6]$$

The reaction [6] has no direct influence on the acidity balance (except for changing the concentration of HSO_4^-), but it considerably limits mineralization.

Processes involved in acidity consumption include the dissolution of base minerals (calcite, dolomite, biotite, plagioclase, orthoclase), cation exchange and redox/precipitation reactions. In the case of lack of base minerals in soils pH may reach an extremely low value, even below 1. It accelerates the rate of silicates weathering and releases aluminium ions into the water. Redox processes with the participation of organic matter usage as a reducing agent result in ferric and sulphate reduction contrary to pyrite weathering. The processes of sulphate and iron reduction are controlled by electron donor availability (Blodau et. al., 2000), temperature and pH (Koschorreck et al., 2003).

4.3 Acid base balance in AMD

In order to understand AMD chemistry and predicting the results of neutralisation strategies it is crucial to know the acid-base theory by Stumm & Morgan (1996). The theory was adopted to AMD by Uhlmann in Nixdorf & Deneke (2004). On the basis of hydrolysis reactions of weak anions (HSO_4^-) or cations (Fe^{3+}, Al^{3+}) acidity and alkalinity can be described as follows:

$$Aci=[H^+]+3[Fe^{3+}_{(aq)}]+3[Al^{3+}_{(aq)}]-[HCO_3^-]-[OH^-] \quad [7]$$

$$Alk = [HCO_3^-]-[OH^-]-[H^+]-3[Fe^{3+}_{(aq)}]-3[Al^{3+}_{(aq)}] \quad [8]$$

where: $[H^+]$, $[Fe^{3+}_{(aq)}]$, $[Al^{3+}_{(aq)}]$, $[HCO_3^-]$, $[OH^-]$ - the activity of the ion.

Calculations of alkalinity and acidity according to the above equations are difficult due to the occurrence of iron and aluminium in different intermediates (hydrolysis), mainly as hydroxyl complexes. Acidity and alkalinity can also be calculated on the basis of strong anions and cations with the differentiation of oxic conditions. For an anoxic milieu acidity and alkalinity can described according the following equations:

$$\text{Aci}_{\text{anox}} = 2[SO_4^{2-}]+[Cl^-]+[NO_3^-]-2[Ca^{2+}]-2[Mg^{2+}]-[Na^+]-[K^+]-[NH_4^+]-2[Fe^{2+}] \quad [9]$$

$$\text{Alk}_{\text{anox}} = 2[Ca^{2+}]+2[Mg^{2+}]+[Na^+]+[K^+]+[NH_4^+]+2[Fe^{2+}]-2[SO_4^{2-}]-[Cl^-]-[NO_3^-] \quad [10]$$

Therefore under oxic conditions ferrous ions are oxidised to ferric, which is a weak base. And the above equations can be modified to:

$$\text{Aci}_{\text{ox}} = 2[SO_4^{2-}]+[Cl^-]+[NO_3^-]-2[Ca^{2+}]-2[Mg^{2+}]-[Na^+]-[K^+]-[NH_4^+] \quad [11]$$

$$\text{Alk}_{\text{ox}} = 2[Ca^{2+}]+2[Mg^{2+}]+[Na^+]+[K^+]+[NH_4^+]-2[SO_4^{2-}]-[Cl^-]-[NO_3^-]- \quad [12]$$

In order to estimate the aciditity of water from copper mines (high copper concentration), Evangelou (1995) formulated the neutralisation potential (NP), which is defined as follows:

$$NP = 2c_{CO_3^{2-}} + c_{HCO_3^-} + c_{OH^-} - c_{HSO_4^-} - c_{H^+} - 2c_{Fe^{2+}} - 3c_{Al^{3+}} - 2c_{Cu^{2+}} \quad [13]$$

c- ion concentration, mmol/dm^3.

Schoepke (1999) adapted this expression for lignite mine water:

$$NP = 2c_{CO_3^{2-}} + c_{HCO_3^-} + c_{OH^-} - c_{HSO_4^-} - c_{H^+} - 2c_{Fe^{2+}} - 3c_{Al^{3+}} - 2c_{Fe3+} - 2c_{Mn^{2+}} \quad [14]$$

If alkalinity according to Stumm&Morgan (1996) can be described as:

$$Alk = -c_{H^+} - c_{HSO_4^-} - 3c_{Fe^{3+}} + c_{HCO_3^-} \quad [15]$$

Then the NP can be shortened to:

$$NP \approx Alk - 2c_{Fe^{2+}} - 3c_{Al^{3+}} - 2c_{Mn^{3+}} \quad [16]$$

Alk means the results of 0.1N NaOH titration to a pH= 4.3.

In acid water NP is below zero and for base has positive values. The goal of remediation strategies is to achieve an NP above 0.2 mmol/dm^3 to lower acidity to the expected level. The choice of a neutralization strategy depends on the neutralization algorithm, which includes the following reactions:

1. protons neutralisation: $H^+ + NaOH = H_2O + Na^+$ [17]
2. Fe(III) hydrolysis, pH>3: $Fe^{3+} + 3NaOH = Fe(OH)_3 + 3Na^+$ [18]
3. Al(III) hydrolysis, pH>4: $Al^{3+} + 3NaOH = Al(OH)_3 + 3Na^+$ [19]

Tootche et al. (2003) studied the buffering mechanisms of acidic mining lakes in the open cast lignite mining lake (Plessa 111) in Lausitz (Germany). On the basis of laboratory investigations, hydrochemical modelling and mineralogical structure he detected four different buffers in four successive sections of the titration curve:

1. hydrogen sulphate buffering with no precipitates :

$$HSO_4^{-}=SO_4^{2-}+H^+ \quad [20]$$

2. iron buffering with iron hydroxide precipitation and sulphate sorption:

$$\text{precipitation: } Fe^{3+}+3H_2O\rightarrow Fe(OH)_{3(s)}+3H^+ \quad [21]$$

$$\text{sorption: } Fe\text{-}OH + SO_4^{2-}+H^+\rightarrow Fe\text{-}SO_4^{-}+H_2O \quad [22]$$

$$Fe\text{-}OH + SO_4^{2-}\rightarrow Fe\text{-}OHSO_4^{2-} \quad [23]$$

$$Fe\text{-}OH + Ca^{2+}\rightarrow Fe\text{-}OCa^{+}+H^+ \quad [24]$$

$$Fe\text{-}OH + Mg^{2+}\rightarrow Fe\text{-}OMg^{+}+H^+ \quad [25]$$

3. aluminium buffering: aluminium hydroxysulphate precipitation:

$$Al^{3+}+ x\, SO_4^{2-} +(3-2x)H_2O\rightarrow Al(OH)_{3-2x}(SO_4)_{x(s)}+(3-2x)H^+ \quad [26]$$

and also Mn-buffer - manganese oxide precipitation

$$Mn^{2+}+2H_2O\rightarrow FeO_{2(s)}+4H^{+}+2e^{-} \quad [27]$$

4. an additional soil phase buffer which was based on:
a. ion exchange calcium and magnesium on goethite,
b. mineral transformation of schwertmannite into goethite with the release of protons and sulphate:

$$Fe_{16}O_{16}(OH)_{16-2x}(SO_4)_{x(s)}+2xH_2O\rightarrow 16\, FeOOH+xSO_4^{2-}+2xH^+ \quad [28]$$

The first section of the titration curve was in a pH range of 2.5 to 3, the second - 3.5 to 4.5, the third - 4.5 to 5.5 and the last one - above 5.5. The longer the buffer zones are, the more of the neutralization agent must be used.

5. Antrophogenic lake district in Łuk Mużakowski – Poland as a case study of the influence of brown coal mining on water resources

5.1 Location and origin

The "anthropogenic lake district" is located along the Polish-German border between Trzebiel and Łęknica. It comprises about 100 reservoirs of different sizes and chemical properties with a total surface of about 1 500 000 m^2. The reservoirs were generated due to lignite mining which started in the second half of the 19-th century and lasted till the seventies of the 20-th century. In 19-th century seven mining companies were operated in this area. The first one was founded in Raszyn in 1983, the next in Tuplice a year later, then "Wiktor" in Trzebiel in 1898 and in Nowe Czaple in 1900. The best known mine in this area was "Babina" in Łęknica. It was founded in 1922 and operated (as the last one) till 1973 (Jędrczak, 1992). Exploitation was done by means of two methods, at the beginning underground and later open cast. Both techniques resulted in unfavourable changes in the landscape, such as pits and sinks, which after being filled with water created pit and hollow lakes.

Reservoirs are divided into three groups depending on their origin and location (Fig. 1.):

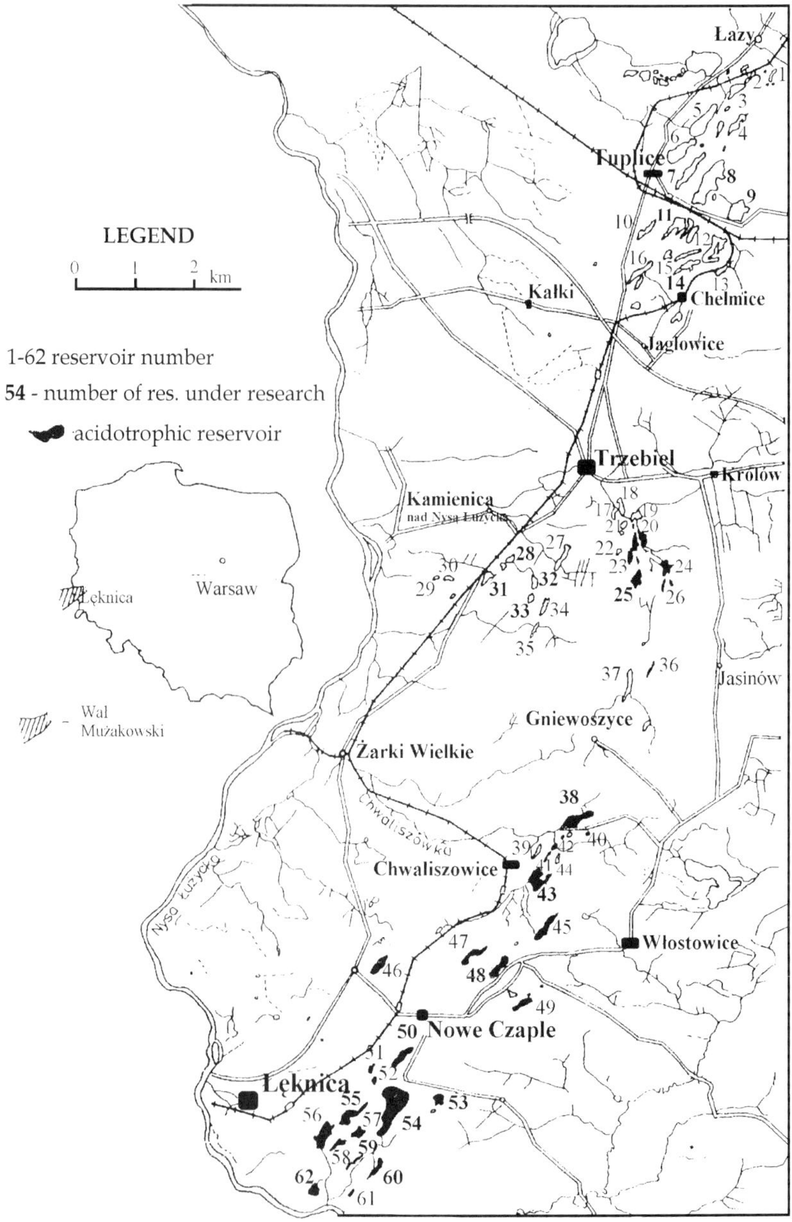

Fig. 1. Map of the "antrophogenic lake district"

1. ca. 30, the oldest hollow reservoirs situated near Tuplice; they have elongated shape with axis NE-SW, their surface varies from <1 to 15 ha and maximal depth from 1.5 to 5.5 m.
2. 33 hollow reservoirs situated between Trzebiel and Kamienica; they are rather small, with a surface lower than 5 ha, but relatively deep to 9.5 m.
3. 33, the youngest and biggest pit reservoirs situated near Łęknica; their surface varies from 1.8 to 22 ha, and the maximal depth from 3.5 to 21.5 m.

The reservoirs have never been remediated. One of them was used as a domestic waste receiver and due to eutrophisation it disappeared. Some of the smallest reservoirs are used as fish farms. The rest of them are not exploited economically nor touristically.

5.2 Methods

The reservoirs of the "antrophogenic lake district" were monitored irregularly. The first data come from the early eighties, about five years after the youngest reservoir came into being. The chemistry of some reservoirs was described by Jędrczak (1992) and Jachimko (1998). In the years 2006-2009, mainly meromictic reservoirs were investigated (Jachimko, 2007; Jachimko&Kołodziejczyk, 2008). All research was based an analysis of water samples taken up along the vertical profile (from the surface and then every two meters in depth), from the deepest place of each reservoir. The water analyses were supplemented by sludge chemistry. The hydrogeochemical simulator PHREEQC (Parkhurst&Apello, 1999; Apello&Postma, 2006) was used to calculate saturation indexes with the use of constants included in the program. NP according to the reaction [14] was calculated to evaluate the acidification.

5.3 Chemistry of reservoirs

The reservoirs of the "antrophogenic lake district" have different chemical properties (Table 3). According to the pH value and redox potential the reservoirs were classified into three groups:

1. acidotrophic with a pH ranging between 2.6 and 3.7 and redox potential of 605 to 755 mV,
2. intermediate with a pH ranging between 2.6 and 3.7 and redox potential of 605 to 755 mV
3. "others" with a pH ranging between 5.2 and 7.4 and redox potential of 380 to 600 mV.

Acidotrophic reservoirs are located in the south of the district. They are mainly the youngest reservoirs of pit origin. "Others" occur in the north and belong to the oldest group. There were only two intermediate reservoirs, no 11 and no 21, but their chemistry varied from one another. No 21 was similar to the acidotrophic class and no 11 to "others". The pH value and redox potential were different in the profiles of most reservoirs. The biggest changes of the factors were found in acidotrophic reservoirs Re 25, 48 and 54. Surface waters were very acidic (pH<3.4) and had a high redox potential (>750 mV). However, waters close to the bottom had a pH of above 4 and a redox potential of less than 400 mV. Within the acidotrophic group the changes of the pH value in the vertical profile were insignificant. The pH values in vertical profile in the group of "others" reservoirs were almost constant or increased from the surface to the bottom (with the exception of very shallow reservoir no 8) (Fig. 2).

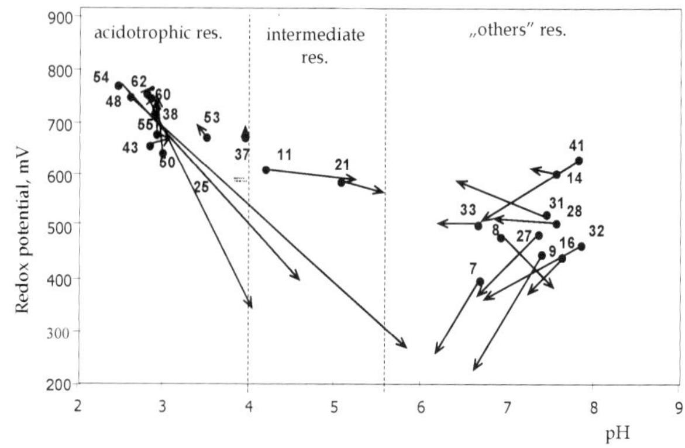

Fig. 2. pH-value versus redox potential in vertical profiles of reservoirs (points represent surface, arrows - bottom layer)

The chemical composition and biocenosis of the "others" reservoirs were similar to the natural lakes. The evidence of mine origin are a high concentration of iron in the deeper part of the water body, the iron to sulphur ratio and a low neutralisation potential. In five reservoirs of this type, (no 7, 9, 27, 33 and 32) iron to the sulphur ratio in the bottom layer meets the values which result from pyrite weathering. The iron concentration in uppers layers is very small, lower than 0,05 mmol/dm^3, due to the low solubility of ferric compounds in this pH region. Close to the bottom due to ferric reduction the concentration of iron remained on the primary, high level. The sulphate concentration changed in a very small range with depth and ranged from 0,15 to 2 mmol/dm^3. In some reservoirs the sulphate concentration decreased with depth, probably as a result of reduction, but in the others it increased as a result of continuous AMD interaction. The bottom layer of reservoirs 25 and 48 contained much more ferrous irons than resulted from pyrite weathering. It was probably connected with meromictic type of the reservoirs.

parameter	unit	"others" reservoirs	intermediate reservoirs	acidotrophic reservoirs
pH	-	5,96 7,85	4,18 5,64	2,41 5,85
calcium	mg/dm^3	10,0 79,6	19,5 50,6	13,9 334
sulphate (VI)	mg/dm^3	10,8 186,3	37 171,6	75,5 3935
tot. iron	mg/dm^3	0,14 39,2	0,87 1,73	0,82 1618
aluminum	mg/dm^3	0,013 0,431	0,114 0,542	0,195 20,59
potassium	mg/dm^3	1,7 11,6	2,3 3,1	2,0 36,9

Table 3. Minimal and maximal values of water quality parameters (own results)

The water of "others" is of the calcium bicarbonate or calcium sulphate type. The percentage of calcium in the "others" reservoirs ranged from 34 to 71% of total cations and decreased with depth. In acidotrophic reservoirs it was smaller and varied from 16 to 59%. The iron content ranged from 0,1 to 30 % of total cations in the "others" reservoirs and reached 68% in the acidotrophic ones. Sulphate averaged from 34 to 72% of anions in the "other" reservoirs and dominated in almost all of them. The percentage of bicarbonate ranged from 9 to 50 % and chloride from 5 to 20% of total anions. Sulphate dominated in all acidotrophic reservoirs and ranged from 81 to 99,6%.

The chemistry of the reservoirs was interpreted with means of SI of ferric, aluminium and calcium minerals. Figure 3 presents the relationship of SI of the selected minerals and pH. The iron content was much higher than calculated from goethite or jarosite K solubility. The reservoirs were oversaturated to goethite in the whole range of pH, and to ferric hydrooxides in a pH over 4. Jarosite K was oversaturated in a pH below 3.5 and over 6.5 (with the exclusion of water with very little iron (concentration below 0.2 mg/dm^3).

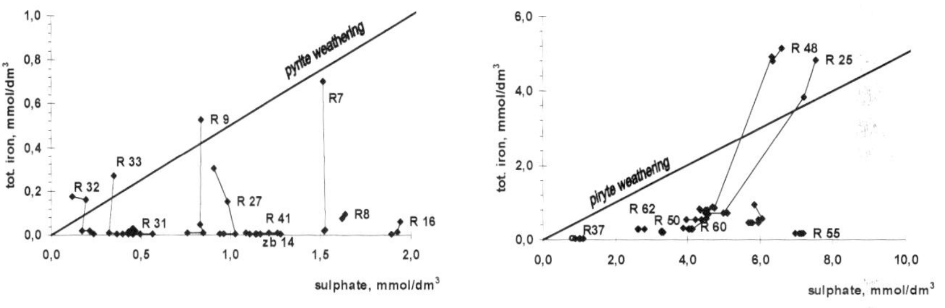

Fig. 3. Coexistence iron and sulphate in "others" (left) and aciditrophic reservoirs of the "antrophogenic lake district".

The saturation of waters by aluminium compounds depended on pH in the following way:

- pH<4.7 - undersaturation to aluminium hydroxide, and below 4 also alunite, SI value incereased with pH,
- pH>4.7 - oversaturation to gibbsite, and in the range between 4 to 7.5 also alunite,
- pH>7.5 - undersaturation to aluminium hydroxide and alunite.

Most of the reservoirs, mainly "others" were slightly undersaturated to calcium sulphate, as gipsum or anhydrite. In the "others" reservoirs (no 14 i 16 i 27) and a pH above 7.5 hydroxapatite could precipitate. The saturation state existed only in acidotrophic reservoirs. The content of aluminium was limited by gibbsite in a pH over 4.7.

Based on the correlation between pH and neutralisation potential, sensibility on accidental acidification of "others" reservoirs and reclamation processes of acid reservoirs can be distinguished, as follows (Fig. 4. and 5.):

1. among the "others" reservoirs no 33 and 7 are the most sensible on acidification,
2. among the acidotrophic reservoirs no 37 and 53 should be reclaimed in first,
3. high pH of the deeper part of reservoirs no 25 and 48 does not indicate their real neutralisation.

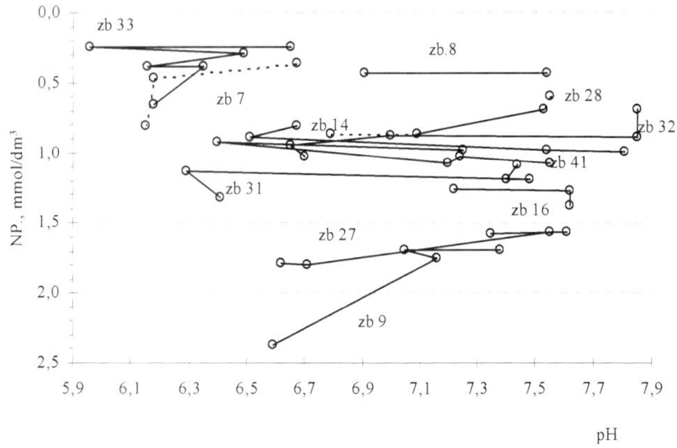

Fig. 4. Correlation between the pH value and the neutralisation potential in vertical profiles of the "other" reservoirs

The reservoirs of "the antrophogenic lake district", despite the different chemical composition and acidity, exhibit features typical of mine waters influenced by pyrite weathering. On the basis of the above analysis it is possible to conclude, that the most crucial factors creating its chemistry is the way of origin. It limits the amount of released pyrite and load of acidity. The inflow of weathered pyrite to the "others" reservoirs was limited by oxygen and ferric accessibility. So, very little amounts of base minerals occurring in the soil were sufficient for neutralisation. In pit lakes, especially in overburden and bottom sludge huge amounts of acidity were released. The following chemistry of acid reservoirs depended on the mixing type. In particular circumstances for meromixis occurred and the water body was divided into two layers of a different chemical composition.

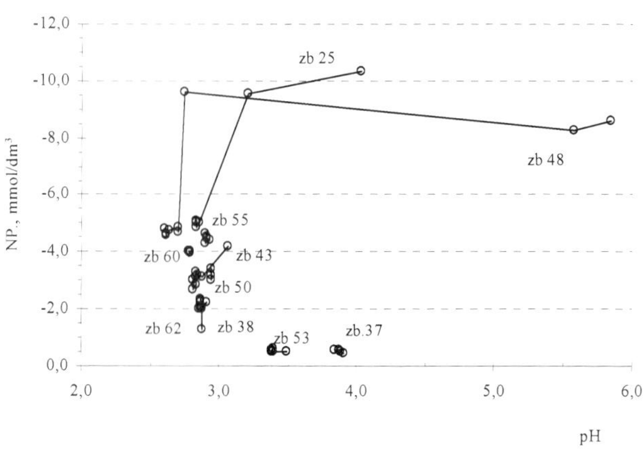

Fig. 5. Correlation between the pH value and the neutralisation potential in vertical profiles of acidotrophic reservoirs

5.4 Influence of the mixing type on the chemistry of lakes – Meromixis reservoirs

Meromictic lakes are lakes which do not mix completely. This phenomenon is not common in natural lakes, but it often occurs in mine lakes. The following mechanism creates meromixis in the development of pit lakes: a) formation of a less mineralized mixolimnion by the inflow of less mineralized water to the lake surface, b) enrichment of iron and sulphate due to the transport of secondary minerals from mixolimnion to monimolimnion, c) influence of sea water d) accumulation of substances in the monimolimnion due to biological decay, e) evaporation f) influence of groundwater with high TDS (Boehrer&Schultze 2006 and references there in). The deepest part of the meromictic lake water body - monimolimnion - has often a very specific chemical composition: strong anoxia, rich in hydrogen sulphide and products of microbial decay. Among the acidotrophic reservoirs of the "anthropogenic lake district" three of them are of the meromictic type with a pH of monimolimnion over 4.6. The pH increase had a great influence on lake chemistry and caused iron and sulphate accumulation in monimolimnion and turned back the stratification of aluminium.

The research presented below focused on the biggest lake - no 54 which comes from 1973 (there are no exact data about the rate and duration of filling up). The reservoir is located about 2 km east of Łęknica, 2.3 km west of Przewoźniki and 2 km south of Nowe Czaple (Fig. 1). The reservoir is surrounded by strongly eroded heaps of excavation. The surrounding area is covered with forest. The morphometry of the reservoir is presented in Table 4.

water level m above see level	surface area m^2	max length, m	max width, m	extension coefficient	max depth m	shore line length m	shoreline development
132.0	202,000.0	896	468	1.91	21.5	2625	1.65

Table 4. Morphometry of reservoir no 54

The first data about composition of the lake come from February 1981, eight years after coming into existence. From the study it is known that the lake is extremely acidic with a high concentration of iron and sulphur. The reservoir has no surface inflows such as streams or other lakes. It results in the fact that the chemical composition of its waters depends mainly on the composition of groundwater and precipitation, and reactions within the water body.

Chemical composition of reservoir modified significantly during the 27 years of observation (Fig. 6). The pH value changed in different ways in mixo and monimolimnion. In 1981 pH was nearly the same at all depths. A small increase of its value, by about 0.2 units, was observed between 18 and 20 m. Since 1988 a difference between the pH values of the upper and lower strata has been observed. The pH of bottom layer has increased systematically and reached values of about 4.8-4.9 in 2008. Also the thickness of miksolimnion was growing. Nowadays, the region of the greatest pH increase is located between 7 and 8 m below the water surface.

The changes in iron concentration in vertical profile had two patterns. The first one, in the years 1981 and 1988, showed gentle increase in iron concentration from the surface to the bottom. Later a sharp boundary between iron concentration in the upper and lower part of the water body appeared. Iron concentration in the mixolimnion layer was almost steady during the following research period. Iron concentration in monimolimnion increased with depth and ranged from about 100 to more than 800 mg/dm^3 in the years 1981-88 and, since 1993, to more than 1500 mg/dm^3.

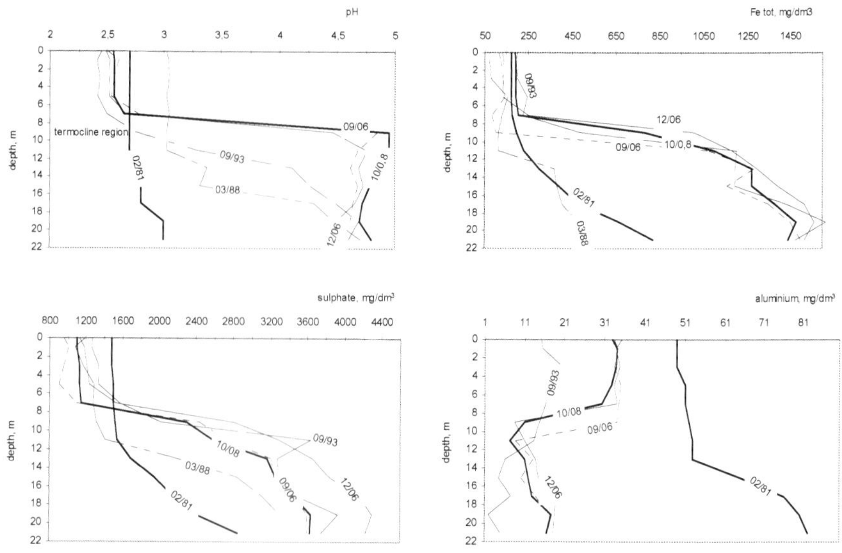

Fig. 6. Chemical composition of reservoir no 54 in the years 1988-2008

Sulphate concentration did not change as rapidly as iron concentration. In the years 1981-88 sulphate concentration was almost steady in the upper part of water body (from the surface to 11 m in depth). Deeper, the sulphur content increased with depth. The research done later, showed a low concentration of sulphate in the layer 0-7 m, a rapid concentration increase between 7 and 11 m to values of about 3000 mg/dm^3, and then a further gentle increase with depth to a maximum concentration of about 4300 mg/dm^3.

Aluminium concentration had two different profiles. The first one comes from 1981, when aluminium concentration was the highest and increased with depth. Later the concentration of aluminium decreased significantly with depth from the amount of about 35 to lees than 10 mg/dm^3.

In 1981 potassium concentration was low and varied from 3 in miksolimnion to 7 mg/dm^3 close to the bottom. The potassium amount in mixolimnion did not change significantly during the following period. The maximum concentration in miksolimnion (5.8 mg/dm^3) was stated in 2008. In monimolimnion potassium concentration increased gradually and exceeded 30 mg/dm^3 in 2008.

Solid name	formula	data	log IS on the depth, m below water surface											
			0	1	3	5	7	9	11	13	15	17	19	21
	$Fe(OH)_{3a}$	1981	-1.15	-1.15	-1.16	-1.17	-1.17	-0.76	-0.72	-0.72	-0.69	-0.67	-0.64	-0.64
		2007	-1,32	-1,34	-1,29	-1,11	-1,18	*)						
goethite	FeOOH	1981	4.50	4.50	4.50	4.40	3.97	3.39	3.39	3.67	3.17	2.85	2.59	3.16
		2007	4.30	4,18	3,99	4,11	4,01							
jarosite K	$KFe_3(SO_4)_2(OH)_6$	1981	2.81	2.66	2.61	2.61	1.7	3.39	3.67	0.75	-0.64	-1.44	-2.14	-0.97
		2007	2.83	2.61	2.28	2.57	2.44							
alunite	$KAl_3(SO_4)_2(OH)_6$	2007							3.33	3.92	2.06	2.68	5.36	6.29
gibsite	$Al(OH)_3$	2007							-0.82	-1.15	-1.59	-1.31	-0.05	0.43

Table 5. SI for selected solids in 1981 and 2007 (lack of value means very low level)

Based on the chemical properties and SI values (table 5) of the lake two periods of its development can be distinguished:

1. "the iron period" with the intensive processes of ferrous precipitation in miksolimnion (termocline was then located between 9 and 11m below the water surface).

The sedimentation of jarosite K and probably be also goethite caused transport of ferrous, potassium and sulphate to the deeper part of the water body. On the boundary of oxic and anoxic zones dissolution of minerals and the release ions into the water body occurred. We can also suppose that the main reason for the pH change in monimolimnion was ferrous reduction. These processes affected the pH values by consuming protons, but didn't reduce the acidity of water (Schoepke 2001). The potassium stratification confirms the main role of jarosite K precipitation in miksolimnion.

2. "the iron-alum period" - from 1988 up till now, with two buffer zones, ferrous precipitation in miksolimnion and alum in monimolimnion. The processes in the mixolimnion over that period were similar to the "iron period". Also the chemical composition of miksolimnion did not show any significant changes and were similar to the initial composition. The only difference is alum concentration which decreased by about 20 mg/dm^3, which means almost 50% of initial value. Due to the fact that monimolimnion acted as a sink for ferrous precipitates, the amounts of precipitate dissolution products increased. The concentration of almost all (excluding aluminium) ions analysed increased by orders of magnitude. The pH values increased from 3 to 4.5-4.9. It created the conditions for aluminium salts precipitation and lowered aluminium concentration in monimolimnion.

It seems that the future of reservoir 54 depends on the dynamics of pyrite oxidation products discharge. The total load of iron and sulphate has begun to increase for a few years. It allows the mixolimnion to reach equilibrium with iron compounds. Monimolimnion will still act as a sink for ferrous and sulphate ions. Lack of organic compounds inhibits sulphate reduction and secondary sulphides precipitation. Such a situation will stabilise meromixis for a long time.

6. Summary

The "anthropogenic lake district" located in Łuk Mużakowski is a drastic example of the destruction of the environment by lignite mining. The pollution of acid reservoirs is very stable and until now no diminishing tendency has been observed. The chemistry of the group of "others" reservoirs balances between the constant inflow of acid groundwater and neutralising processes as reduction with the use organic matter or the dissolution of base minerals. The composition of acidic reservoirs depends on their mixing pattern. Meromixis, which was created in three of the reservoirs in the area caused changes of chemical properties of monimolimnion. The monimolimnion acted as a sink for products of the dissolution of ferrous salts and affected water density and stabilized meromixis. Intensive precipitation of ferrous salts in the oxic zone and the dissolution of the precipitations resulting from ferrous reduction in monimolimnion cause an increase in pH to a value over 4. It created the next buffering aluminium zone in monimolimnion.

7. References

Appelo, C.A.J.; Postma, D. (2006). *Geochemistry, groundwater and pollution*, 2ND edition, ISBN 04 1536 421 3 Amsterdam, the Netherlands

Bigham, J.M.; Schwertmann, U.; Traina, S.J.; Winland, R.L., Wolf, M. (1996). Schwertmannite and the chemical modelling of iron in acid sulphate waters. *Geochim Cosmochim Acta* Vol 60., No.12, pp. 2111-2121

Blodau C., Peine A., Hoffman S., Peiffer S. (2000). Organic matter diagenesis in acidic mine lakes. *Acta Hydrochim Hydrobiol* Vol. 28, pp.123-35

Blodau, C. (2006). A review of acidity generation and consumption in acidic coal mine lakes and their watersheds. *Science of the Total Environment* No.369, pp. 307-332

Boehrer B, Schulze M., (2006). On the relevance of meromixis in mine pit lakes. IMWA/02000/pp 200-213

Brown, T. J.; Idoine, N. E.; Bide T.; Mills, A. J.; Hobbs S. F. (2010) .*European Mineral Statistics 2004-08* A product of the World Min. Statistics database Nottingham British Geol. Survey

Cornell, R., Schwertmann U. (2000). *The Iron Oxides: Structure, Properties, Reactions, Occurrences and Uses.* ISBN 40-471-97056-5 Wiley-VCH

Evangelou, V.P. (1995). *Pyrite oxidation and its control,* ISBN 0-8493-4732-7, CRC Press, New York

Fyson, A.; Nixdorf, B.; Kalin, M. (2006). The acidic lignite pit lakes of Germany-Microcosm experiments on acidity removal through controlled eutrophisation. *Ecological Engineering*, Vol 28, Issue 3, pp. 288-295

Jachimko, B., Kołodziejczyk, U. (2008). Effect of lignite mining on quality of water resources, *Przemysł Chemiczny* Vol. 87, nr 5, pp. 482--484

Jachimko, B. (2007).The influence of heap weathering on the iron and calcium release into water. *Zeszyty Naukowe Uniwersytetu Zielonogórskiego* - (Inżynieria Środowiska ; 13) No133, pp 181 – 187, ISBN 978-83-7481-082-1 (in polish)

Jędrczak, A. (1992). *Chemical composition of water of anthropogenic lake district in the "Mużakowski Łuk"*. Wyd. WSInż. w Zielonej Górze, Seria Monografie;, No 5, ss 135 (in polish), ISSN-0239-7290

Kasztelewicz, Z., Polak, K., Zajączkowski, M. (2008). The methods of excavation and processing of brown coal in the first half of XIX centaury, *Kwartalny Biuletyn Informacyjny Węgiel Brunatny*, No. 4/65 (in polish)

Kosschorek, M., Wendt-Potthoff, K., Geller, W.(2003). Microbial sulphate reduction at low pH in sediments of acidic lake in Argentina. *Environ Sci Technol* Vol. 37, No 6, pp.1159-62

Kwiatkowski, W. (2006). In the laboratory of coal quality *Kwartalny Biuletyn Informacyjny* „Węgiel brunatny", Vol 4 no 57, http://www.ppwb.org.pl/wb/57/11.php (in polish)

Langmuir, D. (1997). *Aqueous Environmental Geochemistry*. ISBN 0-02-367412-1 Prentice-H.Inc.

Lenk, S.; Wisotzky, F. (2007). Chemische Beschaffenheit und mmodellierte Genese von Grundwassern in Braunkohlenabraumkippen des Tagebaues Inden *Grundwasser* Vol 12,pp.301-313

Nixdorf, B.;Deneke, R. (2004). *Grundlagen und Massnahmen zur biogenen Alkalinisierung von sauren Tagebauseen*. ISBN 3-89998-038-7, Wesissensee Verlag, Berlin

Nixdorf, B.; Hemm, M.; Schlundt, A.; Kapfer, M.; Krumbeck, H.; (2001) *Tagebauseen in Deutchland - ein Ueberblick. Umwaltforschungsplan des Bundesministeriums fuer Umwelt, Naturschutz und Reactorsicherheit Forschungsbericht* 298 22 240 UBA-FB 000146

Nordstrom, D.K. (1982). The effect of sulfate on aluminum concentrations in natural waters: Some stability relations in the system Al_2O_3-SO_3-H_2O at 298 K. *Geochim. Cosmochim Acta*, Vol 46, pp.681-692

Parkhurst, D.; Appelo, C.A.J. (1999). *User's guide to PHREEQC (Version2)- a computer program for speciacion, batch-reaction, one-dimensional transport, and inverse geochemical calculations*. Denver, available from http://wwwbrr.cr.usgs.gov/projects/GWC_coupled/phreeqc/

Schoepke, R. (1999). *Erarbeitung einer Methodik zur Beschreibung hydrochemischer Prozesse In Kippengrundwasserleitern* BTU Cottbus LS Wassertechnik, Schriftenreihe Siedlungswasserwirtschaft und Umwelt, 1999, Vol.2

Schöpke, R.; Koch, R.; Querfelli, I.; Striemann, M.; Preuß, V.; Regel, R. (2001). Anwendung des Neutralisationspotentials bei der Bilanzierung von Saeure-Base-Reaktionen im Umfeld des Braunkohlebergbaues. *Grundwasser* Vol 1, pp. 23-29

Stumm, W.; Morgan, J.J. (1981). *Aquatic chemistry - an introduction emphasing chemical equilibria in natural waters*, 2nd ed.; ISBN 0-471-51185-4 New York

Tootche, O.; Pothig, R.; Uhlmann, W.; Buttchr, H.; Steinberg, Ch. (2003). Buffering mechanisms in acidic mining lakes-A model-Based Analysis. *Aquatic Geochemistry* No.9, pp 343-359

Wilk, Z. (2003). *Hydrogeology of Polish mineral deposits and water problems of mining*. ISBN: 83-89388-20-0, Wyd. AGH (in polish)

Wolkersdorfer, Ch., Bowell, B. (2005). Contemporary reviews of mine water studies in Europe. Mine Water and the environment Vol. 24, No1, pp.2-37

Wolska, E.; Schwertmann, U. (1993) The mechanism of solid solution formation between goethite and diaspore. Neues Jahrb. Miner. Monatsh, Vol 5, pp. 213-233

17

Relationship Between Water Quality and Oil-Shale Mines in Northern Estonia

Aare Selberg[1] and Malle Viik[1,2]
[1]*Institute of Chemistry, University of Tartu*
[2]*Centre for Limnology, Estonian University of Life Sciences*
Estonia

1. Introduction

Different parameters and their combinations are used to evaluate and characterize the quality of the surface water: a) chemometric techniques (Brodnjak-Voncina et al., 2002; Kowalkowski et al. 2006); b) weekly chemical analysis of a wide range of major, minor and trace elements (Neal et al., 2006); c) multiple indicators of conditions (concentrations of compounds, intensity of agriculture and industry) (Cuffney et al. 2000); d) calculated water quality index (Pesce e al., 2000; Sanchez et al., 2007; Štambuk-Giljanovic, 1999).

The concentration of various constituents including different forms of nutrients (nitrogen and phosphorus) (Mainstone et al., 2002; Withers et al., 2002) and organic carbon is used to evaluate of surface water quality. Origin of organic matter in water is related to different sources (Peuravuori and Pihlaja, 2007; Zsolnay, 2003) and natural organic matter controls the geochemical processes in aqueous systems as a donor and acceptor of protons as a buffer solution (Aiken and Cotsaris, 1995) and it is known to play important role in the fate and transport of many toxic organic or inorganic chemicals and in nutrient cycling (Dawson et al., 2009; Steinberg, 2008). It is used for the classification of surface water bodies (streams, rivers and lakes). On the other hand the fluxes of nutrients (nitrogen, phosphorus and carbon) are used to evaluate the ecological quality of lakes. The fluxes of nutrients are important for the evaluation of risks of eutrophication of rivers and lakes, but the flux depends on the concentration of compounds and flow velocity of water (Dakova et al., 2000; Rätsep & Liblik, 2001; Tamm et al., 2008; Withers et al., 2002).

The objective of the present paper is to examine the chemical characteristics of the Purtse and Pühajõgi Rivers during the closing of oil shale mines and in comparison the water quality of the Kunda River was evaluated. The Purtse River is located in the centre of mining area and the Pühajõgi River is located between the oil shale mines and Gulf of Finland. The Kunda River is located in the similar geological conditions, but it is not directly connected with the oil shale mines.

2. Mining as one of the significant influence on water quality

The mining deteriorates directly the lithosphere, but the mining has significant influence on the atmosphere (generation of dust and gaseous emissions) and hydrosphere (discharge of mine water) (Equeenuddin et al., 2010; Mighanetara et al., 2009; Tiwary, 2001). The underground mines have significant impact of hydrosphere as large amounts of groundwater is pumped out into the bodies of surface water generating changes in hydrological and hydrogeological conditions.

The environmental impact of mines can be evaluated by the analysis of water quality during mining in comparison of water quality before the mining started. It means that it is important to measure many values of different parameters before constructions of mines. It will be good to analyze quite well the surface water sampled in the different places and at different time for the better comparison (Finkelman et al., 1999; Kolker et al., 2009; Natarajan et al., 2006).

If we have already working mines then it is quite problematic to get additional information about the surface water bodies which are under the direct impact of mines. It is possible to analyze the water sampled from the water bodies which are located in the similar climatic and hydrogeological conditions, but this situation has some uncertainty due to the differences of water sources, composition of soil and bedrock and biological diversity of water bodies. On the other hand the closing of mines gives the alternative opportunity to evaluate the environmental impact of mining during the restoration of natural conditions in the mining area.

3. Specific influence of oil-shale mining on the water quality

In certain mines of base metals and iron where ores have high sulfur content, drainage from mine workings and waste heaps can become highly acidic and can contain high concentrations of dissolved heavy metals. The acid generation from the oxidation of sulfur-bearing minerals like pyrites in coal mines occurs only when the minerals react with water and oxygen in the presence of *Thiobacillus* bacteria to produce sulphuric acid and iron hydroxide or iron sulfate (Banks & Banks, 2001; Equeenuddin et al., 2010; Natarajan et al., 2006; Tiwary 2001).

Common characteristics of acid mine drainage include highly acid water with pH values as low as 2.0 and high sulfate concentration (up to 2 g L^{-1}) and the high metal loading typically including iron (up to 100 mg L^{-1}) and manganese (up to 5 mg L^{-1}) (Banks & Banks, 2001; Equeenuddin et al., 2010; Tiwary 2001). The high acidity results in further dissolution of minerals and the release of toxic metals and other constituents into waterways (Equeenuddin et al., 2010; Tiwary 2001).

The composition and properties of oil shale mine water is quite different in comparison with coal mines. The water quality in closed oil shale mines was studied by Rätsep (Rätsep & Liblik, 1998) and Reinsalu (Reinsalu et al., 2006) with coworkers. For the period of 2000-2004 Reinsalu et al. found the following mean values of characteristics for oil shale mine water: pH = 7.1, sulfate 342 mg L^{-1}, total iron 0.69 mg L^{-1}, calcium 174 mg L^{-1}, total hardness 13,7 mge L^{-1} and conductivity 1095 μS cm^{-1}. The mean values of concentrations of different forms of nitrogen are presented for the illustration: [NO_3^--N] = 11.7 mg L^{-1} (836 mmol m^{-3}), [NO_2^--N] = 0.015 mg L^{-1} (1.1 mmol m^{-3}), [NH_4^+-N] = 0.017 mg L^{-1} (1.2 mmol m^{-3}) (Reinsalu et al., 2006). In 1991-1992 the content of phosphates in mine water did not rise above 0.04 mg L^{-1} and during the following years (1993-1996) their content fell practically to zero level (below 0.001 mg L^{-1}) (Rätsep & Liblik, 1998).

Oil shale mining is accompanied by lowering of the water level and discharge of mine water into bodies of surface water (Rätsep & Liblik, 2000, 2001; Rätsep et al., 2002; Perens at al., 2006; Vaht & Rätsep 2009). For example, during the period of 1990-1998 the average amount of mine water directed into the Purtse catchment was 104 million m^3 per year (Rätsep & Liblik, 2000). An important problem is prognostication of hydrological processes after the shutting down of mines. The water in the mining area filtrates from one mine to another and feeds the waters of operating mines (Perens et al., 2006). There is also evidence that the decreasing amount of pumped-out mine water has a negative influence on the quality of the water of some rivers due to the decreasing amount of water in them (Punning et al., 1997).

4. Spatial changes in the water quality in the Kunda River, directly unconnected with oil-shale mines

The Kunda River is located in Lääne-Viru County in North-East Estonia (Figure1.). The length of the main river is 64 km, and the catchment covers an area of 530 km^2 [Järvekülg, 2001]. The Kunda River is located in similar geological conditions with oil shale mines having limestone ground as sedimentary rock, but the Kunda catchment is not directly connected with oil shale mines. Water was sampled in different points of the main river and streams in 1995, 2000 and 2005, and the sampling points were chosen to monitor the main river and the larger tributaries evenly (Fig. 1 A).

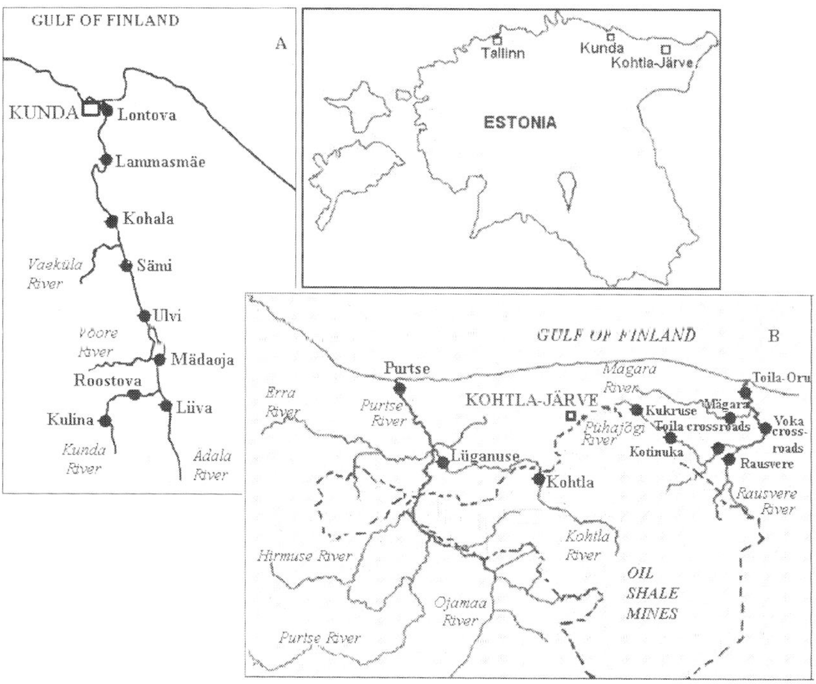

Fig. 1. Location of sampling points (•) of the Kunda (A), Purtse and Pühajõgi (B) catchments and oil shale mines in Northern Estonia.

Sampling was made in July when the water level is usually at the lowest; water quality parameters are the most stable and well comparable. The summaries of results of physico-chemical analysis are presented in Tables 1 and 2. The measurements of conductivity and alkalinity were included into the program of the complex hydrobiological field works after 2000, and because of that the values of conductivity and alkalinity of water samples are presented for 2005 only. Methodology of sampling and analysis was described in detail in our publication (Selberg et al., 2009) and it is not repeated here.

The Kunda catchment is located on the limestone base, and it means that the water samples are rather rich in carbonate ions (HCO_3^- + CO_3^{2-}) resulting in high alkalinity (Table 1). In the present study alkalinity of water samples was calculated in millimoles of bicarbonate ions per m^3 of water (HCO_3^- mmol m^{-3}) for better comparison of the amounts of different ions in the water samples.

Sampling point	pH			Conductivity in 2005, μS cm^{-1}	Alkalinity [HCO_3^-] in 2005, mmol m^{-3}	Total dissolved solids (TDS) in 2005, mg L^{-1}
	1995	2000	2005			
Kulina	7.6	7.2	7.2	506	4899	339
Roostova	8.1	7.4	7.3	523	4999	350
*Liiva	7.7	7.2	7.7	474	4399	318
Mädaoja	8.0	7.4	7.6	497	4498	333
Ulvi	7.9	7.8	8.1	505	4398	338
Sämi	7.9	7.8	7.9	532	4599	356
Kohala	8.2	7.8	7.9	534	4598	358
Lammasmäe	8.0	7.6	8.2	530	4398	355
Lontova	8.1	8.1	7.8	542	4699	363

*-Sampling points of a tributary

Table 1. Values of pH, conductivity, alkalinity and concentration of total dissolved solid in the water samples from the sampling points of the Kunda catchment in July.

The values of pH of water samples from Kunda River (Table 1) were mainly around 8 as typical for the surface water in northern and western Estonia due to the limestone. The alkalinity of water samples from the sampling points of the Kunda catchment was on an average 4700 mmol m^{-3}, independent of sampling points. Conductivity of the water samples of the Kunda River was on the average 500 microsiemens per centimeter (μS cm^{-1}), practically independent of the sampling point.

The concentrations of total nitrogen and nitrate nitrogen of water samples from different sampling points of the Kunda River had quite similar values (Table 2), and during ten years (1995-2005) the concentration decreased on the average only ca 20%. In the water samples of the Kunda catchment the concentration of nitrate nitrogen formed more than 70% of the total nitrogen concentration (Selberg et al., 2009).

The water samples of Kunda catchment had quite steady concentration of nitrite (below 1.2 mmol m^{-3}) over all sampling area and period (1995-2005) (Table 2). The calculated mean values of concentration of NO_2^--N decreased during the ten years (1995-2005) as the mean values were 0.53±0.34, 0.48±0.18 and 0.29±0.1 mmol m^{-3}, respectively.

Sampling point	N_{tot}	$[NO_3^-\text{-}N]$	$[NO_2^-\text{-}N]$	$[NH_4^+\text{-}N]$
	Concentration in 1995,mmol m^{-3}			
Kulina	103	84	0.36	0.71
Roostova	131	121	0.29	0.79
*Liiva	174	152	0.50	0.36
Mädaoja	137	116	1.0	0.43
Ulvi	158	141	1.2	0.36
Sämi	150	138	0.44	0.43
Kohala	146	142	0.36	1.1
Lammasmäe	134	122	0.29	3.1
Lontova	127	110	0.29	0.93
Change of concentration in 2000,+/- %				
Kulina	-41	-31	-61	**41**
Roostova	**1**	**8**	-28	-10
*Liiva	-26	-50	-14	**261**
Mädaoja	-19	-29	-36	**202**
Ulvi	-15	-38	-58	**233**
Sämi	-11	-38	**30**	**179**
Kohala	-5	-35	**39**	**36**
Lammasmäe	-16	-26	**97**	-55
Lontova	**20**	-30	**145**	**83**
Change of concentration in 2005,+/- %				
Kulina	-35	-35	-61	**224**
Roostova	-20	-18	-28	-6
*Liiva	-42	-46	-28	**206**
Mädaoja	-23	-18	-36	0
Ulvi	-26	-43	-70	**139**
Sämi	-19	-35	-34	-33
Kohala	-21	-39	-42	**109**
Lammasmäe	-13	-30	0	-70
Lontova	**6**	-19	-52	**8**

*-Sampling points of a tributary

Table 2. Concentrations of different forms of nitrogen in the water samples from the sampling points of the Kunda catchment in July 1995 and their changes in July 2000 and 2005.

The concentrations of ammonia in water sampled from the Kunda catchment had also quite steady (Table 2). The highest concentration of NH_4^+-N up to 3.1 mmol m^{-3} was determined in water sampled from sampling point of Lammasmäe in 1995, but it was below 1 mmol m^{-3} in the other sampling points. In 2000 the concentrations of ammonia were below 1.7 mmol m^{-3}, while in 2005 the highest concentrations of ammonia up to 2.3 mmol m^{-3} were determined in water sampled from sampling points of Kulina and Kohala. The calculated variations of concentrations of different forms of nitrogen showed that concentrations of ammonia even were grown during the ten years and it can be connected with development in agriculture and settlement.

The fluxes of inorganic nitrogen formed more than 85% from the fluxes of total nitrogen in water of the Kunda River in 1995 and fluxes of toxic NO_2^--N and NH_4^+-N formed less than 1% of inorganic nitrogen (Table 3). In 2005 the fluxes of nitrogen were lower than in 1995, but the discharge of water was also lower in the upper reach of the Kunda River. The inorganic nitrogen formed more than 85% from the fluxes of total nitrogen in the upper reach, but it was below 75% in the lower reach of the Kunda River in 2005. The fluxes of toxic NO_2^--N and NH_4^+-N formed less than 2% of inorganic nitrogen. The fluxes of different forms of nitrogen increased further downstream of the Kunda River indicating the continuous infiltration of nitrogen from the untreated or purely treated wastewater of settlements and agricultural sources into the Kunda River in 1995 and 2005.

Sampling point	N_{tot}, mg s^{-1}	[NO_3^--N], mg s^{-1}	[NO_2^--N], mg s^{-1}	[NH_4^+-N], mg s^{-1}
			1995	
Kulina	nd	nd	nd	nd
Roostova	1190	1099	3	7
*Liiva	1458	1273	4	3
Mädaoja	2293	1956	17	7
Ulvi	nd	nd	nd	nd
Sämi	4204	3874	12	12
Kohala	4112	3980	10	30
Lammasmäe	nd	nd	nd	nd
Lontova	5174	4481	12	38
			2005	
Kulina	337	278	0.7	12
Roostova	797	747	2	5
*Liiva	990	799	4	11
Mädaoja	2059	1856	13	8
Ulvi	nd	nd	nd	nd
Sämi	nd	nd	nd	nd
Kohala	nd	nd	nd	nd
Lammasmäe	11410	8288	28	91
Lontova	13230	8757	14	98

*-Sampling points of a tributary
nd - not determined

Table 3. The fluxes of dissolved nitrogen in the sampling points of the Kunda River

The concentrations of total phosphorus (P_{tot}) and phosphate phosphorus [PO_4^{3-}-P] in the water of the Kunda River were quite low in 1995 (Table 4) and the concentrations of phosphorus were increased in the most sampling points in 2000.

Sampling point	Concentration in 1995, mmol m^{-3}		Change in concentration in 2000, +/- %		Change in concentration in 2005, +/- %	
	P_{tot}	$[PO_4^{3-}\text{-P}]$	P_{tot}	$[PO_4^{3-}\text{-P}]$	P_{tot}	$[PO_4^{3-}\text{-P}]$
Kulina	0.35	0.16	**37**	**63**	-34	-19
Roostova	0.90	0.19	-28	0	-86	-68
*Liiva	1.1	0.42	**9**	**24**	-47	-45
Mädaoja	0.84	0.55	**79**	**35**	-12	-65
Ulvi	0.84	0.32	**67**	**31**	-38	-90
Sämi	0.90	0.68	**22**	-34	-32	-76
Kohala	0.97	0.74	**343**	**441**	-24	-61
Lammasmäe	1.3	0.90	-23	-32	-43	-71
Lontova	1.1	0.71	**527**	**393**	0	-37

*-Sampling points of a tributary

Table 4. Concentrations of different forms of phosphorus in the water samples from the sampling points of the Kunda catchment in July 1995 and their changes in July 2000 and 2005.

In 2005 the concentrations of phosphorus were reduced in all sampling points of the Kunda River. The variations of phosphorus concentrations can be connected with weather condition of sampling time, because rain water can lower the concentration due to dilution. The calculated fluxes of phosphorus were quite similar in the water of the Kunda River in 1995 and 2005 and fluxes of PO_4^{3-}-P formed about 50% of the flux of total phosphorus (Table 5).

Sampling point	1995		2005	
	P_{tot}, mg s^{-1}	$[PO_4^{3-}\text{-P}]$, mg s^{-1}	P_{tot}, mg s^{-1}	$[PO_4^{3-}\text{-P}]$, mg s^{-1}
Kulina	nd	nd	3	2
Roostova	18	4	2	1
*Liiva	22	8	13	5
Mädaoja	31	20	32	8
Ulvi	nd	nd	nd	nd
Sämi	56	42	nd	nd
Kohala	60	46	nd	nd
Lammasmäe	nd	nd	161	56
Lontova	107	64	238	98

*-Sampling points of a tributary

Table 5. Fluxes of dissolved phosphorus in the sampling points of Kunda River

5. Changes in the water quality in the Pühajõgi and Purtse Rivers, directly connected with oil-shale mines

5.1 The Purtse River

The Purtse River is located mainly in Ida-Viru County in North-East Estonia (Fig. 1). The total length of the Purtse River is 51 km and the catchment covers an area about 816 km^2. The whole catchment area is located in the western part of the Estonian oil shale deposit and the catchment area is rich in tectonic faults and karst (Rätsep & Liblik, 2001).

The Purtse River is located in the center of mining area and therefore the water quality and hydrological conditions of the Purtse catchment are quite well studied (Kaasik & Sõukand, 2000; Karhu & Põllumaa, 2006; Rätsep& Liblik, 1998, 2000, 2001, 2004; Rätsep et., 2002). Besides the mine water, the Kohtla River (tributary of the Purtse River, Fig. 1) is also polluted with toxic ash-dump waters of the Kohtla-Järve oil shale processing plant, which contain in addition to sulfates (above 600 mg L^{-1}) and chlorides (up to 500 mg L^{-1}) also phenols (above 50 mg L^{-1}) and sulfides (above 200 mg L^{-1}) (Rätsep & Liblik, 2001). In 1990 the amount of mine water directed into the Purtse River catchment was about 137 million m^3 per year (Rätsep & Liblik, 2000), while the amount of mine water going into the Purtse River decreased to 26 million m^3 per year in 2001 as the Kohtla mine was closed (Rätsep et al., 2002).

Water samples of the Purtse River were sampled and analyzed by different scientists and because there are some problems in comparing different results. The sampling points of the Purtse River are presented in Fig. 1A and results are presented in Tables 6 and 7.

Sampling point	N_{tot}	$[NO_3^- -N]$	$[NO_2^- -N]$	$[NH_4^+ -N]$
	Concentration in 1995, mmol m^{-3}			
Kohtla	22	12	0.14	0.71
Lüganuse	67	2.8	0.93	43
Purtse	121	107	0.71	5.7
Change of concentration in 2000, +/- %				
Kohtla	-72	-65	-67	**300**
Lüganuse	-76	-78	-69	-92
Purtse	**42**	**32**	**150**	-38
Change of concentration in 2005 (or 2011),+/- %				
Kohtla	**300**^{x}	**371**^{x}	nd	-83^x
Lüganuse	**39**^{x}	**700**^{x}	nd	**12**^{x}
Purtse	-18	-33	-70	-62

nd - not determined
^{x}-Samples from Kohtla and Lüganuse were analyzed in 2011.

Table 6. Concentrations of different forms of nitrogen in the water samples from the sampling points of the Purtse catchment in July 1995 and their changes in July 2000 and 2005(or 2011).

In 1995 the concentrations of total, nitrate and nitrite nitrogen (Table 6) in water sampled from the Purtse River were lower than concentrations of respective fractions of nitrogen of water sampled from the Kunda River (Table 2) which can be connected with huge amounts of discharged mine water. In 2000 their concentrations were lowered more than 60%, but in sampling point of Purtse the concentrations were increased more than 30%. Only the concentrations of ammonia were higher in water sampled from the Purtse River in comparison with the results of Kunda River. In 2000 the concentration of ammonia was increased in sampling point of Kohtla (Table 6) and it was lowered in the other sampling points in comparison with results in 1995. In 2005 the concentrations of different forms of nitrogen were significantly increased in water sampled from the Kohtla River. It can be connected with leachate from the ash-dumping heaps.

The concentrations of different forms of phosphorus are presented in Table 7 and in 1995 their values were quite similar to results for the Kunda River (Table 4). There is no the clear trend for the concentrations of phosphorus in the water samples of the Purtse River in 2000 and 2005. In the downstream of the Purtse River (sampling point Purtse) the concentration of phosphorus was increased, but it was lowered in 2011.

Sampling point	Concentration in 1995, mmol m^{-3}		Change of concentration in 2000, +/- %		Change of concentration in 2005 (or 2011),+/- %	
	P_{tot}	$[PO_4^{3-}-P]$	P_{tot}	$[PO_4^{3-}-P]$	P_{tot}	$[PO_4^{3-}-P]$
Kohtla	0.32	0.32	0	0	**370**^{x}	nd
Lüganuse	1.3	1.3	**84**	-50	**27**^{x}	nd
Purtse	0.77	0.45	**25**	**86**	-17	-23

nd - not determined
^{x}-Samples from Kohtla and Lüganuse were analyzed in 2011.

Table 7. Concentrations of different forms of phosphorus in the water samples from the sampling points of the Purtse catchment in July 1995 and their changes in July 2000 and 2005 (or 2011).

5.2 The Pühajõgi River

The Pühajõgi River is located in Ida-Viru County in North-East Estonia (Figure 1B). The length of the main river is 28 km, and the catchment covers an area of 196 km^2 (Järvekülg, 2001). The whole catchment is located in the western part of the Estonian oil shale deposit area and large volumes of mine water were discharged into the upper reach of the Pühajõgi River (Vaht & Rätsep, 2009). The Pühajõgi River is not so well studied (Selberg et al., 2009, Vaht & Rätsep, 2009), but it was included into the hydrological field works. Water was sampled in different points of the main river and streams in 1995, 2000 and 2005, and the sampling was made in July. Summary of results is presented in Tables 8-12 (Selberg et al., 2009, Selberg, 2010).

The analyzed water samples were mostly weakly alkaline (pH 7.3-8.0) and they had quite high alkalinity (Table 8). The results show that the alkalinity of river water samples decreased on the average to 66% of that of sampling point of Kukruse (6000 mmol m^{-3}) to Toila-Oru (4200 mmol m^{-3}) in the Pühajõgi catchment. At the same time the alkalinity of water samples from the sampling points of the Kunda catchment was on an average 4700 mmol m^{-3}), independent of sampling points (Table 1).

Sampling point	pH			Conductivity in 2005, $\mu S\ cm^{-1}$	Alkalinity [HCO_3^-] in 2005, mmol m^{-3}	Total dissolved solids (TDS) in 2005, mg L^{-1}
	1995	2000	2005			
Kukruse	7.2	7.0	7.3	984	6097	659
Kotinuka	7.2	7.2	7.3	814	5195	545
Toila crossroads	7.3	7.3	7.5	991	5392	664
*Rausvere	8.0	7.5	7.3	1860	4786	1246
Voka crossroads	7.6	7.6	7.6	1720	4884	1152
*Mägara	7.6	7.7	8.0	564	4196	378
Toila-Oru	8.0	7.8	7.9	1528	4294	1024

*-Sampling points of a tributary

Table 8. pH, conductivity, alkalinity and concentration of total dissolved solid in the water samples from the sampling points of the Pühajõgi catchment in July.

Although the concentration of carbonate and bicarbonate anions was quite similar in the water samples of both catchments, the measurement of conductivity showed a significant difference. Conductivity of water samples of the Pühajõgi River was between 800 and 1800 $\mu S\ cm^{-1}$ which was 1.5-3 times higher than the conductivity of water sample from Mägara sampling point (tributary of the Pühajõgi River, Fig.1). Conductivity of the water samples of the Kunda River was on the average 500 $\mu S\ cm^{-1}$, practically independent of the sampling point (Table 1).

The determined concentrations of nitrate nitrogen in water samples of the Pühajõgi catchment (Table 9) had quite different values (variation up to 10 times), and the highest concentration of nitrate nitrogen (252 mmol m^{-3}) was determined in the sampling point of Toila-Oru in 1995. The concentrations of nitrate nitrogen were very low (up to 1% from the concentration of total nitrogen) in the sampling points of Kukruse and Toila crossroads which means that nitrogen occurred in other forms (ammonia and organic compounds). The concentrations of total nitrogen and nitrate nitrogen were lowered in 2000 and 2005 mainly in the downstream of the Pühajõgi River.

The analysis of water samples showed that variations of concentrations of ammoniacal nitrogen were significant (between 1.1 and 520 mmol m^{-3}) and the higher concentrations of NH_4^+-N up to 340 mmol m^{-3} (up to 99% from N_{tot}) and 520 mmol m^{-3} (up to 81% from N_{tot}) were determined in the water sampled from Kukruse and Toila crossroads in 1995, respectively. It confirmed that nitrogen occurred in form of ammonia. The concentration of ammonia decreased significantly during the next ten years indicating the reduction of pollution of the Pühajõgi River.

Concentrations of nitrite were quite low (up to 7.8 mmol m^{-3}) in the water samples from the catchment of the Pühajõgi, whereas variations of concentrations of ammoniacal nitrogen were significant and the highest concentration of NH_4^+-N up to 520 mmol m^{-3} was determined in 1995. The concentration of nitrite nitrogen was increased almost in all the sampling points of the Pühajõgi River in 2000, but it decreased during the next five years (Table 9).

Sampling point	N_{tot}	$[NO_3^-\text{-N}]$	$[NO_2^-\text{-N}]$	$[NH_4^+\text{-N}]$
	Concentration in 1995,mmol m^{-3}			
Kukruse	345	3.5	0.43	340
Kotinuka	100	54	0.79	2.6
Toila crossroads	645	4.9	0.14	520
*Rausvere	186	186	2.1	14
Voka crossroads	327	180	2.5	98
*Mägara	31	2.1	0.14	1.1
Toila-Oru	327	252	1.2	18
Change of concentration in 2000,+/- %				
Kukruse	-60	**1186**	**1714**	-94
Kotinuka	-24	-67	**141**	-58
Toila crossroads	-82	**165**	**900**	-97
*Rausvere	**3**	-67	**224**	**321**
Voka crossroads	-50	-71	**172**	-79
*Mägara	0	-83	-57	-67
Toila-Oru	-58	-70	**42**	-77
Change of concentration in 2005,+/- %				
Kukruse	-50	**177**	**714**	-72
Kotinuka	-36	-98	-63	-4
Toila crossroads	-90	-43	**157**	-99
*Rausvere	-59	-80	-62	-63
Voka crossroads	-69	-87	-68	-90
*Mägara	**84**	**614**	0	**64**
Toila-Oru	-79	-92	**50**	-48

*-Sampling points of a tributary

Table 9. Concentrations of different forms of nitrogen in the water samples from the sampling points of the Pühajõgi catchment in July 1995 and their changes in July 2000 and 2005.

Calculated fluxes of nitrogen showed that 80% of nitrogen had infiltrated into the Pühajõgi River as NH_4^+-N in 1995. The fluxes of NO_2^--N and NH_4^+-N were reduced in further downstream of the Pühajõgi due to oxidation as the fluxes of nitrate increased (Table 10). In 2005, the fluxes of nitrogen had reduced 3-5 times in comparison with 1995 and the fluxes of organic nitrogen constituted more than 50% of the fluxes of total nitrogen in the Pühajõgi River.

Sampling point	N_{tot}, mg s^{-1}	[NO_3^--N], mg s^{-1}	[NO_2^--N], mg s^{-1}	[NH_4^+-N], mg s^{-1}
			1995	
Kukruse	39	0.4	0.05	38
Kotinuka	56	30	0.4	1
Toila crossroads	903	7	0.2	728
*Rausvere	1304	969	15	95
Voka crossroads	3993	2186	31	1188
*Mägara	13	0.9	0.06	0.5
Toila-Oru	4131	3177	15	225
		2005		
Kukruse	48	3	1	26
Kotinuka	18	0.3	0.08	0.7
Toila crossroads	57	2	0.3	4
*Rausvere	431	152	4	29
Voka crossroads	630	153	5	59
*Mägara	48	13	0.1	2
Toila-Oru	1460	390	39	195

*-Sampling points of a tributary

Table 10. The fluxes of dissolved nitrogen in the sampling points of Pühajõgi River

The results of water analysis showed significantly different concentrations of total phosphorus and phosphate phosphorus in Pühajõgi and Kunda rivers (Tables 4 and 11). The concentrations of total phosphorus were even close to 100% lower during the ten years (1995-2005) which was a good sign for the improvement of water quality in the Pühajõgi River.

The comparison of P_{tot} and [PO_4^{3-}-P] concentrations showed that the phosphorus occurred mainly as phosphate anion (up to 90%) in the samples of the Pühajõgi River. Water samples from sampling point Mägara (the tributary of the Pühajõgi River) and the Kunda catchment contained a lot of organic phosphorus compounds, and the content of phosphate phosphorus was less than 50% of that of total phosphorus. During the 10-year monitoring period (1995–2005), the concentration of total phosphorus was decreased by 93 to 98% in the water samples of the Pühajõgi River and this is related to the closing of oil shale mines.

In 2005 the situation of the fluxes of phosphorus was much better and PO_4^{3-}-P formed about 50% from the fluxes of total phosphorus as it was in sampling point of Mägara (the tributary is not directly connected with oil shale mines) in 1995. The fluxes of phosphorus increased further downstream of Pühajõgi River indicating the continuous infiltration of phosphorus into the Pühajõgi River in 1995 and 2005.

The fluxes of PO_4^{3-}-P formed about 50% from the flux of total phosphorus in water of the Kunda River (Table 12) and similar situation was observed for Pühajõgi River in 2005. It means that the Pühajõgi River can achieve the natural conditions after the closing of oil shale mines.

Sampling point	Concentration in 1995, mmol m^{-3}		Change in concentration in 2000, %		Change in concentration in 2005, %	
	P_{tot}	$[PO_4^{3-}$-P]	P_{tot}	$[PO_4^{3-}$-P]	P_{tot}	$[PO_4^{3-}$-P]
Kukruse	50	45	-91	-93	-82	-82
Kotinuka	3.6	0.97	-56	-30	-61	-10
Toila crossroads	80	74	-95	-97	-98	-99
*Rausvere	15	14	-36	-37	-93	-96
Voka crossroads	27	24	-70	-75	-96	-98
*Mägara	1.8	0.97	-66	-99	-44	-67
Toila-Oru	30	25	-74	-76	-92	-93

*-Sampling points of a tributary

Table 11. Concentrations of different forms of phosphorus in the water samples from the sampling points of the Pühajõgi catchment in July 1995 and their changes in July 2000 and 2005.

Sampling point	1995		2005	
	P_{tot}, mg s^{-1}	$[PO_4^{3-}$-P], mg s^{-1}	P_{tot}, mg s^{-1}	$[PO_4^{3-}$-P], mg s^{-1}
Kukruse	12	11	6	5
Kotinuka	5	2	0.9	0.5
Toila crossroads	247	228	4	2
*Rausvere	240	210	14	8
Voka crossroads	720	635	14	5
*Mägara	2	0.9	2	0.6
Toila-Oru	846	693	108	81

*-Sampling points of a tributary

Table 12. Fluxes of dissolved phosphorus in the sampling points of Pühajõgi River

6. Conclusion: Environmental impact of closing of oil-shale mines on the surface water quality

Four oil shale mines were closed in 1999-2002 and the discharge of mine water into the rivers was reduced. The results of water analysis give information about the restoration of natural conditions in the rivers. The chemical status of the Purtse River is moderate on the basis of data obtained from the supplementary monitoring of 2003 and the national monitoring. For the drainage of mine waters in the Purtse catchment area, a new hydrological scheme is needed to restore gradually the natural hydrochemical conditions of the Purtse River water. However, the inlet of the ash-dump waters into the Kohtla River must be fully stopped. In connection with the closing of mines the impact of mine water on the hydrologic state of the rivers in the Purtse catchment area decreased significantly, whereas its changed balance has set more favorable conditions for development of phytoplankton.

The determined concentrations of nitrate nitrogen in water samples of the Pühajõgi catchment had quite different values (variation up to 10 times), and the highest concentration of nitrate nitrogen (252 mmol m^{-3}) was determined in 1995. The concentrations of total nitrogen and nitrate nitrogen were lowered in 2000 and 2005 mainly in the downstream of the Pühajõgi River. The analysis of water samples showed that variations of concentrations of ammoniacal nitrogen were significant (between 1.1 and 520 mmol m^{-3}) and the higher concentrations of NH_4^+-N were determined in 1995. The concentration of ammonia decreased significantly during the next ten years (1995-2005) indicating the reduction of pollution of the Pühajõgi River. During the 10-year monitoring period (1995–2005), the concentration of total phosphorus was decreased by 93 to 98% in the water samples of the Pühajõgi River and this is related to the closing of oil shale mines.

The analysis of water quality showed that the Pühajõgi River could achieve natural conditions after closing oil shale mines, but it needs longer time as additional pollutants can be dissolved from the sediments of rivers.

7. Acknowledgment

Supports for this study were provided by the target-financed projects of the Estonian Ministry of Education and Research Nos. SF0170011s08 and SF0180135s08.

8. References

Aiken, G.., Cotsaris, E. (1995) Soil and hydrology: their effect on NOM. *J. AWWA*, 87, 36–45.

Banks, S.B., Banks, D. (2001) Abandoned mines drainage: impact assessment and mitigation of discharges from coal mines in the UK. *Engineering Geology*, 60, 31-37.

Brodnjak-Voncina, D., Dobcnik, D., Novic, M., Zpan, J. (2002) Chemometric characterization of the quality of river water. *Analytica Chimica Acta*, 462, 87-100.

Cuffney, T. F., Meador, M. R., Porter, S. D., Gurtz, M. E. (2000) Responses of physical, chemical, and biological indicators of water quality to a gradient of agricultural land use in the Yakima River Basin, Washington,. *Environmental Monitoring and Assessment*, 64, 259–270.

Dakova, Sn., Uzunov, Y., Mandadjiev, D. (2000) Low flow - the river's ecosystem limiting factor. *Ecological Engineering*, 16, 167-174.

Dawson, J.J.C., Malcolm, I.A., Middlemas, S.J., Tetzlaff, D., Soulsby, C. (2009) Is the composition of dissolved organic carbon changing in upland acidic streams? *Environmental Science and Technology*, 43, 7748–7753.

Equeenuddin, Sk.Md., Tripathy, S., Sahoo, P.K., Panigrahi, M.K. (2010) Hydrogeochemical characteristics of acid mine drainage and water pollution at Makum Coalfield, India. *Journal of Geochemical Exploration*, 105, 75-82.

Finkelman, R.B., Gross, P.M.K. (1999) The types of data needed for assessing the environmental and human health impacts of coal. *International Journal of Coal Geology*, 40, 91-101.

Järvekülg, A. (Ed.) (2001) *Estonian Rivers*. Estonian Agricultural University, Institute of Zoology and Botany. Tartu [in Estonian].

Kaasik, M., Sõukand, Ü. (2000) Balance of alkaline and acidic pollution loads in the area affected by oil shale combustion. *Oil Shale*, 17(2), 113-128.

Karhu, A., Põllumaa, L. (2006) Environmental hazard of the waste streams of Estonian oil shale industry: an ecotoxicological review. *Oil Shale*, 23(1), 53-93.

Kolker, A., Panov, B.S., Panov, Y.B., Landa, E.R., Conko, K.M., Korchemagin, V.A., Shendrik, T., McCord, J.D. (2009) Mercury and trace element contents of Donbas coals and associated mine water in the vicinity of Donetsk, Ukraine. *International Journal of Coal Geology*, 79, 83-91.

Kowalkowski, T., Zbutniewski, R., Szpejna, J., Buszewski, B. (2006) Application of chemometrics in water river classification. *Water Research*, 40, 744–752.

Mainstone, C.P., Parr, W. (2002) Phosphorus in rivers - ecology and management. *The Science of the Total Environment*, 282-283, 25-47.

Mighanetara, K., Braungardt, C.B., Rieuwerts, J.S., Azizi, F. (2009) Contaminant fluxes from point and diffuse sources from abandoned mines in the River Tamar catchment, UK. *Journal of Geochemical Exploration*, 100, 116-124.

Natarajan, K.A., Subramanian, S., Braun, J.-J. (2006) Environmental impact of metal mining - biotechnological aspects of water pollution and remediation - an Indian experience. *Journal of Geochemical Exploration*, 88, 45-48.

Neal, C., House, W. A., Jarvie, H. P., Neal, M., Hill, L., Wickham, H. (2006) The water quality of the River Dun and the Kennet and Avon Canal. *Journal of Hydrology*, 330, 155–170.

Perens, R., Punning, J.-M., Reinsalu, E. (2006) Water problems connected with oil shale mining in North-East Estonia. *Oil Shale*, 23, 228-235.

Pesce, S.F., Wunderlin, D.A. (2000) Use of water quality indices to verify the impact of Cordoba City (Argentina) on Suquia River. *Water Research*, 34, 291-2926.

Peuravuori, J., Pihlaja, K. (2007). Characterization of freshwater humic matter. In: Nollet, L.M.L. (Ed.), *Handbook of Water Analysis*. CRC Press, pp. 435–448.

Punning, J.-M., Boyle, J.F., Alliksaar, T., Tann, R., Varvas, M. (1997) Human impact on the history of Lake Nõmmejärv, NE Estonia: geochemical and palaeobotanical study. *The Holocene*, 7, 91-99 [in Estonian, summary in English].

Reinsalu, E., Valgma, I., Lind, H., Sokman, K. (2006) Technogenic water in closed oil shale mines. *Oil Shale*, 23, 15-28

Rätsep, A., Liblik, V. (1998) Changes in physico-chemical composition of mine water in closed Kiviõli oil shale mine. *Oil Shale*, 15(4), 341-352.

Rätsep, A., Liblik, V. (2000) Technogenic waterflows generated by oil shale mining: impact on Purtse catchment rivers. *Oil Shale*, 17(2), 95-112.

Rätsep, A., Liblik, V. (2001) The influence of polluted water flows on hydrological and hydrochemical conditions of Purtse catchment rivers, NE Estonia. *Nordic Hydrology,* 32(3), 215-226.

Rätsep, A., Rull, E., Liblik, V. (2002) Impact of oil shale mine water discharges on phytoplankton community of Purtse catchment rivers. *Oil Shale*, 19(3), 297-306.

Rätsep, A., Liblik, V. (2004) Impact of oil shale mining and mine closures on hydrological conditions of north-east Estonian Rivers. *Oil Shale*, 21(2), 137-148.

Sanchez, E., Colmenarejo, M.F., Vicente, J., Rubio, A., Garcia, M.G., Travieso, L., Borja, R. (2007) Use of the water quality index and dissolved oxygen deficit as simple indicators of watersheds pollution. *Ecological Indicators*, 7, 31-328.

Selberg, A., Viik, M., Pall, P., Tenno, T. (2009) Environmental impact of closing of oil shale mines on river water quality in north-eastern Estonia. *Oil shale*, 26(2), 169-183.

Selberg, A. (2010) *Evaluation of environmental quality in Northern Estonia by analysis of leachate.* PhD Thesis, University of Tartu.

Steinberg, C.E.W. (2008) *Ecology of Humic Substances in Freshwater*. Springer.

Štambuk-Giljanovic, N. (1999) Water quality evaluation by index in Dalmatia. *Water Research*, 33, 3423-3440.

Tamm, T., Nõges, T., Järvet, A., Bouraoui, F. (2008) Contributions of DOC from surface and groundflow into Lake Võrtsjärv (Estonia). *Hydrobiologia*, 599, 213-220.

Tiwary, R. K. (2001) Environmental impact of coal mining on water regime and its management. *Water, Air, Soil Pollution*, 132(1-2), 85-199.

Vaht, R., Rätsep, A. (2009) Impact of oil shale mine water on hydrology and runoff of a small river. The Pühajõgi River case study. *Oil Shale*, 26(1), 84-93.

Withers, P.J.A., Lord, E.I. (2002) Agricultural nutrient inputs to rivers and groundwaters in the UK: policy, environmental management and research needs. *The Science of the Total Environment*, 282-283, 9-24.

Zsolnay, A. (2003) Dissolved organic matter: artifacts, definitions, and functions. *Geoderma*, 113, 187-209.

18

Study of the Factors Influencing the Shallow Groundwater Quality in Two Settlements with Different Characteristics

György Szabó, Tímea Vince and Éva Bessenyei
University of Debrecen
Hungary

1. Introduction

Recently, clean water became one of the most important natural resources. It is also proven by the fact that the fight for water often causes conflicts, even wars as well (Sultana, 2011). Although the water resources of the Earth are large, most of the water can be found in oceans (96.6%), and this water is hardly available for the humanity since the desalination is a very expensive process and the transportation of the cleaned water to the inner continent is also difficult. Only 3.4% of the hydrosphere is fresh water; most of it is also hardly available such as the water stored in the polar ice caps or the water resources found in the lithosphere deeper than 5 km (Rakonczai, 2008). Easily available water resources are the surface waters of the continents (rivers and lakes), and the groundwater such as shallow groundwater, karstic water and deep groundwater.

Unfortunately, in the last decades the condition of the fresh water resources has worsened. Especially the contamination of the surface water resources is significant in many places, but the groundwater resources are also often in danger (Howden et al., 2009; Oprean et al., 2009; Pál et al., 2009; Szalai, 2004). The mostly endangered groundwater resource is the shallow groundwater since the contaminants can easily get into it due to its little depth. Karstic water resources are also relatively sensitive since contaminants can get into the water through the cracks of the rock. Deep groundwater is less sensitive to contaminants due to the greater depth, especially when there is a confining layer (or layers) above it. Unfortunately, this water reserve of strategic significance has already become contaminated in several countries of the Earth.

Shallow groundwater is one of the most important water resources in several countries, although the use of it is decreased in the developed countries in the last decades. For example in Hungary, in the 1950s shallow groundwater was the most significant water resource, but recently only 6% of the drinking water derives from shallow groundwater (Steiner, 2010); the decrease is owing to the serious contamination. In the settlements several pollution resources can endanger the condition of shallow groundwater, but contaminants can also derive from agriculture in the outer areas.

In the settlements, one of the most important contaminants is domestic sewage that can cause significant pollution in settlements where the sewer network is not constructed.

Contamination occurs since the septic tanks, where domestic sewage is collected, are not supplied with adequate insulation therefore sewage can infiltrate into the soil (Szabó et al., 2010). Besides, significant pollution sources are the out-of-date landfills and sewage disposals that are not supplied with adequate insulation, but the industrial activities can also cause serious contaminations. The residues of fertilizers and insecticides used in the agriculture can also appear in the shallow groundwater. Moreover, the manure deriving from livestock farming, if the storage is not proper, can also endanger the shallow groundwater.

The most common contaminants of the shallow groundwater are mainly the different compounds of nitrogen (ammonium, nitrite and nitrate). Additionally, phosphates are also important contaminants; they can derive from sewage, soil or the decomposition of the organic materials. Phosphates usually appear as inorganic orthophosphate ions in shallow groundwater. Bacterial pollution also causes problems; the sources of infection can be the domestic sewage and the livestock farms. Near industrial areas, shallow groundwater can often be contaminated with heavy metals and different hydrocarbon derivatives (Steiner, 2010).

Several papers have been published regarding the pollution of shallow groundwater, the factors influencing the contamination and the risks resulted from contamination. Several authors emphasized the determinant role of the agricultural contaminants. Jang and Liu (2005) examined the nitrate and ammonium contamination of the water resources situated in different depth in the alluvium of the Choushui River, Taiwan. They proved that the main source of pollution was agriculture. Using isotope analyses, Widory et al. (2004) investigated the nitrate pollution sources of the shallow groundwater in the sample sites in the watershed of the River Arguenon, France. They determined that mainly organic manure and sewage infiltrating into the soil cause nitrate pollution. The authors emphasize that in spite of the more and more increasing national and European efforts (EC Directive 91/676/EEC) nitrate deriving from the intensive agriculture is still the most significant contaminant of the shallow groundwater. Cey et al. (1999) also examined the nitrate concentration of the shallow groundwater in agricultural areas. They proved that nitrate mainly derives from organic manure and chemical fertilizers. The authors also showed that when the shallow groundwater leaves the agricultural areas, the nitrate concentration quickly decreases mainly due to denitrification. Goss et al. (1998) studied the pollution of the shallow groundwater in some farms of Ontario, Canada. The authors also examined the factors influencing the degree of contamination and proved, based on the examinations of 1292 wells, that the degree of nitrate pollution principally depends on the depth of the shallow groundwater. The concentration was significantly lower in the deeper wells. Warner et al. (2008) drew similar conclusion, they carried out water quality examinations in Kathmandu valley, Nepal. They proved that regarding nitrate and bacterial contamination shallow groundwater, that is closer to the surface, was the most polluted. This means serious health risk since in the examined area 50% of the inhabitants used shallow groundwater as drinking water. Chettri and Smith (1995) also studied the pollution of shallow groundwater in Nepal. They experienced that nitrate contamination derived mainly from septic tanks and waste disposal, but agricultural pollution was not significant.

Muñoz-Carpena et al. (2005) examined the nitrate, orthophosphate and total phosphate concentration of the shallow groundwater in an agricultural watershed near the Everglades

National Park. They proved that the temporal changes of the concentration of the examined water quality parameters are significantly influenced by the rainfall and the depth of the shallow groundwater. It was also determined that the concentration of the contaminants increases after intensive rainfalls due to the increased leaching.

Other authors drew similar conclusions regarding the nitrate contamination of the shallow groundwater wells examined in the rural area of SW Nigeria (Adekunle et al., 2007). They observed that the water of the wells was more polluted in the wet season than in the dry season. In the area the most important pollution sources are the waste landfills and the sewage disposal. Moreover, the authors also proved that the contamination occurs mainly within the 200 m radius of the pollution sources. In this zone the nitrate concentration of the water exceeded the limit value determined for drinking water by the WHO. This means serious health risk since the population consume mainly the water of the wells. In Wisconsin, USA, Knobeloch et al. (2000) examined the effect of the nitrate content of drinking water on infants. Two cases were demonstrated when the infants have the so-called "blue baby" syndrome. In both cases the babies consumed the water of the private well created by the families, and this resulted in their sickness. In the surroundings of the polluted wells there were agricultural areas, and the high nitrate concentration of the water presumably derived from nitrogen-rich fertilizers. It is also mentioned in the study that in the USA, between 1979 and 1996, 6 children died in methemoglobinemia caused by the nitrate contamination of the shallow groundwater. Since 1945, 3000 deaths are known in the world, almost half of them occurred in Hungary (Barzilay et al., 1999). However, the high nitrate concentration of the shallow groundwater is dangerous not only for infants but for the adults as well; higher concentrations also have harmful effects on adults. Similar to children, ruminant domestic animals such as goat, sheep and cow are also sensitive to methemoglobinemia since their digestive system quickly transforms nitrate to nitrite (Barzilay et al., 1999).

In this chapter two settlements with different characteristics were examined. We supposed that regarding the quality of shallow groundwater there is significant difference between the two settlements due to the different natural and social-economic characteristics. Thus, our objective was to survey the contamination of the shallow groundwater, to examine the factors influencing the quality of shallow groundwater, and to determine the most determinant factor in terms of water quality. At last, we also intend to survey the health risks related to the use of shallow groundwater.

2. Sample sites

Our examinations were carried out in a small town of Eastern Hungary, Mikepércs, and in a Western Ukrainian town, Beregszász (Fig. 1).

Mikepércs is situated in the eastern part of Hungary, about 5 km south from Debrecen. Mikepércs has diverse pedological conditions as the settlement is situated in the boundary of the Hajdúság and the Nyírség. The loess area of the Hajdúság extends to the western part of the settlement where chernozem soils were formed. Soils of sandy texture can be found in the greater part of the settlement that belongs to the area of the Nyírség. These soils are much more sensitive than the chernozems: in addition to the coarser grain size distribution the less organic matter content and the weaker buffering capacity of the sandy soils also contribute to the considerable sensitivity to contaminants. The depth of the water table in Mikepércs is 1-3 meter but in the higher reliefs more than 5 meter deep water table can occur. The shallow groundwater flows from the north-east to the south-west, towards the Kondoros stream running near the west part of the village.

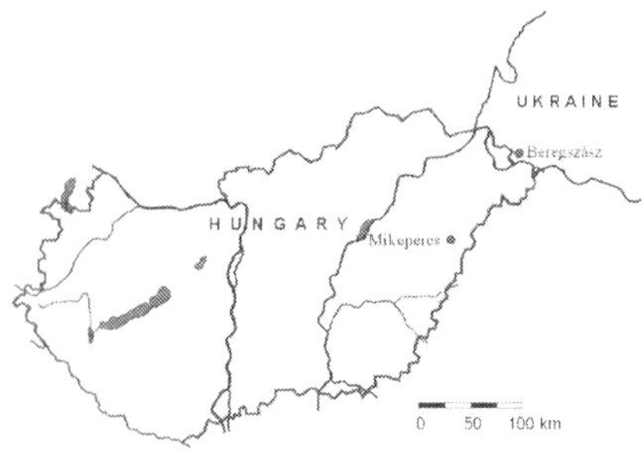

Fig. 1. The localization of the examined settlements

In the settlement of 4000 inhabitants the water supply system is almost 100% constructed but the construction of the sewer system has begun only in the second half of 2006, and the installation was executed only after the examinations, thus the positive effects of the sewerage cannot be demonstrated in this examination series. The inhabitants use the water of the dug wells to water gardens and domestic animals, even to wash cars, but sometimes, mainly elder people drink water from the wells.

Beregszász can be found in the western part of Ukraine, Transcarpathia, almost 50% of its inhabitants are Hungarian. The town is situated in the landscape border of the inner volcanic belt of the Eastern Carpathians, and the Great Hungarian Plain. The Bereg Plain is very fertile since the River Tisza and its tributaries regularly flooded the area over thousands of years, and humus-rich soils of good quality could be formed on the sediment.

The rivers that can be found in the surroundings of Beregszász are lower-course rivers. The River Tisza flows approximately 9 km from the town along the Ukrainian-Hungarian border. The Vérke canal flows through Beregszász and connects the Borzsa and the Latorca rivers. Similarly to most of the plain areas in Transcarpathia, the shallow groundwater is in the depth of 3-7 m in Beregszász; above the water there are fine-grained sediments, often clayey confining layers that make the infiltration of the contaminants into the shallow groundwater from the surface more difficult (Molnár, 2009).

Beregszász is the centre of Beregszász district and the population of the town was 25200 in 2008 (Molnár, 2009). The sewer network is partially constructed in the town, 62.7% of the residential area is supplied with sewer system. Unfortunately, the treatment of the collected sewage is not adequate; it is observable in the Vérke Canal since it is the receiver. The water supply system is also not entirely constructed in the settlement, therefore the inhabitants do not use shallow groundwater only to water gardens and animals but they cook and wash with it and also drink it. The inhabitants regularly drink from the water of 13 wells of the examined 16 ones. According to our observations, the inhabitants occasionally used shallow groundwater even if they can use tap-water. The main explanation is that they have to pay for tap-water, and on the other hand the taste of the water from the wells is considered to be better than tap-water.

3. Methods

In our research the water quality of dug shallow groundwater wells was examined. In Mikepércs 14 and in Beregszász 16 wells were chosen; we tried to cover the whole area of the settlements (Fig. 2 and 3).

Water samples were examined monthly: we have a two-year data set regarding Mikepércs (from June 2005 to June 2006 and from July 2007 to July 2008), and in the case of Beregszász the examinations lasted a year (from April 2009 to March 2010). The samples were transported in hermetically closed plastic flacons to the geography laboratory of the University of Debrecen.

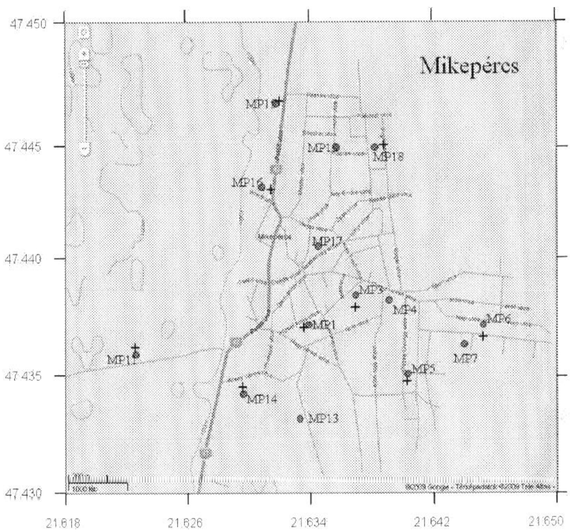

Fig. 2. The location of the groundwater sample wells in Mikepércs (The soil samples are marked by the crosses)

Electric conductivity and temperature were measured in situ with a Schott electric conductometer. The depth of the water table was also determined during the sampling. The determination of nitrite, nitrate, orthophosphate, ammonium and organic matter, and the measurement of the pH were carried out in the laboratory a day after the sampling (Literáthy, 1973).

Among the natural factors influencing the shallow groundwater quality, soil characteristics were examined. Soil samples were collected from the surroundings of 13 wells, from approximately a meter below the water table. Samples were taken from every 20 cm layers and were examined in the geography laboratory of the University of Debrecen, according to the Hungarian standards. We determined the grain size distribution with Köhn-pipette

(MSZ, 1978a), the humus content after Tyurin's scheme (MSZ 1977), and the pH with an electric pH-meter (MSZ 1978b). The determination of the $CaCO_3$-content was carried out with Scheibler-calcimeter (Ballenegger & Di Gléria, 1962). Additionally, the rainfall and the depth of shallow groundwater were also considered.

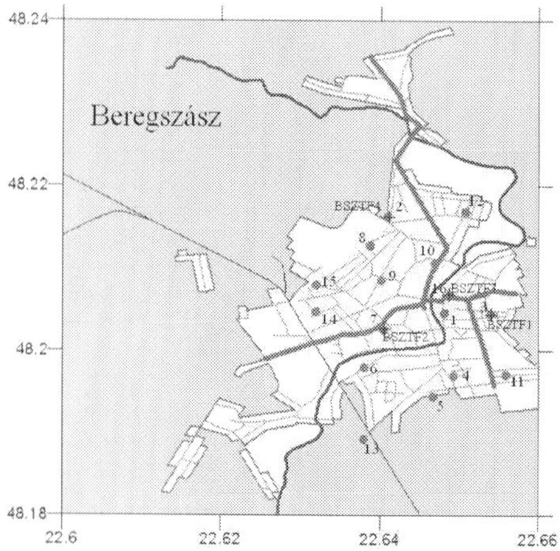

Fig. 3. The location of the groundwater sample wells in Beregszász (The soil samples are marked by the crosses)

Among the social-economic factors we examined the sewer network, the condition of the septic tanks, the construction of the water supply system, the use of water and the sources of pollution near the wells.

The results were saved in an Excel database, and the diagrams were also made with this software. In the course of the statistical analyses normality test were executed with Kolmogorov-Smirnov test. Since most of the data were not normal distribution data, Spearman correlation coefficient was applied during the correlation analysis. SPSS 8.0 software was used in the statistical analyses and to make further diagrams. A questionnaire survey was carried out in order to map the sources of pollution, and regarding the use of the water deriving from the wells.

4. Results and discussion

4.1 The evaluation of the factors influencing the shallow groundwater condition

First, the natural factors are studied, and then the role of the anthropogenic factors is surveyed as influencing factors.

Among the natural factors, the depth of the shallow groundwater is important since if the shallow groundwater can be found deep, then the infiltration of the contaminants is more difficult. There is significant difference between the two examined settlements because in Mikepércs the water table was between 1.5 m and 2.5 m during the examinations (Fig. 4).

However, in Beregszász the depth of water was between 3 m and 5 m, but in some wells there were deeper water tables than 7 m in the dry periods. Based on this we can say that regarding depth Beregszász is in better situation.

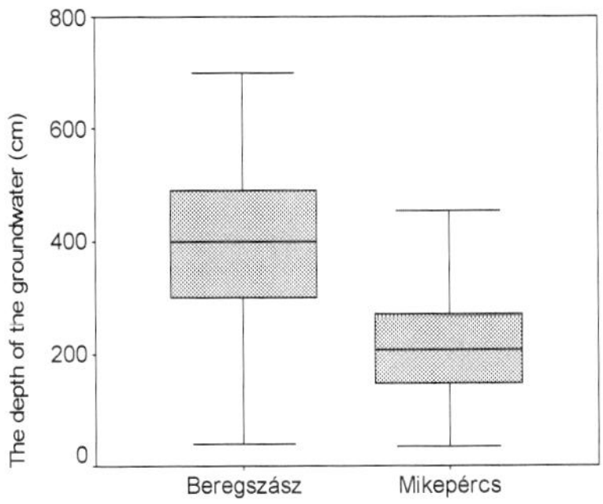

Fig. 4. The depth of shallow groundwater in the examined settlements, based on the data of all the examined wells. (Middle line: median, box: interquartile range, between upper and lower edges: interquartile range by 1.5.)

The role of depth is also proven by the fact that there are significant negative correlations between the concentration of all the examined contaminants and the depth of the shallow groundwater. This means that the deeper shallow groundwater cannot be polluted easily.

Grain size distribution is also an important factor in the contamination of the shallow groundwater. If the soil consists of finer grains, the infiltrating water and dissolved materials, and colloids cannot easily get into the deeper layers of the soil (Nyizsalovszki & Szabó, 2003). If a soil layer is enriched with smaller grains than 0.002 mm and becomes clayey, it can be a confined layer that can prevent the transportation of the contaminants into the deeper layers. In order to determine the grain size distribution we collected soil samples from 9 boreholes in Mikepércs and from 4 boreholes in Beregszász above the water table. Based on the results we can say that in Mikepércs, in the cases of the layers above the water-table fine sand (0.02 0.2 mm) dominates since the percentage of fine sand was more than 80% in every sample. Therefore, the value of the infiltration coefficient is high. The results of the sampled layers near the wells MP6 and MP11 can be seen in Figure 5. The other samples in the settlement showed similar grain size distribution. Campbell's infiltration coefficient was calculated in every layers of the soil above the water table (Campbell, 1985), and we proved that for example in the surroundings of the well MP6 the infiltrating water reaches the water table in 3 m depth during 27 hours. In the surroundings of the well MP11 the water from the surface reaches the water table during 33 hours due to the higher percentage of the silt and clay fractions. In the surroundings of the other wells we determined similar infiltration rates.

Examining the soil layers in Beregszász, the results are totally different. It is observable in Figure 6 that the finer grain size dominates in the layers: the joint percentage of the clay (<0.002 mm) and silt (0.02-0.002 mm) fractions exceeds 50%. There were soil layers in two sample points where the percentage of the clay fraction was more than 60%, that almost made the given layer confined. For example in the case of BSZTF4 in the layer of the highest clay concentration (200-240 cm) the infiltration coefficient did not reach 2 cm/day (1.9 cm/day). Based on the infiltration coefficients determined from the sampled layers, in the surroundings of BSZTF3 water needs 18 days to reach the water table in 4 m depth. Due to the significantly higher clay concentration, water needs 54 days to reach the water table in 3 m depth near BSZTF4. Thus, it can be proven that in Beregszász it is more difficult for the contaminants to get to the water table, compared to Mikepércs.

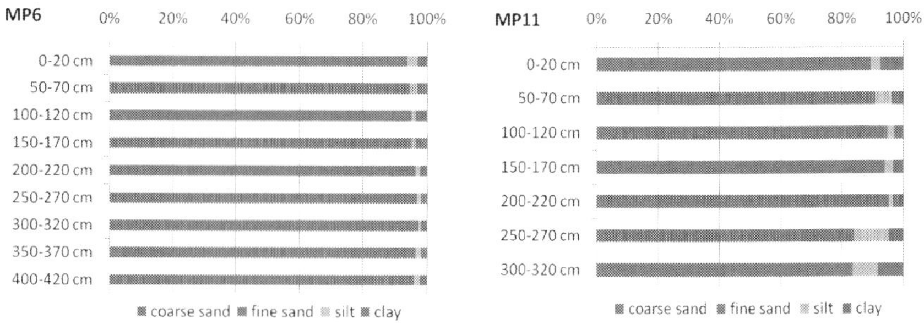

Fig. 5. The grain size distribution diagramms of the soil layers near the wells MP6 and MP11, Mikepércs

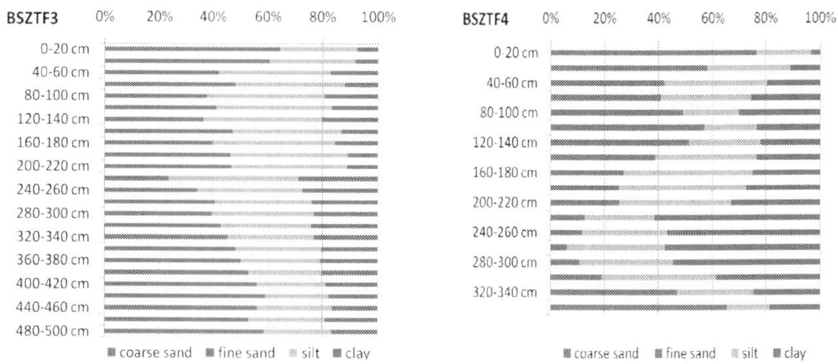

Fig. 6. The grain size distribution of the soil layers near the wells BSZTF3 and BSZTF4, Beregszász

The condition of shallow groundwater can be influenced by the climate of the area. Rainfall has an important role among the climatic factors since contaminants move mainly with the infiltrating water inside the soil layers. In the rainiest areas, leaching is typical of soils: the direction of the water movement in the soil layers is mostly from the surface to the deeper layers. Rainfall also has effects on the depth of water table: when the water table rises in the rainy periods, it comes closer to the contaminants on the surface, and then the leaching pollutants moves the most intensively toward the water table. However, in the cases of the dug wells the dilution of the contaminants can occur due to the water abundance.

In the surroundings of Mikepércs and Beregszász, the leaching processes that also facilitate the leaching of contaminants into the deeper layers become determinant above the precipitation amount of approximately 500 mm/year. The precipitation conditions were studied in both settlements in the research periods, and we proved that in Mikepércs, the annual precipitation was 830 mm in the first research year, and 677 mm in the second one. These values exceeded the average of many years (600 mm) that is typical of the region. In Beregszász, the amount of precipitation was 820 mm in the year of the research, and it is also more than the average of many years (671 mm) measured in the town.

The problem of sewage is the most important one among the anthropogenic factors; it is the most dangerous factor for the shallow groundwater under the settlements. The proportion of the constructed sewer network has significant effect on the condition of shallow groundwater, but the lack of sewer system by itself does not necessarily mean that the sewage gets into the soil and the shallow groundwater since if the sewage is collected in adequately insulated septic tanks and transported properly, the pollution of shallow groundwater do not occur. Unfortunately, according to our experiences, in the settlements where sewer network is not constructed the septic tanks are not supplied with adequate insulation in many cases, so the collected sewage can infiltrate into the soil, then to the shallow groundwater. In Mikepércs where the sewer network had not been constructed by the time of the examination, the inhabitants collected the sewage in septic tanks. Comparing the quantity of water consumption and the transported sewage we observed that more than 90% of the sewage infiltrated into the soil. The main cause of this was mainly that the septic tanks were not supplied with adequate insulation, and sewage can easily infiltrate into the soil since the soil consists of coarse grains and its water permeability is very good. In Beregszász, the situation is better since almost the two-third of the settlement is supplied with sewer network but in the non-sewered parts only a little amount of sewage can infiltrate into the shallow groundwater due to the high clay content of the soils. However, we made a questionnaire survey and it turned out that in 10 cases of the 16 examined households the sewage was not collected in septic tanks but the inhabitants pour it into the drainage channel, or in the backyard and the garden. Obviously, these processes also do not endanger significantly the shallow groundwater since the soils in the settlement have bad water permeability so they can provide some protection against the contamination of the shallow groundwater.

The shallow groundwater wells are endangered not only by the domestic sewage but other pollution sources must also be considered. Manure and liquid manure from livestock farming mean considerable contamination, especially when the proper storage of them is not assured. Livestock farming and the related problems are rather typical of Mikepércs since there was livestock farming in the direct surroundings of 12 of the 14 examined wells, while in Beregszász livestock farming could be found in the close surroundings of only 6 of the 16 examined wells. However, manure was collected in proper storage tanks in neither of the settlements, thus the manure and liquid manure could infiltrate into the soil in both towns, and they could easily reach the shallow groundwater in Mikepércs. In Beregszász, contamination deriving from livestock farming was not considerable due to the reasons mentioned above.

Besides livestock farming, gardens can also be pollution sources if the fertilizer scattered in the gardens and the insecticide infiltrate into the soil and reach the water table. In

Mikepércs, gardens can be found in the surroundings of 86% of the wells, but fertilizers and insecticides are used in only 25% of the gardens. In Beregszász, gardens can be found in the close surroundings of only half of the wells, and the use of fertilizers and insecticides are not typical: fertilizer was used in only one garden and insecticide was applied in three gardens. Summarizing the experiences we can say that gardens cause problems rather in Mikepércs than in Beregszász, but their significance is inconsiderable comparing to the problems related to livestock farming and the lack of sewer network.

4.2 The pollution level of groundwater

4.2.1 pH

There is significant difference between the two settlements regarding pH. In Mikepércs, the pH of the examined water samples was slightly alkaline, while in Beregszász most of the samples were neutral or slightly acidic, but the measured values (except some of them) were within the interval of the (B) limit values (pH > 6.5 and pH < 9.0) determined in the joint decree no. 6/2009 IV. 14. KvVM-EüM-FVM (Fig. 7).

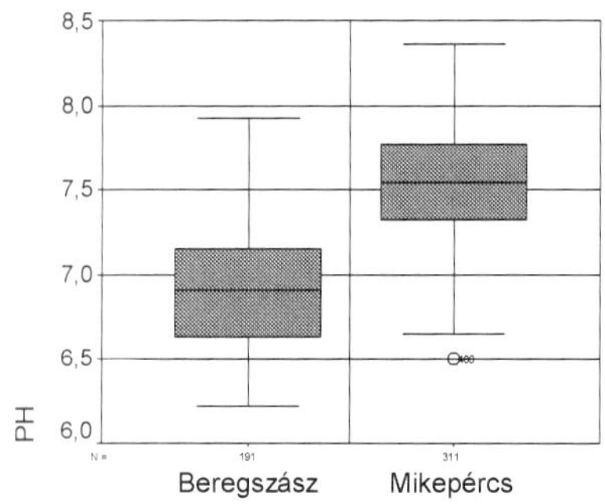

Fig. 7. The pH of the water samples from Beregszász and Mikepércs regarding the entire research period

In both settlements a maximum in the autumn-winter months and a minimum in the spring-summer months can be observed due to the acidic products deriving from the decomposition of organic matter and increasing in the warmer periods (Fig. 8).

4.2.2 Electric conductivity

The electric conductivity of the water samples gives the total ion content of the samples (Szabó, 2008; Szalai, 2008). It is observable in Figure 9 that the electric conductivity of the wells was higher in Mikepércs than in Beregszász. In Mikepércs the values were varied between 1000 and 2500 μS/cm but in some wells we measured values above 4000 μS/cm. In Beregszász electric conductivity values were generally between 1000 and 1300 μS/cm but in the case of the well 4 we usually measured values above 3000 μS/cm that indicates serious pollution.

There are significant differences between the studied areas inside the settlements. However, if the temporal changes of the contamination of a well are examined, we can say that contrary to most of the examined water chemistry parameters the values of electric conductivity are not too variable as it can be seen in Figure 10.

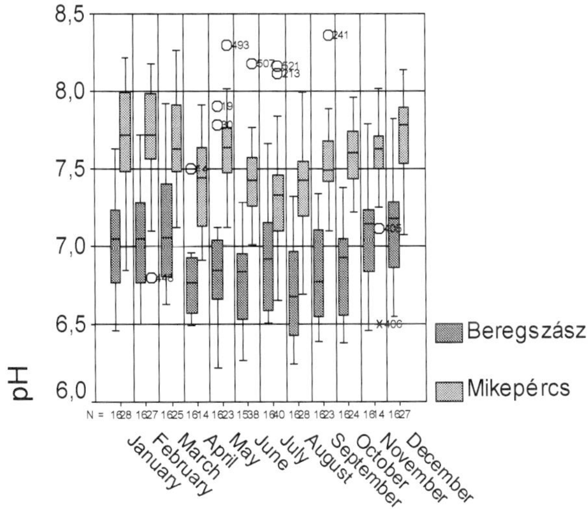

Fig. 8. The seasonal changes of the pH in the studied settlements

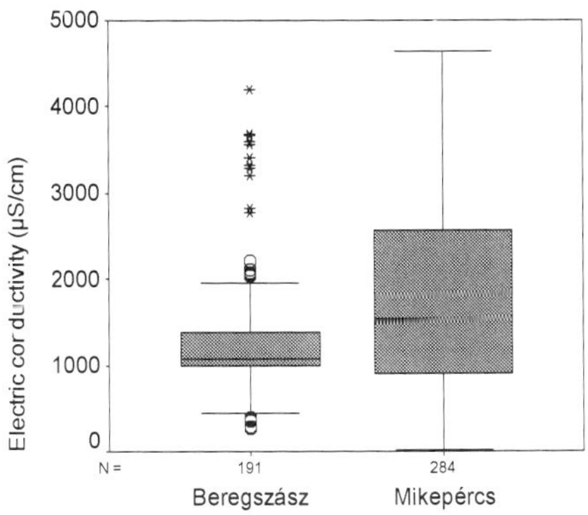

Fig. 9. Electric conductivity in the studied settlements regarding the entire research period (extreme values are not represented)

Following pH, this parameter has the least relative deviation, it slightly exceeded 20%. Although, based on the high value of electric conductivity we cannot determine which ion is mainly responsible for the high value; correlation analyses showed that there is a relatively strong positive correlation between electric conductivity and nitrite content ($r=0.62$, $p<0.01$). Earlier, significant correlation was proven between electric conductivity and nitrate concentration in the cases of several settlements (Szabó et al., 2010).

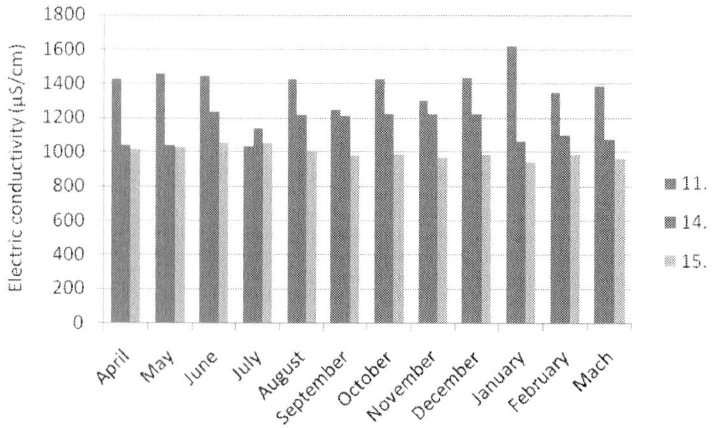

Fig. 10. The temporal changes of electric conductivity in the cases of the wells 11, 14 and 15, Beregszász

4.2.3 Ammonium ion

The ammonium ion concentration of the shallow groundwater originated from the biodegradation of organic matter, so it is one of the most important indicators of the organic contaminations. In water, ammonia can take up and also release protons. The percentage of the ammonia and ammonium ion in water bodies depends on the temperature and the pH of the water (Barótfi, 2000). With the increase of the pH and the temperature, the concentration of the toxic ammonia also increases. Most of the samples from Mikepércs are neutral or slightly alkaline, but the samples from Beregszász were mainly neutral or slightly acidic. The temperatures of the water samples varied between 4°C and 18°C. In this range of pH and temperature the percentage of ammonium ion is 97-100%, thus the percentage of the toxic, free ammonia is below 3%, in the cases of the samples from Beregszász it does not even reach 1% due to the lower pH.

It is observable in Figure 11 that the wells in Mikepércs are more polluted with ammonium. However, if we examine what percentage of the samples is below the critical contamination (B) limit (0.5 mg/l; according to the joint decree no. 6/2009), there is only a small difference between the two settlements. 34.5% of the samples in Beregszász, and 28.3% in Mikepércs contained less ammonium than the contamination limit value.

There is difference between the two settlements in the rate of contamination. In Mikepércs, there are more samples that exceeded many times the contamination limit, 3% of the samples exceeded more than a hundred times the contamination limit. In Beregszász, only one well contained ten times more ammonium than the contamination limit, but in Mikepércs half of the wells did.

Especially high ammonium ion concentration means anthropogenic contamination, and the extreme contamination values indicate direct sewage infiltration. Besides the domestic sewage, the improper treatment of the manure deriving from livestock farming could also contribute to the extreme values.

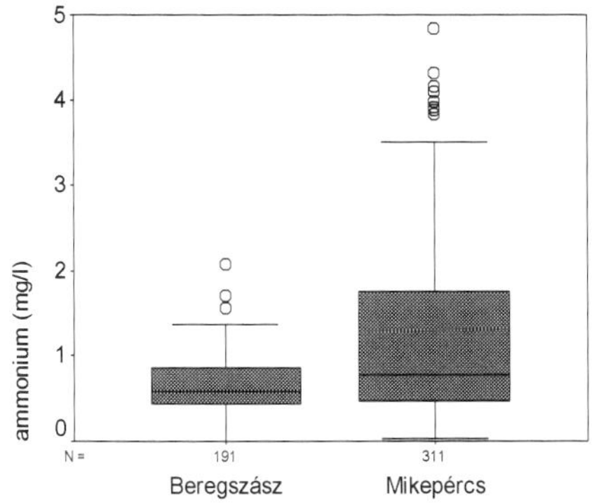

Fig. 11. The ammonium concentration of the water samples in the examined settlements (excluding extreme values)

The highest ammonium concentrations were measured in the autumn-winter months (Fig. 12). In these seasons the degradation of the organic nitrogen occurs, but the oxidization of the ammonium to nitrite is detained by the cold since the activity of the nitrifying bacteria slows down below 10°C, thus ammonium ions accumulates in the water (Bíró et al., 1998).

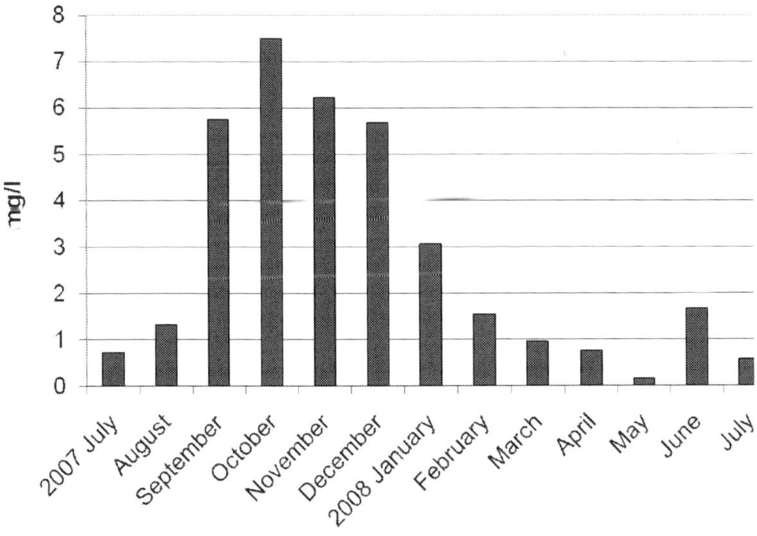

Fig. 12. The ammonium concentrations in Mikepércs in the second research period, based on the averages of 13 wells. (The figure does not contain the data of the well MP19 since extreme high concentrations were measured in this well.)

4.2.4 Nitrite

The next step of the organic matter degradation is the ammonia-nitrite transformation that is a pH-dependent process; it is the fastest between pH 8 and pH 9.5. In Mikepércs, the conditions are better for the transformation due to the higher pH of the water. Water temperature also plays an important role since the nitrifying bacteria do not stand cold, their activity slows down below 10°C, but with the increase of the temperature the nitrite concentration also increases (Barótfi, 2000). It is observable in Figure 13 where the results of the examinations in Mikepércs can be seen. The nitrite concentration of the shallow groundwater decreases significantly in the colder autumn and winter months, the maximum values are measured in the summer months. Among the examined water quality parameters nitrite showed the greatest temporal variability. In some wells of Mikepércs the relative deviation of the nitrite content was generally above 100%, and in Beregszász the mean relative deviation was 83.5%. Nitrite does not accumulate in high concentrations since it immediately oxidizes to nitrate under aqueous conditions.

The joint decree no. 6/2009 does not determine the contamination limit of nitrite in underground water. In surface water bodies 0.3 mg/l nitrite concentration indicates extremely contaminated water (MSZ 12749/1993). Regarding nitrite contamination, the situation is worse in Mikepércs, although nitrite concentration was below 0.3 mg/l in 78.4% of the samples. Extreme nitrite contamination was observed in only one well (MP19) where ammonium concentration was also extremely high. In Beregszász, 92.6% of the samples contained less nitrite than 0.3 mg/l. In fact, we measured extreme values only in the wells where ammonium concentration was also high.

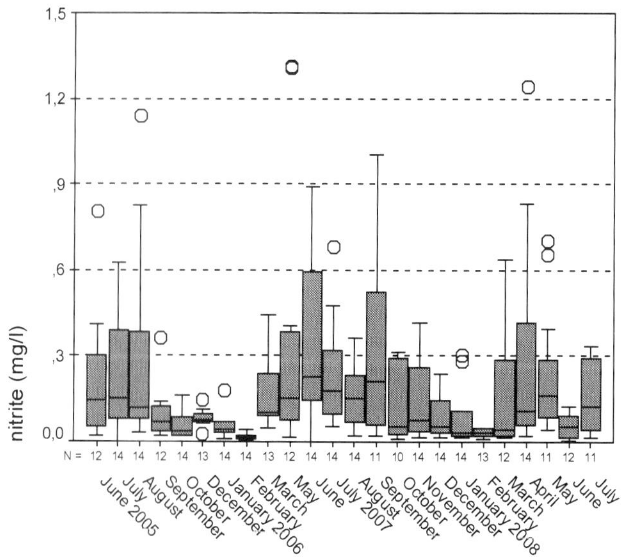

Fig. 13. The nitrite concentration of the water samples from shallow groundwater wells in Mikepércs, in the two research period (Extreme values are not represented)

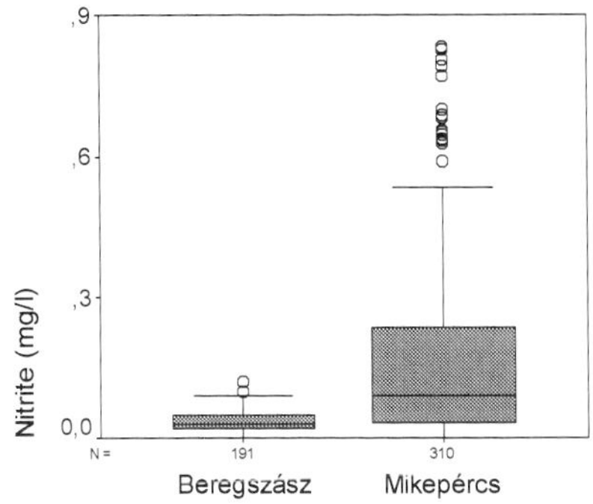

Fig. 14. The nitrite concentration of the water samples in the examined settlements (excluding the extreme values)

4.2.5 Nitrate

After the ammonium-nitrite transformation the next step of the organic matter degradation is the oxidization of nitrite to nitrate. During this process, the concentration of the dissolved oxygen in the shallow groundwater decreases considerably (Barótfi, 2000). In Figure 15, the changes of nitrogen forms in the shallow groundwater can be seen in the well no. 2, Beregszász. It is common that ammonium and nitrite concentrations vary inversely to each other since ammonium oxidizes to nitrite, then to nitrate so it is natural that when the concentration of ammonium decreases, the quantity of nitrate increases.

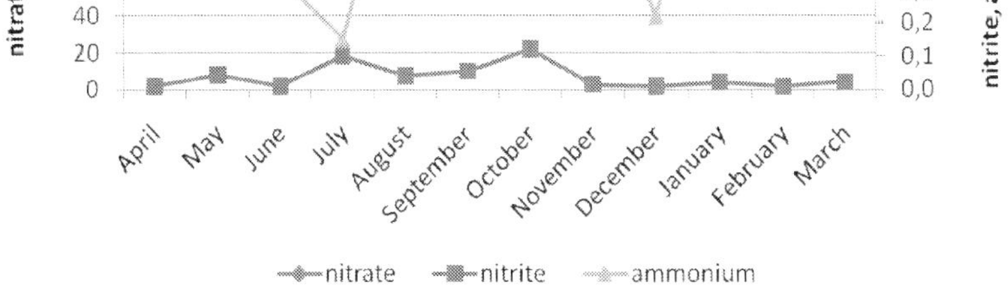

Fig. 15. The changes of the nitrate, nitrite and ammonium concentrations in the well no. 2, Beregszász

Similarly to the other examined water quality parameters, the situation is worse in Mikepércs regarding nitrate (Fig. 16). The mean nitrate concentration of the shallow groundwater is 154.4 mg/l here, while in Beregszász it was only 57.2 mg/l. However, this relatively low value also exceeds the contamination limit (50 mg/l) determined by the joint decree 6/2009 and the directive on nitrates no. 91/676/EEC. In Mikepércs, nitrate concentration exceeded the limit value in 83% of the water samples, and more than 20% of the samples contained more than 250 mg/l nitrate. Moreover, there were values over 500 mg/l. However, in Beregszász only 27% of the samples contained more nitrate than the limit value, and more concentration than 170 mg/l was not measured in any sample.

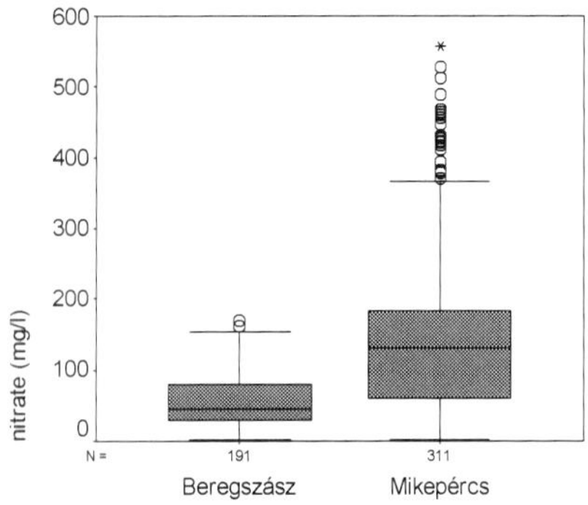

Fig. 16. The nitrate concentration of the water samples in the examined settlements

In the examined settlements the nitrate contamination of the shallow groundwater derived from the domestic sewage, but in Mikepércs fertilizers containing nitrogen scattered in the vegetable gardens and ploughlands near the wells can also contribute to the higher concentrations (Szabó et al., 2010).

4.2.6 Orthophosphate

Although, the phosphorus content of the shallow groundwater can derive from natural sources, higher concentrations are always due to anthropogenic effects. The most significant pollution source is domestic sewage, but the residuals of the fertilizers containing phosphorus can also cause contamination (Szabó et al., 2007).

There are great differences between the two settlements regarding orthophosphate contamination (Fig. 17).

In Beregszász the mean of the results is 0.56 mg/l, and only 27.7% of the samples exceeded the 0.5 mg/l contamination value (B) determined in the decree no. 6/2009. However, in Mikepércs the mean value was 3.21 mg/l and almost 94% of the samples contained more orthophosphate than the contamination limit. The situation is bad since 1/5 of the samples contained ten times more orthophosphate than the contamination limit. Although the situation is much better in Beregszász, more than ¼ of the wells is considered to be polluted but it is far less than the contamination experienced in Mikepércs.

The unfavourable conditions in Mikepércs occurred because the sewer network was not constructed in the research period. On the other hand, the water table is very close to the surface, and it occurred in several times during the research period that the bottom level of the sewage tanks in the surroundings of the wells was under the water table, therefore sewage could directly mixed with the shallow groundwater.

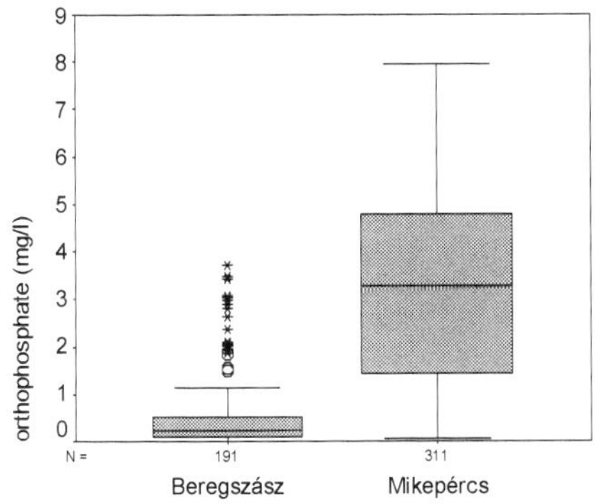

Fig. 17. Orthophosphate concentration of the water samples in the examined settlements

4.3 The health risks related to the use of groundwater

Health risks must be examined depending on the use of shallow groundwater. Based on our questionnaire survey in the settlements, it is proven that in Beregszász 81% of the inhabitants, who live near the wells, generally consume the water of the wells; but in Mikepércs it does not happen (except some cases). The inhabitants in Beregszász are in greater danger in terms of human health. However, Mikepércs is more endangered in terms of animal heath since the contamination of the dug wells is higher than in Beregszász, and water from the wells is used for watering animals in almost every household. In Beregszász only 1/3 of the wells are used for this purpose.

Human and animal health risk can be estimated based on the comparison of the values measured in the examined water samples and the threshold limit values related to drinking water and water for animals. (The evaluation was carried out based on the Hungarian threshold limit values.) The threshold limit values of drinking water are clear, but the law is not obvious regarding water for animals.

4.3.1 Human health risks

According to the part B of the Appendix 1 of the Government Decree No. 201/2001. (X. 25.) on the water quality requirements, the threshold limit value is 50 mg/l for nitrate and 0.5 mg/l for nitrite. In the part C of the Appendix 1 of the government decree the threshold limit values for indicator water quality characteristics can be found. It is 2500 µS/cm for electric conductivity, and the range between 6.5 and 9 can be considered acceptable regarding pH. The D part of the Appendix 1 of the decree describes threshold limit values for karstic water, shallow groundwater and bank-filtered water resources. According to this, if drinking water is provided from these water resources, the threshold limit value for ammonium is 0.2 mg/l, and it is 0.1 mg/l for nitrite.

Regarding the threshold limit values it is proven that none of the examined wells in Mikepércs meets the requirements. In the cases of the examined water quality parameters we measured values that exceeded the threshold limit values many times, therefore the water of the wells are absolutely inappropriate for human consumption.

In Beregszász, the water quality of the wells was much better, but 4/5 of the inhabitants regularly consume water from the wells. It is also observable that we can often find values that exceed the threshold limit value. Electric conductivity was higher than the limit value only in 6.8% of the samples, the percentages of the values above the limit value were 10.2% regarding pH, 15.7% for nitrite, 27.7% examining orthophosphate and 45.2% regarding nitrate. In case of ammonium, 92.1% of the samples contained higher concentration than the limit value, if we consider the threshold limit value determined for shallow groundwater (0.2 mg/l). However, if we would consider the threshold limit value regarding deep groundwater (0.5 mg/l), the percentage would even be 65.4%. If we examine the wells in Beregszász one by one, it is observable that there was not a single well where at least one value of the examined parameters did not exceed the limit value. Regarding all the examined water quality parameters, the values of only 2.6% of the 191 water samples from Beregszász were below the threshold limit values. These data are alarming, even if the water of the wells in Beregszász is less contaminated than the water of the wells in Mikepércs, since if someone regularly drinks water that contains harmful components in higher concentration, they must reckon with the risks of illnesses.

The effects of nitrate and nitrite on health are the mostly studied issues of the special literature. In some cases such as methemoglobinemia the process of the illness and the effects of the intoxication are well-known, but in other cases (such as the formation of stomach cancer) the role of nitrate and nitrite is not clear (Barótfi, 2000; Du et al., 2007; Kerényi, 1995; Milkowski et al., 2010; Yang et al., 2007).

Methemoglobinemia (or "blue baby" syndrome) occurs mostly in infants. The pH of the stomach is almost neutral in infants - contrary to the acidic pH of the adults' stomach -, and this is beneficial for the bacteria that transforms nitrate to nitrite. Nitrite reacts with haemoglobin forming methemoglobin, thus it blocks the oxygen carrying capacity of blood and can cause suffocation (Barzilay et al., 1999). In adults, there is a specific enzyme that transforms methemoglobin back to haemoglobin so it can detoxify the body. This enzyme appears later during the growth of the child, so in infants the detoxification do not occur (Barótfi, 2000). In the case of nitrate concentration above 50 mg/l this illness can occur in infants, therefore in Beregszász people must not give the water of dug wells to infants since less than 50 mg/l nitrate concentration was measured only in 3 of the examined 16 wells.

4.3.2 Animal health risks

The decree no. 41/1997. (V. 28.) states that "As far as possible water of drinking-water quality should be used for the watering of animals." Thus, the decree do not make the use of drinking water obligatory, just says that as far as possible water of drinking water must be used. Unfortunately, it was not carried out in the cases of the examined households (possessing dug wells) in Mikepércs since the inhabitants water the animals everywhere with the water from the dug wells, but the quality of this water is not adequate regarding the directions. On the other hand, they had the possibility to use water of drinking water quality as the water supply system is entirely constructed in the settlement. However, they do not use drinking water due to financial and comfortable causes. The situation is similar in Beregszász: people also used the water of the dug wells to water animals, but in several cases tap-water was not available, and it is also true that where the water supply system is constructed people also do not give tap-water to the animals.

Knowing the water quality of the wells in Mikepércs, we can say that the use of water from the dug wells mean serious risk to animal health since we measured values that exceeded the threshold limit values many times regarding several water quality parameters. It is often proved that, due to some contaminants, in animals the same illnesses can arise as in humans, such as methemoglobinemia. In the case of ruminant animals symptoms similar to the infants' symptoms appear due to the consumption of water containing nitrate in high concentration (Barzilay et al., 1999). An international research group examined the effect of nitrate on human and animal organisms. They proved that besides ruminant animals, pigs are also endangered by methemoglobinemia since they cannot produce enough MetHb-reductase enzyme that transforms methemoglobin to haemoglobin (Cockburn et al., 2010). The research group also studied what risks are caused if nitrite indirectly gets into the human body through animal source foods such as milk, eggs, or meat. According to the examinations only 2.9% of the accepted nitrite load per day gets into the human body with animal source foods. Therefore, the authors state that the consumption of animal source foods containing nitrite does not mean serious health risk to human health. However, regarding several other contaminants (e. g. heavy metals) we must consider the process of bioaccumulation since in the higher levels of the food chain such accumulation can occur that can mean serious risk to human health.

Since in Beregszász water of drinking water quality is not ensured for animals, some risks must be considered, especially in the cases of the wells where significant contamination was experienced regarding some parameters (wells no. 2, 7, 10 and 16). However, comparing to Mikepércs, we can say that the risks to animal health are far less in Beregszász.

5. Conclusion

The water quality of dug wells and the factors influencing water quality were studied in two settlements with different natural and social-economic characteristics. Beregszász is in better situation in terms of the natural characteristics and the anthropogenic effects influencing water quality since the water table can be found deeper, and the water permeability of the soils is worse here. The sewer network is partly constructed in Beregszász, and the role of livestock farming and gardens, that endanger the quality of shallow groundwater, is less.

Our presumption that the water quality of the dug wells is better in Beregszász (that have more advantageous characteristics) is proven. In the cases of all the examined contaminants (ammonium, nitrite, nitrate, orthophosphate) we proved that the contamination of the water of the wells was much more significant in Mikepércs. Based on the results we can say that among the factors influencing shallow groundwater quality the grain size distribution, then the depth of the shallow groundwater play the most important role, so mainly natural characteristics affect the water quality of the dug wells.

Studying the risks of the use of water from the dug wells it is proven that there are health risks in both settlements. Mainly risks to human health are significant in Beregszász since

80% of the inhabitants living in the surroundings of the examined wells regularly consume the water of the wells, while there was not a single well where at least one value of the examined parameters did not exceed the limit value. In Mikepércs, the inhabitants do not consume water from the wells, but 86% of the wells are used to water animals. This means extremely serious risk to animal health since the water of the wells in Mikepércs is absolutely inappropriate for animal consumption (due to the serious contamination).

6. Acknowledgement

The research was supported by the No. 68566 Hungarian Research Found (OTKA), the MERIDIÁN Landscape and Environmental Geography Foundation and the TÁMOP-4.2.2/B-10/1-2010-0024 Hungarian Found

7. References

Adekunle, I.M.; Adetunji, M.T; Gbadebo, A.M. & Banjoko, O.B. (2007). Assessment of Groundwater Quality in a Typical Rural Settlement in Southwest Nigeria. *Int. J. Environ. Res. Public Health*, 4(4), pp. 307–318.

Ballenegger, R. & Di Gléria, J. (1962). *Soil and manure examination methods,* Mezőgazdasági Kiadó, Budapest, Hungary (in Hungarian)

Barótfi, I. (2000). *Environmental technology* - Mezőgazda Kiadó, ISBN 963-286-009-8-9, Budapest, Hungary (in Hungarian)

Barzilay, J.; Weinberg, W.G. & Eley, J.W. (1999). *The Water We Drink,* Rutgers University Press, ISBN 0-8135-2673-6, New Brunswick, USA

Bíró, T.; Thyll, Sz. & Tamás, J. (1998). Risk assassment of nitrate pollution in lower watershed of the Berettyó River. In: Filep, Gy. (ed.): *Soil water environment relationships.* Wageningen, Holland - Debrecen, Hungary

Campbell, G. S. (1985). *Soil Physics with Basics, Transport Models for Soil-Plant Systems,* Development in Soil Science 14, Elsevier, 149 p. ISBN 0-444-42557-8

Cey, E.E.; Rudolph, D.L.; Aravena, R. & Parkin, G. (1999). Role of the riparian zone in controlling the distribution and fate of agricultural nitrogen near a small stream in southern Ontario. *Journal of Contaminant Hydrology* 37, pp. 45–67.

Chettri M. & Smith G.D. (1995). Nitrate pollution in groundwater in selected districts of Nepal. *Hydrogeology Journal* 3, pp. 71–76.

Cockburn, A; Brambilla, G.; Fernández, M.L.; Arcella, D.; Bordajandi, L.R.; Cottrill, B.; Peteghem, C. & Dorne, J.L. (2010) Nitrite in feed: From Animal health to human health. (Aricle in press, August 2011) *Toxicology and Applied Pharmacology.* Available from http://www.sciencedirect.com/science

Council Directive of 12 December 1991 concerning the protection of waters against pollution caused by nitretes from agricultural sources (91/676/EEC)

Decree no. 41/1997. (V. 28.) FM on Veterinary regulation, Available from http://www.magyarorszag.hu (in Hungarian).

Du, S.T.; Zhang, Y.S. & Lin, X.Y. (2007) Accumulation of Nitrate in Vegetables and its Possible Implications to Human Health. *Agricultural Sciences in China*, 6 (10), pp. 1246-1255.

Goss, M.J.; Barry, D.A.J. & Rudolph, D.L. (1998). Contamination in Ontario farmstead domestic wells and its association with agriculture: 1. Results from drinking water wells. *Journal of Contaminant Hydrology* 32, pp. 267–293.

Government Joint Decree No. 6/2009. (IV. 14.) KvVM-EüM-FVM on limit values established for the protection of groundwater and the geological medium, 2009. (in Hungarian)

Howden, N.J.K.; Bowes, M.J.; Clark, A.D.J.; Humphries, N. & Neal, C. (2009). Water quality, nutrients and the European union's Water Framework Directive in a lowland agricultural region: Suffolk, south-east England. *Science of the Total Environment* 407, pp. 2966-2979.

Hungarian Standard no. MSZ 12749 1993, The quality of surface water, qualitative characteristics and qualification, 1993 (in Hungarian)

Jang, C.S. & Liu, C.W. (2005). Contamination potential of nitrogen compounds in the heterogeneous aquifers of the Choushui River alluvial fan, Taiwan. *Journal of Contaminant Hydrology* 79, pp. 135–155.

Knobeloch, L.; Salna, B.; Hogan, A.; Postle, J. & Anderson, H. (2000). Blue babies and nitrate-contaminated well water. *Environ Health Perspect* 108, pp. 675-678.

Milkowski, A.; Garg, H.K.; Coughlin, J.R. & Bryan, N.S. (2010). Nutritional epidemiology in the context of nitric oxide biology: A risk-benefit evaluation for dietary nitrite and nitrate. *Nitric Oxide* 22, pp. 110–119.

Molnár, J. (2009). General information and geographic position. In: *Kárpátalja – A Kárpát-medence régiói 11.* Baranyi, B. (Ed.) Dialóg Campus Kiadó, ISBN 9789639899186, Pécs-Budapest, Hungary (in Hungarian)

MSZ 'Determination of organic carbon content of the soil' *Hungarian Standard* No. 08-0210; 1977 (in Hungarian).

MSZ 'Examination of some chemical characteristics of the soil' *Hungarian Standard* No. 08-0206-2; 1978a (in Hungarian).

MSZ 'Examination of the physical and water management characteristics of the soil' *Hungarian Standard* No. 08-0205; 1978b (in Hungarian).

Muñoz-Carpena, R.; Ritter, A. & Li, Y.C. (2005). Dynamic factor analysis of groundwater quality trends in an agricultural area adjacent to Everglades National Park. *Journal of Contaminant Hydrology* 80, pp. 49–70.

Nyizsalovszki, R. & Szabó, Sz. (2003). The temporal and spatial changes of water table in a piedmont sample area. *Hungarian Geographical Bulletin* 52, (1-2) pp. 23-36. (in Hungarian)

Oprean, L.; Poplăcean, M.; Oancea, S.; Oprean, C. & Lengyel, E. (2009). Monitoring the microbiological and physico-chemical pollution degree of the river Visa. *Studia Universitatis "Vasile Goldiş", Seria Ştiinţele Vieţii*, Vol. 19, issue1, pp. 203-207.

Pál, Z.; Aczél, M. & Pál, K. (2009). Nitrate contamination of the groundwater in rural settlements - example of Imecsfalva. *Collegium Geographicum* 6. sz. pp. 43-51. (in Hungarian with english abstract)

Rakonczai, J. (2008). *Global environmental issues*. Universitas Szeged Kiadó, ISBN 9639416525, Szeged, Hungary (in Hungarian)

Steiner, F. (Ed.) (2010) *The state of the environment in Hungary*, Prospektus Nyomda, Budapest, Hungary (in Hungarian)

Sultana, F. (2011). Suffering for water, suffering from water: Emotional geographies of resource access, control and conflict. *Geoforum* 42, pp. 163–172.

Szabó Gy.; Szabó Sz.; Szabó A. & Szemán B. (2007). Spatial and time variations of the groundwater quality of two different landscapes, In: *Implementation of Landscape Ecology in New and Changing Conditions,* Boltižiar, M. (ed.), ILE Slovak Academy of Sciences, pp. 421-427. ISBN 978-80-89325-03-0

Szabó, Gy.; Angyal, A.; Csikós, A.; Bessenyei, É.; Tóth, E.; Kiss, P.; Csorba, P. & Szabó, Sz. (2010) Examination of the groundwater pollution at lowland settlements - *Studia Universitatis "Vasile Goldiş",* Seria Ştiinţele Vieţii Vol. 20, issue 4, pp. 89-95. ISSN: 1584-2363

Szabó, Sz. (2008). *Examination methods of environmental protecion* - monitoring - távoktatási tankönyv, Debrecen, Hungary (in Hungarian)

Szalai, Z.; Jakab, G. & Madarász, B. (2004). Estimating the vertical distribution of groundwater Cd and Cu contents in alluvial sediments (River Danube). Saturated and Unsaturated Zone; In: *Integration of process knowledge into effective models.* (Eds. Per Aagard et al.) pp. 303-312. La Gordialica Pavese, Rome, Italy

Szalai, Z. (2008). Spatial and temporal pattern of soil pH and Eh and their impact on solute iron content in a wetland (Transdanubia, Hungary) *AGD Landscape and Environment* 2 (1). pp. 34-45. HU ISSN 1789-4921

Yang, C.Y.; Wu, D.C. & Chang, C.C. (2007) Nitrate in drinking water and risk of death from colon cancer in Taiwan. *Environment International* 33, pp. 649-653.

Warner, N.R.; Levy, J.; Harpp, K. & Farruggia, F. (2008). Drinking water quality in Nepal's Kathmandu Valley: a survey and assessment of selected controlling site characteristics. *Hydrogeology Journal* 16, pp. 321–334.

Widory, D.; Kloppmann, W.; Chery, L.; Bonnin, J.; Rochdi, H. & Guinamant, J.L. (2004). Nitrate in groundwater: an isotopic multi-tracer approach. *Journal of Contaminant Hydrology* 72, pp. 165–188.

19

Determination and Speciation of Trace Heavy Metals in Natural Water by DPASV

Amra Odobasic

University of Tuzla, Faculty of Technology
Bosnia and Herzegovina

1. Introduction

The aim of water preservation within each country seeks to be achieved through the increased concern about the water, which often means controlling of the water quality. Chemical analysis and speciation of traces of heavy metals such as lead, cadmium, zinc, copper, etc. is of great importance.

The heavy metals also includes number of physiologically important elements such as Cu, Fe, Zn; than the highly toxic Pb, Mn, Cd, hg, As, Sb and less toxic as Au, Ag and Cr, etc. [1]

The most important anthropogenic sources of metals in aquatic ecosystem are certainly wastewaters, that are being discharged untreated or with different levels of purity, so they can cause many changes in the stream / recipient. Heavy metals pass go through biogeochemical process with different retention time in different parts of atmosphere. They do not decompose and have the ability of bioaccumulation, because they are being retained in plants, animals and nature in general. This is an extremely heterogeneous group of elements in terms of biological and ecological effects. Large numbers of metals is essential for proper functioning of the human body and appertain to the group of essential elements. A deficiency of heavy metals on one side can lead to serious symptoms; and on the other, their presence in slightly elevated concentrations can lead to serious illnesses. Their toxicity depends on concentration, and the allowed concentration range varies from metal to metal. Concentrations in which heavy metals can occur depend from source of pollution and features of system in which they are found, so they can range from traces to very high concentrations. Water, as well as air can receive large amounts of pollutants, but beside ability of self-cleaning, some pollutants among which are also heavy metals leads to modification of water quality to that level that it becomes useless for many purposes. From that reason most of European countries have issued maximum allowed amount of heavy metals in industrial waters that are being discharged into natural flow. Some of those values are given in table 1.

According to the data in table 1, Belgium has the most rigorous regulations regarding maximum allowance of heavy metal concentration.

In water, heavy metals are being rapidly degraded and sediment in the form of hard soluble carbons, sulphats and sulphids on the bottom. At the time when the absorptive capacity of sediment is exhausted, the concentration of metal ions in water is increasing. (Goletić 2005).

Features of these dissolved metals in water will depend on nature of other metal components dissolved in the water. Heavy metals in water appear in different chemical forms and in different oxidizing conditions, so their toxicity is changing depending on chemical form in which they are found. Therefore, the knowledge of total concentration of metals in contaminated water is often insufficient for the proper information regarding their harmful effect. Toxicity, biodegradability, bioaccumulation, mobility, solubility, as well as number of other important characteristics in particular depend on specific physical - chemical form in which the metal is found. Most studies on the topic of toxicity of heavy metals shows that the free hydrated metal ion is the most toxic form,[2] because the free metal ion is significantly easier and faster absorbed on suspended particles or colloids.

Metal	Germany	Slovenia	Hungary	Italia	Belgium
Pb	0,5	0,5	0,05-0,2	0,2	0,05
Cd	0,1	0,5	0,005-0,01	0,02	0,005
Cu	0,5	0,5	0,5 -2,0	0,1	0,05
Zn	-	1,0	1,0-5,0	0,5	0,3
As	-	0,1	0,05	0,5	0,05
Fe	-	2,0	10-20	2,0	-
Cr	0,5	1,0	0,2-1,0	2,0	0,05

Table 1. Maximum allowed concentrations (mg/l) in industrial wastewaters of some European countries.

Heavy metals can also appear in the form of inorganic and organic complexes, tied to colloid fraction (as polyhidroxid complex and polysilicate), as polymers and pseudocolloids (metal hydroxides, humic acid) and tied on suspended particles or live organisms (e.g. Microorganisms). Appearance of different heavy metal chemical forms will depend also from chemical nature of given heavy metal, conditions of environment, presence of complexationsubstances, colloid dispersion, etc.[3] To what extent will heavy metals in water be mobile, depends from number of parameters:

- pH of water
- Presence of carbonates and phosphates
- Hydrated oxides of iron and manages
- Content of organic mater
- Sulphideions and pirit(significant for mobilization process of heavy metals in water environment).

Migration of metals and way of their allocation in water systems is controlled primarily by sediment character in water mass. Intensity of reaching metals from sediment into water depends from its physical structure and chemical nature, since they determine power (strength, binding) force of chemical elements. This means that transport of pollutants in hydrosphere goes through physical - chemical processes in water and sediment.

Chemical characteristics of heavy metals offer information regarding their mobility, relative to the extent of their availability in live organisms. In order to determine it, it is necessary to examine kinetic, thermodynamic, chemical balance, stability constant of metal complex, and do the chemical speciation.

Several methods are appropriate for the speciation of metals. From electrochemical methods those are anodic and cathodic stripping voltammetry and potentiommetry; and from analytical methods, beside separation technique based on ionic exchange, extraction, dialysis and ultrafiltration are being used, as well.

In this chapter, importance of monitoring of heavy metals in natural waters and their speciation using differential pulse anodic stripping voltammetry (DPASV) are going to be pointed out. The DPASV provides the capacity of determining very low concentrations, even up to 10^{-12}mol/l.

Natural water contains various organic and inorganic species, where heavy metals build complexes with both. Today, more and more experiments related to formed mixed ligand complexes are being performed. These complexes contain more than one type of ligand, which differs from solvent molecule, and which is being located in inner coordination sphere around the central metal ion.

In this chapter, the results of examination of Modrac Lake water, used as a source of drinking water, is going to presented. Examination is conducted in order to determine concentration and speciation of heavy metal traces (Zn, Cu and Cd). The DPASV was applied for determination of the total concentrations and stable chemical forms, while for determination of labile chemical forms software programs MINTEQ, CHEAQES and HIDRA were used.

Analyzed water samples of lake Modrac where taken through period of four season and from four different characteristic locations. The experiments have shown that concentration of Cd was very low($<10^{-11}$mol/l), while concentrations of Cu and Zn were around 10^{-8}mol/l. The constant stability and capacity of complexation of Cu was determined, in order to observe ecological status of the water from Lake Modrac from chemical analysis aspect.

2. Importance determination of content copper, zinc and cadmium in natural water

The pollution increase, caused by heavy metals and organometallic compounds is one of the today's most serious problems. Conditionally said, the examined heavy metals are incorporated in drinking water and in entire food chain. Heavy metals, as toxic and hazardous pollutants are also present in the pesticides which are used in agriculture. The common characteristic is that all are insoluble and have tendency to accumulate in vital organs. Zinc and copper fall into metalloids category, ie. metals of essential importance necessary for different biochemical and psychological processes in the organism and toxic only above the "exactly" determined concentration. The "concentration window" of these heavy metals is somewhere around toxic and maximum allowed limit, and it is very slight. The toxicity level of metals significantly differs and depends on their physical-chemical form. According to the World Health Organization (WHO), about 400 million people in developing countries suffer from the diseases which are caused by use of contaminated drinking water.[4] In table 2.1. the critical concentrations of heavy metals in natural waters by EPA.[6]are given.

Metal	Maximum allowed concentration (μgr/l)
Mercury	0.002
Arsenic	0.5
Lead	0.5
Copper	0.6
Zinc	5
Cadmium	0.04

Table 2.1 Critical concentrations of heavy metals in natural water by EPA

2.1 Copper

Copper belongs to IB group of periodic system of elements, which is characterized by electronic configuration with one unpaired element. As a result, the two compounds have a high melting and boiling point. In natural waters, copper appears during dissolution of minerals and in the range of 1-10μg/l. The main sources of copper are wastewaters from process of manufacturing copper, refining, copper processing, and manufacturing of steel.

In water environment, copper is found in three main forms: suspend, colloid and as dissolved. Dissolved form implies free copper ions and complexes with organic and inorganic ligands. Neutral ligands, such as ethilendiamin or pirid form complexes with copper that are typical for four-coordinated compounds. Moreover, they form complexes with humic substances (in fresh water over 90% is tied to these complexes, and in salt water only 10%). This low percentage of binding in sea water is result of its replacement with calcium and magnesium. The amount of copper tied to solid particles can range up to97% of total content in fresh water. Intensive sorption of copper is the cause of its high content in the sediment, where the intensity of sorption depends from content of humic acids, fulvic acids, pH, iron oxide and manganese, etc. When the chloride content is increased, the sorption of copper is reduced due to formation of chloride complexes, results in higher solubility and mobility. Concentrations of copper and bioavailability to aquatic system also depends on many parameters such as hardness of water, alkalinity, ionic strength, pH, redox potential, amount of present complex ligands, suspended matters and interaction of water and sediment.

2.1.1 Eco-toxicity of copper

It is generally known that copper does not represent big eco-toxicity problem. However, taking into account that copper is very widespread, and given to its big exposition from exhaust gases, it is very important is structuring of ecosystem. In small quantities copper represents an essential heavy metal necessary for proper functioning of organisms, and for activity of different enzymes.

In aquatic environment, copper accumulation results in primary exposure of aquatic organisms. Aquatic organisms can accumulate dissolved copper by direct absorption through the skin, while particles are absorbed through the food chain. When it comes to plant aquatic life, copper is absorbed in ionic form and in helatnoj form. Normal concentrations of copper in plant are generally very low, ranging from 2-20μg/g of dry mass. In higher concentrations they are toxic, and toxicity is more expressed with low values of pH. However, their main source comes from wastewaters resulting from the process of copper manufacturing, refining, procession go copper and manufacture of steel. It is essential element, but in large quantities causes, so called, Wilson disease. In natural waters copper appears due to dissolution of minerals, at around of 1 - 10μg/l.[5]

2.2 Cadmium

Cadmium represents one of the most toxic metals. It is usually brought into organism through an thropogenic source. It belongs to the IIB group of transition elements, which are characterized by availability of *s* and *d* orbitals. This availability explains stability of elements of this group, as well as their different physical features, compared to other transition elements.

The standard redox potential of cadmium has negative value which indicates that cadmium relatively easy oxidizes in water solutions. Especially interesting are halogen complexes of cadmium, whose formation is caused by higher cadmium concentration. These complexes are not very stable, and the most stable is tetrajodocadmiat. whose constant of stability is $10^6 dm^{12}/mol.$,[6]. In the organism, Cadmium accumulates and causes weakness or total loss of kidney, liver and pancreas functions. Furthermore, it destroys DNA and leads to damage of the red blood cells. The basis of psychological influence of cadmium has its similarity with zinc in so that cadmium replaces zinc as microelement in some enzymes which then leads to complete change of enzymes stereo-structure and its characteristic features. The researches have shown that on the bioavailability of cadmium influences also other constituents in water and food; like calcium, phosphor and steel.

2.3 Zinc

Zinc, like cadmium, has negative redox potential and ability to easy oxidize in water solution. It is essential element and has important role in biological process in organism. In water solutions, zinc behaves as very strong acidic metal, binding halogenid ions in series F^- $>Cl^->Br^->J^-$. At pH 6.7, zinc in water is in divalent form which is available for sorption and forming complexes with organic substances. Same as lead and cadmium, zinc also has ability of bioaccumulation to some extent, especially in fish. Level of bioaccumulation also depends from the exposure time and conditions that are present in aquatic environment. Conditions that may influence to the toxicity of zinc in aquatic environment are water hardness (content of Ca, Mg), alkalinity, and content of humic substances. In all concentrations, higher than 100 ppm, zinc is fitotoxic.

2.4 Complex compounds of heavy metals in natural waters

Natural and synthetic organic ligands and chloride complex of metal poor sorption and increase time of deposition in water layer. Biological systems have significant impact on chemical reactions of heavy metals in water. Water plants and animals are using nutrients, excreted metabolic products, while in their organism occurs bio-concentration of heavy metals. Moreover, during microbiological degradation of organic substances, complexing agents are formed, which directly influence mobility of heavy metals in aquatic environment.

Solubility of metals in natural waters are determined and in some way controlled by three factors:

1. pH
2. type and concentration of ligand and chelating complex
3. oxidizing state of mineral components and redox environment of system

It is already stated above that metal in traces can be in suspended, colloid or in dissolved form. Suspended particles are in fact those whose diameter is higher than 100μm, dissolved are those whose diameter is lower than 1μm, and colloid are those whose diameter is in range from 1-100 μm. Colloid and suspended particles can consist of 1) heterogeneous mixture of metals in form of hydroxide, sulfide oxide or silicates or 2) organic matters which are tied to metal ion by absorption, ionic exchange or complexation. Besides these forms, metals in water can be also in the form of free metal ions, surrounded by coordinative molecules of water, in form of anions (OH^{-}, CO_4^{2-}, Cl^{-}) or in a form of organic or inorganic complexes with hydrated metal ion that substitutes coordinative water.

Variations in speciation of heavy metals, as well as change of their oxidation state can significantly influence their biodegradability, and therefore toxicity. Thus, for example, As (III) is much more toxic than As (V). Significant changes in nutrition can also lead to decline in biodegradability of organic compounds of heavy metals, so for example, high fiber food and proteins absorb essential elements and can cause lack of minerals. In case of copper, hydroxyl complexes are also toxic even though in lesser measures. Strong complexes and particles associated on colloid particles are usually nontoxic.

Unpolluted sea or fresh water contains very low concentrations of heavy metals like Cu, Cd, Zn, Pb, etc, mostly dissolved and absorbed on organic and inorganic colloid substances. Namely, human by itself makes toxic form, and as such release directly into waterway or causing decrease of pH transform it into reactive metal form. The table 2.4.1 provides for possible physical-chemical forms of Zn, Cu and Cd.

Physical-chemical form	Possible complex	Average diameter (nm)
Particles	Held on the filter 0,45μm	>450
Simple hydrated metal ions	$Cd(H_2O)_6^{2+}$	0.8
Simple organic complex	Cu -glycinat	1-2
Stabile inorganic complex	$ZnCO_3$	1-2
Stabile organic complex	Cu -fulvate	2-4
Absorbed on inorganic colloid	Cu^{2+}-Fe_2O_3	10-500
Absorbed on organic colloid	Cu - hum. Acid	10-500
Absorbed on compound of organic/inorganic colloid	Cu - hum. acid/Fe_2O_3	10 -500

Table 2.4.1. Possible physical-chemical forms of Zn, Cu and Cd. .[7]

As seen from the table, metal ions can complex with organic and inorganic ligands through chemical reactions similar to reaction of hydrolysis of metal ions. As a result, soluble and insoluble complexes can occur, which depends on concentration of metal, ligand concentration and pH value of water.

Inorganicligands are usually present in natural waters in much higher concentrations than from metal with which will form complexes. Every metal ion has specific behavior model in aquatic system which depends from 1) tendencies of metal ion to form complexes with inorganicligands 2) stability of products incurred with hydrolyze reaction. Ions like Cd or Zn ions form series of complexes in presence of Cl- and/or SO_4^{2-} in conditions when concentration is approximate to concentration of sea water. Metal ions can be tied to artificial organic substances and form organometal compounds, than carboxyl groups forming salts of organic acids, electron donor atoms O,N,P,S forming coordinative complexes.

Nowadays, researches related to metal behavior test in natural waters usually includes calculation of thermodynamic balance in model systems of metal and ligands, inside volume of pH which is characteristic for natural water (5 - 9,5).

Taking into account that natural water is complex and unbalanced redox environment to the behavior of metal, significant impact has also redox conditions. Impact of these conditions can be expressed through 1) direct change of oxidation stage of metal ion and 2) change ofredox in availability and competition of ligands.

Nowadays inorganic complex of heavy metals is being the most studied one. However for proper comprehension of biodegradability, toxicity and geochemical reactivity, the smallest particles present in the water must be taken into account and knowledge of organic complexion of heavy metal traces is necessary. Taking this into account, this chapter will provide mechanism of metal concentration.

2.5 Differential Pulse Anodic Stripping Voltammetry (DPASV)

Electroanalytical methods are widely used in scientific studies and monitoring of industrial materials and the environment. One of the most widespread electroanalytical methods is voltammetry (polarography).

Various voltammetry techniques have simple laws and theoretical relations that in detail describe and represent different aspects of dependence i - E such as the shape of curve, height and position of peak.

Nowadays, voltammetry unite large number of instrumental techniques such as anodic striping voltammetry (ASV), catodic striping voltammetry (CSV), adsorbed voltammetry (DPAV), differential pulse anodic stripping voltammetry (DPASV), which due to its large sensibility more and more suppresses classic techniques.

Stripping methods belongs to the electroanalytical methods group with pre-concentration, and these methods have the lowest detection border, so they are applicable for the analysis of media with very low concentrations ranging from 10^{-10} to 10^{-12}M,[8]. All stripping methods are based on two-stage process.

I-stage is pre-concentration of analytes from the tested sample solution on the surface of the working mercury electrode by oxidation or reduction, at a constant potential deposition E_d.This step can include anode or cathode process. Pre-concentration stage usually lasts 5 to 30 minutes, and with longer disposition time leads to higher signal and lower detection border.

II-stage is dissolution of tested electro-active species from surface of working electrode with potential scan. Process of dissolution can be done by changing potential of working electrode or by applying controlled electric pulses. During this stage, solution is in phase of inaction. As response, electrical signal caused by dissolution process is registered. It can be either current or potential, and depending on the type of signal, we distinguish several types of stripping methods.

The electrical signal (response) obtainedin this way is directly proportional to the concentration of metal, and the position of the half-wave peak potential is identified by tested metal.

On the other side, by using pulse methods, it is possible to overcome the deficiency of voltammetry (polarography) with linear change of potential, like complex apparatus or a low detention border. Pulse techniques are intended to reinforce Faraday's current in relation to charging current, which results in increasing detection limit of ion species.

The excitation signal in differential pulse polarography represents linearly increasing potential pulse, on which is superimposed small square pulse potential with amplitude of 10 to 100mV.

As result, curve i - E is obtained, with the height of power peak directly proportional to the concentrations and potential is directly proportional to the half-wave potential.

On figure 2.5.1. voltamogram appearance in linear change of potential is given.

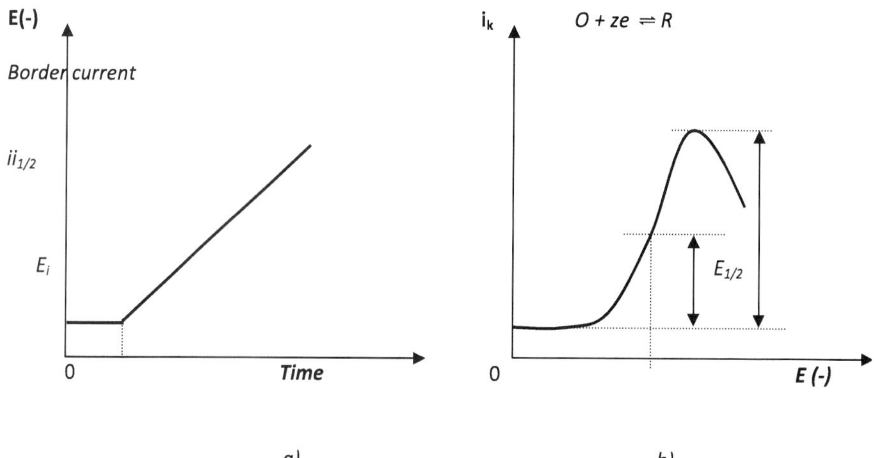

Fig. 2.5.1. (a) Excitation signal, (b) Response signal (voltamogram) in linear change of potential

Potential in maxim is given by equatation:

$$E_{\max.} = E_{1/2} - \frac{\Delta E}{2}$$

In this potential, current of diferential pulse voltamogram is :

$$(\Delta i)_{max.} = \frac{z \cdot F \cdot A \cdot D^{1/2} \cdot c_0}{p^{1/2} t_m^{1/2}} \bullet \frac{1-s}{1+s} \quad 2.5.1$$

where:

- s - function of amplitude of pulse potential;
- D - coefficient of difusion;
- A - surface of electrode.

Quantiative parameter is maximum height and can be changed by this method in order to improve sensitivity. With change of height current peak, sensitivity can be adjusted, which is important analitical mean.

The source of information in voltammetryc technique is always an electrochemical reaction. If reaction is reversible, the peak of potential is equal to standard potential of reaction. In the case of irreversible electrode reaction, position of maximum current peak is shifted toward negative values in relation to formal electrode potential of tested redox system.

Anode striping voltammetry is also known as *inverse voltammetry,* and given the high sensitivity, it is suitable for analysis of low metal concentrations in different matrices. Anode striping voltammetry is carried out through two stages:

The first stage is cathode deposition from solution to the surface of working electrode with appliance suitable negative potential [9]. If mercury electrode is used as working electrode and if metal in mercury is soluble, amalgam will arise. Concentration of metal ions in mercury film depends from concentration of Mn^{+} ions in the solution, time of electrolyze and speed of mixing.

$$M^0 \rightarrow M^{n+} + ne$$

In this phase, anodic current of dissociation in function of the amount or concentration of metal in tested sample is registered.

Catholic deposition is usually performed at potential which is 300 – 400 mV more negative than polarographic half-wave potential of tested metal. Maximum potential of deposition depends on the type of material of working electrode. Mercury electrodes have shown to be most suitable, since they allow simultaneous determination of a larger number of different metal ions in one tested sample.

Reproducibility of measurements depends from hydrodynamic conditions within the solution. Process of cathode deposition is carried out through inert conditions without presence of oxygen.

In «stripping» phase or in anodic dissolution phase, the response is obtained, i.e. curve i - E whose shape depends from used electrode and the speed of change of potential. If the speed of change in potential is small enough, then the height of the peak/spade signal is directly proportional to the speed of change of potential. Current in peaked wave is then given by proposition 2.5.2:

$$i_p = \frac{z^2 F^2 v \cdot l \cdot A \cdot c}{2{,}7 \cdot R \cdot T} \quad 2.5.2.$$

Where: F - Faraday's constant, v - speed of change of potential,l - thickness of mercury, A - surface of electrode, c- concentration of separated metal in mercury, z-charging.

While processing obtained voltamograms, graphical methods are used, where the height of current peak/spade $i_p$is directly proportional to the concentration of deposited metal.If the mercury- film electrode is used as working electrode, then the integration of current (surface) bellow the voltamogram is performed. The relation between height of maximum and concentration for tested metal is determined by method of standard supplement/addition or by method of standard sample.

The great variety of electrode reactions, electrode materials (including solids studies), electrode designs, and conditions of electrode polarization determines a wide field of stripping electroanalytical methods application. Essential advantages of stripping electroanalytical methods over other methods used to determine trace concentration of inorganic and organic substances in solutions and to analyze and study solids are:

- High selectivity, good accuracy and reproducibilty
- The posibility of determining a considerable number of chemical elements and many organic substances
- Low detection limits (10^{-10}M)
- The possibility od determining correlation of the „ composition - property for the solid substances and materials analyzed and of studying structural peculiarities of this entities
- Relative simplicity and low price

Beside Anodic Striping Voltammetry (ASV), Differential Pulse Anodic Stripping Voltammetry (DPASV), is the most applied technique nowadays for determination of different heavy metals chemical forms. With these techniques, it is possible to separate species based on their redox potential and determine labile or electro active complexes and free ions. DPASV falls into electro analytical methods which are based on direct dependence of measured electro chemical magnitude of tested system concentration.[10] By applying DPASV, it is possible to get three types of information regarding tested system:

- Regarding thermodynamic magnitudes
- Regarding setting up mechanism of electrode process and reactions which precede reduction
- Regarding qualitative and quantitative analytical parameters

It is important for pulse measurements that their sensitivity depends from usage of pure solvents and chemicals used as base electrolytes, and whose concentration is generally about 10 times lower than in the field of electrolytes, used in DC - polarography. Beside water, as solvents in differential - pulse analysis can bealso used non-aqueous solvents, with which it is possible to determine electro active species that are infusible in water. The selection by itself, of carrying electrolyte, influences to the selectivity of method. On figure 2.5.2 use of voltamogram polarographic techniques is given.

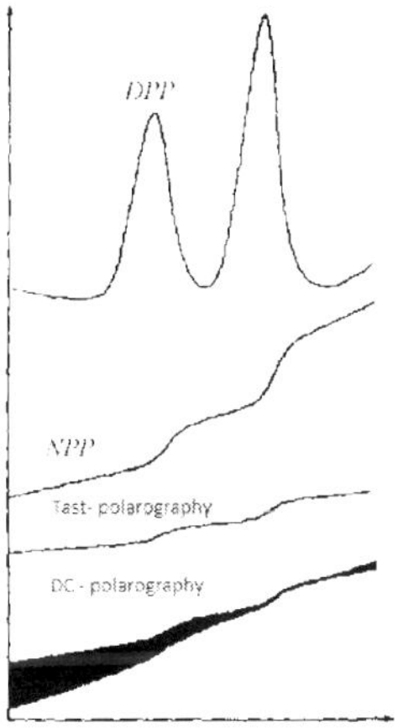

Fig. 2.5.2. Use of voltamogram polarographic techniques for Pb and Cd in 0,1 M HNO_3.

The intense sensitivity of differential pulse anode striping voltammetry is great advantage in analysis of natural waters. The possibilities of this technique lie in the fact that the amount of metal (M) which is being sedimented on electrode during level of deposition depends among other things on deposition rate (k_{-1}) complex metal species (ML_n) on intermediate stages of electrodes - solution:

$$ML_n \underset{k_1}{\overset{k_{-1}}{\leftrightarrow}} M + nL$$

The amount of metal M that is being sedimented depends also from parameters like: total metal concentration, time of deposition, surface of electrodes and temperature.

However, contribution of metal complexes in height of the ASV- peak does not depend from these parameters, but only from k_{-1} and thickness of diffuse layer. This fact has lead to the widespread use of term "ASV - labile metal" in analysis of water, i.e. concentration of metals that are determined with ASV in natural pH of water or in slightly acidic solutions (acetic puffer pH 4,7); which makes ASV and DPASV methods unique in monitoring of metal traces speciation.

The ability of DPASV method to measure labile metals, as well as great sensibility (around 10^{-10}M) for four most common toxic heavy metals: Cu, Cd, Pb and Zn , constitutes the powerful instrument in metal traces speciation.

It is evidently that in the near future, legislative for heavy metal in water will also include need for determination of speciation. Besides that, capacity of complexion can also be included into criteria of water quality.

There are various procedures by which stripping analysis can be used in order to obtain information about metal speciation using method of titration. When metal is used as titration means, than "capacity of complexion" for given metal is being determined. Titrimetric stripping procedures include direct determination of conditioned stability constants for metal complex - ligand. Direct stripping procedure can be set to measure total metal concentration or labile fraction.

Labile metal fraction represents part of total dissolved metal determined by stripping analysis - in well defined experimental conditions, in natural pH of water. This fraction also includes free hydrated metal ions, labile metal complexes.

Labile metal complexes represent bio - available toxic fraction. The advantage of measurements by striping labile metal fraction is in minimal pretreatment of sample.

Non - labile metal fraction also include metals that are tied in complexes or absorbed on colloid mater with speed of dissociation that is slower than stripping measurements time scale. This method is characterized by adequate «detection window» for monitoring and determination of speciation of metal traces, as well as constants of complex stability. The scope of detection window depends from interaction of metal - ligand in solution of sample and is determined by stability and adsorptivity of these surface-active metal complexes. Detection window depends from potential of disposition and detention time of inert complex molecules in diffusive layer during electrochemical deposition.

Opportunities and accuracy of DPASV - method, as well as reproducibility to the extent of concentration of metal traces of 10^{-10}M, is completely satisfactory for all three tested metals, as given in table 2.5.1..[11].

Cu	*Cd*	*Zn*
0,2 - 0,02ppb	0,2 - 0,02ppb	0,2 - 0,02ppb
±5-10%	±2,5%	±2,5%

Table 2.5.1. Scope of concentrations and accuracy

2.6 Chemical speciation

One of the most important voltammetry appliances in environment analysis is for speciation testing. Speciation analysis is defined as determination of concentration of different chemicals forms of elements that make up the total concentration in sample.

Since toxicity of every metal ion depends on physical - chemical form and that the most toxic form of metal ion is hydrated or free metal ion, it is important to do the chemical speciation.

Metal speciation, either in sea or in natural water is different because of:

- Different ionic strength
- Different concentration of metal ion
- Different concentrations of cations and anions
- Higher concentration of organic ligand in natural (clean) water
- Lower content of adsorbent in sea water

The least toxic forms are stable metal complexes and metal absorbed on colloid particles. Metal complexes with highly lipophylic organic ligands are also very toxic.

Concentration of electro-active oxidation state can be determined with measurements of total concentration of metal ion following conversion from electro-inactive to electro-active stage because it is possible to measure concentrations of metals only in ionic form, and not individual species like $PbCl^-$, $CuCl^-$ and others.

Labile species are those that can be detected with DPASV, like hydrated metal ion or metal ion that is dissociated from poorly tied inorganic complex or weakly absorbed on colloid particle. This is the first (I) - phase of speciation.

Chemical speciation of water in lake Modrac refers primarily to determination of chemical forms of heavy metals, copper, cadmium and zinc; whereby copper is separated because of the lake surroundings and high exposure, and its capacity of complexations determined.

In the first (I) phase, based on the obtained experimented results and with application of software programs MINTQ[61,62] and HIDRA chemical forms of tested metals are determined; taking into account pH of the lake, ionic strength and water temperature.

In second (II) phase, concentration of inert metal complex is determined. The content of inert complex is such that the particles does not dissociate under given conditions. That is why the water sample is subject to extended UV-radiation in duration of three hours in acidic environment that accelerates the dissociation from all types of ligands and colloids. The assumption is that complexed metal does not reduce directly during deposition.

During chemical speciation, it must be taken into account presence of organic substances that are naturally present in lake water, and which can lead to error in results of analysis and determination of total concentrations of heavy metals by DPASV. For this reason, pre-treatment of samples is necessary. For elimination of undesired effects surface - active substances and complexing ageneses, different methods can be used. One of these methods involves modification of media, namely the change of pH during deposition and stripping. There are three versions:

- Deposition and stripping at pH 7
- Deposition and stripping at pH 2
- Deposition at pH 7, and stripping at pH 2

In these experiments, when it comes to copper, deposition and stripping are performed at water pH 2, since some comparative studies have shown that change of media and pH leads to significant increase of concentrations of labile fraction. If it is known that metal ions which are incorporated into organic complexes are usually not toxic, than it is obvious why chemical speciation is important. The capacity of Cu complexion is determined by applying ASV titration of water samples with increasing concentrations of Cu^{2+} ions. End point of titration indicates the presence of free Cu ion in water. Based on obtained experimented data using method by Ruzic and equitation 2.6.1.capacity of complexion is calculated.

$$\frac{[M]}{(M_T - [M])} = \frac{[M]}{(C_M + \Delta_M - [M])} = \frac{([M] + 1/K)}{C_L} \qquad 2.6.1.$$

Where:

- C_M i C_L - concentrations of metal and ligand present in the system

- M_T - total metal concentration
- Δ_M - concentration of added metal
- [M] -concentration of non-complexed metal
- K - constant of stability of ML complex

Speciation of copper is interesting and important because of:

- Very wide application that inevitably makes it a component on water surface in concentrations that are significant for water ecosystem
- Known for its affinity toward natural complexing ageneses
- Organic Cu compounds have shown to be significantly less toxic than from free Cu ions or inorganic complexes [12].

Accuracy can be tested by using larger volume of deposition potentials and drawing curves $I = d(E_d)$. If ionic form of metal is only reduced, than current peak rise from 0 to the upper limit in small volume of potential; and if metal complex is directly reduced, than current peak continuously grows with potential of deposition. With these analyses it is recommended to use minimal deposition potential while measuring each metal, but sufficiently negative to give maximal anodic current, after which metals can be determined in individual experiments.

The advantage of chemical speciation using DPASV is that the labile fraction can well correlation with toxic fraction.

This method can also be combined with other methods of separation for the separation of labile and inert fraction, such as leakage through ion - exchange column.

3. Results

3.1 Method applied for analysis of results

For quantitative polarographic determinations in practice, usually three methods are used.[13]

- Method of calibration diagram
- Method of direct comparison
- Method of standard addition.

These methods allow calculating concentration of electro-active species in tested solution, based on the polarographic wave height. Most commonly applied method is method of standard addition since it is suitable for determination of small concentrations in small amounts of sample. In our researches, standard addition method is used.

Method of standard addition is often used for analysis of complex samples. It is based on successive addition of several portions of the standard solution of known concentration into tested sample, always in same amount. This process is known as sample "spiking". It measures the current response for tested sample before addition of standard solution and after each adding. Volume of V_s standard addition shall not significantly influence to the volume of sample V_x. The obtained current response id proportional to the concentration of tested sample, so that:

$$S = \frac{k \cdot V_s \cdot c_s}{V_t} + \frac{k \cdot V_x \cdot c_x}{V_t} \quad 3.1.$$

Where:

- c_x - concentration of tested sample
- k - constant of proportionality
- c_s - concentration of standard addition
- V_s - volume of standard addition
- V_x - volume of tested sample.

Drawing $S = f(V_s)$ equitation of direction is obtained:

$$S = mV_s + b \quad 3.2.$$

Where m is slope of direction, and b is section on y - axis so that:

$$m = \frac{k \cdot c_s}{V_t}; \text{ and } b = \frac{k \cdot V_x \cdot c_x}{V_t} \quad 3.3.$$

Rearranging, concentration of tested sample is obtained

$$c_x = \frac{b \cdot c_s}{m \cdot V_x} \quad 3.4.$$

The figure 3.1.gives a graphical display of standard addition method.

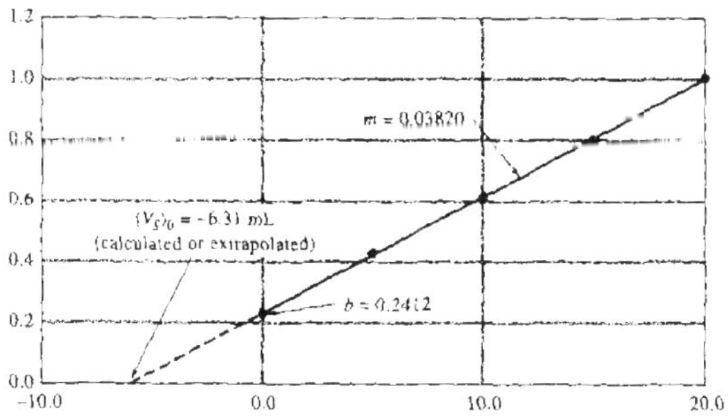

Fig. 3.1.1. Standard addition method

In the process of preparing and analysis of samples, used chemicals where of high purity level (Merck); HNO_3, $HClO_4$, HCl i KNO_3.

All samples were prepared in redistilled water, obtained by triple distilled process whereby second and third level was performed in rock ristaly aparatures.

Standard analyzed metal solutions (Cu, Cd and Zn) were prepared in "concentrations" from 10^{-4} to 10^{-5}mol/l from supra pure salts of nitrate metal ions.

- Copper - standard solution $Cu(NO_3)$ in HNO_3, c(Cu) = 1g/l
- Zinc - standard solution $Zn(NO_3)$, in HNO_3 c(Zn) = 1g/l
- Cadmium - standard solution $Cd(NO_3)$ in HNO_3 c(Cd) = 1g/l

Glassware and electro analytical cells were kept for 24 hours in 6 m HNO_3, after which they were flushed with distilled and redistilled water. As carrying electrolyte 0.1M KNO_3 was chosen which was prepared from salts of KNO_3 supra pure. All metals were determined individually.

The water samples of lake Modrac were taken from four characteristic places during four seasons, from the depth of 2,5 m.

- Place A - area around entrance into area around dam
- Place B - area on the middle of lake
- Place C - area around the mouth of river Tinja
- Place D - area around the mouth of river Spreca

Instrumental parameters for determination of concentrations of Cu, Cd and Zn in water were as follows:

- Working electrode HMDE
- Auxiliary electrode - platinum wire
- Reference electrode Ag/AgCl (E=0,222V)
- Cleaning time (nitrogen deaeration) 120 sec
- Sleep time 15 sec
- Scanning rate 2 mV/sec
- Time of deposition 300 sec

Half-wave potential for each metal is given in table 3.1.1.

Metal	Initial potential (mV)	Final potential (mV)
Cu	-0,45; -0,35	0; 0,05
Cd	-0,9	-0,6
Zn	-1,2	-0,9

Table 3.1.1. Potential of deposition

Results of water analysis samples for free/labile metal ion

In first phase are the concentrations of Cu, Zd and Cd and their labile chemical forms present in the water of lake Modrac. One part of water samples was acidified on the spot with concentrated HCl. In these samples, total concentrations of free metal ion or metal ion that was tied to in labile inorganic complex, were determined. The second part of samples was frozen to -20^0C and analyzed in second phase of experimental work in order to determine total concentrations of metals that were tied in stabile organic complex with ligand; which is naturally present in water lake Modrac, as well as capacity of complexation of Cu. Applying standard addition method in Mathcad program, concentrations of heavy metal traces (Cu,Cd and Zn) in tested samples were determined. On figure 3.2., 3.3.and 3.4. are given voltamograms obtained for Cu,Cd and Zn applying standard addition method.

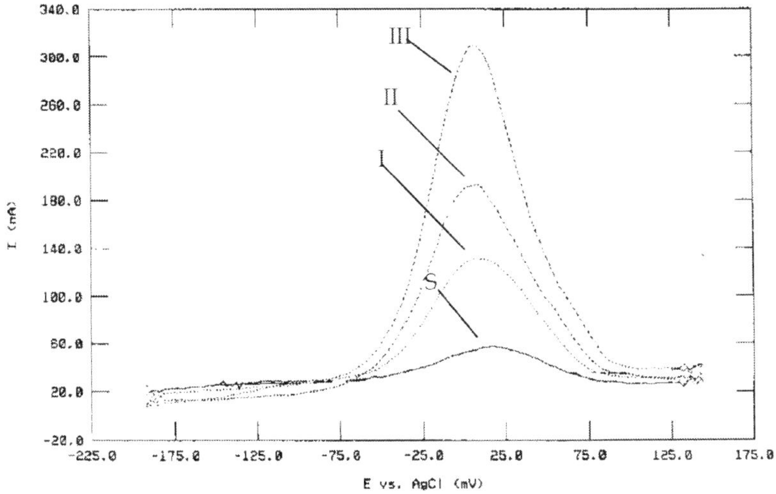

Fig. 3.2. Copper in water of lake Modrac - standard addition method (I - first addition, II- secound addition, III - third addition, S - sample)

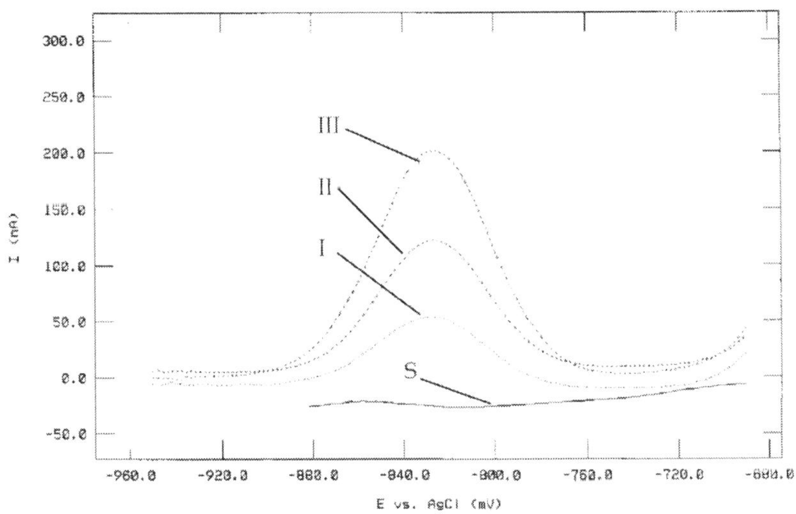

Fig. 3.3. Cadmium in water of lake Modrac - standard addition method(I - first addition, II- secound addition, III - third addition, S - sample)

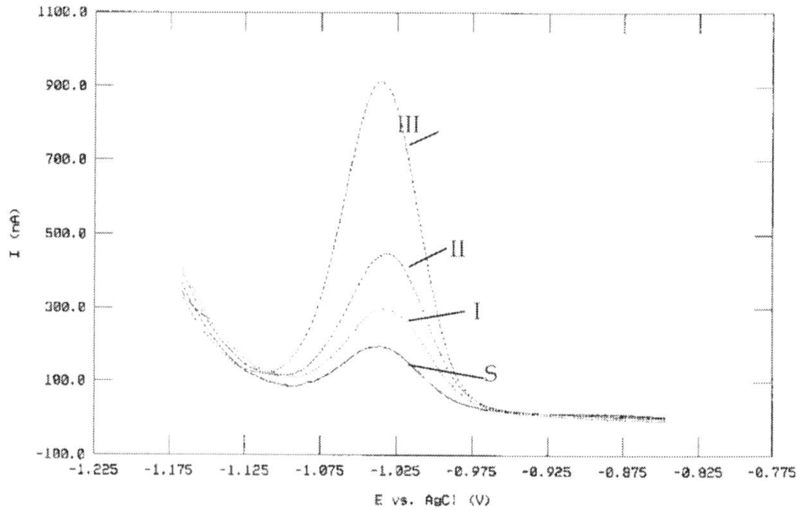

Fig. 3.4. Zinc in water of lake Modrac - standard addition methods(I - first addition, II- secound addition, III - third addition, S - sample)

On figure - diagrams 3.5., 3.6., and 3.7., are given average calculated concentrations of Cu, Cd and Zn during four seasons in Lake Modrac.

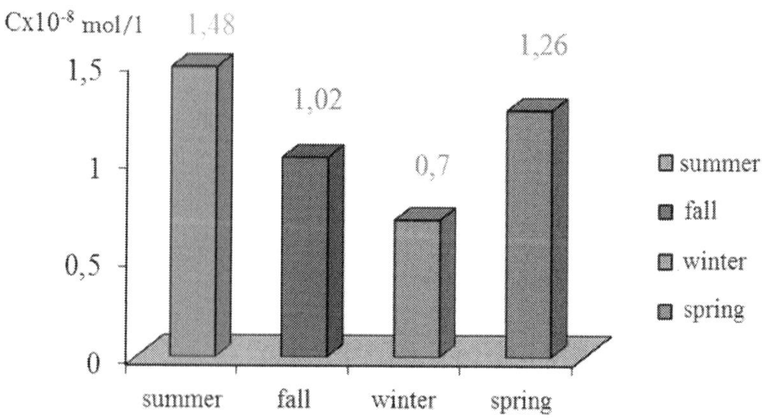

Fig. 3.5. Average amounts of copper during four seasons in Lake Modrac

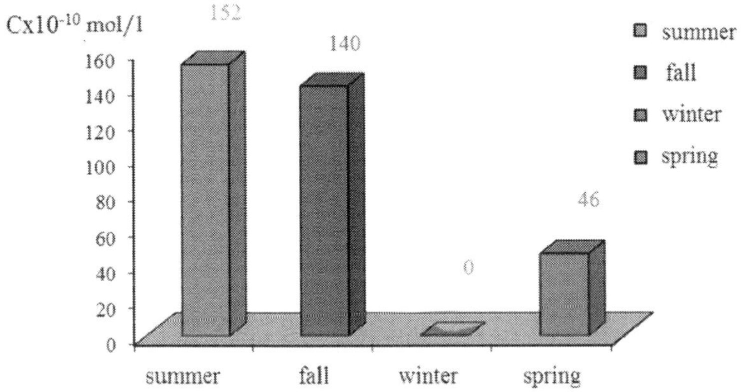

Fig. 3.6. Average amounts of cadmium during four season in Lake Modrac 10^{-10}

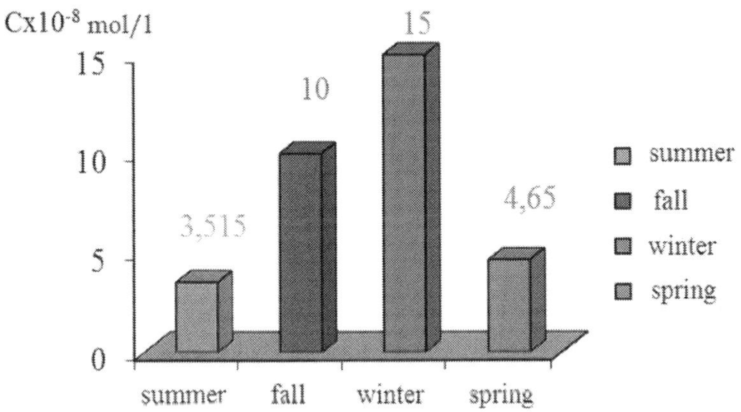

Fig. 3.7. Average amounts of zinc during four season in Lake Modrac

Metal	Season			
	Summer	Autumn	Winter	Spring
Cu	94•10-5 (mg/l)	64,81•10-5 (mg/l)	44,48•10-5 (mg/l)	80,06•10-5 (mg/l)
Cd	1,78•10-3 (mg/l)	1,573•10-3 (mg/l)	<10-6 (mg/l)	5,17•10-4 (mg/l)
Zn	2,29•10-3 (mg/l)	6,54•10-3 (mg/l)	9,811•10-3 (mg/l)	3,041•10-3 (mg/l)

Table 3.2. Summary of calculated average concentrations of Cu, Cd and Zn

In table 3.3. Percentage representation of certain metal complexes of copper obtained by application of software programs HIDRA and MINTQ.

Chemical form	Season			
	Summer %	Autumn %	Winter %	Spring %
Cu^{2+}	13,4	75	64	12
CuII(OH^{+})	17,9	18,5	16	30
CuII$(OH)_{aq}$	-	5,4	14	5
Cu II CO_3	37,12	-	14	-
$CuSO_4$	-	5,4	4	-

Table 3.3. Representation of chemical forms of Cu during four season expressed in percentages.

The table shows that chemical form of copper in summer and spring time varies more in relation to autumn, and especially to the winter. In fall and winter time, copper is mainly found in ionic form, which means in the most toxic form. This is explained with the higher pH value in this two seasons (7,5 - 8), since with higher pH values copper is easily dissociated and from label inorganic complexes goes into ionic form.

In table 3.4.balance constant of found copper complexes, calculated with software program CHEAQES are given.

Unorganic complexes	Real constante of balance (K)
CuII$(OH)^{+}$	6,5
CuII$(OH)_{2aq}$	11,8
CuII$(SO_4)_{aq}$	2,36
CuII$(CO_3)_{aq}$	6,7
CuIIHCO_3)	12,12

Table 3.4.

Cadmium is not found in water samples in winter period, not in any location. More accurate, those concentration were above 10^{-12}mol/l. Inorganic speciation has shown that cadmium is found same as zinc, mostly as free metal ion, even 90% and on all locations and during all four seasons.

Test results of some other authors have shown that cadmium requires higher pH value than copper; in order to occur adsorption to organic or inorganic particles [14]. This is probably reason why cadmium in lake Modrac is usually found in most toxic form. In table 3.5, representation of Cd chemicals forms during four seasons by percentage, obtained by application of software programs HIDRA and MINTQ, are given.

In the table 3.7 are given results obtained for Zinc. During all four season and on all places, zinc is found almost completely dissociated, i.e. in ionic form and one smaller portion in form of zinc-sulfate whose determined equilibrium constant (K=2,3) indicates that it is very unstable complex from which zinc easy dissociate and transforms into ionic form. This goes with results obtained by Figure and McDuffie by which Zn in natural waters is found mostly in very labile form. By Florenc, zinc is also found as free ion in fresh water for about

50%, while in the sea water is usually found in form of chlorine-complexes for about 47% and 27% as free ion.

Test for inorganic chemical speciation of zinc are controversial and with very different results which can be explained by the fact that zinc is component that is present in variety of materials that are being used in lab work, human skin, so that contamination of samples is much bigger problem than in other heavy metals.

Chemical form	Season			
	Summer %	Autumn %	Winter %	Spring %
Cd^{2+}	85	97	-	91
$CdCl^-$	9,6	-	-	1,4
$Cd(SO_4)_{aq}$	7,0	6,0	-	7,5

Table 3.5. Representation of Cd chemical forms during four seasons expressed in percentages.

Inorganic complexes	Real constant of balance (K)
$Cd(SO_4)_{aq}$	2,37
$CdCl^+$	1,98

Table 3.6. Equilibrium constants of found cadmium complexes, calculated using the software program CHEAQES.

Chemical form	Season			
	Summer %	Autumn %	Winter %	Spring %
Zn^{2+}	90	92,7	93	92
$ZnHCO_3^-$	-	-	-	-
$Zn(SO_4)_{aq}$	7,5	6,4	5,1	6,9

Table 3.7. Representation of Zn chemical forms during all four seasons expressed in percentages, obtained by application of software programs HIDRA i MINTQ, are given.

Determination of complex capacity of copper in lake water Modrac

In second (II) phase complex capacity of copper was determined, as one of the possible parameters of natural water quality.

In second phase of speciation, water samples were first defrosted to room temperature, filtrated through sterile membrane filter of 0,45 μm pore diameter. Filtration was carried out in order to remove possible presence of colloid particles. After filtration samples were acidified to pH 2 with s.p. HCL and ventilated with UV radiation intensity 100 W during three hours. Radiation was performed in order to release metal ions tied into stabile organic complexes. Previously acidification was performed to achieve a faster release of metal ions from these complexes. Figure 3.8.representsvoltamogram for copper before and after radiation.

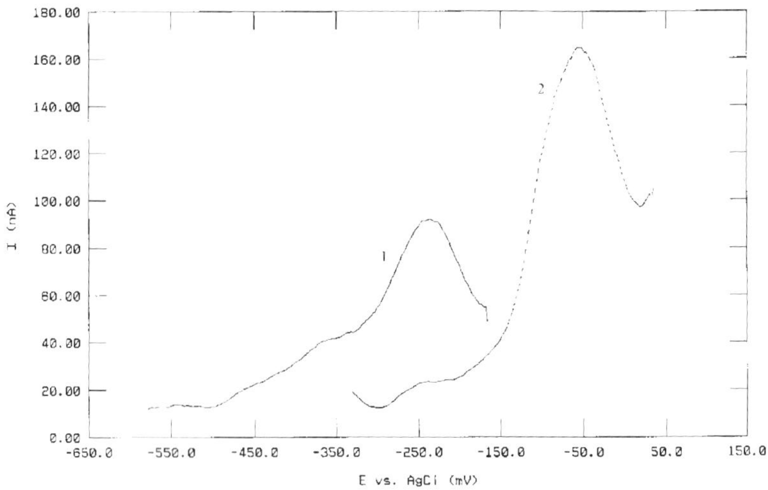

Fig. 3.8. Voltamogram of copper before (1) and after (2) radiation

The results obtained after radiation have shown that there was an increase in copper concentrations in all season, indicating a higher presence of copper, but in the form of stable organic complexes. Movements of half-wave potentials toward positive values are caused by change of pH values.

Results for copper with EDTA

In order to determine the complex capacity of copper, e.i. concentration of organic ligand naturally present in water that can tie copper into stabile complexes, first were done test with known ligand, EDTA, which is by characteristic very similar to natural ligand. This was performed only for data reproducibility and greater accuracy.

Primary the calibration diagram was documented for known concentrations of copper and to the extent in which the concentrations of this metal ion was found in lake Modrac (1,6; 5; 8; 10; 14; 180 and 22)-10^{-6}mol/l. Based on calibration diagram and from obtained equitation of direction, concentration of non-complex copper ion of copper was calculated.

Capacity of complexation and constant of stability are calculated based on Ruzic method. On figure 3.2.1.obtained calibration diagram for copper is shown.

To determine the capacity of complexation of copper with EDTA in 6 copper samples of 10 ml, of above stated concentrations, EDTA concentration of 2,2·10-6 mol/l was added. The samples were left in the sterile plastic bottles overnight, after which concentrations of non-complexing copper with EDTA was recorded, based on the differences in height of current peak for copper before and after EDTA adding. On figure 3.2.2.are given results obtained by linearization of data for height of current peaks for copper and EDTA. The ratio between concentrations of non-complex and complex copper in function of concentration non-complexing copper is given. From the slope of directions, capacity of CuEDTA complexion was determined, and stability constant of resulting complex from section.

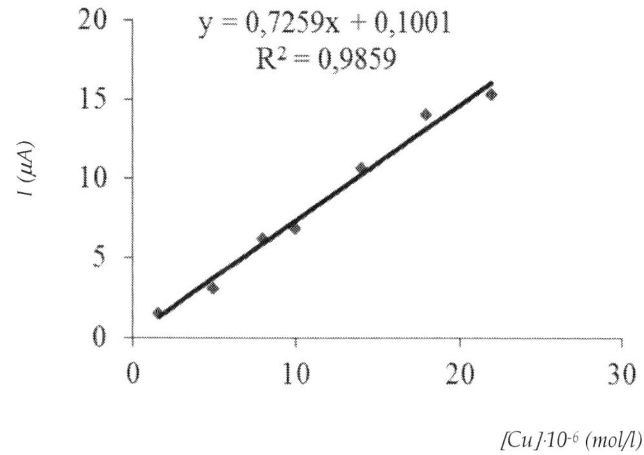

Fig. 3.2.1. Calibration diagram for Cu

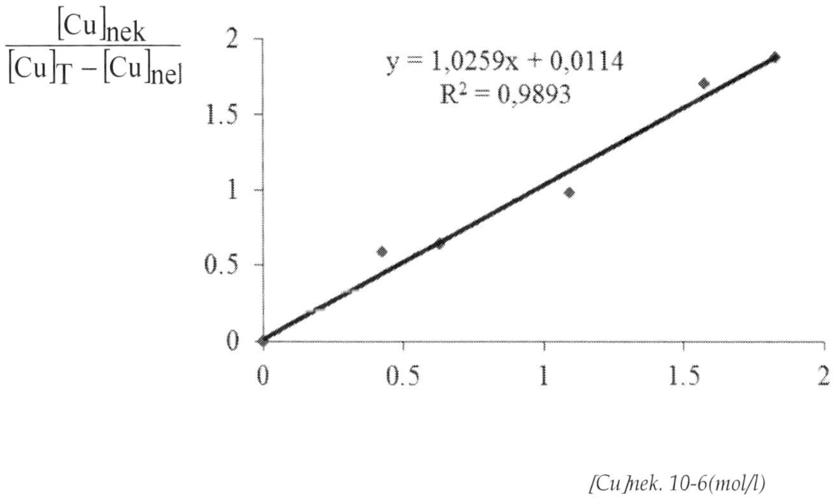

Fig. 3.2.2. Mathematical data processing for copper + EDTA

$$1/C_L = 1{,}0259 \cdot 10^{-6} mol/l \ ; \ C_L = 1/1{,}0259 = 0{,}974 \cdot 10^{-6} mol/l; \ 1/K \cdot C_L = 0{,}0114;$$

$$K = 1/0{,}0114 \cdot 0{,}974 \cdot 10^{-6} = 1/0{,}0111 \cdot 10^{-6} \ ; \ K = 9{,}009 \cdot 10^{7}; \ \log K = 7{,}95$$

Obtained values capacity of complexation and stability constant explain good conformity with literature data.

Results for capacity of complexation of copper in water of lake Modrac

Due to results reproducibility, capacity of complexation for copper in water of lake Modrac was not performed at natural pH value, rather the samples were acidified with 0.1 $MHNO_3$ right before recording so that the pH was 2,3.

First, the calibration diagram was recorded for concentrations that match the concentration of copper in lake water. Following copper concentrations was used (0.1; 0,5; 0,7; 1; 1,5; 2,5; 3; 5,5; 8,5;)·10^{-8}mol/l.

On figure 3.2.3.calibration diagram for copper is given.

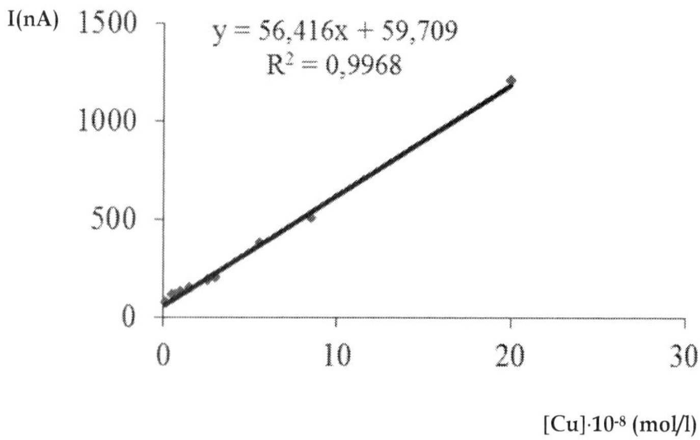

Fig. 3.2.3.Calibration diagram for copper

For analysis, 100 ml of untreated lake water is taken and divided into 10 sterile bottles of 10 ml. into samples increasing concentrations of copper was added; the same concentration by which calibration diagram was constructed. Samples were left overnight to make the maximum absorption of copper into water, after which concentrations of non-complexing copper with natural ligand from water were recorded. From calibration diagram, Fig. 4.3.5. concentrations of non-complexing copper were calculated. From the difference of total concentration of copper $[Cu]_T$ and concentration of non-complexing copper $[Cu]_{nek,}$ concentration of complexed copper $[Cu]_{kom}$ was calculated.

$$[Cu]_T = C_{Cu} + \Delta Cu \quad 3.2.$$

C_{Pb} - concentration of copper found in lake; ΔCu – added concentration of copper.

Linearization of data, from the relationship between concentrations of non-complexing and complexing copper in function of non-complexing copper, the complex capacity in water of lake Modrac from the slope of direction was calculated; and stability constant of resulting complex was calculated from the section. On Figure 3.2.4 diagram $[Cu]_{nek.} / [Cu]_T - [Cu]_{nek.} = f([Cu]_{nek.})$ is given.

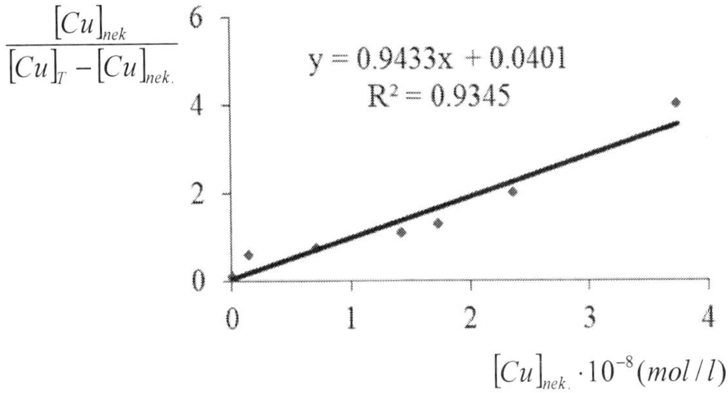

Fig. 3.2.4. Data obtained using mathematical processing of data for complex capacity of copper into water of lake Modrac

From equitation $y = 0,9433 \cdot x + 0,0401; (R^2 = 0,9345)$ capacity of complexation and stability constant for copper were calculated:

$$1/C_L = 0,9433;\ C_L = 1/0,9433 = 1,06 \cdot 10^{-8} mol/l.$$

$$1/K \cdot C_L = 0,0401;\ K = 1/0,0401 \cdot 1,06 \cdot 10^{-8} = 2,5 \cdot 10^{9}$$

$$C_L = 1,06 \cdot 10^{-8} mol/l;\ K = 2,5 \cdot 10^{9},\ \log K = 9,4$$

4. Conclusion

Results of copper testing have shown that this metal in lake water is found mostly in « medium» labile and « poor» labile inorganic form. Copper is usually found in inorganic complexes and smaller percentage in form of free metal ion. The results indicated that dominant chemical form varies depending on season and locality. So that, during summer and spring for which is characteristic neutral pH value, copper is usually found at labile hydroxide complex about 30% and as carbonate about 50%. On the other side, during autumn and winter copper is usually found in the form of free metal ion. Similar results were obtained by Florence.[15] for sea and fresh water, where copper is found about 50% as $CuCO_3$ and 40% as Cu(OH).

The dominance of free Cu ion in fall can be explained with chemical composition of water, which is characterized by total absence of carbonates. Calculated equilibrium constants using CHEAQS program also indicates that it is medium labile complexes, especially when it comes to hydroxide and carbonates (K = 6,5 and 6,7).

On the diagrams 3.5 to 3.8 is given copper distribution and dominant labile inorganic forms in function of pH value, determined using software programs HIDRA and MINTQ.

The diagram shows that with pH increase, portion of copper free metal ion decreases and adsorption on inorganic molecules occurs, so that during the winter and spring there is more in form of inorganic complexes while during autumn in the form of free ion. Chemical form of copper mostly varies during summer period from place to place. High percentage of free metal ion on place B, which represents middle area of the lake, can partly be explained with water currents which are expressed on minimum level, as well as lower concentrations of dissolved oxygen and decreased quantities of organic and inorganic matters as result of hydrologic characteristics. [16]

The determined Cu constant of stability and calculated capacity of complexation addresses that it is very stabile/inert Cu organic complex, and that significant amount of organic active substance is present in the water of Lake Modrac.[16]

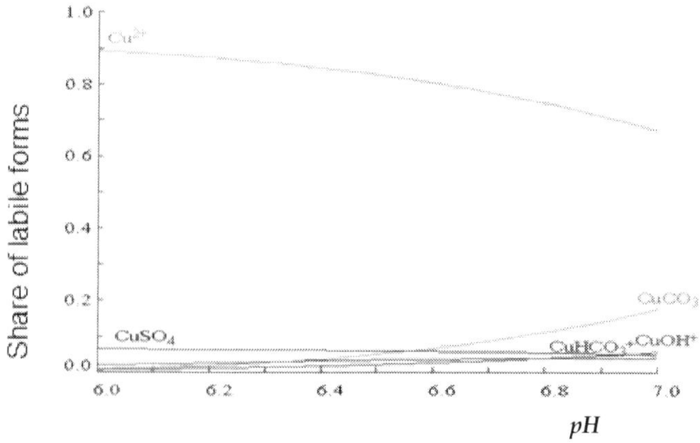

Fig. 3.5. Dominant labile copper forms in summer period

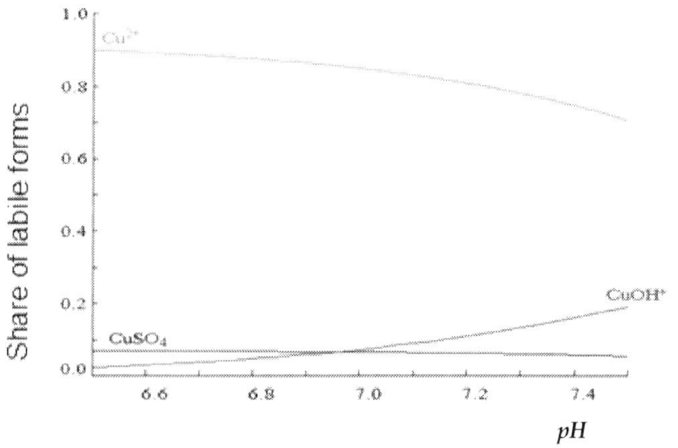

Fig. 3.6. Dominant labile copper forms in autumn period

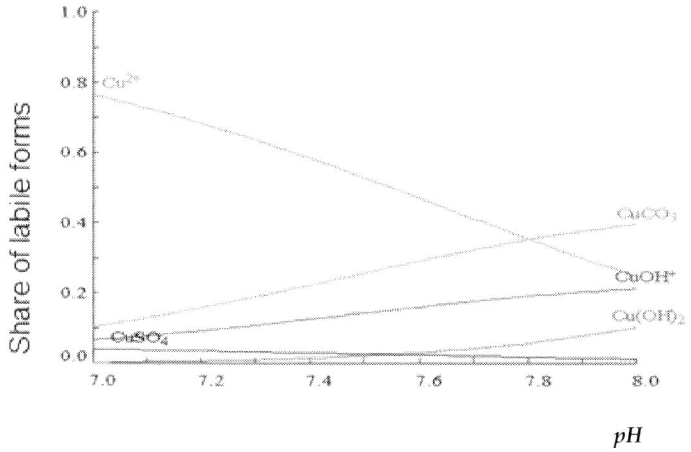

Fig. 3.7. Dominant labile copper forms in winter period

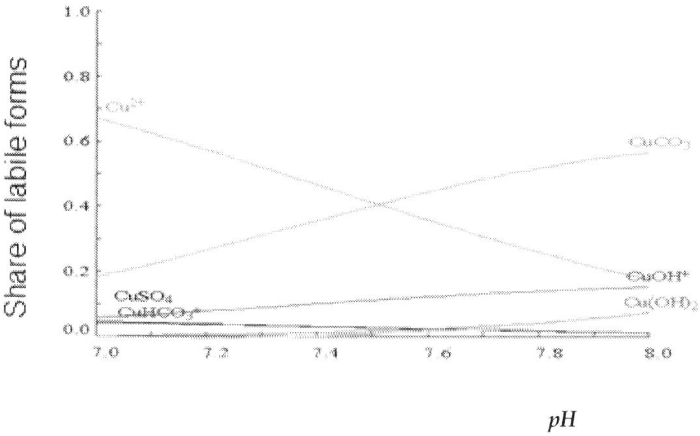

Fig. 3.8. Dominant labile copper forms in spring period

Performed testing have shown that DPASV method that was used as basic polarographic technique, with varying conditions and electrochemical parameters by type of determination, proved to be extremely accurate and sensitive and definitely economically most suitable.

Application of modeled polarographic DPASV technique offers possibility to determine speciation and capacity of complexation of natural waters. Moreover, it allows determination of total concentrations of tested metals in natural waters, sediments and aquatic ecosystem, by which this method, as optimal method, is recommended by all research cycles for this kind of testing.

5. References

[1] E.Kahrović „Anorganska hemija", Univerzitetska knjiga,Sarajevo, 2005

[2] R.M.Sterrit, J.N.Lester " Journal of Science Total Environment, 14 (1985)

[3] J. Agbaba, S. Ugarčina, J. Trifković "Toksične supstance i njihovi efekti, Prirodno-matematičkifakultet Novi Sad,2008

[4] S.E.Manahan : Environmental Chemistry, (Sixth Edition, Lewis Pub. London,1994)

[5] J.Đuković, B.Đukić, D.Lazić, M.Marsenić: Tehnology of water, (Beograd 2000)

[6] W.B.Mills et al.: Part I (Revise 1995), USA EPA Office of Research and Development, Washington DC.20460,EPA

[7] T.M.Florence:Talanta,29,(1982), pp. 345-364

[8] Kh.Brainina,E.Neyman:Electroanalytical Stripping Methods, JohnWiley&Sons,(1993)

[9] I.Piljac: Elektranalitičke metode, Zagreb(2005)

[10] S.P.Kounaves: VoltammetricTechniques,Handbook of Instumental Techniques for Analytical of Chemistry,TuftsUniversity,Department of Chemistry,Ch .37, (2000)

[11] B.Raspor,H.W.Nurnberg,P.Valenta,M.Branica:MarineChemistry,15 (1984),317-327.

[12] M.Gardner,E.Dixon,S.Comber: Chemical Speciation and Bioavailability,1 (2000),121-129.

[13] D. A. Skoog, D. M. West, F. J. Holler: Fundamentals of Analytical Chemistry, Saunders College Publishing, Philadelphia, 1996

[14] I. Šanko, J. Doležal: Electrochemical Chemistry and interfacial electrochemistry, 25, (1970), 299-306.

[15] M.G.C.Campbell: Speciation and Bioavailability Aquatic System,John Wiley&Sons (1995),pp.45-102.

[16] A.Odobašić: Speciation of heavy metals with polarographic techniques in natural water, Doctoral thesis, University Tuzla,2005.

20

Evaluation of Drinking Water Quality in Three Municipalities of Romania: The Influence of Municipal and Customer's Distribution Systems Concerning Trace Metals

Gabriela Vasile[1], Liliana Cruceru[1], Cristina Dinu[1],
Epsica Chiru[2], Daniela Gheorghe[3] and Aurel Ciupe[4]
[1]*National Research and Development Institute for Industrial Ecology ECOIND, Bucharest*
[2]*Apa Nova Company, Bucharest*
[3]*Aquatim Company, Timisoara, Timis County*
[4]*Aquaserv Company, Targu Mures, Mures County*
Romania

1. Introduction

The purpose of every water utility is to provide consumers with drinking water that presents no risk to public health. Safe drinking water is generally obtained by complying with specific water quality standards such as European Union Drinking Water Directive (Council Directive 98/83/EC, 1998).

Corrosion can affect public health, public acceptance of a water supply and the cost of providing safe drinking water. The release of corrosion by-products of household plumbing systems can be a significant source of trace metals found in tap waters (Agatenos & Okato, 2008). Pollution with metals may originate mainly from old and poor quality distribution networks and piping systems.

When drinking water is distributed through pipelines, biofilms will grow on the inner surface of the pipes and soft deposits (organic and inorganic matter) and several metals will accumulate to the pipelines (Lehtola & al, 2004a). Discoloration of drinking water is one of the main reasons customers complain to their water company. An elevated concentration of iron or increased turbidity, affect taste, odor and color in drinking water. Unlined iron pipes in drinking water distribution networks develop extensive internal corrosion scales as the time of use increases. These corrosion scale deposits reduce the hydraulic capacity of the pipes and more energy is required to deliver water at a desire flow rate (Sarin et al., 2004).

The heaviest corrosion was observed mainly in steel and cast iron pipes (Nawrocki et al., 2010). Corrosion of cast iron and iron products is ofen seen as the main source for discolouration of drinking water, iron higher content and may can also induce a chemical decay of the residual chlorine (Lehtola et al., 2004b; Sarin et al., 2004).

High contents of Cu can be released in drinking water as an effect of using the copper or copper alloys pipes for cold-water domestic installation. In copper pipes, the concentration of copper in water incresed with increasing stagnation time (Lytle & Schrek, 2000; Merkel et al., 2002). Merkel et al found that during stagnation, the concentration of copper reached its maximum after 10 hours, and then it started to decline. Water stagnation for more than 4 hours significantly increased the copper concentration in water (Lehtola et al., 2007; Lytle & Nadagouda, 2010). In Romanian legislation, the maximum admissible value for Cu in drinking water (100 µg/L) is lower than the limit imposed by EU Drinking Water Directive (2,000 µg/L).

The major source of lead in drinking water was identified to be plumbing materials. Lead pipes, lead - based solder, brass fittings and plumbing fixtures such as pipe's jointing faucets are known to be dominant lead sources in public water supply systems (Kimbrough, 2001; Lasheen et al., 2008).

Lead pipes were replaced with other types of pipes such as polymer materials like polyvinyl chloride (PVC), polyethylene (PE) and polypropylene (PP). Plastic pipes currently make up about 54% of all pipes installed worldwide. PVC makes up 62% of this demand and PE in its various forms about 33.5% (Raynand, 2004).

PVC polymer is mixed with a number of additives including stabilizers in order to provide the range of properties needed in the final products. Stabilizers are often composed of salts of metals like lead and cadmium (Kim, 2001). Unlike PVC, other plastics including PE and PP do not require metallic heat stabilizers. In addition, galvanized iron pipe can release significant amounts of lead into drinking water, as the zinc coating contains about 1% lead impurities (AWWA, 1996).

Nickel can be leached from the internal surface of some components of domestic installations such as: Cr/Ni plated devices, alloys containing nickel (Gari & Kozicek, 2008).

In the process for drinking water production, usually, the raw water is treated with aluminium sulphate as coagulation agent, for this reason is important to control aluminium. The role of Aluminium in Alzheimer illness is not well known but is certainty known that this metal has a toxic effect on the nervous system (Flaten, 2001). Aluminum has been shown to play a causal role in dialysis encephalopathy (Alfrey et al., 1976) and epidemiological studies suggest a possible link between exposure to this metal and a higher prevalence of AD (Flaten, 2001; Becaria et al., 2006). This association is dependent on the duration of Al exposure and only becomes significant if an individual has resided in an area with high Al in drinking water (>100 µg/L) for several years (Becaria et al, 2006).

Drinking water produced in water plants almost invariably fulfils the water quality requirements set in European Union Drinking Water Directive (Council Directive 98/83/EC, 1998). However, the DWD requires that water quality should also meet the requirements at the consumer's tap.

In some European countries such as Romania or Germany, the water distributors must ensure that microbial and chemically clean water reaches water meters. After that, the owner of the building is responsible for the water quality. Up to the water meter, the drinking water quality is very good, but the drinking water collected from the customer tap could have a lower microbial and chemical quality than the water produced and distributed. (Volker at al., 2010).

In most European countries, drinking water quality is not monitored routinely at household level but rather directly in the distribution system, as waterworks and authorities have limited access to private homes, as well as limited control over household plumbing and operation.

Significant levels of trace metals could be detected after stagnation of the water in distribution system, especially during night - time (Haider et al., 2002; Vasile et al., 2009, 2011; Zietz et al., 2003, 2007). All these studies reported increased concentration of lead, cadmium, copper, iron and nickel after stagnation in household tap water in Austria, Germany and Romania.

2. Aim

Tap water from the municipal supply system is the source of drinking water for majority of consumers in Romania. The main sources of drinking water in Romania include rivers (about 60%), drillings and much less, lakes.

According to Romanian legislation, the last segments checked by the Water Companies are branch pipe and water meter. Less than 0.1% of domestic network of customer is included in monitoring plan of drinking water; usually, the tap water is controlled only at customer request or complaint.

The aim of the study was to identify issues that may affect public health and the risk prevalence of relevant metals in in-building installation systems, in three important municipalities from Romania. Thus, it is possible to take measures to increase security of water systems, replacement of the pipelines at risk of metal corrosion, improve drinking water quality and protect human health against adverse effects caused by contamination of drinking water.

In order to get an overview of the overall current contamination levels of drinking water at the point of consumption were collected and analysed more than 600 samples from cold line-pipe with three different sampling procedures (first draw, fully flush and randome daytime).

A monitoring plan was developed and the content of metals (Al, As, Cd, Cr, Cu, Fe, Mn, Ni, Pb, Se, Sb and Zn) was analyzed from drinking water samples collected directly from consumers and from the municipal distribution network.

3. Methodology of the study

3.1 Cases under study – Municipalities of Bucharest, Timisoara and Tg. Mures

A large distribution system delivering drinking water to about 1.7 million citizens, supplied by three water treatment plants and two different surface water sources was investigated. Bucharest, Romania's capital, is the largest city, the industrial and commercial center of the country, the sixth city of UE population. The Capital is located in south - east part of Romania, in Muntenia Region, in the Romanian Plain. Bucharest has the same administrative level as a county and is divided into six districts.

Three Drinking Water Plants, located outside the city perimeter, supply the city. Arges River, raw water source for Crivina and Rosu Water Plants, is affected by anthropogenic pollution; the river collects wastewater discharges from industrial areas located downstream near Curtea de Arges and Pitesti cities. In case of accidental pollution of both rivers uses as raw water (Arges and Dambovita Rivers), the operator, APA NOVA Bucharest, has in place necessary means to eliminate the pollutants.

Second city selected was Timisoara, Timis County capital, situated in West Part of Romania, in historical region of Banat, the third largest city in Romania. The most important raw water source in collected from Bega River, which is clogged and the water is cloudy for most of the time. The water and sewerage operator, which provide drinking water in Timisoara city is AQUATIM Company.

Targu Mures is the sixteenth largest city in Romania and the sixth in Transylvania, located in central - northern Romania and represent Mures County municipality residence. As only source of raw water is used Mures River. The river collects discharges from several locations upstream of Tg. Mures, most important being Reghin and Toplita cities.

In Tg. Mures municipality, AQUASERV Company is the producer and distributor of drinking water.

In table 1 are presented characteristics of the utilities under study, such as population served, source of raw water, supplied flow rate, water plants, and most important, treatment processes.

	BUCHAREST			TIMISOARA			TARGU MURES
Population served	1725,000 inhabitants (December 2009), 81% from Bucharest population			330,000 inhabitants (December 2010), 99% from Timisoara population			143,000 inhabitants (December 2010), 95,25% from Tg. Mures population
Source of raw water	*Arges River*		Dambovita River	Bega River	Groundwater (63 drilling points)		Mures River
Water Plants	*Crivina*	*Rosu*	Arcuda	Bega	Urseni	Ronat	Tg. Mures
Supplied flow rate (m^3/day)	259,200	400,000	610,000	69,676	26,131	1,311	46,000
Treatment processes	*pre-ozonation*→coagulation (aluminum sulphates) → flocculation → decantation → inter-ozonation**→ sand filtration → disinfection with chlorine gas* *, ** only in Crivina Water Plant			*Coagulation (aluminum sulphates, sodium aluminates, aluminum polychloride) → flocculation → decantation → sand filtration → disinfection with chlorine gas*			*Pre-oxidation with $KMnO_4$ → pre-decantation → coagulation (aluminum poly hydroxyl chloride) → flocculation → decantation → sand filtration → ozone treatment → adsorption on granular active carbon → disinfection with chlorine gas*

Table 1. General characteristics of the utilities under study

3.2 Sampling

3.2.1 Sampling techniques

In order to obtain a large database, the samples were collected from customer's cold line-pipe with three different sampling techniques: first draw sampling (from kitchen); fully flushed sampling procedure after flushing five minutes same tap; random daytime procedure (within office hour, without previous flushing of the tap).

The definition of a *first draw sample* is a sample that is taken first in the morning before the tap in the premise has been used for other purposes. During the stagnation period no water should be drawn from any outlet within the property (this includes flushing of toilets). If any water is drawn during the stagnation period the result will be invalid. It is common practice for such samples to be taken by consumers. There is no control over the quality of the samples. When the sample is taken, the tap should be fully opened or as open as possible without losing sample (WHO & COST Action 637, 2009; Karavoltos et al., 2008).

The definition of a *fully flushed sample* is a sample that is taken after prolonged flushing of the tap in a premise in such way that stagnation of water in the domestic distribution system does not influence the quality of tap water. In practice a sample is taken after flushing at least three plumbing volumes. In case that the temperature of the water from the distribution network is cooler than the ambient temperature, an alternative method is monitoring the temperature of the water during flushing until it stabilises (WHO & COST Action 637, 2009; Karavoltos et al., 2008).

The definition of a *random daytime sample* is a sample that is taken at a random time of a working day directly from the tap in a property without previous flushing. When the sample is taken the tap should be fully opened or as open as possible without losing sample (WHO & COST Action 637, 2009; Karavoltos et al., 2008).

Each type of sampling technique provides different informations about influence of materials used in both distribution systems (municipal and domestic) to the tap water quality. Thus, the results obtained with fully flushed procedure indicates the influence of municipal distribution system to the tap water supplied by the Operators.

The first draw results shows in principal the influence of materials used in domestic distribution system.

In Europe, random daytime (RDT) sampling (1st liter taken during office hours, without fixed stagnation) and sampling after 30 minutes of stagnation (30 Ms) (1st and 2nd liter) were identified as the best approaches for estimating exposure and detecting homes with elevated lead concentration in tap water (Deshommes, 2010; Hayes, 2009; Hayes et al., 2010).

3.2.2 Sampling points

In order to control drinking water quality, in the study were collected samples from producer via distribution system to consumer's tap, from different points such as:

- Water Plants - drinking water produced by the Water and Sewerage Operators;
- monitoring points of the Operators, situated in different locations in the municipalities, such as elementary and high schools, kindergartens, markets, fountains, public institutions, pumping stations (fully flushed samples);

- selected customers of the Operators with the residence in different districts of the municipalities (first draw and fully flushed samples);
- medical centres, pharmacies, schools, private companies, public institutions, food markets, fast foods, restaurants situated in old buildings from the center of the cities – (random daytime samples).

In the study were collected and analyzed also surface and groundwater used as raw water by the Operators.

The drinking water samples from Water Plants and monitoring points of the Operators were collected by the specialists of the NR&DI ECOIND and the Operators, in a common programme. The customer's tap water were collected after an adequate instruction by the customers, which were workers from NR&DI ECOIND in Bucharest and Timisoara, AQUATIM (Timisoara) and AQUASERV (Targu Mures) Companies.

The random daytime points were selected and drinking waters were collected by the scientists from NR&DI ECOIND.

In Timisoara and Tg. Mures were collected samples in two campaigns, one in winter season, the other one in summer time. In addition, in summer, from old, historical buildings situated in the center of the cities, were collected drinking water samples with Random Daytime procedure, in order to control lead concentration in tap water.

The technique used for sampling of drinking water from Operator's monitoring points was fully flushed procedure.

In table 2 are presented data regarding number of samples collected in the study for each municipality and also the date when the samples were collected.

Municipality	Samples									
	Water Plant		Operator Monitoring points		Customer points					
					First Draw		Fully Flush		Random Daytime	
	No	Date	No	Date	No	Date	No	Date	No	Date
Bucharest	3	22.06.09	23	23.06.09	71	22-24.06.09	71	22-24.06.09	-	-
Timisoara	3	16.02.10	15	16 - 18.02.10	30	17.02.10	30	17.02.10	-	-
	3	17.02.10								
	3	18.02.10								
	3	23.06.10	15	23 – 24.06.10	30	24.06.10	30	24.06.10	32	24.06.10
	3	24.06.10								
Tg. Mures	4	16 - 18.02.11	30	16.02.11	18	16.02.11	18	16.02.11	-	-
			30	17.02.11	17	17.02.11	17	17.02.11		
	1	30.06.11	30	30.06.11	17	31.06.11	17	31.06.11	45	30.06.11

Table 2. Number of samples collected in the study

3.2.3 Maps of sampling points distributed in the cities

The map of supply zones with their monitoring points in the water distribution system of the Bucharest Water and Sewerage Operator, APA NOVA, is presented in figure 1. In the monitoring plan were included eighteen points on the treatment flows, in hydro technical nodes, at the six pumping stations (Drumul Taberei, Preciziei, Grivita, Grozavesti, Nord, Sud) and in the distribution network.

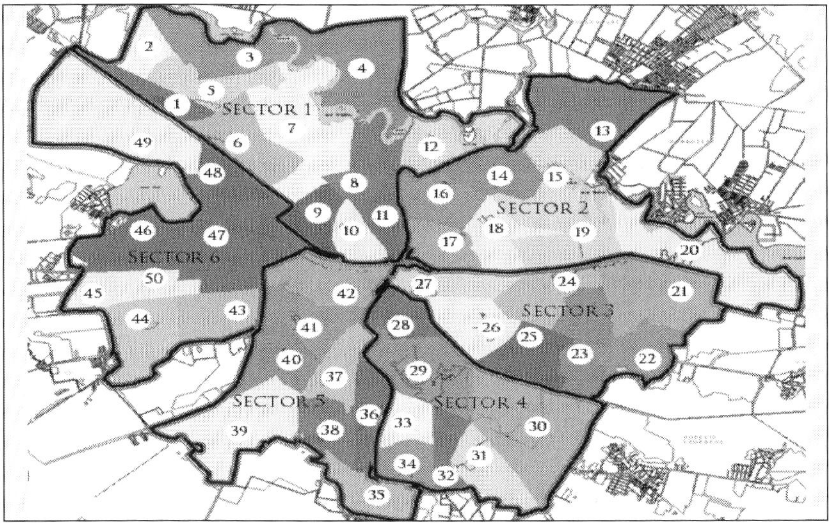

Legend: Points: 1-5, 9, 10, 12-15, 19, 24-26, 28, 29, 35, 41, 43, 47, 49, 50.

Fig. 1. Bucharest APA NOVA monitoring points (50 points)

Figure 2 present the municipal distribution system from Timisoara. On the map are marked sampling points for all type of samples.

On the Targu Mures map (figure 3) were marked sampling points included in monitoring plan, without random daytime points.

3.3 Materials and methods

3.3.1 Parameters

The parameters Al, As, Cd, Cu, Cr, Fe, Mn, Ni, Pb, Se, Sb and Zn were analysed using inductively coupled plasma atomic emission spectroscopy ICP-EOS technique.

In tables 3, 4 are presented the main performance parameters (limit of detection, limit of quantification, accuracy, precision, the uncertainty of measurement) obtained with the equipment and analytical methods used in the study and the maxim admissible value for the metal concentration according to Romanian Legislation (Romanian Law 458, 2002).

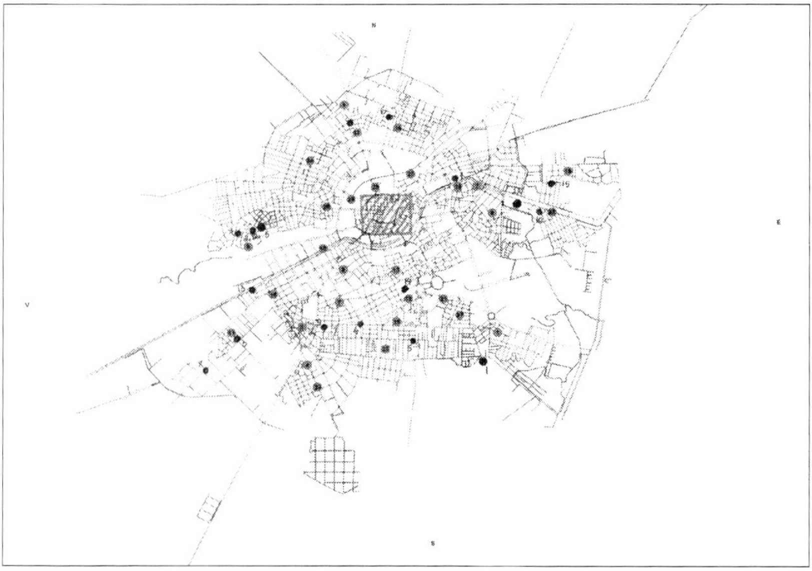

Legend: Indigo dots - Water Plants (Bega 1, Urseni 2/4, Ronat 5); Green dots (15 points) - Aquatim monitoring points; Red dots (30 points) - Customer monitoring points (fully flushed and first draw samples); Orange Square (32 points) - Random Daytime samples, historical center of Timisoara.

Fig. 2. Municipal network system in Timisoara City

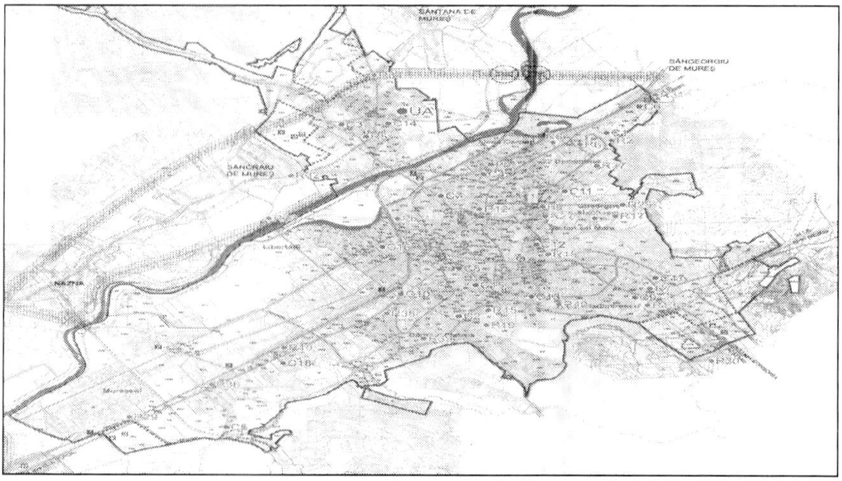

Legend: Indigo dot - Tg. Mures Water Plant; Blue dots (30 points) - Aquaserv monitoring points; Red dots (18 points) - Customer monitoring points (fully flushed and first draw samples)

Fig. 3. Targu Mures Map

Indicator	Al	As	Cd	Cu	Cr	Fe
Max. Value (µg/L)	200	10	5	100	50	200
LOD (µg/L)	1.00	0.05	0.13	0.60	0.50	0.31
LOQ (µg/L)	3.35	0.17	0.43	2.00	1.50	1.00
Accuracy (µg/L)	0.61	0.15	0.04	0.43	0.77	4.50
Precision (µg/L)	4.29	0.83	0.40	4.18	1.93	6.75
U* (%)	12.24	9.09	9.46	10.37	4.58	3.48
Analytical technique	ICP-EOS	ICP-EOS-FIAS	ICP-EOS	ICP-EOS	ICP-EOS	ICP-EOS

* the value represents the expanded uncertanty for a coverage probability of 95%, the coverage factor is k=2.

Table 3. Performance parameters, maxim admissible value and analytical techniques applied in the study (Al, As, Cd, Cu, Cr and Fe)

Indicator	Mn	Ni	Pb	Se	Sb	Zn
Max. Value (µg/L)	50	20	10	10	5	5,000
LOD (µg/L)	0.10	0.95	0.30	0.10	0.10	0.50
LOQ (µg/L)	0.35	3.20	1.00	0.33	0.33	3.70
Accuracy (µg/L)	0.75	0.04	0.03	0.46	0.20	6.20
Precision (µg/L)	1.80	0.55	0.39	0.95	0.67	3.50
U* (%)	4.31	6.64	9.90	10.06	11.21	4.05
Analytical technique	ICP-EOS	ICP-EOS	ICP-EOS	ICP-EOS-FIAS	ICP-EOS-FIAS	ICP-EOS

* the value represents the expanded uncertanty for a coverage probability of 95%, the coverage factor is k=2.

Table 4. Performance parameters, maxim admissible value and analytical techniques applied in the study (Mn, Ni, Pb, Se, Sb and Zn)

3.3.2 Equipments and materials

- Inductively coupled plasma optical emission spectrometer ICP-EOS type Optima 5300 DV Perkin Elmer with Flow Injection Hydride Generation System FIAS 400;
- Simplicity UV System Millipore for Ultra pure water;
- Bi-Distillate System of drinking water type GFL 2104;
- Electric hot places;
- Fume Hood Laborbau System GmbH;
- Suprapure nitric acid 65%, Merck quality;
- Multi-element Standard Reference Material type Quality Control Standard 21, Perkin Elmer, 100 mg/L As, Cd, Cr, Cu, Fe, Mn, Pb, Sb, Se, Zn;

- Uni-element Certified Reference Materials for ICP, 1000 mg/L, CertiPUR, traceable to SRM from NIST, Merck, one CRM for each element;
- Ultra pure water.

3.3.3 Pre-treatment of the samples

The samples were digested with suprapure nitric acid (5 mL for each sample) and concentrated from 150 mL (sample's volume) to 25 mL (Cvijovic et al., 2010; Dumbrava and Birghila, 2009).

3.3.4 Quality Control of the analytical results

For all the analyzed parameters, the laboratory uses standard method (ISO 11885, 2007), reference materials and certified reference materials. The standard method was verified for all metals and the main performance parameters (limit of detection, limit of quantification, linearity, accuracy, precision, selectivity, the uncertainty of measurement) were established with the existing equipment from the laboratory (Tables 3 and 4).

In the study were prepared recovery samples in the same way as real samples, on three different level of concentration, using Standard Reference Material (Quality Control Standard 21), nitric acid and ultra pure water. The recovery percents were situated in the range 94.5% ÷ 114.5%.

The Standard Reference Material was used also for the calibration curve.

The internal control of results were performed using uni-element CRM for each element.

The samples were analyzed in Pollution Control Department of NR&DI ECOIND, which is accredited by RENAR (Romanian Accreditation Association) and follows the requirements of ISO/IEC 17025:2005 standard. The Department has certification with BVQI (Bureau Veritas Quality International) in accordance with ISO 9001/2008 standard and has periodical participation to internal and external audits.

The Department participates every year at tests for the evaluation of its capability by inter-laboratories comparisons (IMEP Belgium, IAWD Germany, CALITAX Spain, IELAB Spain, AQUACHECK LGS, Quality Infrastructure Denmark) for different groups of pollutants (metallic elements, organic compounds, inorganic compounds) from complex matrixes (surface water, waste water, drinking water, soil, sediment, sludge) and the results were included in the accepted range ($-2 \leq$ score $Z \leq 2$).

4. Results and discussions

The quality of raw water used by the Companies in the monitoring campaigns indicates some problems.

In winter season, were detected relatively high contents of organic load (COD - 33÷43 mgO_2/L, BOD - 13÷15.5 mgO_2/L) and Kjeldahl nitrogen (3.1÷4.2 mg N/L) in surface water from Bega River. In summer time, were measured high load of iron (3.6÷4.2 mg/L) and Kjeldahl nitrogen.

In addition, the groundwater used by the Urseni Water Plant has high contents of iron (2.9÷3.5 mg/L), manganese (0.35÷0.45 mg/L) and Kjeldahl nitrogen (3.2÷4.5 mg N/L) in both seasons. The raw water used by the Ronat Plant has also high contents of manganese 0.29÷0.83 mg/L) and Kjeldahl nitrogen (3.1÷7.7 mg N/L).

Mures surface water has in summer season high loads of suspended maters, Kjeldahl nitrogen and phosphorus compounds.

In all investigated periods, the quality of drinking water provided by the Operators (APA NOVA Bucharest, AQUATIM and AQUASERV Companies) was situated in the limits imposed by the Romanian Legislation at their responsibility limit (water connections).

The results recorded for metals in customer's tap water were compared with maximum admissible values according to Romanian Legislation (Romanian Law 458, 2002).

The experimental data shows that concentrations of As, Cd, Sb, Se were situated below the detection limit of the method used (ISO 11885, 2007) to all drinking water samples analyzed from the producer - via distribution system- to the customer's tap.

The data for total chromium shows very low contents, under the limit of detection of the method used (ISO 11885, 2007) or situated close to the limit. For these reasons, only concentrations for seven metals: aluminium, copper, iron, manganese, nickel, lead and zinc, are presented in the next section.

All the results for Cu concentrations reported in this study were situated under the EU Drinking Water Directive (2,000 μg/L), but some of them were higher than the Romanian Legislation limit (100 μg/L).

4.1 Statistical data

In order to obtain more informations, the experimental data were statistically procesed using Excel 2003 Microsoft Programme. Thus, for each set of results (ascendent arrangement) were calculated minimum, maximum, mean, median values, standard deviation of the results, percent of non-compliance samples.

In order to compare the results obtained in the study in all three municipalities, the data were organized in accordance with sampling procedure: first draw, fully flushed and random daytime results.

4.1.1 First draw data

In table 5 are presented experimental data obtained from 71 first draw samples collected in June 2009 in Bucharest. The data shows that the limits, were exceed in some samples by several metals: Al, Cu, Fe, Mn, Ni and Pb. A high percent (around 34%) of samples has Fe content above the limit. High percent of non-compliance samples were recorded for Ni (14%) and Cu (10%).

The tables 6 and 7 present first draw data obtained in winter time, respectively in summer in Timisoara. 30 samples were analized and the data show similarly results, such as 14 - 15 samples (around 50%) has metals (Cu, Fe, Ni, Pb) which exceed the limits.

The tables 8 and 9 shows the data obtained in Targu Mures City in winter (35 samples) and summer time (17 samples). In first draw samples were detected high contens of Cu, Fe, Ni (in winter) and Pb (in summer) and the percentage of non-compliance samples was situated in the range 29-43%.

Parameter / Element	Al	Cu	Fe	Mn	Ni	Pb	Zn
Minim value	3.05	< 0.6	11.1	1.22	< 0.95	< 0.30	1.73
Maxim value	335	621	5,676	520	80.1	39.0	2,194
Median value	42.4	16.0	122	6.73	1.95	< 0.30	250
Mean value	57.3	49.7	271	15.9	9.30	2.81	444
Standard deviation	58.3	112	682	61.1	15.6	6.32	479
Maxim admissible value	200	100	200	50	20	10	5,000
No non-compliance samples / element	3	7	24	1	10	4	0
Total non-compliance samples	34						
Total non-compliance samples (%)	47.89						

Table 5. Bucharest, First Draw Data, June 2009 (μg/L)

Parameter / Element	Al	Cu	Fe	Mn	Ni	Pb	Zn
Minim value	22.4	4.72	10.7	0.10	< 0.95	< 0.30	0.50
Maxim value	735	397	1633	49.7	23.2	22.4	1,881
Median value	82.1	24.3	65.5	4.50	< 0.95	1.54	88.2
Mean value	101	80.7	186	7.23	5.21	4.20	321
Standard deviation	118	111	365	9.32	5.03	5.32	449
No non-compliance samples / element	1	9	5	0	1	5	0
Total non-compliance samples	15						
Total non-compliance samples (%)	50.0						

Table 6. Timisoara, First Draw Data, February 2010 (μg/L)

Parameter / Element	Al	Cu	Fe	Mn	Ni	Pb	Zn
Minim value	17.8	5.43	7.52	1.0	< 0.95	< 0.30	0.50
Maxim value	199	1029	589	20.1	32.7	36.1	1,570
Median value	28.4	35.0	60.1	3.92	< 0.95	< 0.30	340
Mean value	40.5	125	106	6.73	4.82	4.51	470
Standard deviation	37.1	210	130	5.31	8.64	9.52	398
No non-compliance samples / element	0	9	4	0	2	3	0
Total non-compliance samples	14						
Total non-compliance samples (%)	46.67						

Table 7. Timisoara, First Draw Data, June 2010 (μg/L)

Parameter Element	Al	Cu	Fe	Mn	Ni	Pb	Zn
Minim value	19.2	2.51	<0.31	<0.1	<0.95	<0.30	5.22
Maxim value	78.5	235	732	24.1	35.5	7.2	1,224
Median value	31.6	21.8	62.5	4.14	<0.95	<0.30	245
Mean value	33.8	47.0	154	7.21	5.04	2.36	355
Standard deviation	11.7	59.2	201	5.80	8.74	1.13	339
No non-compliance samples / element	0	4	9	0	3	0	0
Total non-compliance samples	15						
Total non-compliance samples (%)	42.86						

Table 8. Targu Mures, First draw Data, February 2011 (μg/L)

Parameter Element	Al	Cu	Fe	Mn	Ni	Pb	Zn
Minim value	21.5	3.11	18.9	1.41	< 0.95	<0.30	8.22
Maxim value	300	244	958	22.0	19.3	19.5	365
Median value	38.9	19.5	58.4	3.80	1.35	<0.30	126
Mean value	60.3	49.8	189	5.73	3.32	3.91	169
Standard deviation	65.9	73.3	259	5.40	4.76	4.30	120
No non-compliance samples / element	1	2	4	0	0	1	0
Total non-compliance samples	5						
Total non-compliance samples (%)	29.41						

Table 9. Targu Mures, First Draw Data, June 2011 (μg/L)

4.1.2 Fully flushed data

When the tap was flushed approximately 5 minutes and then the samples were collected, were recorded lower metal concentrations and the percentage (4.2 - 13.3%) of non-compliance samples decreased (tables 10 -13).

Parameter Element	Al	Cu	Fe	Mn	Ni	Pb	Zn
Minim value	11.2	<0.6	7.25	<0.1	< 0.95	< 0.30	< 0.5
Maxim value	174	63.2	1,387	96.9	17.6	5.23	1,453
Median value	36.1	2.52	27.5	2.95	< 0.95	< 0.30	24.8
Mean value	41.8	5.65	63.2	5.67	2.66	0.51	75.8
Standard deviation	22.8	10.7	168	11.6	4.25	0.92	194
No non-compliance samples / element	0	0	3	1	0	0	0
Total non-compliance samples	3						
Total non-compliance samples (%)	4.23						

Table 10. Bucharest, Fully Flushed Data, June 2009 (μg/L)

Parameter Element	Al	Cu	Fe	Mn	Ni	Pb	Zn
Minim value	22.7	2.52	15.2	3.90	< 0.95	< 0.30	0.5
Maxim value	329	52.8	294	12.5	3.03	11.6	873
Median value	82.9	6.32	36.1	4.31	< 0.95	< 0.30	3.62
Mean value	90.1	8.80	68.6	6.22	< 0.95	2.21	47.3
Standard deviation	53.7	9.21	68.2	4.63	0.61	3.04	152
No non-compliance samples / element	1	0	3	0	0	2	0
Total non-compliance samples	4						
Total non-compliance samples (%)	13.33						

Table 11. Timisoara, Fully Flushed Data, February 2010 (μg/L)

In table 14 were presented fully flushed results from Targu Mures obtained in summer campaign. The data shows o relatively high percentage of non-compliance samples (around 29%), but the number of analyzed samples was lower than in winter (17 samples, respectively 35 samples).

The metals who exceeded the limits are usually iron and lead and only occasionally aluminum, copper and manganese (each, one sample).

Parameter Element	Al	Cu	Fe	Mn	Ni	Pb	Zn
Minim value	15.1	1.81	7.13	1.32	< 0.95	< 0.30	0.52
Maxim value	63.5	69.5	159	20.8	< 0.95	12.1	151
Median value	31.7	5.60	26.1	3.10	< 0.95	< 0.30	15.2
Mean value	34.9	11.2	41.8	4.24	< 0.95	1.61	32.8
Standard deviation	13.5	15.1	45.2	4.02	< 0.95	3.20	38.1
No non-compliance samples / element	0	0	0	0	0	2	0
Total non-compliance samples	2						
Total non-compliance samples (%)	6.67						

Table 12. Timisoara, Fully Flushed data, June 2010 (μg/L)

Parameter Element	Al	Cu	Fe	Mn	Ni	Pb	Zn
Minim value	20.4	1.73	<0.31	1.42	<0.95	<0.30	< 0.50
Maxim value	74.3	53.8	241	18.6	15.7	4.23	1,124
Median value	32.2	4.74	25.1	4.35	<0.95	<0.30	25.4
Mean value	35.4	11.2	50.1	6.04	2.17	2.10	101
Standard deviation	9.89	12.9	56.7	4.46	3.44	0.41	215
Maxim admissible value	200	100	200	50	20	10	5,000
No non-compliance samples / element	0	0	2	0	0	0	0
Total non-compliance samples	2						
Total non-compliance samples (%)	5.71						

Table 13. Targu Mures, Fully Flushed Data, February 2011 (μg/L)

Parameter Element	Al	Cu	Fe	Mn	Ni	Pb	Zn
Minim value	28.2	2.32	32.2	1.23	< 0.95	<0.30	12.3
Maxim value	112	156	429	15.4	3.72	12.6	648
Median value	44.1	4.63	63.1	3.28	< 0.95	<0.30	26,3
Mean value	49.5	24.4	113	5.64	1.58	2.65	96.4
Standard deviation	19.6	40.3	117	6.49	< 0.95	<0.30	13.2
Maxim admissible value	200	100	200	50	20	10	5,000
No non-compliance samples / element	0	1	3	0	0	1	0
Total non-compliance samples	4						
Total non-compliance samples (%)	23.53						

Table 14. Targu Mures, Fully Flushed Data, June 2011 (μg/L)

Fully Flushed procedure was used for the samples collected from monitoring points of the Operators. Therefore, the tables 15 to 19 present the statistical data for these samples.

The data show a low percentage of non-compliance samples for some of the campaigns, such as Bucharest, Timisoara in summer and Targu Mures in winter monitoring programs (4.4 - 13.3%). The concentrations of Fe and Pb were reported in same samples above the limits. Other metals that exceed the limit were aluminum (1 sample, Timisoara, winter), copper (3 samples, Targu Mures, summer) and nickel (1 sample, Targu Mures, summer).

Parameter Element	Al	Cu	Fe	Mn	Ni	Pb	Zn
Minim value	27.3	<0.6	<0.31	<0.1	< 0.95	< 0.30	< 0.5
Maxim value	106	23.6	138	18.7	11.1	31.2	347
Median value	48.5	1.75	12	5.7	< 0.95	< 0.30	6.25
Mean value	51.2	4.11	29.1	5.88	1.34	1.81	48.2
Standard deviation	18.8	5.1	36.3	4.5	2.10	6.46	93.2
Maxim admissible value	200	100	200	50	20	10	5,000
No non-compliance samples / element	0	0	0	0	0	1	0
Total non-compliance samples	1						
Total non-compliance samples (%)	4.35						

Table 15. Bucharest, June 2009 (μg/L), 23 samples (monitoring points of Operator)

Parameter Element	Al	Cu	Fe	Mn	Ni	Pb	Zn
Minim value	22.7	2.51	15.2	3.93	< 0.95	< 0.30	0.50
Maxim value	329	52.8	294	12.5	3.00	11.6	873
Median value	82.9	6.32	36.1	4.33	< 0.95	< 0.30	3.64
Mean value	90.1	8.84	68.6	6.22	< 0.95	2.21	47.3
Standard deviation	53.7	9.21	68.2	4.63	< 0.95	3.04	152
Maxim admissible value	200	100	200	50	20	10	5,000
No non-compliance samples / element	1	0	3	0	0	1	0
Total non-compliance samples	4						
Total non-compliance samples (%)	26.67						

Table 16. Timisoara, February 2010 (μg/L), 15 samples (monitoring points of Operator)

Parameter Element	Al	Cu	Fe	Mn	Ni	Pb	Zn
Minim value	17.8	1.81	19.5	2.04	< 0.95	<0.30	5.80
Maxim value	65.5	80.4	290	22.8	2.23	12,9	2,681
Median value	26.1	7.41	47.5	5.10	< 0.95	<0.30	83
Mean value	29.3	15.5	81.4	7.31	< 0.95	2.52	338
Standard deviation	12.8	19.7	75.9	6.53	< 0.95	3.72	672
Maxim admissible value	200	100	200	50	20	10	5,000
No non-compliance samples / element	0	0	2	0	0	1	0
Total non-compliance samples	2						
Total non-compliance samples (%)	13.33						

Table 17. Timisoara, June 2010 (µg/L), 15 samples (monitoring points of Operator)

In Targu Mures, in summer time, it was student's holiday. In some monitoring points located in public schools and high schools, were recorded high contents of metals (Cu, Fe and Pb), because drinking water remains a longer time in internal distribution system (table 19).

Parameter Element	Al	Cu	Fe	Mn	Ni	Pb	Zn
Minim value	23.2	<0.6	<0.31	<0.1	< 0.95	<0.30	< 0.5
Maxim value	144	18.3	311	20.1	18.7	35.5	275
Median value	41.3	2.21	60.9	6.20	1.63	<0.30	8.92
Mean value	44.9	3.32	78.1	6.61	3.10	2,54	31.6
Standard deviation	21.2	3.65	56.7	3.31	3.35	4.25	47.7
Maxim admissible value	200	100	200	50	20	10	5,000
No non-compliance samples / element	0	0	2	0	0	1	0
Total non-compliance samples	3						
Total non-compliance samples (%)	5.00						

Table 18. Targu Mures, February 2010 (µg/L), 60 samples (monitoring points of Operator)

Parameter Element	Al	Cu	Fe	Mn	Ni	Pb	Zn
Minim value	18.1	< 0.6	13.5	1.12	< 0.95	<0.30	3.13
Maxim value	106	505	536	31.4	65.6	22.5	818
Median value	24.2	2.10	35.6	8.53	< 0.95	<0.30	18.1
Mean value	27.9	42.3	94.8	7.37	3.15	3.45	104
Standard deviation	15.7	123	138	8.53	11.8	4.50	204
Maxim admissible value	200	100	200	50	20	10	5,000
No non-compliance samples / element	0	3	4	0	1	2	0
Total non-compliance samples	8						
Total non-compliance samples (%)	26.67						

Table 19. Targu Mures, June 2010 (µg/L), 30 samples (monitoring points of Operator)

4.1.3 Random daytime data

In thirty-two samples collected with random daytime procedure from the histhorical center of Timisoara City, 28.13 % are non-compliance samples (highest concentrations than admissible values for Cu, Fe, Mn and Pb) (table 20). These data show that internal distribution systems affect drinking water quality. In order to use a better drinking water, the tap must be washed before the water is collected.

In Random Daytime Program were collected and analyzed 45 samples in Targu Mures (table 21). The experimental data show similarly results with Timisoara campaign, so were detected Cu, Fe, Mn and Pb above the maximum limits in 22% of the samples collected.

The aim of this study was to identify issues related with lead concentrations, but the results show also problems with other metals.

Parameter / Element	Al	Cu	Fe	Mn	Ni	Pb	Zn
Minim value	13.3	0.91	5.90	1.22	< 0.95	0.30	2.11
Maxim value	67.7	1064	602	314	< 0.95	16.5	2,404
Median value	28.0	11.8	38.8	5.03	< 0.95	<0.30	124
Mean value	31.5	57.4	90.3	20.5	< 0.95	2.04	335
Standard deviation	11.9	184	131	56.7	< 0.95	3.22	565
Maxim admissible value	200	100	200	50	20	10	5,000
No non-compliance samples / element	0	3	4	2	0	1	0
Total non-compliance samples	9						
Total non-compliance samples (%)	28.13						

Table 20. Timisoara, June 2010 (μg/L), 32 samples

Parameter / Element	Al	Cu	Fe	Mn	Ni	Pb	Zn
Minim value	15.3	<0.6	3.92	0.71	< 0.95	<0.30	5.92
Maxim value	143	741	1,528	1,310	15.6	16.4	1,585
Median value	32.8	6.75	19.3	2.41	< 0.95	<0.30	33.7
Mean value	44.5	58.9	74.1	32.5	3.67	2.38	128
Standard deviation	27.3	147	229	195	5.92	2.15	261
No non-compliance samples / element	0	7	2	1	0	1	0
Total non-compliance samples	10						
Total non-compliance samples (%)	22.22						

Table 21. Targu Mures, June 2010 (μg/L), 45 samples

In table 22 were presented the percentage of samples that exceed the maximum admissible values for all type of samples collected in Bucharest. The results reported for fully flushed procedure indicate a considerable decrease of metal concentrations in tap water.

A relatively high percentages of non-compliance samples were recorded for the samples collected with random daytime procedure (tables 23, 24), such as Cu and Fe.

Municipality	Indicator	The percentage of samples that exceed the maximum admissible values (%)		
		First Draw	Fully Flush	
			Customers	Monitoring points of Operator
Bucharest	Al	4.23	-	-
	Cu	9.86	-	-
	Fe	33.8	4.23	-
	Mn	1.41	1.41	-
	Ni	14.1	-	-
	Pb	5.63	-	1.41

Table 22. Summary of percentage of non-compliance samples for each element - Bucharest

Municipality	Indicator	The percentage of samples that exceed the maximum admissible values (%)						
		First Draw		Fully Flush				Random Daytime
				Customers		Monitoring points of Operator		
		winter	summer	winter	summer	winter	summer	
Timisoara	Al	3.33	-	3.33	-	6.67	-	-
	Cu	30.0	30.0	-	-	-	-	9.38
	Fe	16.7	13.3	10.0	-	20.0	13.3	12.5
	Mn	-	-	-	-	-	-	6.25
	Ni	3.33	6.67	-	-	-	-	-
	Pb	16.7	10.0	6.67	6.67	6.67	-	3.13

Table 23. Summary of percentage of non-compliance samples for each element – Timisoara

Municipality	Indicator	The percentage of samples that exceed the maximum admissible values (%)						
		First Draw		Fully Flush				Random Daytime
				Customers		Monitoring points of Operator		
		winter	summer	winter	summer	winter	summer	
Targu Mures	Al	-	5.88	-	-	-	-	-
	Cu	11.4	11.8	-	5.88	-	10.0	15.56
	Fe	25.7	23.5	5.71	17.65	3.33	13.3	4.44
	Mn	-	-	-	-	-	-	2.22
	Ni	8.57	-	-	-	-	3.33	-
	Pb	-	5.88	-	5.88	1.67	6.67	2.22

Table 24. Summary of percentage of non-compliance samples for each element–Targu Mures

These results shows that the stagnation time of drinking water in internal distribution system even if it is 1, 2 or 3 hours influence the tap water quality in a worse way.

4.2 Correlations between materials and metal concentrations

In tables 25 to 27 are presented the material types inside the customer's building, primary materials in public network and branch pipe in apartments where the metal concentrations were situated above the maximum limits in first draw samples.

Point	Values higher than limits	Household -primary material	Municipal distribution system - primary material	Branch pipe
C1	Cu, Ni	PexAl, Cu	Cast iron	Pb
C2	Fe	PVC, cast iron	Cast iron	Pb
C4	Cu, Ni	PVC, Cu	Cast iron	Pb
C6	Fe	Cast iron, plastic pipe	PEHD	Pb
C7	Fe	Cast iron	Cast iron	Cast iron
C8	Fe	Cast iron	Steel	Pb
C11	Fe	PVC	Steel	Steel
C12	Fe	PVC, cast iron	PEHD	PEHD
C16	Ni	Metallic pipes	Cast iron	Cast iron
C17	Cu	Cu, PE	Cast iron	Cast iron
C18	Fe, Pb	Pb	Cast iron	Pb
C19	Fe, Pb	Pb	Cast iron	Pb
C21	Fe	PVC	Asbestos-cement, Cast iron	Steel
C23	Al, Fe	PVC	Asbestos-cement, Cast iron	Cast iron
C24	Ni	PVC	Cast iron	Cast iron
C25	Al, Fe, Ni	PVC	Asbestos-cement, Cast iron	Cast iron
C27	Ni	Cast iron	Cast iron	PEHD
C31	Cu, Pb	PVC, Cu	Cast iron, Pb	Pb
C34	Fe	Metallic pipes	PEHD	PEHD
C35	Fe	Cu, Pb, galvanized pipe	Cast iron	Cast iron
C37	Ni	PVC	Cast iron	Cast iron
C38	Pb	Pb	PEHD	PEHD
C40	Fe, Ni	Metallic pipes	Cast iron	Cast iron
C42	Fe, Cu	Cu	Cast iron	Cast iron
C43	Fe	Metallic pipes, PExAl	Cast iron	Cast iron
C47	Fe	PVC	Cast iron	Cast iron
C49	Fe, Ni	Plastic	Cast iron	Pb
C50	Fe, Mn, Ni, Zn	Metallic pipes	PEHD	PEHD, Cast iron
C51	Fe	PExAl, Pb	Cast iron	Cast iron
C54	Fe	Cast iron, Pb	Steel	Steel
C58	Al, Fe	PVC, metallic pipes	PEHD	PEHD
C59	Cu, Fe	Cu, cast iron	PEHD	PEHD
C61	Cu	Cu	PEHD	PEHD
C63	Fe	Metallic pipes	PEHD	PEHD

Table 25. Bucharest, source of metals in first draw samples

In Bucharest, a high percent of first draw samples exceeded the iron concentration limit (table 22). Materials used for pipes and fittings in the domestic distribution system in this points are either metallic (Cu, Pb, cast iron) or a combination of plastic and metallic pipes. In addition, the pipes and fittings of APA NOVA Bucharest distribution system are either cast iron, steel or PEHD (high density polyethylene), that's why the effect observed may be a combination of both distribution systems. As was mentioned above (Lehtola et al., 2004b; Sarin et al., 2004), corrosion process in cast iron is the main source for discolouration of drinking water and iron higher content.

Lead contents above the limit are reported in apartments were either the Pb pipes are inside the building or the Pb branch pipes are used to connect municipal distribution network with domestic system.

Source of high levels of copper in first draw samples is only the Cu pipes used for cold water supply in household. The customers have installed central heating in their apartments and hot water circuit was make with copper pipes. In addition, they have replaced the old cold water pipes with copper pipes and can be observed the results, more obvious in Timisoara (Table 26) and Targu Mures (table 27).

Point	Values higher than limits		Household - primary material	Municipal distribution system - primary material	Branch pipe
	Winter	Summer			
C5	Cu	Cu, Ni	Cu	Cast iron	PE
C6	Cu, Pb	Cu, Pb	Cu	Cast iron	Steel
C9	Cu	Cu	Cu	PE	PE
C12	-	Cu	Cu	PE	PE
C13	Cu, Pb	Cu	Cu	Cast iron	Pb
C18	Fe	Fe	Steel	Cast iron, PE	PE
C19	-	Pb	PVC	PE	PE
C20	Pb	Cu, Pb	Cu	Steel	Pb
C21	Cu	-	Steel	PE	PE
C22	Cu, Pb	-	Cu	PVC	Steel
C25	Ni	Ni	Steel	Cast iron	Steel
C27	Cu, Pb	Cu	Steel	PVC	Steel
C29	Fe	-	Steel	PE	PE
C30	Cu	Cu	Cu	Steel	Steel
C31	Al, Fe	Fe	Steel	Cast iron	Steel
C33	Fe	Fe	Steel	Cast iron	Steel
C35	Cu, Fe	Cu, Fe	Cu	Cast iron	Steel

Table 26. Timisoara, source of metals in first draw samples

In Timisoara (table 26) were reported higher values than the limits in first draw samples for Cu, Fe, Ni and Pb.

The sources for lead contents higher than the limit are either branch pipes or PVC pipes. The stabilizers used for PVC production are often composed of salts of metals like lead and cadmium (Kim, 2001), in order to provide properties needed in the final product.

High content of nickel, was observed in same samples (table 25 to 27), which was generated, probably, by cheap taps with poor quality.

As we can see in table 27, source of Fe in first draw samples could we also unprotected steel, used in both distribution systems.

Point	Values higher than limits		Household - primary material	Municipal distribution system - primary material	Branch pipe
	Winter (2 days)*	Summer			
C1	Fe / 2	Fe, Pb	PP	PE	PE
C2	Fe / 1	-	Cu	Asbestos-cement	Steel
C3	Cu, Ni / 1	-	Cu	Steel	Steel
C4	Fe / 2	Al, Fe	Cu	Steel	Steel
C7	Cu / 2	Cu, Fe	Cu	Steel	Steel
C8	Ni / 2	Cu	PE + Steel	PE	PE
C10	Cu / 1	-	Cu	Steel	Steel
C12	Fe / 2	-	Steel	PE	PE
C16	Fe / 2	-	Steel	PE	PE
C18	-	Fe	Steel	PE	PE

*the monitoring programme includes two consecutive days of sampling.

Table 27. Targu Mures, source of metals in first draw samples

5. Conclusions

This research activity demonstrates that materials used in water distribution systems are part of the overall treatment process that affect the water quality which consumers drink at their tap. The interaction between water and the infrastructure used for its supply are fundamental in producing safety drinking water. Subtle reactions between water and different materials used for its transport can affect the finale quality delivered to consumers.

The study, developed in the period June 2009 - June 2011, in a project regarding safety of drinking water distribution systems in some municipalities from Romania, demonstrated that materials used in drinking water domestic installations have a major contribution in the deterioration of water quality supplied by local distribution operators.

In some locations, in first draw samples collected in the morning from kitchen cold taps were detected high quantity of Cu, Fe, Ni and Pb correlated with materials used in internal distribution system (Cu pipes, Pb pipes, PVC pipes, branch pipes, cast iron and unprotected steel pipes, Ni-Cr plated taps). The main causes are the process of water stagnation and the lack of maintenance of the internal distribution materials.

For samples were collected with fully flushed procedure, the quality of drinking water was better, the number of non-compliance samples decreased with 50%.

The customers were advised that, it be not recommended to use the first draw water for cooking and drinking purpose. Recommendations in cases of exceeding the limit values of metals in drinking water were either flushing water for more than five minutes and then use water for household consumption or replacement of pipes and fittings in both, local or domestic distribution systems.

6. Acknowledgments

The Romanian National Research Program has funded this work. The authors want to thank to staff from APA NOVA Bucharest, AQUATIM Timisoara and AQUASERV Tg. Mures COMPANIES for collaboration in this project.

7. References

Alfrey, A.C., Gendre, G.R., Kaehny, W.D., (1976), The dialysis encephalopathy syndrome. Possible aluminium intoxication, *N. Engl. J. Med.*, 294, 184- 188.

Agatenos, C., O kato, P.O., (2008), Studies of corrosin tendency of drinking water in the distribution system at the University of Benin, *Environmentalist*, 28, 379-384.

American Water Works Association (AWWA), (1996), Internal Corrosion of Water Distribution Systems, second ed., *American Water Works Association*, Denver, Colorado.

Becaria, A., Lahiri, D., Bondy, S., Chen, D.M., Hamadeh, A., Li, H., Taylor, R., Campbell, A., (2006), Aluminum and copper in drinking water enhance inflammatory or oxidative events specifically in the brain, *Journal of Neuroimmunology*, 176, 16-23.

Cvijovic M., Djurdjevic P., Cvetkovic S., Cretescu I., (2010), A case study of industrial water polluted with chromium (VI) and its impact to river recipient in western Serbia, *Environmental Engineering and Management Journal*, 9, 45-49.

Dumbrava A., Birghila S., (2009), Analysis of some metal levels in Danube river water, *Environmental Engineering and Management Journal*, 8, 219-224.

Gari, D.W., Kozisek, F., (2008), Nickel and iron in drinking water in Czech Republic, *2nd International Conference "Metals and Related Substances in Drinking Water"*, Cost Action 637, Proceedings Book, 75-77

Deshommes E., Laroche L., Nour S., Cartier C., Prevost M., (2010), Source and occurrence of particulate lead in tap water, *Water Research*, 44, 3734-3744.

Flaten, T.P., (2001), Aluminum as a risk factor in Alzheimer's disease, with emphasis on drinking water, *Brain Res. Bull.*, 55, 187–196.

Haider T., Haider M., Wruss W., (2002), Lead in drinking water of Vienna in comparison to other European countries and accordance with recent guidelines, *International Journal of Hygiene and Environmental Health*, 205, 399-403.

Hayes R.C., (2009), Computational modeling to investigate the sampling of lead in drinking water, *Water Research*, 43, 2647-2656.

Hayes R.C., Aertgeerts R., Barrott L., Becker A., Benoliel M. J., Croll B., (2010), *Best Practice Guide on the Control of Lead in Drinking Water*, Hayes R.C. (Ed), IWA Publishing, London, 13-23. ISO 11885-2007 (SR EN ISO 11885-2009), Water quality: Determination of selected elements by inductively coupled plasma optical emission spectrometry (ICP-EOS).

Karavoltsos, S., Sakellari, A., Mihopoulos, N., Dassenakis, M., Scoullos, M., (2008), Evaluation of the quality of drinking water in regions of Greece, *Desalination*, 224, 317-329.

Kimbrough, D., (2001), Brass corrosion and the LCR monitoring program, *J. Am. Water Works Assoc.* 1, 81–91.

Kim, S., (2001), Pyrolysis kinetics of waste PVC pipe, *Waste Management*, 21, 609-623.

Lasheen, M.R., Sharaby, C.M., Kinoby, N.G.El., Elsherif, I.Y., Wakeel, S.T.El., (2008), Factors influencing lead and iron release from some Egyptian drinking water pipes, *Journal of Hazardous Materials*, 160, 675-680.

Lehtola, M.J., Juhna, T., Miettinen, I.T., Vartiainen, T., Martikainen, P.J., (2004a), Formation of biofilms in drinking water distribution networks, a case study in two cities in Finland and Latvia, *J. Ind. Microbiol. Biotechnol.*, 31 (11), 489–494.

Lehtola, M.J., Nissinen, T.K., Miettinen, I.T., Martikainen, P.J., Vartiainen, T., (2004b), Removal of soft deposits from the distribution system improves the drinking water Quality, *Water Research*, 38, 601-610.

Lehtola, M.J., Miettinen, I.T., Hirvonen, A., Vartiainen, T., Martikainen, P.J., (2007), Estimates of microbial quality and concentration of copper in distributed drinking water and highly dependent on sampling strategy, *International Journal of Hygiene and Environmental Health*, 210, 725-732.

Lytle, D.A., Nadagouda, M.N., (2010), A comprehensive investigation of copper pitting corrosion in drinking water distribution system, *Corrosion Science*, 52, 1927-1938.

Lytle, D.A., Schock, M.R., (2000), Impact of stagnation time on metal dissolution from plumbing materials in drinking water, *J. Water Supply Res. Technol. Aqua.*, 49, 243–257.

Merkel, T.H., Grob, H.-J., Werner, W., Dahlke, T., Reicherter, S., Beuchle, G., Eberle, S.H., (2002), Copper corrosion by-products release in long-term stagnation experiments, *Water Research*, 36, 1547–1555.

Nawrocki, J., Stanislawiak, U.R., Swietlik, J., Olejnik, A., Svoka, M.J., (2010), Corrosion in a distribution system: Steady water and its composition, *Water Research*, 44, 1863-1872.

Raynaud, M., (2004), A view of the European plastic pipes market in a global scenario, *Proceedings of Plastics Pipes XII*, April 19–22, Milan, Italy.

Romanian Law 458, concerning drinking water quality, (2002), *Official Monitor of Romania*, no. 552, modified by Law 311, *Official Monitor of Romania, Part 1*, 382, 2004.

Sarin, P., Snoeyink, U.L., Bebee, J., Jim, K. K., (2004), Iron release from corroded iron pipes in drinking water distribution systems: effect of dissolved oxygen, *Water Research*, 38, 1259-1269.

Vasile G.G., Dinu C., Chiru E., (2009), Monitoring of metal concentrations in tap waters in Bucharest supply system, *3rd International Conference Cost Action 637 "Metals and related substances in drinking water"*, Proceedings Book, 50.

Volker S., Schreiber C., (2010), Drinking water quality in household supply infrastructure. A survey of the current situation in Germany, *International Journal of Hygiene and Environmental Health*, 213, 204-209.

WHO (World Health Organization), (2008), Guidelines for drinking water quality, 3rd ed. Recommendations. Incorporating 1st and 2nd Addenda, vol. 1, Geneva.

WHO (World Health Organization) and COST action 637, (2009), Guidance on sampling and monitoring for lead in drinking water;

Zietz B.P., de Vergara J.D., (2003), Copper concentrations in tap water and possible effects on infant's health-results of a study in Lower Saxony, Germany, *Environmental Research*, 92, 129-138.

Zietz B.P., Lass J., (2007), Assessment and management of tap water lead contamination in Lower Saxony, Germany, *International Journal of Hygiene and Environmental Health*, 17, 407-418.

21

Water Quality Monitoring and Assessment in a Developing Country

O.A.A. Eletta

Department of Chemical Engineering, University of Ilorin, Nigeria

1. Introduction

Water is very essential for man and, it is being increasingly stressed through the action or inaction of man leading to either a temporary or permanent impairment of water quality. Impairment resulting from the presence of foreign substances (organic, inorganic, bacteriological or radiological) tends to degrade the quality of water. It is needed for virtually all of man's major/ routine activities. Water can be useful for industrial, domestic and recreational purposes (Yerima *et al*, 2008).

Apart from water supplied by the utility board, many have dug either boreholes or wells in their homes and industries in order to have adequate supply of clean and fresh water (Spalding and Exner, 1993). River constitutes the bulk of water being used in Cities and its environs either treated or otherwise. River banks could be very busy with varied activities ranging from farming to industrial activities and a lot more.

Water in shallow wells is closer to the surface and to potential sources of contamination. River on its own has the open abuse of being eroded and abused by many if not all in this part of the World as, waste are dumped in the river without recluse to its effect on the ecosystem.

High loads of nitrate in rivers have been responsible for the eutrophication of rivers. This leads to the depletion of oxygen in the water and, leads to an imbalance in the ecosystem. A River would mean different things to different people. To some, it is used for laundry and recreation purposes, to some industries it is a good supply of cooling water and to others, the river is a convenient point of waste discharge from both the home and industries.

Water could be available through precipitation (rain), surface and ground water sources. In the past, the general belief was that, rain water was the purest of these. However, with the advent of industries and many of the activities of man, this source of water supply has suffered its share of pollution. Particulates left in the air, resulting from the activities of man have been washed into water collections during rains besides, collection of water from roof tops have also suffered some level of pollution as the material used in making the roofing sheets have been eroded and washed into the water collection.

Surface water such as rivers and streams have suffered immeasurable contamination resulting from the activities of man. Many industries and workshops are sited along the bank of rivers with one of the purpose being, to have ease of waste disposal as effluents sometimes in their untreated state are emptied into the water bodies. Individuals have been known to leave their residence just to empty their waste into the streams and rivers nearby.

The waste of those upstream becomes source for those downstream. Many drink water from the taps, rains, wells, borehole and even rivers and streams with little or no treatment.

The physico - chemical parameters useful for water quality assessment are determined by the presence of both organic and inorganic compounds that are either suspended or dissolved in it. While some of these compounds are toxic to the ecosystems, others constitutes nutrients to aquatic organisms and others are responsible for the aesthetics of the water body (Olajire and Impekperia, 2000).

The pH of water determines the solubility and bioavailability of chemical constituents such as nutrients and heavy metals. Metals tend to be more toxic at lower pH as they are more soluble. Though an increase in pH may have no direct impact on the aquatic life, they greatly influence the availability and solubility of all chemical forms in the water and this might aggravate nutrient problems leading to a destabilization of the ecosystem.

The prevalent problem in a country resulting from inadequate water supply for use that has turned every man in the country into a local government is responsible for this study. This becomes necessary when you realize that very little can be achieved without water. Three of the sources open to man are being investigated in this study.

Nitrate has been found in wells and in some cases, shallow bore holes. Location of these have been implicated in the concentration discovered. Studies have shown that excess nitrate is responsible for the blue baby syndrome amongst other diseases.

There is so much water on the surface of the earth and yet, insufficient to meet the needs of man as, most of this resides in the ocean which limits its direct use without prior treatment / improvement

The physical and chemical parameters useful for water quality assessment are determined by the presence of both organic and inorganic compounds that are either suspended or dissolved in it. While some of the compounds are toxic to the ecosystem, some constitutes nutrients to aquatic organisms and others are responsible for aesthetics of the water body (Eletta and Adekola, 2005)

The measure of the hydrogen ion concentration determines the mobility/ bioavailabiity of chemical constituents such as nutrients and heavy metals. Metals tend to be more toxic at lower pH since they are more soluble. Though an increase in pH level may not have direct impact on the aquatic life, they greatly influence the availability and solubility of all chemical forms in the water and this might aggravate nutrient problem leading to a destabilization of the ecosystem. Contaminants such as hydrocarbon, heavy metals and pesticides have been known to have direct toxic effect when released into the aquatic environment and, the sediment constitutes a sink for these pollutants. (Fleeger *et al*, 2003)

The cause of pollution typically lies with man while the spread is generally driven by natural means hence the study looking into the impact of man's activity on water quality.

In the urban environment, sewage discharges are a major component of water pollution. This contributes to the oxygen demand and nutrient loading thereby promoting growth of toxic algae and other aquatic plant hence the destabilization of the ecosystem (Olajire and Imepekperia, 2000).

2. Rain water

Rain water is a form of precipitation in which liquid water falls to the earth surface. Rain water harvesting is a common sight in developing countries. This is a major source of water during the raining season as harvesting is done mainly via roof tops without recluse to the effect of the make of the roof sheet, ware on roof sheets with age and erosion of sheet material. Emissions into the air through vehicular exhaust, industrial discharges amongst others get washed down from the sky during rains. Unfortunately, because of the unavailability of sufficient portable water in many of these areas, rain water harvesting is done and, with little or no processing, consumed domestically.

3. Well/boreholes

In many developing countries wells are sited in homes due to the inadequacy of portable water supply. Boreholes are not readily available in many cases and, the siting of wells are not well monitored. Transport of waste through leaching into shallow wells have resulted in pollution.

Studies have shown that water pollution arising from the presence of foreign substances (inorganic, bacteriological or radiological which tends to degrade the quality of water has become a serious concern today (Ibe et al, 2002, Salami A.W., 2003, Yerima et al, 2008).

Environmental scientists are increasingly aware of the need for reliable methods that can determine the chemical forms of toxic trace metals especially the toxic heavy metals in natural water.

In water, the free and hydrated metal ions are considered to be the most toxic form followed by strong metal complexes, and metal species associated with colloidal particles which are least toxic.

In a developing country, the search for water are explored to complement the limited amount of water available for use for domestic, agricultural, recreational and industrial uses. The biodegradable wastes emptied into the water body leads to eutrophication of the water body, depletion of oxygen in the water and extinction of aquatic animal.

Studies have revealed that the waste of one man eventually flows into the areas downstream and becomes the source for the man living downstream of discharge. Hence the need for a proper monitoring of activities capable of leading to changes in water quality.

4. Surface water

In many of the villages in developing countries, the people adopt the streams and rivers as their means of water supply for all the water need. Unfortunately, many see the river as a good point to empty their wastes. Wastes are indiscriminately dumped along the water course when there is rains with the river as the target to empty such waste (Boukori *et al*, 1999, Adeyemi, 2003).

5. Waste disposal

Many see the river as a good location to empty waste from homes and industry such that, the quality of water at a time might be lowered as a result of closeness to residential areas and industries. Indiscriminate location of dumpsite is another avenue of getting pollutants into the water body. During rains, wastes are leached into the surface and ground water sources. The location of septic tank indiscriminately also gives rise to surface and ground water pollution.

Fig. 1. Fishing and recreation

Fig. 2. The young catching fun in water

6. Recreation

Unregulated, recreational activities also leads to lowering of surface water quality. Washing of clothes and other items in the water, bathing and other such activities have a negative impact on the water quality.

7. Industrial

Many industries located close to water sources are in the habit of disposing their waste into surface water aside from the primary use of the water for industrial purposes. For the small to medium scale industries especially that are barely trying to stand, waste treatment is the least of their concern. In a study carried out in 2005 by Eletta *et al* on, the effect of waste discharges from a beverage industry, it was discovered that the nitrate levels were present in quantities above the standards laid down by the World Health Organization (WHO). The concentration of zinc (Zn) was also found to be high in the discharges. The aquatic toxicity of Zn and especially the toxicity in fishes have been reviewed and, it has been concluded that, it has a high toxicity on fish even though the effect on man is not as lethal (Fatoki et al, 2002, Fadiran and Mamba, 2005).

8. Rain water run off

A study on the rain water runoff from paved roads in two cities by Adekola et al, 2001 showed clearly that trace metals were present in significant amounts in the runoffs which eventually are discharged into water bodies. In fact many use opportunity of such run off. Over the time, accumulation of these has a negative effect on the water quality.

9. AIM

This chapter aims at identifying the various sources of water, Some pollutants and suggestion of some form of regulation of activities that impair the quality of water sources.

Fig. 3. A place to empty waste

Fig. 4. Metal works along river bank

Fig. 5. Rainwater harvesting

10. Methodology

Samples were collected and after prior treatment, analysed following standard methods from literature for some trace metals and nutrient loading.

11. Result

The reconstruction of historical inputs of anthropogenic pollutants is important for improving management strategies and evaluating the success of recent pollution control measures (Santschi *et al*, 2001).

There is a need to identify in relation to a particular water body the possible sources of pollution and, give necessary assessment and evaluation of the water based on the activity along/ around it and, where the need arises, education on impact of such activities and in consultation with necessary enforcement agencies, there might be a need to put some measures in place to make sure that right practices are on there.

Three main sources of water available for man's use (ground, surface and rainwater) are randomly investigated for some trace metal and nutrient loading with a view to monitor the impact of man's activities on the quality of water available for man's use.

12. Discussion

The stuay carried out on randomly selected locations confirmed the impact of some of the activities carried out along the course of the river on the concentration of some of the identified pollutants in the various water bodies.

Concentration of copper (Cu) in the water samples across samples and locations were found to be within allowable concentration levels by the World Health Organization (WHO).

Concentration of lead (Pb) in the rain and surface water samples were determined across locations. There was visibly, no significant changes in the Pb concentrations from the two sources. Figure of 0.08 mg /l was recorded at location 4 which has a metal workshop along the bank of the river under study as shown in fig 4.

Concentration of iron (Fe) in the ground water sample was found to be below detection limit. However for the surface water samples, concentration was found above the WHO allowable limits in locations 3 and 5 with average concentrations of 3.08 and 3.6 mg/l respectively with. Location 3 has a soap and detergent company emptying its waste into the water at this point. It was discovered that, the company treats its waste using ferric chloride and, the high concentration observed will not be unconnected with the input into the river at this point. Location 5 has a metal fabricating industry sited just at the bank of the river as shown in the picture above. Fig 5, shows a rusted corrugated iron roof which is a common sight considering it is the cheapest and readily available roofing material to many. However, because of the inadequacy of water for the use of man, rain harvesting is carried out by many during the rains. As the rain impinges on the roof top, there is the eroding of the materials used in the manufacture of the sheet into water collected. The concentration of Fe in roof tops with longer age of use have been found to be higher than newly installed sheets.

Concentration of nitrate was determined in the surface and the ground water. From the data collected, it was found that, the activities around the sampling points had direct influence on the nutrient loading. Location 4 which had an average concentration of 15.41 mg/l has extensive farming activities going on along the bank of the river and a beverage industry empties its waste into the river at this point. Locations 5 and 6 has average concentrations of 9.74 and 10.45 mg/l respectively also has farm activities and refuse dumps along the bank. Location 7 with an average concentration of 7.81 mg/l has extensive irrigation farming along the bank of the river. The surface of the river is covered with sea weed along these areas and when these die off, there is biodegradation and oxygen is used up and this results in the destabilization of the ecosystem. Fishing activities have been greatly reduced along this stretch of the river. In rain water harvesting, the state of the roof tops is often not taken into account. Water is collected from sheets in deplorable conditions as depicted in fig 5.

The average nitrate concentration of 7.89 and 7.35close to mg/ l observed in locations 5 and 6 in the ground water samples could be traced to the improper location of the wells. These two wells are located close to soakaways that are situated uphill of the wells.

13. Recommendations

13.1 Rainwater

Pollution of this source can come from washing down of particulates in the atmosphere during rains in which case, there is a need to monitor discharges in the form of particulates from industrial wastes and vehicular discharges. Trapping of gaseous wastes from industries and proper treatment and discharge will go a long way in reducing the chances of their reacting with water during precipitation. The trace metal pollutants resulting from impinging on the roof tops can be reduced by enforcing that aged roofing sheets be changed after a stipulated length of years after carrying out an intensive study to ascertain the period when optimum leaching is done.

13.2 Surface water

There are several inputs that goes to pollute the surface water. Animals crossing a river and discharges its waste as it crosses a river at shallow ends, man would swim (fig 2), wash and even empty the waste generated in the homes (fig 2) into the river in other to have it washed downstream aside from the washings into the water bodies from gutters and roads after each rain.

For all of these, there is a need for the government to enforce proper conduct with respect to rivers which mean different thing to different people depending on the point of contact. Farm practices along the bed of rivers should only be encouraged at safe distances from the river. A ban on animals wading through rivers should be put in place and the nomadic should be encouraged to wash their animals from puddles or they should be made to have a ranch where the animals are properly monitored.

Ground water pollution is mainly through leaching. There is a need to educate the builders such that wells are not located in areas where wastes cab be washed in easily.

All that constitutes pollution of water can all be addressed with putting laws in place and ensuring that the laws are followed strictly.

Location	Cu_R	Cu_S	Cu_g
1	0.13	0.06	0.04
2	0.1	0.07	0.02
3	0.08	0.06	0.03
4	0.17	0.05	0.02
5	0.13	0.04	0.02
6	0.1	0.06	0.03
7	0.17	0.07	0.02
8	0.12	0.07	0.03

Table 1. Average concentration of Cu in selected rain surface and groundwater samples (mg/l)

Location	Pbavg®	Pbavg(s)
1	0.03	0.04
2	0.03	0.05
3	0.05	0.06
4	0.03	0.08
5	0.04	0.05
6	0.04	0.04
7	0.06	0.05
8	0.05	0.05

Table 2. Average concentration of Pb in selected rain and surface water samples (mg/l)

Location	Feavg®	Feavg(s)
1	1.77	0.54
2	1.59	0.6
3	2.04	3.08
4	1.35	1.37
5	1.41	3.6
6	2.55	2.09
7	2.03	1.66
8	1.56	1.63

Table 3. Average concentration of Fe in selected rain and surface water samples (mg/l)

Location	NO_3s	NO_3g
1	3.74	0.89
2	3.147	0.68
3	5.277	1.1
4	15.41	0.5
5	9.743	7.89
6	10.45	7.35
7	7.81	1.83
8	1.843	0.4

Table 4. Average concentration of NO_3 in selected ground and surface water samples (mg/l)

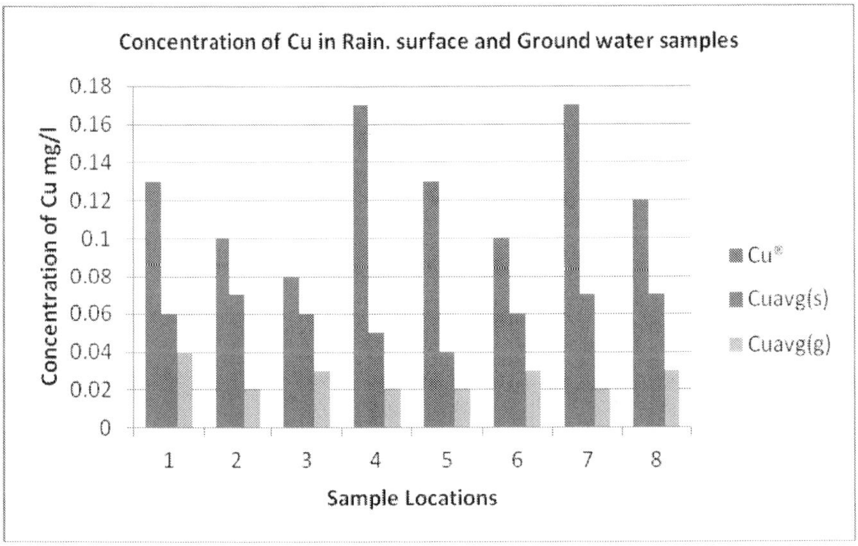

Fig. 6. Average Cu concentration across locations in rain, surface and ground water samples

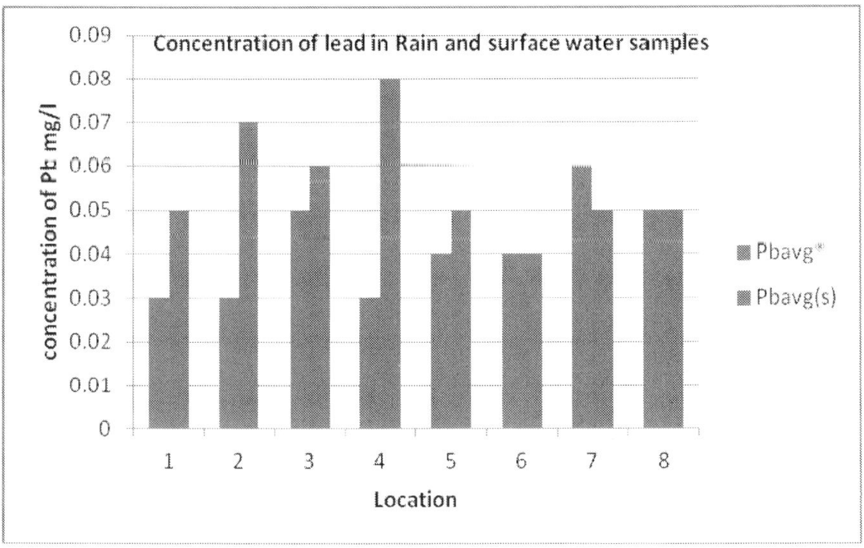

Fig. 7. Average Pb concentration across locations in rain and surface water samples

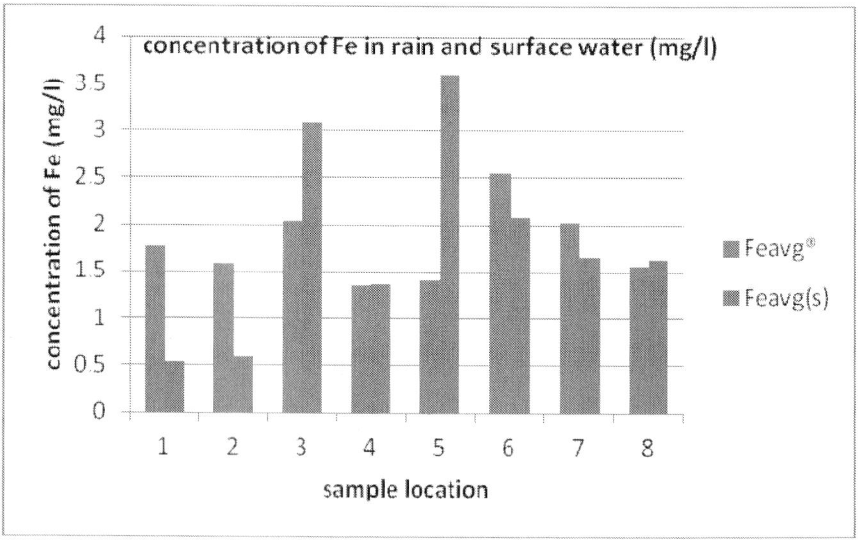

Fig. 8. Average Fe concentration across locations in rain and surface water samples

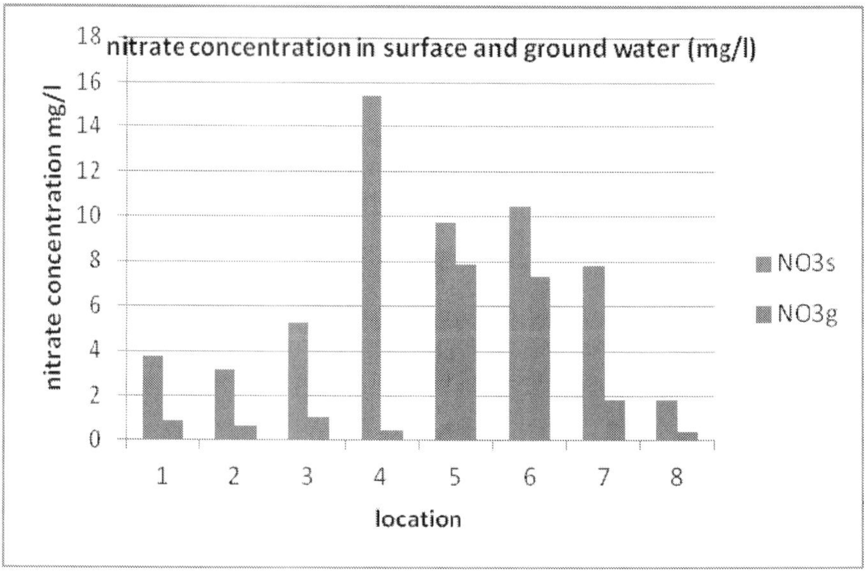

Fig. 9. Average nitrate concentration across locations in surface and ground water samples

14. Appendix

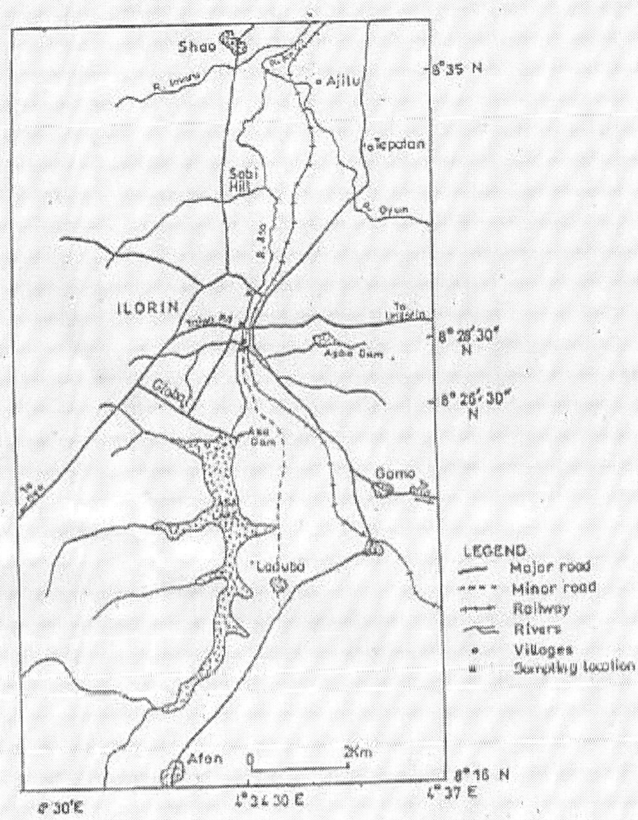

Map of Ilorin and environs showing flow of river Asa with sample locations

Fig. 10.

15. References

[1] Adekola F.A., Eletta O.A.A. and Atanda S.A., 2001, *"Some heavy metals concentration in rain water runoff on paved roads in Lagos and Ilorin Cities, Nigeria"*, Academy Journal of Science and Engineering, 1, 88-96

[2] Adeyemi O.K, 2003, *"Consequences of pollution of Nigerian Aquatic Environment on fisheries resources"* 23(4), 297 - 306

[3] Boukori Y, Bawa I.M., Djareye - Boundjou G., 1999, *"Characterization of some Togo Surface Waters" Bull. Chem. Soc. Ethiopia.* 13(1): 11 - 21

[4] Eletta O.A.A. and Adekola F.A., 2005, *"Studies of the physical and chemical properties of Asa river water, Kwara State, Nigeria"* , Science Focus 10(1), 72-76

[5] Eletta O.A.A., Adekola F.A., and Aderanti M.A., 2005, *" Effect of waste discharge from soft drink plant into Asa river, Ilorin, Nigeria"* J. Appl. Sci. Envron. Mgt. 9(1), 187 -190

[6] Fadiran A.O and Mamba S.M, 2005, *"Analysis Of Nitrates And Nitrites In Some Water And Factory Effluent Samples From Some Cities In Swaziland"*, Bull. Che. Soc. Ethiopia 19(1) 35-44

[7] Fatoki O.S., Lujiza N. and Ogunfowokan A.O, 2002, *"Trace metal Pollution in Umtata River"* water South Africa 28(2), 183 -187

[8] Fleeger J.W., Carmana K.R., and Nisbeth R.M., 2003, *" Indirect effects of contaminants in aquatic ecosystem"* The Science of Total Environ., 317 (1-3), 207 - 233

[9] Ibe K.M., Egereonu U.U and Sowa A.H.O., 2002 *" The impact of hand pump corrosion on water quality in rural areas of west African Sub - Region"* J. Environmental Monitoring and Assessment, 78, 31 -43

[10] Olajire A.A. and F.E. Imepekperia, 2000, *" A study of water quality of the Osun River; metal monitoring and Geochemistry"* Bull. Chem Soc. Ethop 14(1), 1-18

[11] Salami A.W., 2003, *" Assessment of the level of water pollution along Asa River Channel, Ilorin, Kwara State, Nigeria"* Nig. J. Pure Appl. Sci. 18: 1423 -1429

[12] Santschi P.H., Presleyb B.J., Wadec T.L., Garcia - Romeroe B., and Baskaran M., 2001, *" Historical contamination of PAHs, PCBs, DDTs and heavy metals in Missisipi River Delta, Galveeston Bay sediment cores"* Marine Environmental Research, 2, 387 -392

[13] Yerima F.A.K., Daura M.R and Gambo B.A., 2008, *"Assessment of groundwater quality of Bama town, Nigeria"* J. of Sustainable Development in Agriculture and Environment 3(2) 128 -137

22

Assessing Water Quality in the Developing World: An Index for Mexico City

Fabiola S. Sosa-Rodriguez
University of Waterloo, Department of Geography & Environmental Management, Canada

1. Introduction

Safe water supply is one of the most significant worldwide concerns. Since water has become a scarce resource in some regions, over the last several decades efforts have generally focused on supplying more water to people, without considering its quality. Although the quantity of water that people receive influences hygiene and promotes public health, poor quality of water also affects humans and the environment, increasing famine, child mortality, waterborne diseases, environmental deterioration, and social inequalities. The likelihood of being infected by waterborne diseases is not homogeneous among all age groups, indeed the groups most likely to be affected by these diseases are children under five years, the elderly, and people living in poor hygiene conditions (no-running water or proper waste disposal) (World Health Organization [WHO] & United Nations Children's Fund [UNICEF], 2011; WHO, 2008).

Today, freshwater resources (surface and groundwater) are threatened by over-exploitation and pollution. Consequently, they do not necessarily meet the quality standards for ensuring safe human consumption, and must undergo a purification process capable of 1) reducing turbidity, odors and unpleasant taste; 2) removing suspended solids, natural organic compounds (i.e., ammonia nitrogen), synthetic organic compounds (i.e., oil, detergents, pesticides and industrial solvents), and inorganic compounds (i.e., lead, cadmium, mercury, copper and zinc); and 3) eliminating pathogenic microorganisms (bacteria, viruses, protozoa and parasitic worms) that can cause outbreaks of waterborne diseases. Some of these compounds and microorganisms may be present in drinking water, but they must not exceed specific ranges if water safety is to be assured (WHO, 2011; Semarnat, 2011; U.S. Environmental Protection Agency [U.S. EPA], 2001). Nevertheless, some water management practices have increased people's exposure to health risks and insufficient water supply. For example, untreated wastewater discharge to freshwater sources, run-off from agricultural fields with high concentrations of pesticides, or wastewater reuse for crop irrigation.

Use of water polluted with microorganisms and organic and inorganic compounds affects people's health, increasing their exposure to waterborne and other water-related diseases. Waterborne diseases are infectious diseases caused by bacterial, viral or protozoa infections spread primarily through contaminated water (Ashbolt et al., 2001; Grabow et al., 2001;

Payment et al., 2003). The majority of these diseases is transmitted through fecal waste; therefore, lack of or low water quality contributes to their dissemination. The most common waterborne diseases include cholera, typhoid, dysentery, amoebic dysentery, and diarrhea (WHO, 2011). Although water is an important source for the generation and transmission of infectious organisms, waterborne diseases can also be disseminated through different routes, including the ingestion of contaminated food (either due to lack of hygienic conditions for its preparation or because crops were irrigated with untreated wastewater), person to person contact, unsanitary living conditions and/or poor hygiene.

Water-related diseases can as well be caused by high concentrations of organic and inorganic compounds. Many of these compounds exist naturally, but their concentration has increased as a result of human activities. For instance, high concentrations of pesticides, lead, fluoride, nitrates, arsenic, and other heavy metals can affect the nervous system and kidneys; and also cause reproductive disorders, cancer, skin lesions, endocrinal damage, and vascular diseases (WHO, 2011, 2004; Thompson et al., 2007; U.S. EPA, 1998).

People's access to safe water must be assured to guarantee healthy life. To this end, monitoring and assessing the quality of the water that people receive must be an essential component of water management. New methods that support effective water quality management need to be design and implemented. This chapter proposes a method to evaluate the effectiveness of water management in providing safe water, using as an example a study case of Mexico City. A water quality management index (*WQI*) was estimated by principal component analysis. The results were normalized and transformed into decimal units to simplify the data into a language that decision makers can easily understand. Since the quality of the water supply in this city varies among its neighborhoods, it was necessary to identify the distribution and location of the most vulnerable groups using an *ad hoc* Geographical Information System (GIS).

2. Water management in Mexico City

Mexico City is one of the biggest cities worldwide, characterized by important economic, social, environmental and cultural disparities. Its population is around 9 million people distributed in 16 boroughs, but in its metropolitan area are concentrated more than 18 million people. Although many people have access to basic services, including water supply that meet quantity and quality standards, others (mainly those located in irregular settlements) lack access to these services. The city concentrates the highest infrastructure and employment facilities in Mexico.

Mexico City receives 35.1 m^3/s of water from various internal and external sources. Groundwater from the aquifers of the Basin of Mexico is the main internal source, providing 42% of the total supply. Since the amount of water extracted from the aquifers (20.7 m^3/s) is almost three times the natural recharge capacity of this basin (7.9 m^3/s), cracks in the subsoil of these aquifers have occurred (Sosa-Rodriguez, 2010a). These cracks have exposed groundwater to direct pollution caused by wastewater and garbage leaching, with high concentrations of fecal coliform bacteria, chromium, lead, nickel, cadmium, mercury, and arsenic (Mazari, 1996). Another consequence of the city's groundwater over-exploitation is soil subsidence, which has caused the breakage of water and sewage pipelines and wastewater flooding in some areas of the city.

The most important external sources are the Cutzamala and Lerma Systems. These sources provide 43% of the total supply. Water from these external sources must be transported almost 127 km and pumped up nearly 1,100 m to reach the city (Conagua, 2002). This process requires a large amount of electricity, and has increased the dependency of Mexico City on external sources of both power and water. These water management practices are not sustainable; consequently, city authorities face increasing challenges to supply its inhabitants with a volume of water that meet minimum requirement and quality standards.

Most of the city's inhabitants (86% of the total) have access to water directly from pipelines. The other 14 per cent acquire this resource mainly by car tanks (88 per cent of that total), but also via wells, rivers, streams and springs (INEGI, 2005). These alternative sources do not ensure safe water consumption, and sometimes are more expensive. It is estimated that 1.25 million people are exposed to several risks generated by the lack of safe water supply (Sosa-Rodriguez, 2010a).

Mexico City also generates 25 m^3/s of wastewater, but treats only 9% of this volume (SACM, 2008). The remaining 91 per cent of wastewater is discharged without any treatment; thereby further polluting rivers which are used to transfer effluent from the city to the sea. Unfortunately, agricultural areas located in the Valley of the Mezquital and the Valley of Tula use this wastewater for farming activities (Conagua & Semarnat, 2006). This practice has had severe impacts on the health of producers and consumers of these crops, which include sorghum, barley, oat, wheat, corn, tomatoes, carrots, onions and coriander. Furthermore, it also pollutes the soil, groundwater and air of the areas where these crops are grown (Esteller, 2000). Evidence of this pollution has been a documented increase of waterborne diseases. For example, the morbidity rate caused by *Ascaris lumbricoides* in children between zero and four years increases from 2.7 to 15.3 per thousand children in areas where wastewater is used to irrigate crops. Similarly, the morbidity rate caused by *Entamoeba histolytica* increases for individuals between five and 14 years, from 12.0 to 16.4 per thousand (Sosa-Rodriguez, 2010b; Romero, 1994).

2.1 Water quality analysis

To disinfect drinking water, authorities at the beginning of the water distribution network apply an excessive amount of chlorine (approximately 2mg/liter), which oxidizes organic and inorganic materials, in addition to removing pathogens (bacteria, viruses and protozoa). During the process of water distribution, the residual chlorine reacts at full strength at first, but then decreasing in concentration until it disappears in the most distant points of the network. The gradual diminution of chlorine concentration explains why in some parts of the distribution network chlorine is absent, and so is the desired water disinfection.

Residual chlorine concentrations should be within a specific range considered safe since in excess chlorine is toxic, in addition to generating high concentrations of chlorine byproducts, including trihalomethanes (i.e., chloroform [CHCl 3], bromoform [CHBr3], dibromochloromethane [CHBr2Cl]) and haloacetic acids (i.e., monochloroacetic acid, dichloroacetic, trichloroacetic) (Thompson et al., 2007; U.S. EPA, 2001, 1998).

According to the Mexican National Water Law, water provided to residents must be free from microorganisms and any substance that could produce adverse physiological effects and cause harm to human health. Parameters such as residual chlorine or fecal coliform bacteria are the most commonly used indicators for evaluating water quality. Based on the norm NOM-127-SSA1-1994, the amount of residual chlorine accepted as safe fluctuates from 0.2 to 1.50 mg/liter (Semarnat, 2011). This range is wider than the standards established by the U.S. Environmental Protection Agency, in which residual chlorine concentration in drinking water must be between 0.1 and 0.3 mg/liter (U.S. EPA, 2001, 1998). Therefore, the city's inhabitants are not only exposed to health risks caused by low residual chlorine concentration, but also to diseases associated to a greater exposure to chlorine byproducts. These byproducts are classified as carcinogens, and can also affect the functions of vital organs such as the liver or kidneys (Craun et al., 2001; Lindquist, 1999; U.S. EPA, 2001).

Pathogenic microorganisms must not be present in drinking water; nevertheless, the existence of heavy metals (i.e., mercury, cadmium, chromium, lead and zinc), nitrates, nitrites, fluorides, chlorides, sulfates, DDT, pesticides and dissolved solids is allowed if the concentration of these organic and inorganic compounds is below the ranges in which they are considered a threat to people's health (Table 1).

Chemical Substance	Concentration Limit	Chemical Substance	Concentration Limit	Chemical Substance	Concentration Limit
Aluminum=	0.20 mg/l	Zinc=	5.00 mg/l	Cyanide=	0.07 mg/l
Cadmium=	0.005 mg/l	Nitrites=	0.05 mg/l	DDT=	1.00 mg/l
Chromium=	0.05 mg/l	Nitrates=	10 mg/l	Hexachlorobenzene=	0.01 m g/l
Copper=	2.0 mg/l	Chlorides=	250 mg/l	Trihalomethanes=	0.20 m g/l
Iron=	0.30 mg/l	Fluorides=	1.50 mg/l	Pesticides=	0.03 mg/l
Lead=	0.025 mg/l	Sulfates=	400 mg/l	Solids Dissolved=	1000 m g/l
Mercury=	0.001 mg/l	Arsenic=	0.05 mg/l	Residual Chlorine=	0.2-1.50 mg/l

Source: Mexican National Water Law, Norm NOM-127-SSA1-1994

Table 1. Concentration limits for chemical substances in drinking water

The Mexico City Water System (SACM) reported in 2007 that 2 per cent of samples analyzed did not satisfy residual chlorine standards. For fecal coliform analysis, pathogenic microorganisms were identified in 12 per cent of the samples (SACM, 2008). In both cases, the most affected areas were south and southeast of Mexico City. It is important to note that water quality monitoring in this city must be improved since the sampling used is not representative: over 80%of the samples tested are gathered from less-affected areas; thus, 20% or fewer samples are collected from the areas reported to be affected by poor water quality (Sosa-Rodriguez, 2010a).

Due to financial constraints, since 1997 the analysis of samples has been reduced from 160,000 per year to less than 30,000 per year. This has increased the exposure to risks that could adversely affect the heath of the city's residents because local authorities are less capable of identifying water quality problems. Additionally, because of the decline of water source quality, the use of chlorine as the only disinfection mechanism is no longer sufficient for providing safe water (Sosa-Rodriguez, 2010a). For example, bacteria such as *Helicobacter pylori*, total coliforms, fecal coliforms, Streptococci and *Vibrio* spp. have been found in some water samples (Mazari-Hiriart et al., 2005).

2.2 Waterborne diseases mortality

Mexico City is the entity in the country with the highest total mortality rate, with 566 deaths per 100,000 inhabitants. The leading causes of death in the capital are chronic degenerative diseases (i.e., heart attacks, diabetes and tumors), with 489 deaths per 100,000 inhabitants. In the case of intestinal infectious diseases (including typhoid, salmonella, bacterial food poisoning, amebiasis, and dysentery), they are the 19th cause of death, accounting for 0.6% of the total deaths in the city. Although in general the intestinal infectious diseases give the impression that they do not constitute a health problem in the city, this is the main cause of death for infants under four years, explaining 9.7% of the total number of deaths in children under four years. The second cause of death in this group was acute lower respiratory infections (8.9% of total deaths in this age group) (Table 2).

In 2007, the Ministry of Health in Mexico City registered 280 deaths due to intestinal infectious diseases: 93.6% of the total deaths were caused by diarrhea and gastroenteritis, and the remaining 6.4% were attributed to salmonella, *clostridium perfringens* poisoning, typhoid fever, amoebic dysentery and viral intestinal infections (Table 2). The largest number of deaths cause by intestinal infectious diseases occurred southeast of the city, and the borough most affected by these diseases was Iztapalapa, where 22.9% of the total deaths were registered.

Group disease	Rate[1]	Group disease	Rate[1]
Chronic-degenerative	489.2	Total Mortality Rate[1]	565.9
Infectious	62.2	Infant Mortality Rate[2]	20
Injuries	40.57		

Type of disease	Position	Number of deaths	Mortality Rate[1]	Intestinal infectious diseases: Disease	Deaths	Proportion
Cardiovascular	1	10 062	114.1			
Diabetes	2	8 270	93.8	Diarrhea/ gastroenteritis	262	93.6
Cancer	3	6 759	76.7	Enteritis salmonella	6	2.1
Intestinal Infections	19	280	3.2	*Clostridium perfringens*	5	1.8
Respiratory Infections	20	272	3.1	Typhoid Fever	2	0.7
	Total	49 882	565.9	Total	280	100.0

1 Rate/100,000 individuals
2 Rate/10,000 live births
Source: Ministry of Health in Mexico City (SSDF), 2007.

Table 2. Mortality rates in Mexico City

3. Water quality management index (*WQI*)

The water quality management index (*WQI*) was estimated by Principal Component Analysis (PCA) to determine to what extent water management in Mexico City is capable of ensuring safe water supply to the city's inhabitants. This method was chosen due to its capacity to summarize several variables that were measured with different units and to simplify the results interpretation.

Another advantage of PCA is that it avoids problems such as multicolinearity and heteroscedasticity; both problems make the parameters of models biased. In the case of multicolinearity, highly correlated variables can complicate the identification of the individual effect that each variable has on the sample. Regarding heteroscedasticity, which is attributed to non-constant variations in the residuals of a model, the variance of the parameters estimated increases.

The input variables used in PCA were selected based on their capacity to explain the effectiveness of the city's water management in providing a safe water supply—that is, water is free of any pathogenic microorganism, metals and toxic substances so that outbreaks of waterborne and other water-related diseases (i.e., endocrine and reproductive damage, poisoning, skin lesions, and cancer) are prevented. These variables included concentrations of residual chlorine (*RChlor*), the presence of fecal coliform bacteria in water samples (*Fecalbac*), the total number of deaths caused by water pollution (*Totalmort*), and the total number of deaths of children under five years due to waterborne diseases (*Infantmort*). The variables water purification and wastewater treatment plants did not contribute to explain the city authorities' performance in ensuring safe water consumption; therefore, they were not included in the estimated model.

The database and the *ad hoc* GIS built—disaggregated by neighbourhood—were based on information reported by the National Water Commission (Conagua), the National Institute of Statistics and Geographical Information (INEGI), and the Mexico City Water System (SACM)—who together comprise the national and local governmental authorities responsible for managing the water in Mexico and Mexico City.

The index was normalized and transformed into decimal units. By using the 'pass'/'fail' criteria defined by the Mexican Educational System for evaluating the performance of students, and the model outputs were assessed in each neighbourhood. This was done to simplify the results for decision makers, who can easily identify in which neighborhoods water quality problems need to be addressed immediately. Therefore, if $WQI_i = 10$, the water management in the neighbourhood *i* is effective to provide safe water supply. If $WQI_i \geq 6$, then water quality is not problematic and do not endanger people's health in the neighbourhood *i*. However, if this index is smaller than this value ($0 \leq WQIi \leq 6$), water quality problems must urgently be addressed, as they may present great risk to the inhabitants of the neighbourhood *i*.

It is important to consider that *WQI* has been based on 2007 data. Therefore, the accuracy and reliability of the resulting model in other spatial and temporal contexts must still be evaluated.

4. Results

According to results of the *WQI* for Mexico City, the city authorities' performance in providing inhabitants with a volume of water that meets quality standards is unsatisfactory. A score of 6.1 is associated with a performance that needs to be improved; consequently, governmental strategies to guarantee safe drinking water have not had the desired results, and low water quality in Mexico City threatens people's health. The equation that resulted from PCA (Table 3) for assessing authorities' performance in providing safe water supply is (Equation 1)

$$WQI = 0.61705\ Fecalbac + 0.60687\ RChlor - 0.12889\ Totalmort - 0.08876\ Infantmort \quad (1)$$

Component 1 Water quality analysis		Component 2 Waterborne diseases mortality	
Variable	Charge	Variable	Charge
RChlor	0.60687	*Infantmort*	-0.12889
Fecalbac	0.61705	*Totalmort*	- 0.08876

Extraction method: Principal component Analysis.
Rotation method: Varimax with Kaiser Normalization

Table 3. *WQI* component score coefficient matrix

It is important to note that only 69.03% of the variability of the sample used to evaluate the city authorities' performance in ensuring safe water supply was explained by the variables incorporated in the model: 36.38% is attributed to the residual chlorine concentrations and fecal coliform existence (the first component); the remaining 32.65% corresponds to deaths caused by water pollution and waterborne diseases (the second component) (Table 4). These results can be modified if other variables, such as morbidity rates of intestinal infectious diseases, are incorporated in the analysis. Unfortunately, in the case of Mexico City, the information available for morbidity rates is not systematized by the origin of the people infected; it is only reported by kind of disease and the health facility involved in treatment.

	Initial Eigenvalues			Extraction Sums of Squared Loadings		
Component	Total	% of variance	Cumulative %	Total	% of variance	Cumulative %
1	1.345	50.263	50.263	1.068	36.381	36.381
2	1.053	22.455	72.718	1.017	32.653	69.034
3	0.941	15.545	88.263	0.976		
4	0.661	11.737	100.000	0.659		

Extraction Method: Principal Component Analysis
Rotation method: Varimax with Kaiser Normalization

Table 4. *WQI* total variance explained

Although low water quality is a problem widespread in the city, some neighbourhoods are more likely to be affected due to the lack of economic, social and political capacities of their residents to find alternative water sources. These neighborhoods are located south and southeast of Mexico City, and the majority of their residents have low incomes, low education levels, poor housing conditions, and their demands tend not to be a priority of the local government (Figure 1).

In particular, the areas most affected by the lack of safe water supply as a result of low residual chlorine concentrations are located southeast of Mexico City. The presence of fecal coliform bacteria in some samples that met residual chlorine standards in several neighborhoods confirmed that chlorination does not guarantee the elimination of all pathogenic microorganisms that can affect health.

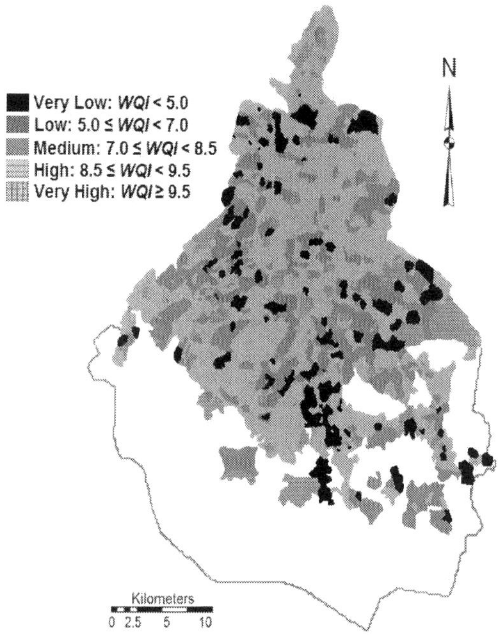

Source: Based on the results estimated by PCA.

Fig. 1. Water Quality Management Index (WQI)

Furthermore, failure to identify in water samples coliform bacteria, whose presence proves fecal contamination, does not guarantee safe water consumption. Regularly, low concentrations of coliform bacteria (i.e., *E. coli*) are eliminated by disinfection with chlorine (or its derivatives); however, some pathogens can resist chlorine disinfection. For example, enteric viruses and some protozoa that can cause viral and parasitic diseases are more resistant to chlorine disinfection than are fecal coliform bacteria; as a result, these pathogens can survive (Spellman, 2008; WHO, 2008; U.S. EPA, 2001) (Figure 2 and 3). Their removal from drinking water requires alternative water disinfection processes that strengthen chlorine effects. Additionally, water quality varies with changes in temperature, precipitation, pipeline pressure, and intermittent supply, and thus, needs systematic monitoring (Ainsworth, 2004).

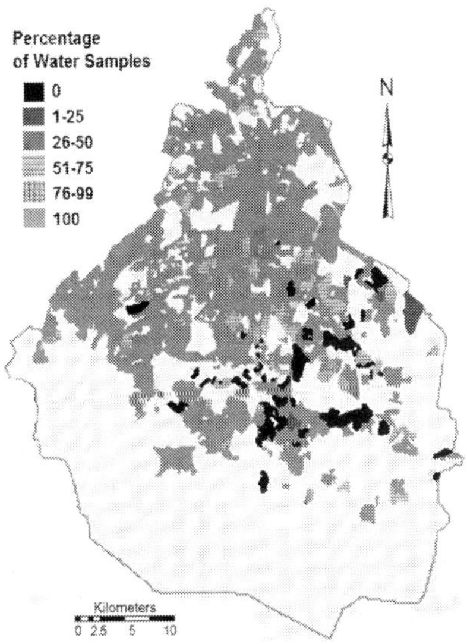

Source: SACM Databases (2007): Water Quality Analysis.

Fig. 2. Water Samples that Meet Residual Chlorine Standards

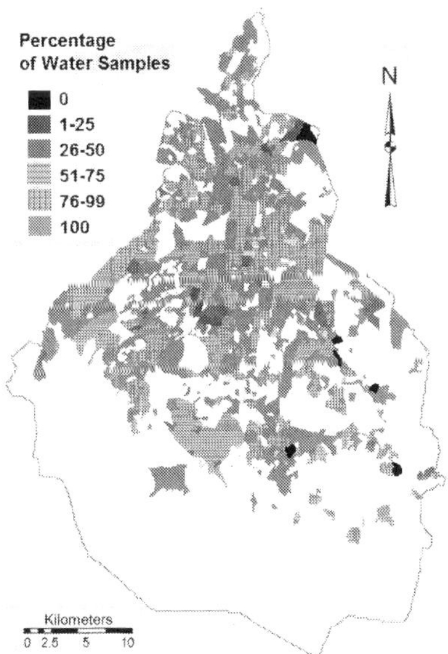

Source: SACM Databases (2007): Water Quality Analysis.

Fig. 3. Water Samples that Meet Fecal Coliform Bacteria Standards

Based on the analysis, improvements in water supply (i.e., infrastructure maintenance and construction, volume supply increase, among others) clearly lead to lower infant mortality rates, but not fewer cases of intestinal infectious diseases. In the case of these diseases, results suggest that advances in sewage disposal and treatment are likely to reduce the number of deaths. This can be explained because greater concentrations of the infrastructure for water supply and sewage disposal are mainly located in the zones that are physically less vulnerable, where live populations with medium and high average income levels (Table 5).

	V1	V2	V3	V4	V5	V6	V7	V8	V9	V10	V11	V12	V13
V1	1.000	-0.668	-0.557	-0.054	-0.481	-0.510	0.102	0.422	0.039	-0.042	-0.055	0.161	0.219
V2	-0.668	1.000	-0.596	0.040	-0.172	-0.086	-0.542	0.090	-0.006	-0.347	-0.432	0.039	0.030
V3	-0.557	-0.596	1.000	-0.039	-0.275	-0.266	0.002	-0.228	-0.553	-0.085	-0.095	0.120	0.084
V4	-0.054	0.040	-0.039	1.000	0.023	0.361	-0.595	0.014	-0.022	0.019	0.035	-0.045	0.032
V5	-0.481	-0.172	-0.275	0.023	1.000	0.928	-0.002	-0.680	-0.130	0.671	0.675	-0.684	-0.618
V6	-0.510	-0.086	-0.266	0.361	0.928	1.000	-0.003	-0.643	-0.092	0.579	0.601	-0.657	-0.596
V7	0.102	-0.542	0.002	-0.595	-0.002	-0.003	1.000	0.034	0.293	-0.038	-0.043	0.014	0.077
V8	0.422	0.090	-0.228	0.014	-0.680	-0.643	0.034	1.000	0.159	-0.194	-0.275	0.492	0.652
V9	0.039	-0.006	-0.553	-0.022	-0.130	-0.092	0.293	0.159	1.000	-0.694	-0.755	0.058	-0.025
V10	0.075	-0.006	-0.057	-0.009	0.066	0.090	0.053	-0.063	-0.036	0.034	0.039	-0.052	-0.067
V11	-0.019	0.051	0.007	0.022	-0.050	-0.032	0.016	-0.020	-0.109	-0.020	-0.014	0.027	-0.018
V12	-0.042	-0.347	-0.085	0.019	0.671	0.579	-0.038	-0.194	-0.694	1.000	0.989	-0.302	-0.316
V13	-0.055	-0.432	-0.095	0.035	0.675	0.601	-0.043	-0.275	-0.755	0.989	1.000	-0.313	-0.225
V14	0.161	0.039	0.120	-0.045	-0.684	-0.657	0.014	0.492	0.058	-0.052	0.027	1.000	0.572
V15	0.219	0.030	0.084	0.032	-0.618	-0.596	0.077	0.652	-0.025	-0.067	-0.018	0.572	1.000

V1= Slope
V2= Land Subsidence
V3= Flooding Areas
V4= Water consumption
V5= Households with piped water
V6= Water supply pipeline network
V7= Leaks
V8= Deaths of children under 5 years
V9= Deaths caused by waterborne diseases
V10= Households connected to the sewerage network
V11= Sewerage network
V12= Iliteracy
V13= Low income (less than 6 US dollars per hour)

Note: Each element of the correlation matrix (a_{ij}), known as correlation coefficient (r), has a value between -1 and 1. If $|r|$ is close to unity, the relationship among analyzed variables is stronger. When $0 < r \le 1$, this relationship is positive; when $-1 \le r < 0$, it is negative; and when $r \approx 0$, there is no relationship among variables.

Table 5. Correlation Matrix

Therefore, people with low income levels are more affected by the lack of these services. Although alternative water sources, such as car tanks, have been used to supply with water the most affected areas of the city, other ways to dispose of sewage are basically to discharge it into rivers, springs, or directly into the soil, polluting the water, air and soil of the affected areas, and thus, increasing waterborne disease occurrence. There was no evidence to conclude that interrupted water supply was related to problems in pipeline network pressure; nevertheless, this water supply problem is shown to be associated with the presence of leaks mainly caused by the city's land subsidence, and lack of pipeline maintenance and their over-capacity operation (Table 5). In this context, water management decision-makers must consider that even though people have access to water supply; this does not ensure safe water consumption since this resource is not always provided continuously and does not always met quality standards.

5. Conclusions

Successful measures to mitigate and prevent negative impacts caused by lack of safe water and adequate sewage discharge and treatment must consider social and spatial differences in people's capacity to cope with low water quality. To this end, alternative methods that support the water management decision-making must be developed, and the use of advanced technological tools, such as Geographical Information Systems (GISs), can improve the identification of those areas most exposed to low water quality, as well as the most vulnerable groups. Thus, the Water Quality Management Index proposed in this chapter is useful for assessing the effectiveness of water management in ensuring safe water consumption. Nevertheless, it is still necessary to evaluate the accuracy and reliability of this index in other spatial and temporal contexts to evaluate its capacity to measure changes in water quality analysis or mortality rates.

On the other hand, the results of the Water Quality Management Index (*WQI*) provide evidence that the use of chlorine to disinfect drinking water is not as effective as is generally thought, so water disinfection needs to be complemented with other methods that can remove a broad spectrum of pathogens such as *E. coli*, *Klebsiella* spp., Faecal streptococci, *Vibriosis*, and *Helicobacter pylori*. These microorganisms have been identified in some water samples in Mexico City, and their presence threatens the health of the city's inhabitants making them more likely to be affected by outbreaks of waterborne diseases. Among methods considered more effective for removing bacteria, inactivate viruses, and protozoa are ozonation, ultraviolet radiation, ion exchange, inverse osmosis, and coagulation; the pathogen target to be removed will dictate the disinfection method selected.

Another measure that can contribute to ensure safe water consumption is to improve the bacteriological water quality analysis by incorporating in the indicators used the presence of enteric viruses (such as *rotavirus* or *adenovirus*), protozoa (Giardia and *Cryptosporidium*) and bacteria (including Fecal Streptococci as *Helicobacter Pylori* and Legionella), in addition to the existence of fecal coliform bacteria. Additionally, it is necessary to evaluate the different routes of waterborne disease transmission, considering both the consumption of polluted water and the ingestion of food irrigated with wastewater or that was prepared in unhygienic conditions.

Finally, it is vital to create an autonomous institution responsible for systematically assessing the physical, chemical and biological characteristics of the water supply to Mexico City's inhabitants. This institution must be independent from the SACM, so it can perform a reliable analysis of the quality of the water–it is difficult to be a judge and defendant simultaneously. This institution should have power to sanction SACM when water quality standards are not met and have the expertise necessary to make recommendations aimed to improving water disinfection.

6. Acknowledgements

This research was conducted with financial support from the Mexican National Council for Science and Technology (Conacyt). Contributions from the Mexico City Water System (SACM) are greatly appreciated. The author thanks Dr. Jean Andrey, Dr. Peter Deadman, and M.Sc. Linda Mortsch the opportunity given to pursue postdoctoral research at the University of Waterloo, and is grateful for the invaluable assistance of Mary McPherson (ELPP—University of Waterloo).

7. References

Ainsworth, R. (2004). *Safe piped water: Managing microbial water quality in piped distribution systems*. World Health Organization, Geneva.

Ashbolt, N.J., Grabow, W.O.K. & Snozzi, M. (2001). Indicators of microbial water quality, In: *Water quality – Guidelines, standards and health: Assessment of risk and risk management for water-related infectious disease*, Fewtrell, L. & Bartram, J. (Eds.), pp. 289-315, IWA Publishing, London.

Conagua (2002). *Plan Hidráulico de la Región XIII, 2002-2006* [Hydraulic Plan of the Region XIII, 2002-2006], Conagua, Mexico City.

Conagua & Semarnat (2006). *Hacia una estrategia de manejo sustentable del agua en el Valle de México y su zona metropolitana [Towards a Strategy for Sustatinable Water Management in the Valley of Mexico and its Metropolitan Area]*, Conagua/ Semarnat/World Water Council (WWC), Mexico City.

Craun, G.F., Hauchman, F.S. & Robinson, D.E. (2001). *Microbial pathogens and disinfection byproducts in drinking water: Health effects and management of risks. Conference Conclusions*, pp.533-545, ILSI Press, Washington.

Esteller, M.V. (2000). Reutilización de las aguas residuales en México [Wastewater reuse in Mexico], *In: Sequía en un mundo de agua [Drought in a World of Water]*, Anton D. & Díaz Delgado, C. (Eds.), pp. 299-320. CIRA/UAEM/Pirigazu, San Jose/ Toluca.

Grabow, W.O.K., Taylor, M.B. & De Villiers, J.C. (2001). New methods for the detection of viruses: Call for review of drinking water quality guidelines. *Water Science and Technology*, Vol. 43, pp. 1-8.

INEGI (2005). II Conteo de Población [2nd Population Count], Database, Aguascalientes, Mexico, INEGI.

Lindquist, H.A.D. (1999). *Emerging pathogens of concern in drinking water*, EPA, Washington, EPA600/R-99/070.

Mazari, M. (1996). *Agua vs población [Water versus People]*, El Colegio Nacional de Ingenieros, Mexico City.

Mazari-Hiriart, M., López-Vidal, Y., Ponce-de-León, S., Calva, J.J., Rojo-Callejas, F. & Gonzalo, C.R. (2005). Longitudinal study of microbial diversity and seasonality in the Mexico City metropolitan area water supply system. *Applied and Environmental Microbiology*, Vol. 71, No.9, pp. 5129–5137.

Payment, P., Walter M. & Dufour, A. (2003). Introducing Parameters for the assessment of drinking quality, In: *Assessing microbial safety of drinking water. Improving approaches and methods*, Durfour, A. (Ed.), IWA Publishing/WHO/OECD, London.

Romero, L.P. (1994). Problemas socio-ambientales en la gestión del agua *[Socio-environmental problems in water management], In: Medio ambiente. Problemas y soluciones [Environment. Problems and Solutions]*, Yunez-Naude, A. (Ed.), Colmex, Mexico City.

SACM (2008). Analysis of Drinking Water Samples, Databases 2007, Water Quality Laboratory, Mexico City.

Semarnat (2007). Catalogo de Normas Oficiales Mexicanas [Mexican Official Standards Catalogue], Semarnat-Conagua, Mexico City.

Sosa-Rodriguez, F.S. (2010a) Exploring the risks of ineffective water supply and sewage disposal: a case study of Mexico City. Environmental Hazards, Vol. 9, pp. 135–146.

Sosa-Rodriguez, F.S. (2010b). Impacts of Water-management Decisions on the Survival of a City: From Ancient Tenochtitlan to Modern Mexico City. *Water Resources Development*, Vol. 26., No. 4, pp. 675–687.

Spellman, F.R. (2008). *The Science of Water: Concepts and Applications*, Taylor & Francis Group, London & New York.

Thompson, T., Fawell, J., Kunikane, S., Jackson, D., Appleyard, S., Callan, P., Bartram, J. & Kingston, P. (2007). *Chemical safety of drinking-water: Assessing priorities for risk management*, World Health Organization, Geneva.

US EPA (U.S. Environmental Protection Agency) (2001). *Controlling Disinfection byproducts and Microbial Contaminants in Drinking Water*, EPA, Washington. EPA 600/R-01/110.

US EPA (U.S. Environmental Protection Agency) (1998). *National Primary Drinking Water Regulations: Disinfectants and Disinfection Byproducts, Final Rule*, EPA, Washington.

World Health Organization (WHO) (2011). Guidelines for drinking-water quality: recommendations, World Health Organization, Geneva.

World Health Organization (WHO) (2004). *Inheriting the World: The Atlas of Children's Health and the Environment*, World Health Organization, Geneva.

World Health Organization (WHO) & United Nations Children's Fund (UNICEF) (2011), *Rapid assessment of drinking-water quality: A handbook for implementation,* WHO & UNICEF, Geneva.

23

Water Quality Degradation Trends in Kenya over the Last Decade

Shadrack Mulei Kithiia
Postgraduate Programme in Hydrology,
Department of Geography and Environmental Studies,
University of Nairobi, Nairobi,
Kenya

"Once the water is polluted, life is polluted; once the water is exhausted, civilization is exhausted; once the water disappears, the human being disappears; protecting the water resources is protecting the human being".
Mingjun, J. (2010).

1. Introduction

Water is described as both a chemical substance vital to life on earth and a means for navigation. Water, is a commodity that is consumed, and carrier of other substances or properties, such as heat, disease vectors, pollutants and energy (Jordaan *et al.* 1993). But, whereas the total quantity of water on earth remains constant, its quality changes in both time and space. The problem of water quality was articulated by Jordaan *et al.* (1993) as causing a great strain on water supply systems, especially in cities along river courses. Similarly, in 1985, at the midpoint of the international water supply and sanitation decade, it was pointed out that although 870 million people lived in urban areas of the developing world, roughly 1.6 million were rural inhabitants and approximately 22% of the urban group was lacking water supply service and 40% were without sanitation. The population represented 64% and about 85% as lacking water supply and sanitation services to the rural population, respectively (Jordaan, *et al.* 1993). In essence, the rural inhabitants are faced with a great threat of using low quality water compared to the urban counterparts. This explains the existent problem of water quality degradation not only in the urban areas but also in the rural areas.

The problem of water pollution and quality degradation in the developing countries is increasingly becoming a threat to the natural water resources. This phenomenon is attributed to the increasing quest of these countries to attain industrialization status and diversification of the national development goals and Kenya is no exception to this phenomenon (Kithiia,S.M & Khroda, G.O, 2011).

Kenya is described as a water scarce country, and yet future projections show that per capita available water currently at 650m^3/year, will likely drop to 359m^3/year by 2020, as a result of population growth. This figure is far much below the global accepted value of 1000m^3

year^{-1} per capita level. Urgent action is therefore needed to increase the capacity of the water sector to improve the availability and accessibility of clean and safe drinking water (GoK, 2007). Many Kenyans especially in the rural areas have limited access to quality water. They walk for long distances in search of this precious commodity and use it raw and untreated from rivers, lakes and dams. The untreated water is not only turbid, but also contain disease causing bacteria and in some cases chemicals. The problem is further compounded by the seasonal fluctuations of water availability which is more correlated to the seasonal patterns of rainfall in most parts of the country.

"Water is essential to sustain life and a satisfactory (adequate, safe and accessible) supply must be available to all. Improving access to safe drinking water can result in tangible benefits to health. Every effort should be made to achieve a drinking water quality as safe as practicable". (**W.H. O; Guidelines for Drinking Water Quality, 1991, 1997**).

In addition, water resources in Kenya are under pressure from agricultural chemicals (fertilizers and herbicides) and urban and industrial wastes, as well as from use for hydroelectric power. Thus, increase in demand for water resources for various uses is likely to cause many constraints and lead to the pollution of the available water resources. Water pollution from urban and industrial wastes poses another environmental problem. Kenya has 20.2 cubic kilometers of renewable water resources with 76% being used in farming activity and 4% used for industrial purposes. Only about 42% of the residents in rural areas and 88% of city dwellers have pure drinking water. The situation is much worse in the informal settlements in the urban areas due to lack of proper sanitation and sewerage system. These settlements are located near or on top of the rivers courses draining the city of Nairobi and contribute significant pollutants amounts and water quality degradation.

In addition, water resources in Kenya are increasingly being polluted by organic, inorganic and microbial matter (Kithiia, 1992 & 1997). The impact of pollution on water resources is manifested by water of poor quality which gives rise to water toxicity to mammals and aquatic life; loss of aesthetic value by becoming unsuitable for recreational activities, high cost of water supply as polluted water is expensive to treat, eutrophication, deoxygenation, acid rain and habitat modification. This is evident in the urban centers and more in the city of Nairobi.

Kenyan urban population which has been growing at a rate of 8% per annum is now more that 27% of the country's total population as described in the state of the Environment report of 2003 (NEMA, 2004). In addition, generation of solid, liquid and gaseous wastes has been increasing at the same level as industrial development and the diversification of consumption patterns. The report by NEMA (2004) indicates that per capita waste generation ranges between 0.29 and 0.66 kg day^{-1} within the urban areas of the country, and that of the municipal waste generated in the urban centres is 21% emanatting from industrial areas and 61% from residential areas.

The city of Nairobi is the heart of industrial production in Kenya and the commercial hub of the East African Community (EAC). It is drained by three tributaries of the Nairobi River, namely; Ngong, Nairobi and Mathare rivers. These rivers collect most of the wastes generated in the city and its waters are heavily polluted by solid, liquid and organic wastes. The polluted water has a strong impact on human beings and other living organisms and the environment in general. Nairobi River has acted as a major depository of waste from both domestic and industrial sources. The riparian reserves of the three main rivers are marked with numerous informal settlements without adequate sewerage and sanitation services and hence discharge their raw sewage into the rivers; and most industries find it cheaper to discharge their effluents into the rivers without any satisfactory treatment (UPIU, 2009).There is therefore, a need to find strategic ways and methods of reducing pollution levels and manage the water quality status within the basin. This is the basis of the investigations explored in the present paper.

2. Background information – Water quality issues in Kenya

Trends in water use and demand are often a guide to the availability of water in terms of both quantity and quality. In many instances, an increase in water demand and use is directly proportional to deterioration in water quality. The amount of waste discharged tends to increase with rising water demand, although the relationship depends in detail on the amount of water and the specific use involved. The amount of water in a river depends on the type and number of water abstraction facilities along the course, the number of tributaries, amount of rainfall and distribution, soil type, temperature and the shape of the drainage basin, as well as the population structure. The nature and extent of human activity, be it industrial, agricultural or both, will in turn determine the magnitude and nature of pollution and the water quality status of the water course. Clearly, in order to avoid the dangerous consequences of serious water pollution, there is a need to properly understand the amount of water required and the volume of effluents to be discharged by any development projects planned for a drainage basin (Kithiia, S.M & Ongwenyi, S.O. 1997).

Water quality degradation problem is not a new phenomenon in Kenya. Initial research reports on the problem in the country dates back in the 1950's. In Kenya, the problem of water quality degradation was first exposed in the MOWD (1976 a&b) containing case studies of three rivers; Nzoia, the Nyando and Kerio. These reports contain the chemical characteristics of water shortly before and after establishment of factories along their courses. Nzoia River which drains into Lake Victoria carries the effluents discharged from Pan Africa Paper Mill in Webuye upstream and from Mumias Sugar Factory downstream; Nyando which also discharges its waters into Lake Victoria receives Molasses from Sugar Factories, Chemilil and Muhoroni. Kerio River drains the Kerio Valley with intermittent flow into Lake Turkana. This is now periodically polluted by effluents from fluorspar factory established three decades ago. All these three reports are a clear indication of the effects of industrial growth on the quality of water courses and their effluents are a major contributing factor to water quality degradation at the extreme vicinity of industrial activities. This was once more confirmed by a recent study done by the Ministry of Land Reclamation and Regional Development (MLRRD, 1993) (now the Ministry of Water and Irrigation). Many studies indicate that groundwater resources are not seriously threatened with water pollution problem but the major concern has been on the surface waters especially in river systems.

Other research reports which indicated water quality deterioration in Kenya are those by Ongwenyi (1979) and Njunguna (1978). Njunguna (1978) also added that pollution in the river (Nairobi-Athi-Sabaki) was chiefly due to domestic waste, industrial wastes and runoff (both rural and urban).

The most recent studies by Kithiia (1992, 2006), Okoth and Otieno (2000) and Mavuti (2003) revealed degradation trends in water quality within the river system due to changes and intensity in land use activities. The study by Kithiia (1992) in the same area revealed high concentration levels of Mercury and Lead surpassing the critical WHO and the Kenya standards guideline values (Mercury 0.03 and Lead 0.1 mg l^{-1}). In the same study pesticide residues notably DDT, *"Ambush"*, *"Ridomil"*, *"Malathion" etc* were all found to surpass the WHO and the Kenya standards for drinking water. DDT was measured at 0.000086 mg l^{-1}, "Ambush "0.1413 mg l^{-1}, "Ridomil" 0.147 mg l^{-1} and "Malathion" 0.00039 mg l^{-1}. In addition, the study by Kithiia (2006) within the same river systems indicated a downstream increase in water pollutants and water quality degradation. Sediment and Heavy metal concentrations were found to increase downstream the river courses. The increasing trends in pollution levels and water quality deterioration are attributed to increased land use activities and population growth rates within the urban area and the basins in general. This calls for stringent measures to address the problem and reduce the ever increasing water pollutants and water quality degradation. This should incorporate measures devoted towards the reduction of the water quality degradation trends within the Nairobi River basins in order to restore the water quality status as of the time when the city was established.

3. Problem discussion

While the amount of water available in a drainage basin normally remains relatively static, the demand for water often rises with changing human activities. The latter include a wide spectrum from changes in the basic way of living to changes in the technology employed to exploit natural resources (Dunne & Leopold, 1978; Kithiia, 1992). Water pollution often arises as a consequence and not only does a low quality supply result, but there may also be a major reduction in its quantity. The sources of water for Nairobi River sub-basins are the Ondiri springs, the Kabete and Karura forests, the Ngong hills, and all the streams which join downstream of Nairobi city to form the main Athi River. The total quantity of water in these streams is 23.6 m^3 s^{-1} which amounts to 36.7 X 10^6 m^3 $year^{-1}$. This amount of water, however, is less than the discharge of waste water from both industry and domestic use in the city. This is a major source of water pollution and hence water quality degradation within the basin and in Kenya generally.

Water demand increase and use is directly proportional to deterioration in water quality in many instances. In most cases, the amount of waste discharged tends to increase with rising water demand, although the relationship depends in detail on the amount of water and the specific use involved. Thus, the nature and extent of human activity will in turn determine the magnitude and nature of pollution as well as the water quality status of the water course/or source.

4. Water quality problems in Nairobi city

The main water pollution problems in the city of Nairobi are closely linked to the various land use systems. Nairobi city and its environs are the focus of the study. The area covers a total of 2435 km^2 and its part of the upper-Athi drainage basin. Its forms a greater part of the conglomeration of the urban centres of Ruiru, Ngong and Mavoko municipalities. The city and its environs are drained by three sub-streams of Ngong, Nairobi and Mathare rivers, which run in a southeast direction towards the Indian Ocean. These streams drain areas of diverse land use activities. The significant contribution of these land use activities to water pollution and quality degradation is quite enormous. Figure 1 shows the three streams and the major land use activities along the river courses from its headwaters downstream through the city of Nairobi (Kithiia & Mutua, 2006).

5. Methods of study and data collection

An attempt to quantify the data that was used in the over-all analysis of the problem is described in this section of the chapter paper. Most of the data used was derived from measurement of water samples from specific river basins and the laboratory results of the various water quality parameters. These formed the basis of the study of water quality degradation trends in the country and the city as well as the resultant environmental and human health impacts. This further informed the researcher on the mitigation strategies to control and abate the problem as proposed in the chapter presentation.

This chapter paper details the results of two studies conducted within the Nairobi River sub-basins in two distinct time span in the years from 1990-1992 and another period of three years from 1999-2002. The study of 1990-1992 basically revealed the existed of water quality degradation problem and its increasing trends. The 1999-2002 study aimed at finding the best water quality management strategies following the identification of the problem in the previous study. The study methods employed involved water sampling and laboratory analysis of the water samples collected. Water samples were collected at designated points to determine the effects of land use systems and in assessing water pollution levels and degradation status as indicated in Figure 1. The river flows (discharges) were at the same time measured using a propeller type current meter. The water samples were obtained using standard methods according to APHA (1995 & 2001). Water samples for water quality determination were collected at the same designated points in a depth integrated manner at the middle of the rivers investigated. Each sample was subjected to the standard laboratory analysis and the concentration of each parameter determined. The basic method applied in the determination of each parameter was the AAS method.

A visual assessment of the water quality status in regard to solid wastes, turbidity, foul smell and colour was carried out. To facilitate the assessment, still photographs were taken during the dry and the wet season to indicate areas of solid waste dumping before and after heavy rainfall within the study area sub-streams. This helped in understanding the impact of storm water in cleaning and restoring the water quality status. These were used as tools for awareness campaign regarding the water quality problem in the Nairobi River sub-basins. The findings of the first study were adequately utilized in the Nairobi River clean-up exercises adopted by Ministry of Environment and UNEP in 2009/2010 fiscal year. This measure only managed to remove partially the huge garbage heaps that had dotted the river profile and thus more investigation would be required to ascertain the chemical pollution of the water in the three river systems.

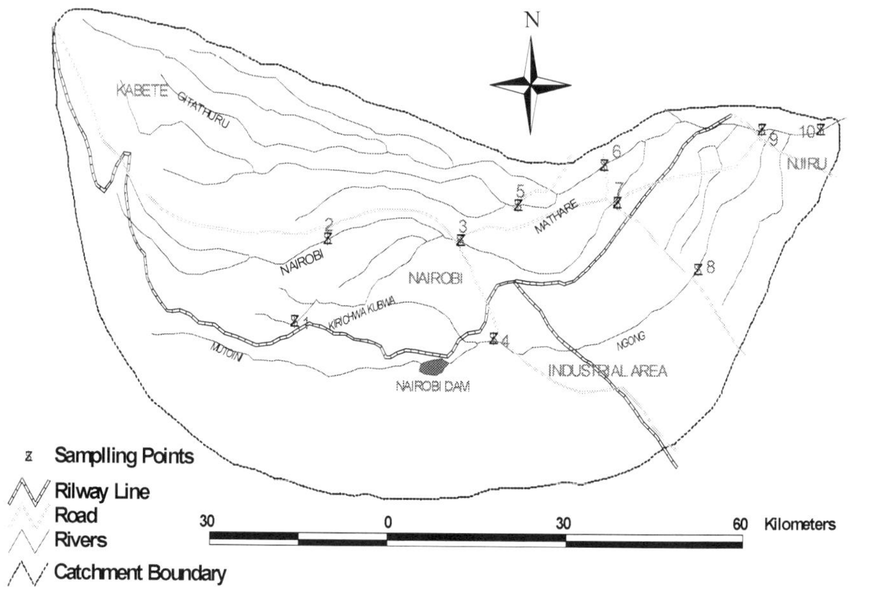

1 - Kirichwa Kubwa D/S Kawangware
2 - Nairobi at Muthangari
3 - Nairobi at Museum
4 - Ngong at Langata Road Bridge
5 - Mathare at Thika Road Bridge
6 - Mathare at Outer Ring Road
7 - Nairobi at Outer Ring Road
8 - Ngong at Embakasi D/S Industrial Area
9 - Nairobi at Njiru 1
10 - Nairobi at Njiru 2 - Ruai

Fig. 1. Distribution of study sampling points (1999-2002)

6. Data analysis and results

Table 1 indicates the distribution of some water quality parameters at various sampling points and the corresponding river discharge changes. The mean concentration of the water quality parameters at each of the sampling point per river illustrates water quality degradation. The table demonstrates an increasing trend in the major metallic ions, possibly due to the effect of leaching from rocks or as a result of effluent discharges into the river courses.

Figure 2 and Table 1 further indicated that increase in river discharge results to decreased water quality physically. Figure 2 illustrates monthly mean river flows per river sub-basin which corresponds well with the bi-modal pattern of rainfall in Kenya. The implication of this is that, storm water can be used to clean up the rivers during the wet seasons (rains), since water can be easily treated if physically polluted. In addition, effects of storm water are to reduce the water quality degradation status thus in essence dilute the water chemically. Plates 1 and 2 demonstrate that water in the Nairobi River can be cleaned through storm water incidents especially after heavy storms.

The two plates demonstrate that runoff can be used to clean up the garbage from the river banks during the rain seasonal. This will only clean up the river physically and therefore further study would be required to establish the chemical pollution of the river basin waters.

Sampling point	River	Mean Concentrations in mg l^{-1}				
		Q m^3/s	TSS Mgl^{-1}	COND $\mu s\ cm^{-1}$	TDS Mgl^{-1}	TUR Mgl^{-1}
Muthangari	Nairobi	0.772	157.6	392.1	239.7	69.4
Museum	Nairobi	1.376	129.4	397.7	244.2	69.3
Outering Rd	Nairobi	2.140	255.7	564.5	290.9	65.5
Njiru 1	Nairobi	5.083	199.2	509.5	310.9	67.8
Njiru 2 (10)	Nairobi	5.341	95.5	474.8	298.8	28.5
Thika Rd	Mathare	0.738	161	352.2	215.5	35
Outering Rd	Mathare	1.371	251.2	527.1	349.9	85
Kibera slums	Ngong	0.110	33	233	88	134
Langata Rd	Ngong	0.305	59	598.8	59	42
Embakasi	Ngong	0.949	180	611.7	174.4	71

Source: Field data 1999-2002; COND=Electric conductivity, TDS=Total dissolved solids, TSS=Total suspended sediments, TUR=Turbidity.

Table 1. Mean measured values of physical water quality parameters at various sampling points (mg l^{-1})

Sampling point	Main river	Mean concentrations of metal ions in mg l^{-1}							
		Pb	Zn	Cu	Fe	Mg	Na	Cl	Ca
Muthangari	Nairobi	0.01	0.00	0.00	0.30	6.4	42.6	53.7	18
Museum	Nairobi	0.02	0.00	0.01	0.73	6.4	44.7	55.5	19.3
Outering Rd	Nairobi	0.05	0.01	0.04	1.99	8.4	59	47	17.9
Thika Rd	Mathare	0.00	0.00	0.01	0.40	6.8	39.6	40	17.1
Outering Rd	Mathare	0.01	0.01	0.02	1.59	9.6	45.8	38.8	19.6
Langata Rd	Ngong	0.05	0.00	0.01	1.10	8.6	58.5	52	27.6
Embakasi	Ngong	0.07	0.02	0.15	1.30	7.8	69.5	46	24.3
Njiru 1	Nairobi	0.04	0.02	0.18	1.44	7.2	46.6	39.5	17.1
Njiru 2 (10)	Nairobi	0.03	0.03	0.20	1.17	4.4	42.5	29	16.4
Donyo Sabuk	Athi (middle)	0.00	0.00	0.00	0.30	4.4	28.8	8.0	10.1
Malindi	Athi/Sabaki (Coast)	ND	ND	ND	ND	29.1	118.8	19.1	21.4

Source: Field data (1999-2002), ND- not detected

Table 2. Mean concentrations of metal ions at various sampling points and rivers

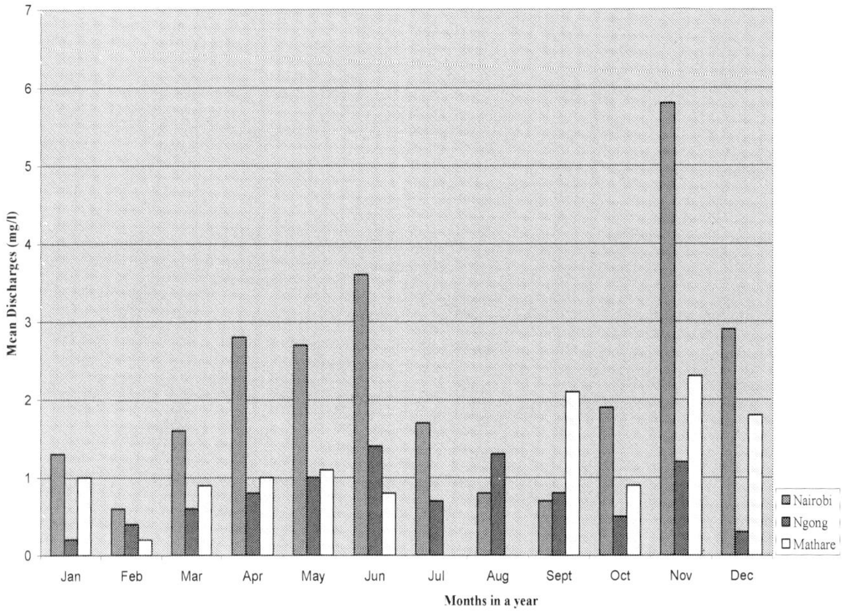

Fig. 2. Mean monthly river flows per sub-basin (mg/l)

Photo 1. Gikomba-Shauri Moyo Bridge D/S Gikomba open-air market-solid wastes, Car shells and *Jua Kali* garages in the foreground.

Source: Field data (1999-2002).

Photo 2. Gikomba-Shauri Moyo Bridge D/S Gikomba open air market-showing clean river bank with all garbage and solid wastes washed away after a storm.

7. Land-use effects on water quality

A range of water quality parameters were investigated for areas under agricultural and industrial land uses in the first study of 1990-1992. The measured concentrations were compared with the recommended WHO and Kenyan guidelines for drinking water quality in order to identify problems of water pollution. The concentrations of chloride, fluoride, manganese and heavy metals were measured in the Ngong River at Embakasi downstream of the Nairobi industrial complex, in the Nairobi River at Museum, in the Kamiti River at Kamiti, in the Kasarani River at Kasarani and in the Nairobi River downstream the city centre at Dandora. Results in Table 3 revealed that most metal concentrations exceeded those recommended by the Republic of Kenya (1985) and by the WHO (1984) as indicated in Table 4, and clearly show the strong polluting effect of industries discharging to these rivers, which is particularly marked for the Ngong River at Embakasi Bridge. There appears to be an increasing trend of industrial pollution along the river courses which presents a potential danger to the people who use water for drinking purposes downstream of the specified sampling stations.

The study also revealed variations of concentrations related to changing river discharge in the two seasons investigated. Problems relating to physical water quality parameters were more significant during the peak flows than in the periods of low flows. Thus, during the period of peak flows, water is highly polluted, has a high content of suspended solids, and is highly coloured, although conductivity is low reflecting the dilution of total dissolved solids (TDS) concentration. During the period of low flows, the water was not highly polluted with respect to physical parameters, but there was a significant increase in the soluble inorganic constituents including the metallic ions, which were present at higher concentrations than during peak flows.

Table 3 also reveals that the concentration of potentially polluting substances varies in the different sub-basins depending on the type of land use that is present. Thus, the Ngong River at Embakasi exhibited high values for most of polluting substances studied, including manganese, lead and mercury, as a result of the industrial activity upstream of this station. The water in the river is therefore of low quality for domestic use and the measured mean concentration values exceed the stipulated Kenyan standards and WHO guidelines for drinking water as depicted in Table 2. The effects of this on human health has not yet reached a critical condition since no incidents have been reported so far, but the water is generally highly polluted and of low quality.

Parameter ($mg\ l^{-1}$)	River and sampling site				
	Nairobi at Museum	Kamiti at Kamiti	Kasarani at Kasarani	Nairobi at Dandora	Ngong at Embakasi
Pb	-	0.1	0.1	0.1	0.1
Zn	-	-	0.4	0.1	0.1
Cu	-	-	-	0.1	0.1
Hg	-	-	-	0.02	0.3
Al	1.6	2.8	2.1	2.5	1.65
Mn	0.4	0.4	1.01	1.0	1.6
Fe	1.5	2.7	2.7	3.0	1.2
Na	43.7	18.9	168.0	55.0	112.8
Mg	6.1	3.1	4.8	7.4	8.4
CI	54.6	14.6	106.0	35.0	64.6
Fl	-	-	-	-	1.84

Source: Field data 1990-1992

Table 3. Mean concentrations of chemical parameters in the Study Rivers.

Substance	Upper limit of concentration (mg l-1)
Zinc as Zn	5.0
Fluoride as Fl	1.5
Copper as Cu	1.0
Iron as Fe	0.3
Aluminium as Al	0.1
Manganese as Mn	0.1
Lead as Pb	0.05
Mercury (total as Hg)	0.005

Source: Field data 1990-1992

Table 4. Concentration limits for chemical substances in drinking water after WHO (1984) and Republic of Kenya (1985).

The relationship between ionic concentration and discharge varied from one sub basin to another. The slope of the concentration-flow relationship was steepest for the Ngong River, followed by the Nairobi River at Dandora, the Kamiti River and the Kasarani River and the Nairobi River at Museum. The different response of ionic concentrations to changes in flow reflects the nature and volume of pollutant discharges which affect different sub-basins.

8. Agricultural pollution and water quality deterioration

By increasing suspended solids, turbidity and colour, nitrates and phosphates from fertilizer application, and animal wastes, agricultural activity causes deterioration in both the hygienic and aesthetic aspects of water quality. Health problems rarely result, but agricultural pollution leads to increased treatment costs and a reduction in the aesthetic enjoyment of the water resources (Kithiia, 1992).

The application of chemicals through agricultural production in the Nairobi river sub-basins was the main subject of the study. The substances investigated were pesticides, which are applied to control crop pests and diseases. Pesticides are considered to be very toxic in water, even in small concentrations, and they render the water harmful for human consumption. The pesticides used in the study area include "Dithane M45", "Ambush", "Malathion", "Ridomil", copper sulphate and DDT, and both soil and water samples were tested for these substances. DDT and "Malathion" were detected in soil samples from the upstream reaches of the Gitathuru and Kamiti Rivers as shown in Table 5. These and other pesticide residues were not detected in the lower reaches of these streams, nor in the Ruaka, Riara, and Gatara Rivers and the main Nairobi River. The presence of DDT residues in the soil reflects its persistence as an organochloride pesticide, while the presence of "Malathion" reflects the relatively high frequency of application. Table 6 indicates that levels of pesticide residues in water varied widely in the study rivers. This variation may be attributed to differences in the amounts of pesticide used and in the intensity of cultivation in terms of land under cultivation per plot. Generally the presence of trace amounts of pesticide residues in water contributed to water pollution and quality deterioration, and measured concentrations were well in excess of the WHO and Kenyan standards for drinking water quality. For example, DDT was detected at a concentration of 0.086 mg l^{-1} in the Gatara River, and "Malathion" at concentrations of 0.14 and 0.63 mg l^{-1} in the Gitathuru and Ruaka Rivers, respectively. The absence of guidelines and/or quality standards for drinking water for most pesticides and herbicides in use in the country as at that time made it complicated to assess the risk to water users from this source of pollution.

Pesticide	Type	Residue (mg kg^{-1})	Sampling site
DDT "Malathion"	Organochloride Organophosphate	0.0678 0.178	Giathuru upstream Kamiti upstream

Source: Field data 1990-1992.

Table 5. Pesticide residues in soil samples.

Sampling site	Pesticide	Residue (mg l^{-1})
Gatara River	DDT	0.086
Gitathuru River	"Ambush"	0.0948
Gitathuru River	"Malathion"	0.14
Ruaka River	"Ambush"	0.255
Ruaka River	"Malathion"	0.63
Riara River	"Ambush"	0.074

Source: Field data 1990-1992.

Table 6. Pesticide residues in water.

9. Recent development in tackling the problem

The present institutional arrangements for the management of the water resources in Kenya can be traced to the launch in 1974 of the National water Master Plan whose primary aim was to ensure availability of potable water, at a reasonable distance, to all households by the year 2000. The Plan aimed to achieve this objective by actively developing water supply systems, sinking of boreholes, construction of catchments dams and provision of the conveyance infrastructure in the form of pipes and furrows. To do so require that the Government directly provide water services to consumers, in addition to its other roles of making policy, regulating the use of water resources and financing activities in the water sector.

The legal framework for carrying out these functions was found in the law then prevailing, the water Act, Chapter 372 of the Laws of Kenya. In line with the Master Plan, the Government upgraded the Department of water Development (DWD) of the ministry of Agriculture into a full ministry of Water which embarked on an ambitious water supply development programme. By the year 2000, it had developed, and was managing, 73 piped urban water systems serving about 1.4 million people and 555 piped rural water supply systems serving 4.7 million people.

In 1988 the Government established the National Water Conservation and Pipeline Corporation (NWCPC), as a state corporation under the State Corporations Act, Chapter 446 of the Laws of Kenya, to take over the Management of Government operated water supply systems that could be run on a commercial basis. By 2000 the NWCPC was operating piped water supply systems in 21 urban centre serving a population of 2.3 million people and 14 large water supply systems in rural areas serving a population of 1.5 million people. Alongside the DWD and the NWCPC, the large municipalities were licensed to supply water within their areas and by the year 2000, ten municipalities supplied 3.9 million urban dwellers. Additionally, about 2.3 million people were receiving some level of service from systems operated by self- help (community) groups who had built the systems, often with funding from donor organizations and technical support from the district officers of the Department of water Development (Government of Kenya, 1999). Persons not served under any of the above arrangements did not have a systematic water service, and had to make do with such supply as they were able to provide for themselves, typically by directly collecting water from a watercourse or some other water source on a daily basis. Indeed, despite the Government's ambitious water supply development programme, by 2000, less than half the rural population had access to potable water and, in urban areas, only two thirds of the population had access to potable and reliable supplies.

In the 1980s the Government begun experiencing budgetary constraints and it became clear that, on its own, it could not deliver water to all Kenyans by the year 2000. Attention therefore turned to finding ways of involving others in the provision of water services in place of the Government, a process that came to be known popularly as "handing over." There was general agreement over the need to hand over Government water supply systems but much less agreement over what it meant for the Government to hand over public water supply systems to others. In 1997 the Government published manual giving guidelines on handing over of rural water supply systems to communities (Ministry of Land Reclamation, Regional and Water Development, 1997). The Manual indicated that "... at the moment the ministry is only transferring the management of the water supply schemes. The communities will act as custodians of the water supply schemes, including the assets, when they take over the responsibility for operating and maintaining them." But, the goal of community management should be ownership of the water supplies, including the associated assets. The Manual stated the criteria for handing over to be the capacity of the community to take over; ability to pay; capacity to operate and maintain the system; involvement of women in management and ability and willingness to form a community based group with legal status. By 2002 ten schemes serving about 85, 000 people had been handed over under these guidelines, focusing on management and revenue collection, not full asset transfer.

Building on this experience, the Government developed a full-fledged policy, the National Water Policy in 1999. It has tackled issues pertaining to water resources management, water and sewerage development, institutional framework and financing of the sector. In each case an attempt has been made to discuss the problems associated with each area and suggest the appropriate strategies and the desired policies that the government will put in place to resolve those problems. The Policy stated that the Government's role would be redefined away from direct service provision to regulatory functions: service provision would be left to municipalities, the private sector and communities.

The Policy also stated that the Water Act, Chapter 372 would be reviewed and updated, attention being paid to the transfer of water facilities. Regulations would be introduced to give other institutions the legal mandate to provide water services and to provide mechanisms for regulation. The Policy justified handing over, arguing that ownership of water facility encourages proper operation and maintenance: facilities should therefore be handed over to those responsible for their operation and maintenance. The Policy stated that the Government would hand over urban water systems to autonomous departments within local authorities and rural water supplies to communities. While developing the National Water Policy, the Government also established a National Task Force to review the Water Act, Chapter 372 and draft a Bill to replace the Water Act. The Water Bill 2002 was published on 15th March 2002 and passed by Parliament on 18th July 2002. It was gazetted in October 2002 as the Water Act, 2002 and went into effect in 2003 when effective implementation of its provisions commenced. The New Water Act Building on this initiative, the current government set out on an agenda of reform with water resources development and management as one of its priorities. In his opening address to parliament, President Mwai Kibaki stated that his Government, " is committed to ensuring that Kenyans have access to clean water." In this context, the Minister for Water Affairs established the new 2002 Water Act, intended to tackle the worsening water services experienced over the earlier decade. This step has given poverty reduction in Kenya a new possibility.

This *Water Act* established an autonomous Water Resources Management Authority, destined to manage and protect Kenya's resources. It also shaped an institutional framework that gave responsibility for providing decentralized services to seven regional Water Services Boards (WSB). These Boards manage water services assets and ensure that they remain in the public realm. An essential aspect of the reform outlined in the *Water Act* is the separation of water and sanitation from the management of resources. Overall supervision of water services will consequently be carried out by the Water Services Regulatory Board (WSRB), an organ in charge of regulating the services supplied by the regional Boards and their providers. From mid-2005, the responsibility for water services provision - originally falling under the Ministry of Water and Irrigation was put entirely to the WSRB. Simultaneously, representatives for the new Area Catchments Advisory Committees are being selected and the Water Users Associations are being encouraged to define their new roles. These efforts were meant to encourage proper use and management of the water resources in the country as well as try to minimize pollution of the meager resources by creating sanitation infrastructure in many of the urban and rural areas.

10. State of water environment: Water pollution in Kenya

Water resources in Kenya are increasingly becoming polluted from both point and non-point sources due to agriculture, urbanization, and industry which contribute to organic, inorganic and aesthetic pollution of water. Ground-water is threatened by intrusion of saline water in the coastal region, leachates from solid waste dumps, and infiltration of fertilizer and pesticide residues. The five principal sources of water pollution in Kenya include, among others, the categories listed below, (JICA/GOK, 1992).

a. Agricultural activities which produce sediments and agro-chemical residues (biocides and fertilizers).
b. Industrial processing of agricultural and forestry products which produce liquid effluents, gaseous emissions and solid wastes.
c. Industrial manufacturing - heavy metals, acids, dyes, oils.
d. Domestic/municipal effluents - sewage and garbage.
e. Sedimentation - soil erosion; and mining which produces tailings and effluents.

Municipalities and industries constitute the largest source of waste water discharges; however, industries discharge an enormous variety of wastes some of which are toxic to human beings and the general environment. Examples of such industrial wastes, which are a critical environmental issue in Kenya, include effluents, sludge and solid waste from sugar, coffee pulping and textile factories, leather tanneries, paper-mills and slaughter- houses.

All these wastes impact on water quality and are manifested through:

i. Deterioration of the quality of the aquatic environment, namely, deoxygenation, eutrophication, siltation, habitat modification and toxicity, among others;
ii. Environmental health effects with the occurrence of water-borne diseases, for example, typhoid and cholera, and chemical poisoning in humans and animals;
iii. aesthetic effects leading to loss of scenic beauty with water becoming unsuitable for recreation;
iv. high cost of water supply as polluted water is expensive to treat; and,

v. Water rights conflicts due to shortage of water of suitable quality for domestic, industrial and irrigation purposes. In order to protect Kenya's water resources from further degradation, water pollution or effluent discharge standards are necessary.

11. Water pollution control guidelines/standards setting

- In practice, standards can be set from either first principles or based on existing guidelines; international or national. For first principles, classification, and prioritization of pollutants is necessary. Pollutant exposure processes and their ecological effects should be determined. Predicted environmental concentrations (PEC) and predicted no effect concentration (PNEC) for the aquatic environment should be derived. Usually, actual analytical data (baseline data) relevant to the locality or country situation are preferred.
- Standards can also be based on existing guidelines such as those of the International Standards Organization (ISO) or other similar country standards. Baseline data collection is then undertaken to improve or adapt the initial standards to own country situation. Thus, baseline data is important for the setting of standards for water pollution as well as:(a) forming a basis for zoning; where general or special standards should apply; (b) assessing the assimilative capacity of various streams; (c) identifying the water courses which require stringent or less stringent application of standards; and, (d) formulating rehabilitation and/or conservation measures.

12. Water pollution effects

Waterborne diseases caused by polluted drinking water may include:

- Typhoid
- Amoebiasis
- Giardiasis
- Ascariasis
- Hookworm

Waterborne diseases caused by polluted beach water may include:

- Rashes, ear ache, pink eye
- Respiratory infections
- Hepatitis, encephalitis, gastroenteritis, diarrhoea, vomiting, and stomach aches

Conditions related to water polluted by chemicals (such as pesticides, hydrocarbons, persistent organic pollutants, heavy metals etc) may include but not exclusive of the following:

- Cancer, incl. prostate cancer and non-Hodgkin's lymphoma
- Hormonal problems that can disrupt reproductive and developmental processes
- Damage to the nervous system
- Liver and kidney damage
- Damage to the DNA
- Exposure to mercury (heavy metal) may further result :
  - *In the womb:* may cause neurological problems including slower reflexes, learning deficits, delayed or incomplete mental development, autism and brain damage.
  - *In adults:* Parkinson's disease, multiple sclerosis, Alzheimer's disease, heart disease, and even death.

Other notes:

- Water pollution may also result from interactions between water and contaminated soil, as well as from deposition of air contaminants (such as acid rain).
- Damage to people may be caused by fish foods coming from polluted water (a well known example is high mercury levels in fish).
- Damage to people may be caused by vegetable crops grown / washed with polluted water (author's own conclusion).

13. Water quality control strategies

Sources of the various pollutants should be identified and mechanisms put in place especially proper land management practices which minimize surface runoff and maximize ground infiltration of surface runoff. An integrated approach to the management of water resources should be adopted to encompass soil conservation and management, forest conservation and re-afforestation, protection of water catchment areas as well as river courses and prohibiting dumping of wastes into river courses and banks. This also should create buffer zones along the river systems to allow for the growth of riverine vegetation (macrophytes) which can take up some of the water pollutants and hence reduce water quality degradation and restore the quality of water. In the slum areas with inadequate drainage and sewerage systems, well constructed pit latrines should be provided far away from river banks, in addition to a better drainage system. Good housing structures with toilets will also reduce direct discharge of raw human wastes into the rivers flowing through these slum areas and to any other water courses in the country.

There are indications that, land use changes and activities have profound effects on water quality. There is therefore, need to review the existing land use policy in the country to guard against encroachment of land use activities into watershed areas. This will encourage land management control and conservation measures as well as riparian use of water resources in the headwater and downstream areas. Use of modern techniques and strategies (Best Management Practices-BMP) to control water quality degradation and maintain the quality of water need be emphasized. These should include Integrated Water Resources Management (IWRM) with emphasis on community participation and policy formulation.

There is also need for public awareness regarding the pollution problem and the consequences arising thereof in the rivers and basins facing water quality degradation. In overall, a sound integrated Environmental Education (EE) programme need be adopted with emphasis on communities' participation and use of green environmental policy.

Research findings and conclusions

- Water quality deterioration within the study area sub-streams was found to vary with increasing stream discharge. Thus, the study revealed that stream discharge and concentration of water pollutants exhibited a discordant relationship. This was found to be true in all the streams investigated.
- Storm water run-off was found useful in cleaning up the river systems by washing away the solid wastes, diluting the chemical industrial effluents and waste water from residential areas and burst sewers. In addition, the water was found to be physically polluted and therefore easily treatable using conventional methods of treating water such as filtration and sedimentation processes.

- Specific plant species were found to take up some of the heavy metals from the water resources and hence contributing to cleaning up the water resources within the basin.

14. Recommendations

- There is need for public awareness regarding the pollution problems and the consequences arising thereof in the rivers and basins investigated. In over all there is need for an integrated Environmental Education (EE) programme within the basins. The programme should focus on the need for people living within the city of Nairobi and its environs to appreciate a cleaner environment. It should try to encourage people to properly manage their domestic raw wastes and avoid dumping these into the rivers. Domestic wastes separation and sorting out would a better option. People should be encouraged to treat the Nairobi River as their friend and therefore protect its water quality for the present and future generations and to enjoy its aesthetic values.
- Integrated water resources management (IWRM) approach would be very integral in addressing the problem of water quality degradation within the city and the country in general. The approach need to involve all Government Ministries, water resources stakeholders and the participation of the people (community) in understanding the implication of declining water quality status and the need to guard against polluting the resources. This will ensure continued safe supply of clean and potable water resources to the people and minimize disease outbreaks.

15. Acknowledgement

The researcher wishes to thank the University of Nairobi who funded the first phase of this project and DAAD (the German Academic Exchange service) for providing substantial funding for research and completion of the project.

16. References

APHA (1995) Standard methods for the examination of water and wastewater-18th Edition. A. Greenber, L.S. Clescer, A.D. Eaton (Ed). American Public Health Association, American Water Works Association, Water Environment, New York.

APHA (2001) American Public Health Association, American Water Works Association, and Water Environment Federation, *Standard Methods for Examination of Water and Wastewater, 20th Ed.*, American Public Health Association, New York, NY.

Dunne, T (1978) Sediment yield of Kenya Rivers. Report to the Ministry of Water Development, Nairobi, Kenya.

Dunne, T. and Leopold, L.B. (1978) Water in Environmental Planning; W.H. Freeman Company. Pp 255-San Francisco.

GOK; Kenya, Republic of, (1985) Kenya Standards for drinking water quality. Kenya Bureau of Standards (KBS), Standards No. KS 05-459.

GOK; Kenya, Republic of, (1986) Sessional Paper No. 1 of 1986 on Economic Management for Renewed Growth, Nairobi, Kenya; p 5.

GOK; Kenya, Republic of, (2003) State of the Environment Report 2003, Kenya; National Environment Management Authority (Nema), 2004.

MOWD (1976a) The Water Chemistry of the Nyando river 1974-1975. Technical report 1; *Ministry of Water Development*, Kenya.

MOWD (1976b) The Chemistry of the upper-Athi river 1975. Technical report 2; *Ministry of Water Development*, Kenya.

MOWD (MLRRWD) (1997) The Athi River Basin Integrated water resources management programme: Phase 1, Management of aquifers underlying the greater Nairobi area and monitoring of industrial pollution problems in the upper-Athi river basin, Draft final report.

http://www.upiu.com/environment/2009/03/26. The untold story of the Nairobi River/ Upiu-5861238090615.

JICA/GOK, Kenya, Republic of, (1992) The National Water Master Plan. Final report., Ministry of Water Development, Kenya.

Jordaan, J., E.J. Plate, E. Prins and J. Veltrop (1993) Water in our common future; IHP Committee on water research (COWAR); A research agenda for sustainable development of water resources, Paris, 1993. UNESCO.

Kithiia, S.M. (1992) Effects of Industries and Other land-use systems on the water quality within the Nairobi river sub-catchments, Kenya. *unpublished M.Sc. Thesis* University of Nairobi.

Kithiia, S.M. (1997) Land use changes and their effects on sediment transport and soil erosion within the Athi river drainage basin, Kenya. In: *Human Impact on Erosion and Sedimentation (proceedings of Rabat Symposium S6,* April 1997). IAHS Publ. No. 245, 1997. 145-150.

Kithiia, S.M., and S.O. Ongwenyi (1997) Some problems of Water Quality degradation in the Nairobi River sub-basins in Kenya. In: *Freshwater Contamination (proceedings of Rabat Symposium S4,* April 1997). IAHS Publ. No. 243, 1997.

Kithiia, S.M. (2006) Effects of land-use types on the Hydrology and water quality of the upper-Athi River Basin, Kenya, *unpublished PhD. Thesis materials,* University of Nairobi.

Kithiia, S. M. and F.M. Mutua (2006b) Impacts of land use changes on sediment yields and water quality within the Nairobi River sub-basin, Kenya. In: *Sediment dynamics and the Hydromorphology of Fluvial systems.* (Ed. by J. Rowan, R. W. Duck & A. Werrity) (Proceedings of a symposium held in Dundee, UK, July 2006), IAHS Publ. 306. IAHS Press, Wallingford, UK, 582-588.

Kithiia,S.M and G.O. Khroda (2011) Sediments yields and Transport within the Nairobi River basins, Kenya: In *River, Coastal and Estuarine Morphodynamics: RCEM2011 © 2011 Tsinghua University Press, Beijing.*

Mavuti, K.M (2003) Nairobi River Basin Project-Phase II, Pollution Assessment and Monitoring Sub-component, UNEP/UON (2003), Final Report, Nairobi, Kenya.

Mingjun, J (2010) Comprensive Introduction on Eco-safety. The Chinese Overseas Publishing House. Beijing.

Nairobi River Clean up Organization (1992) Nairobi River (*unpublished materials*).

Njuguna, G.S. (1978) A study of the effects of pollution on tropical river in Kenya (*unpublished M.Sc. Thesis*) University of Nairobi.

Okoth, P.E. and P. Otieno (eds) (2000) Technical report of the Nairobi River basin project-Africa Water Network, UNEP.

Ongwenyi, G.S. (1979) Water Resources and the Environment in Kenya: Seminar of Environment Engineering, University of Nairobi.

WHO (World Health Organization) (1993) Guidelines for drinking water quality; Vol. 1. Recommendations (2nd Ed.), WHO, Geneva.

WHO (World Health Organization) (1995) Guidelines for drinking water quality; Vol. 1. Recommendations, WHO, Geneva.

24

Water Pollution of Oued Medjerda in Algerian Souk Ahras Region

A. Nait Merzoug[1,2] and H. Merazig[2]
[1]*Laboratory of Science and Techniques of Water and Environnement, University Center Mohamed Cherif Messadia, Souk Ahras,*
[2]*Research Unit of Environnemental Chemistry and Structural Molecular Chemistry, Department of Exact Sciences Faculty, Mentouri University, Route de Ain El Bey, Constantine, Algeria*

1. Introduction

Water has a fundamental impact on the economy and it is at the center of strategic interests. It is is an essential element of daily life for each and every one of us and a vital source of wealth and can become a possible *destabilizing* factor for the countries of the disinherited areas. The water provision is becoming more and more alarming which is due the galloping demography and the increasing droughts during these last years. In 2002, 1.5 million humans did not have access to safe drinking water and 2.5 million individuals in the world did not benefit from cleansing service (Phiri, 2005), (Lefort, 2006).

Algeria is one of the most underprivileged areas in the world with regard to hydrous availability (A. N. B. 2001). Furthermore, the demographic explosion and the economic growth led to a very high water demand which by far exceeds the available resources (Ajayi, 2002), the annual availability of water in east-Algeria is 600 m^3/inhabitant year. This places it in the poor category countries in hydrous resources with respect to the rarity threshold level, fixed by the World Bank, of 1000 m^3/inhabitant year (A. N. B. 2001), (Bouchentouf, 1994).

The strategy of any developing region in a country must necessarily be based primarily on "The control of water resources," they must be translated into a rational, continuous and sustainable water potential in the socio-economic order. Faced with ever-increasing challenges of water resource management in the country, a coherent politic of water at the national level is essential. Where with the development of the urbanization and the industrialization, as well as the evolution of consumption modes the discharges of waste water were considerably evolved in quantity and in quality. So sewer systems collect all the varieties of discharges such as domestic, industrial and commercial discharges with very diverse characteristics. The aim of the wastewater treatment was to reduce the pollution load they carry in order to protect the receiving natural environment such us sea water, oued...etc. and eventually for its future uses (fishing, leisure, food, agricultural or industrial use, etc.).

From this perspective our contribution will be particularly focused on the impact of residual treated wastewater from the sewage treatment plant (STEP) of Hennencha (Souk Ahras) on the water quality of Oued Medjerda, Principal River in the region, and eventually of Ain Dalia Dam.

Souk Ahras city is located at north east of Algeria and extends on a surface of 4360 km^2. It has four dams and an exclusive dependence on Medjerda River for its provision (ANRH, 2003). This natural water source is the principal river of the Maghreb region by its flow length as well as its area and the water volume procured (Derradji, 2004), (Louis, 1956).

2. Materials and methods

2.1 Study area

The low plain of Medjerda is situated in the north east of Algeria, at 680 km from Algiers and at 100 km from Annaba. The Medjerda River crosses the territories of two states: Algeria in its high part and Tunisia in its average and low parts. It takes its source from Khemissa, runs towards the east before discharging its contents in the Mediterranean sea, in the "golf of Tunis" (Figure. 1). It extends on 416 Km, 106 Km of which depend on the authority Souk Ahras city (Athmani, 2005). The river displays a flow that widely varies with the season, ranging from 1000 m^3/s at winter to 1 m^3/s at summer (A.B.H. 2001). The population of the studied area reaches more than 120000 inhabitants.

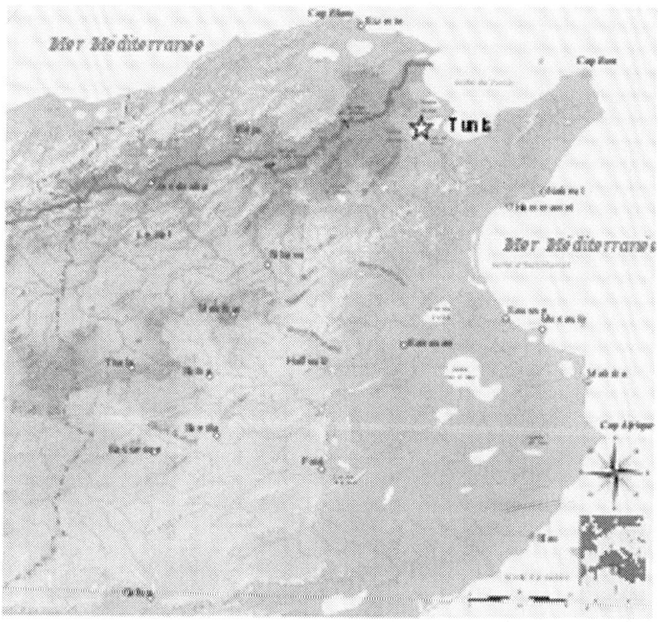

Fig. 1. Localization on map of Medjerda River (Elkenedj, 2009).

Realized in 1991 for a capacity of 200 m^3/day, the treatment plant of waste water (STEP of Henancha) receives currently more than 774 m^3/day causing serious water quality problems in Oued Medjerda which is the first receiver of the STEP rejections and also on the water quality of Ain Dalia dam located at 600 m downstream from STEP.

Geographical location:

The commune of Hennencha is located at 15 km at the West of Souk Ahras city. It is localised with the Lambert coordinates (Northern Algeria): X=96,977Km; Y=340,349Km figureN^0 2.

2.2 Sampling

Samples from the Medjerda River were taken from sites A and E respectively located upstream upstream and downstream the STEP. However, both sites are situated upstream the Ain Dalia dam. The sampling procedure was carried out once a month from Mars to July 2009.

2.3 Sampling procedure and material

The samples were taken in PVC 2-liter bottles. Before use, the bottles were treated with nitric acid, rinsed with distilled water and dripped before being washed three times with the water to be analyzed.

The sampling was carried out manually from the two fixed intake points. The bottle was held with one hand, immersed to a 20 cm depth, filled with water and hermetically closed. Then the samples were carefully and promptly transported to the laboratory after collection at a temperature of 4^{0}C in order to ensure the physico-chemical parameters.

2.4 Chemical and physical analysis

The knowledge of some basic physical and chemical parameters offers a preliminary estimation of the water quality and the extent of eventual contamination. The physico-chemical parameters were determined by standard chemical and instrumental methods.

The suspended Matter (SM) in water after and before STEP rejections was determined using centrifugation. A sample of 100 ml of water is poured into the cup and centrifuged for 20 min at 3000 tr/min. The reaping cullet centrifuge is placed in a porcelain dish and dried to constant weight at 1050 C. After drying, the dish is placed in the desiccator for one hour and then it is weighed quickly.

Chemical Oxygen Demand (COD) is measured by colorimetric method after digestion of the sample for two hours in a COD reactor (DCO Spectrophotometer DREU2010 Adapteur de tube DCO sur DREL12010). When the biochemical oxygen demand (BD05) is defined as the amount of oxygen consumed speaks aerobic micro-organisms for the decomposition of biodegradable organic matter in the water to be analyzed for a period of five days.

The sample is stirred in a flask sealed incubator connected to a mercury manometer During biodegradation of organic matter, microorganisms consume oxygen from the air in the bottle causing a reduction in pressure above the sample. This vacuum is transmitted to the mercury manometer and oxygen consumption and read on the scale gauge.

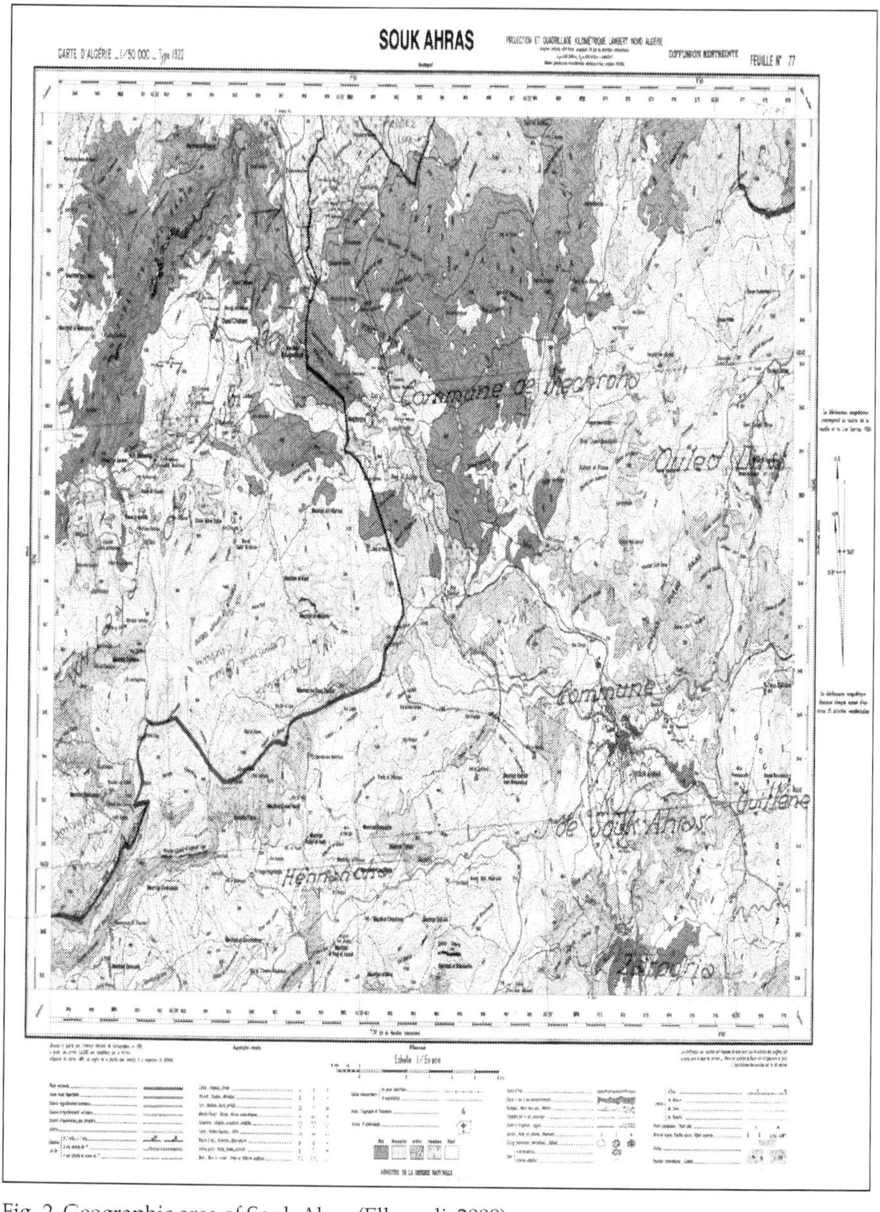

Fig. 2. Geographic area of Souk Ahras(Elkenedj, 2009).

So, we used colorimetric methods in order to quantify the major elements (NH_4^+, NO_2^-, NO_3^-). For each determination, a water volume of 20 ml was mixed in a cell with the suitable reagent HACH. The unit was stirred and then left at rest for a pre-determined time to allow the complete colour-making reaction of the analyte with the reagent. The cell was then placed in a calibrated spectrophotometer, previously set at the wavelength of maximum absorbance. The quantification of the heavy metals lead (Pb) and cadmium (Cd) and selenium (Se) was carried out by a Lovibond spectrophotometer. The methods of analysis of the remaining parameters are gathered on the (Table. 1).

Temperature	Thermometer
pH	Potentiometry
COD	Reflux method
BDO_5	5- days incubation
Suspended matter	Gravimetric
Nitrates	Colorimetric method
Nitrites	Colorimetric method
Ammonium ions	Colorimetric method
Lead	Photometric method
Cadmium	Photometric method
Selenium	Photometric method

Table 1. Water quality test methods.

3. Results and discussion

The results of the physico-chemical analyses of the Medjerda water before and after the STEP effluent input (site E and A respectively) are reported on **Tables 2** and **3** respectively.

Period	T (°C)	pH	SM (mg/l)	COD (mg/l)	BDO_5 (mg/l)	NO_3^- (mg/l)	NO_2^- (mg/l)	NH_4^+ (mg/l)	Pb (mg/l)	Cd (mg/l)	Se (mg/l)
March	18.5	7.60	19	11	1.4	0.006	0.002	0.008	0.01	0.08	/
April	18.5	7.39	25.5	82.2	4.6	5.4	0.04	0.11	0.76	0.09	/
May	19.7	7.77	31.6	9	2	1.9	0.003	0.12	0.85	0.09	/
June	20.3	7.81	10.5	25	15	1.7	0.002	0.12	1.18	0.10	/
July	24.8	7.52	55	37	19	0.7	0.001	0.13	1.44	0.18	/
Limiting values	30	7-9	30	120	40	50	0.1	0.5	0.03	0.01	/

Table 2. Water parameters at site E (upstream STEP).

Period	T (°C)	pH	SM (mg/l)	COD (mg/l)	BDO_5 (mg/l)	NO_3^- (mg/l)	NO_2^- (mg/l)	NH_4^+ (mg/l)	Pb (mg/l)	Cd (mg/l)	Se (mg/l)
March	18.2	7.47	66.5	118	45.8	0.7	0.26	2.64	0.04	0.123	/
April	18.4	7.19	90.5	296	47	9.4	0.68	5.92	1.84	0.133	/
May	20.3	7.50	109.6	132	33	2.6	0.37	6.82	1.86	0.169	/
June	21	7.88	154.5	144	75	2.3	0.14	7.14	1.95	0.172	/
July	25.1	7.72	155	158	145	1.9	0.006	11.92	2.08	0.280	/

Table 3. Water parameters at site A (downstream STEP).

3.1 Temperature

The water temperature plays an important role in the solubility of salts and gases. As reported in (figure. 3) the maximal temperatures were recorded in July, specifically 25,1 °C at site A and 24,8 °C at site E. The temperatures observed at the both was lower than the Algerian standards; we may recall that temperatures higher than 15°C favor the development of micro-organisms and in the same time intensify the organo-leptical parameters like odor and activate chemical reactions (Athmani, A. S. 2005), (Guasmi, 2004), (Botton et all. 1999), (Meinck et all. 1977).

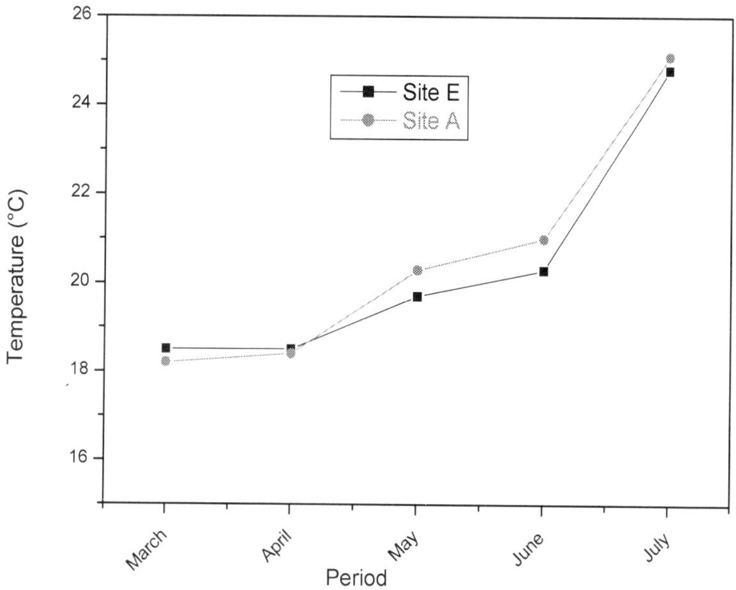

Fig. 3. Temperature variations before and after the intake of STEP effluent as a function of time.

3.2 pH

The pH is also a factor which influences the biological activity of the water micro flora. As seen on (figure 4) all samples from both sites display a slightly basic pH, ranging from 7,4 to 7,9 in the first site and from 7,1 to 7,9 in the second one. Generally, the obtained values were in agreement with the Algerian and the word health organization standards of 7,0 to 8,5. A neutral pH in the interval (6.5 - 8.5) characterizes water where life develops in optimal way (Larpent, 1997), (Guasmi & Djabri, 2006).

3.3 Suspended matters

To appreciate a water quality; it is always necessary to estimate quantitatively its load in dissolved and particular matter. As shown in (Figure. 5) we notice that the contents in suspended matter (SM) are more raised in the site A than in the site E and the values vary from 19 to 55 (mg/g) in the first one and from 66,5 to 155 in the second.

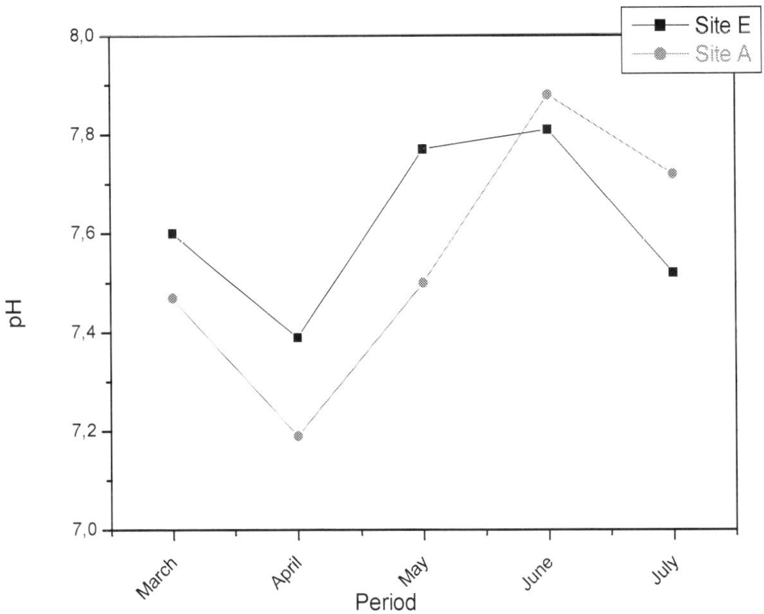

Fig. 4. pH variations before and after the intake of STEP effluent as a function of time.

We also noticed that the suspended matter content varies significantly with the season and the rates are higher in dry period than in pluvial one.

As soon as suspended matter content exceeds the Algerian standards in site A and that can induce the accumulation of higher quantities of toxic substances such as metals, pesticides, mineral oils, polycyclic aromatic hydrocarbons (Tiri et all. 2007), (Derradji et all. 2007), (Abulude et all. 2007), (Bouchenafa et all. 2008)which may have an adverse effect on water quality.

3.4 Chemical oxygen demand (COD) and biological demand of oxygen (BDO_5)

The Chemical Oxygen Demand (COD) is the amount of oxygen consumed in mg/ l, for oxidable materials in an effluent. It is representative of most of organic compounds but also oxidable inorganic salts (sulfides, chlorides...).

As shown in (Figures. 6) we noticed that values of chemical oxygen demand are lower than the Algerian standards in the site E but higher in the site A. We also note that most of the values exceed the Algerian standards in the site A for a maximum of 296 mg/l. This allows to classifying the water of this site as very bad quality. On the other hand, the high temperatures of the warmest months and also the dry periods of the year, seem to increase significantly the COD of these water.

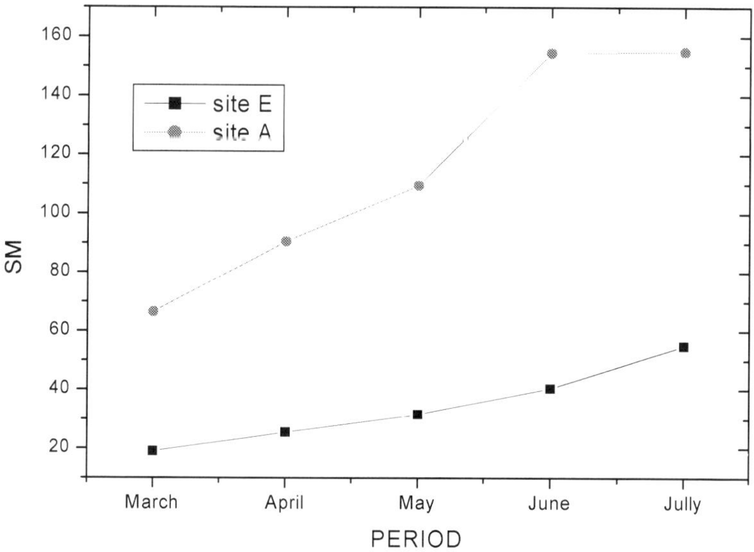

Fig. 5. Total suspended solids variations before and after the intake of STEP effluent as a function of time.

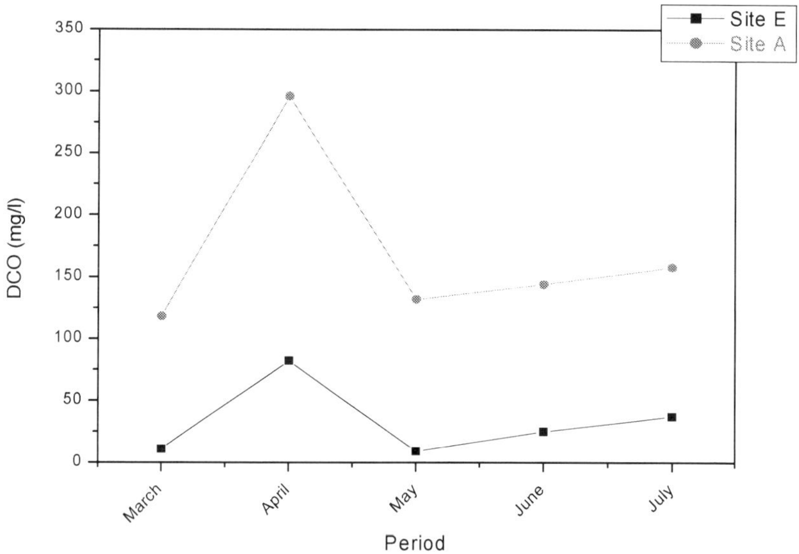

Fig. 6. Variations of chemical oxygen demand before and after the intake of STEP effluent as a function of time.

The biochemical oxygen demand (BDO_5) allows the evaluation of the present biodegradable organic matters in water. (Figure 7) showed that the BDO_5 values are lower than the Algerian standards in the site E but exceed them in the site E only in dry period and this is in agreement with the results obtained by Guasmi et Djabri...etc (Guasmi & Djabri, 2006), (Tiri et all. 2007), (Benzha et all. 2005), (Guasmi & Djabri, 2005).

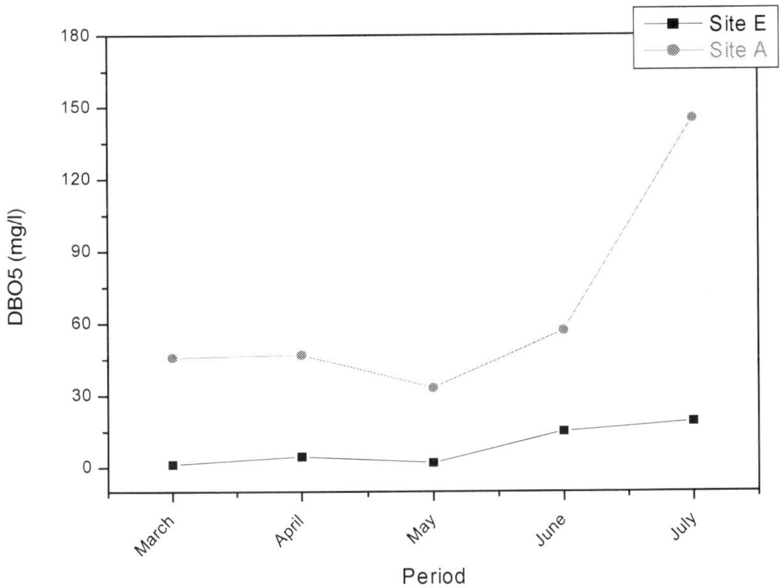

Fig. 7. Variations of biochemical oxygen demand before and after the intake of STEP effluent as a function of time.

3.5 Dissolved nitrogen

Human activities and mainly those related to agriculture are a major cause of the presence of nitrates and nitrites in surface water. These two substances are responsible for many problems not only for environment but also for human health. Indeed, although not directly toxic, they participate in eutrophication phenomena of surface water.

The made analyses allow to notice that the rates of nitrates obtained in the two sites are lower than the standards required which are in the order of 50 mg / l (Figures 8, 9 and 10). On the other hand the results concerning nitrites and ammonium ion show superior rates to the standards required (0,1 mg / l) at the site A. Their presence can be explained by an incomplete oxidation of the ammonia water; or a nitrate reduction reaction. This pollution can be caused by intense agricultural activity (the region of study is known for its agricultural vocation) and misuse of chemical fertilizers around the sewage waste water and Oued Medjerda.

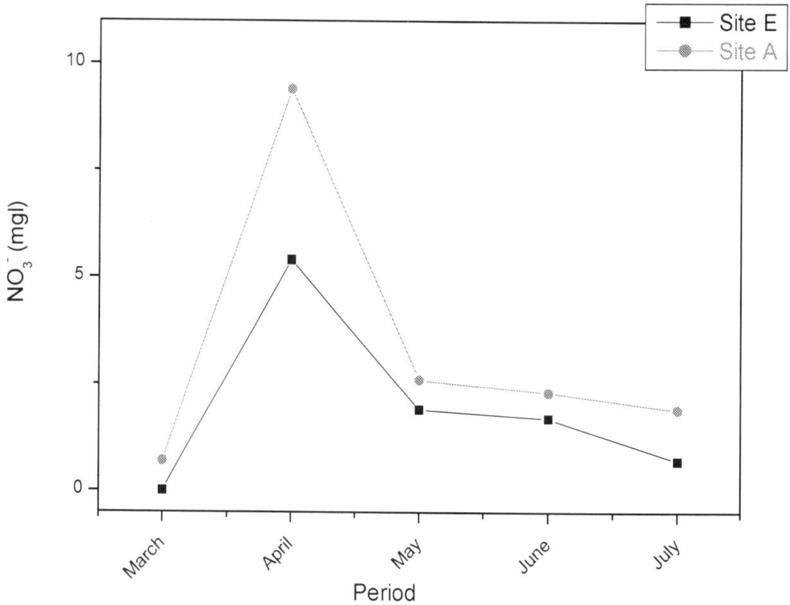

Fig. 8. Nitrates variations before and after the intake of STEP effluent as a function of time.

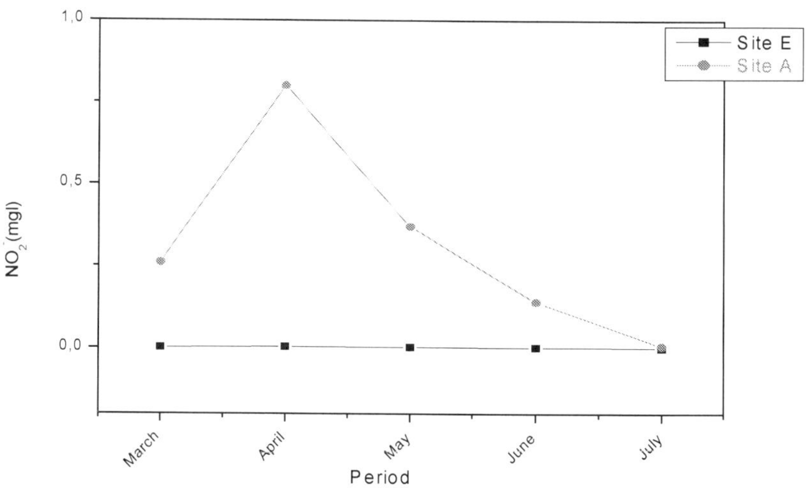

Fig. 9. Nitrites variations before and after the intake of STEP effluent as a function of time.

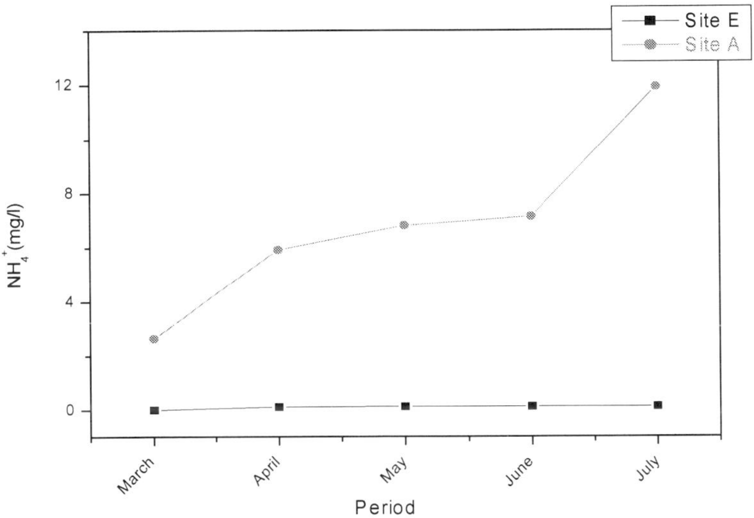

Fig. 10. Ammonium ion variations before and after the intake of STEP effluent as a function of time.

The highest contents of nitrates and nitrites are observed in April and may be explained by the leaching by rainfall observed at this precipitation period. The highest ammonium concentration is observed in July and may be explained by a dilution effect, the water volume flowing in the river being much lower in this dry period (Derradji et all. 2007), (Abulude et all. 2007), (Bouchenafa et all. 2008).

3.6 Metals

It has been clearly demonstrated that serious environmental problems were caused when the waste or the treated water may contain high levels of trace elements. These elements are potentially toxic to plants and human beings. How ever the case of selenium is complex since it is at the same time a trace element; essential to life; and a toxic; and this in a narrow range of concentration.

According to the results obtained (figure 11 and 12) we noticed that lead (Pb) levels Varied between 0,01 to 1,44 mg/l in the first site and from 0,04 to2,08 mg/l in the second one which are within acceptable limit value of 0,03 mg/l. Similarly for cadmium (Cd), its concentration varies from 0,08t o 0,18 mg/l in the site E and from 0,11 to 0,3 mg/l in the site A, whereas the limit value is 0,01 mgl/l. while this is not the case for selenium (Se). For which we noted its non-existence.

So we can say that there is a generated pollution by metals such as lead and cadmium. With high levels in the site A and less in the site E. The high content in of metals is probably due to the direct pollution of the rivers by the industrial wastes; atmospheric precipitation involving a certain part of the pollutants present in the atmosphere and in consequence of the streaming of water on the grounds involving accumulated lead or cadmium at surface (Abulude et all. 2007), (Igbinosa & Okoh, 2009).

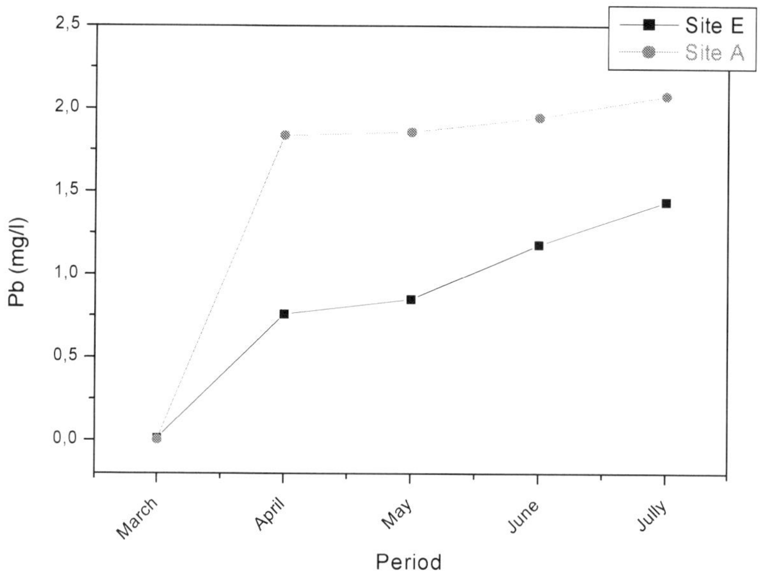

Fig. 11. Lead variations before and after the intake of STEP effluent as a function of time.

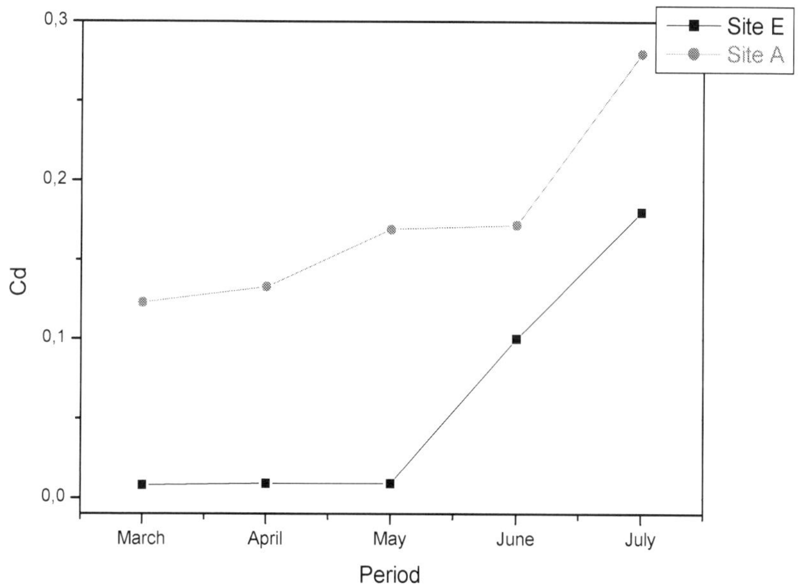

Fig. 12. Cadmium variations before and after the intake of STEP effluent as a function of time.

4. Conclusion

Medjerda is the principal river in the north-east of Algeria. This study examines the impacts of discharges of sewage wastewater (STEP Henencha) on the quality of this river. Monitoring was made upstream and downstream the STEP rejections from March to July 2009. Parameters measured include pH, temperature, total suspended matters, chemical oxygen demand, biological oxygen demand, nitrate, nitrite, ammonium ions and trace elements (cadmium, lead and selenium) using standard methods. Unacceptably, high levels of the assayed parameters were observed downstream the STEP in many cases for chemical oxygen demand (118- 296 mg/l), biological oxygen demand (33- 145 mg/l), nitrite (0,14- 0,68 mg/l), ammonium ions (2,64- 11,92 mg/l); lead (0,04- 2,08 mg/l) and cadmium (0,1- 0,2 mg/l) during the study period and outside the compliance levels of Algerian guidelines. We can also conclude that the lowest levels of metal ions were registered during flood periods and this is certainly due to the high dilution facing the river following rainfall. The results obtained confirmed that there was an adverse impact on the physico-chemical characteristics and this implies that the water of this river has a major pollution problem.

5. Abbreviations

STEP Henancha : Wastewater treatment station of Henancha
COD: Chemical oxygen demand
BDO_5: Biological oxygen demand
SM : Suspended matter
NH_4^+: Ammonium ions
NO_2^-: Nitrites
NO_3^-: Nitrates
Pb: Lead.
Cd: Cadmium.
Se: Selenium.

6. References

Abulude, F. O. Obidiran, G. O & Orungbemi, S (2007). Determination of physico-chemical parameter and trace metal contents of drinking water samples in Akure Nigeria, EJEAF che, Vol. 6, N0 8, pp. 2297-2303

ANRH: Agence Nationale des ressources hydriques (2003). Page Web. www.anrh.dz.

Ajayi, O.O., J. Ogun, and I.I. Oguntimehin. (2002). "The Influence of Waste Dumps on Some Underground Water Sources in Warri". *Journal of Science Laboratory Technology International*. VOL.1, N01. In Press.

Athmani, A. S. (2005). Evaluation de la qualité des eaux de surface, cas du bassin versant Oued Medjerda (Souk Ahras, Algérie), Thèse de Magister, Centre universitaire de Souk Ahras

Benzha, F. Taoufik, M. Dafir, J.E. Kemmou, S. & Loukili, L (2005). Impact of drainage operation on water quality of Daourat Reservoir, Rev. Sci.Eau N018 , pp. 57-74.

Bouchenafa, N. Abdelkader, D. & Maatoug.,M. (2008). Water pollution of Oued Mina under catchments, in Algerian Tiaret region. Ass. Univ. Bull. Environ. Res. Vol.11 N02, pp. 43-51.

Botton, B. Bretton, A. Fevre, M. Gauthier, S. Guy, Ph. Larpent, JP. Reymond, P. Sanglier, J.J. Vayssier, Y.& Veau, P. (1999). Moisissures utiles et nuisibles : importance industrielle. Collection Biotechnologiques. Ed Masson, p p. 97-213-216-219.

Bouchentouf, K. (1994). Les bilans d'eau vus à travers les paramètres physico-chimiques et hydrodynamiques: cas du basin versant de la haute Mina (Tiaret, Algérie), Thèse de Magister. Centre universitaire de Chelef, Algeria.

Derradji, F. (2004). Identification quantitative et qualitative des ressources en eau dans la région d'Annaba (El Tarf, Algérie). Doctorat des sciences, Université de Annaba, Algérie

Derradji, F. Ben snoubra, H. Kherici, N. Romeo, M. Karuba, R. (2007). Impact de la pollution organique sur la qualité des eaux superficielles dans le nord est algérien, Sécheresse Vol.18, N0 1, pp. 23-27

Document inédit: Agence National des barrages, (2001).Algérie

Elkenedj, Y.B. (2009). Etude de l'effet de l'aération naturelle sur la pollution résiduaire des eaux épurées *de la zone de l'Hennencha W. Souk Haras (N.E.Algérien),* Thèse de Magister, Université Badji Mokhtar Annaba, Algérie

Guasmi, I. (2004). Dégradation de la qualité de l'eau dans le bassin versant de l'Oued Medjerda (Souk Ahras, Algérie), Thèse de Magister, Université de Annaba, Algérie

Guasmi, I. Djabri, L (2005). L'évolution des nitrates, nitrites et ammonium dans les eaux de surface de l'oued Medjerda (Souk Ahras). 3 éme journée des géosciences de l'environnement, El Jadida les 8, 9 et 10 juin. Tunis

Guasmi, I.& Djabri, L. (2006). Pollution des eaux de l'Oued Medjerda par les nutriments. Larhyss Journal N05,pp. 113-119.

Igbinosa, E.O., Okoh, A.I. (2009). Impact of discharge wastewater effluents on the physico-chemical qualities of a receiving watershed in a typical rural community. Int. J. Environ. Sci. Tech., Vol.6, N0 2, pp.175-182, Spring 2009, ISSN: 1735-1472

Louis, D. (1956). Etude géologique des monts de la haute Medjerda, Thèse SC. Science de la carte géologique de l'Algérie, Algérie

Lefort, R. (2006). "Down to the Last Drop". *UNESCO Sources.* 84.

Meinck, F. Stoof, H. Kohlschutter. (1977). H. Les eaux résiduaires industrielles, 2eme Ed. Masson, Paris, p 863.

Larpent, J.P & Gourargaud. (1997). M. Mémento technique de microbiologie (3eme édition. Lavoisier Tech & Doc, pp173-217.

Phiri, O., P. Mumba, B.H.Z. Moyo, and W. Kadewa. (2005). "Assessment of the Impact of Industrial Effluents on Water Quality of Receiving Rivers in Urban Areas of Malawi". *Int. J. Environ. Sci. Tech.*. Vol.2, NO 3, pp:237-244.

Tiri, A. Boudoukha, A. Bouguerne, A. & Lahbari, N (2007). Influence of the industrial and urban waste water on the quality of water (Case of Koudiat Medouar Watershed), J. Eng. Applied, Vol.2, N0 11, pp. 1592-159.

25

Water Quality Issues in Developing Countries – A Case Study of Ibadan Metropolis, Nigeria

Adegbenro P. Daso[1,2] and Oladele Osibanjo[2]

[1]*Department of Environmental and Occupational Studies, Faculty of Applied Sciences, Cape Peninsula University of Technology, Cape Town,*

[2]*Department of Chemistry, Faculty of Science, University of Ibadan,*

[1]*South Africa*

[2]*Nigeria*

1. Introduction

To describe water as the engine of life will not constitute an overstatement. This is because water in its various forms, accounts for more than 70 per cent of the entire earth surface and all life forms regardless of their habitat depend on this abundant resource for their continuous existence. However, as huge as this vital resource is, only small percentage of its natural form could be readily used for drinking and sanitation purposes by man. These are normally stored up in repositories and embankments such as the aquifers, lakes, rivers and other surface freshwater bodies.

Due to the increasing influence of natural events and anthropogenic activities on these natural water sources, the pristine characteristics exhibited by these water sources often fade out with time. Today, the understanding of water quality has become conceptualised because of the numerous uses to which different types of water could be subjected to. More so, due to the complexity of several factors determining water quality and the countless choice of variables used to provide quantitative evaluation of this term, it is difficult to adopt a single definition of water quality (Chapman, 1996). In a simple term, however, water quality refers to the composition of any water body as affected by nature and human cultural activities, expressed in terms of both measurable quantities and narrative statements (Novotny, 2003). Depending on the area of application, the criteria for establishing water quality requirements differ in many aspects. Hence, water which is suitable for a particular purpose, for instance, agricultural irrigation might not be useful for other purposes due to differences in water quality requirements.

The causative factors responsible for the deteriorating water quality in most developing countries are quite similar. For instance, the city of Ibadan which is the largest indigenous city in Africa has several inter-related factors which directly or indirectly impact the quality of water bodies within the city. These are largely due to improper waste disposal, poor physical planning and increasing population pressures on the dilapidated infrastructures within the city. Omoleke (2004) also identified the culture of the indigenous people living in the core of the city as a vital factor contributing to these menace. Traditionally, the city has been a commercial centre for local marketers of maize, yam and other food stuffs where heaps of refuse are generated on a daily basis. Due to the clustered distribution of old houses within the interior of the city, the mechanised collection of these refuse becomes virtually impossible. Hence, people resort to dumping their solid wastes into drains and stream channels which often results into clogging and flooding. More so, most of the houses around these areas do not have toilet facilities, as such people defecate indiscriminately on undeveloped plots of land or along the streams and rivers within the city. These uncivilised behaviours have continued to aggravate many dimensions of water pollution problems within the city.

The consequences of these illicit practices are quite enormous. Besides the increasing number of waterborne diseases related cases in hospitals, the destruction of lives and properties caused by catastrophic flood incidences in the city are notable. In the last 60 years, the city had witnessed several flood disasters notably in 1951, 1955, 1960, 1963, 1969, 1978 and most recently in 2011. According to a newspaper report (Businessday, Nigeria) of 5th September 2011, more than 700 lives have been lost during flash flooding incidences between 1980 and 2011. The report further indicated that government's dereliction of duty and insensitivity to the plight of its citizenry, complete disregard to town planning rules by landlords and the awful traditional habits of small-scale business people and their patrons of dumping wastes in drainages as the cause of the sporadic flood disaster recently witnessed in the city. Interestingly, the cumulative effects of the water pollution problems arising from the flood disasters and industrial discharges are not restricted to the residents of the city. Many villagers living several miles away from the city who often depend on water from these polluted streams and rivers for their drinking and sanitation purposes are not exempted from these effects. Incidentally, the extracted water by these villagers, in most cases, is not subjected to any form of chemical treatment prior to their use. Although most of these streams and rivers do exhibit some natural regenerative potential as they flow further downstream, the extent to which these can be achieved generally depends on but not limited to these factors, namely the quality of industrial effluent discharged, the proximity of the village or other extraction points to the city and the hydrodynamics of the stream or river concerned.

As a result of the increasingly difficult nature of managing solid wastes in the city, little attention is paid to the potential impacts of discharges of untreated, partially treated or treated industrial effluents on the water quality of the receiving water bodies. Usually, the impact of industrial discharges is difficult to assess because there is no clear cut between the water pollution caused by industrial activities and that contributed by improper management of solid wastes. This study was therefore aimed to establish whether the pollution of the two major rivers (Ona and Alaro) within Oluyole Industrial Estate, Ibadan was due to industrial discharges or as a result of indiscriminate disposal of municipal wastes and transportation of pollutants into these rivers via urban run-off especially during high precipitation. To achieve this objective, this study was conducted at the peak of the rainy season so that the influence of urban run-off on the water quality could be readily evaluated. In this study, the water quality was assessed in terms of the physicochemical characteristics of the river water and the levels of trace and heavy metal of both the river water and bottom sediment collected from different sampling locations along the stretch of both rivers. Because of the pivotal role that solid waste management played on the subject matter, mention was made on the institutional and regulatory frameworks established to coordinate the proper management of the environment within the city; suggestions for better ways of protecting the environment were also mentioned.

2. Solid waste management in Ibadan: Current approach and practice and the role of institutional and regulatory frameworks

Since the advent of the disastrous flood of April 1978 in Ibadan, there has been an unprecedented public interest on the channelization of Ogunpa river and solid waste management within the city (Areola and Akintola, 1979). The massive accumulation of solid wastes in both Ogunpa and Kudeti rivers made these rivers overflow their banks resulting in severe flooding in parts of the city in the past years. When it became apparent to the city managers that the rivers within the city could no longer cope with the huge run-off generated during rainfall, the government began the channelization of the major river as early as 1970s. Several administrations have expended heavily on this project yet the problem of flooding still persists. The most recent was the flood incidence of August 25, 2011 where several lives and properties were lost. The extent of the damage caused is depicted in Figure 1. A critical look at this problem shows that the flood abatement measures are already in place but the major cause of the flooding which is the obstruction of drains and channels by dumped solid wastes is yet to be completely tackled.

Fig. 1. A picture showing the devastating effects of the flood incidence of 25th August 2011 in parts of Ibadan city

The task of keeping the environment of the city clean is vested in the hands of three government institutions, namely:

i. Ibadan Solid Waste Management Authority,
ii. Local Government Councils, and
iii. Ministry of Environment.

Prior to the enactment of edict No. 8 of 1997 which established the Ibadan Solid Waste Management Authority, the management of environment of Ibadan city was the primary responsibility of the local government councils that evolved from the defunct Ibadan Municipal Council. Upon the implementation of the edict, the functions of these local councils under the 1979 Nigerian constitution which were to collect, transfer and dispose solid wastes were delegated to the newly constituted authority (Omoleke, 2004). The authority has since its inception been fully responsible for the collection, transfer and disposal of solid wastes in the city. Other functions of the authority include the coordination of private practitioners engaged in solid waste collection and the enforcement of regulations related to the proper management of solid wastes within the city.

To further strengthen the decision-making processes and the implementation of relevant environmental policies, Oyo State House of Assembly passed a law to establish the Ministry of Environment and Water Resources which commenced functioning on 1st January, 2001. As highlighted by Omoleke (2004), the responsibilities of the ministry include:

a. formulation and enforcement of policies, statutory rules and regulations on waste collection and disposal, general environmental protection, control and regulation of the ecological system and all activities related thereto;
b. advise the government on the environmental policies and priorities and on scientific and technological activities affecting the environment;
c. coordinate the activities of the Local Governments and government agencies on environmental and ecological matters;
d. establish and take measure to ensure effective environmental structures in the state for flood control, solid and liquid wastes collection and disposal, water and air pollution eradication, noise control and general sanitation;
e. conduct public enlightenment campaign and disseminate vital information on environmental and ecological matters;
f. prepare master plans for drainage, solid and liquid wastes and general aesthetics for development of environmental technology and ensure the implementation of such plans;
g. promote cooperation in environmental science and technology with bodies whose main objective is the protection of the environment;
h. initiate appropriate policy action on the environmental implications of environmental related activities;
i. monitor sources of toxic pollutants into the air, land and water.

The ministry is empowered under Section 5 of this act to disseminate basic scientific data and other information pertaining to eradication of pollution and environmental protection matters through relevant publications; and to establish environmental criteria, guidelines, specifications or standards for the protection of the environment and public health. Besides these laws, there are many other associated legal regulations which specifically addressed diverse environmental problems at the federal, state and local government levels. Based on the aforementioned preview, it is clearly evident that the solid waste management problems commonly encountered in the city, as it is the case in most cities of other developing countries, has nothing to do with lack of appropriate institutional and regulatory frameworks but rather on the inability of relevant government's agencies to translate these policies into meaningful results.

Presently, waste collection methods in most parts of the city differ from place to place. This often depends on the ease of accessibility of mechanised technology adopted for this purpose. In the interior of the city where accessibility of motorised vehicle could be a problem, communal depots and block system (bring and dump) are the commonest means of solid waste collection. The house to house method is commonly practised in areas with better road networks. Since most of these services usually require some forms of financial commitment, the very poor who could not afford these costs mostly resort to sporadic and indiscriminate dumping of their wastes into available plots of land, sidewalks, roadways, streams, channels and drainages. Omoleke (2004) reported that more than 70% of the refuse generated in the city are mostly disposed in this way - a method commonly known as 'bulk loading'. This is partly responsible for the incessant flood disaster being witnessed in the city over the years. Unfortunately, most of the measures adopted by relevant government's agencies to combat these problems could not be sustained due to poor funding, lack of appropriate technology and manpower for effective management of solid wastes and the poor attitude of most of the residents of the city.

3. Methodology

3.1 Sampling sites

A total of ten sampling sites representing the upstream (mainly residential), industrial zones and downstream of both River Ona and River Alaro were carefully selected. The entire sampling locations where both river water and bottom sediment were collected are shown in Figure 2.

3.2 Sample collection

Two sets of water samples were collected at the region of good mixing into thoroughly cleaned polyethylene bottles from different sampling sites. The first set of water samples was required for the determination of the physicochemical parameters while the other set was acidified upon collection for the assessment of the following heavy metals, namely: Pb, Zn, Cd, Ni, Cu and Fe. Similarly, bottom sediment samples were collected from these sites using a hand trowel wrapped with several layers of polyethylene bags to prevent interaction with the metallic surface. These samples were collected into self-sealing polyethylene bags and were kept cool *en-route* to the laboratory. On arrival at the laboratory, all the samples were kept in the refrigerator until analysis. However, the debris in the bottom sediment samples was removed prior to their preservation.

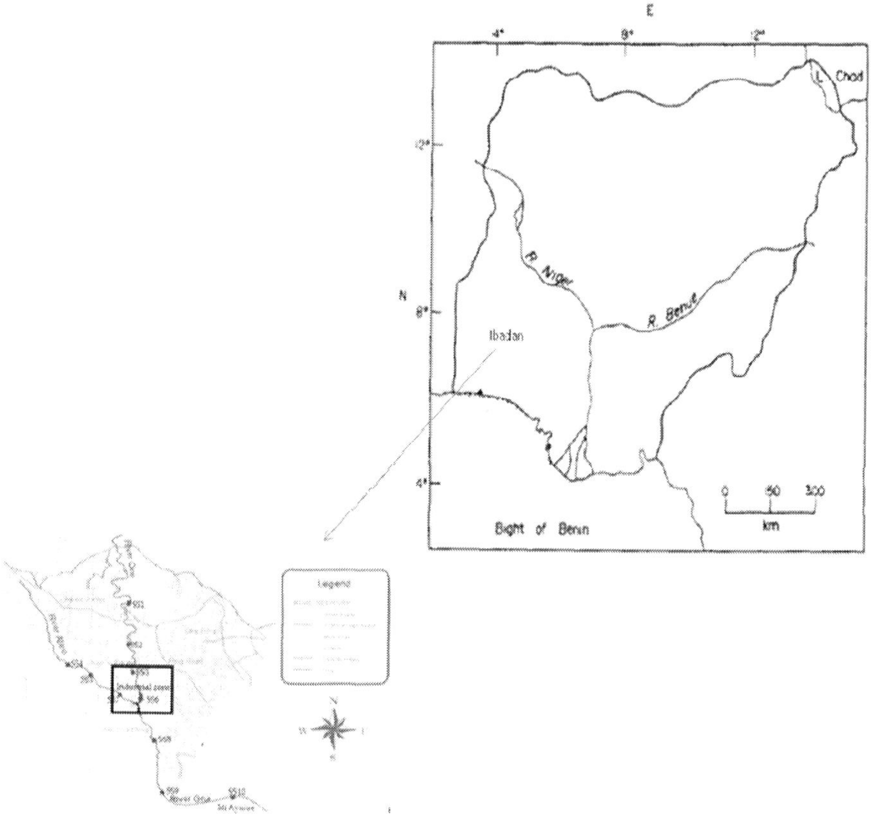

Fig. 2. The map of Nigeria showing its geographical coordinates as well as the sampling sites within the City of Ibadan

3.3 Sample analysis

The pH and temperature of water samples were determined on the field immediately after sample collection. The analysis of water samples for other physicochemical, inorganic and organic parameters was carried out according to the decreasing order of stability of parameters to be determined. In these cases, established analytical protocols were employed for the analysis of these parameters. All the parameters except nitrate were determined using the standard methods for the examination of water and wastewater (APHA, 1985). The analysis of nitrate was done by the spectrophotometric method described by Osibanjo and Ajayi (1980). In this case, nitrate was determined by nitration of 3,4-xynenol in acidic medium. This was followed by the extraction of the nitration product with a suitable alkali solution (NaOH) to form a coloured product whose absorbance was read at 432 nm using CECIL CE202 UV spectrophotometer.

For the heavy metal determination in water samples, an initial pre-concentration of water samples was carried out. In this case, 50 mℓ portion of water sample was measured into a thoroughly clean heat-resistant conical flask and evaporated to a final volume of 2 - 8 mℓ. In each case, the concentration factor was estimated to account for the actual metal concentration in each sample. The concentrated samples were thereafter transferred into 50 mℓ capacity digestion tubes following which 2M HNO_3 was added. The tightly covered tubes were then placed into a 1 ℓ capacity heat-resistant beaker containing water and boiled for 2 h. The digestion tubes were regularly checked at intervals to ensure that the content were intact. After digestion, the tubes were cooled and samples were filtered and made up to 25 mℓ mark. A blank sample was similarly prepared using 10 mℓ of the acid.

In the case of bottom sediment samples, the air-dried samples were ground and mechanically sieved using a 2 mm mesh size sieve. About 2 g of the sediment sample was weighed into a clean digestion bottle. The samples were digested as previously described for the water samples using the same strength and volume of the acid. The heavy metal contents of these digested samples were determined using a Bulk Scientific VGP 210 Model Atomic Absorption Spectrophotometer (AAS). This analysis was carried out at the Central Chemistry Laboratory of the University of Agriculture, Abeokuta, Nigeria. Equations 1 and 2 below were employed for the calculation of the heavy metal content of the river water and bottom sediment samples, respectively. The total organic carbon (TOC) content of the sediment samples was determined using the Walkey-Black method as described by Schumacher (2002).

$$\text{Metal conc. in water } (\mu g/\ell) = \frac{(a - b) * 25 * 1000}{c.f * d} \quad (1)$$

$$\text{Metal conc. in sediment } (\mu g/g) = \frac{(a - b) * 25}{e} \quad (2)$$

where:

- a, b = instrument reading (μg/mℓ) for sample and blank, respectively.
- d = volume of digested water (mℓ),
  c.f = concentration factor.

The TOC content was calculated using equation 3 and the organic matter reported was estimated using the conversion given below.

$$\text{TOC } (\%) = \frac{(meK_2Cr_2O_7 - meFe(NH_4)SO_4) * 0.003 * 100 * f}{\text{mass of air} - \text{dried sediment}} \quad (3)$$

where:

- f = correction factor = 1.33,
- me = Normality of solution * volume of solution used (mℓ),
- % Organic matter in sediment = % TOC * 1.72.

As part of the quality assurance measures taken, all the sampling bottles and glassware employed for sample collection and preparation of reagents were properly cleaned to avoid possible contamination problems. The analysis of procedural blanks during heavy metal determination in river water samples was carried out to eliminate trace levels of target analytes from the reagents used. Since the heavy metal analysis was carried out by an external analyst, several standard solutions of the heavy metals investigated were prepared and sent for analysis prior to the analysis of the real samples by groups of research scientists in the university to ascertain the accuracy of the results obtained. More so, blind samples were prepared and incorporated into the real samples to further establish the reliability of the results obtained from the instrumental analysis.

4. Results and discussion

4.1 Results

The results of the physicochemical, inorganic and organic parameters of river water samples collected from both River Ona and River Alaro are presented in Table 1. These results, except for the industrial zone of River Alaro, generally indicate the average values of each parameter investigated. It is necessary to point out that River Alaro discharges its contents into River Ona where these industries are mostly concentrated along the catchments of the river. Hence, samples were only collected at the upstream and industrial zones of this river. The quality of the surface water is also greatly influenced by the chemistry of the underlying sediment. Therefore, an assessment of the sediment (which serves as a sink for most water pollutants) as well as surface water for heavy metal concentrations was carried out. The results obtained are presented in Table 2.

In this study, the evaluation of the quality of the surface water of both rivers involves the monitoring of inorganic, organic and general water parameters at different sampling sites.

A comparative assessment of the results obtained in this study with those of other similar studies around the world was done. This, however, could only give a clue as to whether a significant pollution has occurred but will not necessarily imply that the pollution is caused by industrial discharges because several other factors have been found to significantly contribute to surface water pollution across the world. Some of these factors would include the geological formation of the study area, urban run-off, agricultural run-off, atmospheric deposition, acid mine discharges amongst others.

4.2 Discussion

4.2.1 Physicochemical characteristics of the rivers

The trend of most of the physicochemical parameters suggests that the rivers were properly mixed. This is not surprising, since the study was conducted at the peak of the rainy season when the rivers had excessive volume of water that flowed over the riverbanks at certain locations along their catchments. The pH of the river water, which ranged from 6.88 to 7.60 showed no inconsistency along the watercourses of both rivers suggesting that there were no significant inflows of substances that abruptly influence the pH of the water system. Generally, the temperature of water from both rivers reflects those typical of a tropical river ranging from 26 to 28°C. However, the higher average of 27°C recorded for River Ona which was higher than that recorded for River Alaro (26.3°C) was indicative of possible thermal pollution from industrial processes.

The evaluation of solids at different sampling sites revealed possible contributions of industrial discharges at certain sampling locations. At sites 3, 8, 9 and 10, the dissolved solids content were found to be greater than the suspended solids levels of these sites, which was expected. In other cases, the suspended solids contents were greater than the dissolved solids which indicated turbulence in the water bodies during sampling and probably because the suspended solids were largely non-settleable. The elevated level of dissolved solids at site 3 may be attributed to the high volume of a combination of partially treated and untreated industrial effluents that are regularly discharged to this sampling site. The industries in this area are mostly food and household products manufacturing industries whose wastewaters may be rich in dissolved organics as reflected in the result obtained. Contributions from both urban run-off and industrial activities may be responsible for the levels obtained at other locations.

4.2.2 Inorganic characterization of the river water

Generally, the trend of the results of most inorganic anions studied agreed with those reported in a similar study (Sengupta *et al.*, 1988). The water samples from both rivers did not show a positive test to phenolphthalein alkalinity determination. By implication, the alkalinity results obtained in this study, therefore, could be attributed to the presence of carbonates (CO_3^{2-}) and bicarbonates (HCO_3^-) in these samples. It further indicates that hydroxide ion is not present in these samples and this claim is supported by the pH values (< 8.00) obtained for these samples. The methylorange alkalinity ranged from 96.9 to 142 mg $CaCO_3/\ell$ while the total acidity, on the other hand, ranged between 24.7 and 43.8 mg $CaCO_3/\ell$ in river water samples from both rivers.

For most of the anions, a somewhat unique pattern in their levels at different sampling sites of the rivers was observed. For instance, the levels of the major anions (Cl^-, NO_3^- and SO_4^{2-}) showed a distinct and regular pattern in its levels along the watercourse of both rivers, particularly before and after the convergence of these rivers. While these anions had a considerable decrease in their levels at the industrial zone of River Ona, all but one of these anions had an appreciable increase in their levels within the same sampling location in River Alaro. The average chloride level of 10.9 mg/ℓ obtained in River Ona was higher than that obtained in River Alaro (8.23 mg/ℓ). Incidentally, these levels were far below the WHO guideline for drinking water. Furthermore, the average chloride level in River Alaro found in this study was significantly lower than the average reported (475 mg/ℓ) in a similar study conducted on the same river during the dry season (Fakayode, 2005). The dilution effect of the rainwater could be responsible for the lower chloride levels obtained in this study. Common sources of chloride in the river water would possibly include the use of mineral acids (e.g. HCl), common salts and other chloride containing compounds often used as inputs in the food and other manufacturing industries.

The decreasing trend in the levels of these anions especially NO_3^- and SO_4^{2-} might be due to the possibility of chemical precipitation and the action of certain micro-organisms on these inorganic species. Nriagu (1978) highlighted several processes through which sulphur can be removed from sub-aqueous biogeochemical cycle to include: (a) the formation of insoluble iron (and other metal) sulphides, (b) the precipitation of sulphate (particularly gypsum and anhydrite) out of solution, which may be engendered by extended desiccation, (c) the removal of organic debris and other sulphur-containing suspended particulates; and (d) the exchange of sulphur across the air-water and sediment-water interfaces.

In view of the above, it follows that in aerobic surface waters; the loss of sulphur should be quite negligible and would primarily entail the sedimentation of suspended particulate and organic matter. The average levels of sulphate in both rivers were 4.94 and 5.64 mg/ℓ for River Ona and River Alaro, respectively. It was observed that the upstream of both rivers generally had higher levels of sulphate. This might be due to the increased utilization of cement for building and other construction works since these areas were mainly residential.

Similarly, the higher levels of nitrate at the upstream of River Ona and in the industrial zone of River Alaro could be attributed to increased usage of nitrogen-based fertilizers and from the decomposition of biodegradable household wastes commonly dumped along the banks of these rivers. Although the loss of nitrogen via the air-water and sediment-water interfaces cannot be ruled out, the reduction of nitrate to ammonia and nitrogen molecules by certain denitrifying bacteria can significantly contribute to the observed trend in the study. Brooks and Effler (1990) noted a rapid depletion of nitrate and nitrite in a typical lake waters an observation attributed to the actions of denitrifying bacteria and possible assimilation of these nitrogen species. The organic constituents of the industrial discharges might also aid the removal of nitrate from solution by facilitating the sedimentation process involved in its removal.

Although the average nitrate level in River Ona (7.91 mg/ℓ) was higher than that found in River Alaro (7.02 mg/ℓ), nitrate levels found at the upstream and industrial zone of River Alaro were generally higher than those found at the same locations in River Ona. This trend is expected because River Alaro is bound to several large-scale farms and industries where increased usage of nitrogen-based fertilizer, agricultural composts as well as poultry and other agricultural and industrial wastes could have significantly contributed to the elevated level of nitrate in the river. The high value of nitrate obtained at sampling site 10 was probably due to the influx of urban run-off into smaller streams that emptied their content into the river downstream.

The total phosphorus levels showed a progressive increase along the sampling sites in both rivers. These generally ranged from 0.71 to 5.71 mg/ℓ and were similar to those previously reported for some Nigerian rivers (Ajayi and Osibanjo, 1981). The discharge of effluents of high phosphate content into these rivers could have resulted into the increased levels observed. Phosphorus-rich detergents are important sources of phosphorus into surface water bodies especially in urban areas. More so, the use of super-phosphate fertilizers and leaching from municipal dumpsite could also contribute to these elevated levels of phosphorus in these rivers. The extremely high level of phosphorus at the sampling site 10 was a reflection of the cumulative effects of industrial discharges and urban run-off conveyed into River Ona via other notable rivers (e.g. R. Ogunpa and R. Kudeti) and streams which converged at a point few metres away from sampling site 9.

Besides certain parameters which gave an indication of good mixing in both rivers, the levels of total hardness found in this study further suggested that the river water were properly mixed during sample collection. These levels were generally uniform at the sampling sites of these rivers. The average level of the total hardness in River Alaro (103 mg $CaCO_3$/ℓ) was slightly higher than that obtained for River Ona (102 mg $CaCO_3$/ℓ). These levels were generally lower than the WHO standard of 500 mg $CaCO_3$/ℓ for drinking. Apart from the economic losses associated with the extremely high levels of hardness in form of poor lather formation with use of soap during washing, there is no known health effects related to this parameter at levels commonly found in natural waters.

Parameters	River Ona			River Alaro	
	Upstream (1,2)	Industrial zone (3,6)	Downstream (8,9,10)	Upstream (4,5)	Industrial zone (7)
pH	7.56	7.57	7.17±0.32	7.55	7.40
Temperature ^{0}C	28.0	27.5	26.0±0.00	26.0	27.0
Total solids (mg/L)	370	420	490±168	290	260
Suspended solids (mg/L)	250	230	140±20.0	250	240
Dissolved solids (mg/L)	120	190	350±150	40.0	20.0
Acidity (mg $CaCO_3$)	29.5	31.4	39.5±4.71	35.6	29.5
Alkalinity (mg $CaCO_3$)	112	108	134±11.4	112	96.9
Total hardness (mg $CaCO_3$)	92.0	102	110±12.2	108	94.0
Sulphate (mg/L)	5.51	3.31	6.25±1.53	6.62	3.68
Total phosphorus (mg/L)	1.07	2.855	4.64±1.48	2.15	2.86
Chloride (mg/L)	11.8	9.63	12.0±1.59	7.59	9.52
Nitrate (mg/L)	3.75	3.32	18.2±15.4	6.26	8.55
Dissolved oxygen mg DO/L	5.08	5.77	5.08±0.21	6.87	5.47
COD mg O_2/L	91.8	71.4	133±20.4	40.8	81.6
Oil and grease mg/L	197	398	379±21.3	407	431

number(s) in parenthesis - sampling sites

Table 1. Mean (±standard deviation) concentrations of physicochemical, inorganic and organic parameters in water samples

Parameters	River Ona			River Alaro	
	Upstream (1,2)	Industrial zone (3,6)	Downstream (8,9,10)	Upstream (4,5)	Industrial zone (7)
River water (µg/ℓ)					
Pb	156	53.5	100±73.7	68.5	88.5
Zn	159	195	151±2.75	228	183
Cu	22.0	31.3	25.7±5.25	21.8	28.5
Cd	5.00	7.50	0.50±0.87	2.50	5.50
Ni	145	ND	65.0±104	65.0	80.0
Fe	802	956	1560±425	1160	1320
Bottom sediment (µg/g)					
Pb	19.4	27.9	31.1±16.9	24.9	24.6
Zn	26.9	156	74.0±37.5	46.7	20.7
Cu	5.88	3.78	6.27±2.38	5.66	7.59
Cd	3.67	11.2	1.60±0.87	5.31	1.58
Ni	8.13	9.13	9.67±6.87	14.4	12.9
Fe	2260	4280	3420±444	4790	2780
OM (%)	0.26	0.74	0.98±0.71	1.48	0.48
pH	7.10	7.28	7.68±0.37	7.81	7.08

OM - organic matter; number(s) in parenthesis - sampling sites

Table 2. Mean (±standard deviation) concentrations of selected heavy metals in river water and bottom sediment samples

4.2.3 Organic pollution indicators

A holistic evaluation of surface water quality will be deficient if the organic constituents are not properly examined. In this study, some gross organic pollution parameters were carefully selected and determined to establish the organic pollution status of the investigated rivers. For instance, a cross-section examination of the dissolved oxygen (DO) content of the river water samples collected at different sampling sites showed that the rivers were not highly polluted with organic matter. It is, however, pertinent to note that the slight drop in DO levels at sampling site 2 could be indicative of influx of organic matter or oxygen demanding substances into the river. Interestingly, this drop in DO levels did not result in appreciable increase in the COD value obtained for the sampling site. The variation in the dissolved oxygen concentrations along the sampling sites in River Ona is presented in Figure 3.

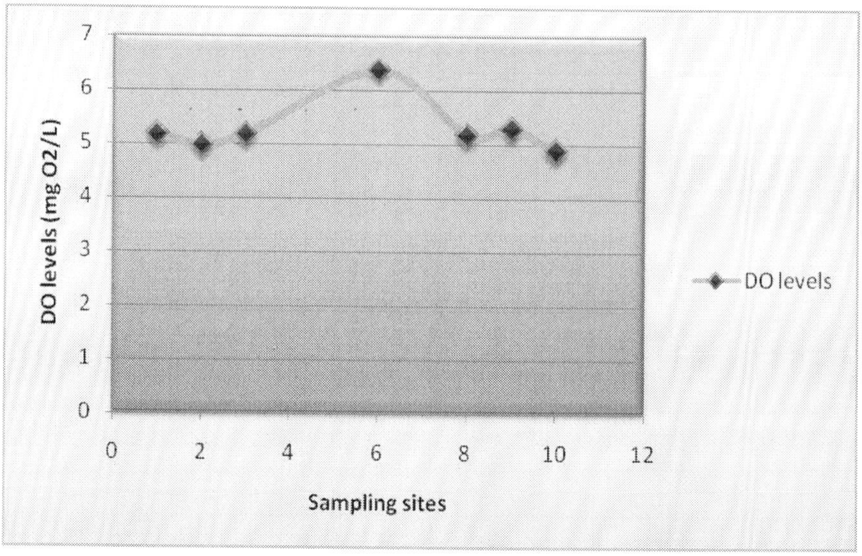

Fig. 3. Variation in dissolved oxygen levels along the sampling sites of River Ona

A seemingly interesting observation was found around the industrial zone of River Ona with relatively high DO level at the sampling site 6 which was contrary to our expectation. The observed level could be due to the additional measures taken by a major agricultural farm along the catchments of the river to further reduce the impacts of its discharges into the river. It was noted that some forms of physical treatments (e.g. aeration) of the river water some distance upstream of the sampling site 6 was undertaken by the farm management. As shown in Figure 3, the slight drop in DO level at sampling site 10 might be due to the influx of high oxygen demanding substances in urban run-off conveyed into the river by other rivers and streams.

The COD values obtained in this study ranged from 20.4 to 143 mg/ℓ with average levels of 99.1 and 54.4 mg/ℓ for River Ona and River Alaro, respectively. The lower COD values obtained in the industrial zone of River Alaro imply that most industries might probably have efficient wastewater treatment plants which in reality is not true. Apart from the dilution effects of the rainwater, the possibility of decreased production rates during the period of sample collection could also be responsible for observed results. Furthermore, the relatively higher COD levels after the convergence of the rivers further corroborate the influence of both urban and agricultural run-offs into River Ona via other streams and rivers as previously mentioned. Based on the classification of surface waters by Prati *et al.* (1971), River Ona with an average COD value of 99.1 mg/ℓ may be categorised as being heavily polluted while River Alaro with an average of 54.4 mg/ℓ may be categorised as being slightly polluted.

The assessment of oil and grease levels in both rivers revealed a high level of oil pollution at most sampling sites. The average levels of this parameter were found to be 338 and 415 mg/ℓ for River Ona and River Alaro, respectively. The relatively higher levels of oil and grease in River Alaro especially at the upstream region could be due to urban run-off which conveyed spent oils from various auto-repair workshops and particularly from the oil depot sited further upstream of the river. The Pearson correlation analysis performed on the three organic pollution indicators revealed that only DO had a significantly strong negative relationship ($r = -0.829$) with the COD at 0.01 level (2-tailed). Furthermore, a weak positive relationship ($r = 0.220$) exist between the DO and oil and grease while a moderately strong negative correlation ($r = -0.580$) was observed between the COD and oil and grease. Going by these statistical analyses, it can be inferred that oil and grease did not contribute appreciably to the total amount of chemical oxidisable organic pollutants in the river water samples. However, the presence of oil and grease in water bodies could significantly impair the normal growth and development as well as the survival of diverse aquatic fauna and flora.

4.2.4 Heavy metals

The evaluation of the quality of the river water with respect to heavy metal concentrations of both rivers before and after their convergence is an essential index of establishing the pollution status of the investigated rivers. A summary of these metals in both rivers before their convergence is presented in Table 3.

Metals	River Ona	River Alaro
Pb	105 (23.6)	75.2 (24.8)
Zn	177 (52.6)	213 (38.0)
Cu	26.6 (4.83)	24.0 (6.30)
Cd	6.25 (7.45)	3.50 (4.07)
Ni	72.5 (8.63)	70.0 (13.9)
Fe	879 (3270)	1210 (4120)
pH	7.57 (7.19)	7.50 (7.56)
OM (%)	(0.50)	(1.15)

values in parentheses represent the heavy metal concentrations in sediment samples; OM – organic matter.

Table 3. Average concentrations of heavy metals, pH and organic matter in river water (μg/ℓ) and sediment (μg/g) before the convergence of the rivers

From this assessment, it is evident that all the heavy metals except Zn and Fe had higher levels in river water samples of River Ona than in River Alaro. Similarly, relatively higher levels of these metals were found in bottom sediment of River Ona except Pb, Ni and Fe which had higher levels in River Alaro.

The concentrations of Lead (Pb) ranged from 38.5 to 184 μg/ℓ and 6.38 to 50.5 μg/g for river water and bottom sediment, respectively. The levels of Pb found in bottom sediment were higher than those obtained for river water samples; hence the bottom sediment could be an important influential factor contributing to the observed levels in river water samples of both rivers. Although the statistical correlation between Pb levels in sediment and water samples showed a weak positive relationship ($r = 0.175$), There is, perhaps, the possibility of sediment contributing to the Pb levels in the river water samples.

In similar studies conducted between 1977 and 1979, the Pb levels found at different sampling sites of River Ona ranged between 0.20 to 21.0 μg/ℓ (Mombeshora *et al.*, 1983), and the levels found in the sediment were from 0.2 to 17 μg/Kg (average of 2.50 - 5.00 μg/Kg) (Mombeshora *et al.*, 1981). Based on these present findings, the accumulation of lead in River Ona has escalated over the past years which are also true for other rivers within the city. Of utmost concern is the higher level of this metal found in the river water samples. Three out of the ten sampling sites had levels exceeding the WHO maximum permissible limit of 0.1 mg/ℓ, hence making the water unsuitable for drinking. This calls for immediate concern as the infants and young children are the most vulnerable to the health effects of lead, especially in rural areas of most developing countries where drinking water are fetched from these rivers without subjecting it to any form of treatment.

The levels of Zinc ranged from 105 to 251 μg/ℓ and 14.3 and 135 μg/g for river water and bottom sediment, respectively. The levels of Zn in river water samples of these rivers were several orders of magnitude lower than those found in four different rivers (1.10 - 3.03 mg/ℓ) within Ibadan city between July - August, 1990 (Fatoki, 1993). However, the levels obtained for sediment samples were comparable to those reported for a typical tropical rivers where Zn levels ranged from 24.9 - 148 mg/Kg (Adekola and Eletta, 2007). Zinc being an essential element for living organisms and due to its relatively low toxicity to humans, a maximum permissible limit of 5 mg/ℓ has been set as guideline value by WHO for drinking and other domestic purposes. Although the levels found in this study were generally lower than the permissible limit, it could however exert its toxic effects on sensitive biota in the aquatic ecosystems.

The levels of Copper in river water samples ranged from 18.5 to 31.5 μg/ℓ while these ranged from 2.28 to 8.24 μg/g for the bottom sediment samples. Appreciable increases in Cu levels were observed along the sampling sites of both rivers particularly in the industrial areas thus suggesting the possible influence of industrial discharges and agricultural run-off on the contamination of the rivers. The levels of Cu found in this study were generally lower than the WHO limit of 0.05 mg/ℓ for drinking water and much lower than those found in Challawa River, Kano (Nigeria) which ranged from 0.32 to 0.57 mg/ℓ (Akan *et al.*, 2007).

Cadmium levels ranged from nd to 7.50 μg/ℓ and 0.65 to 19.3 μg/g for river water and bottom sediment samples, respectively. Although the levels of Cd found in this study were comparatively lower than the WHO maximum permissible limit of 0.05 mg/ℓ, the fact that it is very persistent and toxic especially to aquatic organisms at relatively low concentrations calls for concerns. In fact, it was reported that 95% of its lethal effects in freshwater occur at concentrations > 2 μg/ℓ while in marine waters the corresponding concentration is > 95 μg/ℓ (Taylor, 1981, as reported in Taylor, 1983). Furthermore, the fact that it was not detected in river water samples of four different rivers within the city of Ibadan about two decades ago calls for concerns (Fatoki, 1993). This, however, suggest that industrial discharges as well as improper solid wastes management could be responsible for the observed trend.

A statistical correlation between Cadmium and other heavy metals in water samples revealed that Cd showed a significantly strong positive relationship with Zinc (r = 0.677) and Copper (r = 0.713) at the 0.05 level (2-tailed). This implies that notable sources of both Zn and Cu into the environment may be responsible for the release of Cd into these rivers and *vice versa*. In addition to the natural sources of this metal, other probable sources into the surface water bodies would include leaching from Ni-Cd based batteries, run-off from agricultural soils where phosphate fertilizers were previously used and other metal wastes. With the indiscriminate dumping of solid wastes and the crude techniques of managing solid wastes without proper sorting of electronic wastes from other municipal wastes before disposal, the freshwater systems will continue to be negatively impacted with the toxic effects of Cd and other pollutants contained in these wastes.

Nickel, at low concentrations, can cause allergic reactions and certain Ni compounds may be carcinogenic (Denkhaus and Salnikow, 2002). Nickel occurs naturally in the environment, however, because of its unique physical and chemical properties; metallic nickel and its compounds have found increased usage in modern applications. Several activities related to its handling during production, recycling and disposal can result into serious environmental pollution. In this study, the levels of Ni ranged from not detected (ND) to 185 μg/ℓ and 1.75 to 21.0 μg/g for river water and sediment samples, respectively. These levels were higher than the typical concentration of 0.0005 mg/ℓ in surface water given by the South African Department of Water Affairs and Forestry (DWAF, 1996). All the sampling sites except where the levels were not detected had higher concentrations than this concentration thus suggesting the possibility of water contamination from diverse sources. Possible sources of nickel in surface waters would include combustion of fossil fuels, old battery and other e-wastes, components of automobiles, old coins and many other items containing stainless steel and other nickel alloys. Among the known health-related effects of nickel are skin allergies, lung fibrosis, variable degree of kidney and cardiovascular system poisoning and stimulation of neoplastic transformation (Awofolu *et al.*, 2005).

The levels of Iron (Fe) ranged from 538 to 1920 μg/ℓ and 1900 to 6570 μg/g for the river water and bottom sediment samples, respectively. The maximum permissible limit for Fe in drinking water is 1.00 mg/ℓ, a guideline value set by the WHO. Three out of the ten sampling sites had river water levels higher than the set limit suggesting that the direct use of the river water for drinking can be detrimental. Iron is one of the most abundant elements in the earth; hence the elevated levels found especially in the sediment samples were expected. This assertion does not rule out the possibility of anthropogenic sources of this metal into the freshwater systems as there were several steel industries within the industrial areas of both rivers which could significantly contribute to the observed levels.

The organic matter content of sediment samples ranged from 0.26 to 2.58%. Organic matter is the organic fraction of decomposed plant and animal residues that plays critical roles in water retention, aggregation and soil structure. Organic matter has great influence on the mobility, flux as well as the availability of heavy metals, especially in the aquatic environment. Although all the metals investigated showed positive correlation with the organic matter, only Zn (r= 0.638), Ni (r= 0.771) and Fe (r= 0.811) showed a statistically significant positive correlation with the organic matter. These metals had significant positive correlation at 0.01 level except Zn which showed a significant correlation at 0.05 level.

An attempt was made to compare the results of the heavy metal analysis with similar studies around the world to establish the degree of industrial pollution of surface water of the investigated rivers. The evaluation showed that the average levels of all the metals in river water samples were higher than the 'world average' reported for freshwaters. These were also relatively higher than those reported in other similar studies except for the South African study where higher levels of some of these metals were reported. Of greater concern is the fact that most of these metals had average values which exceeded the WHO guideline limit for drinking water. It is therefore necessary to stress that the exposure of vulnerable groups (i.e. infants and young children) in local communities who depend on this water bodies as source of drinking water would be most affected. The higher average level of Pb is of utmost concern as it is known to exert its toxic effects on the developing brains resulting into a condition termed encephalopathy.

The average levels of these metals in bottom sediment samples were generally lower than those reported in other studies except for the South African study where lower levels were observed. This trend might be due to the significant contribution of bottom sediment to the overall heavy metal content of the river water samples. The high turbulence witnessed in most parts of these rivers could greatly affect the deposition of suspended solids thus influencing the metal distribution in these rivers. Although most of these metals occur naturally in the environment, their detection at levels higher than their natural background levels is indicative of additional sources of these metals into the investigated rivers.

5. Conclusion

The evaluation of the surface water quality at different sampling sites over the stretch of both rivers has been conducted. The pattern exhibited by most of the physicochemical as well as inorganic and organic parameters monitored revealed that the negative impacts of industrial discharges were quite significant in both rivers. However, the detection of relatively high levels of some of these parameters at the upstream of both rivers especially River Alaro suggests that sources other than industrial discharges are also contributing to the water quality deterioration of these rivers.

The heavy metal contamination of these rivers was more pronounced in River Ona than in River Alaro. This is because the average levels of at least four out of the six metals were higher in the industrial zone than in the upstream region for water and sediment samples of River Ona whereas only two of the six metals had higher average levels in the industrial zone than in the upstream region of River Alaro for the sediment samples. Interestingly, the levels of most parameters monitored downstream were significantly elevated than the levels found in the upstream before the convergence of these rivers. The influx of discharges via other notable rivers and streams into River Ona further downstream was responsible for the observed trend.

Sample type/Location	Metals						Reference
River water (μg/ℓ)	Pb	Zn	Cu	Cd	Ni	Fe	
R. Gomti, India	4.00–55.0	30.0–91.0	0.00–35	BDL	73.0–105	NR	Gaur et al. (2005)
Diep River, South Africa	0– 0	100–4400	- 800	NR	0.00–400	100–513000	Jackson et al. (2009)
Niger Delta Rivers, Nigeria	8.08	15.2	11.3	1.61	6.00	383	Kakulu and Osibanjo (1992)
Ona and Alaro Rivers, Nigeria	94.5	180	25.6	3.70	69.5	1180	**This study**
World Average for freshwater	1.00	30.0	10.0	NR	2.20	40	Ikem and Adisa (2011)
WHO limit	10	NGV	2000	3	70	NGV	WHO, 2006
Bottom sediment (μg/g)							
Ottawa River, Canada	26.0	84.0	28.0	NR	22.0	9200	Oliver (1973)
R. Calabar, Nigeria	31.0	184	64.0	0.30	67.0	30,800	Ntekim et al. (1993)
R. Derwent, England	96-3120	82-2760	NR	0.6-13.8	NR	NR	Burrows and Whitton (1983)
R. Gomti, India	5.80–100	14.9–182	3.89–91.2	0.41–17.3	5.32–57.6	NR	Gaur et al. (2005)
Pearl River, China	168–264	268 -426	49.3–69.0	7.2–7.8	42.1–65.1	NR	Cheung et al. (2003)
Tyume River, South Africa	0.04–0.07	0.08–0.49	0.08–0.50	ND–0.005	0.40–0.98	NR	Awofolu et al. (2005)
Ona and Alaro Rivers, Nigeria	26.2	54.7	5.70	4.68	10.5	3570	**This study**

BDL - below detection limit; ND - not detected; NGV - no guideline value given; NR - not reported.

Table 4. Comparison of heavy metal concentrations in water and sediment samples with the other studies around the world

The non-point intrusion of seepage from indiscriminately dumped solid wastes in the drainages and along the banks of major rivers within the city was responsible for the elevated levels of most parameters found at the upstream of the rivers investigated in this study. The perennial flood disaster witnessed in the city in recent times would persist unless measures are taken to address the problems of solid wastes management. The inadequate attention given to environmental issues in the past is being reflected in the form of serious ecological problems. To overcome these challenges, government must be proactive in its attempts to address environmental issues as anything short of this would amount to huge losses in the form of properties and human lives.

Furthermore, there is need to strengthen the government agencies responsible for waste management in the city. Waste management is capital intensive, has the potential to deliver good returns on investment and can provide countless number of jobs if properly managed. The heavy duty machineries required for handling these wastes must be maintained at all times so as to prolong their lifespan. Since one of the factors contributing to the poor environmental conditions in the city is illiteracy, government must embark on vigorous environmental education to enlighten the people about the consequences of untidy environment. This could even be integrated into the school curricula of primary and secondary school students to provide a forum whereby good ethics can be inculcated into these young children.

The roles of the legislative instrument to forestall these problems must not be overlooked. Government must critically review relevant environmental laws and amend them appropriately so as to be able to address immediate environmental challenges. The relevant authorities must ensure that personnel responsible for the enforcement of these laws demonstrate a sound understanding of the laws enacted. Appropriate sanctions in the form of fines must therefore be levied on the violators of these laws regardless of their societal status. The implementation of these recommendations would assist in proffering a lasting solution to solid waste management problems and this would directly or indirectly enhance the quality of water bodies within the city.

6. References

Adekola, F. A. & Eletta, O. A. A. (2007). A study of heavy metal pollution of Asa River, Ilorin, Nigeria; trace metal monitoring and geochemistry. *Environ. Monit. Assess.*, Vol.125, pp. 157 - 163.

Ajayi, S. O. & Osibanjo, O. (1981). Pollution studies on Nigerian rivers II: water quality of some Nigerian rivers. *Environmental Pollution (Series B).*, Vol.2, pp. 87 - 95.

Akan, J. C., Abdulrahman, F. I., Ayodele, J. T. & Ogugbuaja, V. O. (2007). Studies on the effects of municipal waste and industrial effluent on the pollutant levels of River Challawa, Kano State, Nigeria. *Research Journal of Applied Sciences.*, Vol.2, No.4, pp. 530 - 535.

APHA, AWWA, WPCF. Standard methods for the examination of water and wastewater. American Public Health Association, Washington, USA. 1985.

Areola, O. & Akintola, F. O. (1980). Managing the urban environment in a developing country: The Ogunpa River channelization scheme in Ibadan City, Nigeria. *Environment International.*, Vol. 3, pp. 237 - 241.

Awofolu, O. R., Mbolekwa, Z., Mtshemla, V. & Fatoki, O. S. (2005). Levels of trace metals in water and sediment from Tyume River and its effects on an irrigated farmland. *Water SA.*, Vol.31, No.1 , pp. 87 - 94.

Brooks, C. M. & Effler, S. W. (1990). The distribution of nitrogen species in polluted Onondonga lake, NY, USA. *Water, Air & Soil Pollution.*, Vol.52, pp. 247 - 262.

Burrows, I. G. & Whitton, B. A. (1983). Heavy metals in water, sediments and invertebrates from a metal-contaminated river free of organic pollution. *Hydrobiologia.*, Vol. 106, pp. 263 - 273.

Chapman, D. V. (1996). Water quality assessments - a guide to use of biota, sediment and water in environmental monitoring. 2 ed. University Press, Cambridge, UK.

Cheung, K. C., Poon, B. H. T., Lan, C. Y. & Wong, M. H. (2003). Assessment of metal and nutrient concentrations in river water and sediment collected from the cities in the Pearl River Delta, South China. *Chemosphere.*, Vol. 52, pp. 1431 - 1440.

Denkhaus, E. & Salnikow, K. (2002). Nickel essentiality, toxicity and carcinogenicity. *Critical Reviews in Oncology/Hematology.*, Vol.42, pp. 35 - 56.

Department of Water Affairs and Forestry (DWAF). South African Water Quality Guidelines, Field Guide. Vol.8 (1st ed.), DWAF, Pretoria, 1996.

Fakayode, S. O. (2005). Impact assessment of industrial effluent on water quality of the receiving Alaro River in Ibadan, Nigeria. *AJEAM-RAGEE.*, Vol.10, pp. 1 - 13.

Fatoki, O. S. (1993). Levels of dissolved Zinc and Cadmium in some surface waters of Western Nigeria. *Environment International.*, Vol. 19, pp. 285 - 289.

Gaur, V. K., Gupta, S. K., Pandey, S. D., Gopal, K. & Misra, V. (2005). Distribution of heavy metals in sediment and water of River Gomti. *Environ. Monit. Assess.*, Vol.102, pp. 419 - 433.

Ikem, A. & Adisa, S. (2011). Runoff effect on eutrophic lake water quality and heavy metal distribution in recent littoral sediment. *Chemosphere.*, Vol.82, No.2, pp. 259 - 267.

Jackson, V. A., Paulse, A. N., Odendaal, J. P. & Khan, W. (2009). Investigation into the metal contamination of the Plankenburg and Diep Rivers, Western Cape, South Africa. *Water SA.*, Vol.35, No.3, pp. 289 - 299.

Kakulu, S. E. & Osibanjo, O. (1992). Pollution studies of Nigerian rivers: trace metal levels of surface waters in the Niger Delta area. *International Journal of Environmental Studies.*, Vol.41, pp. 287 - 292.

Mombeshora, C., Osibanjo, O. & Ajayi, S. O. (1983). Pollution studies on Nigerian rivers: the onset of lead pollution of surface waters in Ibadan. *Environment International.*, Vol.9, pp. 81 - 84.

Mombeshora, C., Ajayi, S. O. & Osibanjo, O. (1981). Pollution studies on Nigerian rivers: toxic heavy metal status of surface waters in Ibadan city. *Environment International.*, Vol.5, pp. 49 - 53.

Novotny, V. (2003). Water quality: diffuse pollution and watershed management. John Wiley & Sons, Inc., New York, USA.

Nriagu, J. O. (1978). Sulphur in the environment. Wiley, New York, USA.

Ntekim, E. E. U., Ekwere, S. J. & Ukpong, E. E. (1993). Heavy metal distribution in sediments from Calabar River, southeastern, Nigeria. *Environmental Geology.*, Vol. 21, pp. 237 - 241.

Oliver, B. G. (1973). Heavy metal levels of Ottawa and Rideau river sediments. *Environmental Science & Technology.*, Vol.7, No.2, pp. 135 - 137.

Omoleke, I. I. (2004). Management of environmental pollution in Ibadan, an African city : the challenges of health hazard facing government and the people. *J. Hum. Ecol.*, Vol.15, No.4, pp. 265 - 275.

Osibanjo, O. & Ajayi, S. O. (1980). Rapid and sensitive spectrophotometric method for the determination of nitrate in rain water using 3,4 - xylenol. *Analyst.*, Vol.105, pp. 908 - 911.

Prati, L., Pavanello, R. & Pesavin, F. (1971). Assessment of surface water quality by a single index of pollution. *Water Research.*, Vol.5, pp. 741 - 751.

Schumacher, B. A. Methods for the determination of Total Organic Carbon (TOC) in soils and sediments. United States Environmental Protection Agency (USEPA), Environmental Sciences Division. ERASC NCEA-C-1282. EMASC-001, Las Vegas, NV, USA. April, 2002.

Sengupta, B., Laskar, S., Das, A. K. & Das, J. (1988). Inorganic pollutants of Ganga water in the region of Berhampore to Katwa, West Bengal. *Indian J. Environ. Health.*, Vol.30, No.3, pp. 202 - 208.

Taylor, D. (1983). The significance of the accumulation of Cadmium by aquatic organisms. *Ecotoxicology and Environmental Safety.*, Vol. 7, pp. 33 - 42.

WHO.2006. Guidelines for drinking-water quality, 3rd ed. Geneva, Switzerland. 20.09.2011, Available from http://who.int/water_sanitation_health/

26

Groundwater Quality Development in Area Suffering from Long Term Impact of Acid Atmospheric Deposition – The Role of Forest Cover in Czech Republic Case Study

Z. Hrkal[1,2], J. Burda[3], D. Fottová[3], M. Hrkalová[4], H. Nováková[1] and E. Novotná[1]

[1]*T. G. Masaryk Water Research Institute,*
[2]*Charles University, Faculty of Science, Dept. of Hydrogeology,*
[3]*Czech Geological Survey,*
[4]*Czech University of Life Sciences, Faculty of Agrobiology, Food and Natural Resources, Dept. of Water Resources, Czech Republic*

1. Introduction

About 84% of the Czech Republic is covered by so called hardrocks, igneous, metamorphic and strongly cemented rock. Such kind of rock environment is widespread throughout the world, the total extension is estimated at about 20% of present land surface, i.e. approximately 30 millions km^2 (Krásný 1999). From the hydrogeological point of view only near surface aquifer is important. Aquifer thickness is generally in the range of tens of meters and permeability decreases downwards. Nevertheless shallow hardrock aquifer is of major importance in generation of regional groundwater runoff and, consequently, groundwater resources development. Over the large areas hardrock aquifer is only source of drinking water. Problem is that shallow aquifer confined to crystalline rocks is very vulnerable to a decline in atmospheric precipitation and infiltration and exploitable yield should be very unstable.

Hardrock aquifers in Central Europe and in the Czech Republic in particular pose a specific problem consisting in gradual degradation of quality of ground waters caused by long-term acid atmospheric deposition. During the last few decades a dramatic change took place so that the proportion of major cations and anions resulted in change of chemical type of ground waters. Due to significant decrease in pH, the metals bound in the rock massif were mobilized and their enhanced contents in ground waters resulted in damage of mountain ecosystems. In extreme cases there occurred a failure in water supply, which could have been linked with worsening of groundwater quality.

Significant damage to national economy caused by acidification initiated a range of projects focused on the investigation of interaction between atmospheric deposition and biosphere and hydrosphere. The vast majority of these activities were based on the monitoring of the small catchments of which areal extent did not exceed 1 km^2 and in which a mass balance was measured and established. The input of bulk deposition and throughfall in the system and then the output in form of chemical composition of groundwater leaving the catchment were measured. Detailed studies were focused on individual factors influencing the quality of water drained from the catchment. The investigation revealed that besides the character of rock environment, the vegetation cover, the forest in particular, plays an essential role in the quality of ground waters.

The rain washes down into soil dry deposition, dust particles and aerosols caught on leaves and needles of trees. These particles represent in the Czech Republic 2/3 of the total dry deposition (Havel et al. 1996). The intensity of this process is closely connected with the age and type of the forest (Malek & Astel 2007). Old conifers in general exhibit the highest ability to catch dry atmospheric deposition (Likens & Bormann 1995). The acidification is accelerated in particular by spruce conifers that mostly intensify the deposition of H^+ ions (Probst et al. 1992). For instance, Bergkvist and Folkeson (1992) found throughfall in spruce forest in comparison with beech trees and birch to have exhibited three times greater deposition of SO_4 and even eight times greater deposition of H^+. Also Jezeří catchment in the Krusne hory exhibited 49.3 kg S ha/yr under conifers and only 11.6 kg S under deciduous trees in 2000, while bulk deposition in an open area was only 8.1 kg S (Fottová 2003). These data support the argument against planting coniferous monocultures in areas with elevated air pollution. However, apart from this the quality of groundwater is also influenced by internal biological cycle of the forest (Parker 1990).

The atmospheric precipitation only after its passage through vegetation cover is infiltrated into the ground being first in contact with soil, unsaturated and finally with saturated zone of the rock massif. The groundwater gradually changes its chemical composition and the acidity of atmospheric deposition is attenuated by ongoing buffer reactions. During the first stage of acidification, which is more or less discreet, the pH remains stable because the acid deposition is buffered by HCO_3^- ions of which concentrations gradually decrease. Once the bicarbonates are consumed during this process the pH decreases and further progress in acidification is buffered by humic compounds and aluminum (Fottová 2003, Hruška & Krám 2003)

However, changes taking place in groundwater have also a backward effect on the health of the given forest. Low contents of calcium and magnesium and high concentrations of metals mobilized by low pH pose the first problem. Aluminum and beryllium appear to be the most dangerous elements for root system of trees as far as toxicity is concerned (Puhe & Ulrich 2001, Navrátil et al. 2002). According Ebben (1991) particularly spruce forest (*Picea abies*) is extremely sensible to high concentration of toxic metals.

However, the obtained results have only limited informative value for regional hydrogeological considerations as they provide although very exact data, but relating to very small area often not exceeding a few hectares. Moreover, the hydrochemical data from surface streams characterize an overall runoff from the catchment, which represents a mixture of surface and groundwater of which the mutual relationship varies considerably in the course of the year. Consequently, to separate data characterizing only the quality of groundwater from data related to the overall runoff was found very difficult (Hrkal et al. 2002).

So far acquired data became a motivation for the assessment of groundwater quality and its development on regional scale. Consequently, three mountain regions in Bohemia that were most affected by atmospheric deposition were selected for investigation- specifically the Krusne hory Mountains (Erzgebirge), Jizerské hory Mountains and Krkonoše (Giant) Mountains. All these regions suffered in recent past from large ecological damage resulting in extinction of the forest cover due to acid atmospheric deposition. The results of investigation of small catchments revealed close links between the type and/or health of the forest and the quality of groundwater in the catchment. Therefore, the project was focused on the assessment of the role, which the forest plays in the quality of groundwater on regional scale in an environment that is strongly affected by acid atmospheric deposition.

2. Description of pilot sites

Three mountain regions were selected as the pilot sites for investigation - Krusne hory Mts. (Erzgebirge), Jizerské hory Mts. and Krkonoše (Giant) Mts. (Fig.1).

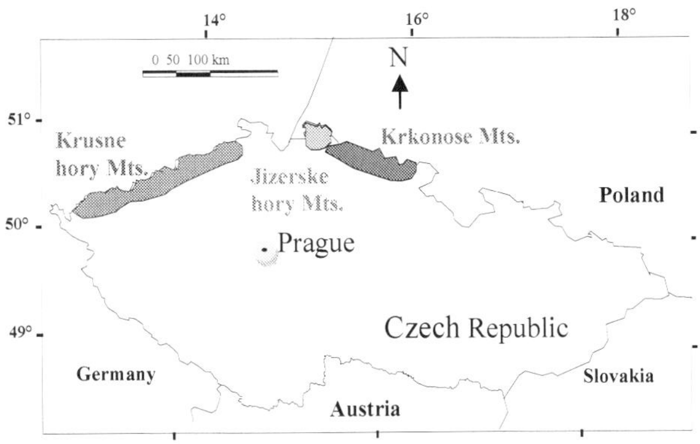

Fig. 1. Situation of pilot areas

All these areas suffered in recent past from large ecological disaster resulting in extinction of the forest cover evidently due to acid atmospheric deposition. All three regions are similar to one another as far as their geology and topography are concerned, but differ significantly from each other by the length and intensity of acid atmospheric deposition. Krusne hory Mts. were exposed to negative impact of acid atmospheric deposition for more than one century of which intensity culminated on the turn of 1970s and 1980s. The peak of negative impacts of acid precipitation occurred ten years later in the Jizerské hory Mts and Krkonoše Mts. Another difference between these regions is the level of environmental protection. While in the Krusne hory Mts only small areas are protected, the Jizerské hory Mts as a whole are Protected Landscape Area and Krkonoše Mts belong among most protected areas having a status of National Park.

2.1 The Krusne hory Mts.

The Krusne hory Mts. were selected as a model region because they suffered from uncontrolled anthropogenic activities. The total area of the investigated part of Krusne hory

Mts. is 1392 km^2. The region is a part of the so-called "Black triangle" close to a junction of borders between Czech Republic, Germany and Poland. The region is characteristic of strong accumulation of heavy industry, chemical plants, power stations and open cast mines for lignite, which were in operation for more than forty years. Huge open cast mines, some of them as much as 300 m deep completely changed the landscape of the piedmont area resembling a "moonscape" exhibiting entirely different hydrological regime. Burning of low-grade lignite high in sulfur (ranging between 8 and 12% - Tyracek et al. 1990) in local power stations resulted in extremely high values of acid atmospheric deposition, which caused acidification of all elements of the environment. As a consequence, the apical parts of the mountain range suffered from mass extinction of the forest cover, which led to stronger erosion, degradation of surface and groundwater quality including negative impacts on hydrological regime in general.

The situation has improved considerably during the nineties due to several factors and imposed measures, which were functioning in the area on parallel lines.

Krusne hory Mts. build a natural boundary between the Czech Republic and Germany. The mountain range is elongate in NE-SW direction with highest peak Klínovec 1224 a.s.l. The area under consideration belongs to slightly cool-to-cool climate with higher total precipitation mostly in areas more than 800 m a.s.l. where long-run rainfall is close to 1000 mm. Due to relatively low average atmospheric temperature (between 5°C and 7°C) the precipitation in form of snow lasts for considerable part of the year.

The potential natural vegetation in the Krusne hory Mts. is characterised by mixed forest of European beech, Silver fir, Norway spruce with a smaller admixture of birch oak and maple (Jankovská 1992). During the last centuries, forests in the Krusne hory have been dominated by Norway spruce. In the 1950s, before air pollution substantially increased, Krusne hory Mts. were predominantly covered by spruce monocultures with a small admixture of beech and larch. The first damage to the forest stands appeared in 1947, when the first symptoms to soil acidification due to enhanced air pollution has been observed (Nemec 1952). Since the 1950s to early 1970s, the area of damaged forest has slowly but continuously spread. Marked acceleration of forest health deterioration was observed after 1979 and was triggered by a temperature drop in 1978. Following figure shows the development of damage to spruce stands over the period 1960-1990.

In 1960 more than 50% of stands were classified as healthy. Later on the healthy stands disappeared and the share of clearcuts increased. The "harvested stands" category is sum of all spruce stands cut due to the forest decline. In 1990 more than 50% of the spruce stand was cut. Extending cutting have occurred, namely in higher elevation of the mountains, e.g. above 700 m a.s.l. where almost all older stands were harvested.

In many places, a site preparation prior to reforestation was done. One of the most common methods was topsoil removal with bulldozers, which was applied on almost 4000 ha. This method was introduced to increase the productivity of reforestation. It was expected to level the soil surface for planting machines and even to improve the soil's physical and chemical properties. However, research results show the rather negative influence of this method on soil fertility and growth of newly planted seedings.

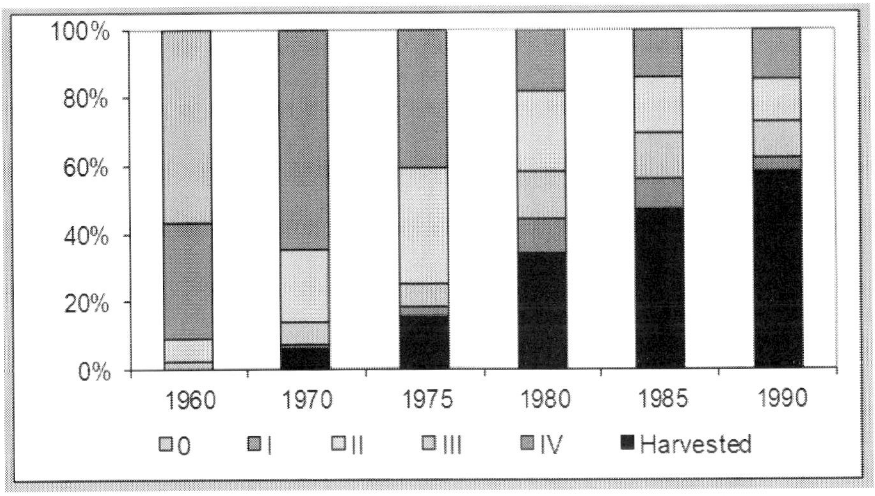

Fig. 2. Damage classes of spruce stands in the Krusné hory Mts., 1960 - 1990 (healthy = 0, dying = IV) after Kubelka et al 1992

The combination of different stress factors limited the use of local tree species i.e. spruce and beech, for reforestation. Forest regeneration was oriented to the establishment of new (so-called "substitute" or "emergency") forest stand composed of free species more tolerant to air pollution.

The dominant species in the "substitute" stands are birch, Blue spruce and Mountain ash. The main goal of establishing such "substitute" stands is to recovery of ecological function of forests in the affected regions, as species used have the most important pioneer function. The large scale usage of these tree species substantially changed the tree species composition of the new forest (Fig. 3.).

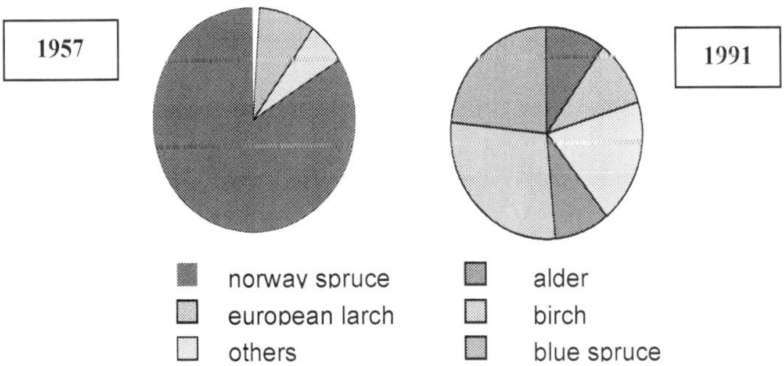

Fig. 3. Tree species composition in the forest of Krusne hory in 1957 (all stands) and 1991 (stands in the age class 1 - 30 years)

2.2 The Jizerské hory Mts.

The Jizerské hory Mts. constitute the northern morphological boundary between the Czech Republic and Poland (see Fig. 1). The altitude of the relatively flat crest of the mountain range slightly exceeds 1000 m above sea level. The Jizerské hory Mts. belong among the wet regions with annual total mean rainfall corresponding to 800 - 1 700 mm. Due to the topography of these mountains there exists a thick drainage pattern. The only natural water surface areas are peat bogs of which thickness varies within a few meters. More than 50 peat bogs of which the total areal extent exceeds 250 hectares exist in the mountains.

Geologicaly the study area is divided into two domains: the majority area is underlain by various types of granite of the Krkonose-Jizerske Mts. massif. Due to its siliceous character, the area of the granite massif contributes to acid character of local groundwaters. Only the northernmost part of the study area, composed of Proterozoic chlorite-sericite phyllite, amphibolite and mica schist, yields small amounts of acid of the groundwater.

The character and type of rock mantle and Quaternary alluvium generally play very significant role in other regions, but in the Krkonose-Jizerske granite is weathering into sandy eluvium showing equal character over the entire area. Only its thickness varies to certain degree. Some more or less random data revealed that greater thickness, as much as about 15 m, are expected to occur in depressions and on slopes, thus in areas of lower altitude.

Forest is the main natural vegetation cover in mountains occupying 274 km^2 of the total area of 368 km^2. Woodless areas are scarce being mostly confined to apical debris covered parts, peat bogs and wetlands. The original forest cover consisted mostly of beech, spruce and fir. The original forest cover remained untouched until the German colonization in the 13th century during which the forested area was reduced. The apical parts of the mountain range maintained their original features until the 16th century and the logging began as late as in the 17th century. Glassworks consumed the majority of wood that was used as a fuel. The plunder of forests started in the 18th century when deciduous and mixed forests became completely extinct at numerous sites. Artificial spruce monoculture planted from the 19th century turned out to be not very suitable because since the very beginning of the 20th century these spruce forests suffered from catastrophic blow-downs and calamities. In the middle of 1960s the spruce (*Picea abies*) formed 85% of the forest while the share of beech (*Fagus sylvatica*) was only 9%. Poor health of the spruce forest was gradually worsening and finally completely damaged by acid atmospheric deposition from power plants in Poland (power plant Turów) and in the former German Democratic Republic (power plant Hirschfelde). Mass extinction of the forest, due to immunodepression of forest species took place on flat apical parts of the mountain range and the dead forest was completely lumbered in the 1970s and 1980s. This area was then gradually forested by planting of a very limited number of tree species. Blue spruce (*Picea pungens*) was introduced in addition to autochtonous Norway spruce.

2.3 The Krkonoše Mts.

The Krkonoše mountain crest is 35 km long constituting a natural boundary with Poland. The apical parts of the mountain range with elevations around 1400 m above sea level are flat but its north-eastern slopes are plunging steeply into Poland. The opposite side of the mountain range is dissected by deep valleys but in general it plunges much more gently southward. Krkonoše Mts. are the highest mountain massif of the Czech Republic with the highest peak Sněžka 1603 m a.s.l.

As concerns the climate the Krkonoše Mts. are the harshest mountain ranges of which the apical parts can be correlated with the climate of Greenland shores. The mean maximum of snow cover attains as much as 2 meters and the average annual temperature fluctuates between 0 and 2°C.

Geological structure consists mostly of metamorphic rocks, mainly phyllites and gneisses accompanied by a granite massif and rare effusive rocks. Rare crystalline limestones occur in the eastern part of the mountain range. Two phenomena participated in creation of the present form (topography, morphology) of the mountain range. Tertiary alpine folding during which a gradual uplift and arching of the mountain range took place and resulted in today's altitude and shape. River erosion and activity of continental glacier in particular are responsible for the current and definite shape of the mountain range.

Krkonoše Mts. forms a natural water-divide between the North and Baltic seas and where numerous rivers rise including the river Labe (Elbe).

Fig. 4. Typical situation of catchment covered by mixture of recently seeded and dead forest (Krkonoše Mts., Kozi hřbety area, June 2010)

The present forest cover underwent a dramatic development and change since the last ice age. Characteristic vegetation cover after the recession of continental glacier consisted of pine, oak, hazel, lime and elm. Spruce, alder and dwarf pine began to grow there approx. 4 500 years ago and two thousand years later beech began to appear in the mountains. The last principal changes in species composition took place roughly before 800 B.C when fir and dwarf form of Norway spruce and dwarf pine in apical parts of the mountain rage began to grow. This kind of forest is considered to be the natural vegetation cover that continued until the 13th and 14th centuries when Krkonoše Mts began to be colonized. During the 16th and 17th centuries much of the forest was cut down mostly in connection with ore mining and production of char coal when beech, ash and elm trees gradually disappeared. The reforestation began as late as in the 18th century when spruce monoculture imported from Austria was planted so that at the end of 19th century. At the first half of twenty century forest in the Krkonoše are dominated by Norway spruce (*Picea abies*), which occupies 87,7% of the forested area. The second most commonly present species is *Pinus mugo* (6,1%) which forms the three border in the higher elevations of the mountains. The share of other species (e.g. beech *Fagus sylvatica*, maple *Acer sp.* and birch *Betula sp.* was very low (2,8%, 0,5% and 0,8% respectively (Moravčík, Černý in Černý, Pačes 1995). Forestation of dwarf pine in tundra parts also took place.

However, the construction of power plants burning coal in the so-called black triangle had fatal consequences on the forest cover in Krkonoše Mts. where gradual extinction of forests in apical parts took place. The first observation of forest damage was made in 1979 (Tesař et al. 1982). Consequently, the forest covering almost 8 000 hectares was completely destroyed during the twenty years due to acid rains. Geographical distribution of spruce defoliation is uneven over the area of the Krkonoše Mts. The defoliation is found at sites in the western part of the mountains in higher elevation exposed to wind (Černý, Pačes 1995).

A managed and gradual forest regeneration consisting of planting fir and beech and to lesser extend even sycamore maple and rowan began in 1990s. Restoration of self-functioning ecosystems specific for the area of National Park is another remedy implemented in the Krkonoše Mts. The dead wood mass is intentionally left in the forest and its decay ensures the return of alkaline substances into soil thus reducing naturally its acidity and also supports species diversity. This step, however, has also some drawbacks because the current forest is not healthy enough and strong to be able to combat the bark beetle calamity since the dead wood left in the forest is a very risk factor.

3. Data available and methodology of the treatment

Three types of data acquired during prolonged periods of time were necessary to meet the designed project objectives. They include data on the development of dry and wet deposition, results of monitoring of chemical composition of ground waters and data on species proportion over the whole area and its development and on the forest health.

The development of ground waters chemistry on regional scale was reconstructed during the first stage of investigation and the results of groundwater quality were compared with the development of atmospheric deposition.

The second stage of studies was focused on the relationship between the quality of ground waters and the state of health of the forest in relevant catchments. The study was based on a network of catchments forming the infiltration areas of the monitored springs and defined within a Digital Elevation Model (grid 50 x 50 m) using ARCVIEW 3D Analyst software. All springs drained a shallow sub-surface aquifer confined to rock mantle and open fissures in metamorphic and magmatic rocks reaching a maximum depth of 30 m. Therefore, the hydrogeological divide was identical with the hydrological divide defined by local topography. A database of factors influencing the groundwater quality and governing changes in it was created for each catchment. The database included altitude above sea level of each spring, the areal extent of the catchment and the basic characteristics of the vegetation cover

The following chapters provide a detailed overview of all data which were used for processing and considerations including description of methods of their acquisition.

3.1 Atmospheric deposition

Data on atmospheric deposition are available for three "pilot sites" that were monitored within the GEOMON project that has been operated by the Czech Geological Survey and is currently supported by the Ministry of Environment of the Czech Republic through the project SP/1a6/151/07. The above project is monitoring a network of catchments of which the Jezeří catchment in Krušné hory Mts. (JEZ), the Uhlířská catchment (UHL) in the Jizerské hory Mts and Modrý potok catchment (MOD) in the Krkonoše Mts were studied in detail. The basic programme of regular monitoring in individual catchments is the following:

- sampling of bulk precipitation - monthly cummulative samples
- sampling of throughfall precipitation - a mix of samples from the regular network of nine sampling stations in each catchment (for estimating the variability in vegetational density) - monthly cumulative samples
- sampling of the runoff from the catchment - basis
- aquiring data on the monthly precipitation amount in the unforested segment of the catchment
- aquiring data on the monthly throughfall
- a continual record of water level in the concluding profile of the catchment and determination a daily average outflow (from the consumption curve)

The sampling and analytical methods are uniform in the network which markedly raises the reliability of data and its mutual comparability (Fottová 1995, Fottová 2003). Accredited laboratories of the Czech Geological Survey run the following tests on all samples:

Na, Ca, K, Mg, S (SO_4), N (NO_3, NH_4), Cl, F, Mn, Fe, Zn, Al, As, Cd, Pb, Ni, pH, conductivity, DOC, DN.

Calculations of the element fluxes - input to the catchment of a defined area by bulk and throughfall precipitation and output from the catchment by surface runoff - is carried out from the data obtained in one hydrological year (from Nov. 1st to Oct. 31st of the following year). This is a question of monthly data on the concentrations of the monitored substances in both types of precipitation and runoff, monthly precipitation amount and the average daily outflow. The deposition of a particular element is counted as a sum of the 12 monthly deposition in units of mg/m^3, calculated by multiplying the concentration in mg/l by the total monthly precipitation (both for bulk and throughfall data) in mm. The results for every element are presented in units of kg/ ha /year.

The assessment of atmospheric deposition has especially been focused on trends of acidifying components so that the bulk and throughfall values of sulfur deposition as sulfates and nitrogen as nitrates are presented in graphic form (Figs. 5 and 6). An increase in pH that was higher in throughfall deposition was recorded in 1994 - 2009 in all three catchments. For instance, pH in the Uhlířská UHL catchment in Jizerské hory Mts. increased during the monitoring period from 3.7-4.5 (bulk) and 3.2 - 3.7 (spruce throughfall) in 1994 to 4.6 - 6.1 (bulk) and 4.3 - 5.7 (spruce throughfall) in 2009. Almost the same situation was recorded in the Jezeří JEZ catchment in Krušné hory Mts. The increase in pH in the Modrý potok MOD catchment in Krkonoše Mts was lower but still significant.

As concerns sulfur (Fig. 5) a significant decrease in bulk as well as throughfall deposition (also including dry deposition) was recorded evidently due to a desulfurization program implemented by the government in 1994 - 1998. The graphs show a marked peak in 1996 in the Jezeří JEZ catchment where a curious situation occurred when, due to the installation of ash-fly separators, the concentration of solid particles decreased, but desulfurization units were not yet installed. As a consequence, the acid constituents were not neutralized by basic components contained in ash-fly which resulted in extremely acid atmospheric precipitation and even glaze ice. This in fact also affected the throughfall precipitation (Krejčí 2001).

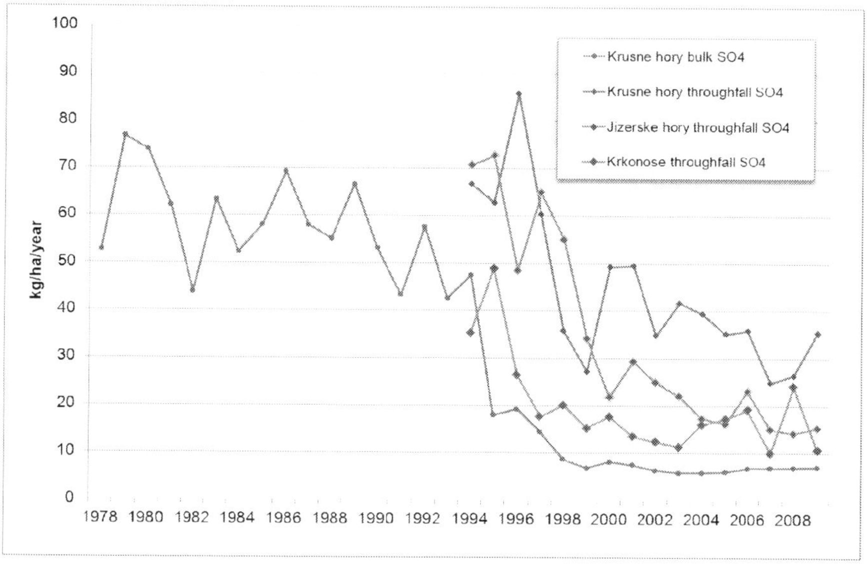

Fig. 5. Sulfate atmospheric deposition on study areas between 1978 - 2009

The deposition of nitrogen in the form of nitrates (Fig. 6) in both the Jezeří JEZ and Modrý potok MOD catchments showed no trend at all during the period 1994 - 2009, while in the Uhlířská UHL catchment some decrease in nitrogen content was recorded which is actually an exception within the whole of the GEOMON network of catchments. The deposition of nitrate nitrogen in the central part of the Czech Republic stagnated and in numerous catchments even decreased which a varying source of this form of nitrogen may be responsible for. This element came from emissions of power plants which were desulfurized but were still emitting nitrogen. Gasification of heating and recently a large increase in motor traffic are the new sources of nitrogen. The effect of aerosols, due to longer dry periods linked with changes in frequency of atmospheric precipitation, appears to be another source of nitrogen. It can be said that nitrogen after the year 2000 has become nationwide a more significant acidification element than sulfur.

The excess of nitrogen is manifested, for instance, by unnaturally high growth of trees. It is one of the primary reasons resulting in damage to the trees in combination with the direct effect of harmful substances on assimilation parts of the trees (including the effect of ozone) and acidification linked with mobilization of toxic metals in the soil. The weakened trees then have difficulties with secondary stressors such as sudden climatic changes, long dry periods in particular, insect pests or harmful fungi - *Ascocalyx abientina* (Hruška, Ciencala 2002).

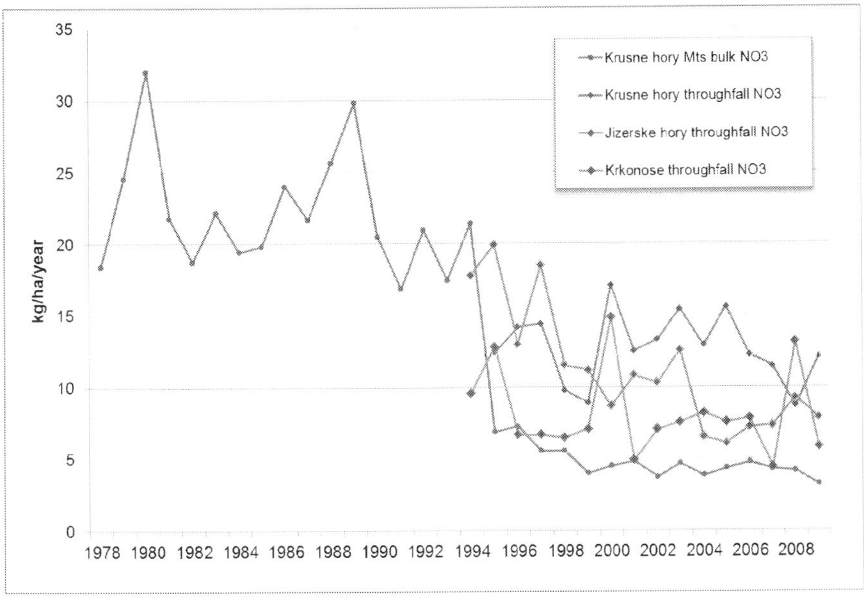

Fig. 6. Nitrates atmospheric deposition on study areas between 1978 - 2009

3.2 Forest health and surface changes

Maps of defoliation and mortality of conifers derived from Landsat-TM/ETM imagery were compiled by the Institute for Economic Development of Forests within the grid 30 * 30 meters. Ten categories (graded in 10% divisions) expressing the degree of damage to conifers and also defining the average defoliation of conifers in the vegetation cover were used to assist interpretation of the analytical results. Category 1 indicates completely healthy forest, and category 10 indicates dead forest.

The oldest data come from 1984 and from continuous series of interpreted satellite images so that a time interval could have been selected which corresponded to a period when samples of groundwater were collected and analyzed.

Series of images enabled a detailed analysis of changes in the areal extent of forest in individual catchments, and could also be used to assess their health. An example of interpretation of the forest health and its development is shown in Fig. 7.

3.3 Quality of ground water

The quality of groundwater is a very dynamic phenomenon so that a very representative data set is needed for unbiased assessment of its areal and temporal variability.

Water samples were analyzed using standard methods in accredited laboratories of the Vodní Zdroje GLS Company and Water Research Institute. The flame atomic absorption spectrometry or optical emission spectrometry with induced plasma were used for the determination of major and minor cations (Na, Mg, Ca, K, Mn, Li, Fe, Al, Zn, SiO_2), whereas trace elements (Cu, Pb, As, Cd, Be, Al, Co, Cr, Mo, Ni and V) were established by means of atomic absorption spectrometry with electrothermic atomization. Liquid ion chromatography was applied to determine SO_4, NO_3 and Cl. A glass electrode combined with pH meter (model Radiometer Copenhagen) was used for measurements of pH.

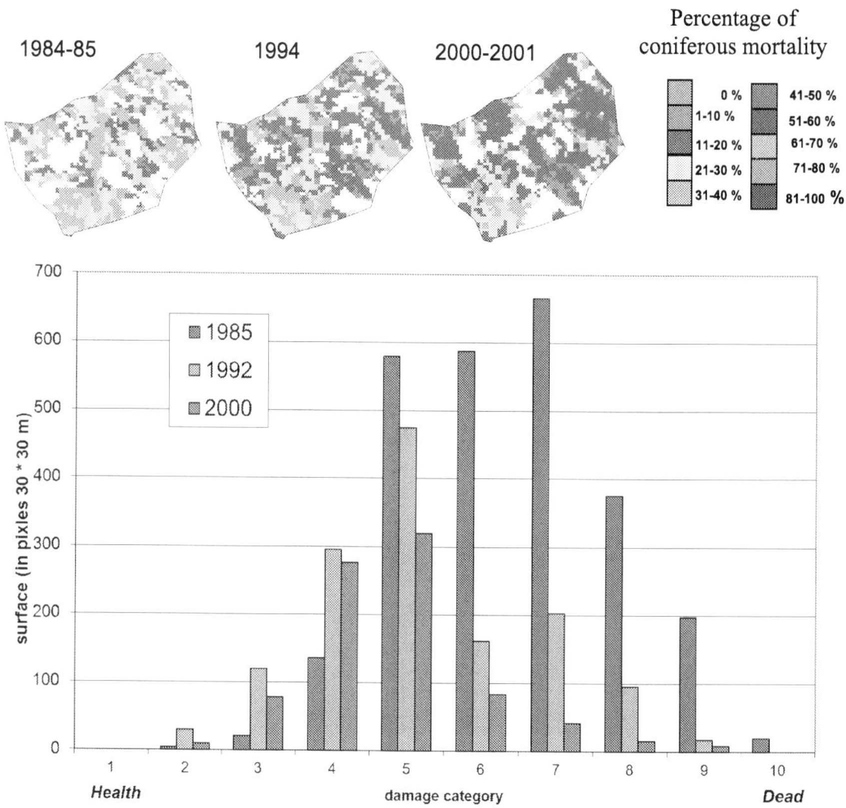

Fig. 7. Example of heath changes of forest cover in catchment No 17 in Krusne hory Mts.

The earliest data on groundwater quality were gathered during the years 1959 - 1965 in the Krusne hory Mts. A database of analyses of single samples from 130 springs was compiled by the Czech Geological Survey during hydrogeological mapping undertaken at that time. The historical analyses gave only contents of the main cations and anions. A total of 62 springs from this group were reanalyzed for the same elements during the years 1965 - 1974 and 1979-1989 (during the same period of the year). Twenty six of these springs were again sampled and analyzed in 2000 - 2002 as part of the LOWRGREP 5. FWP EU project (Hrkal et al. 2003). This time, analyses were made at monthly intervals and trace elements and toxic metals were included.

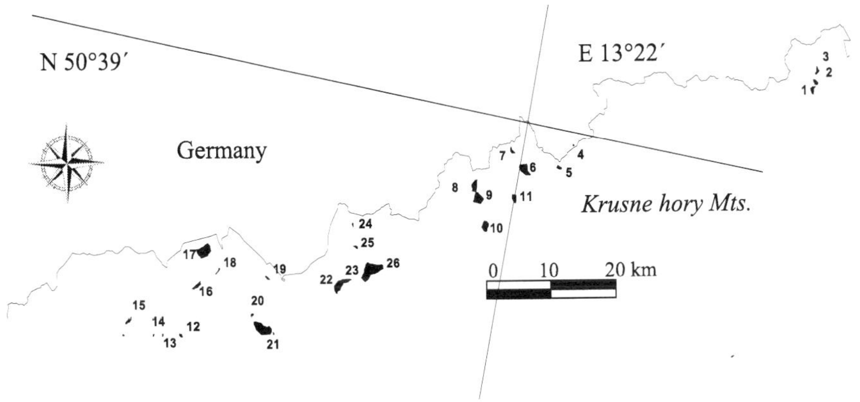

Fig. 8. Situation of experimental catchments in Krusne hory Mts.

The assessment of the Jizerské hory region is based on data collected in 37 small experimental catchments in which the variation in chemical composition of groundwater was monitored since 1971 until the present. The data come from springs which were analyzed in the frame of geological survey undertaken in the years 1971-1975 and 1983 - 1987 and then checked by single sampling in 1997, 2008 and 2009. The sampled springs more or less regularly cover the whole area of the Jizerské hory Mts. and can be considered a representative data set characterizing all phenomena, which may influence the resulting chemistry of local ground waters.

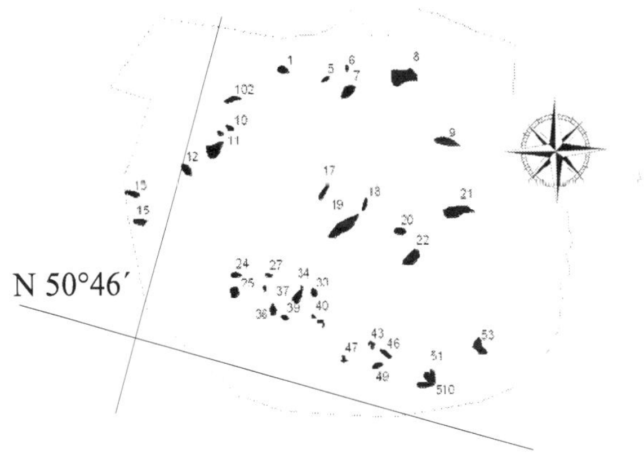

Fig. 9. Situation of experimental catchments in Jizerské hory Mts.

With respect to rather monotonous geology and lithology of the area studied that is built of granite, the resulting quality of groundwater is controlled by the chemistry of atmospheric precipitation and/or deposition, the time of interaction between the rock and water and the land-use. The first two phenomena are in the data set reflected by topography - altitudes above sea level of monitored springs vary between 340 m up to 1040 m. As concerns the land-use, the forestry is the dominating type of exploitation, while meadows without any cultivation occupy only small part of the area. This is also reflected in the character of experimental catchments, which are forested and in many of them the forest covered as much as 100 % of their total area.

Interpretation of the development of groundwater chemistry in Krkonoše area is based on a set of 34 springs. The data come from the years 1972 - 1974, and additional sampling took place in 1984 and 1994, and finally the very last sampling was carried out in 2008 and 2009. Experimental sub-catchments form two morphologically distinct groups which differ from one another by the type of vegetation cover and by climatic conditions.

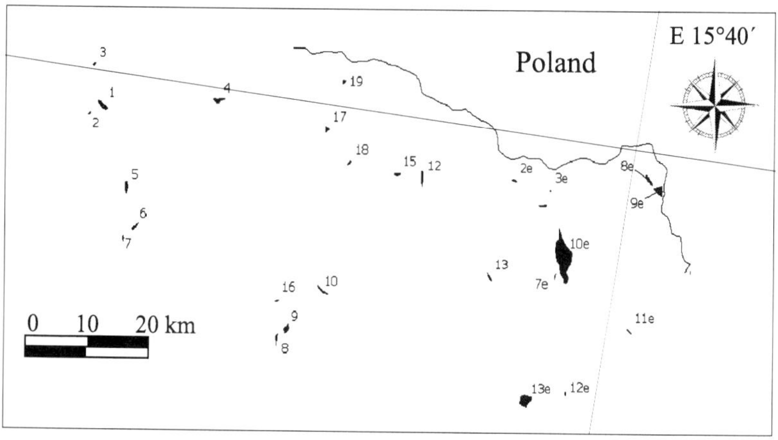

Fig. 10. Situation of experimental catchments in Krkonoše Mts.

The first group includes springs in apical parts of the mountain range with altitudes exceeding 1000 m above sea level of which catchment is mostly bare or only locally covered with dwarf pine, while the second group is located on mountain slopes with forest cover that underwent the above-mentioned development and changes.

4. Results and discussion

4.1 Comparison of the development of groundwater chemistry in pilot areas

The chemistry of ground waters in the Krusne hory area clearly shows the strongest affection by acidification among the three investigated regions. Weakly mineralized ground waters of Ca - SO_4, HCO_3 type were found to have been prevailing during the whole period of their monitoring. Unfortunately, no regional data on ground water quality are available from the 1980s in the Krusne hory area. On the other hand, data from the 1970s clearly point to major differences in average pH values of ground waters in the Krusne hory, Jizerské hory and Krkonoše areas. The contents of sulfates in ground waters from Krusne hory were in 1990s more than fivefold relative to HCO_3 component. A marked decrease in sulfates content was documented only the most recent analyses, but sulfates still remain the dominant component in Krusne hory Mts. ground waters.

The secondary concentration of sulfur in the soil horizon during the recent decades of atmospheric precipitation is believed to be responsible for this situation. At present, this accumulation is being progressively washed out again (Novak et al. 2000, Lischeid 2001). The study (Fottova et al. 2008) has shown that approximately 30% of the total sulfur content in the drained water comes from organic matter in the upper humic layer of the soil horizon. Consequently, the return of groundwater quality to the so-called "natural state" may take quite a long time.

period	Na	K	NH_4	Mg	Ca	Al	Be	Cl	NO_3	HCO_3	SO_4	F	Mineralization	pH
	mg/l						µg/l	mg/l						
1971-1975	6,61	2,85	0,01	4,12	14,7	0,00	?	7,80	1,13	13,57	45,9	0,08	99,9	5,49
1983-1987	?	?	?	?	?	?	?	?	?	?	?	?	?	?
1997	7,05	3,14	0,06	5,50	18,4	0,35	?	7,24	16,2	8,26	54,8	0,26	121,6	5,99
2000-2002	5,73	1,94	0,00	?	12,3	0,1	0,107	8,62	5,55	10,75	29,6	0,07	86,2	5,64
1971-2002	6,27	2,64	0,02	4,56	15,2	0,15	0,107	7,89	7,64	11,46	42,1	0,14	102,6	5,37

Table 1. Development of ground waters chemistry monitored in the years 1971 - 2009 in the Krusne hory area (an average from 26 monitored springs)

The results of monitoring revealed that similarly to Krusne hory the slightly mineralized ground waters of Ca - SO_4, HCO_3 type also prevail in Jizerské hory area. The chemistry of ground waters was found to be more or less stable in long-run. Nevertheless, some trend can be observed in some components, particularly those which indicate acidification. Local springs drain relatively shallow sub-surface zone of hardrocks where the time of interaction between the water and lithology is short so that the resulting chemistry of ground water is significantly governed by the chemistry of rainfall water. With regard to the decrease in acid atmospheric deposition, due to ecological strategy adopted by the Czech administration, the decrease in sulfates concentration from 33 mg/l in the 1970s down to 21 mg/l today is therefore not surprising. Similar trend exhibit also nitrates of which content dropped from 14 mg/l in 1970s to 4 mg/l currently. These values are already close to natural background.

Ground waters in Krkonoše region underwent an interesting development during the last thirty years. The characteristic type of groundwater in the 1970s up to the 1990s was similar to that in the Krusne hory and Jizerské hory areas, thus corresponding to Ca - SO_4, HCO_3. However, the most recent latest analyses undertaken in 2010 - 2011 revealed such a marked decrease in sulfates content so that ground waters can be classified to belong to Ca - HCO_3, SO_4 type, and in some cases even to the distinct Ca - HCO_3 type. This is not a random phenomenon, but a result of long-run trend supported by gradual increase in HCO_3 and decrease in sulfates content (Fig. 11).

period	Na	K	NH_4	Mg	Ca	Al	Be	Cl	NO_3	HCO_3	SO_4	F	mineralization	pH
	mg/l						µg/l	mg/l						
1971-1975	5,21	2,01	0,04	4,85	17,6	?	?	9,40	14,21	16,91	33,1	0,12	113,5	6,52
1983-1987	6,32	1,72	0,12	3,42	17,3	0,19	?	8,81	11,6	14,91	27,6	0,17	90,7	6,06
1997	5,64	1,57	0,05	2,44	13,8	0,21	0,053	4,21	5,12	15,72	27,2	0,19	87,3	6,15
2008-2009	4,48	1,37	0,04	2,14	8,9	0,27	0,082	3,71	4,40	18,43	20,7	0,20	61,9	6,07
1971-2009	5,31	1,51	0,06	2,9	13,6	0,24	0,067	6,01	8,30	24,01	26,2	0,18	97,1	6,13

Table 2. Development of ground waters chemistry monitored in the years 1971 - 2009 in the Jizerské hory area (an average from 37 monitored springs)

Similar trend can be seen in the case of nitrates of which contents vary around 5 mg/l. The ground waters monitored in the Krkonoše area always showed on average higher pH values relative to ground waters in the afore mentioned regions and also markedly lower degree of acidification. This is demonstrated by the fact that in the course of the 20th century, the pH at any of the monitored catchments did not fall below 6.0, not even at sites located at elevations exceeding 1 000 m above sea level.

period	Na	K	NH_4	Mg	Ca	Al	Be	Cl	NO_3	HCO_3	SO_4	F	mineralization	pH
	mg/l						µg/l	mg/l						
1971-1975	3,1	0,6	0,05	1,4	7,1	?	?	4,78	2,9	9,50	15,8	?	45,7	?
1983-1987	3,7	2,4	0,29	3,1	18,1	0,23	?	3,85	11,4	23,1	37,9	0,04	106,3	6,94
1997	2,6	1,0	?	1,9	9,8	?	?	2,45	7,1	20,5	13,7	0,06	68,0	6,86
2008-2011	?	?	0,16	2,9	11,5	0,20	0,02	2,50	5,7	36,4	9,5	0,03	74,2	6,42
1971-2009	3,0	1,3	0,19	2,5	11,3	0,14	0,23	2,86	6,32	27,78	14,23	0,04	69,17	6,62

Table 3. Development of ground waters chemistry monitored in the years 1971 - 2009 in the Krkonoše area (an average from 29 monitored springs)

Regardless of positive trends in groundwater quality it is obvious that some acidification processes still continue on regional scale partly due to persistent nitrogen deposition and partly also due to release of sulfur accumulated in long-run in soil profile. Characteristic chemistry of ground waters in these anomalous areas with prolonged pH lower than 5.5 is demonstrated in the following Tables 4 and 5.

These data clearly demonstrate fundamental differences in historical development of acidification of ground waters in all three compared regions. While in the 1970s almost 80% of springs in the Krusne hory area showed pH lower that 5.5, in Jizerské hory only 30% and in Krkonoše Mts all analyses revealed pH higher than 6.0. Very slow improvement can be demonstrated by analytical data obtained in the years 2008-2009 according to which 57% of local catchments still belong to the category of acidified ground waters. The number of acidified catchments in Jizerské hory area dropped to 19%.

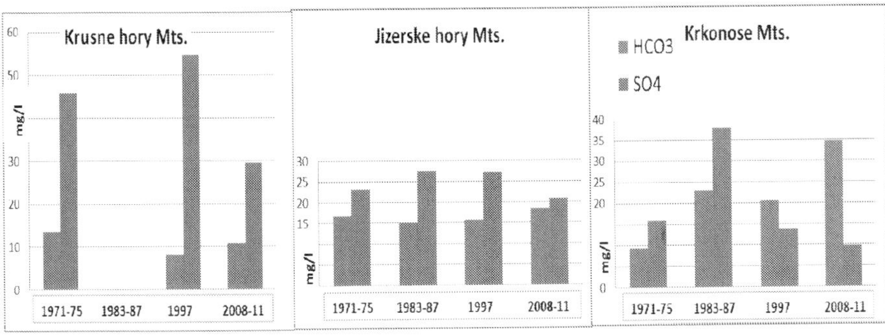

Fig. 11. Comparison of HCO_3 and SO_4 changes during the period 1971 – 2011 in three pilot areas

The common feature of analyses of ground waters from these anomalous catchments is their low alkalinity - contents of HCO_3 are believed to have mostly been eliminated through buffer reactions. Springs showing low pH in both regions, relative to average pH values from the whole of the area, have enhanced contents of Al and Be. The Al contents in Jizerské hory areas are three times higher while concentrations of Be are doubled. In Krusne hory area the aluminum contents in acidified springs are as much as five times higher relative to springs not affected by acidification.

period	percentage from total number	Na	K	NH_4	Mg	Ca	Mn	Fe	Al	Be	Cl	NO_3	NO_2	HCO_3	SO_4	F	minera-lization	pH
		mg/l								µg/l	mg/l							
1971-1975	78%	5,9	2,7	0,01	3,15	13,5	0,0	0,23	?	?	7,14	1,2	0,0	10,65	40,7	0,09	92,97	5,19
1983-1987	?																	
1997	37%	3,7	1,3	0,06	2,61	10,3	0,17	1,17	1,27	?	3,76	7,6	0,0	1,72	41,4	0,29	76,95	4,62
2008-2009	57%	3,4	1,1	0,01	?	7,4	0,0	0,5	0,2	0,07	5,20	5,0	0,0	5,7	19,3	?	62,2	5,1
1971-2009		4,3	1,7	0,03	2,88	10,4	0,06	0,63	0,74	0,07	5,3	4,6	0,0	6,0	33,8	0,19	77,27	4,97

Table 4. Development of groundwater chemistry in acidified springs (pH < 5,5) during the period 1971-2009 (Krusne hory Mts., average from 11 springs)

4.2 Impact of forest on groundwater quality

Forest health is in general an essential indicator of environmental conditions. This particularly applies to mountainous regions of the Czech Republic where the forest has gone through dramatic changes during the last few decades. The following analyses and their interpretation were focused on the assessment of relationship between forest health and the quality of ground waters.

period	percentage from total number	Na	K	NH_4	Mg	Ca	Mn	Fe	Al	Be	Cl	NO_3	NO_2	HCO_3	SO_4	F	minera-lization	pH
		mg/l								µg/l	mg/l							
1971-1975	30%	3,5	0,8	0,26	1,7	11,7	0,10	0,89	0,07	?	3,8	8,4		7,1	32,3	0,26	77,4	5,10
1983-1987	31%	3,9	0,8	0,21	2,5	10,1	1,68	0,73	0,09	?	4,0	7,6	0,01	7,8	29,1	0,22	73,4	5,26
1997	37%	4,4	0,6	0,05	1,5	11,4	0,10	0,49	0,57	0,01	3,2	1,4		7,4	32,0	0,34	68,1	5,29
2008-2009	19%	2,7	0,5	0,03	1,1	4,35	0,07	0,19	0,67	0,01	1,6	2,0	0,01	4,8	22,0	0,33	23,7	5,07
1971-2009	27%	3,4	0,6	0,10	1,8	8,39	0,8	0,57	0,63	0,01	2,8	3,6	0,01	6,6	27,8	0,32	64,5	5,16

Table 5. Development of groundwater chemistry in acidified springs (pH < 5,5) during the period 1971-2009 (Jizerské hory Mts., average from 12 springs)

This close link can be well demonstrated with the example of Jizerské hory Mts. where detailed analysis showed similar dynamics of evolution of forest health as that recorded in the Krušné hory Mts. (see chapter 2.1.). In 1984, forest health in Jizerské hory Mts was at the worst condition when 20% of conifers and 11% of deciduous trees were in a strongly to very strongly damaged category and more than 50% of conifer forest were in worse condition than the medium damaged category. Since 1984, there has progressively existed a better condition or trend in forest health which until 1997 was subjected to massive felling of sick trees. The areal extent of conifer forest decreased during the period 1984 - 1997 by 30%, while the proportion of deciduous trees increased. Only the next decade is marked by coniferous tree planting of which the area currently slightly exceeds its initial areal extent. Simultaneously, the initial proportion of conifer and deciduous trees returned to the original state.

Trends in the evolution of forest health correspond to changes in the quality of groundwater described in chapter 2.2. The close link between forest health and the quality of ground waters is demonstrated by the following two graphs for a coniferous forest.

The graph shows markedly greater proportion of damaged forest in catchments affected by acidification. An improving trend in forest health towards the end of the 1990s, and on the contrary a worsening of the forest health at the onset of new century correlate with the evolution of pH and concentrations of Al and Be.

Similar conclusions demonstrating a close link between the quality of ground waters and forest health can be drawn from the results obtained in the Krušné hory Mts. The following Table 6 demonstrates the close relationship between forest health and the quality of ground waters in experimental catchments in Krušné hory Mts.

In catchments where the forest showed a low level of damage, the pH of the groundwater was up to half an order higher than in areas showing a greater degree of damage. In catchments with severely damaged forest, the groundwater had enhanced concentrations of aluminum and low alkalinity.

Damage degree	Characterstic of the forest
O	Health forest
O/I	First symptoms of damage
I	Slightly damaged forest
II	Medium damaged forest
IIIa	Highly damaged forest
IIIb-IV	Dead forest

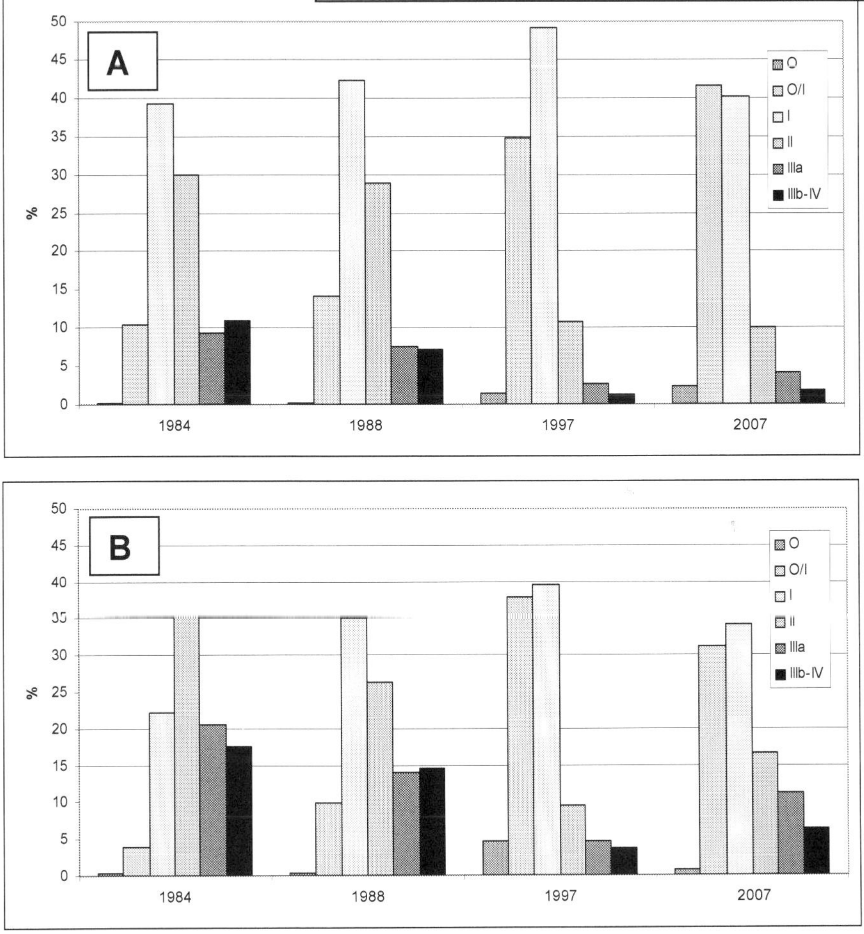

Fig. 12. Comparison of the evolution of coniferous forest health in all experimental catchments in the Jizerské hory Mts (A) and in strongly acidified catchments with pH lower than 5.5 (B).

pH	NO_3	HCO_3	Al	Harm category
	mg/l			
5,37	7,07	8,33	0,40	IIIa + IIIb IV
5,68	3,18	9,86	0,20	I - II
5,82	3,45	19,87	0,20	O+ O/I
5,86	8,40	16,13	0,10	No forest

Table 6. Relationship between mean values of major indicators of groundwater acidification and the state of health of the forest in Krusne hory catchments

The behavior of nitrates corresponds with their use as fertilizers. Only relatively healthy forest is able to eliminate the impacts of atmospheric nitrogen. Therefore, the enhanced concentrations of nitrates occur in areas without forest vegetation or in those with heavily damaged forest.

The fact that the mere existence of the forest as such magnifies the effect of atmospheric deposition on the quality of groundwater is demonstrated by comparison of the chemistry of groundwater in catchments differing from one another by forest density. Catchments with areal extent of forest exceeding or smaller than 50% of the total area were mutually compared. As follows from Table 7 the catchments covered with greater proportion of forest are characteristic of higher degree of acidification, have lower contents of HCO3, lower pH and enhanced contents of Be, Al and sulfates. Forested catchments in Krušné hory Mts. as well as in Jizerské hory Mts. are low in nitrates. In this particular case the forest plays a positive role because it uses nitrogen from atmospheric deposition as a fertilizer. Anthropogenic deposition of nitrogen is directly linked with greater forest growth (Kaupi et al. 1992, Binkley and Hogberg 1997, Aamlid et al. 2000).

Area	Percentage of forest from total surface	Na	K	NH_4	Mg	Ca	Al	Be	Cl	NO_3	HCO_3	SO_4	mineralization	pH
		mg/l						μg/l	mg/l					
Krusne hory Mts.	to 50%	5,8	2,1	0,03	4,4	14,0	0,19	0,029	7,2	6,2	12,6	39,8	82,4	5,7
	above 50%	4,9	0,7	0,04	2,6	9,16	0,25	0,075	7,1	2,2	6,9	36,8	60,9	5,2
Jizerské hory Mts.	to 50%	5,6	2,0	0,1	4,1	15,5	0,11	0,032	5,9	11,0	41,5	23,8	142,2	6,3
	above 50%	5,2	1,5	0,1	2,3	11,8	0,42	0,077	6,3	7,2	13,4	27,0	91,0	6,1

Table 7. Comparison of groundwater chemistry in dense and sporadic forested catchment area

Futher analysis was focused on the assessment of groundwater quality in the regions with damaged forests. Two types of data were compared: catchments where at least 20% of total surface fall in category highly damaged forest and areas where this criterium was not achieved.

The most conspicuous difference between the two data sets compared from both the Krušné hory and Jizerské hory regions are significantly higher concentrations of Al and Be in catchments with damaged forest.

area	forest health status	Na	K	NH_4	Mg	Ca	Al	Be	Cl	NO_3	HCO_3	SO_4	mineralization	pH
		mg/l						µg/l	mg/l					
Krusne hory Mts.	damaged	3,9	1,5	0,01	4,1	8,8	0,34	0,042	4,4	5,8	7,1	29,3	64,2	5,4
	health	3,7	1,3	0,01	1,2	7,8	0,02	0,021	4,4	2,2	7,7	20,3	51,6	5,5
Jizerske hory Mts.	damaged	3,1	0,6	0,07	0,9	5,8	0,58	0,108	2,0	3,0	7,7	15,2	49,4	5,8
	health	5,4	1,5	0,04	2,8	14,1	0,29	0,048	6,1	7,3	16,9	32,1	93,5	6,2

Table 8. Comparison of groundwater chemistry in damaged and health forest catchment area

This analysis raises a question of what is the cause and what is the consequence of such a situation? The primary cause is clear - acid atmospheric deposition that obviously worsened the quality of groundwater which in areas covered by damaged forest exhibits high concentrations of Al and Be and low pH. On the other hand, the forest at the same time intensifies the negative effect of atmospheric deposition.

A paradoxical result can be observed at first glance both in the Krušné hory and Jizerské hory regions when catchments with dead or strongly damaged forest show SO_4 and NO_3 concentrations, the major groundwater components, are lower than in the areas with healthy vegetation cover. This phenomenon can be explained by the reduction of the area (needles and leaves) that can catch the dry atmospheric deposition. Consequently, the dead forest does not play its negative role in intensification of the effect of atmospheric deposition on the quality of groundwater.

A similar approach to the assessment of the relationship between the forest cover and quality of ground waters made in the Krušné hory and Jizerské hory regions turned out to be less applicable in the Krkonoše Mts. because of rather poor results. The first reason appears to be the unsuitable selection of monitored catchments. The monitoring grid was established in the 1970s having been merely aimed at monitoring the quality of ground waters. Although the distribution of sampling sites was suited to hydrogeological requirements, the majority of monitored catchments were not forested or the composition and quality of the forest did not characterize natural conditions of this mountain range. The change in the distribution of monitored objects could eliminate this problem, but on the other hand the unique hydro-chemical data acquired in the past could not be used.

The topography of the Krkonoše Mts. is another phenomenon which is different from other regions. Significantly higher altitude (on average 1 000 m a.s.l.) in the Krkonoše Mts. at which 45% of monitored catchments are located, while in the Jiszerské hory region only one catchment lies above 1 000 m a.s.l. and in the Krušné hory area all monitored springs lie at altitudes below 900 m a.s.l. Consequently, the set of hydrochemical data in Krkonoše Mts characterized mostly an environment without any forest or its proportion in relation to total area was small and mostly did not exceed 25%.

The chemistry of ground waters in the Krkonoše Mts differs more or less significantly from that in the Krušné hory and Jizerské hory regions which could have been due to a very small number of samples collected in forested catchments. So, the current better quality of ground waters in the Krkonoše Mts. can be influenced, besides the primary cause which is their shorter exposure to acid atmospheric deposition, by prevailing tundra with dwarf pine covering the apical parts of the mountain range. This environment must have had limited ability to catch dry deposition in relation to the grown forest, and the bulk deposition was mostly responsible for acidification in the Krkonoše Mts.

5. Conclusion

A comparison of the evolution of groundwater chemistry in three mountain regions of the Czech Republic shows that dramatic changes have taken place during approximately the last 50 years. The regional interpretation of hydrochemical data was based on the assumption that the chemical composition of ground waters at all the localities studied was more or less the same and corresponded to natural background during the period prior to the impact of anthropogenic activities. This assumption was justified by the fact that the geology underlying all three regions investigated is more or less the same. The bedrocks are chiefly metamorphic rocks intruded by granites. The topography and climate of the three regions are similar and until the 1960s the land was covered by spruce monocultures, which gradually became extinct due to acid precipitation.

The results of the investigation revealed that long-term acid precipitation from the atmosphere had a markedly detrimental effect on the quality of local ground waters. The changes common to all regions were the decreases in pH and alkalinity and increases in the contents of SO_4, Al and Be. However, the intensity of these changes varied considerably from one region to another. The period of time during which the regions became subject to the negative effects of anthropogenic activity has been shown to play a decisive role in changes of groundwater chemistry.

The greatest changes in the quality of groundwater, that have persisted until today, occurred in the Krušné hory Mountains where acid atmospheric deposition has taken place more or less continuously for almost a hundred years, peaking in the 1960s and 1970s. The initial Ca-HCO_3 type of groundwater formerly found in the Krušné hory region has clearly been changed to groundwater of Ca-SO_4 type containing a concentration of sulfates five times more than that of the HCO_3 component.

Acid atmospheric deposition in the Jizerské hory and Krkonoše Mountains has taken place only for about 25 years. The groundwater chemistry showed the same trend as in the Krušné hory region but to a much lower extent. Moreover, the Krkonoše area also revealed that these hydrochemical changes are not irreversible because the chemistry of ground waters during the last few years is gradually returning to the preexisting Ca-HCO_3 type.

In contrast, it is believed that the return of groundwater chemistry in the Krušné hory region to the initial quality will take a much longer period of time. Although the problem of acid atmospheric deposition of sulfur has been successfully resolved, the prolonged accumulation of sulfur in the soil profile has created a persistent secondary source of this element that still affects the quality of local ground waters.

The quality of the ground waters in the three regions investigated showed significant heterogeneities and anomalies. Detailed analysis of the factors influencing the quality of ground waters affected by acid atmospheric deposition revealed the significant negative role played by coniferous monocultures in groundwater chemistry. The negative impact of acid deposition on the quality of groundwater in the majority of cases was found to have been more intense in areas covered by forest. This was best demonstrated in the case of spruce forest that has a much greater capacity to trap dry atmospheric deposition in the form of dust particles and aerosols. Forest catchments more often revealed anomalous concentrations of toxic aluminum and/or beryllium. The other aspect of this phenomenon is that the deterioration of ground water quality caused by the forest in turn affects the health of the forest itself. Damaged or completely dead forest was found to occur most frequently in catchments showing high degrees of acidification.

6. Acknowledgment

This work was supported by the projects of Czech Ministry of Environment MZP 0002071101, SP/1a6/151/07 and Ministry of Education MSM 60460709.

7. References

Aamlid, D., Torseth, K., Venn, K., Stuannes, A.O., Solbergs, S., Hylen, G., Christopheresen, N. & Framstad, E. (2000). Changes of forest healt in Norwegian boreal forest during 15 years. Forest Ecology and Management 127, 130-118.

Bergkvist, B. & Folkeson, L. (1995). The influence of tree species on acid deposition, proton budgets and element fluxes in south Swedish forest ecosystems. Effects of acid deposition and tropospheric ozone on forest ecosystems in Sweden, 90-99.

Binkley, D. & Hogdberg, P. (1997). Does atmospheric deposition of nitrogen threaten Swedish forests ? Forest Ecology and Management 92, 119-152.

Černý, J. & Pačes, T. Eds. (1995). Acidification in the Black Triangle Region. Czech Geological Survey, Prague

Ebben, U. (1991). Die toxische wirkung von aluminium auf das wachstum und die elemntgehalte der feinwurzeln von altbuchen und altfichten. Ber forsungzentr waldoksyst Univ, Gottingen A:64.

Fottová, D. (1995). Regional evaluation of mass element fluxes: GEOMON network of small catchments. Environmental Monitoring and Assessment 34, 215-221.

Fottová, D. (2003). Trends in sulphur and nitrogen deposition fluxes in the GEOMON network, Czech Republic, beween 1994 and 2000. Water, Air and Soil Pollution 150: 73-87.

Fottová, D., Hruška, J., Krám, P. & Skořepa, J. (2008). Assessment of the climatic change impact on the hydrological balance and the proposal of practical measures to their mitigation. Final report of the Czech Ministry of the Environment project SP/1a6/151/07, [In Czech].

Havel, M., Krejčí, R., Černý,J. (1996). Decline of atmospheric deposition in Krušné hory Mts. MS, Final rapport of project GACR 205/93/0675, Czech Geological Survey, Prague. [In Czech].

Hrkal, Z., Buchtele, J., Tikkanen, E. & Käpyaho, A. (2002). The role of groundwater in the acidification of the hydrosphere - examples from small catchments in the Bohemian Massif. Nor. geol. unders. Bull. 439. Trondheim. p. 99 – 105.

Hrkal, Z., Bender, S., Sanchez Navarro, J.A., Martin, C. & Vayssade B. (2003). Landscape-use Optimisation with Regards to Groundwater Protection Resources in Hardrock Mountain Areas (LOWRGREP); some results from an European research program (5th PCRD). Proceeding from Water Symposium UNESCO, Cannes, France.

Hrkal, Z., Fottová, D. & Rosendorf, P. (2009). The relationship between the quality of ground waters and the forest cover in regions affected by high levels of acid atmospheric deposition – a case study of the Krušné hory Mts., Czech Republic, Polish J. of Environ. Stud. Vol. 18, No. 6, pp. 1001-1010.

Hruška J. & Ciencala E. eds. (2002). Longterm acidification nad nutrient degradation of forest soil - restrictive factor of contemporary forestry (in Czech). MS Ministry of the Environment Czech Republic

Hruška, J. & Krám, P. (2003). Modelling long-term changes in stream water and soil chemistry in chatchments with contrasting vulnerability to acidification (Lysina and Pluhuv Bor, Czech Republic). Hydrology and Earth System Sciences 7, 525-539.

Jankovská, V. (1992). Evolution of Krusne hory Mts. since the last glaciation [In Czech]. Lesnická práce, 71. pp.73-75.

Kauppi, P.E., Mielkainen, K. & Kuusela, K. (1992). Biomass and carbon budget of European forests, 1971to 1990 Science 256, 70-74.

Krásný, J. (1999). Hard-rock hydrogeology in the Czech Republic. - Hydrogéologie, 2, 25-38. Éditions BRGM. Orléans.

Krejčí, R. (2001). Spruce forest damage in Krušné hory Mts. (in Czech) Vesmír 80, October 2001, 576-578.

Kubelka, L., Karásek, A., Rybář, V. & Slodinčák, M. (1992). Reconstruction of forest in immission endagered area of the Krušné hory Mts. Final Report. North Bohemia State Forestry. Research Office Krušné hory, Teplice, 139 pages. [In Czech].

Likens, G.E. & Bormann H.F. (1995). Biogeochemistry of a Forested Ecosystem. Springer-Verlag, 175 Fifth Avenue, New York,NY 10010, New York Inc.

Lischeid, G. (2001). Investigation short-term dynamics and longterm trends of SO4 in the runoff of forested catchment using artificial neural networks. Journal of Hydrology. 243, 31-42, 20.

Malek, S. & Astel, A. (2007). The effect of stand age on throughfall chemistry in spruce stands in the Potok Dupnianski catchment in the Silesian Beskid Mountains, southern Poland. Thescientificworldjournal. 7, 181-191.

Němec, A. (1952). Contribution to the question of spruce die-back in Krusne hory with the special emphasis on smoke damage. In: Práce výzkumných ústavů lesnických, Brázda, Praha, pp.167-227. [In Czech].

Novák, M., Kirshner, J.W., Groscheová, H., Havel, M., Černý, J., Krejčí, R. & Buzek, F. (2000). Sulfur isotope dynamics in two Central European watersheds affected by high atmospheric deposition of Sox. Geochimica et Cosmochimica Acta. 64, 3, pp. 367-383. Elsevier Science. USA.

Navrátil, T., Skřivan, P., Minařík, L. & Žigová, A. (2002). Beryllium Geochemistry in the Lesni Potok Catchment (Czech Republik), 7 Years of Systematic Study. Aquatic Geochemistry, 8, 2, pp. 121-133.

Parker, G.G. (1990). Evaluation of Dry Deposition, Pollutant Damage, and Forest Health with Throughfall Studies. In: Lucier, A.A., Hanines, S.G. (ed), Mechanisms of Forest Response to Acidic Deposition, 10-61. Springer-Verlag, New York.

Probst, A., Viville, D., Fritz, B., Ambroise, B. & Dambrine, E. (1992). Hydrochemical budgets of a small forested granitic catchment exposed to acid deposition: The Strengbach catchment case study (Vosges massif, France). Water, Air, and Soil Pollution. 62, (3-4), pp. 337-347.

Puhe, J. & Ulrich, B. (2001). Global climate change and human impacts on forest ecosystems. Ecological Studies 143, p. 593.

Tesař, V., Anděl, P., Schwarz, O. & Vacek, S. (1982). Findings on the magnitude of effects on Krkonoše forest accrding to 1979 knowledge (in Czech). Opera corcontica, 19 :79-94.

Tyráček, J., Malkovský, M., & Schovánek, P. (Eds) (1990). Synoptic geological map of the North Bohemian coal basin and its surroundings (1 :100 000). Geological Survey of Prague.

27

Don't Know Responses in Water Quality Surveys

Zhihua Hu and Lois Wright Morton
Iowa State University
USA

1. Introduction

Water quality is a major environmental challenge in the United States (U.S.). The passage of the Clean Water Act in 1972, and subsequent amendments provided for the development of indicators and the monitoring of the quality of all U.S. navigable waters and analyses of the discharge of pollutants and their effects on water conditions (U.S. EPA, 2009). These indicators have focused on tracking physical water conditions and environmental impacts. The U.S. has 3.5 million miles of rivers and streams, 41.7 million acres of lakes, ponds and reservoirs, and 87,791 square miles of bays and estuaries. The 2009 National Water Quality Inventory Report to Congress on 2004 water conditions reported that 44% of all states' assessed rivers and streams were impaired or not clean enough to support at least one of their designated uses (e.g. swimming, fishing) (U.S. EPA, 2009). Sources of impairments were pathogens, habitat alterations and organic enrichment/oxygen depletion. Of assessed lakes and reservoirs 64% were impaired with mercury, polychlorinated biphenyls (PBCs), and nutrients identified as leading sources of impairment. Thirty percent of assessed bays and estuaries were reported impaired with pathogens, organic enrichment/oxygen depletion, and mercury major causes of impairment.

This legislation also specified that states and the national government were to estimate the environmental impacts and the economic and social costs of recommended interventions to achieve a level of water quality which "provides for the protection and propagation of a balanced population of shellfish, fish, and wildlife and allows recreation activities in and on the water" (U.S. EPA, 2009). It is the social aspect of getting to better water quality that this chapter addresses. Social science research and education are necessary to build an educated public that has the capacity to meet the challenges of degraded water bodies and engage in thoughtful problem-solving (Blockstein and Brunette 2008).

Water quality problems, like all other environmental issues, are social problems at root. Thus we need to not only monitor and assess biophysical conditions but also develop indicators that measure the human and social connections to water resources. Population-based water surveys and purposeful geographically focused interviews and focus groups provide useful indicators of public knowledge and willingness to address public goals for water protection (Prokopy and Floress, 2011). Understanding public awareness of water conditions and sources of impairment is a critical aspect of human capacity to solve the problems of water pollution. Knowledge built from surveys of citizens' perceptions of water quality offer useful guidance for the development of appropriate and effective environmental intervention strategies.

Social survey response items often consist of a continuum of positive and negative responses (e.g. highly agree, agree, disagree, highly disagree; excellent, good, fair, poor). Non-substantive responses, responses of "don't know" and "no opinion," are frequently absent from the survey or if present are not analyzed. Traditionally the non-substantive responses were considered as conveying no clear opinion of the respondents and were usually treated as missing. These don't know (DK) responses, however, are very useful in terms of revealing valuable insights about citizens' awareness, knowledge, or the lack thereof regarding their water conditions. Patterns of DK responses can help public policy makers and educators to better understand the public's awareness and knowledge about water quality, and guide future design and development of targeted programs for community engagement in solving water problems. In this sense, analysis of DK responses provides an important piece of information to researchers and community leaders with interest in moving from attitude and knowledge assessment to citizen action and engagement.

In this research, the authors analyze data from a national general population survey on water issues and in particular explore two questions related to DK responses. First, we ask if there are systematic patterns of don't know responses in water quality surveys. Secondly, we search for social factors that may be useful in understanding knowledge/awareness or the lack of knowledge about water quality. We propose that underlying don't know about water quality responses is a lack of visibility of water conditions in the respondent's everyday life. Although water is essential to life and used daily, it can easily be taken for granted by the consuming public. To answer these questions we propose two social factors that might be associated with DK survey responses: type of water supply system the respondent has for drinking water and community size. After describing the methodology, we report our findings and then offer conclusions and implications from our findings.

1.1 Non-substantive responses in survey research

Two core dimensions of a public opinion survey are the respondent's knowledge or awareness of the issue and their interest in the problem or concern about it (Rossi et al., 1983). A challenging issue facing social researchers is the presence of "don't know" (DK) responses in survey data and how to handle these non-opinion responses. No general guidelines exist for handling such responses. A typical practice, what is called the standard question form, is not to include a "don't know" option as part of a question (Schuman and Presser, 1981; Rossi et al., 1983). Researchers holding this position often argue that inclusion of no-opinion options in the surveys may not necessarily enhance data quality and instead may preclude measurement of meaningful opinions (Krosnick et al., 2002). The assumption is that DK is the lazy answer that respondents will choose when given the option (Rossi et al., 1983). As a result, DK responses are typically treated as a form of missing data in the analysis.

DK responses, however, differ from refusal to answer the question in nature, and should therefore be analyzed separately (Shoemaker et al., 2002). First of all, omitting a DK response option risks frustrating the respondent when she or he truly doesn't have knowledge or an interest in the item in question (Rossi et al., 1983). Without a DK option, respondents are forced to state an opinion about something they have no experience with, never thought of before and may not likely consider again (Rossi et al., 1983). Furthermore, DK responses can be indicators of lack of knowledge, low saliency of the issue (it is not important), and/or indifference. The implications of a DK response suggest disengagement from the issue or lack of confident knowledge which can be a deterrent to a readiness to act on a public problem such as water quality. Past studies have identified consistent correlates with DK responses, and respondents with certain characteristics are found to be more likely to give DK responses than others in attitudinal and opinion surveys. In particular, researchers have found that females, nonwhites, low-educated, low-income, and non-involved respondents with feelings of low political efficacy give a predictably high number of DK responses (Francis and Busch, 1975; Faulkenberry and Mason, 1978; Pickery and Loosveldt, 1998; Singer et al., 2000; Krosnick et al. 2002; Stocke, 2006).

1.2 Water supply systems and water quality

Public water supply systems have become increasingly complex, requiring technologies and skilled technicians to implement public safety regulations designed to ensure a safe water supply for communities. Daily monitoring for bacterial and other contaminant levels to determine treatment that assures quality and safety as well as adequate flow levels has made the provision of the public water supply complicated. Further, how the water system works and is managed can exceed the knowledge and expertise of the ordinary citizen.

Anthony Giddens, in his discussion about modernity, writes about expert systems and their implications for everyday life in the modern society. Expert systems are "systems of technical accomplishment or professional expertise that organize large areas of the material and social environments that we live in today" (Giddens, 1991, p.27). There are always experts who know about all the details and who will ensure the whole system goes all right. To a large extent, a water supply system is an expert system. As the system gets more complex, ordinary people usually do not know how their water is treated before it reaches their home for drinking. Instead, ordinary people are more likely to have only some "surrounding" knowledge such as how to turn on their tap to get water. Therefore, because "experts" are taking care of their water, they usually do not need to worry about the quality of their water source. In large communities where residents usually depend on city water supply systems, people may have little idea about where their drinking water comes from, or what is added to their water to make it safe and clean to drink. They simply trust the expert systems of water supply and turn on their tap expecting their water to be of good quality and safe to drink.

In contrast, in many smaller communities and outlying rural places where residents get water from private wells or nearby surface water bodies, the system is much less complicated. Users have more direct experience with their water and a personal responsibility to assure a safe and consistent water supply. Private wells, if monitored at all, require that home residents do their own testing and taking actions to assure safety. The mechanical condition of the well pump, water levels, and water pressure are concerns that the rural resident must pay attention to. Thus, we posit that these users are more likely to be experts on their water supply themselves and relatedly, more likely to be knowledgeable about the water quality in general in their area.

1.3 General environmental and social context

In addition to knowledge and direct experience, the general social and environmental context is thought to also exert influence on an individual's perception process and therefore, have an effect on the non-substantive responses showing up in water quality surveys. Kilbourne et al. (2002) propose a general model for the formation of environmental attitudes and motivation of behaviors that include the following aspects: (1) institutional structures, (2) value systems, (3) general environmental beliefs, (4) specific beliefs and attitudes, (5) behavioral commitments, and (6) behavior. Institutional structures, laws and regulations, cultures and beliefs are often place specific suggesting differences in political, economic, social, and environmental contexts may influence differences in attitudes, perceptions, and awareness towards water quality among residents of different regions. And these differences may translate into different degrees of concern or indifference towards water quality issues. In this sense, the DK response rate to water quality questions may well reflect how important/unimportant a water issue is to the general public within the certain environmental and social context. Based on the arguments of Kilbourne and other scholars (Kilbourne el al., 2002; Stern et al., 1995), we examine the DK response rate across several U.S. regions and states in search of detectable patterns.

1.4 Hypotheses

Based on the above argument about expert systems, we hypothesize that respondents who get their drinking water from public water supply systems (city or rural district) are less likely to have reason to learn or be aware of the water quality conditions in their area. The users of individual water supply on the other hand, are hypothesized to be more knowledgeable and thus concerned with local water quality, and therefore, less likely to give DK responses to water quality questions.

- H1: public water supply users more likely to give DK responses to water quality questions compared with individual water supply users.

Secondly, we expect the size of the community where a person lives to have an effect on water quality awareness/knowledge or the lack of such awareness/knowledge. As a community increases in size, it is more likely that the water supply system becomes more complex, and the citizens more distant from local water management processes and thus less likely to be knowledgeable about water conditions.

- H2: The larger the community size, the more likely a respondent will give a "don't know" response to questions about their water quality.

Other variables controlled for in the study include age, gender, and education. We are using these variables to test whether the previously found patterns about female, less-educated respondents and their association with don't know responses also hold true in water quality surveys.

Finally, we assume that the general environmental and social context exerts influence on one's awareness and interest towards water quality. Although an imprecise measure for the general social context, we use state as a proxy variable to capture the geo-political, regulatory and institutional conditions that people within the same state would experience as water problems are identified and addressed. Although not a formal hypothesis, we posit that the rate of don't know responses to water quality questions varies by state rather than maintains constant across all the sampled states.

The hypotheses are based on the assumption that DK responses were given purely because of lack of knowledge, awareness, or interest on the subject matter. In other words, from the nature of our data (collected via mailed survey), we assumed there were no confounding effects from interviewers' characteristics, sensitivity issues, or general attitudes toward the survey itself.

2. Methodology

Data were collected from a multi-state water issue survey completed in 36 of the 50 U.S. states (2002 through 2009). According to geographic adjacency and regional conditions, the U.S. Environmental Protection Agency divides the fifty states into ten water regions, and our sampled states covered nine of the ten regions.[1] Figure 1 provides an illustration of the sampled 36 states and which regions they belong to. States where no data were available are colored with white.

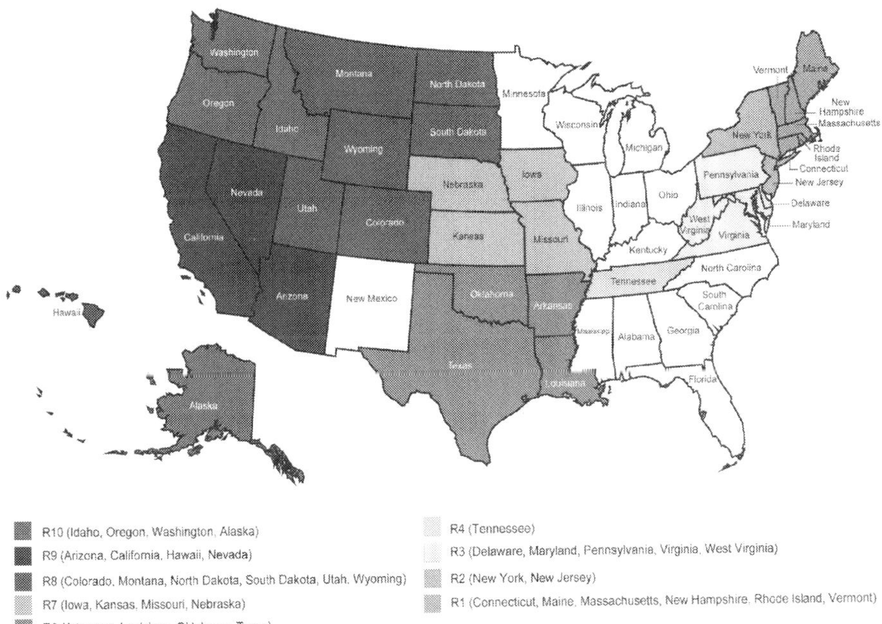

Fig. 1. Sampled States and Regions.

[1] See http://water.epa.gov/type/location/regions/ for more information about water regions and regional information.

The survey was conducted by Dr. Robert Mahler, University of Idaho under a USDA (United States Department of Agriculture) NIFA (National Institute of Food and Agriculture) Integrated Water Quality project, and the data were made available to the authors for analysis. Households were randomly sampled from phone books in each state, and calculation of targeted sample size was based on the total population of the state. Mailed surveys were sent to sampled names and addresses, with any adult in the household, whether or not addressee, invited to complete the survey questionnaire. Question content and wordings in the surveys included identical core questions as well as differing state-specific questions, with the total survey length about 50 questions. It is the core identical questions across states that asked about respondents' perceptions of water quality, water use importance, factors responsible for water pollution, sources of information about water, general environmental attitudes, and demographic information that were of interest to this study. Standard mail survey methods as recommended by Dillman (2000) were followed in each of the surveys with a total of 9332 returned surveys and response rates ranging from 37% (Massachusetts) to 70% (Wyoming).

Survey questions examined in this study include drinking water supply types (individual system, community well, or city/rural public water supply system) which represents the complexity of water supply systems, community size, three demographic items and overall ground and surface water quality. Respondents were asked where they got their drinking water, whether it is from individual system (well or surface water) or community well system (well serving 15 or more residences but not a city system), or public (city or rural) water system[2]. Community size was measured by five increments: population less than 3500; 3500 to 6999; 7000 to 24,999; 25000 to 100,000; and more than 100,000. Demographic variables were age, gender, and education. Age was divided into six increments of adult years (18-29; 30-39; 40-49; 50-59; 60-69; 70 and above). Gender responses were male or female. Educational attainment responses were five increments including less than high school, high school graduate, some college, college graduate, and advanced degrees. Two perceptions of water quality questions asked about ground and surface water conditions: "In your opinion, what is the quality of surface (ground) water in your area?" Possible responses range from "poor" to "good" to "excellent" plus "don't know/no opinion". For the purpose of analysis, responses of "don't know/no opinion" were recoded as 1 and all other substantive responses were recoded as 0. After recoding, the means reflect percentage of DK responses. The final sample with no item missing data resulted in a total number of 6401 cases.

First, Analysis of Variance (ANOVA) was used to examine DK about surface and ground water quality as percentages across groups with different water supply types, residence community size, and demographic characteristics. Post hoc Tukey tests were used to evaluate pairwise differences. This particular statistical test was chosen because of its relative advantage in statistical power and its appropriateness in pairwise comparison (Kutner et al., 2005). Then we looked at the DK variation across states. Next, all the predictor variables were fit into a logistic regression model for prediction of the occurrence of DK about ground water quality and surface water quality responses. The logistic models were then tested separately for each state to detect geographical differences.

[2] The water supply source question also has an option of "purchase drinking water", but this category is omitted in this study and covered in a separate research.

3. Findings

3.1 Water supply systems as expert systems

For both ground and surface water quality questions, the percentage of don't know responses given by respondents who use individual water supply (well or surface water) or community well is significantly less than that given by users of public water supply systems. Especially for the ground water quality question, 30% of respondents who used public water supply systems reported "don't knows", while only less than 6% of the users of individual/community water supply systems said "don't know". The difference between the two percentages is highly significant. This confirms our hypothesis about water supply systems - as the water supply systems get more complex, users of these systems are more likely to say they don't know about their water quality.

	N	DK Mean[3]	Std. Dev.	Std. Error
A. individual/community	1294	.079 B	.269	.007
B. public system	5107	.117 A	.321	.004
Total	6401	.109	.311	.004

Table 1a. DK surface_water by water supply type (ANOVA)[4]

	N	DK Mean	Std. Dev.	Std. Error
A. individual/community	1294	.059B	.235	.007
B. public system	5107	.307A	.461	.007
Total	6401	.256	.437	.005

Table 1b. DK ground_water by water supply type (ANOVA)

3.2 Community size

Findings about community size also support our hypothesis that people residing in larger communities are significantly associated with greater likelihood of responding DK to ground or surface water quality questions (Tables 2a, 2b). For both water quality questions, the percentage of DK responses rises as the respondents' community size increases. In communities with a population of less than 3500, fewer than 15% of the respondents said they did not know about their ground water quality. But in large communities of 100,000 people or more, about 33% of the respondents gave DK responses to the same question. With surface water quality, the differences in the percentages of DK responses are less strong as with ground water quality, but the general pattern is consistent as shown with ground water quality.

[3] The number in this column reflects percent of don't know responses. For example, a mean of .079 represents means that 7.9% of all responses were don't knows. Footnotes 3 and 4 apply to Tables 1 through 5.

[4] The letters behind DK means indicate from which group the mean is significantly different in a post hoc pairwise comparison using Tukey method.

	N	DK Mean	Std. Dev.	Std. Error
A. <3.5 thousand	821	.083[E]	.276	.010
B. 3.5 to 7 thousand	625	.096	.294	.012
C. 7 to 25 thousand	1146	.095[E]	.293	.009
D. 25 to 100 thousand	1794	.107	.309	.007
E. >100 thousand	2015	.133[AC]	.340	.008
Total	6401	.109	.312	.004

Table 2a. DK surface water by community size (ANOVA)

	N	DK Mean	Std. Dev.	Std. Error
A. <3.5 thousand	821	.146[CDE]	.353	.012
B. 3.5 to 7 thousand	625	.189[DE]	.392	.016
C. 7 to 25 thousand	1146	.223[ADE]	.416	.012
D. 25 to 100 thousand	1794	.270[ABCE]	.444	.010
E. >100 thousand	2015	.331[ABCD]	.471	.010
Total	6401	.258	.437	.005

Table 2b. DK ground water by community size (ANOVA)

3.3 Demographics

Respondents were divided into six increments according to their age (Tables 3a, 3b). The percentage of DKs about surface water quality is higher in the youngest age group (18-29, about 10%), and the percentages in three age groups from 30 to 59 are about the same (around 8%). The age group of 60-69 has a slightly higher percentage of DKs (about 10%), and the age group of 70 and above has the highest percentage of DK responses (over 20%). DK responses to the ground water quality question tend to share the same general pattern (Table 3b). However, with ground water quality, the DK percentage within each age category is much higher than that with surface water quality. DK responses show the highest percentage in the youngest and the oldest age groups, and are about the same across all other age groups from 30 to 69.

	N	DK Mean	Std. Dev.	Std. Error
A. 18-29	273	.099[F]	.299	.018
B. 30-39	724	.087[F]	.282	.010
C. 40-49	1155	.078[F]	.268	.007
D. 50-59	1541	.084[F]	.278	.007
E. 60-69	1236	.104[F]	.306	.009
F. 70 and above	1472	.202[ABCDE]	.380	.010
Total	6401	.129	.311	.004

Table 3a. DK surface water by age groups (ANOVA)

	N	DK Mean	Std. Dev.	Std. Error
A. 18-29	273	.293	.456	.028
B. 30-39	724	.253[F]	.435	.016
C. 40-49	1155	.226[F]	.418	.012
D. 50-59	1541	.221[F]	.415	.011
E. 60-69	1236	.261[F]	.439	.012
F. 70 and above	1472	.311[BCDE]	.463	.012
Total	6401	.257	.437	.005

Table 3b. DK ground water by age groups (ANOVA)

About 17% female respondents said they don't know about their local surface water quality, while only about 8.2% among the male respondents reported DKs. With ground water quality, the difference between the two groups of respondents is even more considerable. Almost 35% of female respondents responded don't know compared to 21% of the male respondents (See Tables 4a, 4b).

	N	DK Mean	Std. Dev.	Std. Error
A. Female	1987	.170[B]	.375	.008
B. Male	4414	.082[A]	.274	.004
Total	6401	.109	.312	.004

Table 4a. DK surface_water by gender (ANOVA)

	N	DK Mean	Std. Dev.	Std. Error
A. Female	1987	.348[B]	.476	.010
B. Male	4414	.216[A]	.411	.006
Total	6401	.257	.437	.005

Table 4b. DK ground_water by gender (ANOVA)

Previous studies have reported a correlation between education and DK, with lower education correlating with higher DK responses. Our data, however, did not reveal a similar pattern (Tables 5a, 5b). From our findings, there are no significant differences across any educational achievement groups in terms of their DK responses to the ground water question. For surface water quality DK responses, the pattern is fairly curious - respondents who were high school graduates gave more DK responses than respondents of any other educational levels. No other significant differences were found with regards to educational levels.

	N	DK Mean	Std. Dev.	Std. Error
A. < high school	769	$.087^{B}$	.282	.010
B. high school	1087	$.146^{ACDE}$	.354	.010
C. some college	1678	$.110^{B}$	.313	.008
D. college graduate	1649	$.107^{B}$	.3.10	.008
E. advanced	1218	$.090^{B}$	.286	.008
Total	6401	.109	.312	.004

Table 5a. DK surface water by education (ANOVA)

	N	DK Mean	Std. Dev.	Std. Error
A. < high school	769	.228	.420	.015
B. high school	1087	.278	.448	.014
C. some college	1678	.257	.437	.011
D. college graduate	1649	.258	.438	.011
E. advanced	1218	.254	.435	.012
Total	6401	.257	.437	.005

Table 5b. DK ground water by education (ANOVA)

3.4 DK Percentage by state

Next we examine the patterns of don't know response percentages by state. The average DK percentage for the total sample is 13.07% for surface water quality and 27.71% for ground water quality. Generally there were more people giving don't know responses to the ground water quality question.

In only one state, Arizona, the percentage of don't know responses to the surface water quality question exceeds 20 percent. Other states with higher DK responses to surface water quality include Texas (19.9%), California (17.8%), New York (17.6%), and Utah (17.2%). In thirteen states, the percent of DK to surface water quality is below ten percent. States with the lowest percent of DK responses are Alaska (0%), Vermont (3.9%), Maine (5.9%), Idaho (6.1%), and Oregon (6.6%).

Many more people gave DK responses to the ground water quality question. In eleven states, DK responses exceed thirty percent of all valid responses to that question, and only in one state (Vermont), the DK percentage is below ten percent. States with high percentage of DK on the ground water quality question include Missouri (38.5%), Tennessee (35.2%), and Oklahoma (34.4%). On the lower end, there were Vermont (9.9%), Alaska (11.7%), Montana (13.1%), and Idaho (14.7%).

It seems that some states, states like Alaska, Vermont, Maine, Montana, and North Dakota, tend to have lower DK percentage than others on both of the two water questions, while other states like Oklahoma, Texas, New York, and California have consistent higher percentage of DK on both of these questions.

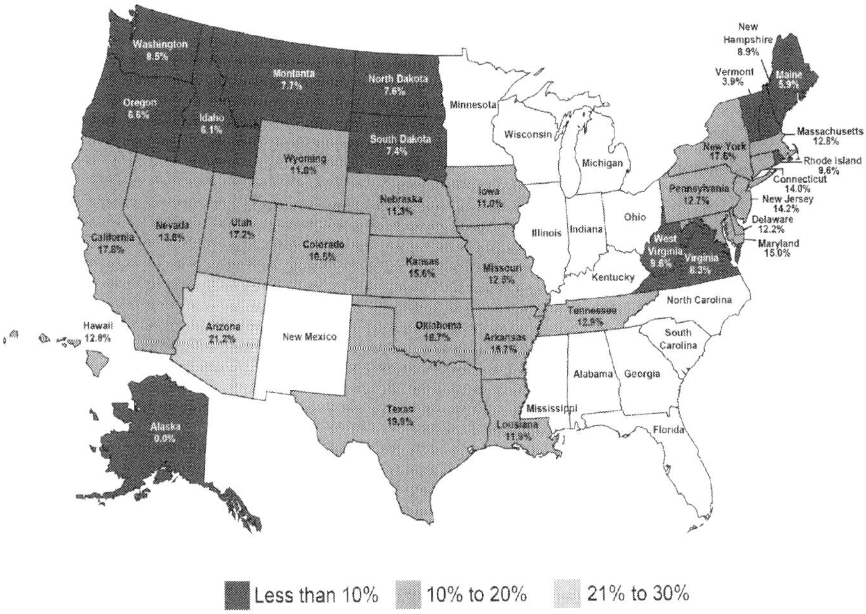

Fig. 2. DK for surface water

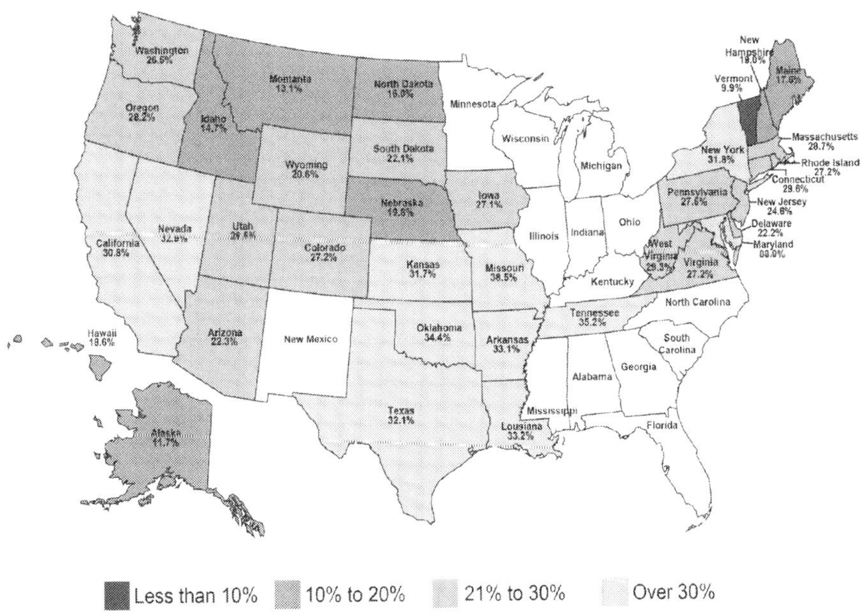

Fig. 3. DK for ground water

3.5 Results of logistic regression

The expert system and user knowledge hypotheses about water quality DK responses were further tested by a logistic regression model, where water supply type and community size were used as predictors of DK responses while demographic variables such as age, gender, and education were controlled for (Tables 6a, 6b). Although these models do not explain a large amount of the variance in DK responses (Pseudo R^2 less than 10% of the total variation in both cases), the statistical significance of tested variables confirm our hypotheses for both surface and ground water quality DKs. Public water supply system and larger community size are associated with an increased likelihood of DK responses to the ground water quality question when respondents' age, gender, and educational achievement are held constant. The logistic model for surface water showed similar results.

	B	Std. Err	Sig.	Odds Ratio
Individual/community supply	-.257	.121	.033	.774
Comm. size	.130	.034	.000	1.139
Age	.022	.003	.000	1.022
Education	-.039	.033	.239	.961
Female	.857	.082	.000	2.355
Cox and Snell Pseudo R^2	.032	Pearson Goodness-of-Fit	.023	

Table 6a. Logistic regression model: don't know about surface water quality (reference group = know)

	B	Std. Err	Sig.	OR
Individual/community supply	-1.816	.126	.000	.163
Comm. size	.149	.025	.000	1.161
Age	.008	.002	.000	1.008
Education	-.051	.025	.037	.950
Female	.692	.062	.000	1.998
Cox and Snell Pseudo R^2	.088	Pearson Goodness-of-Fit	.176	

Table 6b. Logistic regression model: don't know about ground water quality (reference group = know)

As Table 6a shows, the odds ratio of respondents using individual/community water supply systems versus public water supply system users is .774 for DK responses to surface water quality, which means the odds that an individual/community water supply user says don't know to the surface water quality question are 22.6% less than a user of public water supply systems when all other conditions held equal. Community size also showed up as a significant predictor for DK responses. Compared with a resident whose community size is one category below (for example, a resident of community size over 100 thousand people compared to another person whose community size is between 25 to 100 thousand people), the odds for a person from a larger community to report DK about local surface water quality is about 13.9% higher, as long as the two persons have the same water supply type, age, education, and gender. For a person who is older in age by one year, the odds that the person says DK to the surface water quality question are 2.2% more than a person who is one year younger, other variables being equal. Education is not a significant predictor for DK responses to surface water quality. When all other variables are held equal, there is no significant difference between respondents with different educational achievement in terms of their odds of saying "don't know" about their local surface water quality. A female respondent, when other things being equal with a male respondent, is found to be more likely to say DK to the surface water quality question, with her odds more than twice that of the counterpart male respondent.

The logistic regression model for don't knows about ground water follows the same pattern (Table 6b), except that education is a significant predictor. For a person whose education is one category higher (for example high school graduates compared with less than high school), the odds that the person responds with DK to ground water quality question are reduced by 5% when compared with a person with lower education. In addition, the difference in odds between individual/community water supply users and public water supply system users are even more pronounced with ground water quality DK responses. Compared with a city water supply user, an individual/community water supply user's odds associated with ground water quality DK responses are reduced by 83.7%. Findings about community size, age, and gender are very similar to those for surface water quality DK responses.

For logistic regression, there is no straightforward statistic like R^2 in ordinary least square regression which measures the variance explained in the dependent variable. However, several statistics, among which Cox and Snell pseudo R^2 is one, can be used to measure the strength of association between the dependent variables and explanatory variables. Cox and Snell pseudo R^2 for the surface water quality model is .032, and .088 for the ground water quality model. The proposed explanatory variables seem to be more useful predicting DK responses for the ground water quality questions. The ground water quality model also has a better goodness -of-fit, which means that the proposed model well represents the variance structure in the sampled data.

3.6 State as an explanatory variable

We then added the state where the respondent lives as a variable in our logistic model to predict their DK responses. Due to the long list of thirty-six states, we are not presenting the logistic regression results in tables, but the results show that, as hypothesized, it does matter what state the respondent is from. For the surface water quality question, the results show that with all other conditions being equal, if a respondent is from New York, Utah, Arizona, California, or Texas, then the person tends to have a higher chance of responding with DK to the question. Our results also show that with everything else being equal, if a respondent is from Vermont, Montana, Alaska, or Idaho, chances are less that the person is going to respond with DK to the ground water quality question. These results are consistent with what we found from the two DK percent maps.

3.7 Logistic regression models tested on individual states

Based on the previous findings that state is an important factor influencing residents' non-substantive responses to water quality questions, we tested the two proposed models on each individual state to see how the proposed explanatory variables predict DK responses within states when the variance caused by residence is controlled. The following tables (Table 7a and 7b) present a summary of the strength of association between the explanatory variables and the DK response and the model goodness-of-fit.

In some states, the set of explanatory variables show higher association with DK responses than in other states. For example, in Delaware (.223), Rhode Island (.151), Louisiana (.149), Arkansas (.147), Maine (.133), Wyoming (.132), Massachusetts (.108) and New Hampshire (.104), the explanatory variables show higher association with DK responses for the surface water quality questions. And for the ground water quality question, the model has higher Cox and Snell pseudo R^2 in the states of Connecticut (.225), Missouri (.183), Delaware (.170), Louisiana (.170), Montana (.152), Wyoming (.150), and Oklahoma (.150).

State	Cox and Snell	Goodness-of-Fit
Connecticut	0.041	0.173
Maine	0.133	1.000
Massachusetts	0.108	0.995
New Hampshire	0.104	0.000
Rhode Island	0.151	1.000
Vermont	0.052	1.000
New York	0.048	0.794
New Jersey	0.034	0.280
Delaware	0.223	0.434
Maryland	0.032	0.159
Pennsylvania	0.063	0.006
Virginia	0.047	0.659
West Virginia	0.024	0.588
Iowa	0.047	0.644
Kansas	0.066	0.402
Missouri	0.018	0.233
Nebraska	0.073	0.686
Colorado	0.022	0.467
Montana	0.022	0.370
North Dakota	0.058	0.272
South Dakota	0.027	0.603
Utah	0.051	0.683
Wyoming	0.132	0.644
Arizona	0.043	0.326
California	0.025	0.637
Hawaii	0.032	0.401
Nevada	0.063	0.322
Alaska	NA[5]	NA
Idaho	0.091	0.274
Oregon	0.029	0.349
Washington	0.011	0.117
Arkansas	0.147	0.335
Louisiana	0.149	0.999
Oklahoma	0.080	0.454
Texas	0.096	0.491
Tennessee	0.039	0.181

Table 7a. Surface water

[5] Not applicable because there were no DK responses for the surface water quality question in Alaska.

State	Cox and Snell	Model Fit
Connecticut	0.225	0.011
Maine	0.130	0.719
Massachusetts	0.120	0.079
New Hampshire	0.112	0.184
Rhode Island	0.101	0.407
Vermont	0.096	0.995
New York	0.115	0.354
New Jersey	0.110	0.866
Delaware	0.170	0.714
Maryland	0.140	0.391
Pennsylvania	0.128	0.120
Virginia	0.106	0.514
West Virginia	0.110	0.090
Iowa	0.104	0.229
Kansas	0.119	0.491
Missouri	0.183	0.568
Nebraska	0.137	0.831
Colorado	0.130	0.730
Montana	0.152	0.998
North Dakota	0.110	0.018
South Dakota	0.062	0.280
Utah	0.069	0.538
Wyoming	0.150	0.849
Arizona	0.045	0.412
California	0.096	0.181
Hawaii	0.046	0.194
Nevada	0.139	0.488
Alaska	0.067	0.526
Idaho	0.042	0.168
Oregon	0.035	0.296
Washington	0.089	0.315
Arkansas	0.139	0.493
Louisiana	0.170	0.635
Oklahoma	0.150	0.285
Texas	0.071	0.278
Tennessee	0.062	0.529

Table 7b. Ground Water

Overall, the model has better usefulness in predicting DK responses for the ground water quality question than for the surface water quality question. In both cases the model fits the data structure fairly well in most states, except for New Hampshire and Pennsylvania in the surface water quality case, and Connecticut and North Dakota in the ground water quality case.

4. Discussion

Water supply system type is found to be a highly significant predictor for DK responses to water quality questions. Public water supply users are much more likely to give DK responses to the water quality questions than users of individual or community water supply systems. This confirms our hypothesis about the system complexity. Public water supply users are more likely to trust water experts for their safe drinking water, and therefore are found to be less knowledgeable or concerned with the water quality in their area. In contrast, users of individual or small community well water systems are more likely to be experts themselves about their water supply, and they are also more concerned or knowledgeable about their local water quality because it is closely related to their daily drinking water safety. The size of the community where one lives is positively associated with don't know responses to the water quality questions. As the community size increases, there is increased chance that respondents say they do not know the quality of either the ground or surface water in their community. It may be due to the fact that in a smaller community residents have more opportunities to participate in community water activities including conservation and protection activities or decision-making processes regarding their water. They may have a personal well, belong to a cooperative rural water system, and/or personally use the community well or water supply system and know how it functions and what the water quality is like. Furthermore, those in smaller communities may be more likely to observe their lakes, streams, and rivers which are sources of their water supply.

Our findings about water quality questions also confirm that females are more likely to give DK responses. Older respondents are also found to be associated with higher rates of DK responses. The association between education and DK responses to water quality questions is not consistent across ground and surface water. Education shows as a significant predictor for DK responses to the ground water quality question, but not for the surface water quality question. This may be due to the nature of these two water sources. Surface water is more visible and its quality may be perceived as easily judged by its appearance, whereas the quality of ground water is not directly experienced. The public learns through other sources such as media, tech reports, water-related activities, etc. about the condition of ground water which may explain the connection to a person's education.

Generally there are more DK responses to the ground water quality question than to the surface water quality question. As discussed earlier, this may in general be related to visibility and direct experiences (swimming, fishing, viewing) with surface water (e.g. lakes, pond, rivers, and streams) and perceptions of quality seem easier to judge based on clearness, turbidity, algae and other plant growth conditions.

State is found to be an important factor associated with personal DK responses to water quality questions. In some states the DK responses to both surface and ground water quality questions are consistently higher than in other states. In addition, when we tested our proposed model within the individual states, the results show that the predictability and model goodness-of-fit also vary by state. Geographic and climate differences, as well as political, cultural and social settings in the states likely play a role in influencing residents' awareness and knowledge in water quality problems. More research is needed to better understand these variations. Local water programs and educational outreach programs, for example, might be a key factor in promoting the public's concern for their water quality. Also, in arid and rain-rich states, people's awareness of water quantity and quality might also substantially differ because of their collective experiences with flooding, drought, and water shortages. Future research is encouraged to test potentially significant explanatory state level variables, such as annual precipitation, state laws and regulations governing public water supplies, and funding on water programs.

5. Conclusion and policy implications

In everyday life we interact with hundreds of "expert systems": bank systems, computer systems, automobile systems, etc. We give full, almost blind trust to these expert systems, but the trust comes inherently with risk - that the system may not function, or that it functions well but not to our benefit. As our water resources become more limited and their quality being impaired, the risk to human health and well-being increases. Science and technology are valuable and essential resources for responding to the challenges of ensuring sustainable and secure water (Blockstein and Brunette 2008). But ultimately, improving water quality will require an engaged public that is concerned, willing to invest resources and act to protect the environment. This means ordinary citizens must move from not knowing or caring about local water conditions to a personal knowledge about their ground and surface waters, sources of pollution and what actions are necessary to counteract or reverse harmful trends and reinforce positive ones.

A high percent of DK responses on water quality issues signals that much education and communication remains to be done if public unawareness and lack of knowledge patterns are to be altered. It is important that water quality surveys include opportunity for respondents to mark their "don't knows". This response provides valuable information for the design of educational interventions, state-wide and regional media campaigns as well as strategic targeting of specific audiences. Traditional environmental education often focuses on formal school curricula and classroom education, which might explain why people of higher education tend to have lower percentages of don't know responses regarding their ground water quality. In order to reach a wider range of audiences, formal and informal environmental education programs need to be more diversified and age and gender tailored. Awareness of local water quality could be increased by providing people of all ages more opportunities to personally experience their water resource base through volunteer clean up river days and recreational services. Efforts should be made to reach older residents, those with lower education and females, and especially the more distant public water users who live in large communities. Awareness of water quality is a first step in addressing local water contamination and degradation issues. If citizens are to be mobilized to act on water concerns, they must first be motivated to increase their awareness and encouraged to learn about their water resources.

6. References

Blockstein, D.E. &Brunette, D.L. (2008) Environmental Research and Education Needs. National Council for Science and the Environment, Washington D.C.

Converse, J.M. (1976-1977). Predicting "No Opinion" in the Polls. *Public Opinion Quarterly*,Vol. 40, pp.515-530.

Dillman, D.A. (2000). *Mail and Internet Surveys*, 2nd ed. Wiley, New York, USA.

Faulkenberry, G.D. & Mason, R. (1978). Characteristics of Nonopinion and No Opinion Response Groups. *Public Opinion Quarterly*, Vol. 42, pp.533-543.

Francis, J.D. & Busch, L. (1975). What We Now Know about "I Don't Knows". *Public Opinion Quarterly*, Vol.39, pp.207-218.

Giddens, A. (1991). *Modernity and Self-Identity: Self and Society in the Late Modern Age*. Polity Press, Cambridge, UK.

Kilbourne, E. W., Beckmann, S. C., & Thelen, E. (2002). The Role of the Dominant Social Paradigm in Environmental Attitudes: A Multinational Examination. *Journal of Business Research*, Vol. 55, pp.193-204.

Krosnick, J. A., Holbrook, A.L., Berent, M.K., Carson, R.T., Hanemann, W.M., Kopp, R.J., Mitchell, R.C., Presser, S., Ruud, P.A., Smith, V.K., Moody, W.R., Green, M.C. & Conaway, M. (2002). The Impact of "No Opinion" Response Options on Data Quality: Non-Attitude Reduction or an Invitation to Satisfice? *Public Opinion Quarterly*, Vol.66, pp. 371-403.

Kutner, M. H., Necter, J., Nachtsheim, C.J. & Li, W.(2004). *Applied Linear Statistical Models*. Fifth edition. Irwin, Homewood, Illinois, USA.

Pickery, J. & Loosveldt, G. (1998). The Impact of Respondent and Interviewer Characteristics on the Number of "No Opinion" Answers: A Multilevel Model for Count Data. *Quality and Quantity*, Vol. 32, pp. 31-45.

Prokopy, L.S. & Floress, K. (2011). Measuring the citizen effect: What does good citizen involvement look like? Chapter 7 in *Pathways for Getting to Better Water Quality: The Citizen Effect* eds L.W. Morton and S.S. Brown. Springer, NY, NY.

Ribaudo, M. O, Horan, R.D. & Smith, M.E. (1999). Economics of Water Quality Protection From Nonpoint Sources: Theory and Practice. *U.S. Department of Agriculture. Agricultural Economic Report No. 782*. Resource Economics Division, Economic Research Service, USA.

Rossi, P.H., Wright, J.D. & Anderson, A.B. (1983). *Handbook of Survey Research*. Academic Press, Inc., San Diego, California, USA.

Sharpley, A., Meisinger, J.J., Breeuwsma, A., Sims, J.T., Daniel, T.C. & Schepers, J.S. (1997). Impacts of Animal Manure Management on Ground and Surface Water Quality. In: *Animal Waste Utilization: Effective Use of Manure as a Soil Resource*, J.L. Hatfield & B.A. Stewart, (Ed.), Ann Arbor Press, Chelsea, Michigan, USA.

Schuman, H. & Presser, S. (1981). The Assessment of No Opinion. *Questions and Answers in Attitude Surveys*. Academic Press, Inc., Waltham, Massachusetts, USA.

Singer, E., van Hoewyk, J. & Maher, M.P. (2000). Experiments with Incentives for Survey Participation in Telephone Surveys. *Public Opinion Quarterly*, Vol. 64, pp. 171-188.

Shoemaker, P.J., Eichholz, M. & Skewes, E.A. (2002). Item Nonresponse: Distinguishing Between Don't Know and Refuse. *International Journal of Public Opinion Research*, Vol.14, pp. 193-201.

Stern, P.C., Dietz, T., Kalof, L. & Guagnano, G.A. (1995). Values, Beliefs, and Proenvironmental Action: Attitude Formation Toward Emergent Attitude Objects. *Journal of Applied Social Psychology*, Vol. 25, pp. 1611-36.

Stocke, V. (2006). Attitudes Toward Surveys, Attitude Accessibility and the Effect on Respondents' Susceptibility to Nonresponse. *Quality and Quantity*, Vol. 40, pp. 259-288.

U.S. Environmental Protection Agency. (1994). *National Water Quality Inventory: 1994 Report to Congress*. Available from http://www.epa.gov/305b/94report/index.html

U.S. Environmental Protection Agency. (2009). *National Water Quality Inventory: 2004 Report to Congress*. Available from http://www.epa.gov/305b/

U.S. Environmental Protection Agency. (1996). *Watershed Approach Framework*. Available from http://water.epa.gov/type/watersheds/framework.cfm